GEOSPATIAL CON

THE FUNDAMENTALS OF GEOSPATIAL SCIENCE

FIRST EDITION

NICOLAS R. MALLOY | AMY E. ROCK

PUBLISHED BY

GEOSPATIAL INSTITUTE

EUREKA, CALIFORNIA

Geospatial Concepts: The Fundamentals of Geospatial Science, First Edition

Nicolas R. Malloy | Amy E. Rock

Published by Geospatial Institute.

www.geospatial.institute

This book uses ***EPUB 3.0*** specifications

CONTENTS

Preface

It was sometime after the implementation of the Geospatial Concepts course at Humboldt State University that a graduate student approached me. This student wanted to bypass the Geospatial Concepts course prerequisite and jump right into taking the Intermediate Geographic Information Science (GIS) course. This student had never completed a formal geospatial course before, though she asserted that someone in her workplace had shown her how to use "the GIS." She assured me that she was "good with computers" and was confident that she could pass the course. I understood her situation. No graduate student wants to take a 100-level course. When I was a graduate student, I also had to do the same. I knew how she felt, so I considered her request.

However, I also wanted to gauge her understanding of basic geospatial science. After asking a few fundamental questions, it was clear that she just had no idea. This student did not understand how geospatial data is structured, how to use it appropriately, how to communicate effectively, or how to identify sources of error and uncertainty. Worse yet, she was utterly oblivious to the fact that there was a need to understand it. From her point of view, GIS was merely a matter of knowing how to use the GIS software.

I explained that there was more to GIS than knowing how to operate the software. I made the case that when entering a new scientific disciple, understanding the fundamental concepts is a reasonable expectation. She did not quite see it that way and argued that she did not need to know about "all that other stuff." She just wanted to learn how to operate the software. In the end, I did not let her into the course, and the student left, clearly upset. She had come with some pre-conceived notions about what GIS was.

Unfortunately, this was not the last time I encountered this situation. I have the same conversations nearly every semester. Many people have certain expectations when it comes to the geospatial sciences, especially GIS. In part, this is because GIS software is one of the most ubiquitous tools used in the geospatial sciences but commonly used with little understanding of the nature of geospatial data. Throughout this text, you will use GIS software, and later, I will explain more about geographic information systems in detail. However, it is essential to realize that it is not enough to know how to operate a particular GIS software package. While GIS software is handy, all of its capabilities and features will not amount to much if you fail to understand the data that goes into it. This understanding requires knowing where data comes from and how it is measured, recorded, and represented. Most importantly, it requires an understanding of what its limitations are.

My goal is to provide you with the foundational material common to all geospatial science. I hope you find the information presented in the following chapters, both thought-provoking and informative. Most importantly, I hope that by reading these chapters, you will come to see the value in having this knowledge, regardless of your interests, major, or career goals.

— Nicolas R. Malloy

Why a Printed version?

Initially, the authors chose to publish this book in eBook format rather than in print. The eBook format offers several advantages over a traditional printed textbook related to cost, dissemination, accessibility, interactivity, and sustainability. Despite the benefits of the eBook format, some readers still prefer a hardcopy book. Many readers benefit from the tangible aspect of a physical book and find hardcopy text easier to read. If you are one of these readers, then this is the book is for you!

What the reader should know

The concepts and tutorials presented in this book are for readers with little to no experience using geographic information systems (GIS) software. This book is intended for use in an introductory college-level course. Readers are expected to have a sound basis of academic, and personal effectiveness competencies outlined in the U.S. Department of Labor Geospatial Technology Competency Model, including:

» Interpersonal Skills
» Integrity
» Professionalism
» Initiative
» Reading
» Writing
» Mathematics
» Basic Computer Skills
» Communication
» Critical Thinking

Learning Outcomes

The text and tutorials focus on the academic, workplace, and industry sector competencies outlined in the U.S. Department of Labor Geospatial Technology Competency Model, including:

- Reading
- Writing
- Core Geospatial Abilities and Knowledge
- Working with Tools & Technology
- Planning and Organizing
- Creative Thinking
- Problem Solving
- Positioning & Data Acquisition

After completing the text and tutorials, readers should be able to do the following:

- Describe fundamental concepts that are the basis for all geospatial sciences
- Distinguish between sources of geospatial data needed for a particular task
- Critique maps used by the media, politicians, and scientists.
- Develop a professional geospatial project from start to finish, including data acquisition, written reports, and cartographic design.

Each chapter in this book also has additional learning outcomes described within.

Required Hardware and Software

This book assumes that the reader has access to ArcGIS Desktop™ Advanced 10.5.1 or higher. The authors assume that readers have a computer that runs the Microsoft Windows™ 10 operating system and with enough power and memory to meet the ArcGIS Software requirements.

Other software requirements include the following:

- Google Chrome
- 7-Zip (7-Zip is a free, open-source file compression/decompression utility)
- Microsoft Office
- Adobe Acrobat Reader

Additionally, some tutorials need access to a GPS receiver that can connect to a computer for data download. Any hand-held GPS receiver that can connect to a Windows PC to download data should work fine. The base model for the Garmin GPSMAP 64 GPS receiver is the recommended model. However, readers are not required to choose this model. Readers are encouraged to look for used or refurbished models that fit their budget, such as the Garmin eTrex 10.

Internet Connection

For some content referenced in this book, access to a reliable high-speed internet connection is required.

Public Domain Data

Whenever possible, this book uses public domain data acquired from various websites belonging to governmental and non-profit organizations. The goal is to provide readers with data acquisition knowledge and experience. Due to the nature of obtaining data from public sources, there may be times when the data is temporarily unavailable. Additionally, some datasets may change over time, causing the results of a given tutorial to vary from the outcomes described within this book. Whenever possible, the authors will revise the text to account for recent changes to publicly sourced data.

About the Chapters

This book contains seven chapters, each representing approximately two-weeks of work for a three-credit 16-week semester course. Each chapter starts with text related to fundamental concepts related to geospatial science and its sub-disciplines. Each chapter also includes one or more tutorials designed to reinforce the concepts learned. Tutorials may take between one to six hours to complete, depending on their complexity. When possible, the authors provide an estimated time to complete tutorials. Additional references, such as video content and external websites, may also be mentioned throughout the text.

Chapter 1: Understanding Geospatial Data

Chapter 1 introduces the primary and secondary geospatial sciences. These are the roots and branches of geospatial science and determine how one creates, represents, manages, and displays geospatial data. Geospatial data has particular properties, distinguishing it from other types of information. One uses it to solve problems and answer questions related to geographic location, distribution, extent, changes over time, and geographic relationships. Commercial enterprises, government agencies, non-profits, and the average person on a day-to-day basis all use geospatial data.

A phrase familiar to computer science says, "Garbage in, garbage out." It means that the results of one's work depend upon the quality of data that goes into it. This phrase also applies to geospatial science. Understanding geospatial data will ensure that a project, analysis, or procedure will result in producing quality work. This Chapter covers the concepts, structure, data types, file types, and management of geospatial data.

Chapter 2: Communication and Design

In the advent of automobile global positioning system (GPS) navigation, free services like Google Maps, and the availability of computer mapping software for personal computers, maps are part of one's everyday lives in more ways than ever before. Increasing access to public geospatial data, and the availability of free and open source (FOS) software like QGIS and Open Street Maps have placed mapmaking within reach of anyone with computer skills and the willingness to learn. Today everyone can be a mapmaker. However, not every mapmaker is a cartographer. Cartography is the art and science of making maps to communicate geospatial information effectively. The difference between a lay mapmaker and a cartographer comes from an understanding of geospatial data, cartographic conventions, and in recognizing that a map's primary role is to communicate information visually. One might compare this to the writing one does every day on social media or in an email, versus a professional author or journalist.

Chapter 2 presents the fundamental principles of cartographic design and communication. Maps are a medium for communication with a unique set of methods and techniques. Understanding how maps communicate will allow one to view maps in a new light and with a critical eye. One begins by learning the essential map elements and the visual variables of graphic communication.

Chapter 3: Geodesy and Transformation

Since ancient Greece, mathematicians and philosophers have speculated about the size and shape of Earth. The Greek mathematician Pythagoras was one of the first to advocate the idea of Earth as a sphere. Since then, there have been many estimates on the circumference of Earth by those that followed. **Chapter 3** presents the discipline at the root of geospatial science, geodesy. Geodesy is a branch of applied mathematics. It is the science of measuring and representing the size and shape of Earth, the exact position of points on the planet, and the study of Earth's gravitational and magnetic fields as they change over time.

Chapter 4: Map Scale and Spatial Reference Systems

Most people have the idea that coordinate systems are static, unchanging definitions of where they are. One can log on to google maps and look up their latitude and longitude coordinates and feel confident that these numbers have a universal meaning that does not change. In reality, the numbers one sees on google maps are just one of many versions of latitude and longitude coordinates that can define their location. Determining a position on earth in a way that is meaningful to others is a difficult challenge. In part, the difficulty is due to the variations in map projections and datums used across the world, which can change longitude and latitude coordinates in different ways. It may seem like a small detail, yet boundary definitions and positional information can have significant legal, political, and military consequences. Chapter 4 presents how distance and location are defined and communicated using map scale and spatial reference systems.

Chapter 5: Mobile Mapping Fundamentals

Most readers of this book may expect some level of mobile mapping, geospatial fieldwork involving the measurement of a position, an elevation, a perimeter, or an area to define positional information on Earth. One aspect of mobile mapping is a land survey, the direct application of geodesy, linking mathematical models of Earth to physical reality through precise field measurements. While most readers may not expect to achieve a surveyor's level of precision, this book discusses some of the underlying concepts related to land surveying, and to a broader extent, mobile mapping. Chapter 5 presents a series of methods and equipment for mapping data in the field. This chapter differs from others due to the hands-on nature of field collection that is difficult to translate into a digital textbook. The activities included in this chapter have far less focus on software and incorporate some outdoor activities that readers will have to perform.

Chapter 6: Image Acquisition and Interpretation

It is possible that many, if not most, readers were not alive before digital aerial imagery became commonplace. Most also have little direct experience with film-based photographs. Today, anyone with an internet connection and a web browser can view images from aircraft and space satellites. With imagery so commonplace and accessible, many might take it for granted. However, there are still new frontiers emerging in the collection, application, and processing of images. The scientific and educational potential of civilian-operated unmanned aircraft systems (UAS) is just one. Chapter 6 presents the phenomenon, concepts, equipment, and methods behind the science of Remote Sensing.

Chapter 7: Selection, Proximity, and Overlay

Too often, people conduct a geospatial analysis without consideration for uncertainty and error, map projections, and datums. More often, there is little regard for cartographic convention and communication design goals. A geospatial analysis should consider the properties of geospatial data before applying GIS software tools. Chapter 7 introduces the first steps in learning how to conduct a geospatial analysis. The topics presented within should help to prepare readers for more sophisticated uses of GIS.

Acknowledgments

This book is the result of years of development, fact-checking, testing, and revision. The authors would like to thank all of the people that made this possible, including both fellow faculty members and our students. We also thank the reviewers who provided feedback and advice for this publication.

Chapter 1: Understanding Geospatial Data

Chapter 1 introduces the primary and secondary geospatial sciences. These are the roots and branches of geospatial science and determine how one creates, represents, manages, and displays geospatial data. Geospatial data has particular properties, distinguishing it from other types of information. One uses it to solve problems and answer questions related to geographic location, distribution, extent, changes over time, and geographic relationships. Commercial enterprises, government agencies, non-profits, and the average person on a day-to-day basis all use geospatial data.

A phrase familiar to computer science says, "Garbage in, garbage out." It means that the results of one's work depend upon the quality of data that goes into it. This phrase also applies to geospatial science. Understanding geospatial data will ensure that a project, analysis, or procedure will result in producing quality work. This Chapter covers the concepts, structure, data types, file types, and management of geospatial data.

Learning Outcomes

Readers should be able to accomplish the following outcomes by the end of this chapter:

- Explain the relationship between the primary and secondary geospatial sciences
- Describe the properties of geospatial data
- Summarize general geospatial data classification methods
- Discuss the difference between precision and accuracy
- Recognize sources of uncertainty and error
- Compare the differences in geospatial data models
- Demonstrate proper file management practices
- Acquire geospatial data from a public source

The Geospatial Sciences

Geospatial science is a multidisciplinary field of research and application using geographic information and technology to answer scientific questions as they relate to both space and time. Geospatial science incorporates five subfields, categorized into primary geospatial sciences and secondary geospatial sciences. Watch Geospatial Revolution: Episode 1, produced by *Penn State Public Broadcasting*, to learn more (**Figure 1.1**). This episode covers what is involved in the geospatial revolution, the origins of mapping and geospatial technology, and a look at the use of crisis mapping in Haitian earthquake relief efforts.

Figure 1.1: Geospatial Revolution: Episode 1, produced by Penn State Public Broadcasting. URL:*https://youtu.be/poMGRbfgp38*

Primary Geospatial Sciences

The defining characteristic of primary geospatial sciences is that they produce geospatial data. This book defines the primary geospatial sciences as the following:

- Geodesy
- Mobile Mapping
- Remote Sensing

Geodesy is a branch of applied mathematics. It is the science of measuring and representing the size and shape of Earth, the exact position of points on Earth, and the study of Earth's gravitational and magnetic fields as they change over time. Geodesy is the foundation, quite literally, upon which all other geospatial science stands.

This book uses the term **mobile mapping**, geospatial fieldwork involving the measurement of a position, an elevation, a perimeter, or an area to define positional information on Earth. One aspect of mobile mapping is a **land survey**, the direct application of geodesy, linking mathematical models of Earth to physical reality through precise field measurements. While most readers may not expect to achieve a surveyor's level of precision, this book discusses some of the underlying concepts related to land surveying, and to a broader extent, mobile mapping.

Remote sensing is the geospatial science related to obtaining data from a distance using devices that detect emitted or reflected electromagnetic energy. Remote sensing applications produce raw geospatial data, usually as images. Subsequently, other geospatial sciences often use images and data acquired via remote sensing. However, recent developments in the field of remote sensing have broadened its scope. It is now very closely linked with geographic information systems (GIS) and mobile mapping technology, providing highly accurate information about Earth from a variety of sources.

Secondary Geospatial Sciences

Secondary geospatial sciences integrate existing geospatial data to answer specific questions and communicate the results. These sciences include:

- Cartography
- Geographic Information Systems (GIS)

Cartography is akin to reading and writing with maps. It is the art and science of making maps to communicate geospatial information effectively. The difference between a lay mapmaker and a cartographer comes from an understanding of geospatial data, fundamentals of design, cartographic conventions, and in recognizing that a map's fundamental role is to communicate information visually.

A **geographic information system (GIS)** is a combination of software, hardware, data, and scientific methods used to store, analyze, produce, and display geospatial information. Its primary use is for the direct application of geospatial analysis on geospatial data to make decisions and answer questions.

PROPERTIES OF GEOSPATIAL DATA

One of the common ways one obtains information about political views in this country is by using public opinion polls. Pollsters make highly accurate projections using rigorous scientific methods, collecting random samples of information, and using mathematical and computer analysis to create a representative model of public opinion. For polling to be accurate, pollsters need to be sure that the sample group is the right size and that it genuinely characterizes the broader population. Failure to understand the makeup and qualities of the larger population will lead to inaccurate polling. Like pollsters, geospatial scientists often collect a representative sample to model the world around us. They too must consider the size of the sample and the degree to which it represents the greater whole.

Geospatial data has specific properties, distinguishing it from other types of information. Geospatial data must always fit within the framework of **geographic space**, a model of Earth that is continuous, three dimensional, and approximately spherical (**Figure 1.2**).

Figure 1.2: Geographic Space is a continuous phenomenon modeling the shape of Earth.

Geospatial data refers to a collection of information containing non-spatial attributes that correspond to a specific location and extent in space and time. Geospatial data has the following qualities:

» Geospatial data is dependent upon the physical features of geographic space.

» Geospatial data can be represented graphically via a map.

» Geospatial data is a selective and generalized representation of the real world.

Applications for Geospatial Data

Geospatial data helps one understand the environment. The most common use for geospatial data is for answering a geographically related question. There are specific types of questions that geospatial data is particularly suitable for:

» Questions about a location

» Questions about geographic distribution or extent

» Questions about geographic change over time

» Questions about geographic relationships

Watch Geospatial Revolution: Episode 2, produced by *Penn State Public Broadcasting* to learn more (**Figure 1.3**). This episode looks at how local governments and business use geospatial technology to deliver services and run efficiently, keeping a continuing eye on future developments and applications.

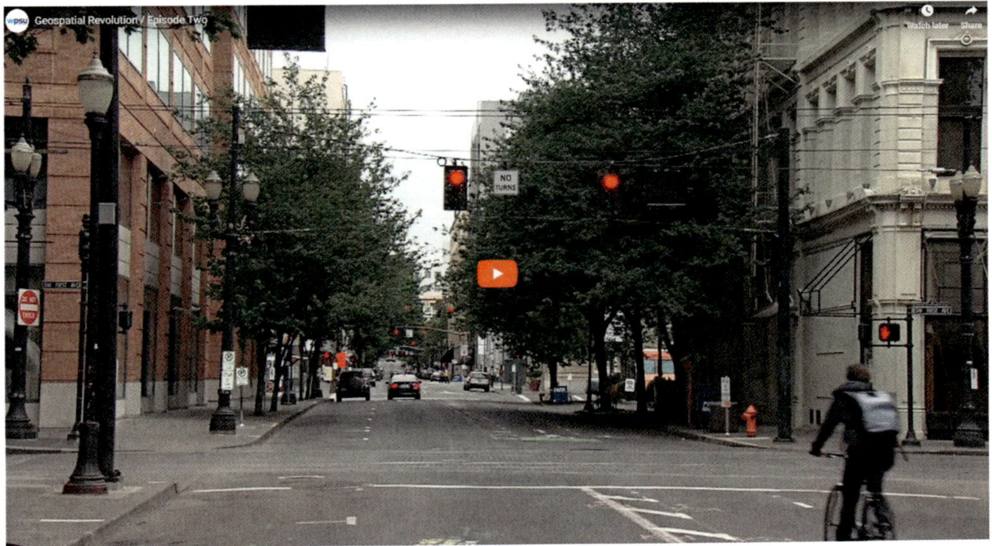

Figure 1.3: Geospatial Revolution: Episode 2, produced by Penn State Public Broadcasting. URL: *https://youtu.be/GXSobsRoe7w*

Having a clear understanding of the type of question one tries to answer is essential in finding, using, and applying geospatial data. Once one sees a solution to a problem, the next step is often to communicate those answers graphically via a map. **Chapter 2** covers the issues and challenges associated with the cartographic communication of information.

Geospatial Data Models

As mentioned previously, geospatial data is a simplified representation of physical features and their attributes. GIS software uses a graphical and structural representation of geospatial data, called a data model, to symbolize each feature. Think of the lines and shapes on a Google map. These shapes are data models of real-world features. There are two primary types of geospatial data models, the **vector data model**, and the **raster data model**. The types of data models used are dependent upon the spatial nature and geometry of the feature they represent. The vector data model usually represents discrete features, such as buildings and roads. The raster data model usually represents continuous features, such as elevation, aspect, and slope, but it also includes aerial imagery.

Vector Data

The raster data model does not adequately represent discrete features, such as buildings, roads, or fire hydrants. These features often call for precise dimensions and positional information. A raster cell substantially generalizes these features, losing much of the information. A much more suitable data model for discrete features is the **vector data model** (**Figure 1.4**).

Figure 1.4: In this image, polygons represent building and landscape features on the map.

Vector Object Types

The vector data model has three fundamental objects types, a point, a line, and a polygon. Vector data relies on a series of x y coordinate pairs to represent these shapes (**Figure 1.5**).

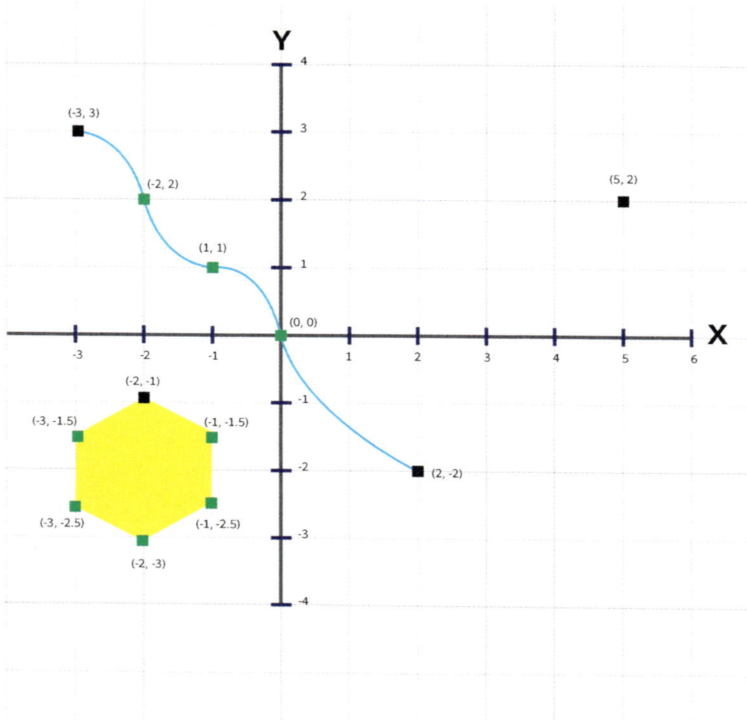

Figure 1.5: The vector data model has three basic object types: a point, a line, and a polygon.

A single XY coordinate pair called a **node** represents the point feature seen here in the upper right quadrant (**Figure 1.6**). Five is the X value, and two is the Y value. In geospatial data, another coordinate system such as latitude and longitude would identify these x and y values.

When using vector data models, a **linear feature**, sometimes called an **arc**, is made up of nodes and vertices connected by line segments. The term **node** refers to the beginning and ending points of the line feature. Any points in between help to define the shape of the line. These intermediate points are called **vertices**. In the example here, the black points represent nodes, and the green points represent vertices (**Figure 1.7**).

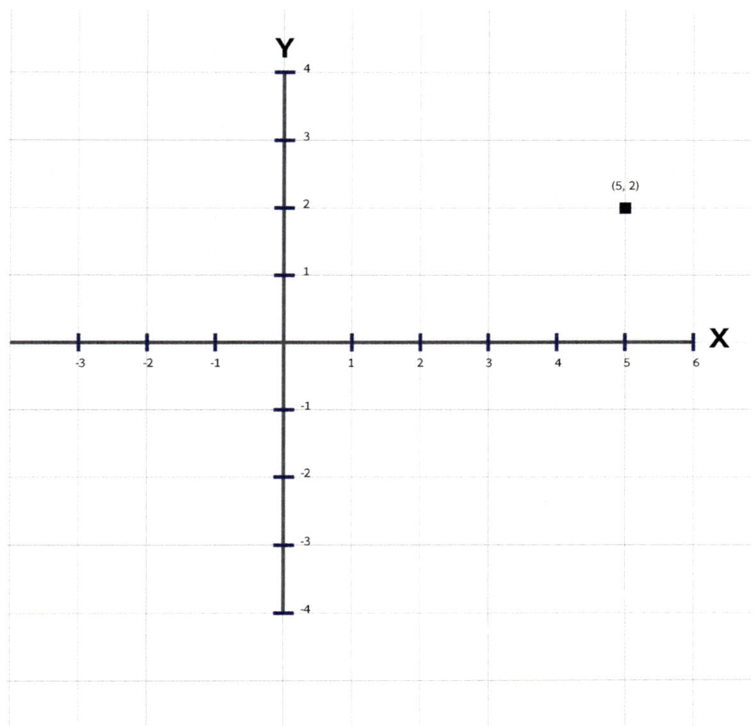

Figure 1.6: X Y coordinate pairs define the location of a node.

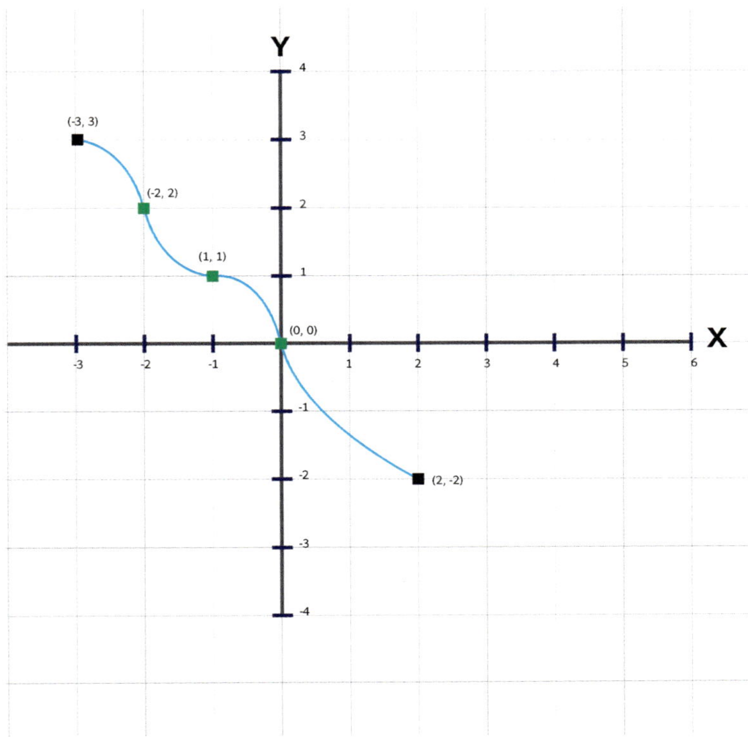

Figure 1.7: The nodes and vertices help to define the blue line in this image, with the curves calculated mathematically.

In a similar manner as a linear feature, nodes, vertices, and arcs make a **polygon**. However, in this case, the polygon is a closed feature. The same point defines the location of the starting node and the ending node. A polygon also has an interior region and can contain other features within it. It is likewise possible for polygons to share a boundary with other polygons. In the example here, the black point (-2, -1) represents the starting and ending node. The green points represent the vertices in between (**Figure 1.8**).

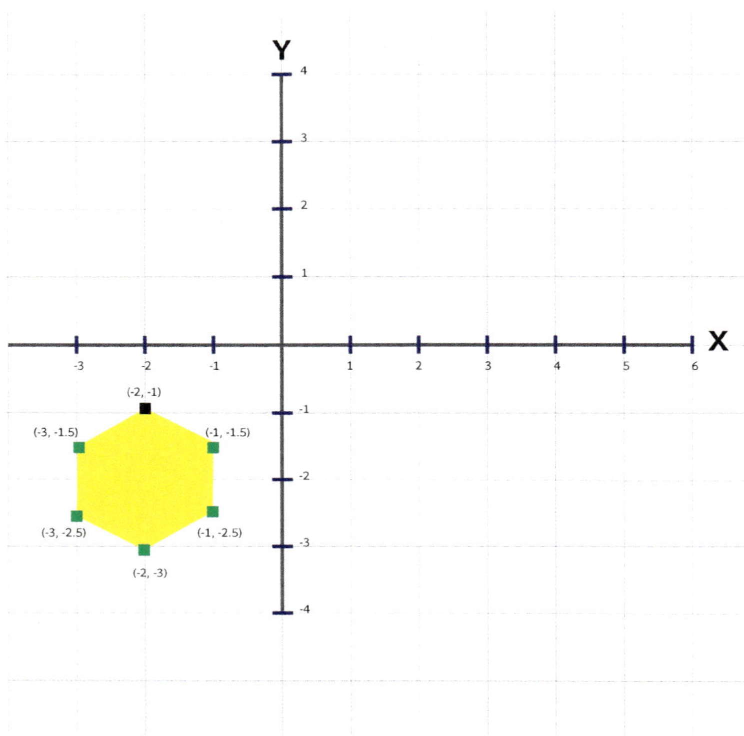

Figure 1.8: A closed set of nodes and vertices make up this yellow polygon.

Raster Data

Although they may not realize it, most people have already used raster data. A digital photograph is an example of a raster data model (**Figure 1.9**). Look very carefully at any digital photograph. Notice that it consists of many individual squares called **pixels** (**Figure 1.10**). Each square pixel has only one color. Each pixel in a digital image holds an alphanumeric value. A computer translates this value into a color that one sees on the screen. In the raster data model, each pixel, or **cell**, is assigned a single data value, called an **attribute**. In the case of a photograph, that data is the color of the pixel. In a geospatial raster, that value might represent something like elevation, or precipitation.

Figure 1.9: Many rows and columns of square pixels comprise a digital image. When viewed together, they form an image.

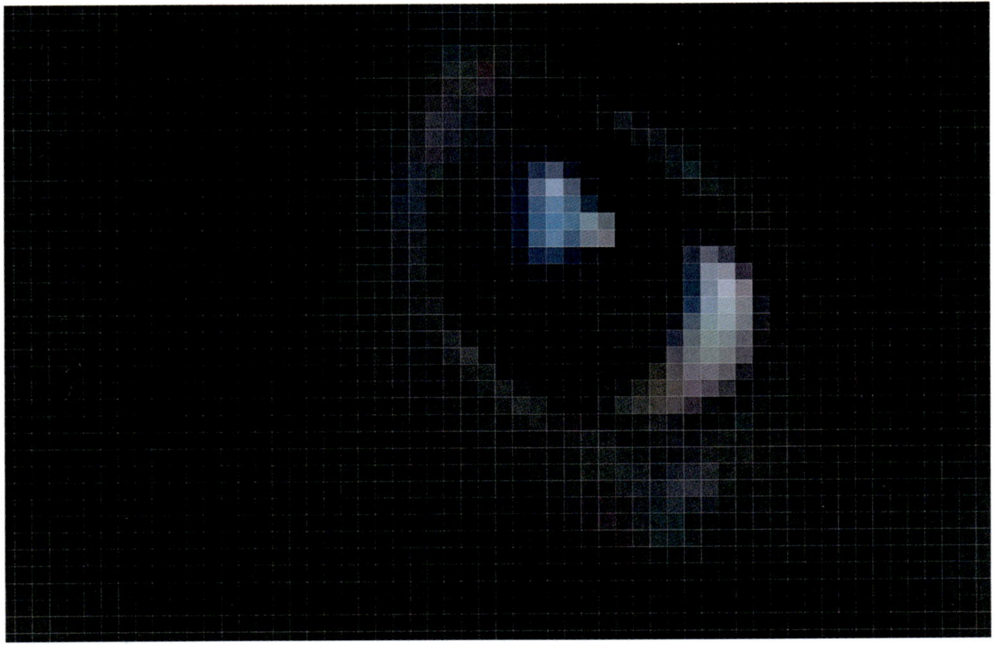

Figure 1.10: If one zooms in closely on a digital image, they can see that it is made up of many rows and columns of square pixels.

There can be no gaps in raster data, so even if a cell does not have data, it must be assigned a placeholder value such as **NoData**. Raster data is most often used to store attributes that represent a continuous phenomenon, such as snowfall, temperature, slope, or elevation. The elevation values in a **digital elevation model (DEM)** are an excellent example of a continuous phenomenon since there is no point on the surface where elevation does not exist (**Figure 1.11**).

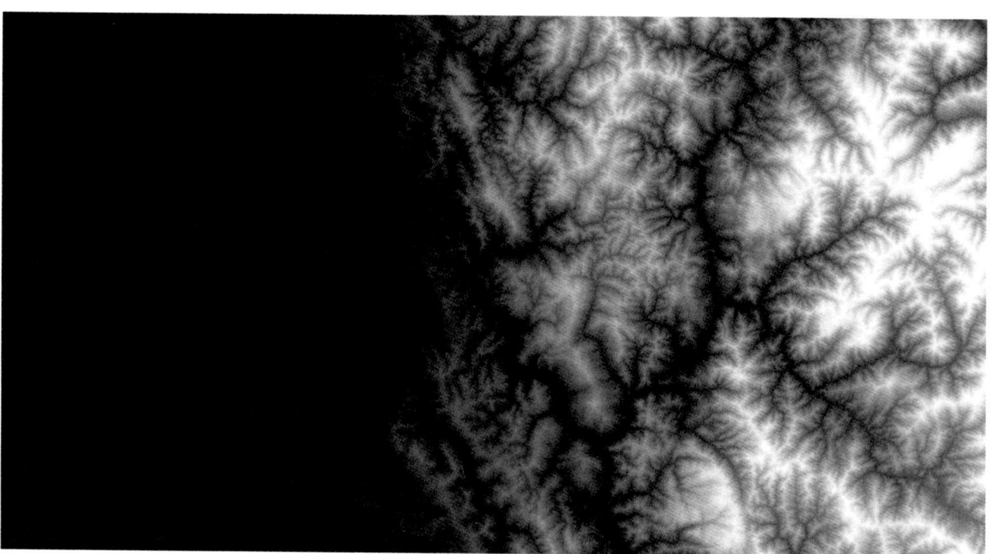

Figure 1.11: Each cell in this digital elevation model (DEM) stores a number representing elevation in meters. The lighter colors on the DEM represent, the higher elevations, while the darker colors indicate, the lower elevations.

The **raster data model** works very well for storing types of data that have a lot of different values for one attribute. Each cell records its value of that attribute. In the example here, each cell stores a number, and that number represents an elevation value in meters (**Figure 1.11**). A raster dataset can record the elevation values across the region, constrained only by the size and the number of cells. On the screen, a computer translates those values into colors, such as shades of gray. To the computer, a raster file looks more like the image shown here (**Figure 1.12**).

Raster Cell Size

One also uses rasters in geospatial science, though the terms used are slightly different. For example, one often refers to a pixel as a **cell**. When looking at raster data, each cell represents a distance and an area on the ground. The length on the ground that each side of a cell represents is called the **cell size**, also commonly referred to as **resolution** (**Figure 1.13**).

Figure 1.12: Rasters store a single value for each cell.

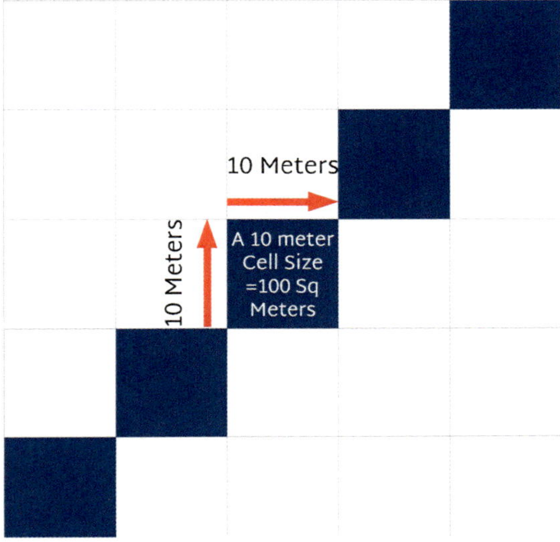

Figure 1.13: A ten-meter cell size represents an area of 100 sq. meters. Each side of the cell is 10 meters long.

Raster Data Considerations

There are several limits one must be aware of when dealing with raster data. As mentioned previously, each cell can only hold a single attribute value. The value represents the entire area covered by a cell, which is a simplification of reality. For example, a digital elevation model with a cell size of 30 meters assumes that the elevation across the entire 900 square meter area is equal. In reality, the elevation may rise or dip lower in places at that location. As the cell size increases, information gets lost as it becomes generalized or averaged. Another problem can arise when the cell size is too small. A decrease in cell size for any given area means an increase in the number of cells overall. The increase in the number of cells places a much higher demand for computer processing requirements and data storage (**Figure 1.14**).

Figure 1.14: Aerial imagery is another continuous phenomenon commonly represented by raster data. However, the higher the resolution, the larger the file will be.

In the advent of **unmanned aircraft systems (UAS)** for environmental research applications, it is now possible to collect imagery data at an extremely precise scale (**Figure 1.15**). This increase in resolution poses a difficult challenge for the hardware and software currently available. **Chapter 6** discusses the applications of unmanned aerial systems in more detail.

Figure 1.15: Unmanned aircraft systems (UAS) can capture extremely high-resolution data.

Raster versus Vector

There is a fundamental difference between raster and vector data models. Their construction determines the dissimilarity between the two. As mentioned previously, rasters are a grid of square pixels. The vector data model does not have pixels. Instead, line segments connect nodes and vertices. The computer keeps track of line segments, vertices, nodes through a series of mathematical equations that keep track of length and curvature. In a vector-based image, one can scale infinitely without pixelation occurring (**Figure 1.16**).

Figure 1.16: Both top and bottom images come from the same graphic shown in the center. The picture here shows a raster graphic at approximately 3200% zoom. The image on the bottom is a vector shown at the same zoom level. Notice the difference in edge quality.

FILE MANAGEMENT BASICS

Geospatial file management is different from what most people have learned from working with documents and spreadsheets. The difference is one of the hardest things for new users to remember. The way one uses files in their everyday lives is very similar to muscle memory. People repeatedly perform specific actions like saving, copying and pasting, moving, and deleting files without much conscious thought. A person's file management habits pose a big problem because geospatial data works differently. Geospatial data is almost always made up of multiple files working in unison, whereas a typical computer file is a single entity.

For example, suppose a writer working on a new novel uses a program like Microsoft Word to create and save their work as a word document file (.docx). If the writer works from an office and then decided to work from home, they could easily copy and paste a single file onto a flash drive and take it with them. When they opened up that file at home, all of the same information would be there. The same is true for many other types of files such as audio files, image files, and video files.

Geospatial File Management

Take a look at a digital elevation model. While it may look like any other image file when viewed in GIS software, it appears very different from a file management perspective. Many are probably familiar with an image composed of a single file, such as a .jpg, .gif, or .png. However, if one uses Microsoft Windows file explorer to view a digital elevation model (DEM), they would see that instead of a single file, it displays many files (**Figure 1.17**).

These files probably do not make much sense at the moment, and that is okay. Understanding this under the hood level of detail is beyond the scope of this book. All one needs to know, for now, is that the folder and the files work together to construct the digital elevation model. Removing any one of these pieces corrupts the data and makes it unusable.

The most common file format for vector data is the **shapefile**. A shapefile consists of many files linked together (**Figure 1.17**). Some of these files store spatial information, such as geographic location. Some files store attribute data as a database file. Others will save the geometry of a feature. Separating any one of these pieces makes the data unusable. The safest way to manage files is by using a geographic information system software package. When using GIS software, for example, a shapefile and a digital elevation model are represented by one icon (**Figure 1.18**).

← → ↑ This PC › Desktop › working   Search working

Name	Status	Date modified	Type	Size
dem.img		4/7/2019 8:03 PM	Disc Image File	507,764 KB
dem.img.aux.xml		4/7/2019 8:02 PM	XML Document	3 KB
dem.img.xml		4/7/2019 8:02 PM	XML Document	7 KB
dem.rrd		4/7/2019 8:03 PM	RRD File	306,622 KB
Study_area.CPG		10/19/2017 3:21 PM	CPG File	1 KB
Study_area.dbf		10/19/2017 3:21 PM	DBF File	7 KB
Study_area.prj		10/19/2017 3:21 PM	PRJ File	1 KB
Study_area.sbn		10/19/2017 3:21 PM	SBN File	1 KB
Study_area.sbx		10/19/2017 3:21 PM	Adobe Illustrator Tsu...	1 KB
Study_area.shp		10/19/2017 3:21 PM	SHP File	155 KB
Study_area.shp.xml		4/7/2019 8:02 PM	XML Document	8 KB
Study_area.shx		10/19/2017 3:21 PM	SHX File	1 KB

12 items

Figure 1.17: The files that begin with the word "dem" are components of a digital elevation model as seen in Microsoft Windows. The files that begin with the word "Study_area" are components of a shapefile, a common vector file type. Though it looks like many separate files, each must work together for use in a geographic information system.

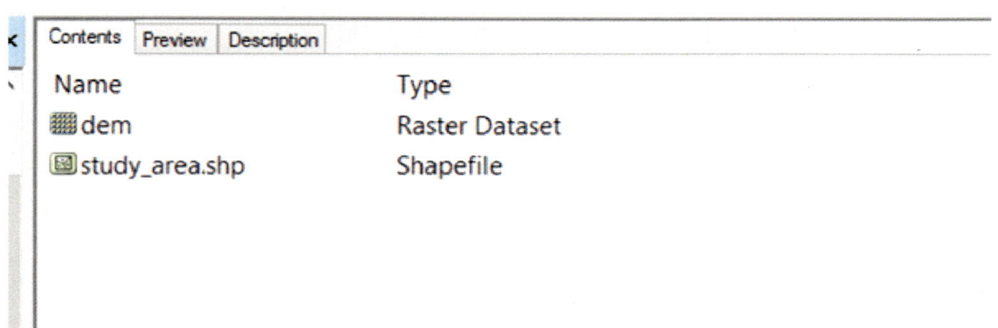

Figure 1.18: This image shows the same two files in the previous image as seen in a GIS software package. As one can see, when using GIS software, the files appear to be single entities. Managing files using GIS software helps to maintain the integrity of one's data.

To the end-user, the GIS software gives the impression that one file comprises the data. This representation allows the end user to manage the data familiarly regarding copying and pasting, editing, and moving files from one directory to another. It makes it seem like one is moving only a single file, and the GIS software keeps everything together. In addition to all of the parts of vector and raster files, geospatial files also have a third component called the **map document (.mxd)**.

When working with geospatial data, many files are linked together.

A central file such as a **map document (.mxd)**, manages the data. However, a map document file *does not* store any of the data. It only *links* to it. The problem occurs when using the Microsoft Windows operating system to manage geospatial data files. Because the Microsoft operating system displays the data as many individual pieces, it is common for geospatial files to become corrupted when moving parts from one place to another. Often a piece is left out or inadvertently deleted. It is, therefore, a good idea to store all of the raster files, vector files and the map document in a single master folder, called a workspace, to keep them all together. Subfolders further organize the data, but the workspace folder keeps everything in one place.

A sound file management process keeps track of original and modified data so that one can retrace their steps if needed. An example of a typical folder structure might look like this:

- » workspace folder
    - » original folder
    - » working folder
    - » final folder

The **original folder** is for storing all of the original downloaded data. The **working folder** is for any modified files, which can include selections and exports from original files, geoprocessing outputs, or manually edited data. The **final folder** is an excellent place to store map outputs and other finished products.

One should expect to use both Microsoft Windows and a GIS software package as part of their regular workflow. The critical thing to remember is that if one is moving a single dataset, be sure to use GIS software. When managing files in Microsoft Windows, the safest way to transfer geospatial data from one directory to another is by moving the entire workspace folder. In this way, one is sure to keep all of the pieces together.

This workflow seems like a straightforward thing to understand, but file mismanagement is a leading cause of problems and frustration.

One's regular habits for file management are so ingrained that mistakes are inevitable. The best way to prepare is to remember to store all of one's data and map document(s) in a master workspace folder. Save work often and always make a backup of the workspace folder using cloud storage.

COMMON GEOSPATIAL DATA SOURCES

The most common question geospatial readers ask is, "Where can I find geospatial data?". The usual answer is, "You can find it everywhere, but you are asking the wrong question." When looking for geospatial data, first consider the goal or purpose for needing a particular dataset. Then amend the question to ask, "Where can I find quality geospatial data produced by experts in the topic related to my data needs?" The answer, of course, is "Go to the experts."

Initially, the military developed and applied many sources of geospatial data and technology. For this reason, government agencies are the leading source of geospatial information. Watch Geospatial Revolution: Episode 3, produced by *Penn State Public Broadcasting* to learn more (**Figure 1.19**). This episode explores geospatial technology in the world of security. It reveals how new technologies help to broker peace, wage war, and fight crime but can also compromise personal privacy.

Figure 1.19: Geospatial Revolution: Episode 3, produced by Penn State Public Broadcasting. URL: *https://youtu.be/OePOK6nzcaY*

Suppose one needed geospatial data related to rainfall. Start by asking, "Who are the experts?" A great start would be to visit the National Oceanic and Atmospheric Administration (NOAA) website and see what one can find there. They are the experts on rainfall, so it is very likely that they would have geospatial data about this topic.

Another critical question in one's search for data is, "What is the geographic extent?" If one needs statewide data, they might start by looking at data resources for their state of interest. Perhaps one might need information about roads and highways in Humboldt County. There are several places one might look including the state transportation department (Caltrans) website or the county website. A Google search would reveal that each of these entities has a GIS data library that one can access. One can expect Caltrans to have statewide datasets, while Humboldt County would likely have localized datasets.

One thing one may notice is that not all geospatial data portals are the same. Some will present information graphically via a digital map while others might present information as a list. The data may also be served using different file formats. For example, a **shapefile**, a vector file format for geospatial data, consists of many separate files, which would make it challenging to download each piece individually. To solve this problem data providers often compress geospatial files into formats such as .zip, .7z, or .tar. Data compression file formats allow one to download a single file that one can decompress when ready to use it.

Below is a list of some commonly used geospatial data sources:

» National Oceanic and Atmospheric Administration (NOAA)[1]
» USDA Forest Service[2]
» USDI National Park Service[3]
» U.S. Geological Survey (USGS)[4]
» California Department of Transportation (Caltrans)[5]
» Los Angeles Geospatial Data Portal[6]
» Humboldt County[7]

If one not sure who the experts are, or what entity houses data for that area, try searching terms like "(thing I want) shapefile" or "(county name) GIS."

1 https://nauticalcharts.noaa.gov/data/gis-data-and-services.html
2 http://data.fs.usda.gov/geodata/
3 http://www.nps.gov/gis/
4 http://www.usgs.gov/
5 http://www.dot.ca.gov/hq/tsip/gis/datalibrary/
6 http://egis3.lacounty.gov/dataportal/
7 http://humboldtgov.org/276/GIS-Data-Download

Geospatial Data Classification

Data classification is used to simplify and understand numeric data. Through classification, it is possible to identify patterns and trends or to alter data by creating a subset of specific classes. Classification is mostly organizing data in groups. One undertakes classification by creating a series of **class intervals**, a range of values that do not overlap. The value that defines the boundary between each class interval is called a **class break**. Think about a series of buckets into which the data gets sorted. Many statistical methods are used to classify data. This chapter examines five basic classification methods:

» Equal Interval
» Quantile
» Natural breaks
» Manual classification
» Standard deviation

A **histogram** is a graph of all the individual values of data, organized from smallest to largest, and displaying the number of times each value occurs. A histogram makes it easier to examine data classification. The X-axis shows the data values, and the Y axis the number of times that value occurs in the data set.

Equal Interval Classification

The **equal interval classification** method divides the range of data values by the number of classes chosen. By doing so, it creates class intervals with equal ranges of values. In the example below (**Figure 1.20**), the data gets divided into four classes. The data values range from 0 to 8188, so one divides the difference (8188-0) by 4 to get 2047. The first class break is 2047. The next is at 4094 (2047 + 2047), and so on. It is possible for the equal interval method to create class intervals with no data. Use the equal interval classification for data with ranges of continuous values that gradually change from low to high (no significant gaps in the histogram).

Range of Data Values

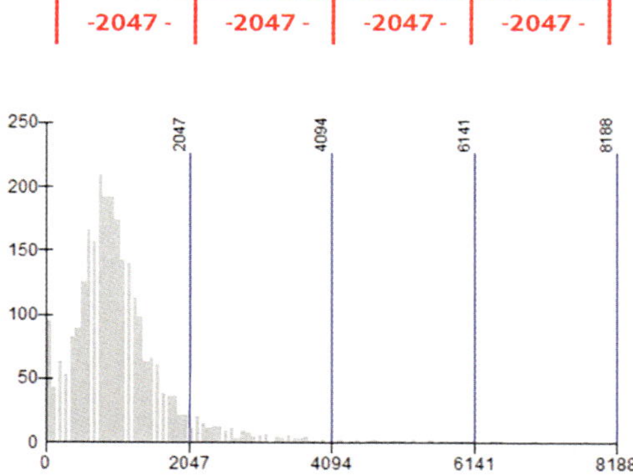

Equal Interval Classification

Figure 1.20: In this graph, the range of population values is from 0 to 8188, and there are four classification intervals. The equal interval methods divide 8188 by 4 to get four equal classification intervals, each with a range of 2047.

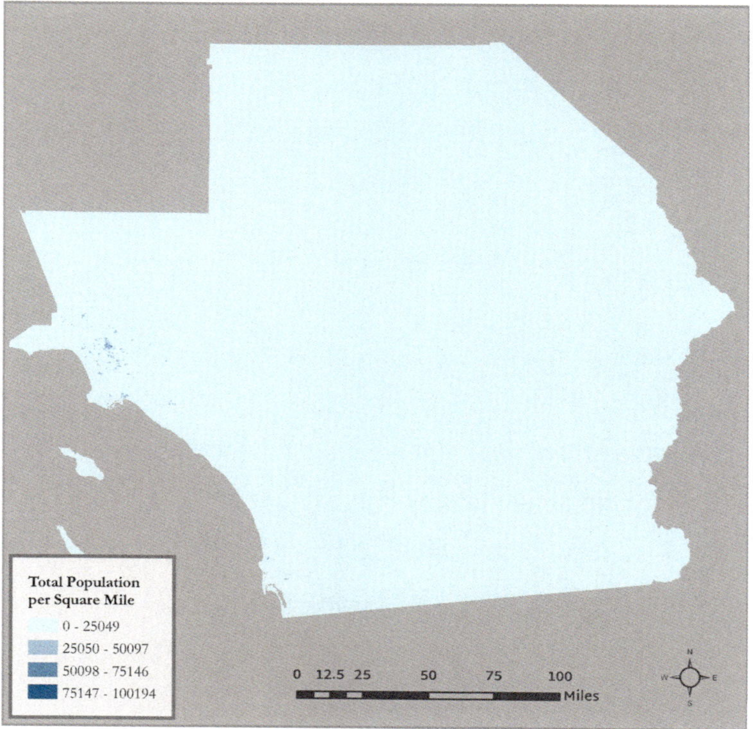

Figure 1.21: This map displays the 2010 population of Southern California using equal interval classification.

Quantile Classification

The **quantile classification method** is often confused with the equal interval method. Rather than dividing the range of data values by the number of classes, the quantile method divides the number of features by the number of classes chosen. If one returns to the bucket analogy, quantiles have the same number of observations in each bucket, which means that the break values may not be evenly spaced. To calculate quantiles, take the total number of observations and divide by the number of classes. The sample data (**Figure 1.22**) has 2,676 observations and four classes, so 2,676 divided by four is 669. The first break falls at the value of observation number 669, the second break at the value of observation 1338, and so on.

The quantile method of classification creates a series of class intervals that have the same number of features. This method works best when applied to linearly distributed data. The quantile method also benefits from a higher number of classes, which easily divide the number of features. This method is excellent if one wants to compare segments of observations, for example, the poorest 20% of counties as compared to the wealthiest 20%.

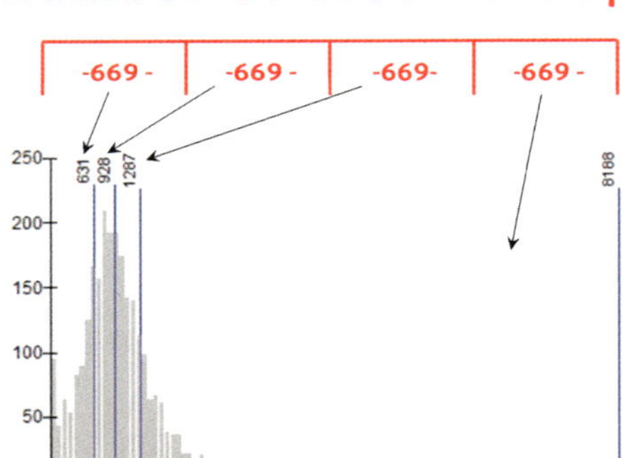

Figure 1.22: In this graph, the range of population values is from 0 to 8188, and there are four classification intervals. Each class has 669 features.

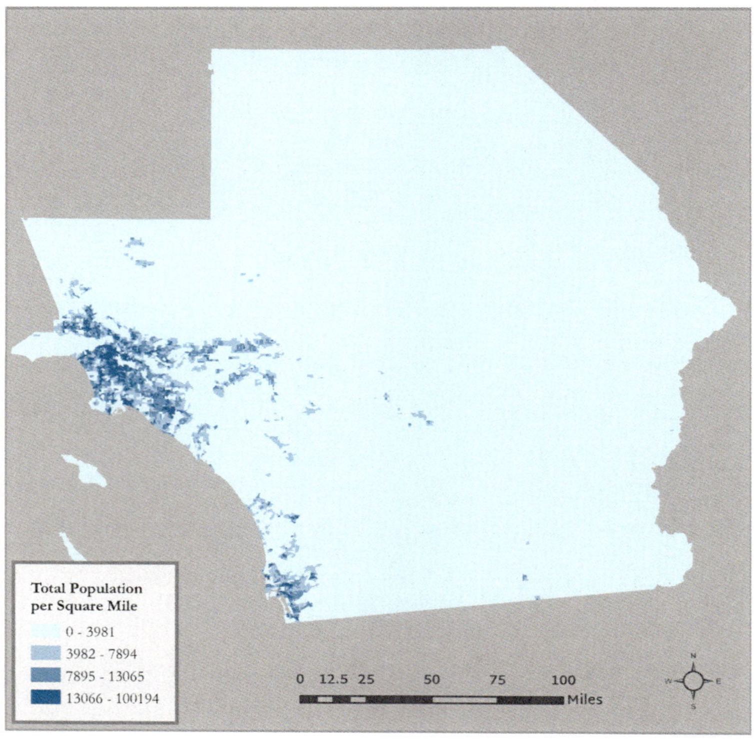

Figure 1.23: This map displays the 2010 population of Southern California using quantile classification.

Natural Breaks Classification

The **natural breaks classification** method utilizes a statistical approach that reduces the differences within class intervals but increases the differences between class intervals. The algorithm looks for natural gaps in the data and identifies clusters of values (**Figure 1.24**). The natural breaks method of classification sets class breaks at each significant gap. It starts with the most significant gap in data values. The number of classes chosen determines how the breaks get defined. Many GIS software packages use natural breaks as the default classification method, but using natural breaks poses a problem when comparing two or more sets of data. An individual dataset defines its class breaks. Therefore, it is unlikely that two or more datasets will use the same breaks to represent the data. The best use for natural breaks is to understand trends in a single dataset with naturally occurring clusters of values. Do not use natural breaks when comparing maps based on different sets of data, or when data is continuous.

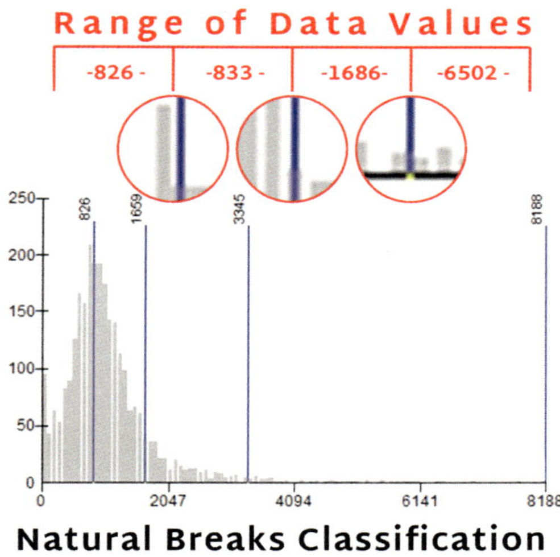

Figure 1.24: In this graph, the range of population values is from 0 to 8188, and there are four classification intervals. Each class has a different number of features and a different range of values.

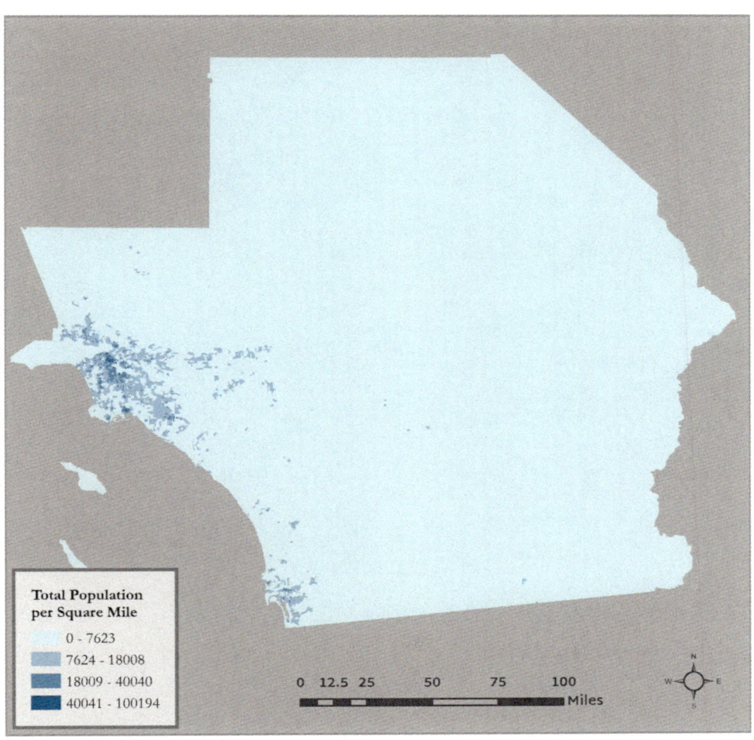

Figure 1.25: This map displays the 2010 population of Southern California using natural breaks classification.

Manual Classification

Manual classification is the arbitrary selection of breaks in class intervals. Use manual classification when trying to separate a specific set of values from the dataset. Manual classification can be subject to bias and error. Avoid manual classification for most circumstances without a compelling reason or a clear understanding of the data. Proper use of manual classification improves the map readers' understanding without substantially distorting the perception of the data. For example, a map author might recognize that slightly adjusting breaks to round numbers, such as changing 2113 to 2100, will increase clarity while still maintaining a high degree of fidelity to the original statistical methods of classification. Manual classification is also quite useful when trying to compare two or more maps so that the class breaks match and a consistent message gets communicated.

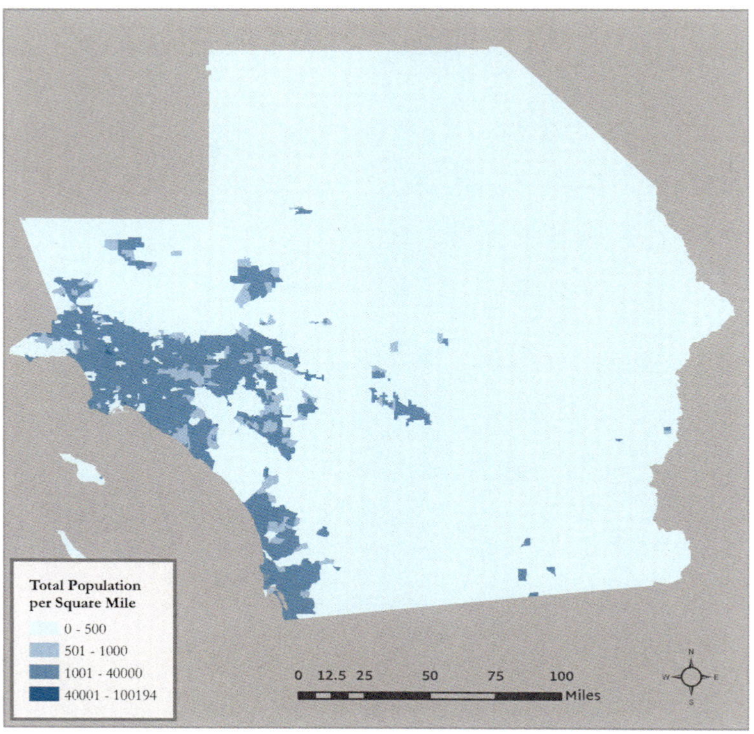

Figure 1.26: This map displays the 2010 population of Southern California using manual classification.

As one can see, the method of data classification can have a significant impact on the message the map communicates (See **Figures 1.21, 1.23, 1.25, and 1.26**). Always take the time to consider which method of classification that suits the data distribution and phenomenon best.

Accuracy and Precision

Confusing the term accuracy with the term precision is common. **Accuracy** is the degree to which information on a map matches true or accepted values. It is a measure of truth or correctness. **Precision** is the exactness of measurement. It represents the level of detail, the level of measurement, the level of repeatability, or the meticulousness of description. Accuracy sometimes improves with precision, but a high degree of precision should not be an automatic indicator of accuracy.

When working with geospatial data, one may need to make a comparison between a trusted dataset known to be accurate and the data in question. Ideally, the trusted dataset should be from a different source or organization.

For example, the California coast is not a straight line. There are vast stretches of the California coast made up of many sinuous inlets and bays. If one were able to measure the length of each of these using a yardstick, they would find the coastline was a great deal longer. Even if one had the resources to measure the coastline foot by foot or even inch by inch, they would still not have a measurement of the actual length. There is always a smaller feature that can be measured around.

What one would find is that the smaller (more precise) unit of measurement one used, the longer the length of the coast. In a way, a coastline is infinitely long. Many refer to this as the coastline paradox, an observation made by the famous mathematician Lewis Fry Richardson. The coastline paradox serves to demonstrate that geographic space, for all practical purposes, is a continuously unbroken phenomenon. One's ability to measure an absolute location in space is limited only by one's methods and technology.

Uncertainty and Error

One of the most poorly understood qualities about geospatial data is the inherent uncertainty and error that is always present. The uncertainty comes from the very nature of geospatial data. Geospatial data is incomplete, generalized, and selective. The error is the degree of uncertainty that one can justifiably measure. Understanding uncertainty and error is key to making geospatial data useful. Like public opinion polls, if the information is understood, gathered, and applied correctly, it can have highly accurate results.

There are two common sources of error that occur during the production or use of geospatial data.

» Human error
» Spatial Error

The easiest to understand and account for is **human error**. Every measurement, record entry, or procedure is subject to human error. Even when scientists operate in a controlled environment, accounting for human error is always a consideration. The fact is, geospatial data is rarely if ever, produced in a controlled environment. Thus, the probability of human error is a significant consideration. Luckily, for many decision-making purposes, minor errors do not have substantial impacts. Because one expects a human error, it often becomes easier to identify. However, human error is sometimes deliberate. Personal or political bias may motivate deliberate errors. Some errors result from methods of classification or develop as a consequence of cartographic representation.

Spatial error, sometimes called **positional error**, is a far more insidious error source. It introduces error in many ways that can go unnoticed by the amateur mapmaker. As discussed previously, geographic space is nearly spherical. A transformation needs to occur to make use of geographic data. The spherical three-dimensional model of Earth transforms into a flat, two-dimensional rectangular map. This process leads to significant distortions in some properties such as area, distance, direction, and shape.

There are methods for minimizing spatial error. However, many novice mapmakers frequently misunderstand or ignore the spatial error. Understanding why and where the spatial error occurs and how it affects the data is one of the primary differences between a trained geospatial professional and someone who only knows how to operate a software package. **Chapter 3** covers this transformation process and the sources of spatial error in more detail.

Accounting for Error and Uncertainty

There are many ways to communicate error and uncertainty. One method includes a mathematical formula called the root mean square error (RMSE). To obtain the root mean square error, one starts with a survey of a small sample of points. Then one compares the surveyed points and tests them against the map. The degree of spatial error is determined mathematically using the RMSE formula (**Figure 1.27**).

$$RMSE = \sqrt{\sum \left(\frac{\left({}^{Y}map\ position - {}^{Y}measured\ position \right)^2}{N} \right)}$$

Figure 1.27: The root mean square error formula is one way to measure error and uncertainty.

This formula calculates the square root of the sum of mean errors. The mean error is the averaged difference between the positions of data on the map and the actual position of data determined by measurements using highly accurate surveying methods. The difference is divided by the total number of errors (N) to get an average error value. The root mean square error provides geospatial scientists with a standardized mechanism for evaluating the degree of spatial error.

Another way to communicate uncertainty and error is through the use of metadata. **Metadata** refers to data about data. The documentation about the source of data, methods of data collection, times of data collection, and data authorship are all part of metadata. Thorough and well-documented metadata can be a valuable resource for any company, agency, or government.

Uncertainty and error can also be communicated using standardized **quality assurance and quality control (QAQC)**procedures. These are standardized procedures established by specific organizations to assess error and uncertainty, completeness, and overall quality of datasets. Conventional methods include documentation of existing metadata quality, inspecting spatial accuracy and resolution, and providing related notes or comments about the data to coworkers and colleagues. Always perform QAQC procedures for both attribute data and spatial data at the start of any project during the data collection phase.

The following are some common questions to ask about the data for QAQC documentation:

» How would one describe the quality of the metadata summary and description? (e.g., detailed, sparse, non-existent)
» Does the metadata provide information about who created the data? If so, who is the author of the data?
» Does the metadata describe the way the data was collected or assembled? If so, what was it?
» Are there any use limitations? If so, what are they?
» What type of spatial reference does the data have? Geographic or Projected?
» Upon visual inspection, how precise does the data appear?
» When compared to a known or trusted dataset, how accurate does the data appear?

Tutorial: Managing Geospatial Data Using ArcGIS

The goal of this activity is for readers to learn how to acquire data from a public source and how to practice essential file management as it relates to geospatial data. In this exercise, you download GIS data from the Humboldt County website. You create and organize a workspace folder using a standardized folder structure. You then download and decompress the data. The files included four shapefiles representing roads, fire hydrants, the Humboldt County boundary, and election precincts. Once you are done managing and arranging your data using Esri ArcGIS software, you compress your entire workspace folder with the data intact.

ESTIMATED TIME TO COMPLETE THIS TUTORIAL: 4 HOURS

Learning Outcomes

Readers should be able to accomplish the following outcomes by the end of this tutorial:

- » Summarize the steps for creating and organizing a project workspace folder structure
- » Describe best file management practices for transferring geospatial data
- » Demonstrate the ability to download data from a public source
- » Practice decompressing zipped files into a specific folder
- » Compare workflows for geospatial data in Microsoft Windows File Explorer and Esri ArcGIS
- » Examine data using ArcGIS software
- » Identify the main features of the ArcMap user interface
- » Locate and add additional toolbars in ArcMap
- » Practice looking up properties for both a data frame and a shapefile in ArcMap
- » Apply direct selections to map features in a data frame
- » Demonstrate the relationship between the map in the data frame and records in the attribute table
- » Repair data sources in ArcMap
- » Practice compressing a project level folder with contents intact

Setting up Your Workspace

In a typical workflow, you work on geospatial data using a local hard drive. When done, you compress your data and back up your work to your cloud storage so that you can retrieve the files from anywhere. When referring to a **local hard drive**, it means you are working on data physically located on the computer in front of you.

In contrast, some computers also include networked drives. **Networked drives** link to cloud storage and save the data elsewhere. Examples include services like OneDrive or Google Drive. For this tutorial, use the **desktop** as your local hard drive location. You may also use an external USB drive if you plan to work in multiple places.

You must avoid using networked drives while you work. They increase the processing time and can cause technical glitches.

In this tutorial, you use a particular folder structure. Start by creating your **workspace folder** on the local hard drive. A **workspace** is a folder or series of folders that contain all of your project files. The top-level folder in your workspace should indicate the lab assignment or the project. Organize all of your work within the workspace folder. On your desktop, create a new folder and give it a descriptive name, such as *GSP101_Activity1*. Be sure there are *no spaces*. You may use underscores instead of spaces. Inside this folder, create the following three subfolders: original, working, and final. Having a standardized folder structure helps to keep a project organized, primarily when you are working with multiple partners. The folder structure you see here (**Figure 1.28**) is the standard used in each of the tutorials presented in this book.

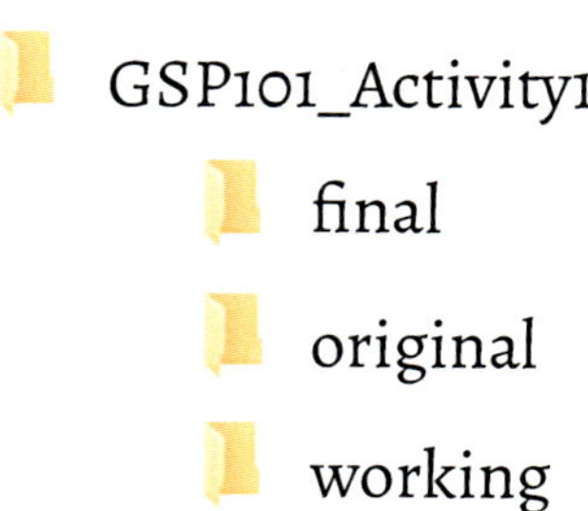

Figure 1.28: This diagram represents a basic folder structure used in this book.

As the name indicates, use the **original folder** for storing original, unaltered data. As you are working on a project, if for some reason your working version of the data gets lost or corrupted, you can go back to your *original folder* and find a fresh copy of the data. Use the **working folder** for data that you create or alter while working on your project.

Use the **final folder** for storing any output you produce as a result of your work such as images, maps, tables, or reports. In this activity, you primarily use the *original folder* and the *working folder*. However, setting up the standard folder structure for a project is good practice and a habit you want to develop.

Downloading Data from the Humboldt County Website

Humboldt County's Planning Division collects and develops GIS data and software to support the *General* Plan Update. The goals of the division have expanded to meet the GIS needs of federal, state, and local agencies. This data is readily available for free download through the ***Humboldt County GIS Data Download***[1] web page.

Open the Chrome browser and use the link above to navigate to the official Humboldt County GIS download page.

If you do not have the Chrome browser, you can download it here[2].

Take a moment to read through the disclaimer about the data provided on this website (**Figure 1.29**). This section gives you some information about uncertainty and error when using this data.

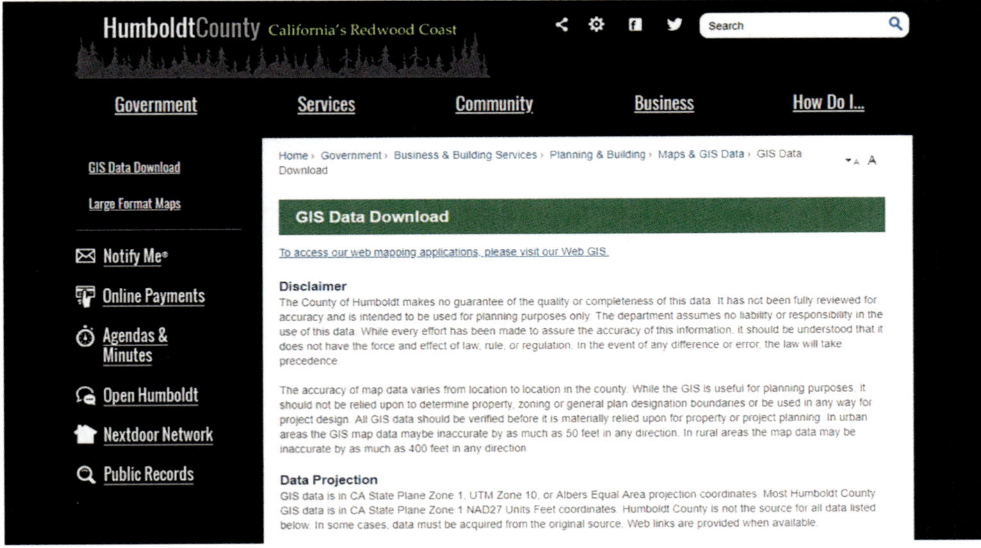

Figure 1.29: The Humboldt County website provides free GIS data related to Humboldt County.

---

1 http://humboldtgov.org/276/GIS-Data-Download

2 https://www.google.com/chrome/browser/desktop/index.html

There is also some further information about data projections and spatial reference information. You don't have to worry about understanding this now. **Chapter 3** covers projections and spatial reference systems in detail. There are additional notes, including information about the metadata and file compression format. Take a moment to read these as well.

You download four shapefiles from this website, a road layer, a fire hydrant layer, a county boundary layer, and a layer for the election precincts in Humboldt County. One of these layers, the roads, is located near the top under Frequently Requested Data Sets. Right-click on the link that says "Roadway Centerline Shapefile (ZIP)" (**Figure 1.30**). Select *Save link as*, then navigate to your *original folder*. Save the compressed file inside your *original folder*.

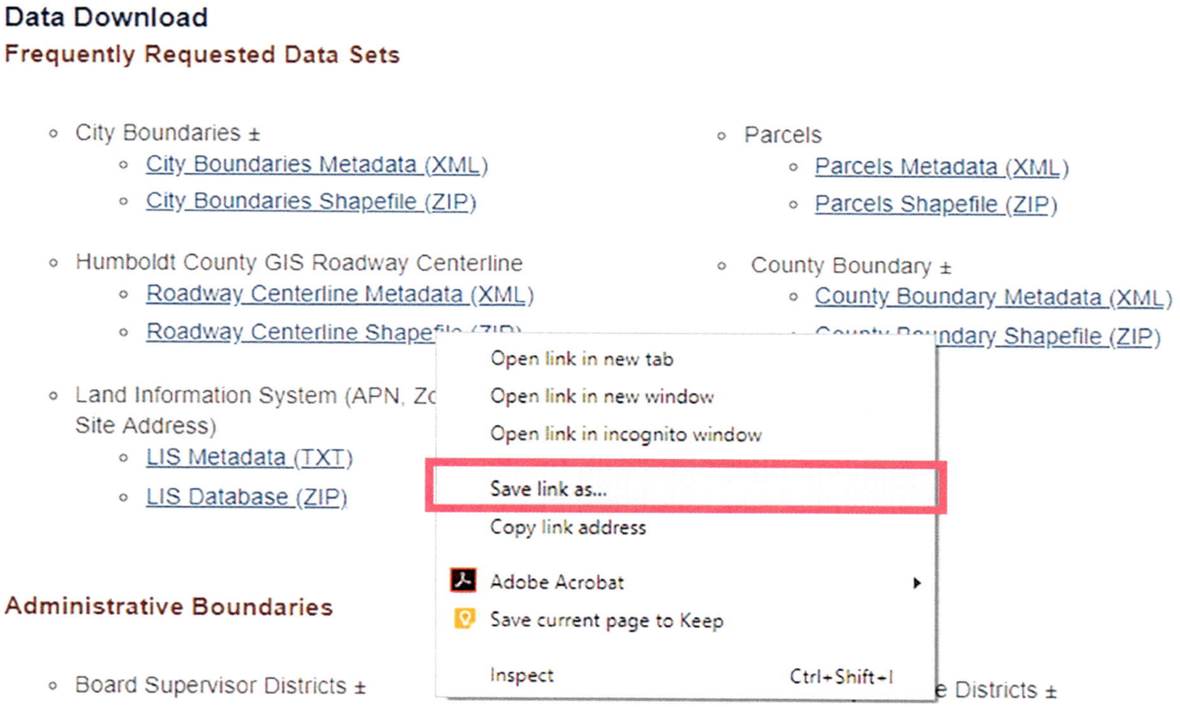

Figure 1.30 In Chrome, you can right-click on a link to save a downloaded file to a specific location.

Scroll down until you see the data located under administrative boundaries. Repeat these steps for the Humboldt County boundary shapefile and the Election Precincts shapefile. Be sure to save the files to your *original folder* (**Figure 1.31**).

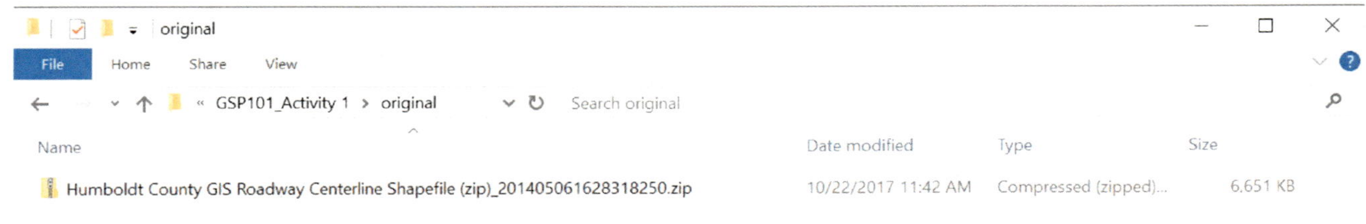

Figure 1.31: Be sure to save the compressed data to your *original folder* and not a different location.

Once you are ready, scroll down further and download the Fire Hydrants shapefile located under fire plan data.

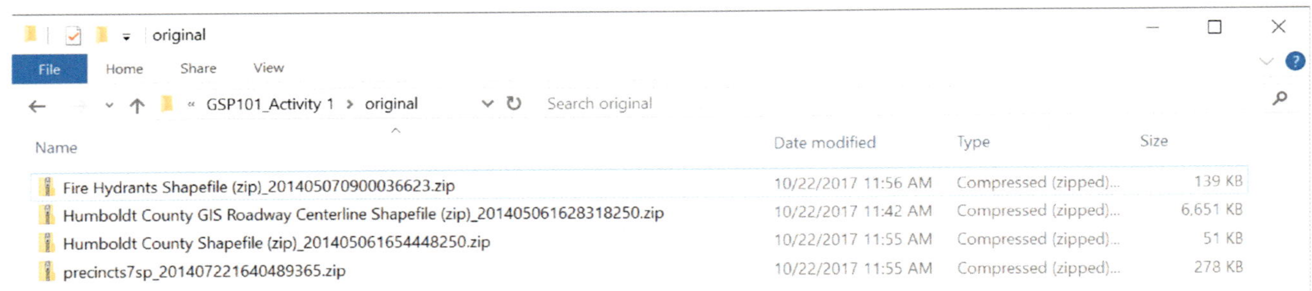

Figure 1.32: When done, you should have four compressed shapefiles located in your *original folder*.

Decompressing the files using 7zip

In this tutorial, you are required to use *7zip*[3], a file decompression software that handles multiple file compression formats commonly used for geospatial data. If you are working from home, you may need to install 7zip on your personal computer. Watch *Installing 7zip on Windows 10*, produced by Nicolas R. Malloy, to learn more (**Figure 1.33**).

Figure 1.33: Watch the video above to learn how to install 7zip on your personal computer. URL: *https://youtu.be/VnezWqRxwRg*

3 http://www.7-zip.org/

Navigate to your *original folder*. Right-click on the Fire Hydrants zip file, select 7zip, then *extract here* (**Figure 1.34**).

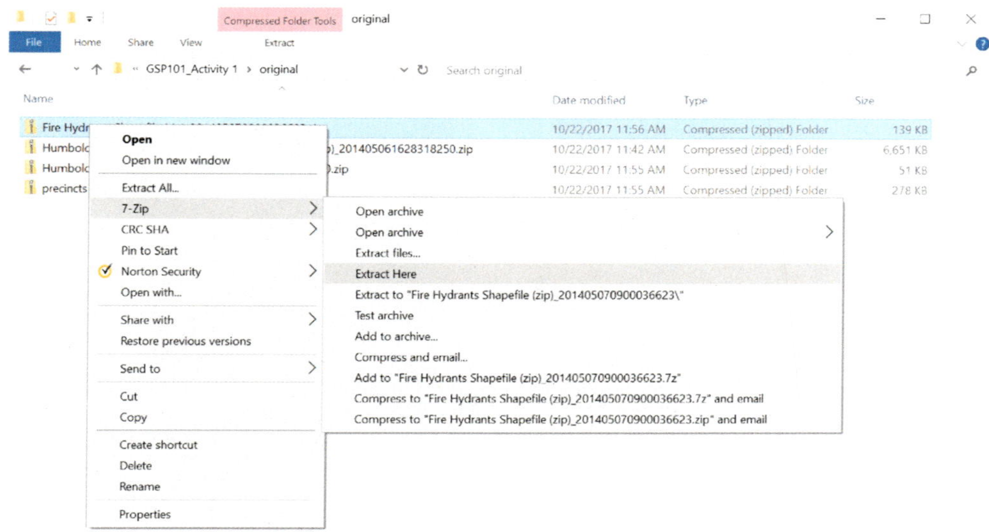

Figure 1.34: You can access the 7zip software by right-clicking on a compressed file.

The fire hydrant zip file decompresses, and the contents appear in the *original folder* (**Figure 1.35**). Once the shapefile decompresses, delete the fire hydrant zip file from your *original folder*. You won't need it anymore. Removing it saves space and helps to prevent confusion later. Repeat these steps for each of the compressed shapefiles. Be sure to delete the zip files once you are done decompressing them (**Figure 1.36**).

original

File Home Share View

GSP101_Activity 1 › original › Search original

Name	Date modified	Type	Size
Fire Hydrants Shapefile (zip)_201405070900036623.zip	10/22/2017 11:56 AM	Compressed (zipped) Folder	139 KB
Humboldt County GIS Roadway Centerline Shapefile (zip)_201405061628318250.zip	10/22/2017 11:42 AM	Compressed (zipped) Folder	6,651 KB
Humboldt County Shapefile (zip)_201405061654448250.zip	10/22/2017 11:55 AM	Compressed (zipped) Folder	51 KB
hydrant8_sp.dbf	6/8/2005 2:29 PM	DBF File	1,235 KB
hydrant8_sp.met	6/8/2005 2:47 PM	MET File	2 KB
hydrant8_sp.prj	6/8/2005 3:07 PM	PRJ File	1 KB
hydrant8_sp.sbn	6/8/2005 3:07 PM	SBN File	29 KB
hydrant8_sp.sbx	6/8/2005 3:07 PM	Adobe Illustrator Tsume File	1 KB
hydrant8_sp.shp	6/8/2005 3:07 PM	SHP File	84 KB
hydrant8_sp.shx	6/8/2005 3:07 PM	SHX File	25 KB
precincts7sp_201407221640489365.zip	10/22/2017 11:55 AM	Compressed (zipped) Folder	278 KB

Figure 1.35: As you can see, many parts make a shapefile. It must use these files together to work correctly.

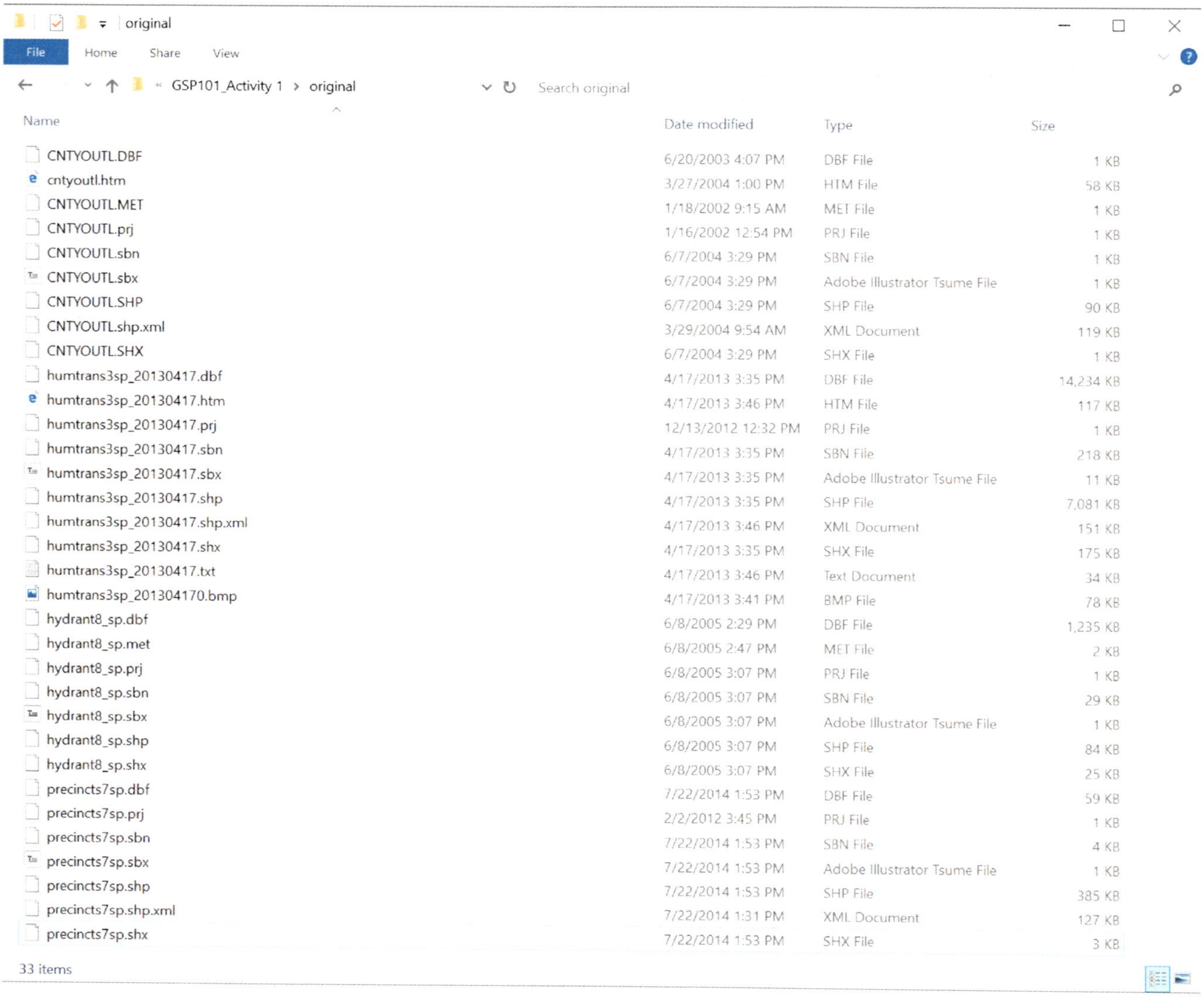

original

File Home Share View

GSP101_Activity 1 > original   Search original

Name	Date modified	Type	Size
CNTYOUTL.DBF	6/20/2003 4:07 PM	DBF File	1 KB
cntyoutl.htm	3/27/2004 1:00 PM	HTM File	58 KB
CNTYOUTL.MET	1/18/2002 9:15 AM	MET File	1 KB
CNTYOUTL.prj	1/16/2002 12:54 PM	PRJ File	1 KB
CNTYOUTL.sbn	6/7/2004 3:29 PM	SBN File	1 KB
CNTYOUTL.sbx	6/7/2004 3:29 PM	Adobe Illustrator Tsume File	1 KB
CNTYOUTL.SHP	6/7/2004 3:29 PM	SHP File	90 KB
CNTYOUTL.shp.xml	3/29/2004 9:54 AM	XML Document	119 KB
CNTYOUTL.SHX	6/7/2004 3:29 PM	SHX File	1 KB
humtrans3sp_20130417.dbf	4/17/2013 3:35 PM	DBF File	14,234 KB
humtrans3sp_20130417.htm	4/17/2013 3:46 PM	HTM File	117 KB
humtrans3sp_20130417.prj	12/13/2012 12:32 PM	PRJ File	1 KB
humtrans3sp_20130417.sbn	4/17/2013 3:35 PM	SBN File	218 KB
humtrans3sp_20130417.sbx	4/17/2013 3:35 PM	Adobe Illustrator Tsume File	11 KB
humtrans3sp_20130417.shp	4/17/2013 3:35 PM	SHP File	7,081 KB
humtrans3sp_20130417.shp.xml	4/17/2013 3:46 PM	XML Document	151 KB
humtrans3sp_20130417.shx	4/17/2013 3:35 PM	SHX File	175 KB
humtrans3sp_20130417.txt	4/17/2013 3:46 PM	Text Document	34 KB
humtrans3sp_201304170.bmp	4/17/2013 3:41 PM	BMP File	78 KB
hydrant8_sp.dbf	6/8/2005 2:29 PM	DBF File	1,235 KB
hydrant8_sp.met	6/8/2005 2:47 PM	MET File	2 KB
hydrant8_sp.prj	6/8/2005 3:07 PM	PRJ File	1 KB
hydrant8_sp.sbn	6/8/2005 3:07 PM	SBN File	29 KB
hydrant8_sp.sbx	6/8/2005 3:07 PM	Adobe Illustrator Tsume File	1 KB
hydrant8_sp.shp	6/8/2005 3:07 PM	SHP File	84 KB
hydrant8_sp.shx	6/8/2005 3:07 PM	SHX File	25 KB
precincts7sp.dbf	7/22/2014 1:53 PM	DBF File	59 KB
precincts7sp.prj	2/2/2012 3:45 PM	PRJ File	1 KB
precincts7sp.sbn	7/22/2014 1:53 PM	SBN File	4 KB
precincts7sp.sbx	7/22/2014 1:53 PM	Adobe Illustrator Tsume File	1 KB
precincts7sp.shp	7/22/2014 1:53 PM	SHP File	385 KB
precincts7sp.shp.xml	7/22/2014 1:31 PM	XML Document	127 KB
precincts7sp.shx	7/22/2014 1:53 PM	SHX File	3 KB

33 items

Figure 1. 36: When done, you should only have the components of the four shapefiles. Remove the zip files from the *original folder*.

Managing Data Using the Catalog Window

ArcMap is part of the ArcGIS software suite. In this tutorial, you primarily use ArcMap to manage data and create maps. Locate ArcMap on your computer and launch the software. If you are using Microsoft Windows 10, click the windows button and type *ArcMap* to find the desktop application (**Figure 1.37**). Launch the ArcMap software.

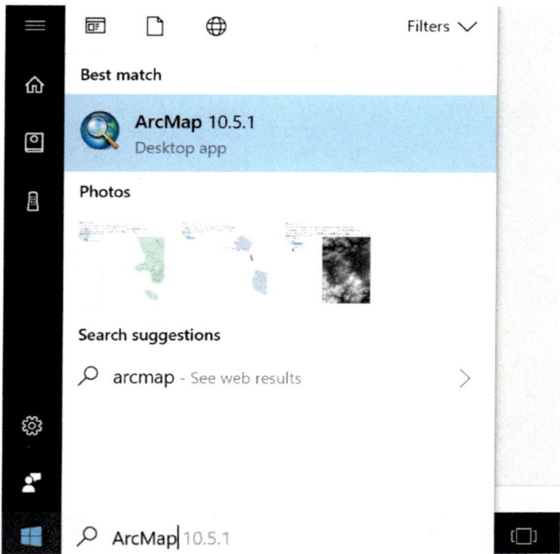

Figure 1.37: The version of ArcMap may vary over time.

When you first launch ArcMap, a window appears that gives you the option to either open a blank map or to open any of your recent map documents (**Figure 1.38**). Choose to open a blank map document.

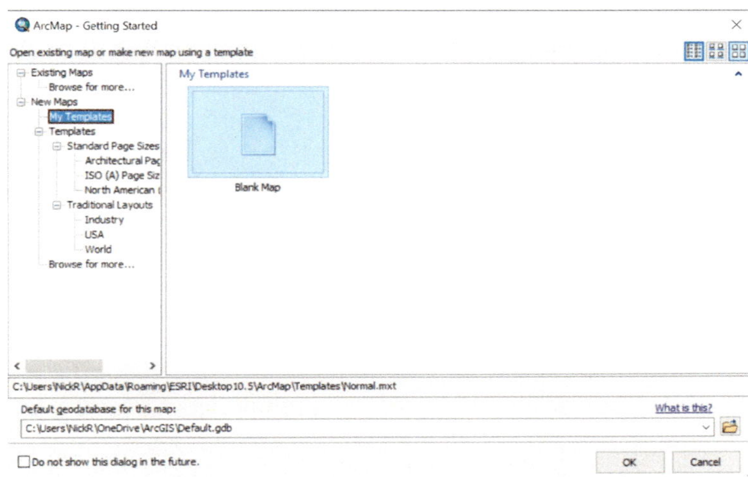

Figure 1.38: The ArcMap Getting Started window provides several options on startup.

The ArcMap user interface has three main windows that you use regularly, the *Table of Contents*, the data frame, and the Catalog Window (**Figure 1.39**). The **Table of Contents** displays a list of data frames and map layers loaded into the map documents. The **data frame**, sometimes spelled dataframe, represents the layers visually as a map and defines the map extent. The **Catalog Window** displays a hierarchical view of folder connections and data as a **Catalog Tree**.

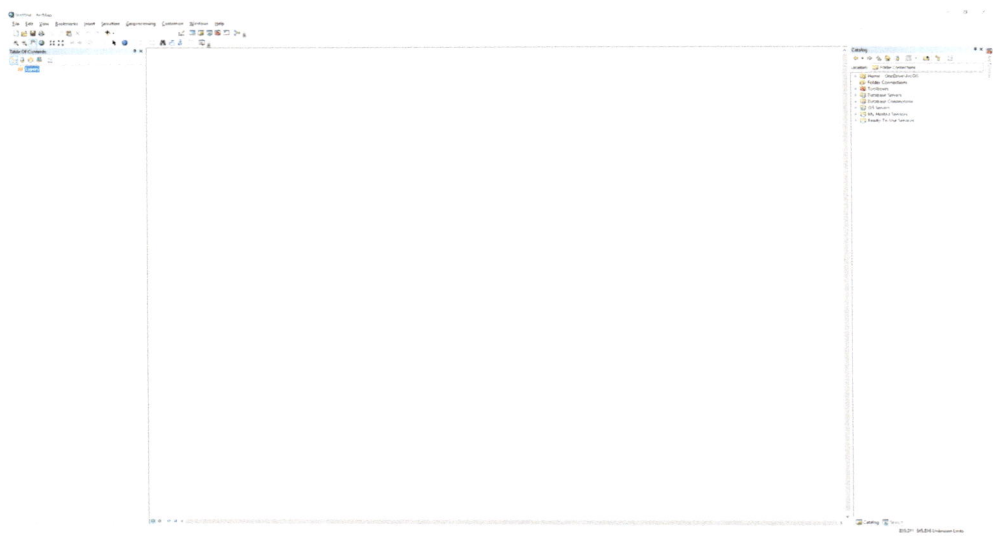

Figure 1.39: From left to right, the three main windows are the Table of Contents, the data frame, and the Catalog Window.

The icon for the catalog window looks like a yellow file cabinet. If your catalog window is missing, you can find the icon on the toolbar across the top. Click to open it.

Figure 1.40: Click the yellow file folder icon if your Catalog Window is missing.

In this step, you start by working with the Catalog Window. The catalog window is where you manage your geospatial data in ArcMap. Start by clicking on the Connect to Folder button (**Figure 1.41**). Navigate to your workspace folder on your local hard drive. In this step, it is essential to select the primary workspace folder, *GSP101_Activity1*, to add it to the Catalog Window (**Figure 1.42**). When ready, click *OK*.

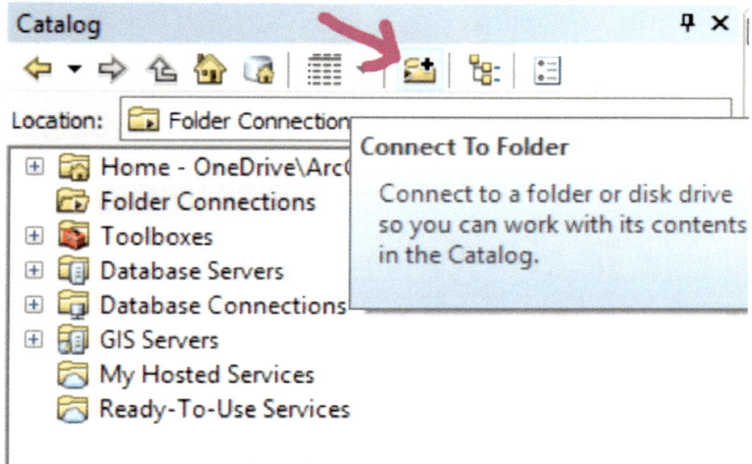

Figure 1.41: You connect to a folder so that the contents appear in the Catalog Window.

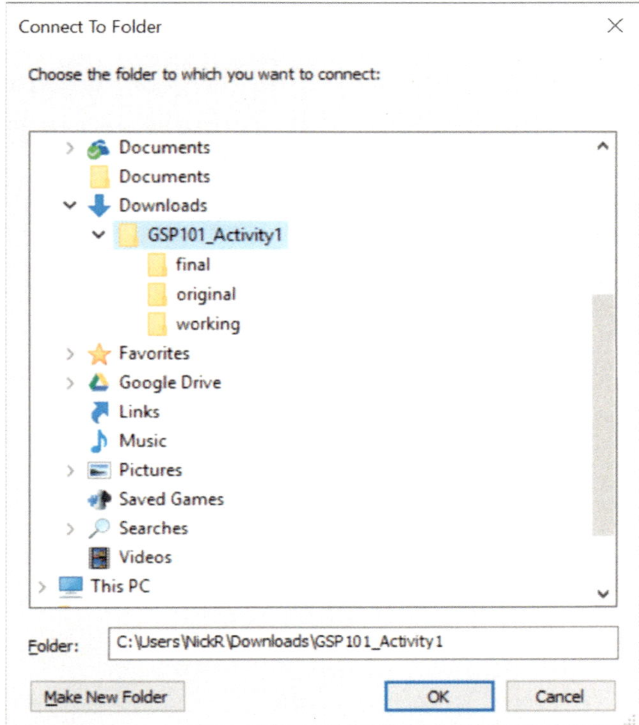

Figure 1.42: The Connect to Folder window displays a directory of folders on your computer. Individual files do not show up here.

Once you add your workspace folder to the Catalog Window, expand the folder by clicking on the plus sign. You should see your three subfolders inside (**Figure 1.43**). Expand the *original folder* to view the contents. This display of folders and files within the Catalog Window is sometimes called the **Catalog Tree**.

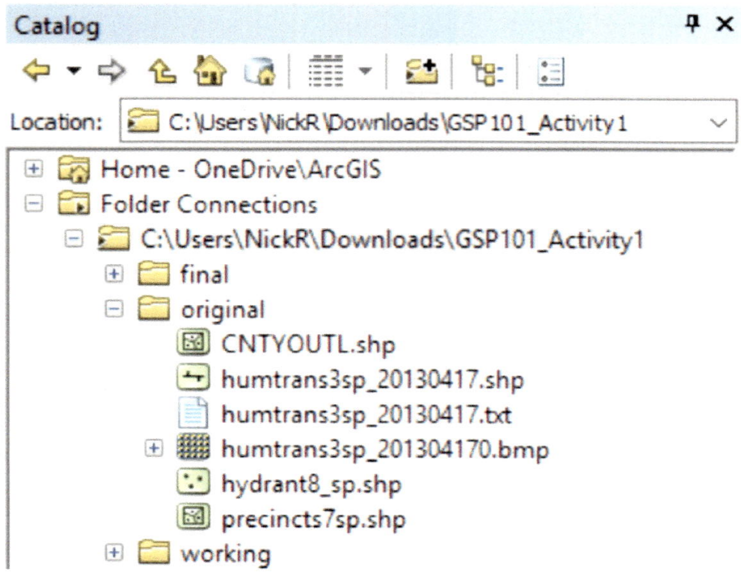

Figure 1.43: The contents of the workspace folder are displayed and organized in the Catalog Window. This display of folders and files is sometimes called the Catalog Tree.

Notice the difference between the appearance in Microsoft Windows File Explorer and the way the Catalog Window displays the data (**Figure 1.44**). The Catalog Window reduces the number of files that you see to make moving and copying data easier. However, it keeps the component files together when moving or copying data. You demonstrate this feature in the next step. In the Catalog Window, right click on the Humboldt County boundary shapefile. In this example, it is called CNTYOUTL.shp. Copy the file. Paste the file into the *working folder* (**Figure 1.45**). When done, you should see a copy of the Humboldt County boundary shapefile under the *working folder* in the Catalog Tree (**Figure 1.46**).

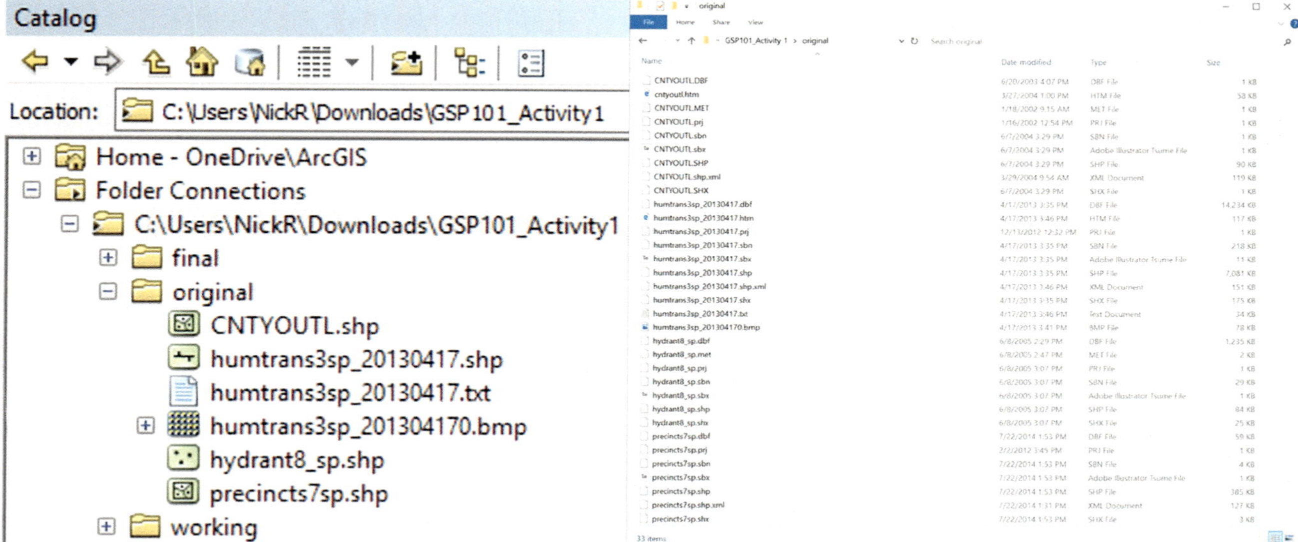

Figure 1.44: On the left, the image displays the contents of the *original folder* in the Catalog Window. On the right, the same folder is viewed using Microsoft Windows File Explorer.

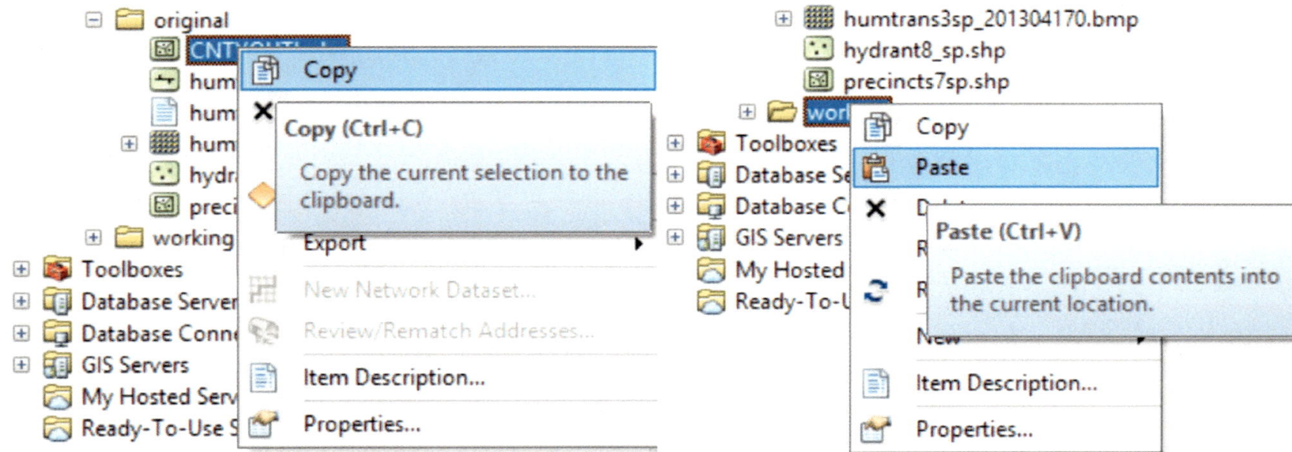

Figure 1.45: You can access contextual menus for copy, pasting, and renaming by right-clicking on a folder or file.

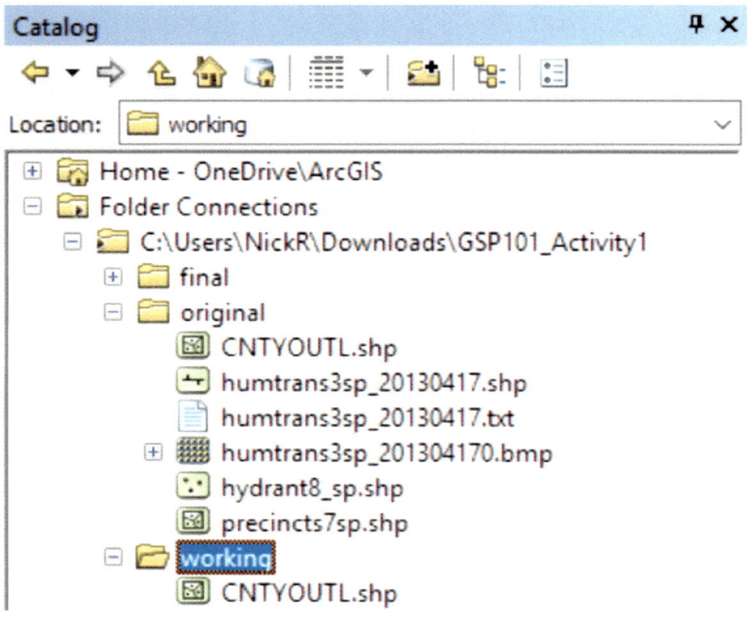

Figure 1.46: The Catalog Tree should now display the CNTYOUTL.shp file in both the *original folder* and the *working folder*.

Though it may appear that you copied and pasted only one file in the Catalog Window, the ArcGIS software also made copies of all the parts that make up a shapefile (**Figure 1.47**). To verify that this is true, open your *working folder* using Microsoft Windows File Explorer and examine the contents.

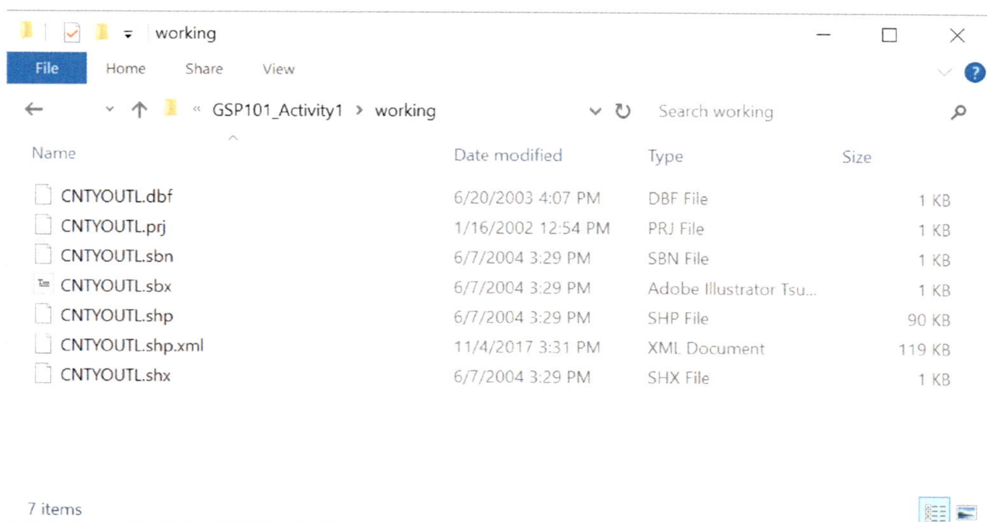

Figure 1.47: The *working folder* now contains all of the parts that make up the Humboldt County boundary shapefile.

Remember! When working with shapefiles, never delete any of the component files, or it breaks.

The Catalog Window also makes it easy to edit geospatial data. To demonstrate this feature, return to ArcMap and rename the shapefile in the *working folder*. Right click on CTNYOUTL.shp and select Rename (**Figure 1.48**). Change the name to "Humboldt_County_boundary." Once again, examine the changes in the *working folder* using the Microsoft Windows File Explorer.

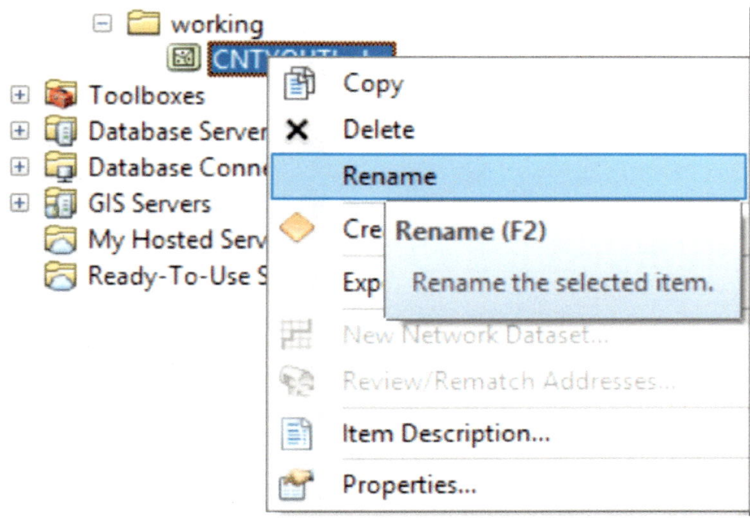

Figure 1.48: You can access contextual menus for copy, pasting, and renaming by right-clicking on a folder or file.

As you can see, the change to the name of the shapefile in the Catalog Window extends to the component files (**Figure 1.49**).

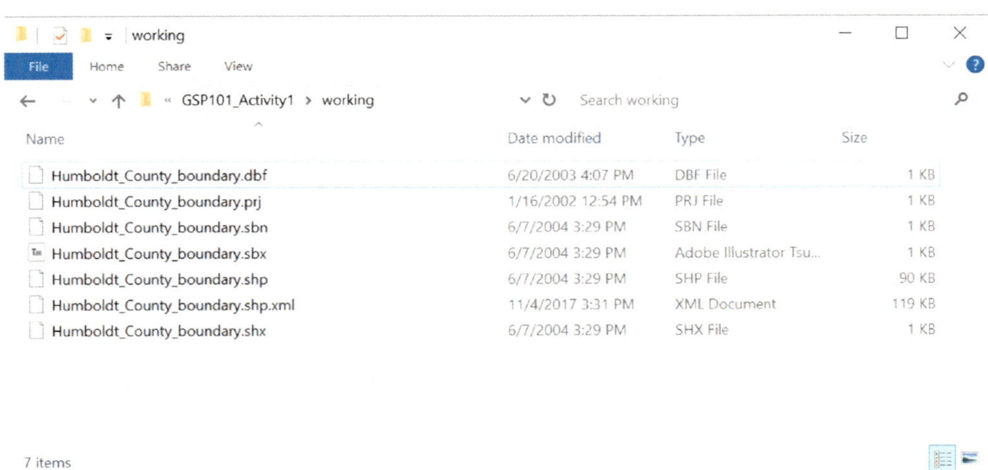

Figure 1.49: The name in each of the parts of the shapefile changed.

Imagine that someone assigned you to manage dozens or even hundreds of shapefiles. If you tried to rename or transfer individual files using Microsoft Windows File

Explorer, the task would be both time-consuming and prone to errors. The advantage of using GIS software to manage geospatial data is that it can save time and reduce the possibility of corrupting the data.

Practice what you have learned. Copy each of the remaining shapefiles from the *original folder* to the *working folder* using the Catalog Window. The shapefiles each have an icon that looks like a green square (**Figure 1.50**). You can ignore the rest of the files for now. Give the copies located in the *working folder* more human-friendly names such as "roads," "election_precincts," and "fire_hydrants."

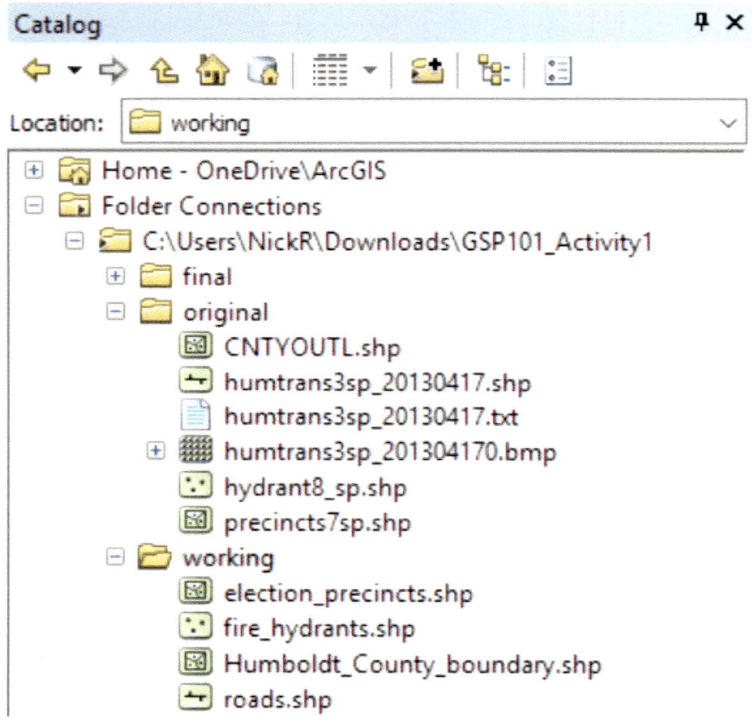

Figure 1.50: The icon for shapefiles looks like a green square. The details vary depending on whether the shapefile is a point feature, a linear feature, or a polygon feature.

Remember! Avoid using blank spaces in the file names. You can use an underscore instead of empty spaces. Blank spaces can cause some tools in ArcMap to stop working.

Inspecting the Metadata

Each shapefile can store metadata. **Metadata** refers to data about data. The documentation about the source of data, methods of data collection, times of data collection, and data authorship are all part of metadata. Thorough and well-documented metadata can be a valuable resource for any company, agency, or government. As a GIS analyst, inspecting the metadata should be the first step you take when beginning a project as part of **quality assurance and quality control (QAQC)**.

Before inspecting the metadata, you need to make sure ArcMap is set to read it correctly. There are different metadata formats available. In this example, you use the **ISO 19139 Metadata Implementation Specification**, a commonly used format. In ArcMap, locate the *Customize* menu at the top. Select *ArcMap Options* (**Figure 1.51**). When the ArcMap Options window opens, select the *Metadata* tab.

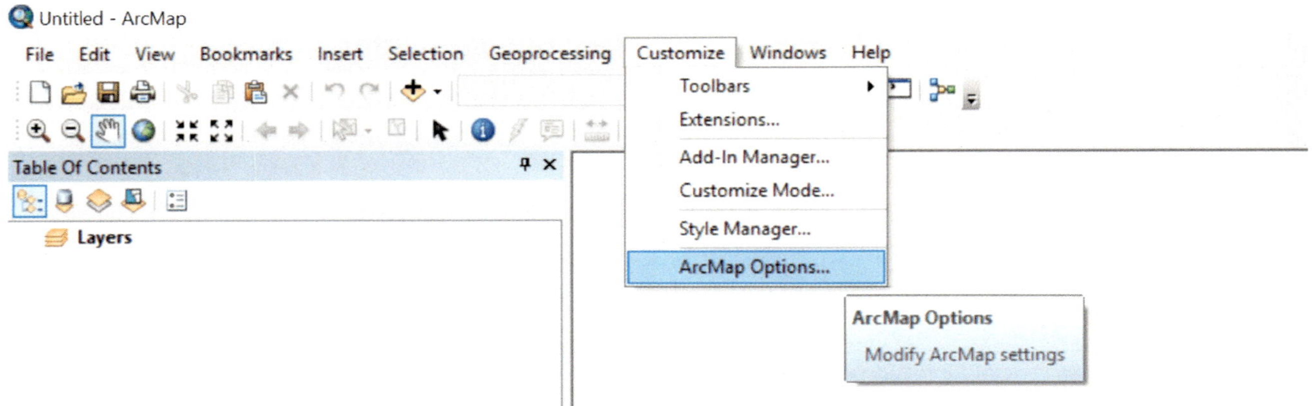

Figure 1.51: Locate the *Main menu* across the top of ArcMap.

Under the Metadata Style, choose the *ISO 19139 Metadata Implementation Specification* from the drop-down menu (**Figure 1.52**). Leave all other default settings and click *OK*. Return to the Catalog Window and right click on the roads shapefile located in the *original folder*. The file name starts with "humtrans." Select, *Item Description* to open the metadata stored in the shapefile (**Figure 1.53**).

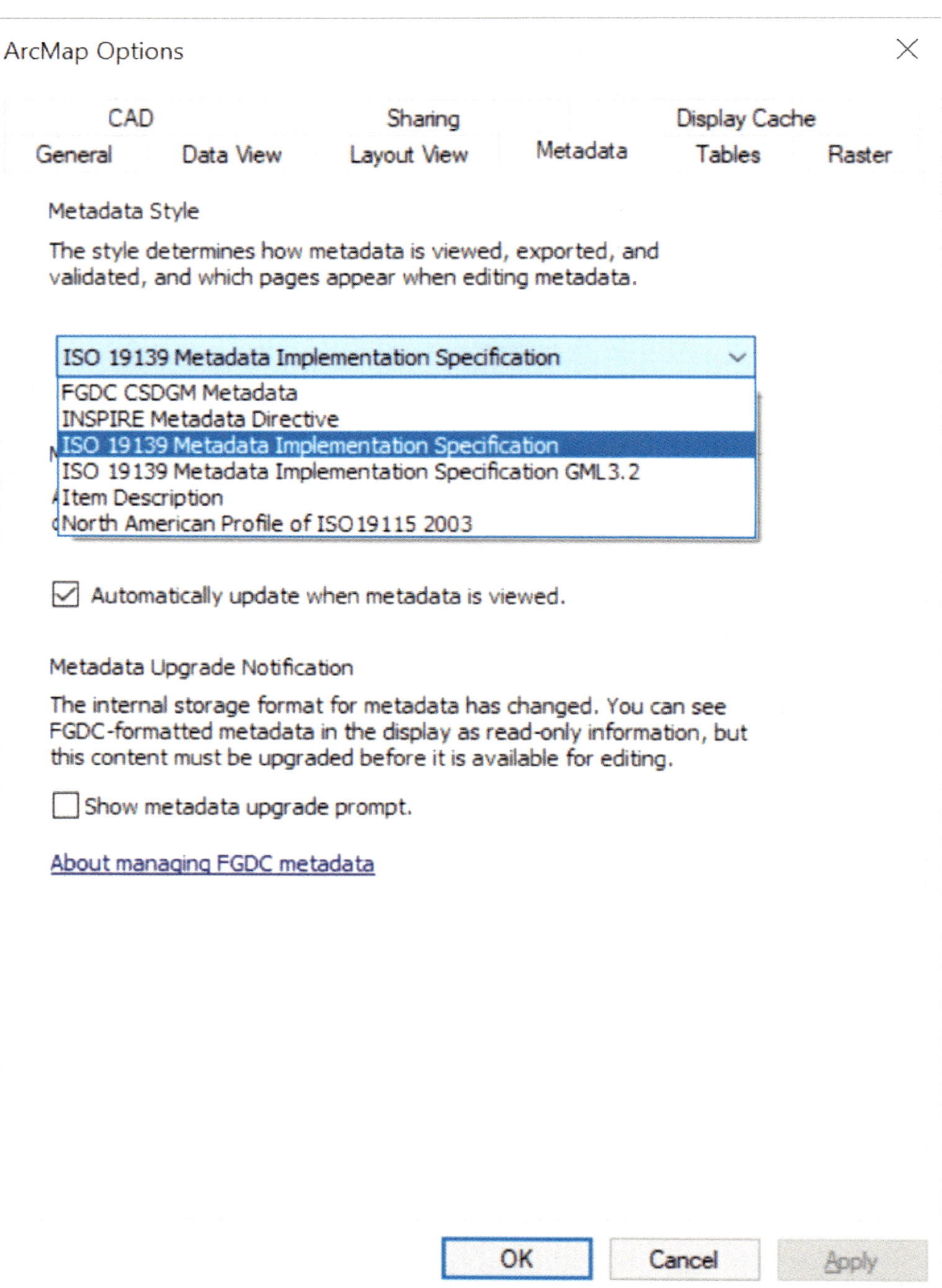

Figure 1.52: ArcMap Options allow you to change many of the global software settings.

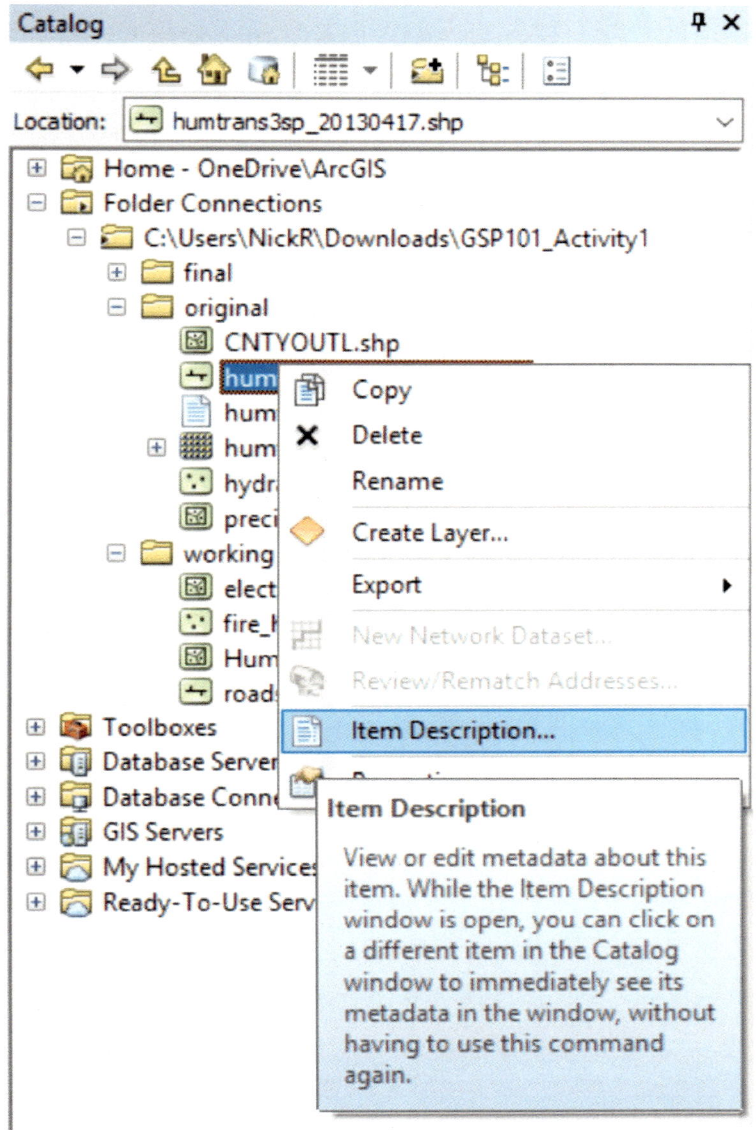

Figure 1.53: You can access contextual menus in the Catalog Window by right-clicking on files and folders.

As you can see, this shapefile includes some metadata such as a summary, a description, credits, and the use limitations (**Figure 1.54**). Take a moment to read these first four sections.

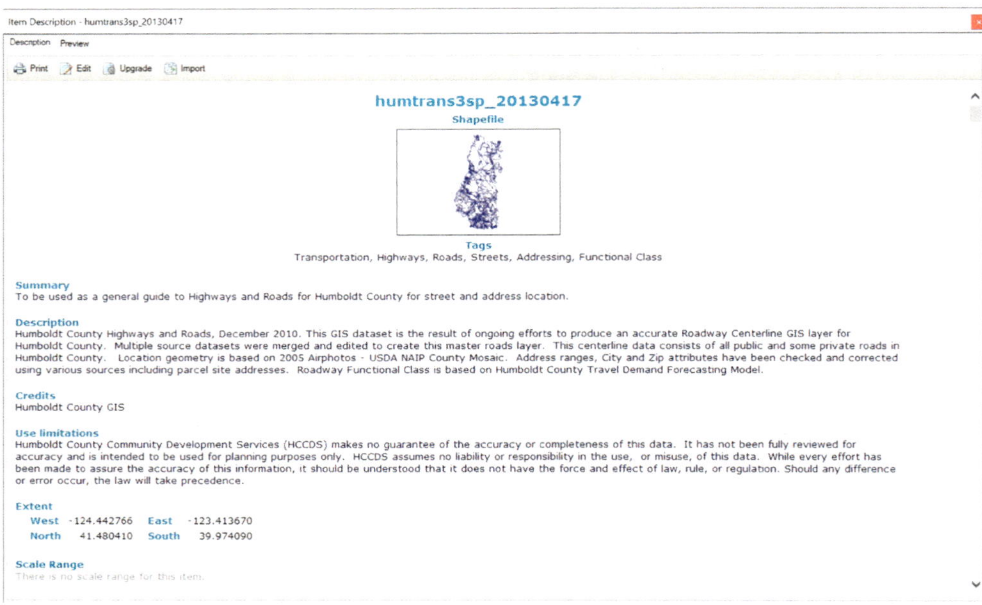

Figure 1.54: The Item Description window has two tabs, Description, and Preview. The Description tab displays the metadata.

Scroll down to see some additional information, such as the spatial reference. This information may not make much sense to you. However, as you learn more about geospatial science over time, you will begin to understand all the information presented here.

With the Item Description window open, click on the remaining shapefiles, one at a time. The Item Description window updates automatically. You may discover that some of the shapefiles do not store detailed metadata. For quality assurance and quality control purposes, the lack of metadata is also valuable information. It tells you that there is a potential for error and uncertainty. It might even indicate that the metadata is stored elsewhere, such as on a website or in a separate file.

Be cautious when using geospatial data which lacks any metadata documentation.

When you are ready, close the Item Description window.

Adding Data to ArcMap

There are several ways to add data to ArcMap. The most direct method is to drag and drop files from the Catalog Window. In this step, you add the Humboldt County boundary to the map document. From the Catalog Window, select the Humboldt County boundary shapefile from the *working folder*. Click and drag the file over the data frame, the interior window pane in ArcMap, then release. The Humboldt County Boundary appears in the data frame window (**Figure 1.55**). The Humboldt County boundary is an example of a polygon feature.

Figure 1.55: The Humboldt County Boundary layer is an example of a vector data model representing a polygon feature.

Another way to add data is to use the *Main menu*. From the menu across the top, select *File*, then *Add Data* (**Figure 1.56**). When the flyout menu appears, several options appear for adding data. In future activities, you may use each of these options. For now, select *Add Data* from the flyout menu.

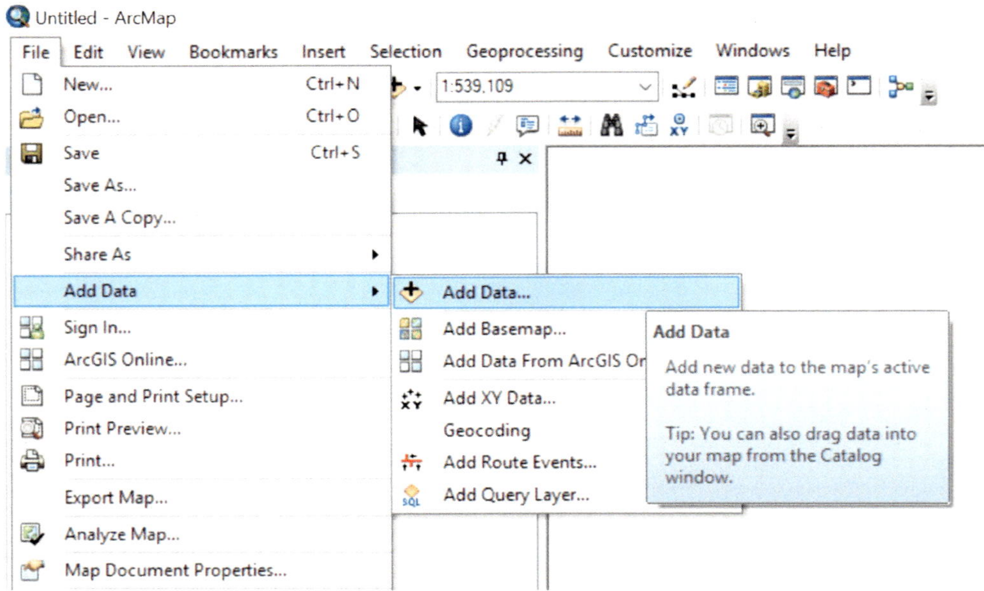

Figure 1.56: The File menu provides several options for adding data.

Using the drop-down menu, navigate to your *working folder* and add the roads shapefile (**Figure 1.57**).

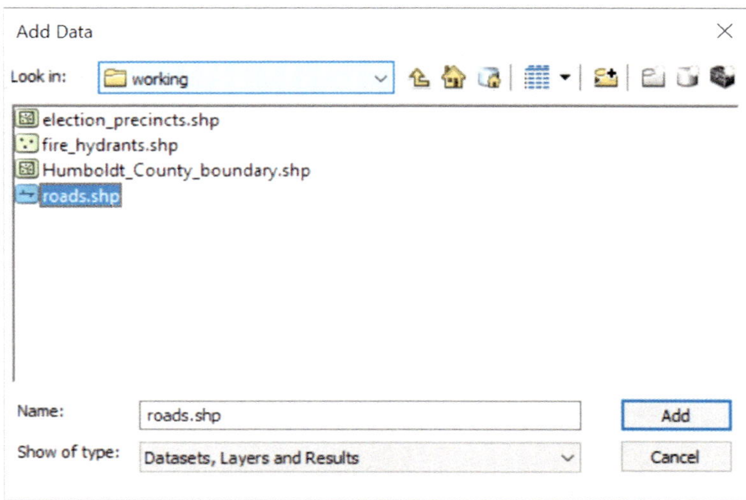

Figure 1.57: Be sure to add the data from the *working folder*.

The roads in Humboldt County appear in the data frame window (**Figure 1.58**). The roads are an example of a linear feature.

Figure 1.58: The roads layer is an example of a vector data model representing a linear feature.

The *Standard toolbar* also provides a means to add data. Unlike the *Main menu*, toolbars can be moved around, removed from view, and added back in. Currently, there are two active toolbars, the *Standard toolbar*, and the *Tools toolbar*. To demonstrate which one is which, select *Customize* from the *Main menu*. Then, select Toolbars. A long list of toolbars appears (**Figure 1.59**). Near the bottom of the list, a check mark appears next to *Standard* and *Tools*. Uncheck the *Standard toolbar*, and it disappears from ArcMap. Check it, and it appears again. In a typical workflow, you customize ArcMap to display the toolbars that you use the most. Under some circumstances, toolbars may become visible when you switch views or activate specific tools.

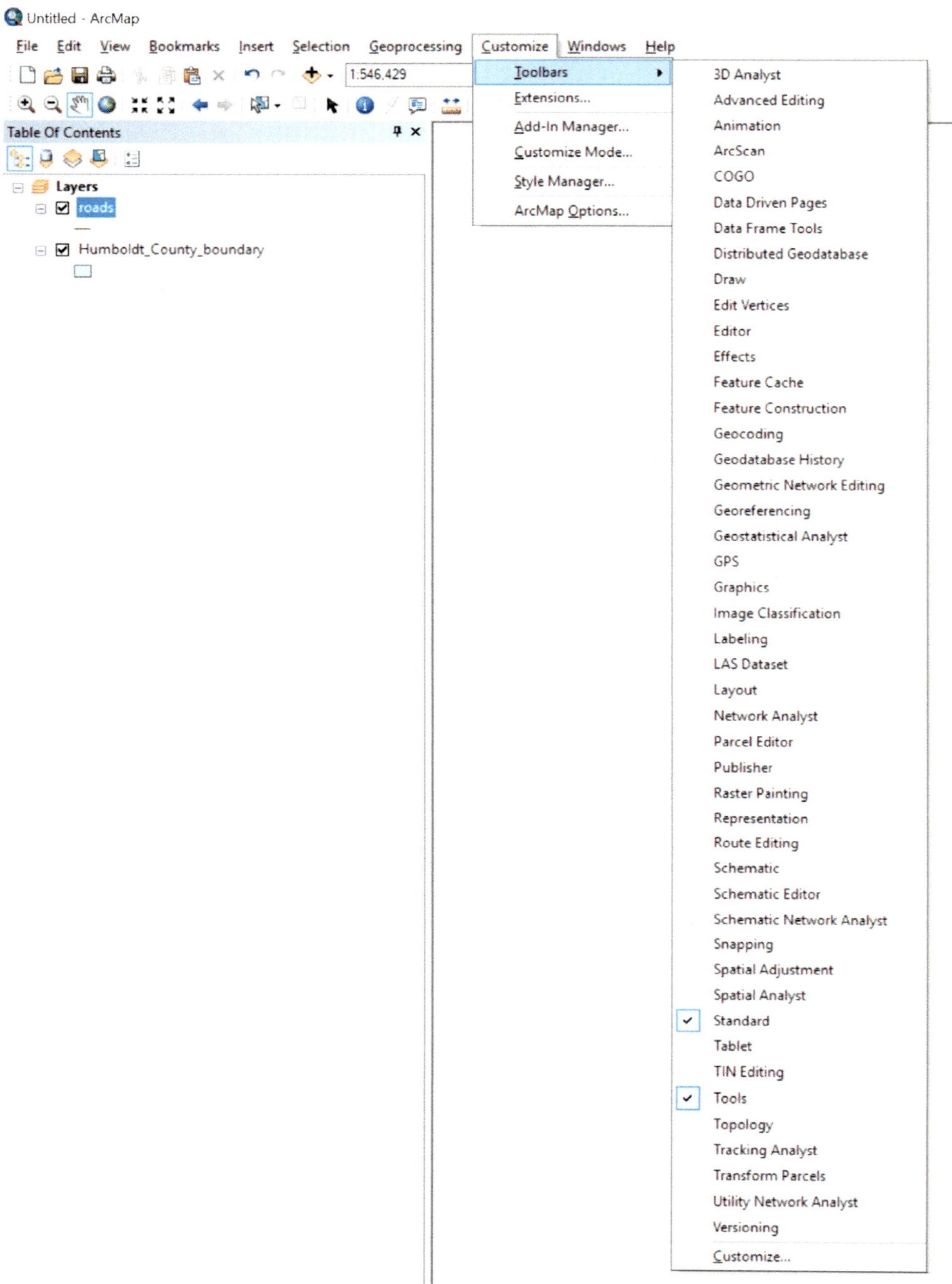

Figure 1.59: ArcMap provides a wide selection of toolbars from which to choose.

On the *Standard toolbar*, click the Add Data icon (**Figure 1.60**). It looks like a black plus sign over a yellow diamond.

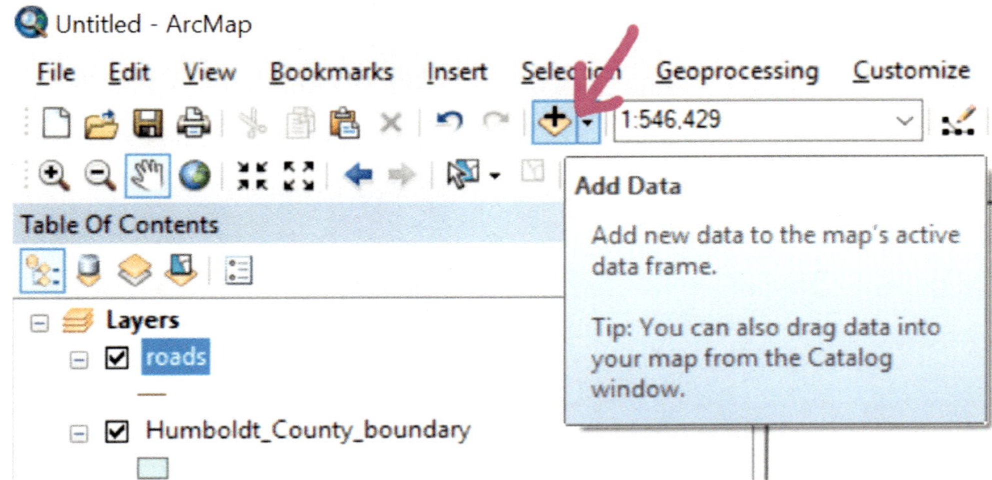

Figure 1.60: The *Standard toolbar* contains many commonly used tools.

Using the drop-down menu, navigate to your *working folder*. Add the fire hydrant shapefile. The location of the fire hydrants in Humboldt County appears in the data frame window (**Figure 1.61**). The fire hydrants are an example of a point feature.

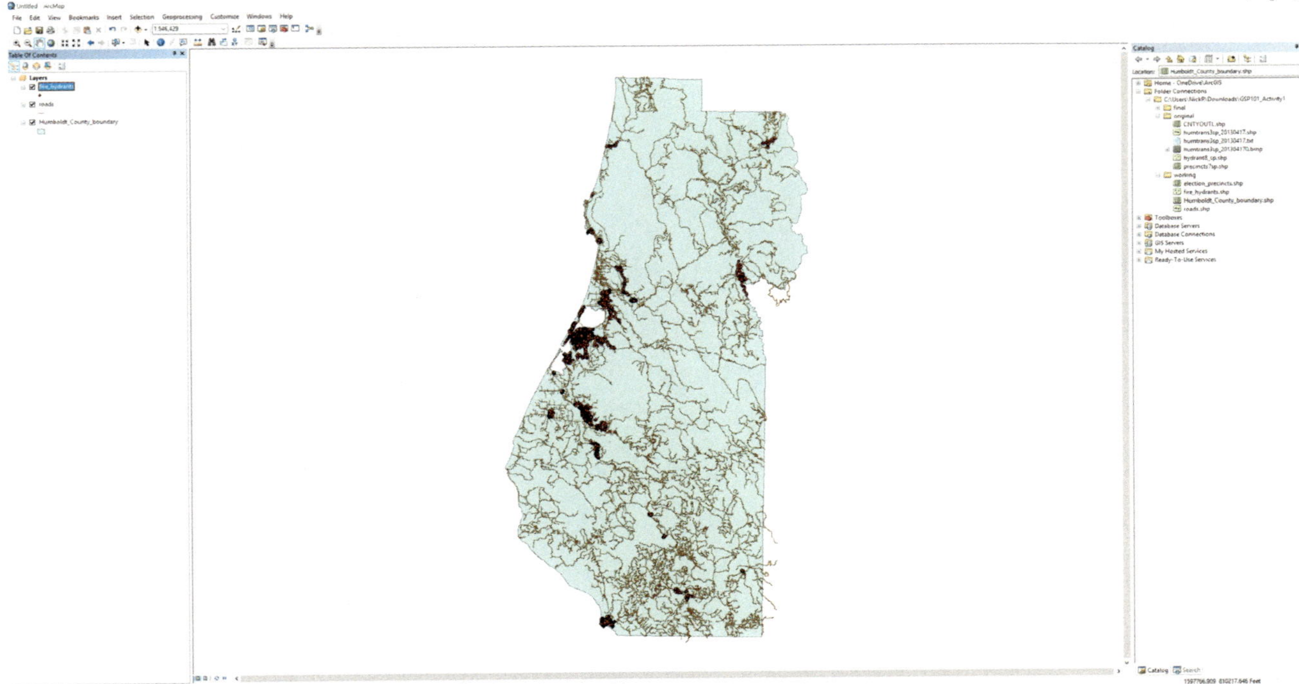

Figure 1.61: The fire hydrant layer is an example of a vector data model representing a point feature.

Exploring The ArcMap User Interface

In addition to the layers appearing as maps in the data frame window, the *Table of Contents* lists the layers on the left. There are several ways the *Table of Contents* displays the list of data. By default, the view setting for the *Table of Contents* is *List by Drawing Order*. In this view, the ArcMap draws the layers on the map in the order the layers get listed. To demonstrate this feature, drag the fire hydrants layer to the bottom of the table of contents (**Figure 1.62**). On the map, the fire hydrants disappear (**Figure 1.63**). You cannot see them because they get drawn under the roads and the Humboldt County Boundary. The Humboldt County boundary is an extensive polygon feature which covers up anything underneath.

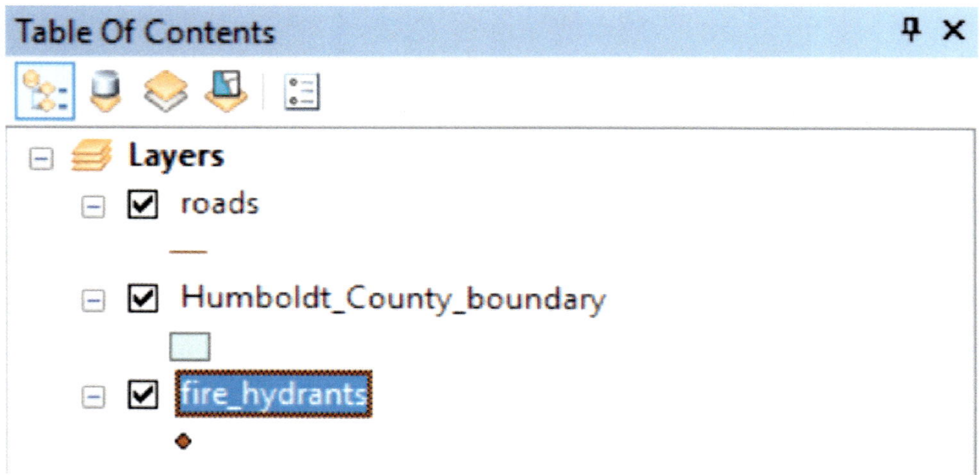

Figure 1.62: The order listed determines the order drawn on the map.

Drag the roads layer to the bottom of the *Table of Contents*. You should see the same thing happen to the roads as they become hidden by the Humboldt County Boundary, which is on top (**Figure 1.63**). Some of the roads extend beyond the Humboldt County Boundary. You might see some peeking out of the sides. Drag the Humboldt County Boundary back to the bottom of the table of contents. Another commonly used view in the *Table of Contents* is *List by Source*. Near the top of the *Table of Contents*, click the second icon from the right (**Figure 1.64**). It looks like a silver cylinder over a yellow diamond. The *Table of Contents* groups the individual layers by file location.

Figure 1.63: The order listed determines the order drawn on the map. The roads and fire hydrants appear under the County boundary.

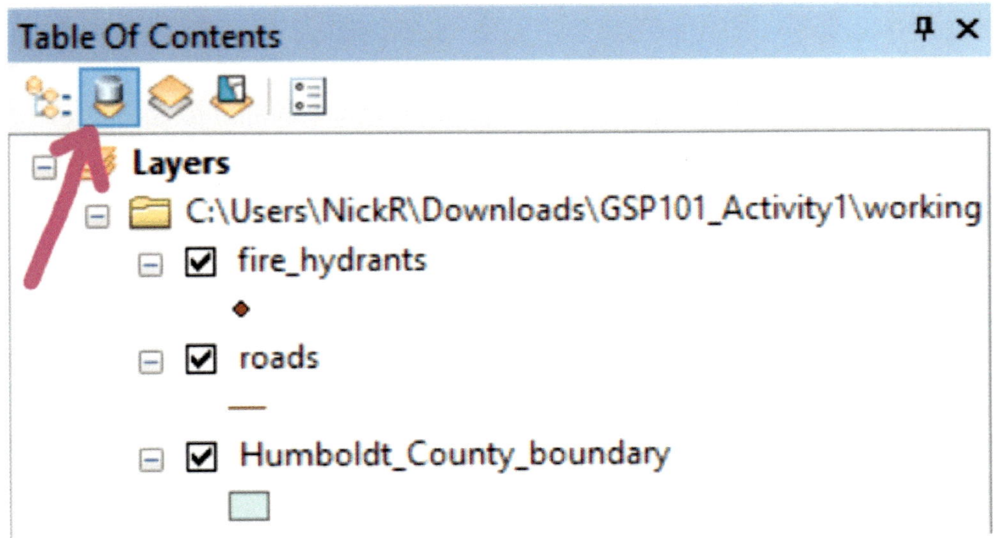

Figure 1.64: *Layers* are grouped based on a file location.

To demonstrate this feature, open the *original folder* from the Catalog window and drag and drop the initial County boundary onto the data frame. The layer appears in the *Table of Contents* under a different group (**Figure 1.65**). Notice that each group indicates the file path.

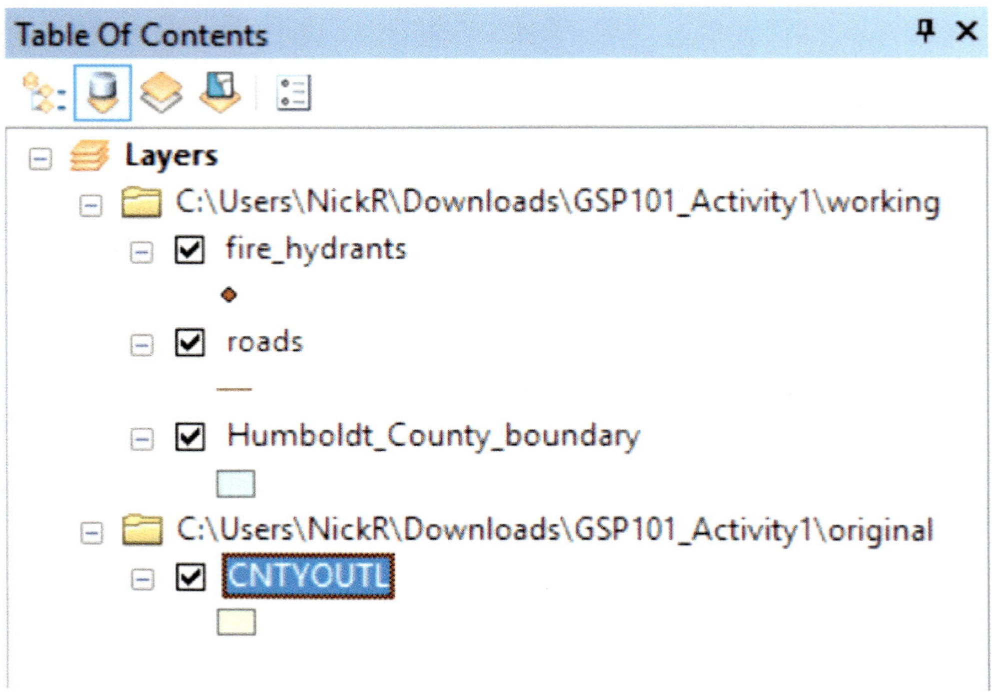

Figure 1.65: The original county outline shapefile appears in a different group because of its location.

Try to drag the newly added County boundary to the top of the *Table of Contents*. As you can see, it does not work while the *Table of Contents* is in the *List by Source* view. The *Table of Contents* contains two additional views, *List by Visibility* and *List by Selection*. These last two views rarely get used in a typical workflow. However, they might prove useful under specific circumstances. Feel free to explore these views at a later time. For now, return to *List by Drawing Order*. Right-click on the County boundary layer from the *original folder* and select *Remove* (**Figure 1.66**). You do not need a duplicate County boundary on the map.

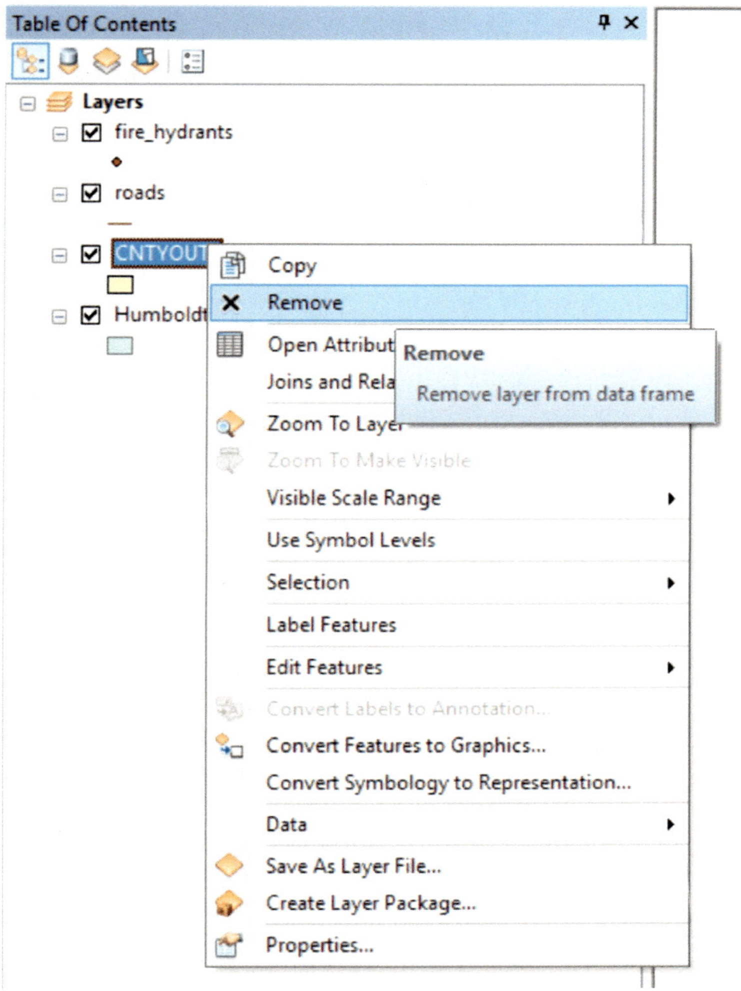

Figure: 1.66: You can access contextual menus in the Table of Contents by right-clicking on layers and data frames.

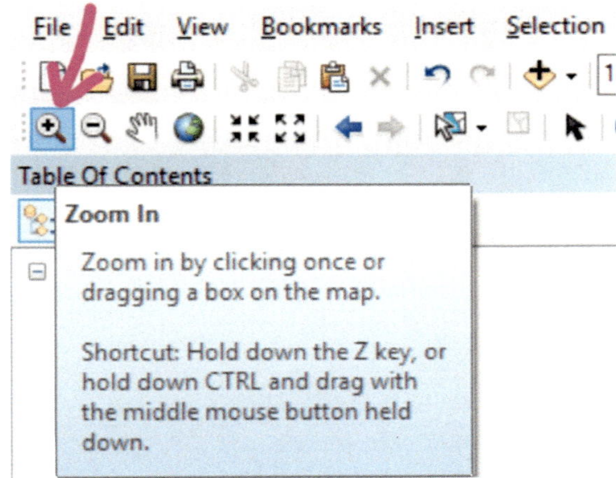

Figure 1.67: The *Tools* toolbar provides several ways for changing the scale of the map, including the *Zoom In* and *Zoom Out* tool.

The data frame changes the map extent closer to the bay (**Figure 1.68**). Use the Zoom tool again click and drag a box near downtown Arcata, just north of the Bay.

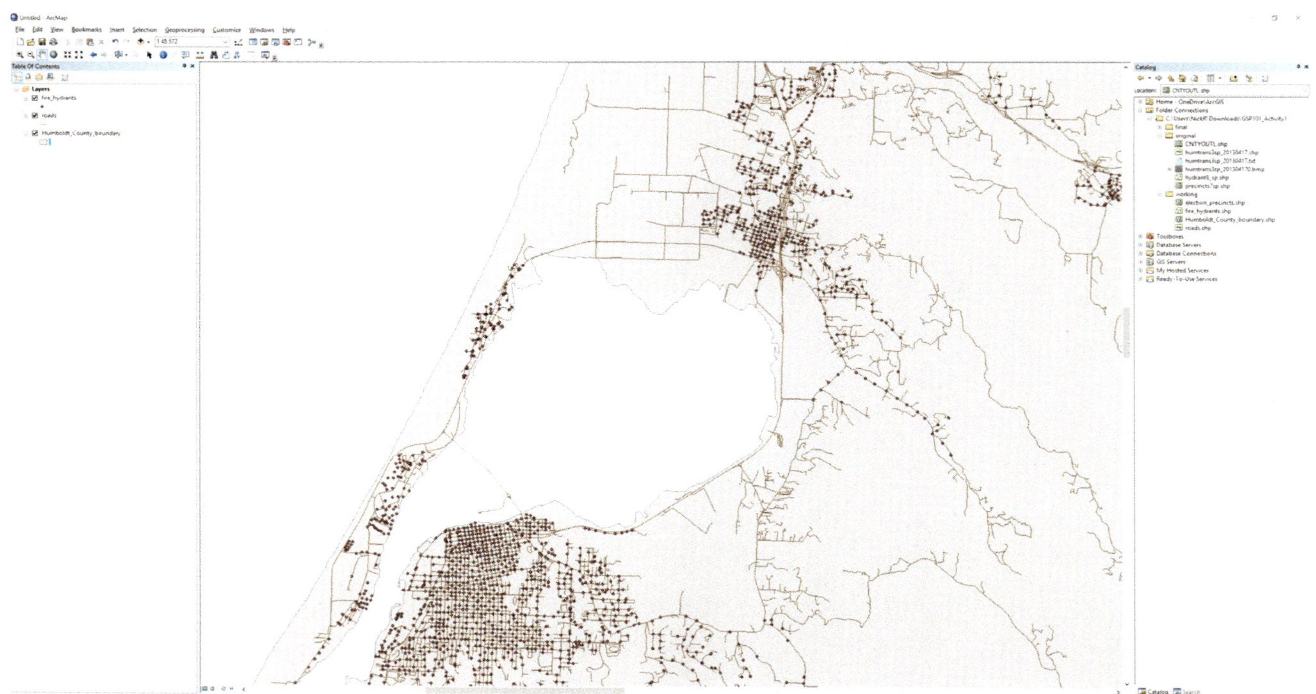

Figure 1.68: The individual roads and fire hydrants should be visible.

Three additional zoom tools exist on the *Tools toolbar*. The *Zoom Out* tool looks like a magnifying glass with a minus sign and works similar to the *Zoom In* tool. Also included on the *Tools toolbar* is the *Fixed Zoom In* and the *Fixed Zoom Out* tools (**Figure 1.69**). They appear as four arrows either pointing inward or outward. You can use the *Fixed Zoom In* and the *Fixed Zoom Out* tools to zoom in and out in controlled jumps from the center of the data frame. Take a moment to experiment with each tool.

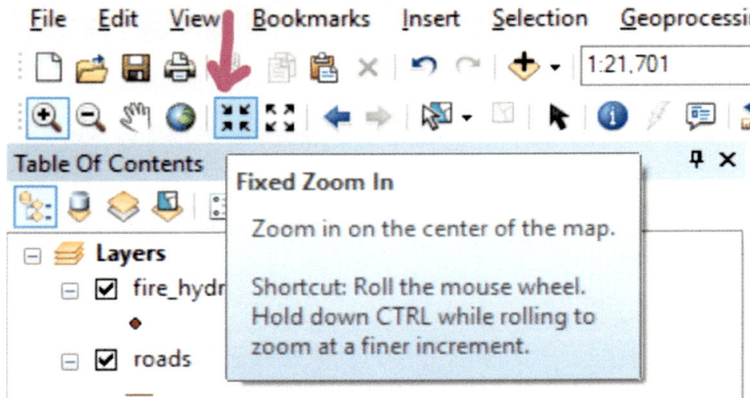

Figure 1.69: The *Tools* toolbar provides several ways for changing the scale on the map, including the *Fixed Zoom In* and *Fixed Zoom Out* tools.

Next to the *Zoom Out* tool is the *Pan* tool. The icon looks like a hand. Click the *Pan* tool and move the map until the Humboldt Bay is visible just between the City of Arcata and the City of Eureka. You may need to zoom in or out a little. The Humboldt County boundary extends around the Bay, but the bay is not explicitly part of the data. What you see as the bay is the background color of the data frame, currently set to white (**Figure 1.70**). You can represent water in the bay by changing the data frame background color to blue.

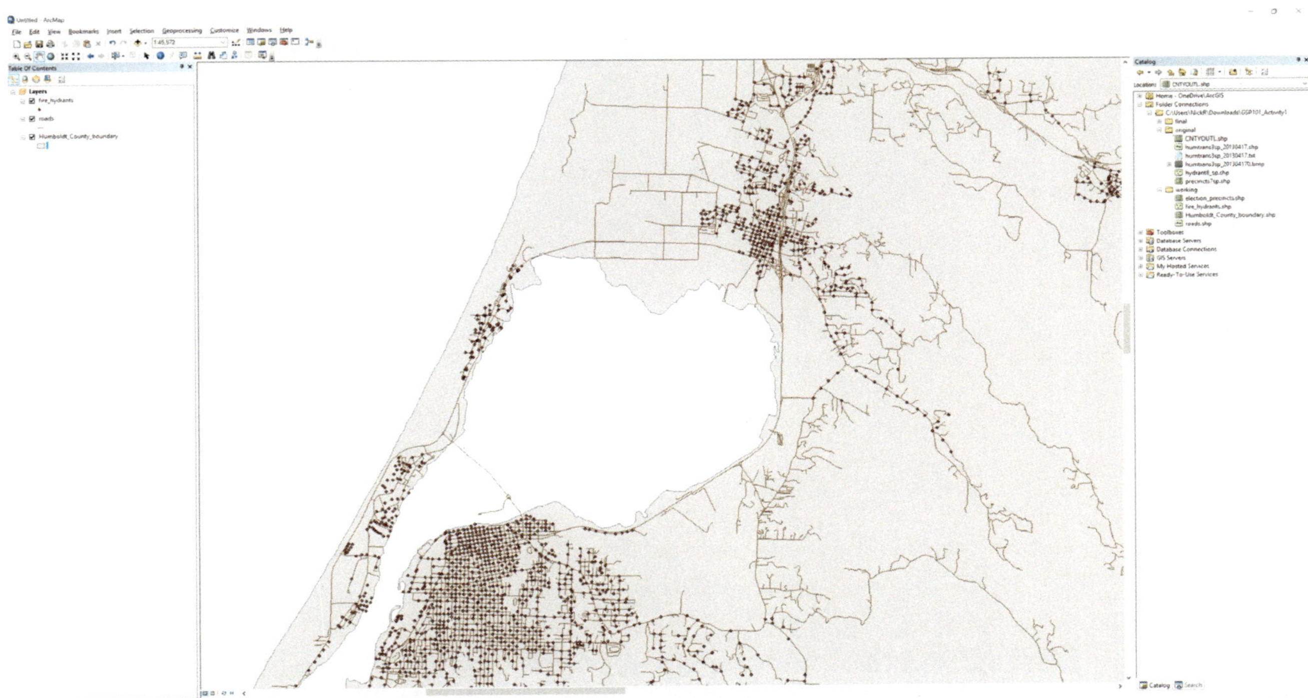

Figure 1.70: The water features appear white because of the background color of the data frame.

In the *Table of Contents*, the word *Layers* appears above the three map layers. I believe Esri chose very poorly when naming this element in the *Table of Contents*. The name consistently leads to confusion. This item on the *Table of Contents* is not a layer. It represents the **data frame**. Access the properties for the data frame by right-clicking on the word *Layers* in the *Table of Contents*. Select, Properties from the contextual menu (**Figure 1.71**).

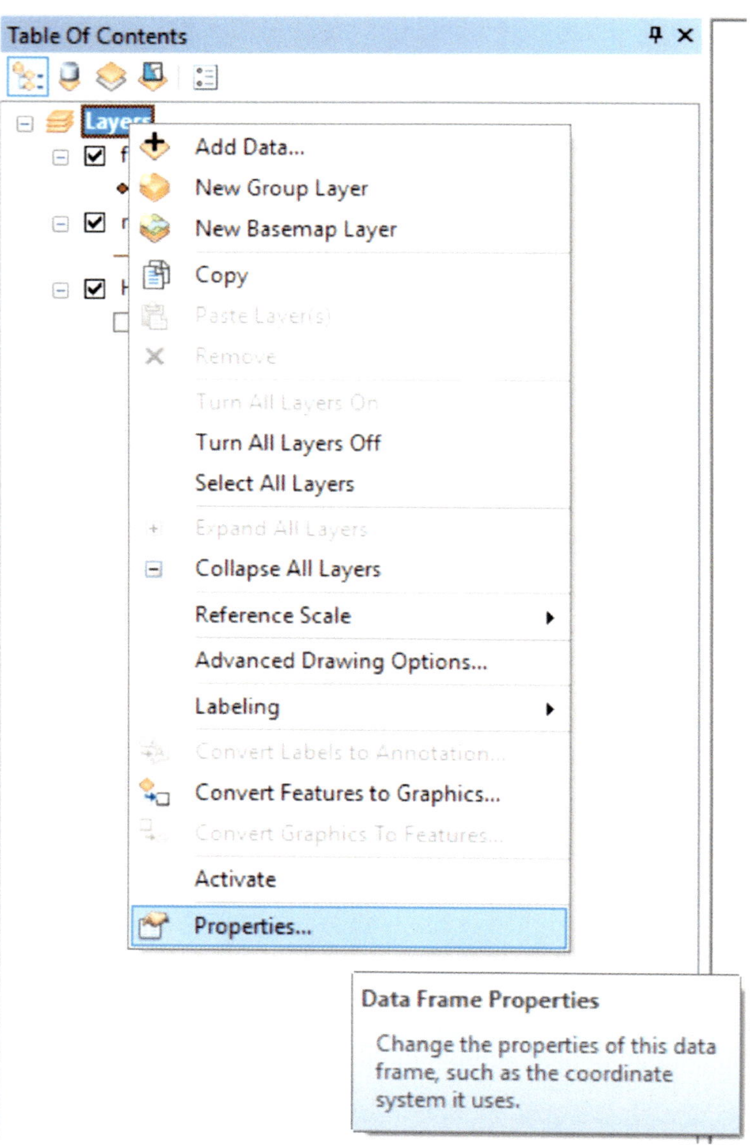

Figure 1.71: You can access contextual menus in the Table of Contents by right-clicking on layers and data frames.

When the Data Frame Properties window opens, click the *General* tab (**Figure 1.72**). Change the name of the data frame to something more descriptive, such as "Humboldt County Fire Hydrants." Click *Apply*. The name updates in the *Table of Contents*.

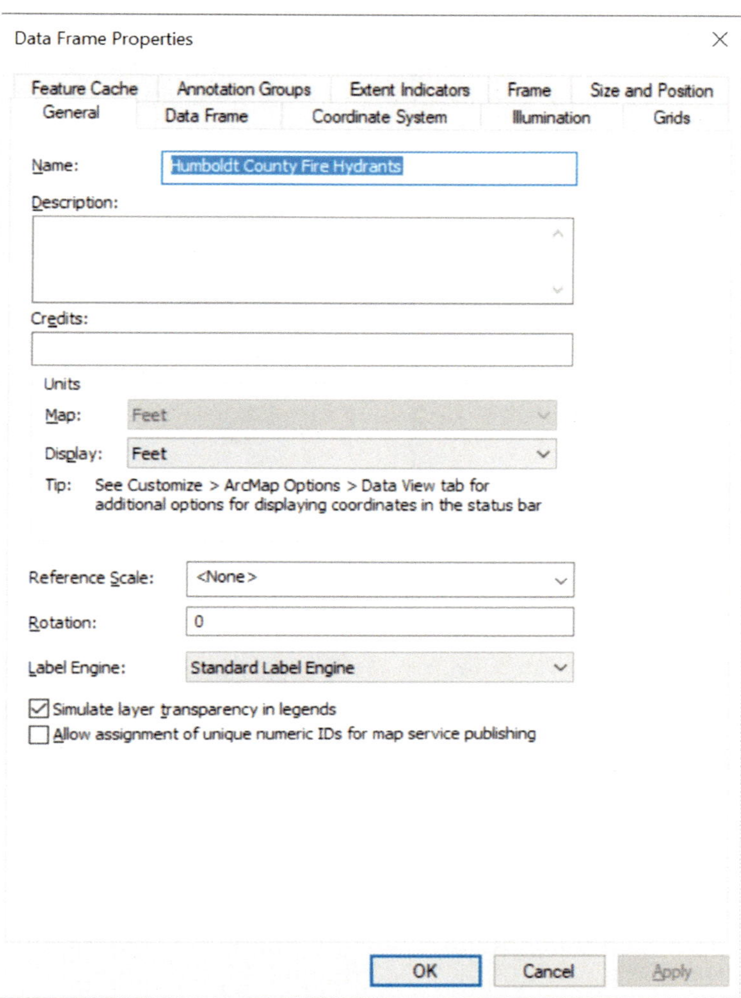

Figure 1.72: Changing the name of the data frame helps to add clarity to the Table of Contents.

In the Data Frame Properties window, Click the *Frame* tab (**Figure 1.73**). Do not confuse it with the *Data Frame* tab, which has a similar name. Click the drop-down menu under Background and change the color to your choice of a shade of blue. When you are ready, click *OK*.

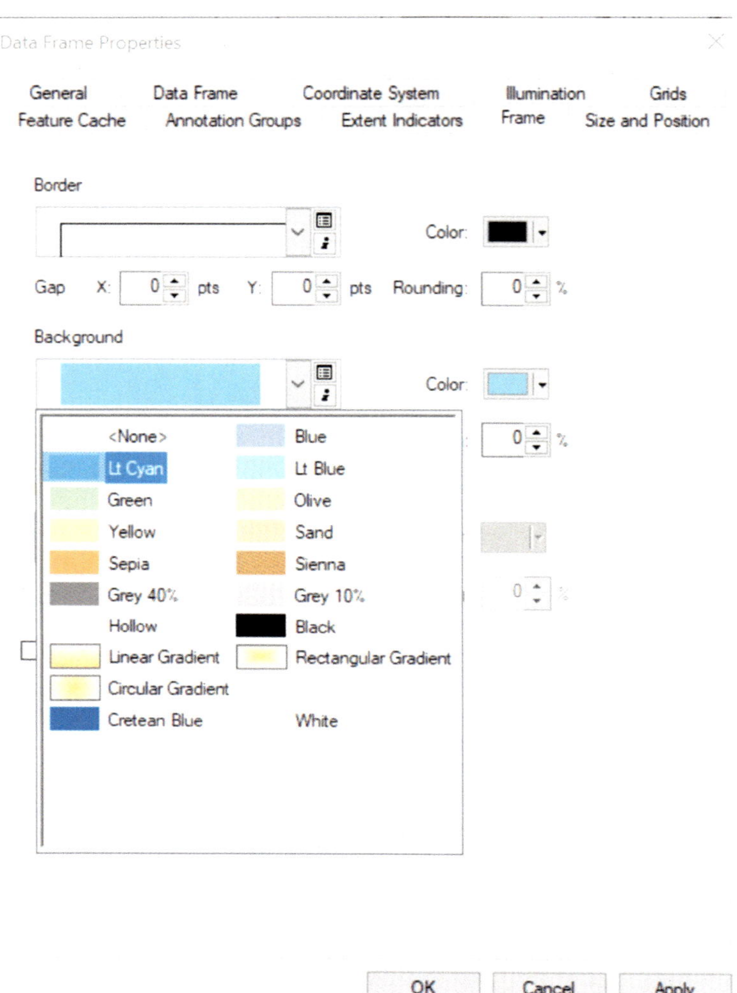

Figure 1.73: The *Frame* tab has options for changing the map border and the background.

As you can see, the background on the map changes to blue (**Figure 1.74**).

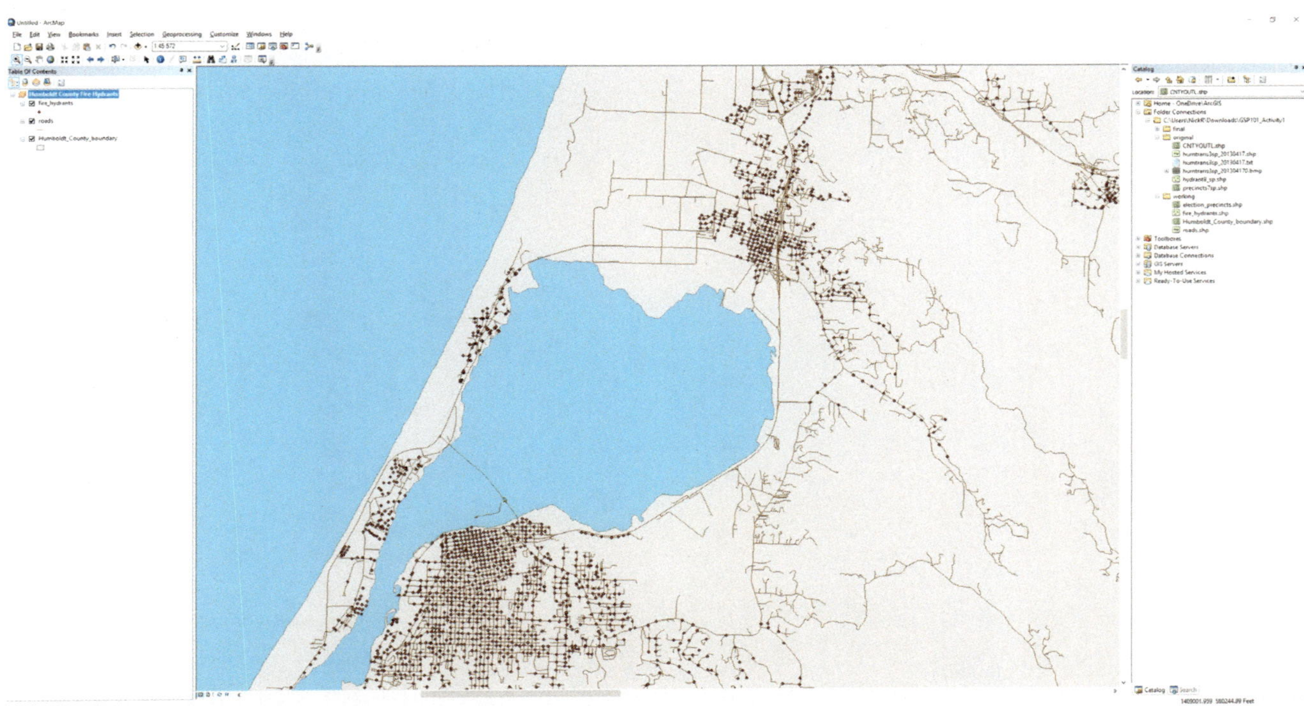

Figure 1.74: Adding a blue background to the data frame simulates water features, such as the Humboldt Bay and the Pacific Ocean.

Adding a Second Data frame to the Map Document

In ArcMap, it is also possible to have multiple data frames in a single map document. From the *Main menu*, select *Insert*, then *Data Frame* (**Figure 1.75**).

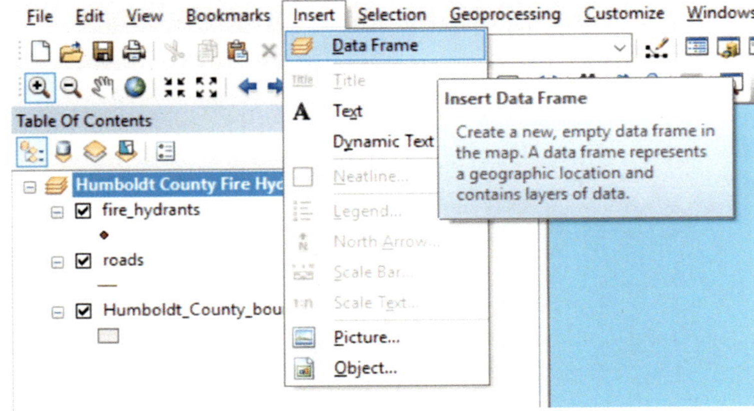

Figure 1.75: The *insert* menu provides several options for inserting items to ArcMap.

The data frame window turns white, and the map disappears (**Figure 1.76**). Currently, the new data frame contains no data. In ArcMap's current view setting, called the **data view**, only one data frame at a time is visible.

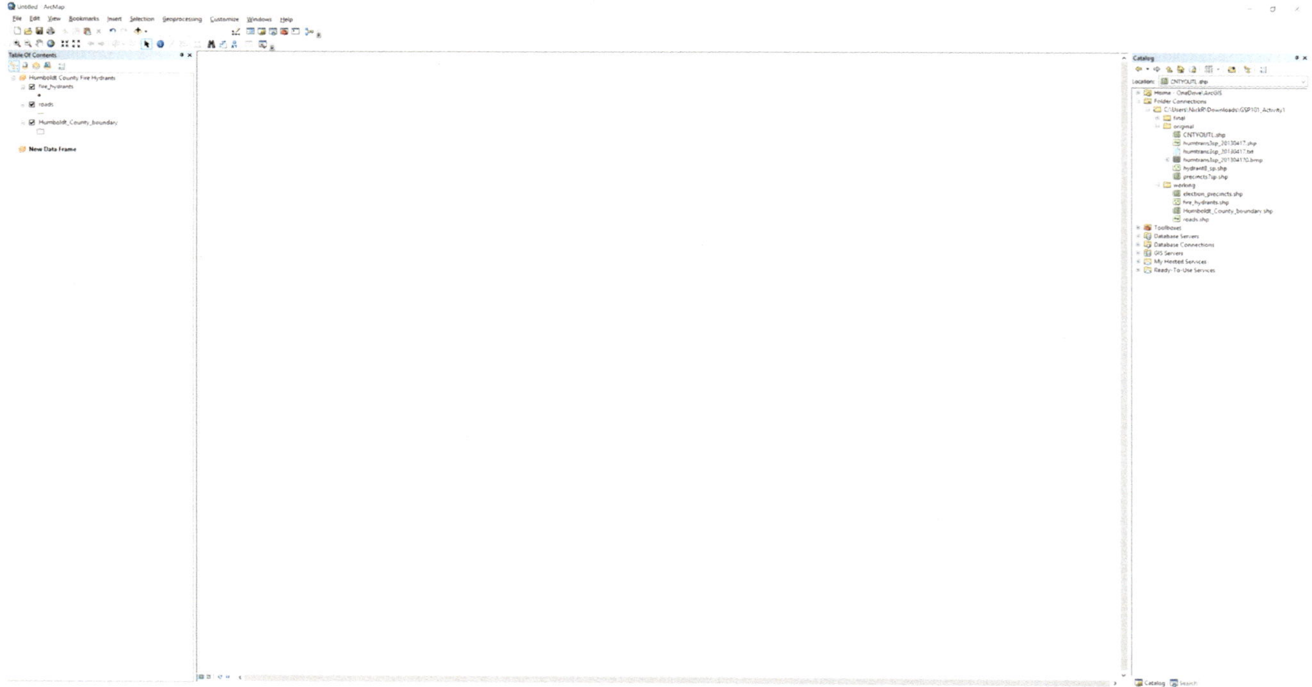

Figure 1.76: With no data added, the map appears blank.

You can switch back and forth between multiple data frames by *activating* them. To demonstrate this feature, right-click on the data frame titled *Humboldt County Fire Hydrants* and select *Activate* (**Figure 1.77**). Switch back to the New Data Frame by right-clicking on it. Then, once again, choose *Activate*.

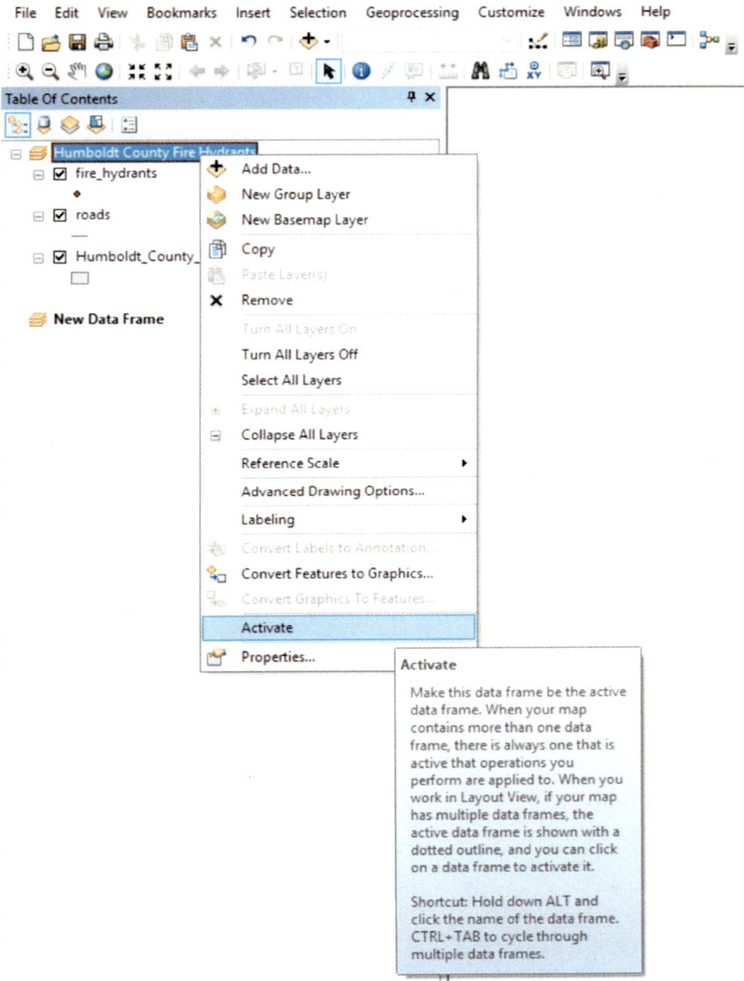

Figure 1.77: You can access contextual menus in the Table of Contents by right-clicking on layers and data frames.

Rename the new data frame. Call the data frame, "Election Precinct Map." Add the election precinct shapefile from the *working folder* to the new data frame (**Figure 1.78**).

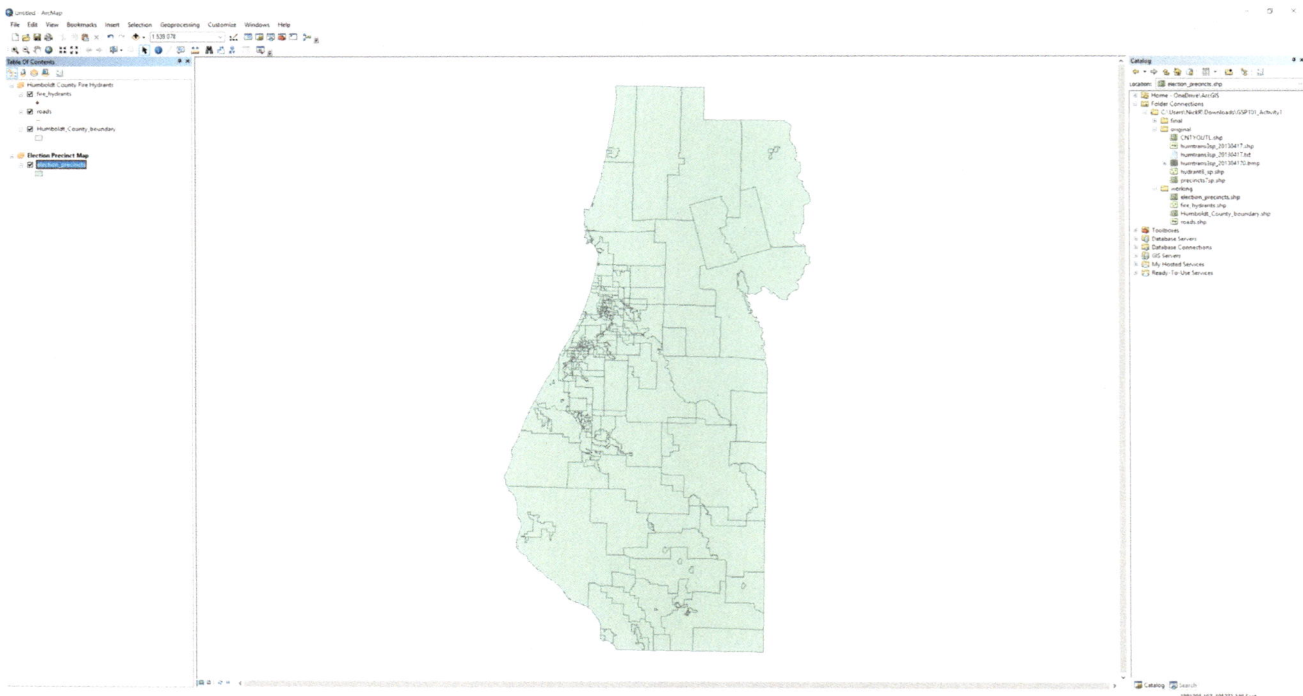

Figure 1.78: The election precinct is the only layer in the new data frame.

One of the most powerful features of GIS software is the connection between a database and the map. Each shapefile comes with a database table. In ArcMap, you refer to the database table as an **attribute table**. To view the attribute table, right-click on the election precincts in the *Table of Contents* and select *Open Attribute Table* from the contextual menu (**Figure 1.79**).

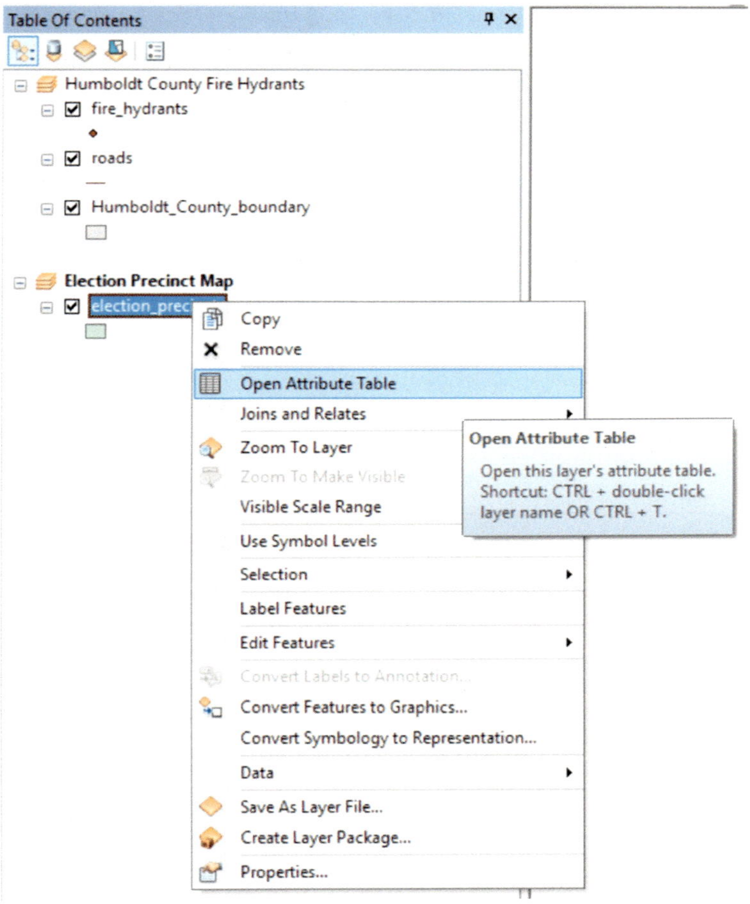

Figure 1.79: You can access contextual menus in the Table of Contents by right-clicking on layers and data frames.

The first time you open the attribute table, it appears floating above the map. Click and drag your attribute table towards the bottom of ArcMap. As you drag, position your cursor over the blue arrow that appears near the bottom (**Figure 1.80**). When you place your cursor over the blue arrow, the attribute table snaps to the bottom of ArcMap.

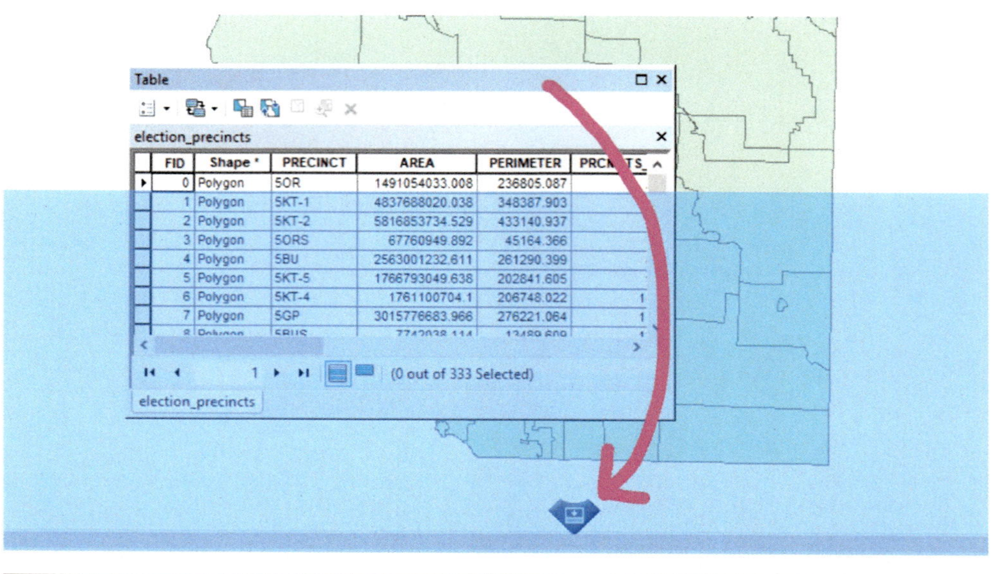

Figure 1.80: Snapping the attribute table to the bottom of the map makes it easy to read.

The attribute table and the map are directly related. The attribute table represents each point, line, or polygon feature on a map as a **record**, or row, in the table. If you select a feature on the map, a corresponding record gets highlighted in the attribute table. If you select a record on the attribute table, ArcMap highlights the corresponding feature in the map. To demonstrate this behavior, click the gray box on the left side of the attribute table for the record that says *5OR* under the *Precinct* field (**Figure 1.81**). ArcMap highlights the polygon representing precinct *5OR* on the map.

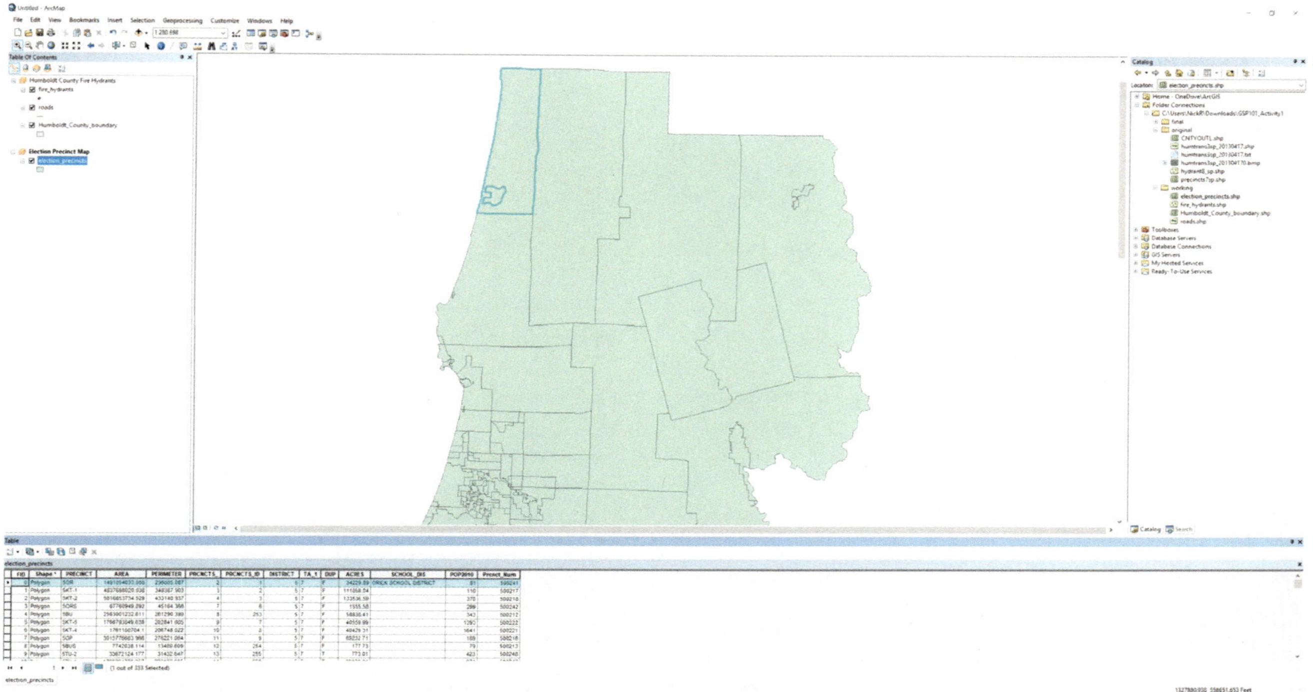

Figure 1.81: Selecting a record on the attribute table selects a feature on the map.

On the *Tools toolbar*, click the *Select Features* tool. The icon looks like a white arrow over a blue and white square (**Figure 1.82**). On the map, click on the polygon just right of precinct *5OR*. ArcMap highlights the polygon on the map and the record for precinct *5KT-1* in the attribute table.

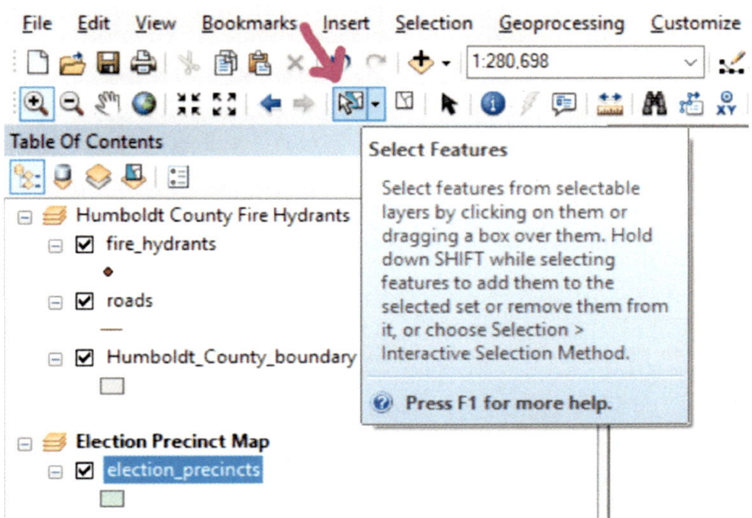

Figure 1.82: You can select features by clicking on them or dragging a box over them.

Close the attribute table. On the *Tools toolbar*, click on the *Clear Selected Features* icon (**Figure 1.83**). The icon looks like a white square.

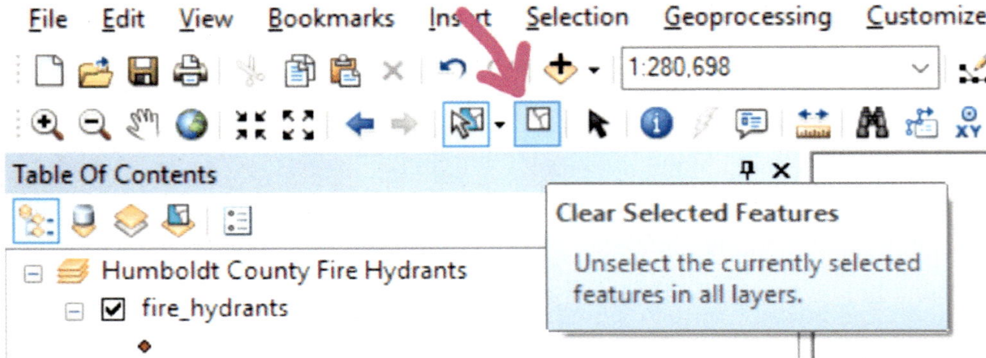

Figure 1.83: It is always a good habit to clear any selected features as you work.

Repairing Broken Links

Take a moment to save your map document. From the *Main menu*, select File, then Save. Always give your files meaningful names.

Untitled is the default file name for a map document. Never use the default name.

Call the map document file, "Humboldt County Maps." Save the file inside your main workspace folder, GSP101_Activity1 (**Figure 1.84**).

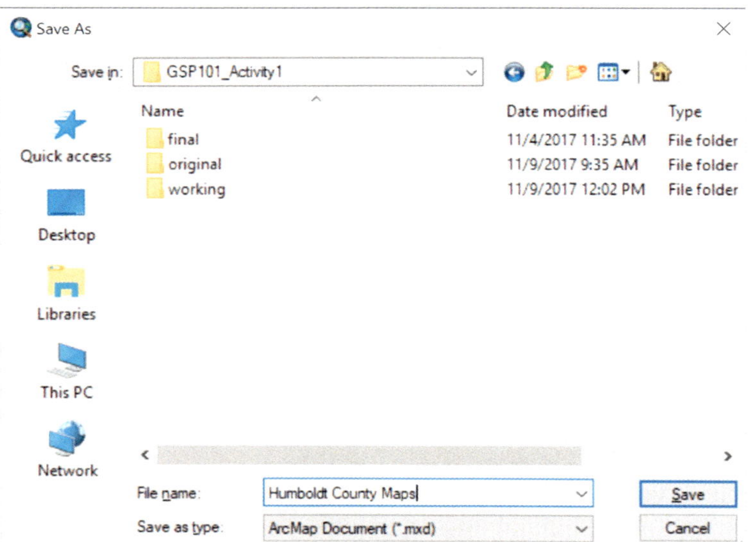

Figure 1.84: Generally, you should avoid spaces. However, empty spaces are allowed when naming map document files (.mxd)

The most challenging fact for novice GIS users to remember is that a map document file (.mxd) *does not* store the data layers. To represent the information on the map, it creates a link to the shapefiles or any other data you add to the map document. If the shapefiles get deleted, corrupted, or moved, the map document will no longer be able to display the data on the map. The link to the data gets broken, and ArcMap loses track of the data. To demonstrate this behavior, close the ArcMap window. Temporarily move your *original*, *working*, and *final folder* to a different folder. If you are working from the desktop, try moving them to the *downloads* folder. The location does not matter, as long as it is different. The Humboldt County Maps.mxd should now be by itself inside the folder GSP101_Activity1 (**Figure 1.85**). Double click the Humboldt County Maps.mxd file to open the map document.

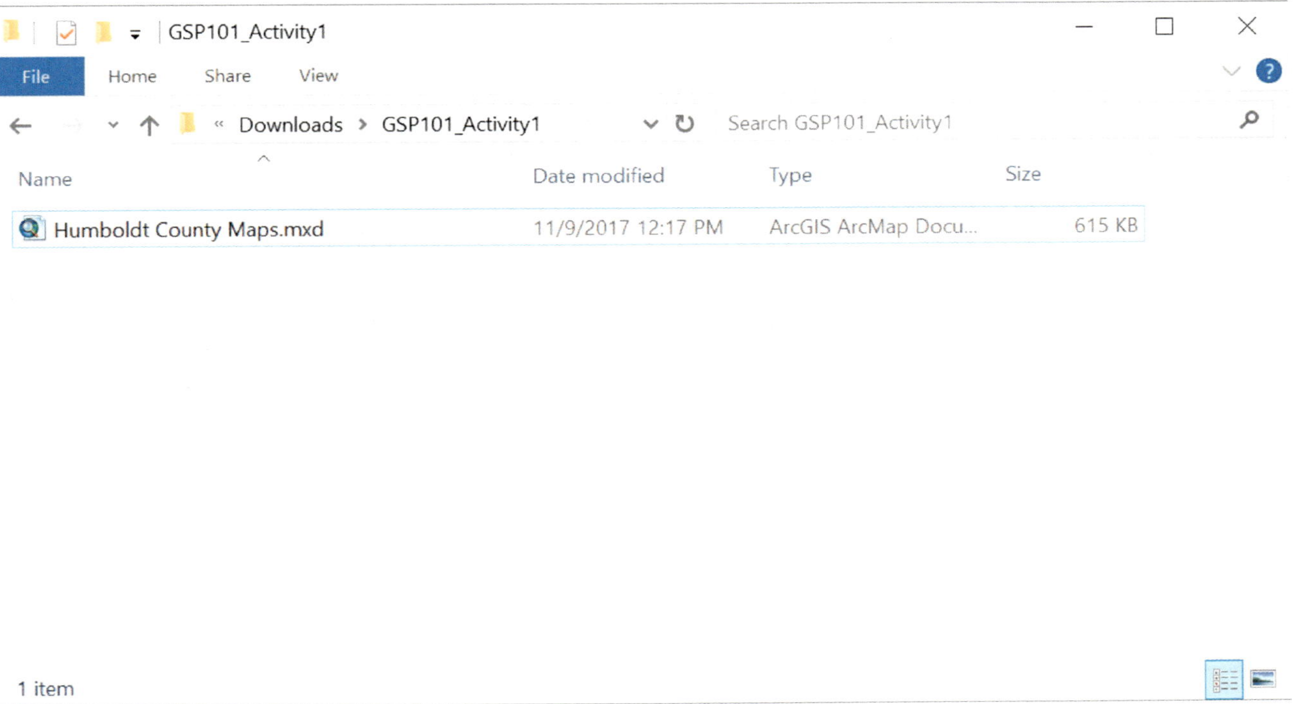

Figure 1.85: A map document (.mxd) by itself is useless.

The chances are that you may see an empty data frame instead of your map. In the *Table of Contents*, a red exclamation point appears next to each layer (**Figure 1.86**). Because the folder moved to a new location, the links broke.

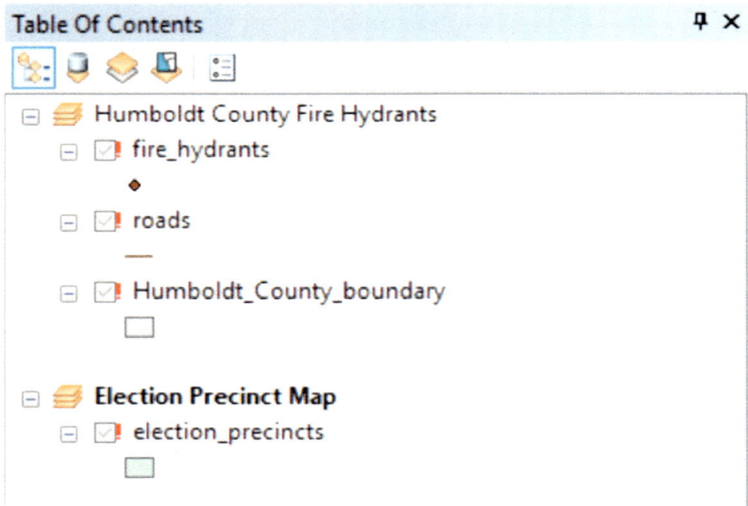

Figure 1.86: Red exclamation points in the Table of Contents indicates missing data.

You can repair the broken link by clicking on the red exclamation point. When the *Set Data Source* window opens, navigate to the location of the data (**Figure 1.87**). You may need to connect to the Desktop to locate the *working folder*. Select the missing data and click *Add*. ArcMap should automatically repair the link for the remaining data layers in that folder.

Close your map document and *do not* save. Instead, return the *original*, *working*, and the *final folder* to GSP101_Activiy1. Then, reopen the map document. Make sure there are no broken links. If you encounter any, be sure to repair them before moving on.

Moving your workspace folder from one location to another is standard practice. To help to avoid broken links, you change the map document properties. From the *Main menu*, select File, then Map Document Properties. Check the box that says, store relative pathnames to data sources (**Figure 1.88**). Click *OK* when you are ready. On the *Standard toolbar*, click the Save icon. The icon looks like an old fashion floppy disk. This setting allows ArcMap to remember the location of the data sources relative to its current position. Now when you move the entire workspace folder, ArcMap keeps track of the contents.

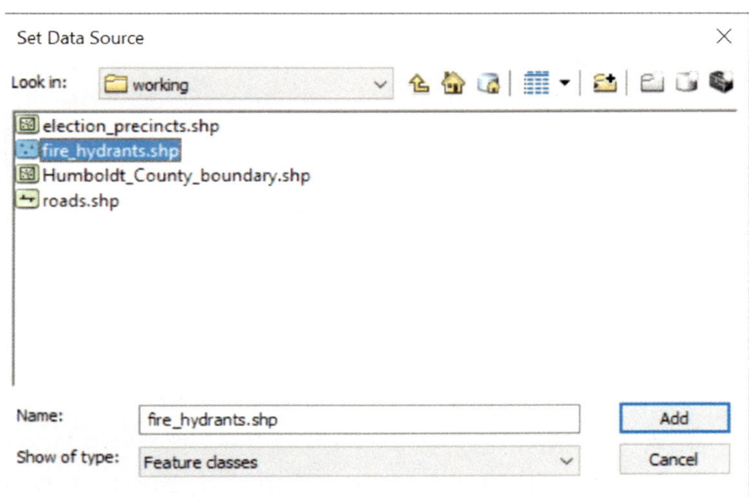

Figure 1.87: Be sure the file you select matches the data you are repairing. In this example, I am repairing the fire hydrant layer.

Map Document Properties

General

File: vnloads\GSP101_Activity1\Humboldt County Maps.mxd

Title:

Summary:

Description:

Author:

Credits:

Tags:

Hyperlink base:

Last Saved: 11/9/2017 12:17:54 PM

Last Printed:

Last Exported:

Default Geodatabase: C:\Users\NickR\OneDrive\ArcGIS\Default.gdb

Pathnames: Store relative pathnames to data sources

Thumbnail: Make Thumbnail Delete Thumbnail

OK Cancel Apply

Figure 1.88: The store relative pathnames to data sources setting allows ArcMap to remember the location of the data sources relative to its current position.

The store relative paths setting does have some limitations. It only remembers the location of the map document relative to its position to the data. This setting means that if you rearrange or move anything out of the workspace folder, GSP101_Activiy1, then the links may get cut.

To safely back up your data, you must move or copy the entire workspace folder, with the subfolders and map document inside.

Compressing the Workspace folder as a 7z file

Sometimes when working with geospatial data, it is necessary to move the data from one location to another. Compressing your workspace folder with the sub-folder structure intact is a safe way to do so. The zip file format is a universal file compression type. However, other file compression types, such as 7z, sometimes work better. Save and then close your map document. Compress your GSP101_Activiy1 folder as a 7z file (**Figure 1.89**). Navigate to your workspace folder, named after this lab activity. Right-click on the folder, select 7zip, then Add to "GSP101_Activity1.7z." Be sure the file extension is **.7z** and not .zip.

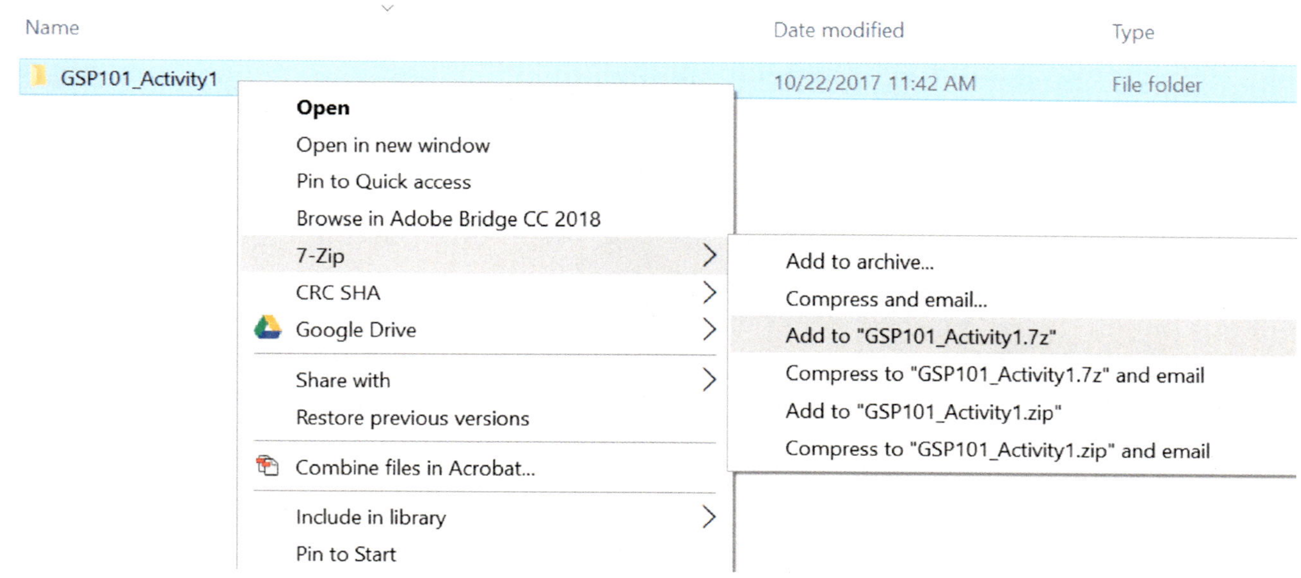

Figure 1.89: You can access the 7zip software by right-clicking on a folder.

When done, you should see your workspace folder compressed as a 7z file (**Figure 1.90**).

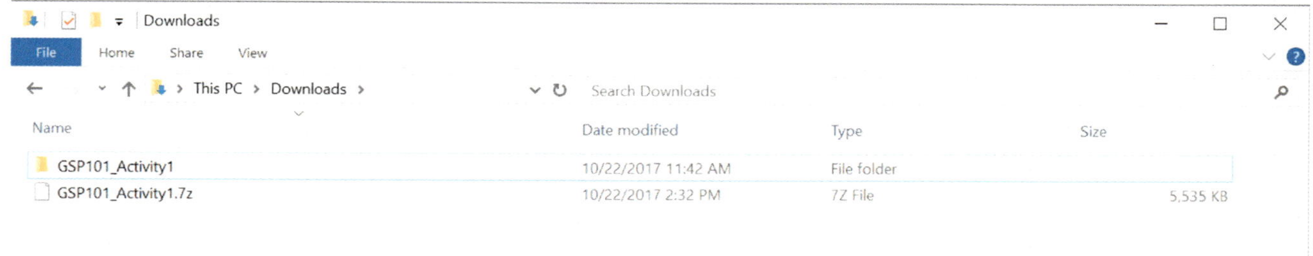

Figure 1.90: Check the file type to be sure it is a 7z file.

Tutorial: Basic Quality Assurance and Quality Control (QAQC)

The goal of this activity is to document a basic review of quality assurance and quality control for geospatial data. As you learned in this Chapter, you can communicate uncertainty and error using standardized **quality assurance and quality control (QAQC)** procedures. Organizations, such as governmental agencies, tribal governments, or non-profit organizations often establish standardized QAQC procedures to assess error and uncertainty, completeness, and overall quality of geospatial datasets. Standard methods include the documentation of existing metadata quality, inspecting spatial accuracy and resolution, and providing related notes or comments about the data to coworkers and colleagues. Always perform QAQC procedures for both attribute data and spatial data at the start of any project during the data collection phase.

ESTIMATED TIME TO COMPLETE THIS TUTORIAL: 2 HOURS

Learning Outcomes

Readers should be able to accomplish the following outcomes by the end of this tutorial:

» Summarize the steps for creating and organizing a project workspace folder structure

» Describe best file management practices for transferring geospatial data

» Demonstrate the ability to download data from a public source

» Practice decompressing zipped files into a specific folder

» Examine data using ArcGIS software

» Document a basic review of Quality Assurance and Quality Control (QAQC)

» Practice compressing a project level folder with contents intact

Scenario

In this scenario, you are working for a non-profit organization as a GIS technician. You have a co-worker that helps to manage geospatial data for the organization. You are about to start a project that uses U.S. Census data to answer questions about different social issues related to poverty and housing across the United States. Your coworker has asked you to collect county boundary shapefiles for the project and perform a basic QAQC evaluation of the data. Your company is still in the process of developing a standardized QAQC form. For now, record the results of your QAQC review using a Microsoft Word Document.

Setting up Your Workspace

In a typical workflow, you work on geospatial data using a local hard drive. When done, you compress your data and back up your work to your cloud storage so that you can retrieve the files from anywhere. When referring to a **local hard drive**, it means you are working on data physically located on the computer in front of you.

In contrast, some computers also include networked drives. **Networked drives** link to cloud storage and save the data elsewhere. Examples include services like OneDrive or Google Drive. For this tutorial, use the **desktop** as your local hard drive location. You may also use an external USB drive if you plan to work in multiple places.

You must avoid using networked drives while you work. They increase the processing time and can cause technical glitches.

In this tutorial, you use a particular folder structure. Start by creating your **workspace folder** on the local hard drive. A **workspace** is a folder or series of folders that contain all of your project files. The top-level folder in your workspace should indicate the lab assignment or the project. Organize all of your work within the workspace folder. On your desktop, create a new folder and give it a descriptive name, such as *Basic_QAQC*. Be sure there are *no spaces*. You may use underscores instead of spaces. Inside this folder, create the following three subfolders: original, working, and final. Having a standardized folder structure helps to keep a project organized, primarily when you are working with multiple partners. The folder structure you see here (**Figure 1.91**) is the standard used in each of the tutorials presented in this book.

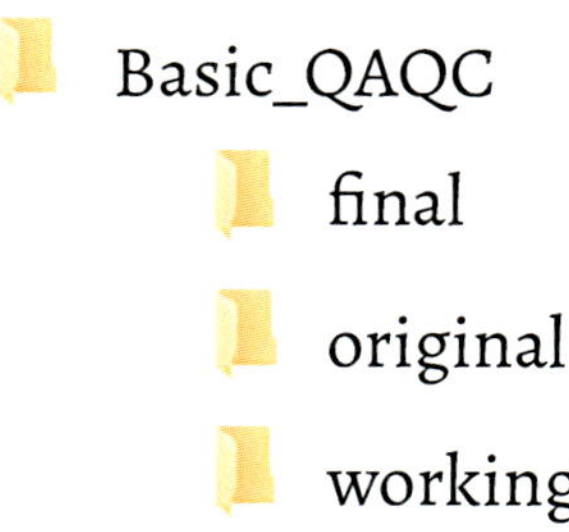

Figure 1.91: This diagram represents a basic folder structure used in this book.

As the name indicates, use the **original folder** for storing original, unaltered data. As you are working on a project, if for some reason your working version of the data gets lost or corrupted, you can go back to your *original folder* and find a fresh copy of the data. Use the **working folder** for data that you create or alter while working on your project. Use the **final folder** for storing any output you produce as a result of your work such as images, maps, tables, or reports. In this activity, you use the *original* and the *final* folder. However, setting up the standard folder structure for a project is good practice and a habit you want to develop.

Download Data from the United States Census Bureau

In a previous tutorial, you learned how to download data from the Humboldt County GIS Data Download web page. In this exercise, you download the data from the United States Census Bureau (**Figure 1.92**).

Open the **Chrome** browser and navigate to the ***United States Census Bureau Cartographic Boundary Shapefile***[1] page.

If you do not have the Chrome browser, you can download it here[2].

Click the link that says County (**Figure 1.92**).

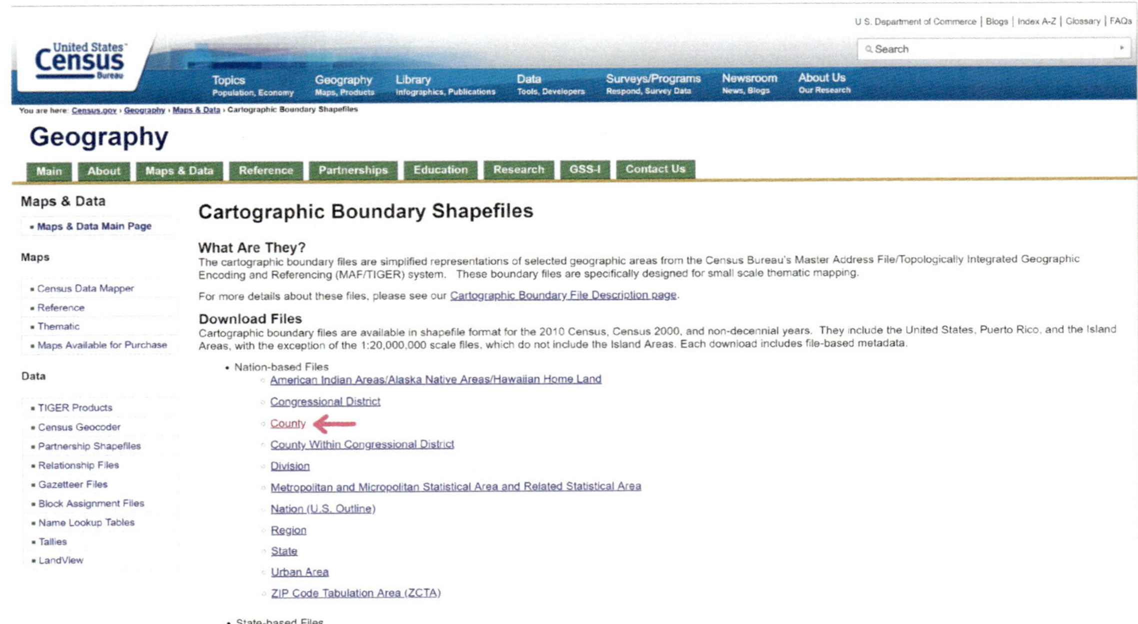

Figure 1.92: Click the link under the header Nation-based Files that says County.

1 https://www.census.gov/geo/maps-data/data/tiger-cart-boundary.html

2 https://www.google.com/chrome/browser/desktop/index.html

On the Counties page, right-click the link to download the 500K resolution (**Figure 1.93**). Select, *Save Link As*, then browse to your *original* folder and click *Save*.

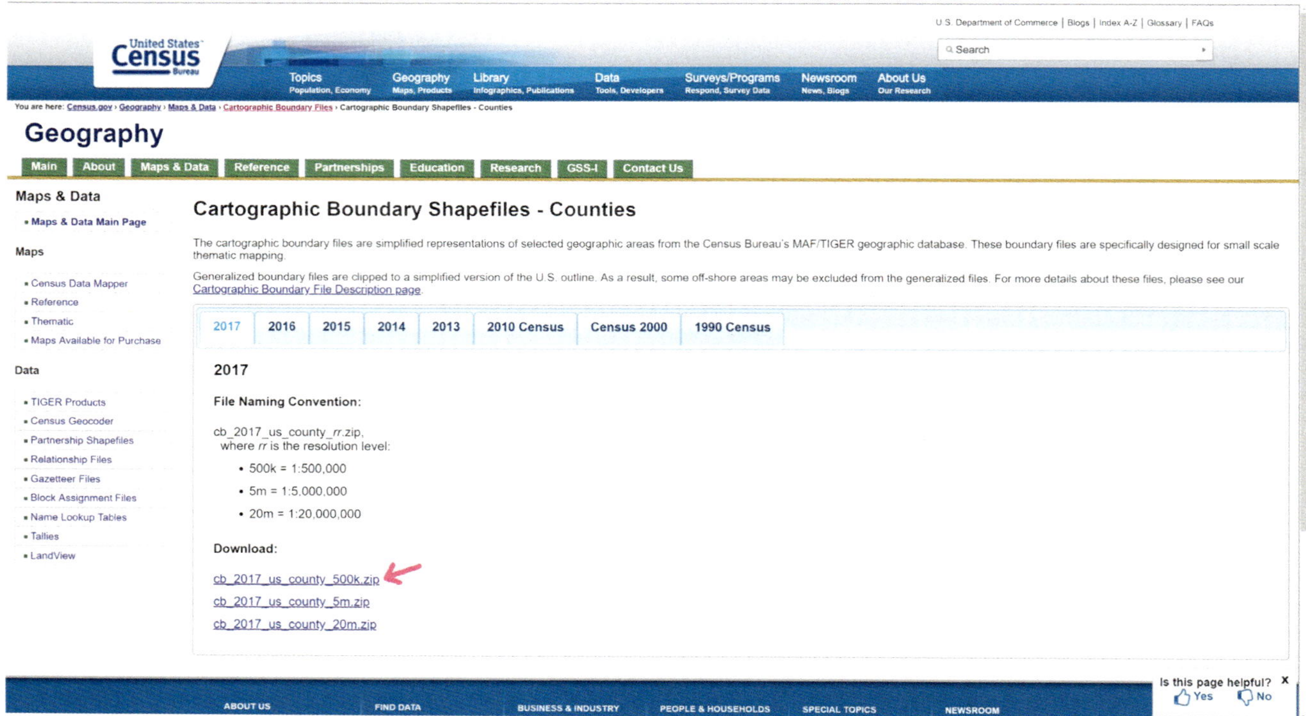

Figure 1.93: The county-level data comes in several resolutions.

In a previous tutorial, you learned how to decompress a file using 7zip. In Microsoft Windows, navigate to your original folder. You should see the zip file for the county 500k data. Right-click on the file, select 7zip, then *Extract Here*. Be sure to delete the zip file when you are done decompressing it. You won't need it anymore. Eliminating the zip file saves space and helps to avoid confusion later.

Skill Drill: Download Census Data

Using what you learned in the previous step, download the cartographic boundary shapefiles for county_5m and county_20m (**Figure 1.94**).

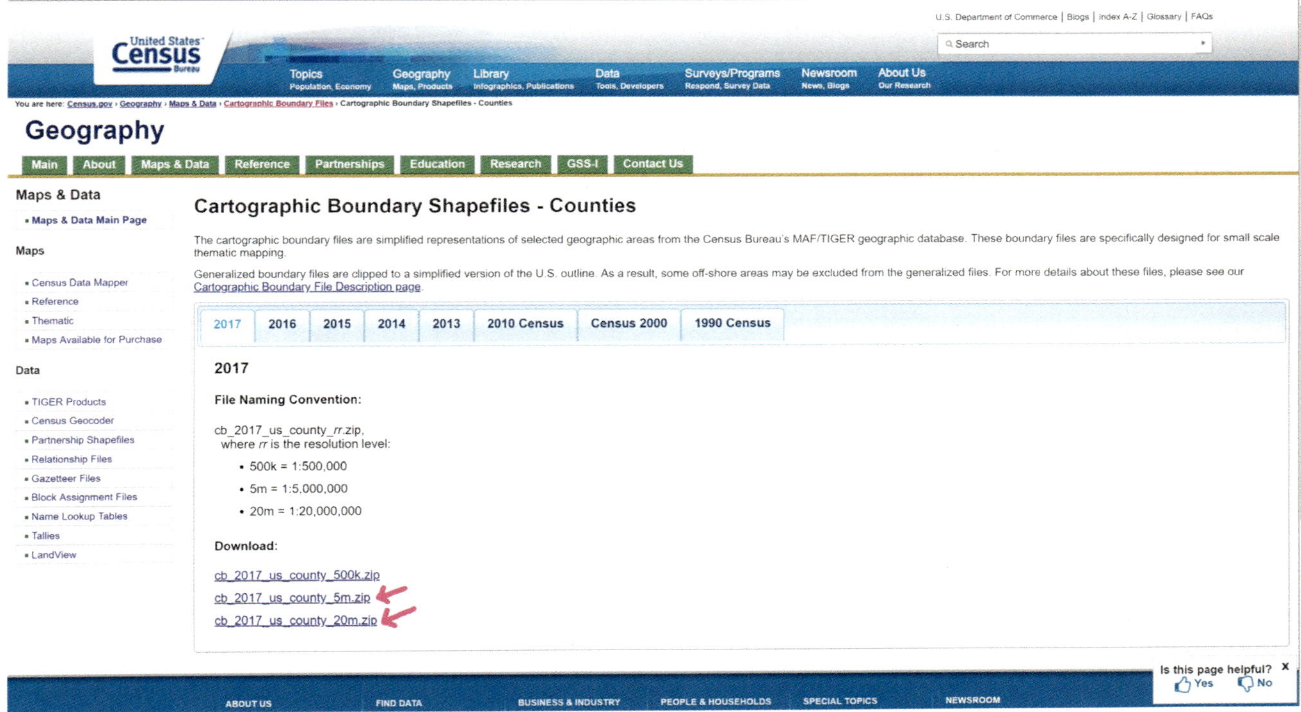

Figure 1.94: The county-level data comes in several resolutions.

Be sure to save the files to your original folder. Decompress each zip file using 7zip. Then, delete the zip files after decompressing them.

Adding County Data to ArcMap

Locate ArcMap on your computer and launch the software. If you are using Microsoft Windows 10, click the windows button and type *ArcMap* to find the desktop application (**Figure 1.95**). Launch the ArcMap software.

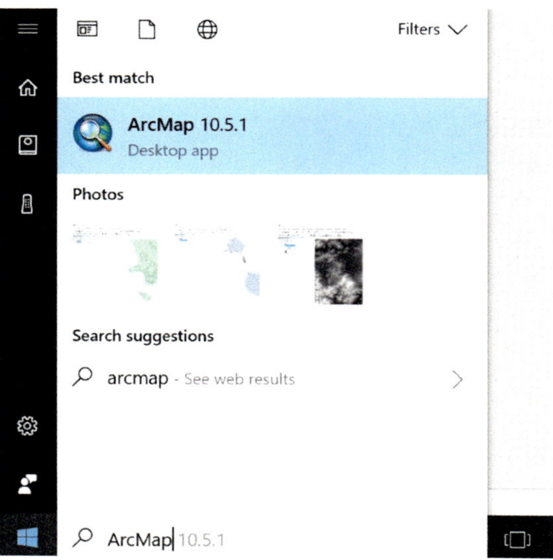

Figure 1.95: The specific version of ArcMap may vary over time.

When you first launch ArcMap, a window appears that gives you the option to either open a blank map or to open any of your recent map documents (**Figure 1.96**). Choose to open a blank map document.

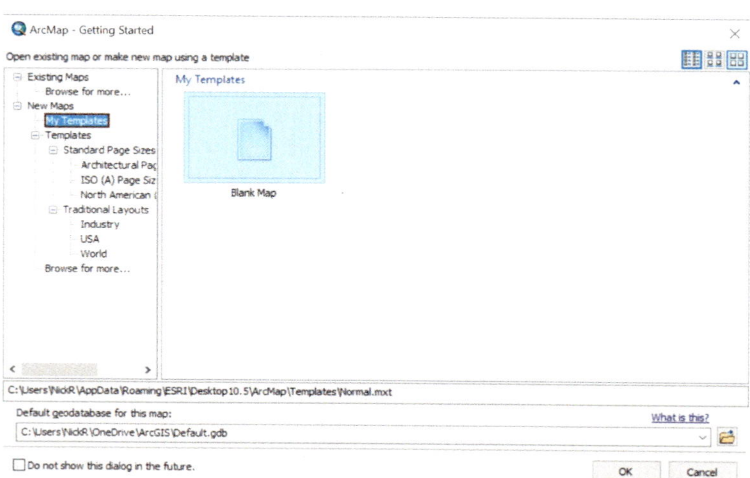

Figure 1.96: The ArcMap Getting Started window provides several options when starting up.

The ArcMap user interface has three main windows that you use regularly, the *Table of Contents*, the data frame, and the Catalog Window (**Figure 1.97**). The **Table of**

Contents displays a list of data frames and map layers loaded into the map documents. The **data frame**, sometimes spelled dataframe, represents the layers as a map and defines the map extent. The **Catalog Window** displays a hierarchical view of folder connections and data as a **Catalog Tree**.

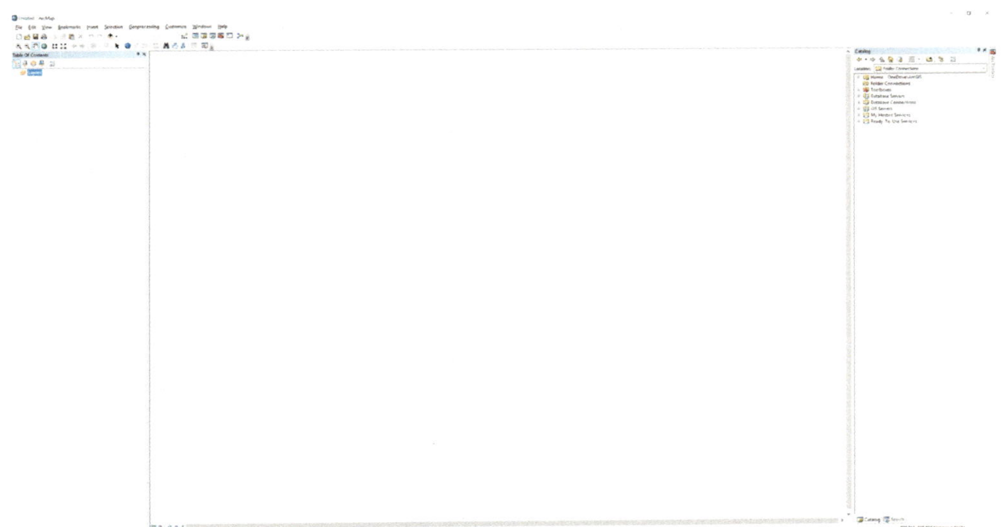

Figure 1.97: From left to right, the three main windows are the Table of Contents, the data frame, and the Catalog Window.

> ***The icon for the catalog window looks like a yellow file cabinet. If your catalog window is missing, you can find the icon on the toolbar across the top. Click to open it back up.***

Figure 1.98: Click the yellow file folder icon if your Catalog Window is missing.

In this step, you start by working with the Catalog Window. The catalog window is where you manage your geospatial data in ArcMap. Start by clicking on the Connect to Folder button (**Figure 1.99**). Navigate to your workspace folder, *Basic_QAQC*, on your local hard drive.

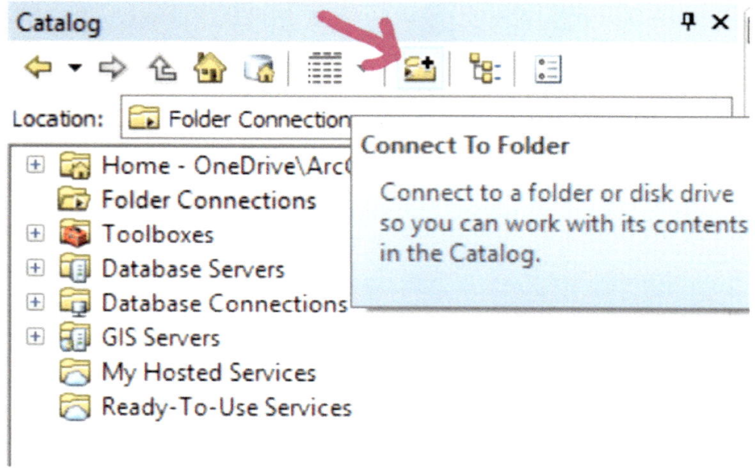

Figure 1.99: You connect to a folder so that the contents appear in the Catalog Window.

In this step, it is essential that you select your primary workspace folder, *Basic_QAQC*, to add it to the Catalog Window (**Figure 1.100**). When ready, click *OK*.

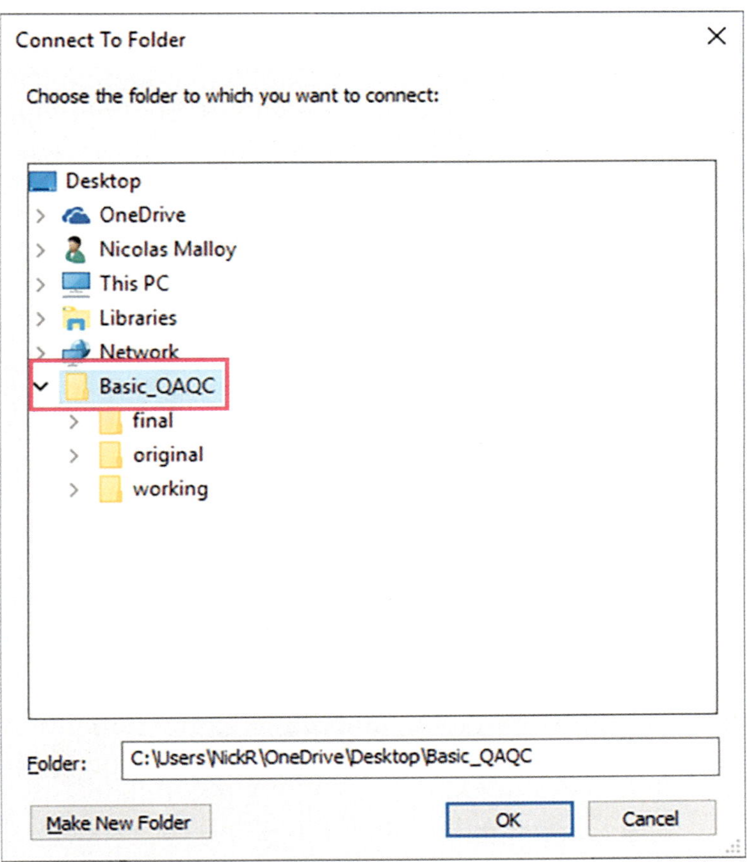

Figure 1.100: The Connect to Folder window displays a directory of folders on your computer. Individual files do not show up here.

Once you add your workspace folder to the Catalog Window, expand the folder by clicking on the plus sign. You should see your three subfolders inside (**Figure 1.101**). Expand the original folder to view the contents.

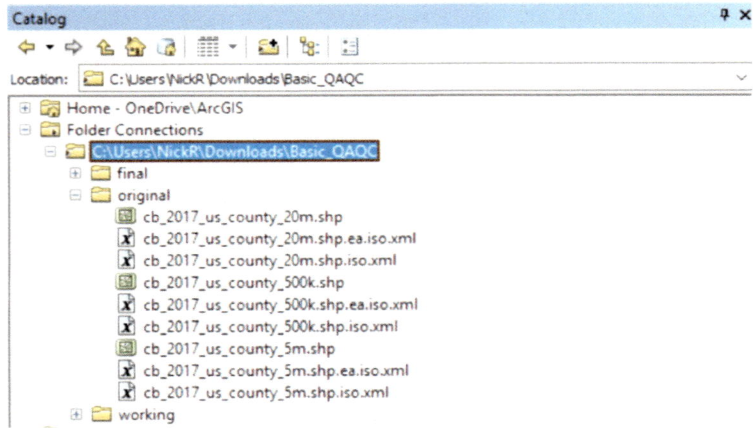

Figure 1.101: The contents of the original folder are displayed and organized in the Catalog Window. This display of folders and files is sometimes called the Catalog Tree.

There are several ways to add data to ArcMap. The most direct method is to drag and drop files from the Catalog Window. In this step, you add the county boundaries to the map document. From the Catalog Window, select the 500k county boundary shapefile from the original folder. In this example, it is called "cb_2017_us_county_500k.shp," represented by a green square icon. Click and drag the file over the data frame, the interior window pane in ArcMap, then release. The county layer appears in the data frame window (**Figure 1.102**).

Figure 1.102: The county shapefile contains boundaries for all U.S. States and Territories.

In a previous tutorial, you learned how to use various zoom tools. In ArcMap, zoom to the west coast of the continental United States (**Figure 1.103**). Try to set your map so that the West Coast takes up most of the space.

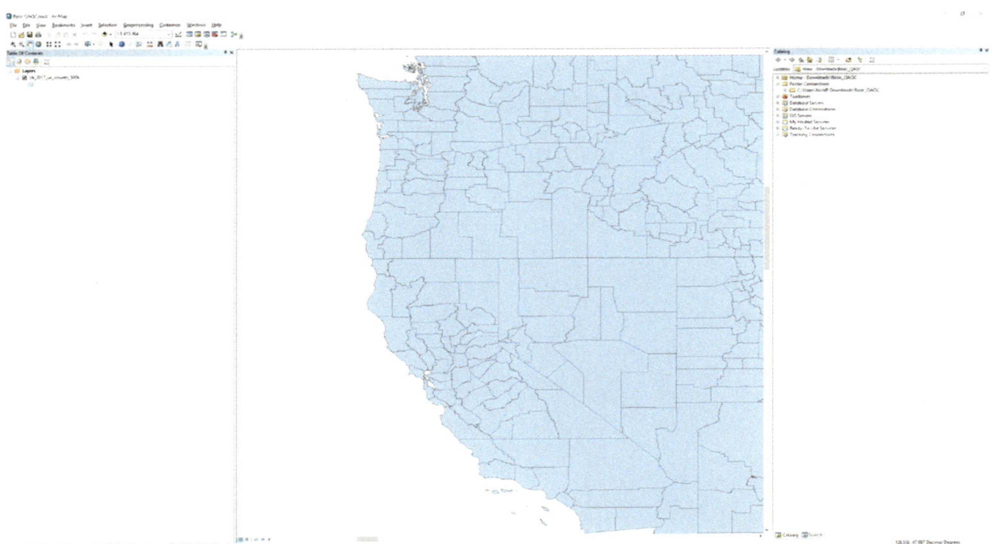

Figure 1.103: The map extent displays the U.S. West Coast.

Skill Drill: Adding Data to ArcMap

Using what you learned in the previous step, add the shapefiles for county 5m and county 20m to the map document (**Figure 1.104**). When done, you should have three map layers in the *Table of Contents*.

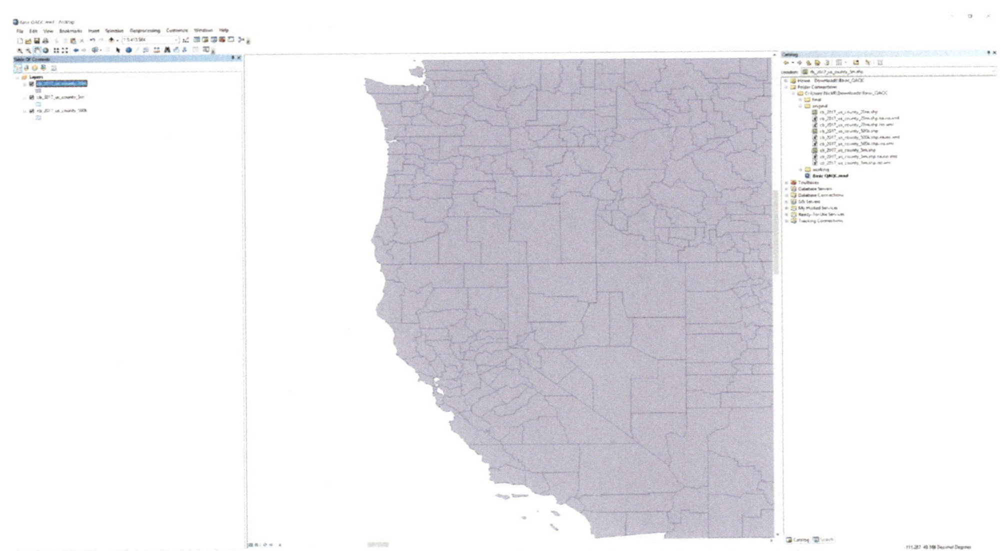

Figure 1.104: The three versions of the U.S. counties shapefile get added to the map document.

Documenting the Quality of the Metadata

Each shapefile can store metadata. **Metadata** refers to data about data. The documentation about the source of data, methods of data collection, times of data collection, and data authorship are all part of metadata. Thorough and well-documented metadata can be a valuable resource for any company, agency, or government. As a GIS analyst, inspecting the metadata should be the first step you take when beginning a project as part of quality assurance and quality control (QAQC).

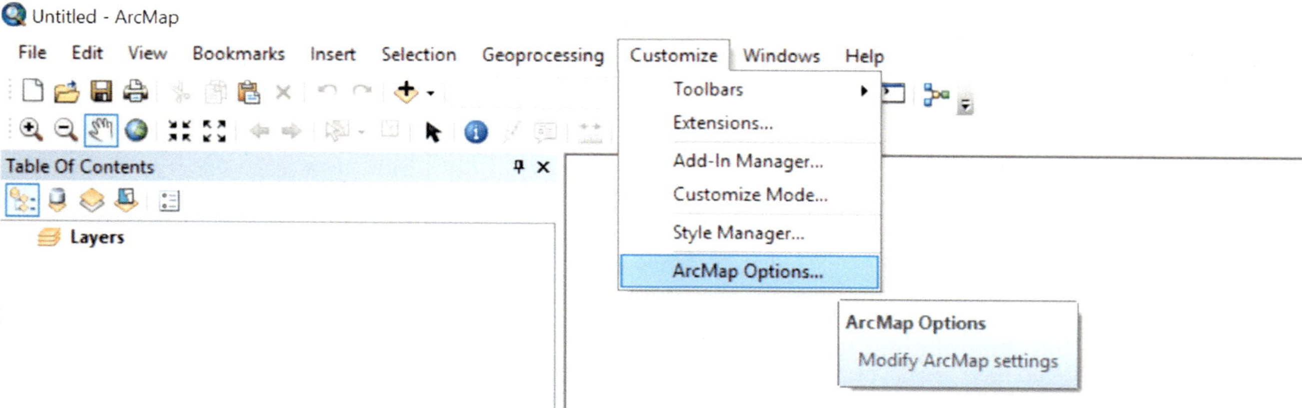

Figure 1.105: Locate the *Main* menu across the top of ArcMap.

Before inspecting the metadata, you need to make sure ArcMap is set to read it correctly. There are different metadata formats available. In this example, you use the **ISO 19139 Metadata Implementation Specification**, a commonly used format. In ArcMap, locate the Customize menu at the top. Select ArcMap Options (**Figure 1.105**). When the ArcMap Options window opens, select the Metadata tab. Under the Metadata Style, choose *ISO 19139 Metadata Implementation Specification* from the drop-down menu (**Figure 1.106**).

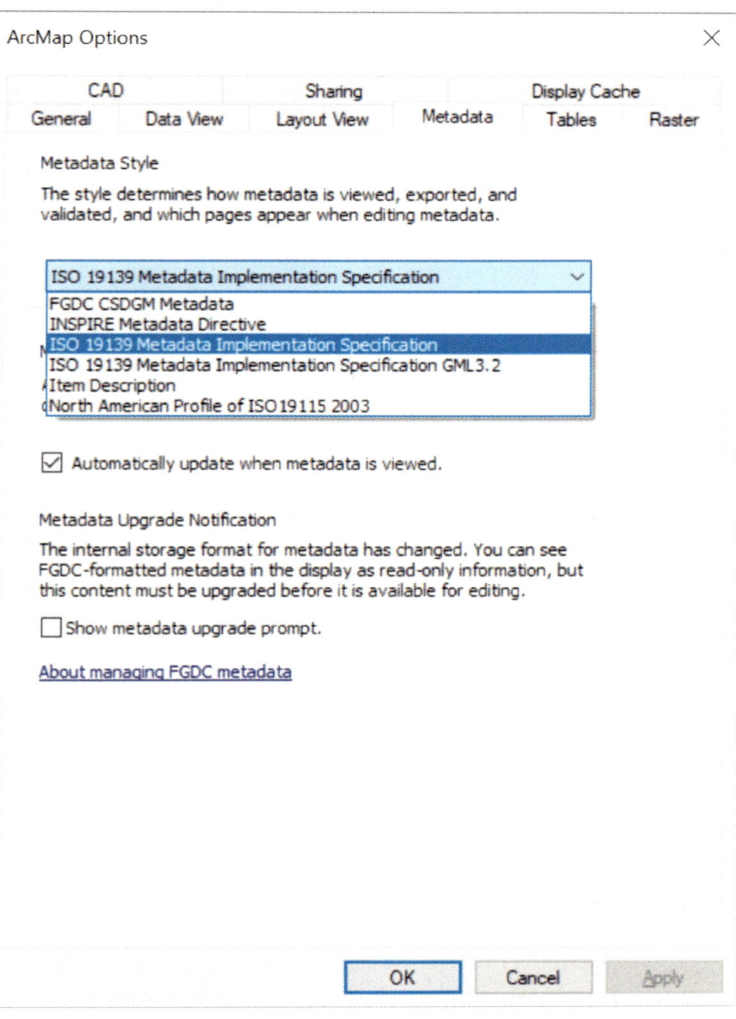

Figure 1.106: ArcMap Options allow you to change many of the global software settings.

Leave all other default settings and click *OK*. Return to the Catalog Window and right click on the county 500k shapefile located in the original folder (**Figure 1.107**). Select, *Item Description* to open the metadata stored in the shapefile.

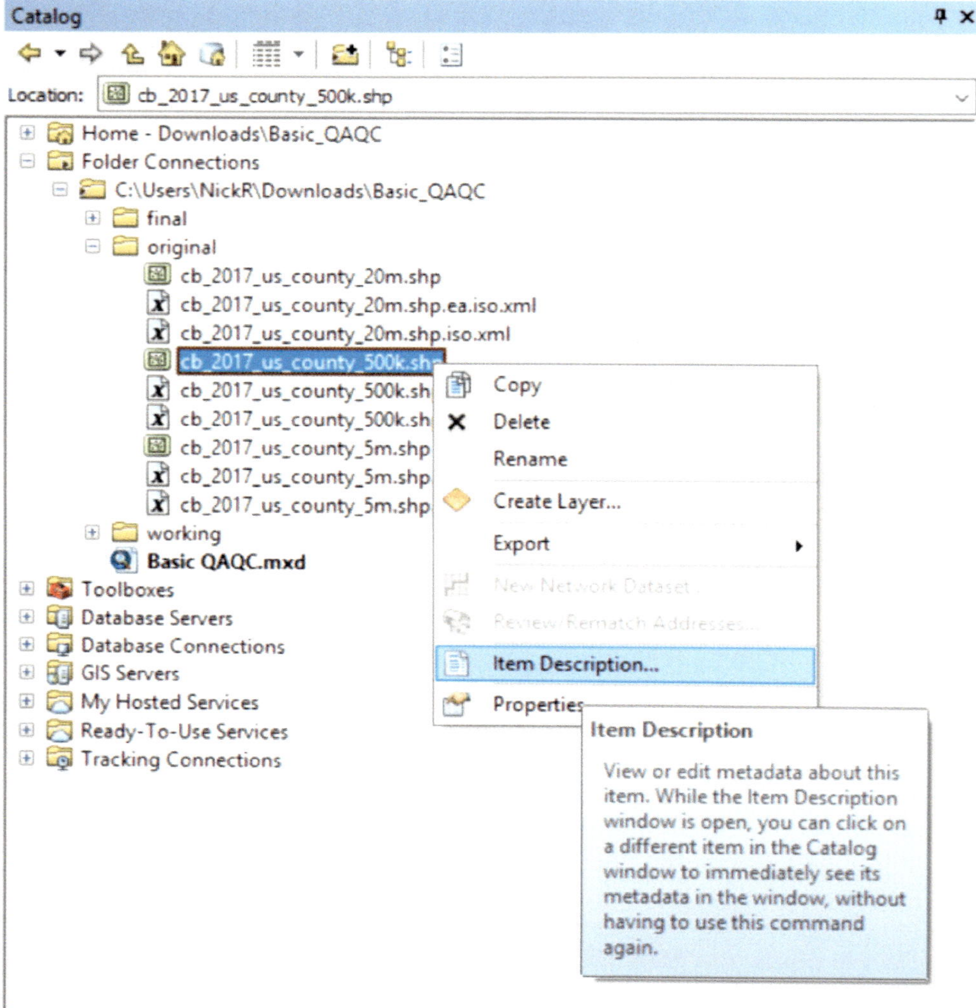

Figure 1.107: You can access contextual menus in the Catalog Window by right-clicking on files and folders.

Enlarge the metadata window so that you can effortlessly read the contents (**Figure 1.108**). As you can see, this shapefile includes some metadata such as a summary, a description, credits, and the use limitations. Take a moment to read these first four sections.

Figure 1.108: The Item Description window has two tabs, Description, and Preview. The Description tab displays the metadata.

Open a new Microsoft Word document. Add a Heading 1 style heading that says, "Metadata Inspection" (**Figure 1.109**). Below the Heading 1, add a Heading 2 style heading that says, "County 500K."

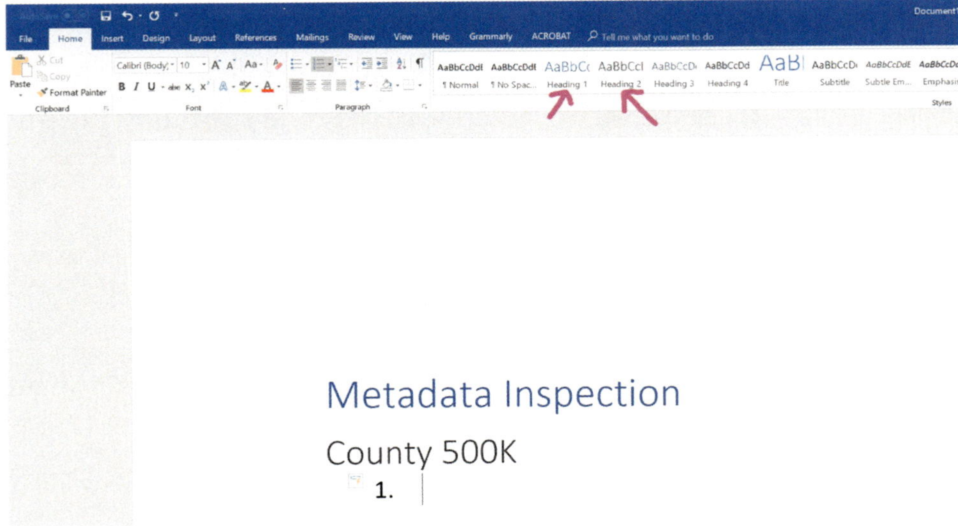

Figure 1.109: You can locate the heading styles on the Home tab.

Below the heading, record the answers to the following questions:

» How would you describe the quality of the metadata summary and description? (e.g., detailed, sparse, non-existent)
» Does the metadata provide information about the organization or person that created the data? If so, who is the author of the data?
» Does the metadata describe the way the data was collected or assembled? If so, what was it?
» Are there any use limitations? If so, what are they?
» What type of spatial reference does the data have? Geographic or Projected?

For this last question, do not worry about the meaning of the terms Geographic or Projected. **Chapter 3** and 4 cover these concepts in detail. For now, try to find the answer in the metadata.

When done, save the Microsoft Word document to your *final* folder. Name the file "Basic QAQC" and save as a .docx file.

Keep the Microsoft Word document open. In the next steps, you add additional documentation.

Skill Drill: Document the Metadata Quality

In this step, you add additional metadata quality documentation to your Microsoft Word document. Using what you learned in the previous step, document the metadata for the county 5m and county 20m shapefiles.

Below your documentation for the County 500K shapefile, add an H2 style heading that says "County 5m." Below the heading, record the answers to the following questions:

» How would you describe the quality of the metadata summary and description? (e.g., detailed, sparse, non-existent)
» Does the metadata provide information about the organization or person that created the data? If so, who is the author of the data?
» Does the metadata describe the way the data was collected or assembled? If so, what was it?
» Are there any use limitations? If so, what are they?
» What type of spatial reference does the data have? Geographic or Projected?

Repeat these steps for the 20m shapefile. When done, re-save your Word document.

Keep the Microsoft Word document open. In the next steps, you add additional documentation.

Precision vs. Accuracy

In this chapter, you learned about the difference between accuracy and precision. **Accuracy** is the degree to which information on a map matches true or accepted values. It is a measure of truth or correctness. **Precision** is the exactness of measurement. It represents the level of detail, the level of measurement, the level of repeatability, or the meticulousness of description. Accuracy sometimes improves with precision, but a high degree of precision *should not* be an automatic indicator of accuracy.

In ArcMap, zoom to the San Francisco Bay (**Figure 1.110**). In the *Table of Contents*, uncheck all of the boxes next to each layer. Then, turn them on and off, one at a time. Make a comparison between the three layers regarding accuracy and precision.

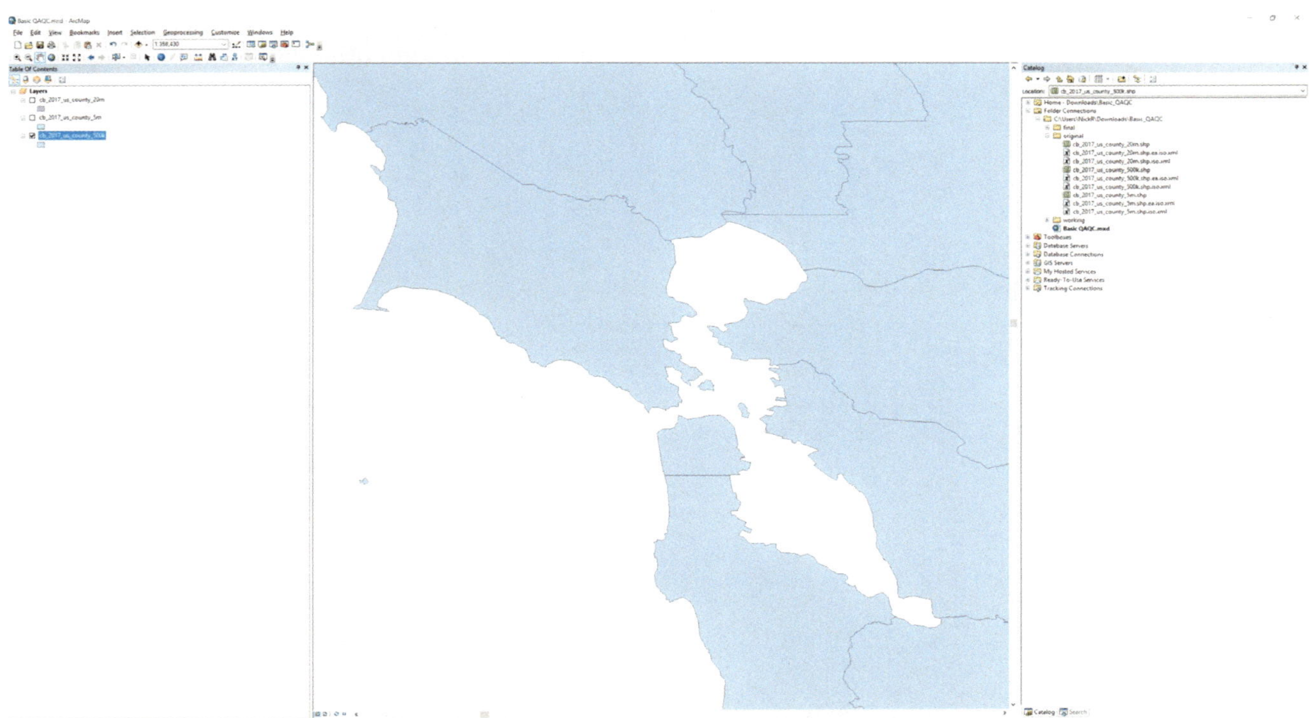

Figure 1.110: You can view one layer at a time by checking and unchecking the boxes.

In the same Microsoft Word document, add a Heading 1 that says, "Visual Inspection." Below the heading, record the answers to the following questions.

» Which dataset has the highest level of precision?
» Based on your understanding of precision and accuracy, is the dataset with the highest level of precision the most accurate? Explain your answer.

Compressing the Project Folder as a 7z File

Sometimes when working with geospatial data, it is necessary to move the data from one location to another. Compressing your workspace folder with the sub-folder structure intact is a safe way to do so. The zip file format is a universal file compression type. However, other file compression types, such as 7z, sometimes work better.

Save and close your Microsoft Word Document. Then, save and close your map document. Compress your Basic_QAQC folder as a 7z file. Navigate to your workspace folder, named after this lab activity. Right-click on the folder, select 7zip, then Add to "Basic_QAQC.7z" (**Figure 1.111**). Be sure the file extension is .7z and not .zip.

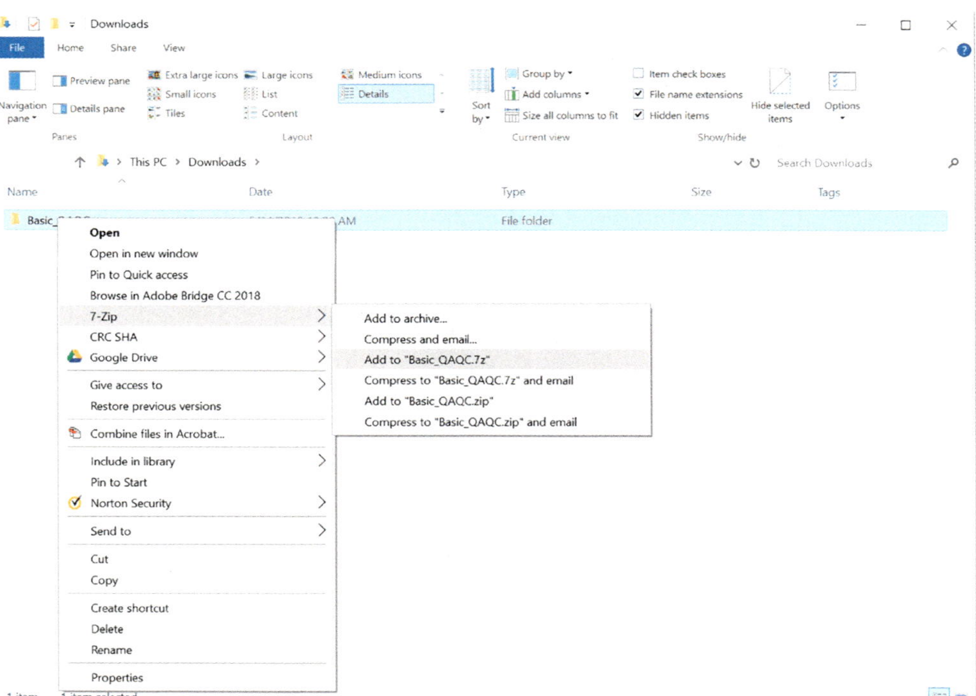

Figure 1.111: You can access the 7zip software by right-clicking on a folder.

When done, you should see your workspace folder compressed as a 7z file (**Figure 1.112**).

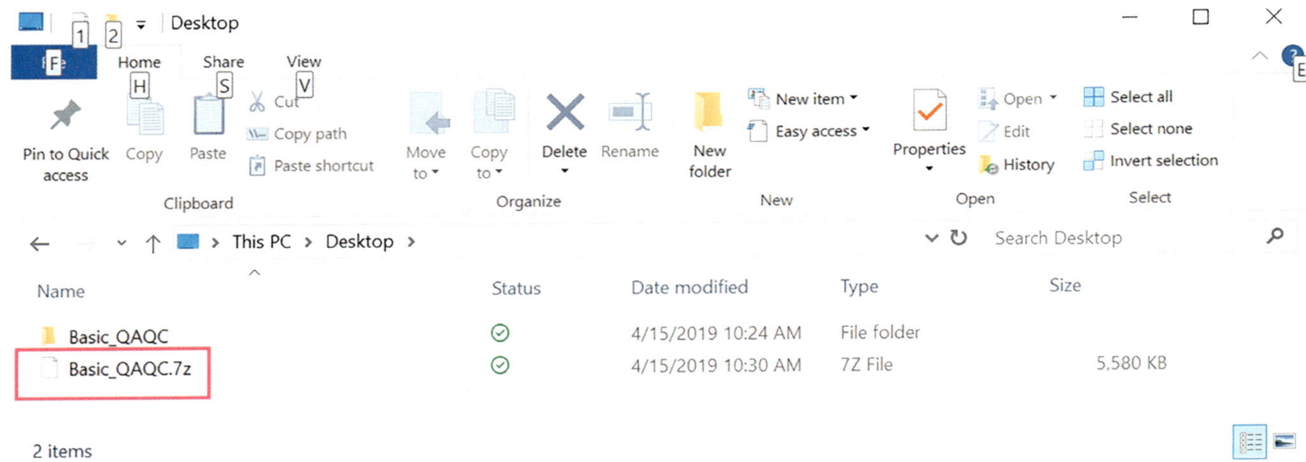

Figure 1.112: Check the file type to be sure it is a 7z file.

Principal Terms

accuracy
arc
attribute
attribute table
cartography
Catalog Window
Catalog Tree
cell
cell size
class break
class intervals
data frame
digital elevation model (DEM)
equal interval classification
final folder
geodesy
geographic information system (GIS)
geographic space
geospatial data
histogram
human error
ISO 19139 Metadata Implementation Specification
land survey
linear feature
local hard drive
manual classification
map document (.mxd)
metadata
mobile mapping
natural breaks classification
networked drives
NoData
node
original folder
pixels
point feature
polygon
positional error
precision
quality assurance and quality control (QAQC)
quantile classification method
raster data model
record
remote sensing
resolution
shapefile
spatial error
Table of Contents
unmanned aircraft systems (UAS)
vector data model
vertices
working folder
workspace folder
workspace

Chapter 2: Communication and Design

In the advent of automobile global positioning system (GPS) navigation, free services like Google Maps, and the availability of computer mapping software for personal computers, maps are part of one's everyday lives in more ways than ever before. Increasing access to public geospatial data, and the availability of free and open source (FOS) software like QGIS and Open Street Maps have placed mapmaking within reach of anyone with computer skills and the willingness to learn. Today everyone can be a mapmaker. However, not every mapmaker is a cartographer. **Cartography** is the art and science of making maps to communicate geospatial information effectively. The difference between a lay mapmaker and a cartographer comes from an understanding of geospatial data, cartographic conventions, and in recognizing that a map's primary role is to communicate information visually. One might compare this to the writing one does every day on social media or in an email, versus a professional author or journalist.

Chapter 2 presents the fundamental principles of cartographic design and communication. Maps are a medium for communication with a unique set of methods and techniques. Understanding how maps communicate will allow one to view maps in a new light and with a critical eye. One begins by learning the essential map elements and the visual variables of graphic communication.

Learning Outcomes

- Explain how maps are a medium for communication
- List basic guidelines for cartographic design and communication
- Describe three primary types of maps
- Demonstrate correct use of the map elements
- Discuss techniques for establishing visual hierarchy and visual balance
- Identify the visual variables for graphic communication
- Express a basic understanding of typography
- Illustrate the visualization techniques for classification
- Critique the communication and design aspects of a published map

Cartographic Conventions

Cartography is a subjective discipline and is often confused with graphic design. Both are forms of visual communication. However, some differences distinguish cartography from graphic design. For example, graphic design might primarily focus on catching the eye for use in advertisement and art while cartography focuses on communicating information. One main difference between graphic design and cartography is the adherence to cartographic conventions. **Cartographic conventions** are a loosely defined set of rules and traditions established over hundreds of years of mapmaking. Think of these practices as the essential grammar for maps. Poor grammar can distract the reader and obscure the message. Proper use of grammar goes unnoticed and allows the author's message to come first in the reader's mind. Like grammar, appropriate use of cartographic conventions is invisible.

While there are no hard and fast rules to cartographic design, cartographic conventions give cartographers some guidelines to follow. Learning them all takes time and practice. This book focuses on the fundamental standards for cartographic design and communication.

Here are some general guidelines to consider:

- A map's design must reflect the purpose and the audience
- Clear communication takes precedence in all design decisions
- A map's design should suit the intended media and the viewing resolution, either electronic or print
- A map's production size and dimensions ought to take place at the same size and dimensions as the final output
- Establish a visual hierarchy among symbols, text, and map elements using visual variables including color, size, and position.

Many cartographic conventions connect to cognitive expectations. **Cognitive expectations** are what the reader expects to see. For example, water is typically blue, and the direction north is generally the top of the map. When cartographers deviate from these expectations, the map reader does more work to interpret the map. One should only deviate from these conventions when there is a good reason.

Map Purpose and Audience

First, consider the following reasons for making a map:

- What data or information does the map convey?
- Is the map meant for navigation, to inform, or to persuade an audience?
- How much detail does the map need?
- Is the map intended to be distributed in print or using electronic media?
- What are the expected size and viewing distance of the map?

A map planned for research, business, or navigation will have different design aesthetics than a map designed for tourism. Having a clear understanding of the map's purpose at the start of the design process will save time and money. There are two basic categories of maps, reference maps, and thematic maps. Both often have a specific purpose and a set of information to communicate. How they do so can be quite different.

Reference Maps

A **reference map** displays a variety of information about the physical features of a geographic region with no special emphasis on any particular one (**Figure 2.01**). The primary purpose of the reference map is to provide information about feature location, distance, and direction. Typical information found on a reference map includes cities, political boundaries, roads and highways, parks, and water bodies. A reference map usually shows these features together without any emphasis on any particular one.

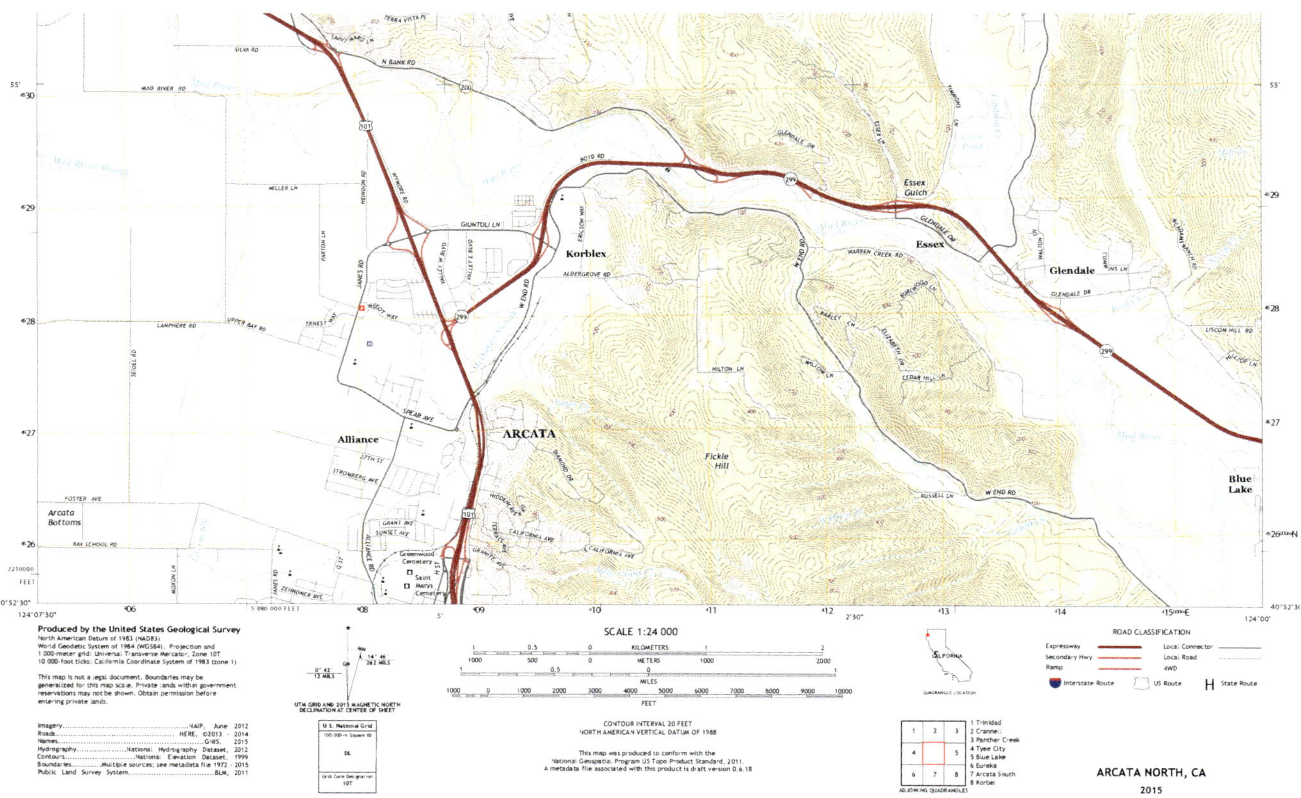

Figure 2.01: The USGS Arcata North Topographic Quadrangle serves as a reference map with a focus on physical features and transportation. Source: USGS.

Navigation Maps

A specialized type of reference map is a **navigation map**. These maps focus on particular features (roads, water bodies, obstacles, environmental hazards, etc.) and help us travel by land, water, or air. The primary map purpose is to assist in route planning and navigation. These types of maps often highlight specialized features and include elements specific to a mode of navigation. For example, cartographers create a nautical chart specifically for water navigation (**Figure 2.02**). Such a map emphasizes information such as shipping lanes, ocean currents, wind patterns, water depth, and navigational hazards. One would find a different set of elements on a map designed for air travel or travel across land.

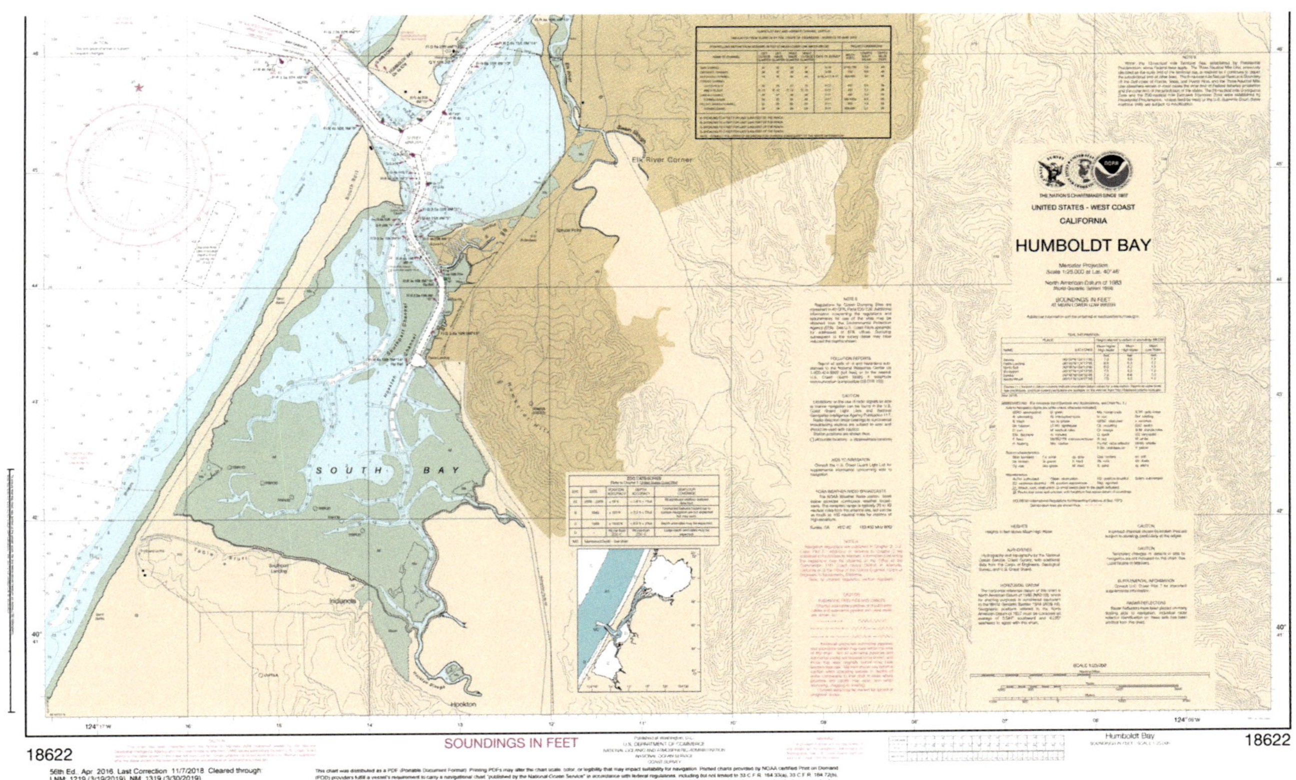

Figure 2.02: This portion of a nautical chart from NOAA depicts many specialized features specific to water navigation. Source: NOAA.

Thematic Maps

A **thematic map** has a very different purpose when compared to reference maps. A thematic map represents conceptual information that is not part of the physical landscape, such as cultural or socioeconomic data (**Figure 2.03**). A map displaying crime rates in the United States is an example of a thematic map. Other common thematic map topics include poverty, disease, election results, population density, age, and race. Thematic maps are often designed to inform or persuade the map audience.

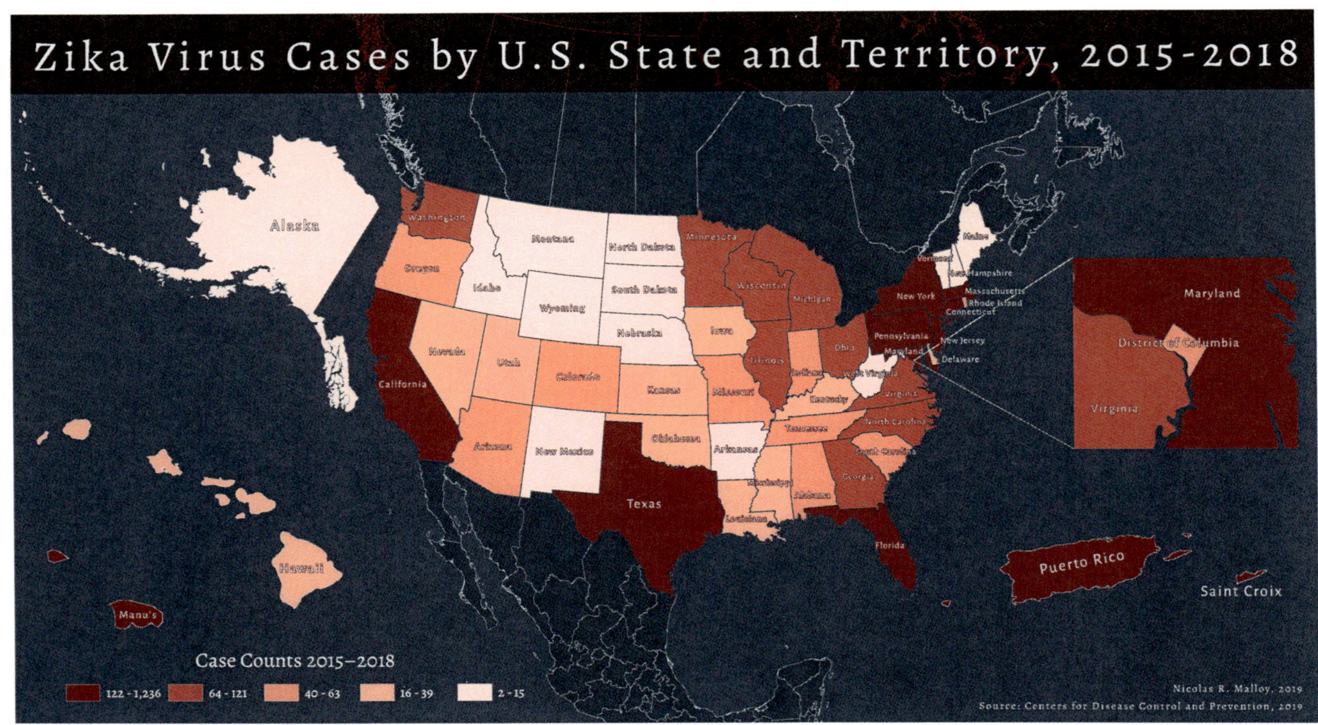

Figure 2.03: This thematic map illustrates a conceptual phenomenon not physically part of the landscape.

Understanding the Audience

To communicate effectively, a cartographer must next consider the audience.

- Are they children?
- Fellow scientists?
- The general public?
- Is the audience familiar with the topic?
- Does the intended audience have any particular areas of expertise?
- What type of map best suits the purpose and audience?

- What type of visual aesthetic does the audience expect?
- Are there any special considerations such as a visual impairment?

Planning and executing maps with a specific audience in mind is a characteristic of an experienced cartographer.

Common Map Elements

Most maps share a standard set of elements that are universally recognized by the map-reader. While cartographers often use these map elements, it is not always necessary to include each one in every map. The decision of whether or not to include any specific map element falls back on the map's purpose and the audience. One develops an understanding of alternate ways to convey these elements as cartographic design experience increases.

Here are a few of the most common map elements:

- Title
- Legend
- Scale indicator
- Directional indicator
- Neatline
- Source note
- Map author & date

Additional elements may include the following:

- Descriptive or explanatory text
- Inset map
- Locator map
- Charts
- Graphs
- Images
- Spatial reference system information

Map Title

A **map title** should be concise and informative. In many cases, the title acts as a gateway to understanding the map's purpose and leads the map reader to the most prominent features (**Figure 2.04**). It helps to think of the map title like a newspaper headline, designed to quickly grab attention and give the reader a sense of what is coming. It is appropriate for a map title to include the subject, place, or time related to the map's purpose.

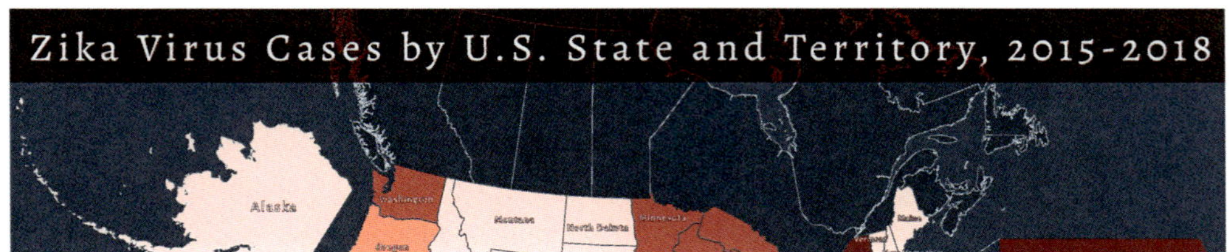

Figure 2.04: The map title is concise and communicates the theme and related time period.

Map Legend

A **map legend**, sometimes called a map key, helps to explain the symbology on the map. A map legend sometimes includes a legend title. The word "Legend," "Key," or similar title should *never* appear in the legend. Using these words as a legend title is redundant since the map reader already knows that they are looking at a legend. A legend title should be explanatory and serve to help the map-reader understand the information included in the legend. If the cartographer feels that a legend title will not serve a purpose, then it should be left out. These conventions are especially crucial in thematic maps where the symbols may otherwise be difficult to interpret.

In many cases, the word "Legend" is the default title inserted by the GIS software. One should always change or remove this wording.

The legend should contain only information visually present on the map. By default, GIS software may incorporate legend items representing features in the data not currently visible on the map. These symbols should be removed unless one creates an interactive map where those items may reappear. Legend symbols should exactly match the size and color of their counterparts on the map.

Legend labels should be concise, relevant, and easy to understand (**Figure 2.05**). For numeric legend labels, insert commas where appropriate and use only the necessary number of decimal places. For example, if the data consists of whole numbers, change the legend labels so that they display whole numbers. If the data has only one or two decimal places, change the legend labels so that they display only one or two decimal places.

By default, some GIS software may assign six places after the decimal. Unless there is a reason to accept this setting, failing to make the proper edits makes one look inexperienced or careless.

Figure 2.05: Instead of using a generic word like "Legend" or "Key," the legend title helps to communicate the meaning of the legend items.

Scale Indicators

There are three common types of scale indicators:

» Verbal scale
» Representative fraction (RF)
» Graphic scale

Most often, cartographers use a **graphic scale**, as it is the most flexible for reproduction and measurement. A graphic scale is a graphic symbol, such as a scale bar, representing ground distance. A **representative fraction (RF)** is the relationship between the map distance over the ground distance, expressed as a ratio or fraction. Cartographers often use representative fractions on topographic and other maps produced in series, so the user can quickly see that the maps are at the same scale. A **verbal scale** uses a word statement to express scale. Cartographers typically reserve the verbal scale for more casual maps, such as tourist maps, where the measurements need not be precise. **Chapter 4** covers the concepts related to scale indicators in greater detail.

Always consider the map's purpose to help determine when and how to include scale indicators appropriately.

Scale indicators supplement the map content and therefore should not compete with the map's purpose, message, or data. They should be legible, but not overly conspicuous. Scale units should be in round numbers that the map user can easily use and understand. In Figure 2.06, the miles, meters, and feet end at thousands, rather than hundreds or smaller units. A good rule of thumb for determining the end value for a graphic scale, such as a scale bar, is to determine which number is most natural to add together repeatedly (e.g., 1500 rather than 1450, or 50 rather than 49). It's also essential to make sure the size of the divisions and the size of the digits are selected such that they don't overlap and confuse the map reader.

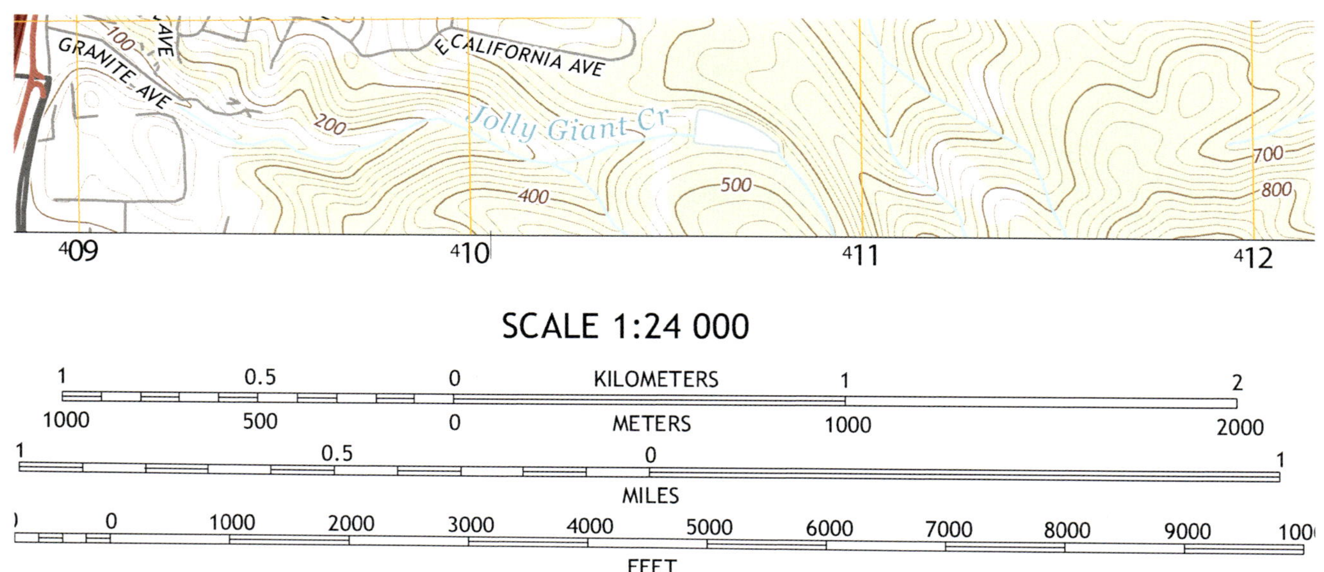

CONTOUR INTERVAL 20 FEET
NORTH AMERICAN VERTICAL DATUM OF 1988

This map was produced to conform with the
National Geospatial Program US Topo Product Standard, 2011.
A metadata file associated with this product is draft version 0.6.18

Figure 2.06: This map uses a graphic scale indicator (scale bar) in three different units, a representative fraction, and a word statement. Source: USGS.

Directional Indicators

A **north arrow** is a directional indicator which shows the direction of north. It is frequently the most abused map element because it is often tempting for novice mapmakers to choose elaborate and decorative styles (**Figure 2.07**). Doing so is considered bad practice since it draws attention away from the map content. Ornate north arrow styles may be appropriate in some cases, such as a tourist map. However, for most purposes, they are not. If in doubt, consider the purpose and audience of the map when deciding on the style of the north arrow.

Figure 2.07: Consider that elaborate north arrows are often one of the things map readers are drawn to in historical or tourist maps, more so than the map content, and one may see why this is not always a good idea.

A graticule is a grid of lines, called **parallels** and **meridians**, draped over the shape of the earth (**Figure 2.08**). **Chapter 3** discusses parallels and meridians in more detail. Like a north arrow, a **graticule** is also a directional indicator. When displayed on a map, a graticule should be drawn underneath large landmasses and remain inconspicuous compared when with the principal map content. Because the graticule serves as a directional indicator, maps typically do not have both a graticule and a north arrow. There are some exceptions to this rule on reference maps where the graticule may serve multiple purposes, such as determining location and scale.

Figure 2.08: A graticule serves as a directional indicator, making a north arrow unnecessary.

Borders and Neatlines

A **map border**, typically a rectangle, outlines the entire contents of a map, including all elements and supporting information. A **neatline** is a specific type of border, which delineates the geographic extent of the map content. It is acceptable practice to have map elements reside between the primary border and the neatline (**Figure 2.09**). Neatlines may also enclose other elements, such as the legend. This practice should not be overused as it may draw too much attention away from the primary map content.

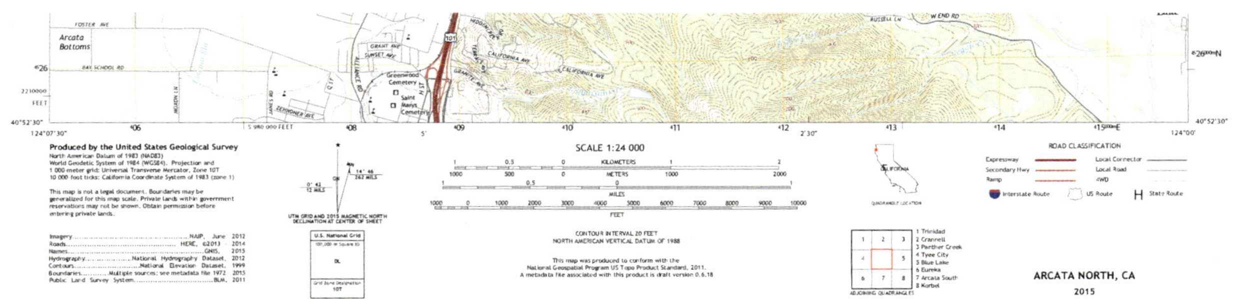

Figure 2.09: The map elements seen here lie just outside the neatline, which delineates the geographic extent of the map. Source: USGS.

Sources, Credits, and Descriptive Text

A **source note** often refers to the person or agency that produced the data and information depicted on the map. Including a source note is similar to citing one's sources for a written paper, although in a more compact manner. For example, the citation for U.S. Census data in a research paper bibliography may have something similar to the following:

» U.S. Census Bureau; Census 2010, Summary File 1, Table P001; generated by Sue Mapmaker; using American FactFinder; <*http://factfinder2.census.gov*>; (8 February 2017).

In a source note, one would likely list the source as U.S. Census along with the date:

» Source: US Census, 2010.

An accomplished cartographer will always cite the data sources when it is appropriate to do so. Some maps may have multiple data sources (**Figure 2.10**). For less formal maps, one may list them together (e.g., Sources: US Census, GNIS, Natural Earth Data), but in the case of a more precise map, such as the topographic map shown below, it may be necessary to be more specific.

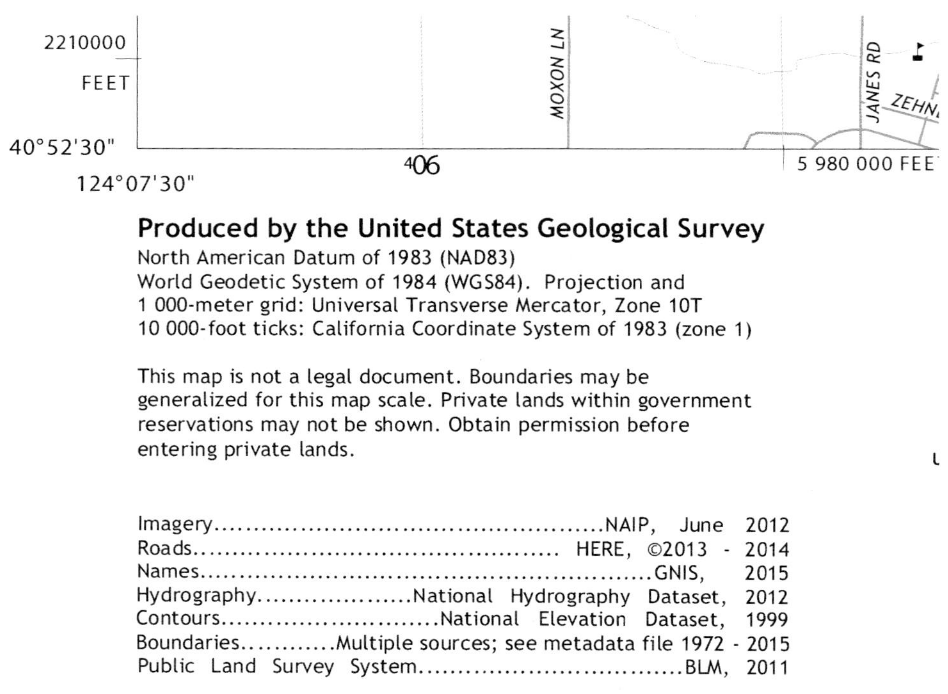

2210000
FEET
40°52'30"
124°07'30"
406
MOXON LN
JANES RD
ZEHN
5 980 000 FEE
Produced by the United States Geological Survey
North American Datum of 1983 (NAD83)
World Geodetic System of 1984 (WGS84). Projection and
1 000-meter grid: Universal Transverse Mercator, Zone 10T
10 000-foot ticks: California Coordinate System of 1983 (zone 1)
This map is not a legal document. Boundaries may be generalized for this map scale. Private lands within government reservations may not be shown. Obtain permission before entering private lands.
Imagery..........NAIP, June 2012
Roads.......... HERE, ©2013 - 2014
Names..........GNIS, 2015
Hydrography..........National Hydrography Dataset, 2012
Contours..........National Elevation Dataset, 1999
Boundaries..........Multiple sources; see metadata file 1972 - 2015
Public Land Survey System..........BLM, 2011

Figure 2.10: Here a map reader can find the agency that produced the map, the data sources, and dates, as well as some supplemental information. Source: USGS

It is common practice to include the map author name and date someplace inconspicuous for reference and credit. An experienced cartographer generally wants to take credit for the maps that they make. However, it may not always be possible depending on the map's purpose, audience, client, or employer. Including a date makes it easy to determine the age and applicability of the map. If one is allowed to include one's name, it should be clear and legible yet inconspicuous so that it does not compete with the map content and purpose.

Inset and Locator Maps

Another map element often used to support the main content is an inset map. An **inset map** shows more detail for a specific region than the primary map. It is usually a small map set within the neatline of a larger map (**Figure 2.11**). A **locator map** provides a broader view, and thus less detail, of the main map's geographic extent. Its purpose is to provide a geographic context for the primary map and used to orient the map reader to an unfamiliar region (**Figure 2.11**).

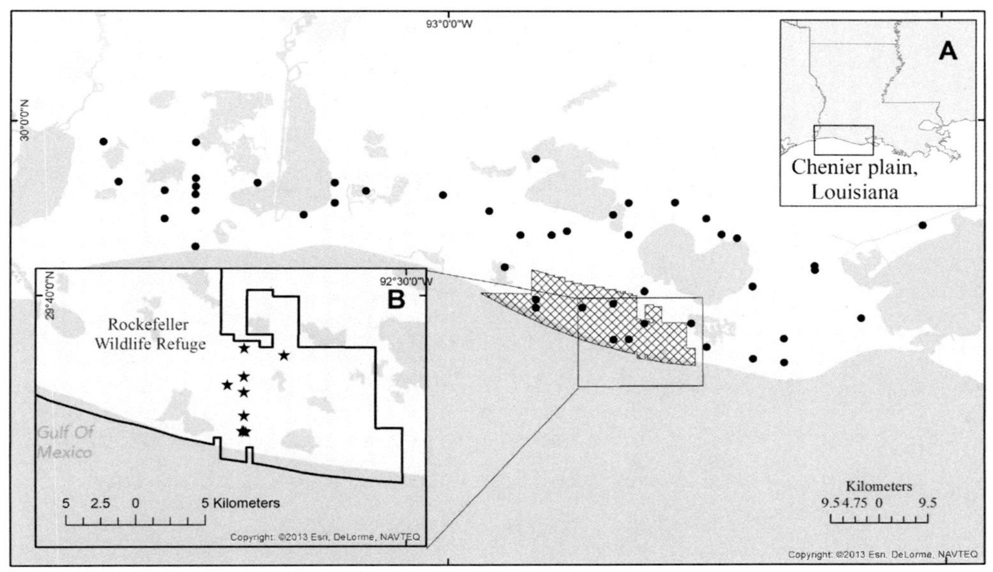

Figure 2.11: This map from the USGS website uses both a locator map (A) to familiarize the map-reader with the location and geographic extent displayed on the map, as well as an inset (B) to provide the map-reader a detailed view of a smaller region. Source: USGS.

Supporting Information

Supplemental information often includes charts, graphs, and images. These elements should always support the purpose and audience of the map. Never use these elements for decorative purposes or to fill up empty space unless it also enhances the information cartographers want to convey to the map-reader. Sometimes cartographers may also want to include a map's spatial reference system. This detail is especially important if the map is used for measuring distances. **Chapter 4** covers spatial reference systems in greater detail. When this information is placed on a map it should be legible but inconspicuous (**Figure 1.12**). **Descriptive text** is used to communicate additional information to the map-reader. Usually, this information is difficult to communicate graphically. Some maps are self-explanatory and do not need descriptive text. Although, if one produces a thematic map one will almost always need some descriptive text to assist the map-reader in understanding the information.

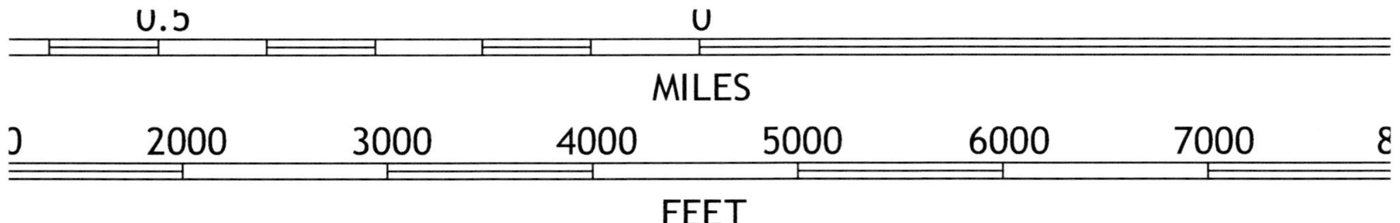

CONTOUR INTERVAL 20 FEET
NORTH AMERICAN VERTICAL DATUM OF 1988

This map was produced to conform with the
National Geospatial Program US Topo Product Standard, 2011.
A metadata file associated with this product is draft version 0.6.18

Figure 2.12: The spatial reference information such as coordinate systems, map projections, and datum, as well as a legal disclaimer are all supporting information.

Visual Hierarchy

How does a map reader know what first to look at on a map? Writers are able to lead their readers through a story from beginning to end by virtue of sequential placement of paragraphs. Even comic books have a structure to the panels that guide us through the story in a particular way. In a map, however, it is much harder to direct the order in which the map reader views the various components. Cartographers use visual hierarchy to accomplish this, by establishing the relative importance of features on a map. **Visual hierarchy** is the logical arrangement of map elements along a wide spectrum between figure and ground. Features high and low in the visual hierarchy have a figure-ground relationship. The contrast between the shape of an object and its background defines the fundamental principles behind the **figure-ground phenomenon** (**Figure 2.13**). Visually, the figure seems to rise above, and the ground appears to recede below. Map elements high in the visual hierarchy *rise to figure*. Map elements low in the visual hierarchy *recede to ground*.

A well-established visual hierarchy places map elements gradually between the figure and the ground in a logical fashion relative to importance. By adjusting the relative weight of features, or density of text or symbols, one can move things up in the hierarchy or down, and thus adjust the relative importance of elements in the map. When deciding which features should rise in the visual hierarchy and which should diminish, consider the map's purpose and audience. The map title is frequently highest in the visual hierarchy, followed by the primary content related to the map's purpose. Auxiliary map elements such as a north arrow, a scale bar, source notes, and credits, should fall *lowest* in the visual hierarchy. Maps executed with a well-defined visual hierarchy are read with minimal effort. Maps lacking visual hierarchy are often unclear and tough to read. Dr. Aileen Buckley, Mapping Center Lead at Esri, offers some tips for improving figure-ground contrast in her blog post, ***Graphic Design Principles for Mapping Figure-Ground Organization***[1].

Figure 2.13: Figure-ground relationships help us to separate shapes and objects (figures) from the shapeless area behind (ground). The figure-ground relationship in this artwork is somewhat ambiguous, making it tricky for the eye to resolve onto one specific shape. Do you see a candlestick? Or do you see two faces?

1 https://www.esri.com/arcgis-blog/products/product/mapping/graphic-design-principles-for-mapping-figure-ground-organization/

Visual Balance

When planning a map layout, the position of an element on a map will imply relative importance to the map-reader. Two things first to consider are the visual center and the visual balance. The **visual center** is the point on the page where the eye is naturally drawn. The visual center is located slightly above the geometric center of the page (**Figure 2.14**).

> ***Elements placed near or above the visual center are more prominent than features below and will rise in the visual hierarchy. For this reason, map titles are typically placed near the top of the map, while supporting elements such as a scale bar, are placed near the bottom.***

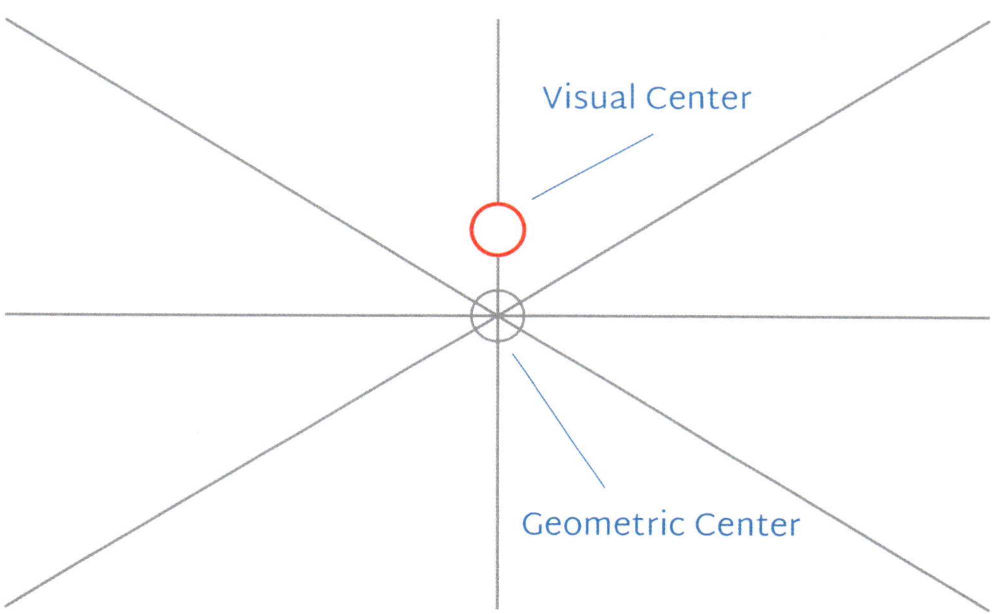

Figure 2.14: The eye will naturally fall just above the geometric center.

Visual balance is a design principle based on using the visual weight of map elements to balance the page around the visual center. Like visual hierarchy, visual weight depends on how much a map element attracts one's eye. The heavier the weight, the higher in the visual hierarchy. Visual balance attempts to leverage elements with heavier weight against elements with lighter weight using visual variables in order to produce a balance across the page. Elements that appear out of balance will draw the map-reader's eye and rise higher in the visual hierarchy, which can be undesirable.

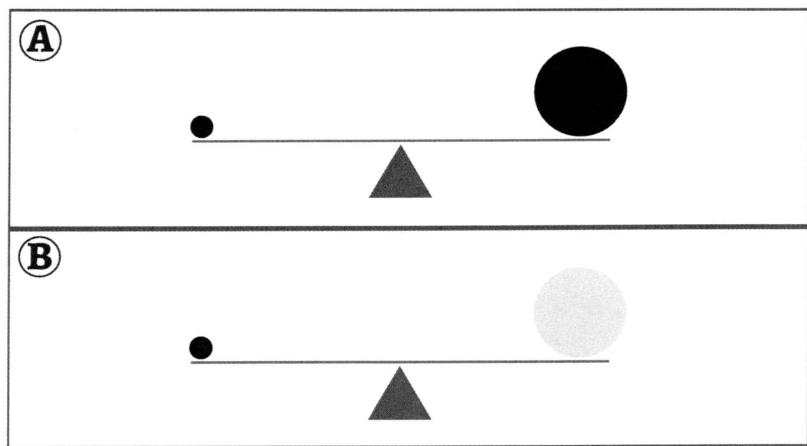

Figure 2.15: Larger objects have more weight than smaller objects, meaning they draw the eye (A). Here the larger circle is more dominant. The reader is left with the impression that the page is about to tip to the right. Darker and more saturated colors have more weight than unsaturated light colors. Here the unbalance in size is countered through the use of color (B).

Visual Variables

Careful implementation of visual variables creates visual hierarchy. **Visual variables** are the graphic representation of information using visualization techniques that clarify symbols and features from one another. Jacques Bertin, a French cartographer, described the visual variables in his 1967 book *Sémiologie Graphique*. Since then, there have been other works expanding the number and types of visual variables. This book discusses some basic visual variables. What is learned here will establish a strong foundation for best practices.

The following are the visual variables of which this book will discuss:

» Shape
» Size
» Orientation
» Hue
» Saturation
» Value
» Position

The visual variables of hue, value, saturation, shape, position, and size all affect weight and hierarchy.

Shape

Cartographers use **shape** to symbolize attributes on a map. Symbols with solid forms or bold, crisp outlines tend to rise higher in the visual hierarchy, while finely detailed symbols may fall lower in the hierarchy (**Figure 2.16**). This can be countered with other variables, such as color, size or position. Highly detailed symbols that more accurately represent the corresponding feature rise higher. Most maps contain a number of different symbols. If each symbol is of relatively equal importance they should have the same level of detail or abstraction.

Figure 2.16: Map symbols should use generalize shapes with just enough detail to communicate the required information. Avoid overly detailed images as this may cause them to rise to figure.

Size

Cartographers also use **size** as another way to establish a visual hierarchy (**Figure 2.17**). The amount of space a feature, a label, or map element takes on a map influence the map-reader's perception of importance. Size can also denote numeric value, so it is important to use it mindfully. A common mistake for the novice mapmaker is to exaggerate one of the secondary map elements such as a North arrow. Inappropriately exaggerating the size of map elements directs the attention of the map-reader away from more important features.

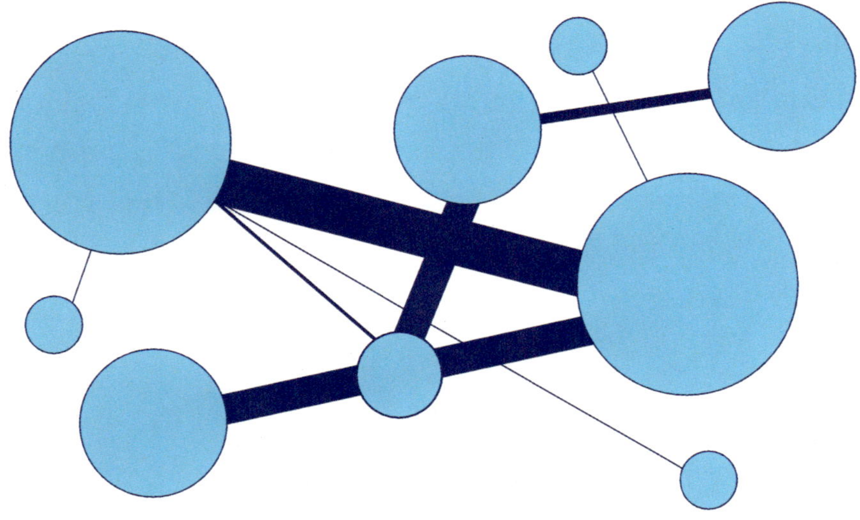

Figure 2.17: Without any labels, this abstract diagram can suggest importance among and between the shapes through the use of variable sized lines and points.

Orientation

Orientation describes the direction labels, symbols, or features are facing on a map. Use orientation patterns to distinguish between similar features or to communicate directional phenomena, such as wind or ocean currents (**Figure 2.18**). The horizontal orientation of shapes and patterns recedes in the visual hierarchy while vertical orientation tends to rise. The diagonal orientation of patterns and shapes will tend to rise higher than both horizontal and vertical in the visual hierarchy.

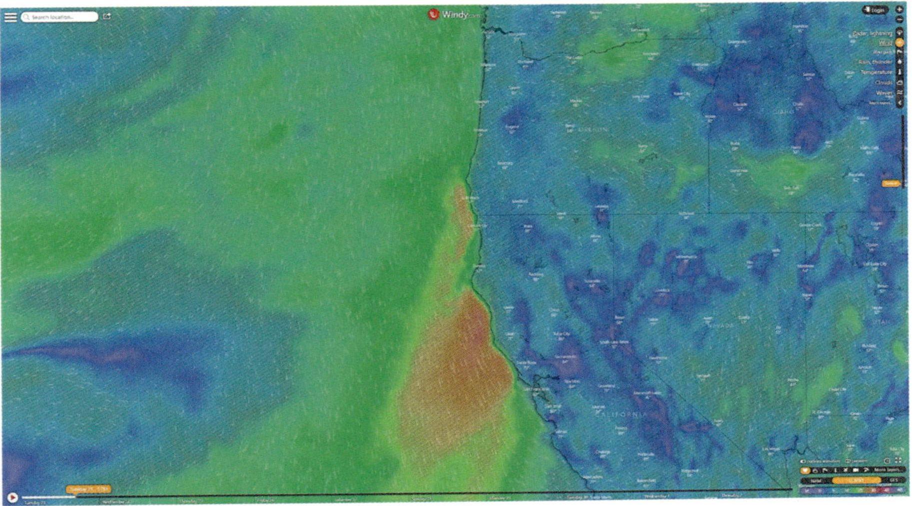

Figure 2.18: In this map, symbols are oriented to correspond with wind direction. Source: *windy.com*.

Color

Color is frequently the most difficult visual variable to master due to its complexity. It is also one of the most powerful visual variables for establishing visual hierarchy. To understand color, cartographers use a color space. A **color space** is a means to organize and reproduce color. Some common color spaces include RGB (red, green, blue) or CMYK (cyan magenta yellow and black). In terms of visual variables, the following three dimensions characterize color:

- Hue
- Saturation
- Value

These qualities are collectively referred to as **HSV color space** (**Figure 2.19**). Though there are other color spaces, these are topics better suited for a discussion on map distribution, such as digital (RGB) and print (CMYK). The HSV color space is ideal for describing qualities of color that directly affect *visual hierarchy*.

Figure 2.19: HSV color has three dimensions. The hue changes along the circumference. Saturation decreases down at the outer edges, while value decreases inward toward the center. Colors have both the highest value and saturation about the circumference.

Hue

The names one commonly associates with color, such as orange, purple, or red, define the **hue** (**Figure 2.20**). The dominant wavelengths of visible light distinguish one hue from another. **Chapter 6** covers the properties of different wavelengths of light.

Figure 2.20: The colors in the wheel maintain roughly the same value and saturation, though each is different based on hue.

Use hue to establish a qualitative organization of map data. A qualitative organization is achieved when features are grouped into different categories sharing similar characteristics. Cartographic convention suggests the use of a specific set of hues to correspond to certain land cover features. For example, map users associate blue with water features and green with vegetation (**Figure 2.21**). There may be times when it is okay to use colors with strong geographic connotations for other purposes. This is especially true for thematic maps. Always consider purpose and audience when choosing colors.

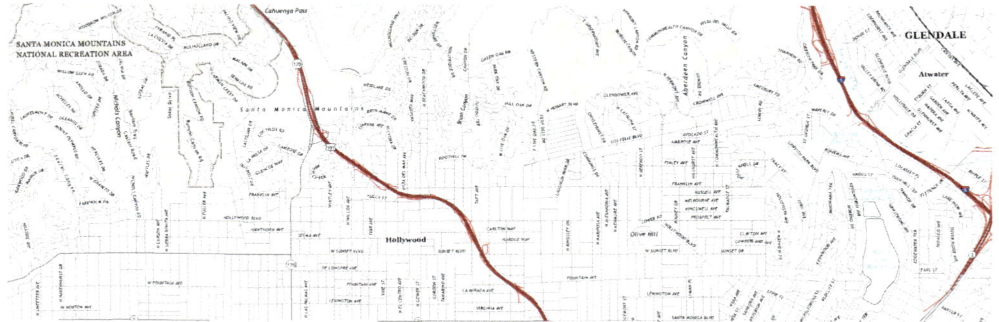

Figure 2.21: In this example, notice how important features, such as the roads, are given bold hues that leap off the page, while other items, such as forests and water features, are softer, receding to ground. Source: USGS.

Value

The lightness or darkness of a particular hue defines its **value** (**Figure 2.22**). Visually, colors with a high value appear to reflect a higher amount of light. Use changes in value to symbolize quantitative data. On thematic maps representing quantitative data, a feature with a larger numeric attribute should appear darker than a feature with a smaller numeric attribute (**Figure 2.23**).

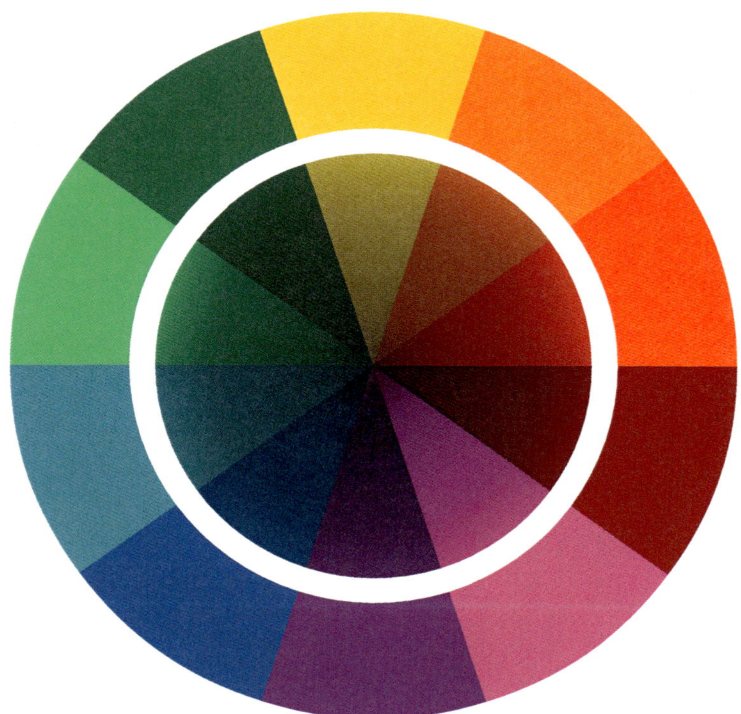

Figure 2.22: The outer ring represents the original colors. The inner circle represents a decrease in value towards the center.

In the map below, the violet hue ranges from light to dark to represent the population per square kilometer of indigenous persons living in the United States (**Figure 2.23**). People often interpret darker values as larger numbers, so this not only draws focus to higher concentrations but also fits well within the map reader's expectations. For this reason, using darker values to represent greater intensity is another of the cartographic conventions one uses to make it easier for the audience to understand the map.

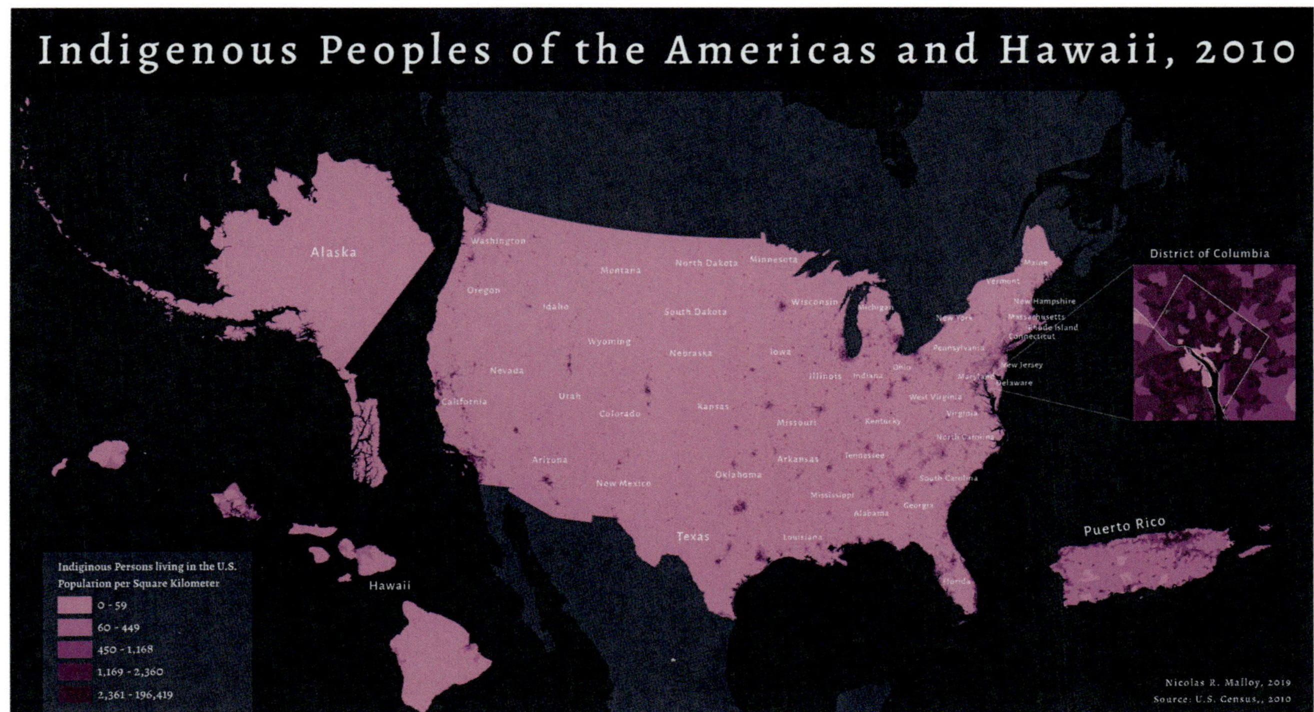

Figure 2.23: This map uses value to represent population density. For this example, changes in saturation were intentionally avoided. Normally, a cartographer uses both saturation and value together.

Saturation, sometimes referred to as chroma, represents the purity or intensity of the color (**Figure 2.24**). For example, candy-apple red is a highly saturated color, but pastel pink is a desaturated color. Saturation is also sometimes described as the amount of gray added to a color, so a saturated color may look bright and vibrant, while a desaturated version may look somewhat muddy or dull. Saturation not only affects the visual hierarchy but, like value, can be used to represent quantitative data. More saturation means higher numbers, just as with darker values.

Figure 2.24: The outer ring represents the original colors. The inner circle represents a decrease in saturation towards the center. Pastels, like pink, have very low saturation. Highly saturated colors rise higher in the visual hierarchy and should be used with deliberation. The overuse of saturated colors on a single map is usually a sign of inexperience.

The map depicting malaria diseases of the United States in 1870 uses increasingly saturated red to indicate increased mortality from the disease (**Figure 2.25**). The combination of red, a high-impact color, and deep saturation makes the high-risk areas jump to the top of the hierarchy and grab the viewer's attention.

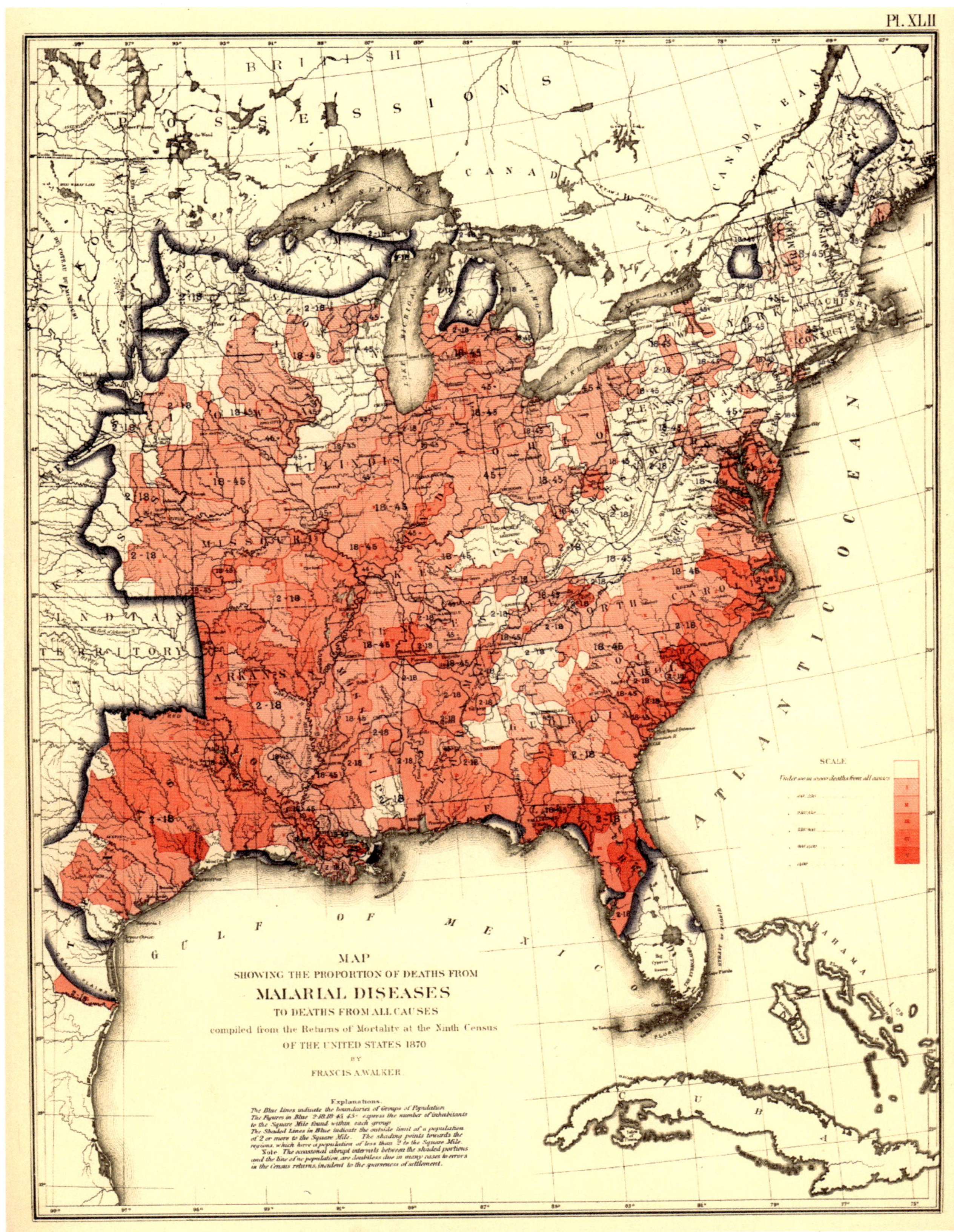

Figure 2.25: This map uses differences in saturation to indicate quantity. Source: Library of Congress.

Classification and Visualization

Classification is the method used to group something based on shared qualities, magnitude, or quantity. Through classification, one simplifies a phenomenon to make it easier to understand. Be aware that this will suggest to the map-reader that these features are geographically homogenous even though in reality this will not be the case. There are two broad categories of classification, quantitative and qualitative. As discussed earlier, **qualitative classification** groups map features together based on the similarity of attributes. Use shape, orientation, and hue to create a distinction between qualitative classes. The changes in hue displayed in the top color scheme make it suitable for qualitative classification (**Figure 2.26**). The changes in value presented in the middle color scheme make it ideal for **quantitative classification** such as changes in magnitude, rate, or quantity. A diverging color scheme is also suitable for quantitative data, though the middle regions indicate a significant value or break, such as an average, or the break between positive and negative values.

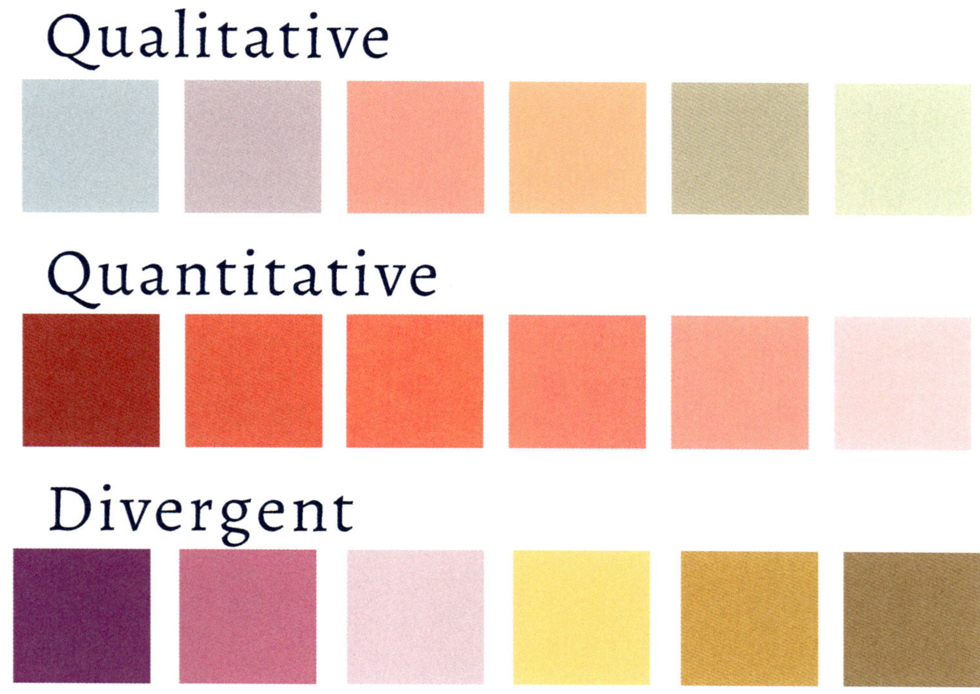

Figure 2.26: The top row is an example of changes in hue. The middle row demonstrates changes in saturation and value. The bottom row shows a divergent color scheme, which uses changes in hue, saturation, and value.

Typography

Because label placement is a major component of every map, cartographic proficiency requires a basic understanding of typography. **Typography** is the art and technique of type placement, arrangement, and design for the purpose of aesthetics, communication, and readability of written language. A skilled cartographer views the typography on a map as both a means to communicate language and as a visual variable. Typography is useful in establishing an effective visual hierarchy to reduce initial confusion and enhance communication. This chapter covers type structure, measurement, and classification to establish a common framework of concepts and vocabulary for effective use of type in cartographic design.

Typefaces and Fonts

Most people are familiar with the term font. Yet in most cases, the term font is often confused with the term typeface. A **typeface** encompasses style and symbology across an entire alphabet including letters, numbers, and punctuation. A typeface is also an entire family of fonts, such as Tahoma, Book Antiqua, or Courier New (**Figure 2.27**). A **font** is a subset of a typeface. It describes a specific size and style of a typeface. Tahoma, Book Antiqua, and Courier New are all *typefaces*. Tahoma bold 10-point, Book Antiqua italic 12-point, and Courier New bold italic 8-point are all *fonts* because they describe a style (italic and bold) and size (10, 8, and 12-point) (**Figure 2.28**).

Tahoma

Book Antiqua

Courier New

Figure 2.27: Each of these represents a different typeface, Tahoma, Book Antiqua, and Courier New.

Tahoma

Book Antiqua

Courier New

Figure 2.28: A font is more specific than a typeface. The letters at the top use the font Tahoma bold 10-point. The letters in the middle use Book Antiqua italic 12-point and the letters at the bottom use the font Courier New bold italic 8-point.

Type Size

Type size is measured in points. A point is 1/72 of an inch, so the 72-point font is roughly one inch tall. However, **point size** is commonly misunderstood because point size does not refer to the size of a character, but rather the space that contains the character. Historically, characters carved or cast into the faces of wooden or metal blocks were used to transfer ink onto paper using a printing press. Making all of the blocks the same size considerably sped up the typesetter's ability to set up a page for printing. Point size referred to the size of this block on which the type was cast and allowed for consistency and predictability in printing (**Figure 2.29**).

In modern times, point size refers to an invisible bounding box that surrounds each character. Characters of different typefaces will occupy more or less space within its bounding box. For this reason, different typefaces may appear larger or smaller from each other even though they are set at the same point size. Some typefaces may also be rounder, while others are more upright and oval in forms, which can affect one's perception of size. In the example below (**Figure 2.30**), both fonts are set at 12-point, which means they would have been cast onto the same size blocks. However, one quickly sees that the variations in the individual characters make one appear smaller and take up less space.

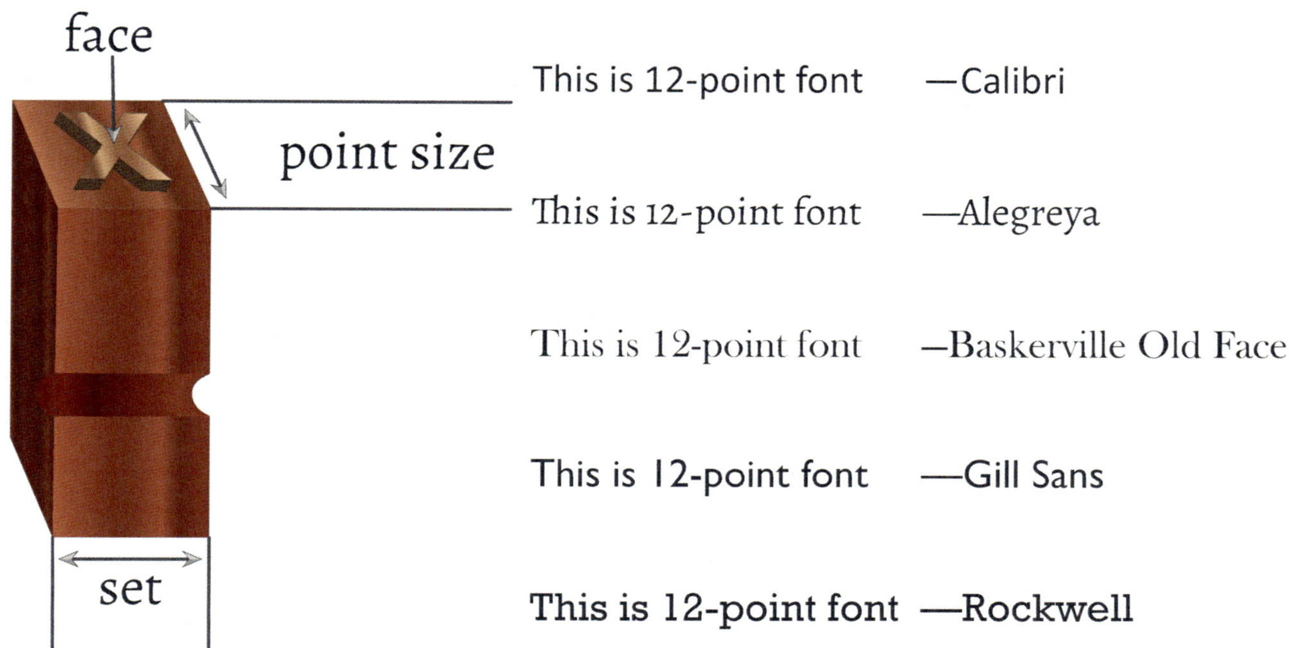
face
point size
set
This is 12-point font —Calibri
This is 12-point font —Alegreya
This is 12-point font —Baskerville Old Face
This is 12-point font —Gill Sans
This is 12-point font —Rockwell

Figure 2.29: Point size refers to the size of the block onto which the letters are carved or cast.

Lorem ipsum dolor sit amet, a et phasellus at massa ut amet, felis dui gravida ac suscipit, ut volutpat vitae pulvinar phasellus. A sit, eu morbi viverra consequat vitae ante a, morbi risus at gravida, amet vel consectetuer elementum suspendisse, ac sed viverra aenean ut. Nec condimentum, aenean erat quisque, neque ligula, sapien metus sem. Tellus cursus faucibus pretium, wisi eget aliquam in, justo augue vestibulum augue sodales. Diam sem vestibulum sit, bibendum eum posuere proin, torquent fusce et nec, vitae libero tempor at donec felis wisi. Feugiat ornare, faucibus at vel quis nam varius wisi, lectus nullam ac lacinia non pulvinar.

Times New Roman 12 Point

Lorem ipsum dolor sit amet, a et phasellus at massa ut amet, felis dui gravida ac suscipit, ut volutpat vitae pulvinar phasellus. A sit, eu morbi viverra consequat vitae ante a, morbi risus at gravida, amet vel consectetuer elementum suspendisse, ac sed viverra aenean ut. Nec condimentum, aenean erat quisque, neque ligula, sapien metus sem. Tellus cursus faucibus pretium, wisi eget aliquam in, justo augue vestibulum augue sodales. Diam sem vestibulum sit, bibendum eum posuere proin, torquent fusce et nec, vitae libero tempor at donec felis wisi. Feugiat ornare, faucibus at vel quis nam varius wisi, lectus nullam ac lacinia non pulvinar.

Arial 12 Point

Figure 2.30: Each of these fonts uses the same point size.

As a rule of thumb, the smallest legible point size for use in print is usually a 6-point font. Digital type read onscreen may appear slightly larger than print. There are too many unknown variables to account for when the intended map audience reads onscreen, including a range of devices and resolutions that can adversely impact legibility.

Never rely solely on point size. Instead, one should use their eyes and judgment when choosing the point size. If possible, print a draft of the map if it is designed for print or view the map on a variety of electronic devices if it is designed for the screen.

The Letterform

A **letterform** is a term used to describe the shape of the letter glyph. Professional type designers have an extensive list of terms to describe the details of letterform anatomy (**Figure 2.31**). This book covers a few of the basic terms.

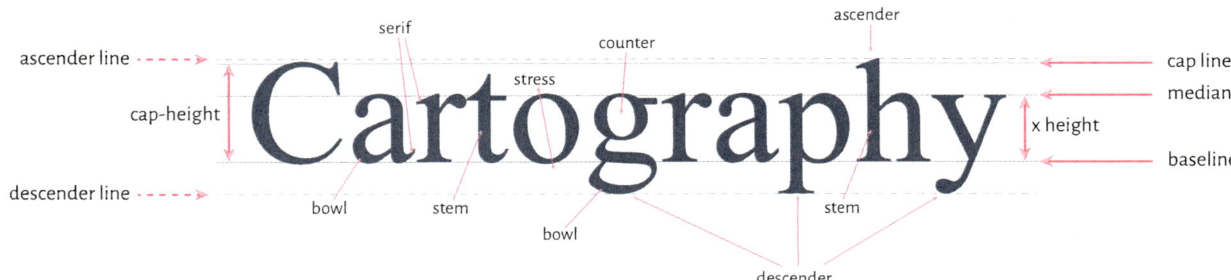

Figure 2.31: The letterform has many characteristics that determine the shape of the letter glyph.

The following are the letterform structures of which this book will discuss:

- x-height
- Median
- Baseline
- Cap-height
- Ascender line
- Descender line
- Ascender
- Descender
- Stem
- Counter
- Bowl
- Stress
- Serif

Letterform Boundaries

Letterform boundaries are imaginary lines that define the upper and the lower limits of letterform anatomy. The **x-height** defines the height of the main body of lowercase letters. The height of the lowercase letter x is used for this purpose (**Figure 2.32**). Since type on maps is often very small, choosing a typeface with a greater x-height can increase legibility. The lower limit of the x-height, called the **baseline**, is the imaginary line on which all the characters sit. The **cap height** is like the x-height, though it refers to the height of an uppercase X. The upper boundary of the cap-height is called the **cap line**. The **ascender** is the part of the lowercase letterform that rises above the x-height. The uppermost limit of the ascender is called the **ascender line**. Sometimes the ascender line rises above the cap line. The **descender** is the part of the lowercase letterform falls below the baseline. The **descender line** is the lower limit of the descender. Exaggerated ascenders and descenders can give fonts an elongated look and can be difficult to use as labels on maps because they can interfere with nearby features and other labels.

Figure 2.32: The baseline, cap line, ascender and descender lines are used to demarcate the boundaries of the letterform components.

Letterform anatomy

Letterform anatomy are specific parts of the letterform. The **stem** is the main part of the letterform that is straight and either vertical or diagonal (**Figure 2.33**). The **counter** is the negative space on the interior of the letterform. A **bowl** is the rounded part of a letterform, including the portion that encloses a counter. **Stress** is the orientation of a letterform's curved strokes from thin to thick. A **serif** is a small tick, flourish, or decorative line at the end of a stem. As mentioned previously, the **ascender** and the **descender** are parts of the lower case letterform that either rise above the x-height or fall below the baseline.

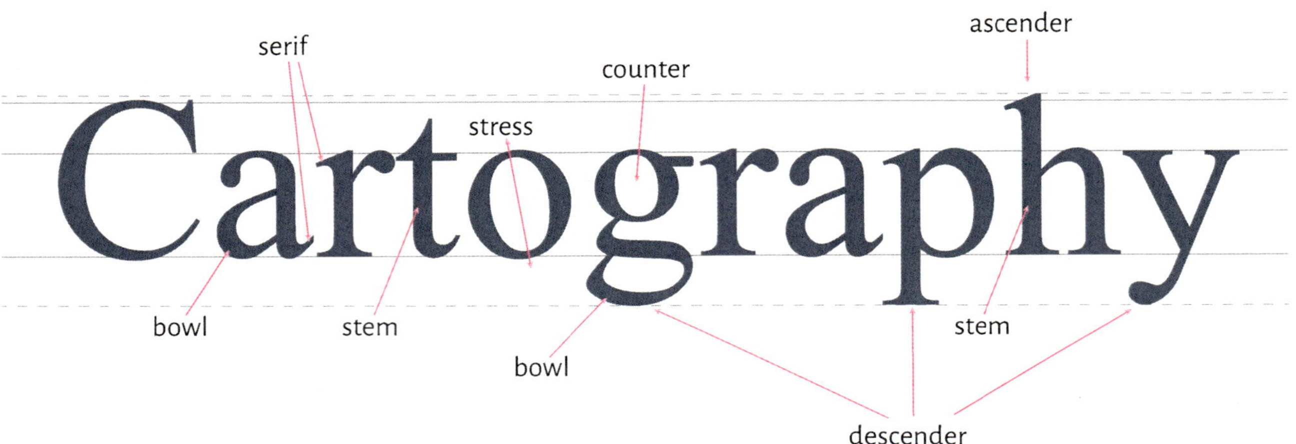

Figure 2.33: Understanding the parts of the letterform allows cartographers to make informed decisions when choosing a typeface.

Serif versus Sans Serif

Most typefaces fall within two broad categories, serif, and sans serif (**Figure 2.35**). Sans means without in French. **Sans serif typefaces** do not have any serifs at the end of stems, while **serif typefaces** do.

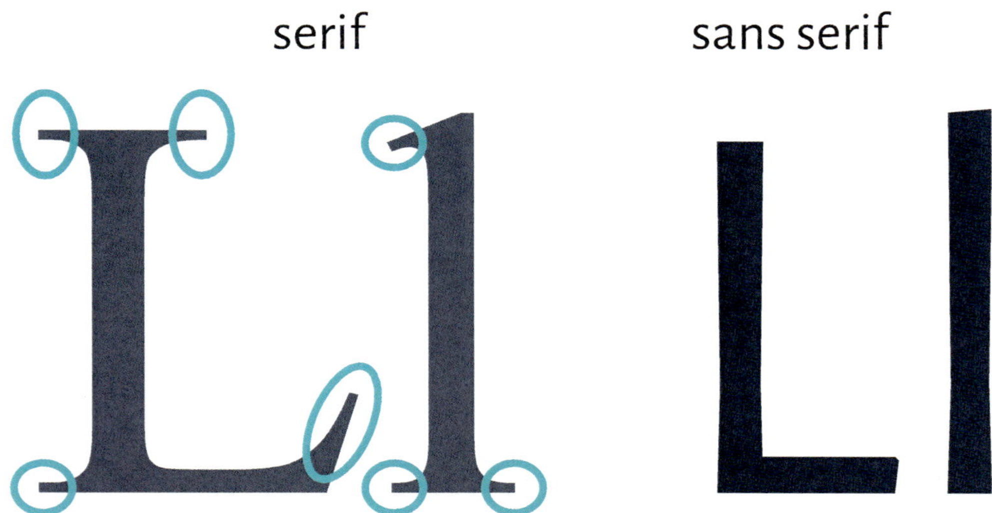

Figure 2.34: The blue circles indicate the parts of the letterform which are serifs.

Professional type designers classify typefaces into many more distinct categories, often based on historical periods and geographic origin. This book covers three: serif, sans-serif, and display. A **display typeface** is a catchall category, highly stylized to catch the reader's attention (**Figure 2.35**). Typefaces in this group are often difficult to read and should be used sparingly. Keep things simple and try to limit the total number of typefaces on a map. A good starting point is to choose one serif and one sans serif typeface, each in balance with the other. If one is new to coordinating fonts, a fun resource is *Type Connection*[2], a fun game that helps one pair a serif and a sans serif font. Like a discordant note, typefaces that clash can be distracting. Traditionally, cultural features such as the names of political boundaries, towns, and roads use sans serif typeface. Physical features such as mountains, forests, and water bodies use a serif typeface. Use display typefaces with caution, as they can render one's map illegible. They can sometimes safely be used in a map title to convey a personality, elicit an emotion, or to catch the map-reader's attention. Never use display type for main map content.

Serif Typefaces	San Serif Typefaces	Display Typefaces
Alegreya	Arial	ALGERIAN
Baskerville Old Face	Calibri	Broadway
Bookman Old Style	Century Gothic	Colonna MT
Century	Corbel	Harrington
Georgia	Gill Sans	Hobo Std Medium
Palatino Linotype	Myriad Pro	Jokerman
Times New Roman	Veranda	Snap ITC

Figure 2.35: Here are a few examples of serif, sans serif and display typefaces.

2 http://www.typeconnection.com/

Typographic Design and Communication

Cartographic use of typography includes three design and communication goals:

» Hierarchy

» Organization

» Geographic association

Cartographers use the same visual variables to establish hierarchy and organization for type as they do for symbols and features. One may have noticed in this book that words in bold type tend to stand out more, as well as the larger fonts used for headings. In addition to size and weight, there are four additional typographic visual variables:

» Spacing

» Leading

» Style

» Case

Spacing is the adjustment of space between letters and words. Space between characters can increase or decrease visual hierarchy. On complex maps, a tight spacing is sometimes necessary to fit many labels and text. Tracking and kerning are two distinct spacing techniques.

Lorem ipsum dolor sit amet, a et phasellus at massa ut amet, felis dui gravida ac suscipit, ut volutpat vitae pulvinar phasellus. A sit, eu morbi viverra consequat vitae ante a, morbi risus at gravida, amet vel consectetuer elementum suspendisse, ac sed viverra aenean ut. Nec condimentum, aenean erat quisque, neque ligula, sapien metus sem. Tellus cursus faucibus pretium, wisi eget aliquam in, justo augue vestibulum augue sodales. Diam sem vestibulum sit, bibendum eum posuere proin, torquent fusce et nec, vitae libero tempor at donec felis wisi. Feugiat ornare, faucibus at vel quis nam varius wisi, lectus nullam ac lacinia non pulvinar.

Lorem ipsum dolor sit amet, a et phasellus at massa ut amet, felis dui gravida ac suscipit, ut volutpat vitae pulvinar phasellus. A sit, eu morbi viverra consequat vitae ante a, morbi risus at gravida, amet vel consectetuer elementum suspendisse, ac sed viverra aenean ut. Nec condimentum, aenean erat quisque, neque ligula, sapien metus sem. Tellus cursus faucibus pretium, wisi eget aliquam in, justo augue vestibulum augue sodales. Diam sem vestibulum sit, bibendum eum posuere proin, torquent fusce et nec, vitae libero tempor at donec felis wisi. Feugiat ornare, faucibus at vel quis nam varius wisi, lectus nullam ac lacinia non

Figure 2.36: In this image, the font size and line height are the same, but a different spacing is used.

Tracking, sometimes called letterspacing, is the uniform adjustment of space between words and letters across the entire typeface (**Figure 2.37**). Use tracking to modify larger bodies of text, such as a paragraph, to improve readability and balance negative space on a map. In cartography, tracking is often used to imply the extent of a feature across an area such as mountain range, state, or country. Though this makes the label larger, it has the effect of making the text feel less dense, and therefore lowering the visual hierarchy.

New York City

N e w Y o r k C i t y

Tracking

Figure 2.37: Tracking is the uniform adjustment of space for a small range of text such as a single word or group of words.

Kerning is the adjustment of space between letters (**Figure 2.38**). Normally, the space between letters is precisely uniform, as if there was a gap between the type blocks. With kerning, space is adjusted according to the letters on either side, creating an aesthetically pleasing result. Manual kerning is exceptionally time-consuming and should be reserved for titles and similar small blocks of text on a map. Many applications support auto kerning, which can significantly reduce the time spent. Advanced applications such as Adobe InDesign™ and Adobe Illustrator™, typically have several options for fine-tuning the kerning.

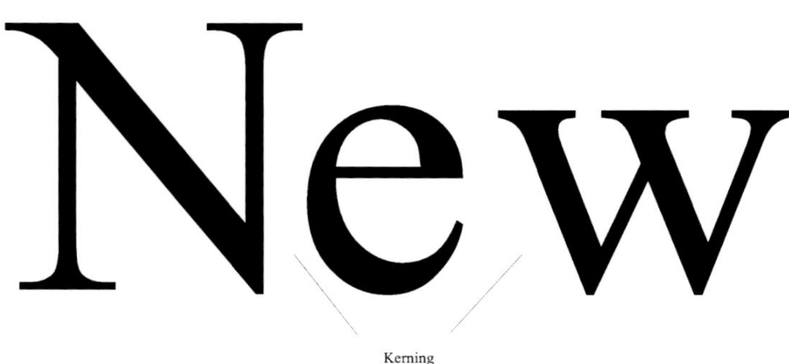

Figure 2.38: Kerning is the adjustment of space between individual typeface characters.

Leading, also referred to as line height or line spacing, is the space between lines of type, measured in points from one baseline to the next (**Figure 2.39**). This term also comes from the world of handset type and refers to the thin strips of lead placed between rows of type. Increasing leading in a body of text meant for continuous reading, such as a paragraph, can improve readability and has the effect of lowering visual hierarchy. A compact line spacing increases density and causes text to rise in visual hierarchy. Leading is used more commonly between items in a legend, or in blocks of supporting text, but can also be helpful when dealing with long feature labels that wrap onto multiple lines. Too much leading, however, can make rows of text appear to be disconnected, so cartographers must use discretion when increasing the leading.

Lorem ipsum dolor sit amet, a et phasellus at massa ut amet, felis dui gravida ac suscipit, ut volutpat vitae pulvinar phasellus. A sit, eu morbi viverra consequat vitae ante a, morbi risus at gravida, amet vel consectetuer elementum suspendisse, ac sed viverra aenean ut. Nec condimentum, aenean erat quisque, neque ligula, sapien metus sem. Tellus cursus faucibus pretium, wisi eget aliquam in, justo augue vestibulum augue sodales. Diam sem vestibulum sit, bibendum eum posuere proin, torquent fusce et nec, vitae libero tempor at donec felis wisi. Feugiat ornare, faucibus at vel quis nam varius wisi, lectus nullam ac lacinia non pulvinar.

Lorem ipsum dolor sit amet, a et phasellus at massa ut amet, felis dui gravida ac suscipit, ut volutpat vitae pulvinar phasellus. A sit, eu morbi viverra consequat vitae ante a, morbi risus at gravida, amet vel consectetuer elementum suspendisse, ac sed viverra aenean ut. Nec condimentum, aenean erat quisque, neque ligula, sapien metus sem. Tellus cursus faucibus pretium, wisi eget aliquam in, justo augue vestibulum augue sodales. Diam sem vestibulum sit, bibendum eum posuere proin, torquent fusce et nec, vitae libero tempor at donec felis wisi. Feugiat ornare, faucibus at vel quis nam varius wisi, lectus nullam ac lacinia non pulvinar.

Figure 2.39: In this image, the font and letter spacing are the same, but different leading is used.

Cartographers use font style and case to indicate hierarchy and organization. The most common styles used in cartography are regular, italics, and bold. **Regular**, sometimes called roman, refers to a font that does not slant. This font style is appropriate for most circumstances and especially for long passages of text. **Italics** are a supplemental set of fonts with an oblique or slanted appearance designed to complement their regular counterpart. Traditionally italics are used to denote categories or for emphasis. As with serif and sans-serif fonts, the choice of bold, italic or regular can help the map reader recognize similar types of features or understand the relative importance.

A common cartographic convention is to label water features in italics, which can sometimes convey a sense of motion but also distinguishes rivers from roads, lakes from forests, and generally reduces confusion, particularly in black and white maps where the use of blue to indicate the water is not an option. The oblique slant of italics increases the density of the font, which tends to rise to figure. Italics are typically lighter in weight and appear smaller than a regular font, which tends to recede to ground. The net effect is often neutral which makes italics useful for denoting a category without having a significant impact on figure-ground balance. Adding weight to a font above regular creates a **bold style**, which causes the label to rise in the visual hierarchy. This can be great for calling out significant features within a category (such as larger cities or political capitals) or developing a general hierarchy in a label-dense map.

Regular *Medium Italic*

Italic **Bold**

Medium ***Extra Bold Italic***

Figure 2.40: Font style can help develop the hierarchy and build map grammar, and, depending on the font, move a label toward figure or ground

Label Placement

Label placement establishes a geographic association. GIS software generally comes with features that include automatic label placement. Automatic label placement saves a considerable amount of time but should never be relied upon solely. Cartographers use automatic label placement as a starting point and then proceed to fine-tune the typography on a map. Effective label placement is legible and clearly associated with a single geographic feature. Proximity is the most reliable indicator of geographic association, while the position in or around a feature serves to reinforce legibility and association (**Figure 2.41**). Label placement on complex maps is often about compromise. One must try to find the optimal label position while maintaining balance and clarity among other labels and features nearby.

Figure 2.41: The labels on the left have ambiguous geographic association. On the right, the labels are clearly associated with the point symbols using location and proximity.

The general guidelines for label placement depend upon the shape of features labeled and tend to fall into three categories:

» Point feature labels
» Areal feature labels
» Linear feature labels

Point Feature Labels

There are four optimal locations for point feature label placement, each approximately 45° off the point from horizontal (**Figure 2.42**). One may place point labels at 90° or 270° if necessary but never set them at 0° or 180° off the point. Doing so may decrease legibility or cause the point symbol to be confused with its label. Often the best location depends upon geographic context and the location and number of other features and labels nearby. As a rule, a point label should not overlap other features or labels. It also should be closer to the point it represents than any other point. Additionally, it should not be confused with the point symbol itself. Labels for coastal point features should be entirely in the ocean or water body. Labels for non-coastal point features should be entirely on land.

Areal Feature Labels

Areal feature labels should reside inside the extent of the area feature whenever possible. For large areas, use tracking to extend the label across the area rather than increasing font size (**Figure 2.43**). Do not increase the font size unless the areal feature important and one intends for the label to rise in the visual hierarchy.

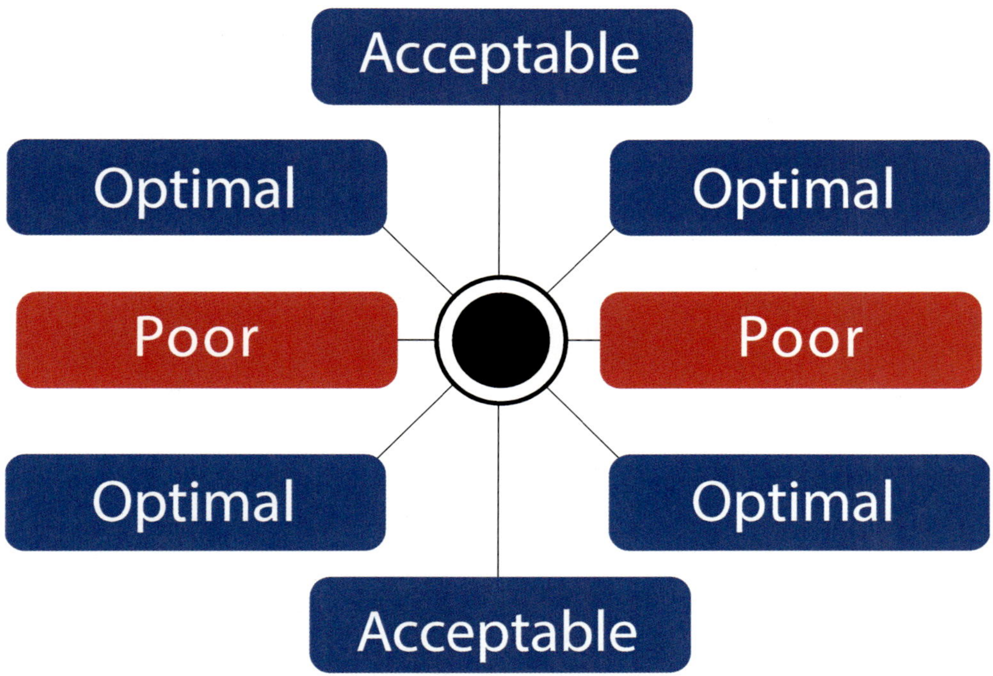

Figure 2.42: Optimal, acceptable and poor label placement for point features

Figure 2.43: In this example, the area label is stretched across the extent of the feature but is quite bold. If there were other features which needed to be labeled, this label might need to be smaller or have less contrast with the background to drop it lower in the visual hierarchy.

Stagger nearby areal labels to avoid unintentional horizontal alignment (**Figure 2.44**). Doing so may imply a multi-word label. In Figure 2.44A, the labels for several states are aligned, implying connection like words in a sentence. In this instance, the labels could be staggered within their extents to improve balance within the feature and relieve the alignment issue.

Figure 2.44: In this image, the labels are horizontally aligned, giving the impression of a multi-word label (A). A better choice would be to stagger the labels so that they rest on differing planes (B).

Linear feature labels

Linear feature labels should follow a smoothed curve mimicking the shape of the feature, parallel and slightly above the linear feature when possible (**Figure 2.45**). For very long linear features, labels may be repeated rather than trying to use tracking to extend the label across the entire length. Avoid touching or crossing the linear feature with the label. Never place any part of a linear feature label upside down. Avoid placing labels below the linear feature unless necessary.

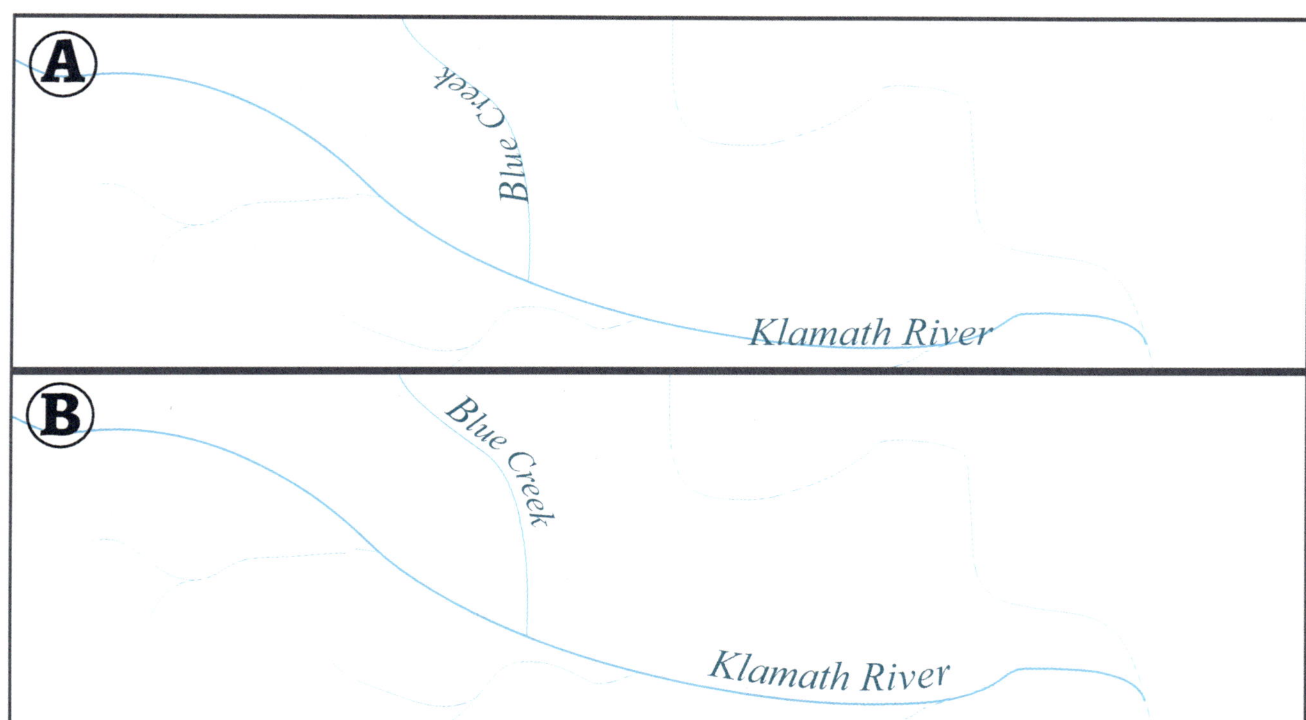

Figure 2.45: The Blue Creek label should be adjusted because it is both touching the linear feature and running upside down. The Klamath River label is a better use of linear feature labels.

Tutorial: Mapping Earthquakes in Northern California

The goal of this activity is for readers to learn how to create a basic map layout. In this activity, you create and organize a workspace folder using a standardized folder structure. You then download and decompress several datasets from public sources. Using the data, you create a map of earthquake magnitude in the State of California, with an emphasis on Northern California.

ESTIMATED TIME TO COMPLETE THIS TUTORIAL: 5 HOURS

Learning Outcomes

Readers should be able to accomplish the following outcomes by the end of this tutorial:

- » Summarize the steps for creating and organizing a project workspace folder structure
- » Illustrate the ability to download data from a public source
- » Demonstrate decompressing zipped files into a specific folder
- » Perform data preparation in Microsoft Excel
- » Convert tabular XY Data to GIS data using ArcGIS software
- » Practice changing map output size in ArcMap
- » Apply symbology and color choices to map features
- » Insert essential map elements using ArcMap
- » Create an inset map
- » Export a map as a PDF file

Setting up Your Workspace

In a typical workflow, you work on geospatial data using a local hard drive. When done, you compress your data and back up your work to your cloud storage so that you can retrieve the files from anywhere. When referring to a **local hard drive**, it means you are working on data physically located on the computer in front of you.

In contrast, some computers also include networked drives. **Networked drives** link to cloud storage and save the data elsewhere. Examples include services like OneDrive or Google Drive. For this tutorial, use the **desktop** as your local hard drive location. You may also use an external USB drive if you plan to work in multiple places.

You must avoid using networked drives while you work. They increase the processing time and can cause technical glitches.

In this tutorial, you use a particular folder structure. Start by creating your **workspace folder** on the local hard drive. A **workspace** is a folder or series of folders that contain all of your project files. The top-level folder in your workspace should indicate the lab assignment or the project. Organize all of your work within the workspace folder. On your desktop, create a new folder and give it a descriptive name, such as *Earthquake_Map*. Be sure there are *no spaces*. You may use underscores instead of spaces. Inside this folder, create the following three subfolders: original, working, and final. Having a standardized folder structure helps to keep a project organized, primarily when you are working with multiple partners. The folder structure you see here (**Figure 2.46**) is the standard used in each of the tutorials presented in this book.

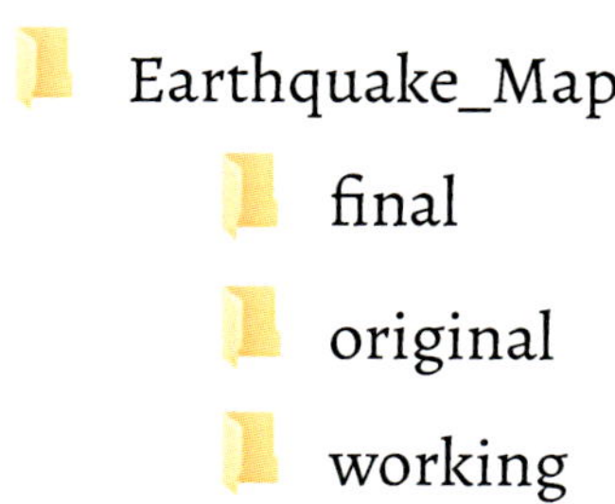

Figure 2.46: This diagram represents a basic folder structure used in this book.

As the name indicates, use the **original folder** for storing original, unaltered data. As you are working on a project, if for some reason your working version of the data gets lost or corrupted, you can go back to your *original folder* and find a fresh copy of the data. Use the **working folder** for data that you create or alter while working on your project. Use the **final folder** for storing any output you produce as a result of your work such as images, maps, tables, or reports. Setting up the standard folder structure for a project is good practice and a habit you want to develop.

Downloading Data from Natural Earth

Natural Earth is a website created to provide free vector and raster data to meet the needs of cartographers using a variety of software applications. The data on this website is free to use without restrictions. Open the Chrome browser and navigate to the ***Natural***

Earth website[1] (**Figure 2.47**). Click the *Downloads* link. Under *Large-scale data*, click the *Cultural* link.

> ***If you do not have the Chrome browser, you can download it here***[2].

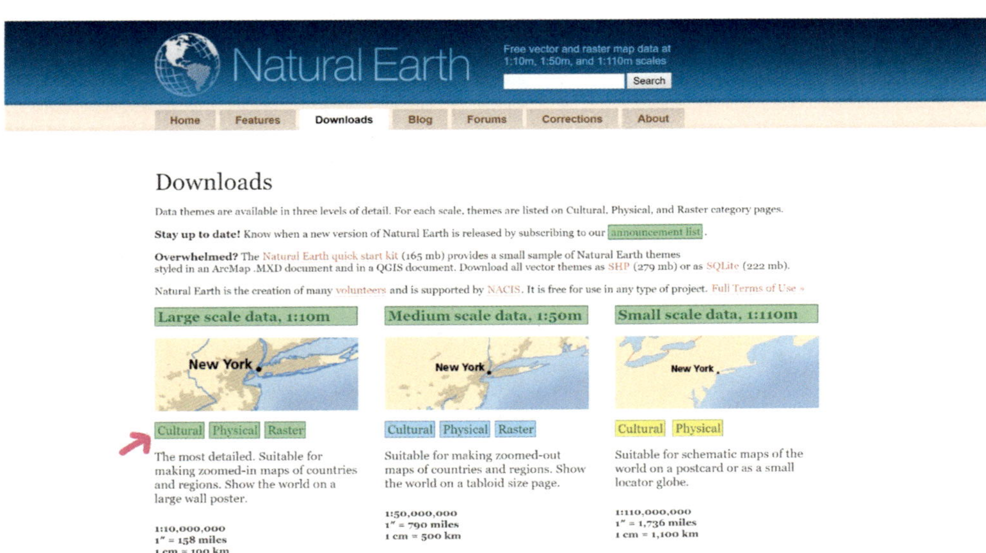

Figure 2.47: Natural Earth provides physical and cultural data in large, medium, and small scale.

Under *Admin 1 States, Provinces*, right-click *Download states and provinces* (**Figure 2.48**). Select, *Save Link As*, then navigate to your original folder and save.

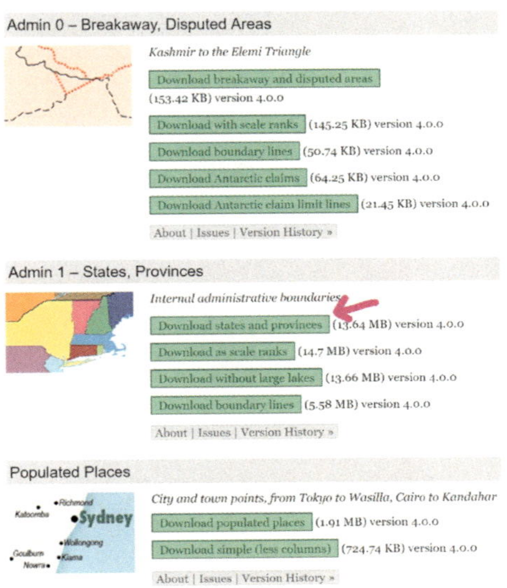

Figure 2.48: The image here shows the Natural Earth Cultural Vectors download page for large-scale data.

1 http://www.naturalearthdata.com/

2 https://www.google.com/chrome/browser/desktop/index.html

In a previous tutorial, you learned how to decompress a file using 7zip. In Microsoft Windows, navigate to your *original* folder. You should see the zip file for the states and provinces. Right-click the zip file, select *7zip*, then *Extract Here*. Be sure to delete the zip file when you are done decompressing it. You won't need it anymore. Eliminating the zip file saves space and helps to avoid confusion later.

Download Data from the Northern California Earthquake Data Center (NCEDC)

UC Berkeley Seismological Laboratory created the Northern California Earthquake Data Center. Using the **Chrome** browser, navigate to the ***Northern California Earthquake Data Center (NCEDC) website***[3]. The catalog has quite a few search settings which at first may seem confusing. In this tutorial, we do not cover all of the details related to the settings. To find out more about how to use this website and to learn about the options available, you can click the ***help link***[4] to read the additional documentation. For the input catalog, choose USGS NCSN Catalog 1967 to Present (**Figure 2.49**). For the output format, pick NCSN catalog in CSV format.

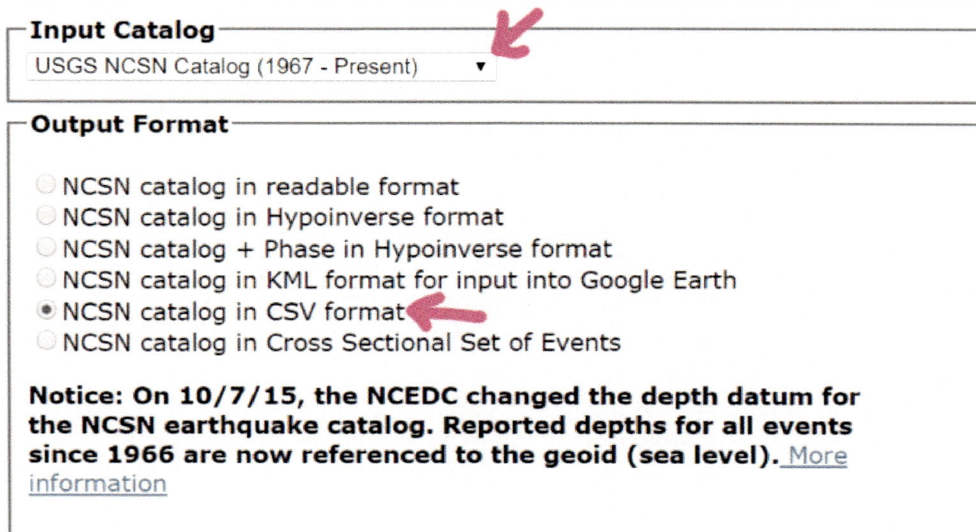

Figure 2.49: The image here shows the NCEDC Input and Output Catalog settings.

3 http://ncedc.org/ncedc/catalog-search.html
4 http://www.ncedc.org/ncedc/catalog-search-help.html

For the earthquake parameters center the values below:

» **Start Time:** 2017/01/01

» **End Time:** 2017/12/31

» **Min Magnitude:** 3.0

» **Min Latitude:** 32

» **Min Longitude:** -114

» **Max Latitude:** 47

» **Max Longitude:** -130

» **Event Types:** Earthquakes

Earthquake Parameters

Start Time:	2017/01/01	End Time:	2017/12/31
Min Magnitude:	3.0	Max Magnitude:	
Min Depth (km):		Max Depth (km):	
Min Latitude:	32	Max Latitude:	47
Min Longitude:	-114	Max Longitude:	-130

Event Types: ● Earthquakes ○ Blasts (Quarry or Nuclear) ○ All Events

☐ Include Events with no reported Magnitude

Figure 2.50: The image here shows the NCEDC Earthquake Parameters.

Accept all other default settings for *Earthquake Parameters* not mentioned here. Under *Output Mechanism*, select *Send output to an anonymous FTP file on the NCEDC* (**Figure 2.51**).

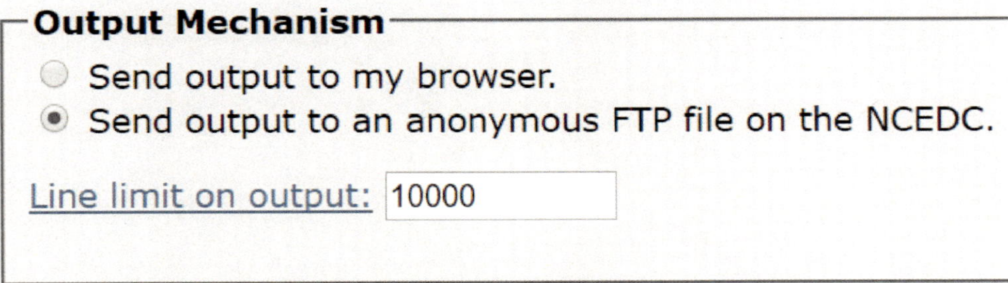

Figure 2.51: The image here shows the NCEDC output mechanism settings.

When you are ready, click *Submit request*. In the **Chrome** browser, the results load in a new tab. Under *output can be downloaded from*, click the link next to the word *URL* (**Figure 2.52**).

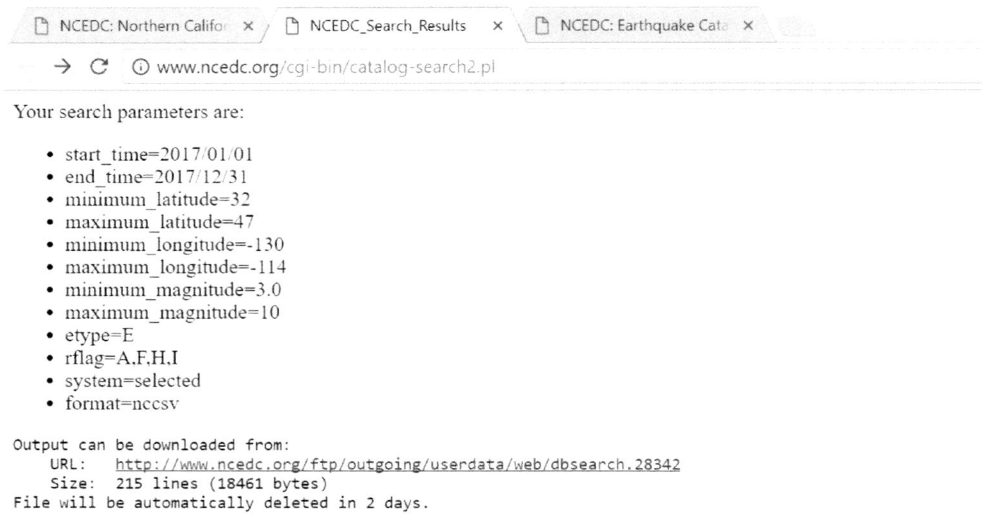

Figure 2.52: Access the data products for this tutorial through the Northern California Earthquake Data Center (NCEDC).

The next page displays the data in CSV format (**Figure 2.53**). You need to save the results to your *original* folder.

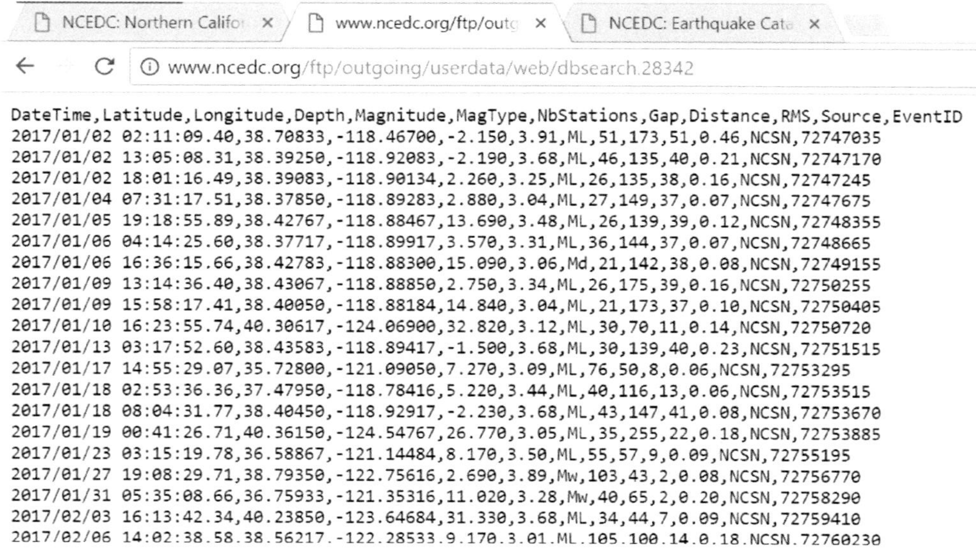

Figure 2.53: CSV stands for comma separated values . As you can see on the page, each of the values gets separated by a comma.

In the **Chrome** browser hit *Ctrl S* to save. Browse to your *original* folder. For the filename, enter earthquake2017.csv. Next to *Save as type*, choose *All files*. When you are ready, click *Save*. In Microsoft Windows navigate to your *original* folder and open the earthquake 2017 CSV to view the results. By default, a CSV file should open in Microsoft Excel (**Figure 2.54**).

If for some reason, the CSV file does not open in Microsoft Excel, right-click and select Open with, then choose Microsoft Excel as the default program.

When opened in Microsoft Excel, the CSV file appears as a table (**Figure 2.54**). The commas in the data create columns, also called **fields** in ArcMap. A row, also called a **record** in ArcMap, represents each earthquake. In this instance, the top row is different from the remaining records. It is made up of column headers, called **field names** in ArcMap. When working with geospatial data, this format is critical. The first row of any geospatial data table must contain the field names. **Chapter 7** covers these types of tables in greater detail.

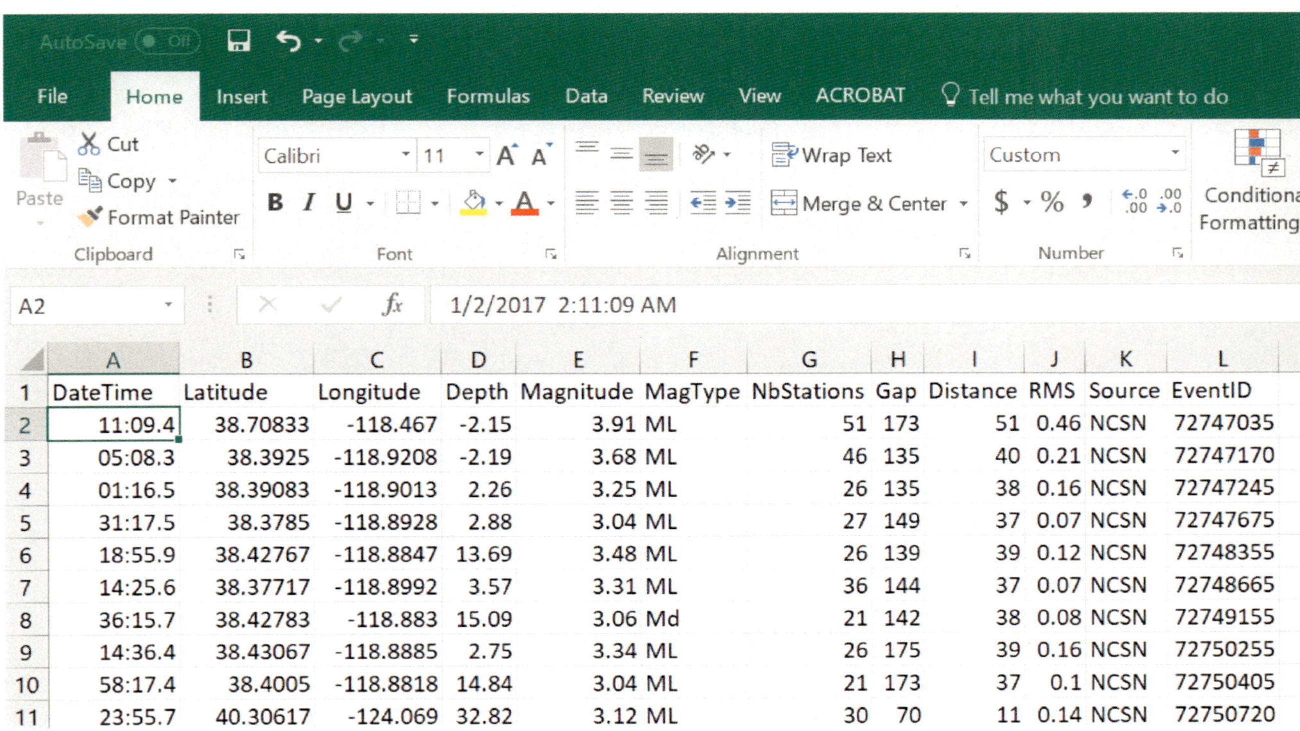

	A	B	C	D	E	F	G	H	I	J	K	L
1	DateTime	Latitude	Longitude	Depth	Magnitude	MagType	NbStations	Gap	Distance	RMS	Source	EventID
2	11:09.4	38.70833	-118.467	-2.15	3.91	ML	51	173	51	0.46	NCSN	72747035
3	05:08.3	38.3925	-118.9208	-2.19	3.68	ML	46	135	40	0.21	NCSN	72747170
4	01:16.5	38.39083	-118.9013	2.26	3.25	ML	26	135	38	0.16	NCSN	72747245
5	31:17.5	38.3785	-118.8928	2.88	3.04	ML	27	149	37	0.07	NCSN	72747675
6	18:55.9	38.42767	-118.8847	13.69	3.48	ML	26	139	39	0.12	NCSN	72748355
7	14:25.6	38.37717	-118.8992	3.57	3.31	ML	36	144	37	0.07	NCSN	72748665
8	36:15.7	38.42783	-118.883	15.09	3.06	Md	21	142	38	0.08	NCSN	72749155
9	14:36.4	38.43067	-118.8885	2.75	3.34	ML	26	175	39	0.16	NCSN	72750255
10	58:17.4	38.4005	-118.8818	14.84	3.04	ML	21	173	37	0.1	NCSN	72750405
11	23:55.7	40.30617	-124.069	32.82	3.12	ML	30	70	11	0.14	NCSN	72750720

Figure 2.54: GIS data table must be in a particular format. The top row must contain the field names.

Go ahead and close the CSV file in Microsoft Excel. In a subsequent step, you use the CSV file in ArcMap. It is essential not to have it opened in two places at once.

Skill Drill: Download Data from the United States Census Bureau

In a previous tutorial, you learned how to download data from the United States Census Bureau. Open the Chrome browser and navigate to the ***United States Census Bureau Cartographic Boundary Shapefile page***[5].

Click the link under the header *Nation-based Files* that says *County*. On the Counties page, right-click the link to download the *500K* resolution. Select, *Save Link As*, then browse to your *original* folder and click *Save*. Decompress the file using 7zip. Be sure to delete the zip files when you are done decompressing them. You won't need them anymore. Eliminating them saves space and helps to avoid confusion later.

Adding California as a Basemap Layer

In geospatial science, a **basemap** refers to a collection of data layers that are used to create a background for the map. The purpose of a basemap is to provide context for the primary data, both spatially and thematically. Typical basemap layers include roads, administrative boundaries, and aerial imagery. In this activity, the state and county boundaries serve as the basemap layers.

Locate ArcMap on your computer and launch the software. If you are using Microsoft Windows 10, click the windows button and type ArcMap to find the desktop application (**Figure 2.55**). Launch the ArcMap software.

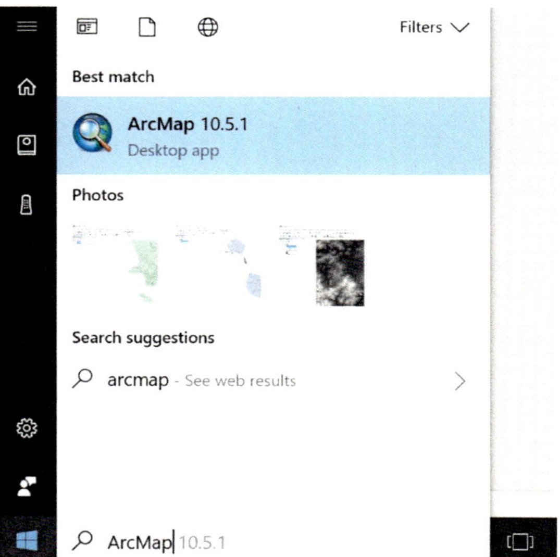

Figure 2.55: The specific version of ArcMap may vary over time.

5 https://www.census.gov/geo/maps-data/data/tiger-cart-boundary.html

When you first launch ArcMap, a window appears that gives you the option to either open a blank map or to open any of your recent map documents (**Figure 2.56**). Choose to open a blank map document.

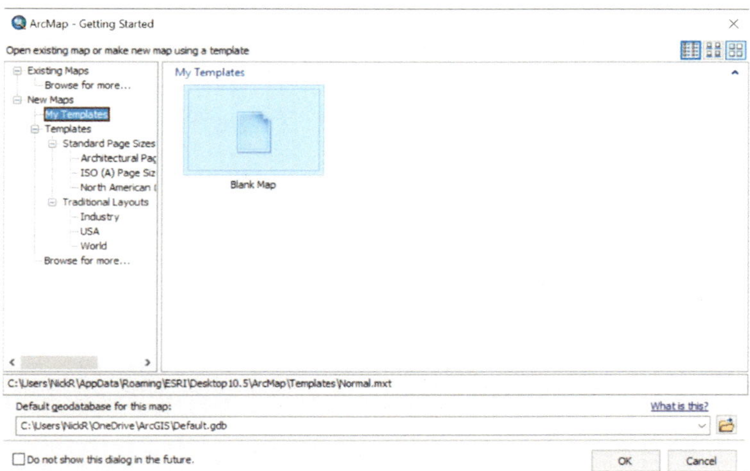

Figure 2.56: The ArcMap Getting Started window provides several options.

The ArcMap user interface has three main windows that you use regularly, the **Table of Contents**, the **data frame**, and the **Catalog Window** (**Figure 2.57**). The *Table of Contents* displays a list of data frames and map layers loaded into the map documents. The data frame, sometimes spelled dataframe, represents the layers as a map and defines the map extent. The Catalog Window displays a hierarchical view of folder connections and data as a **Catalog Tree**.

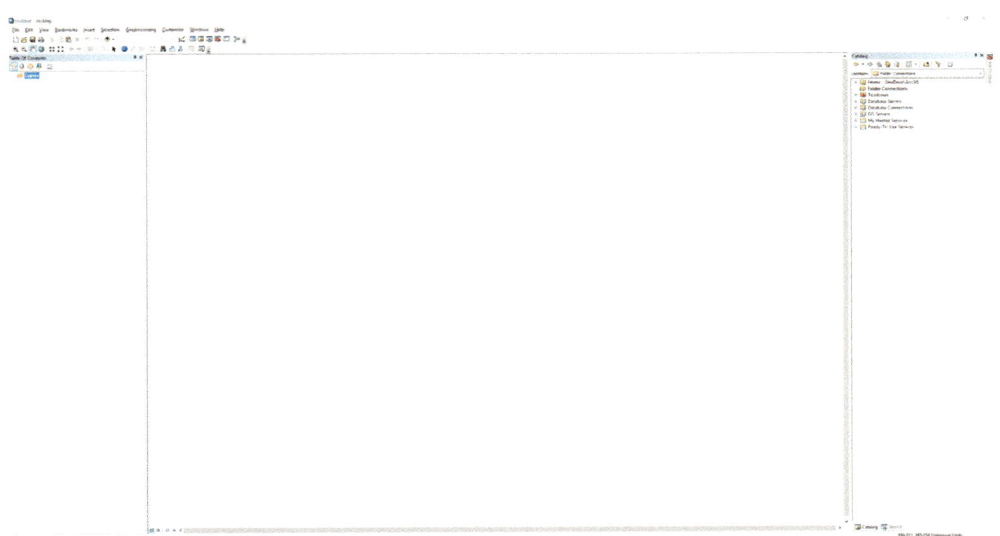

Figure 2.57: From left to right, the three main windows are the Table of Contents, the data frame, and the Catalog Window.

The icon for the catalog window looks like a yellow file cabinet (**Figure 2.58**). If your catalog window is missing, you can find the icon on the toolbar across the top. Click to open it back up.

Figure 2.58: Click the yellow file folder icon if your Catalog Window is missing.

In this step, you start by working with the Catalog Window. The catalog window is where you manage your geospatial data in ArcMap. Start by clicking on the *Connect to Folder* button (**Figure 2.59**). Navigate to your workspace folder, *Earthquake_map*, on your local hard drive.

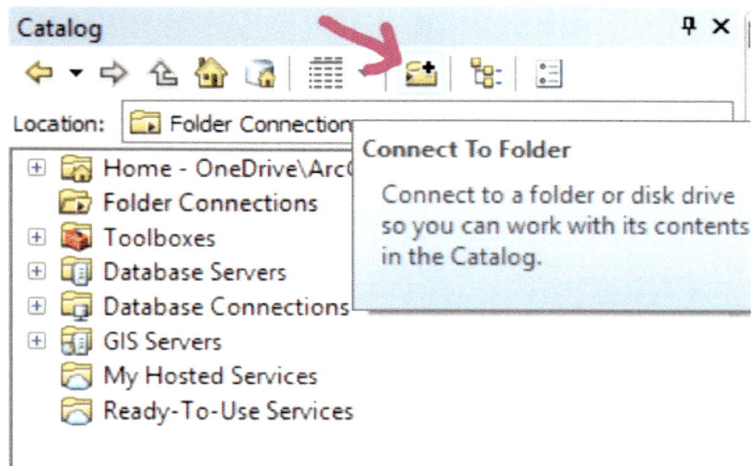

Figure 2.59: You connect to a folder so that the contents appear in the Catalog Window.

In this step, it is essential to select your primary workspace folder, *Earthquake_map*, to add it to the Catalog Window (**Figure 2.60**). When ready, click *OK*.

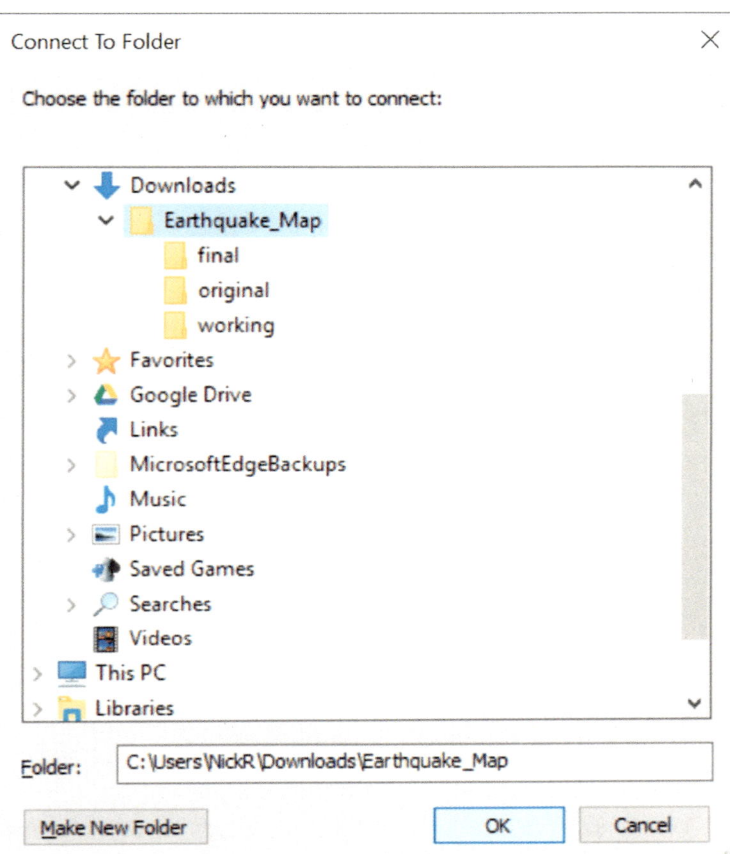

Figure 2.60: Connect to Folder adds the folder to the Catalog Window.

Once you add your workspace folder to the Catalog Window, expand the folder by clicking on the plus sign. You should see your three subfolders inside. Expand the original folder to view the contents. This display of folders and files within the Catalog Window is sometimes called the **Catalog Tree**.

In a previous tutorial, you learned how to add data to ArcMap. Add the states and provinces layer to the map (**Figure 2.61**).

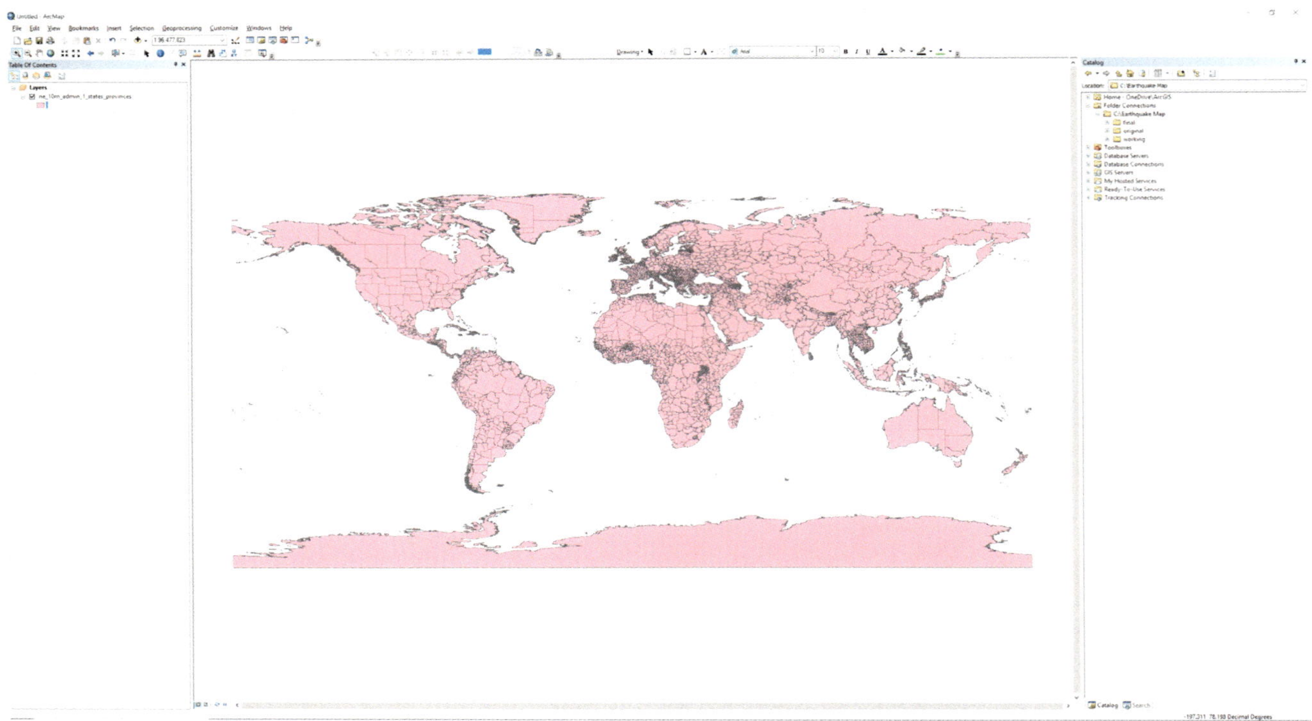

Figure 2.61: Natural Earth provides data for the entire globe.

In a previous tutorial, you learned how to use various zoom tools. In ArcMap, zoom to the State of California and surrounding areas (**Figure 2.62**). Try to set your map so that California takes up most of the space.

Because California is the primary geographic extent for your earthquake maps, you want to be able to distinguish California from the other states. You do this by creating a new shapefile with California as the only polygon feature.

In a previous tutorial, you learned how to directly select features on the map using the *Select Features* tool. Use the tool to select the State of California. On the *Tools* toolbar, click the *Select Features* tool. The icon looks like a white arrow over a blue and white square. Then click on California. When you have a feature selected in ArcMap, you have the option to export it as a new shapefile (**Figure 2.63**). With California selected, right-click on the states and provinces layer in the *Table of Contents*. Select, *Data*, then *Export Data*.

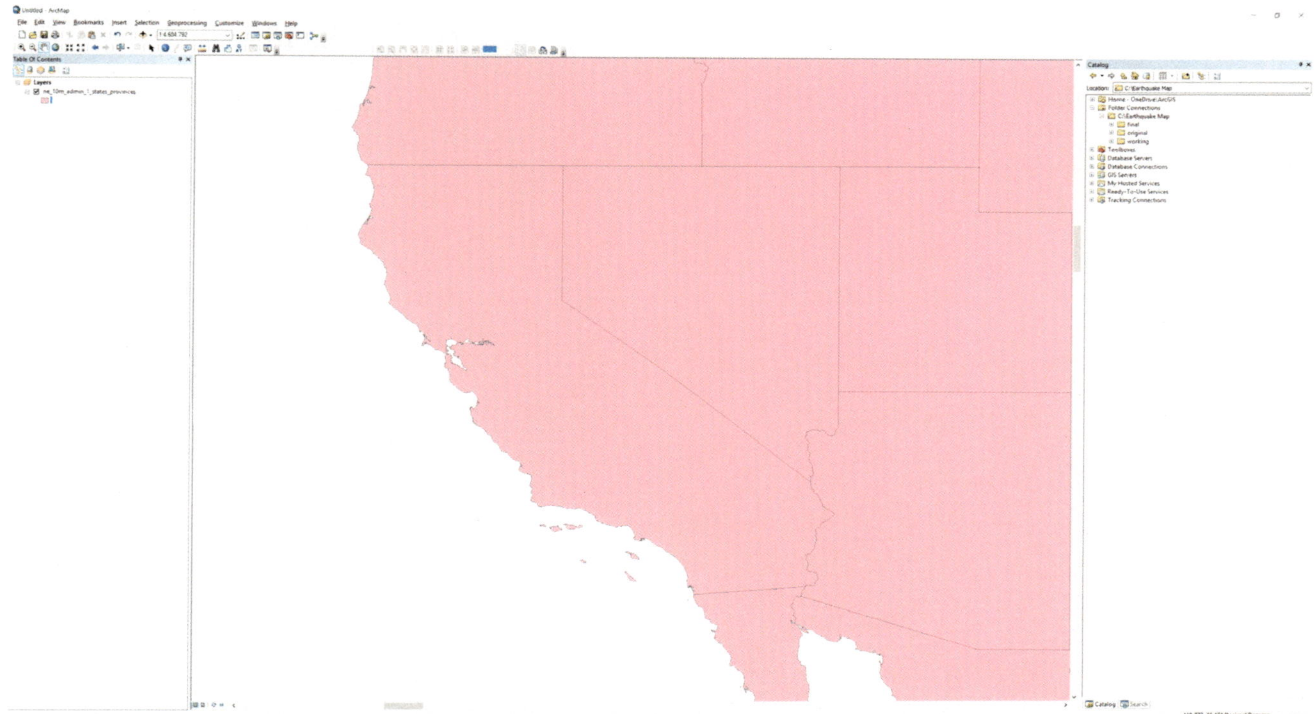

Figure 2.62: The map extent displays the State of California and surrounding areas.

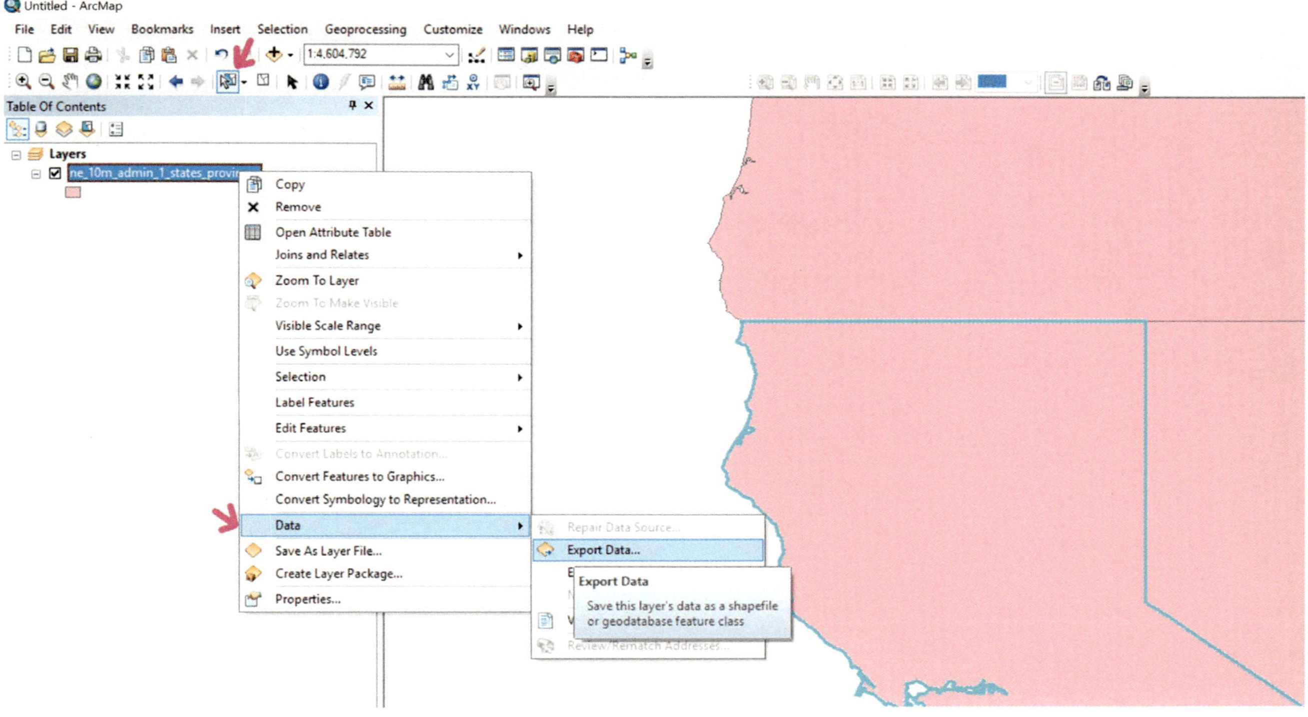

Figure 2.63: You can access contextual menus by right-clicking.

When the *Export Data* window opens, you see some options (**Figure 2.64**). Next to Export, be sure it says *Selected features*. Next, click the yellow file folder icon and browse to your *working* folder.

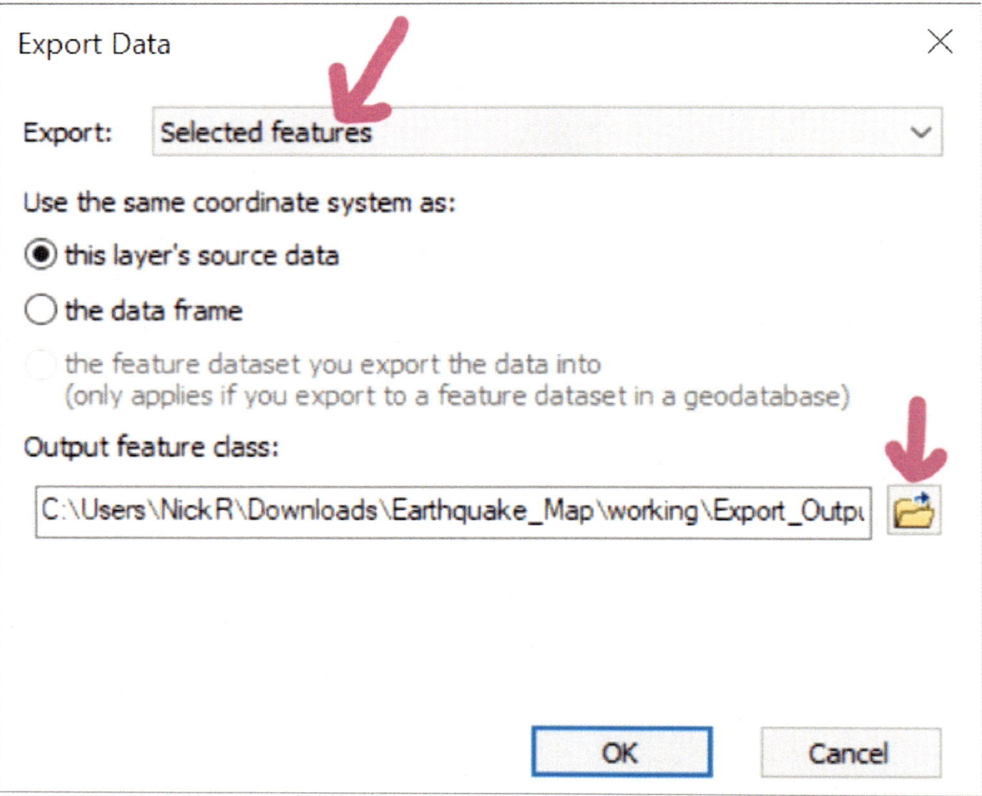

Figure 2.64: You have several options to choose from when exporting data.

Never accept the default output location in ArcMap. Always browse to the folder location you intend to use. This step helps to prevent lost data.

Next to *Save as type*, choose *Shapefile*. Name the file California.shp and click *Save*, then click *OK* (**Figure 2.65**). When ArcMap asks if you want to add the exported data as a map layer, click *Yes*.

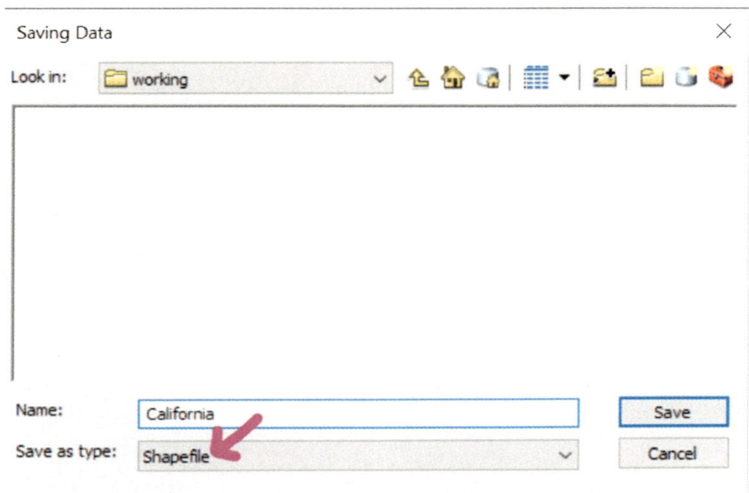

Figure 2.65: When saving data, make sure you are in the correct folder. In this example, it is the *working* folder.

The California layer gets added to the *Table of Contents*. On the map, California remains selected. Clear the selection by clicking on the *Clear Selected Features* button. Your map document now has two layers, one representing the State of California and one that currently displays the surrounding States and Mexico (**Figure 2.66**).

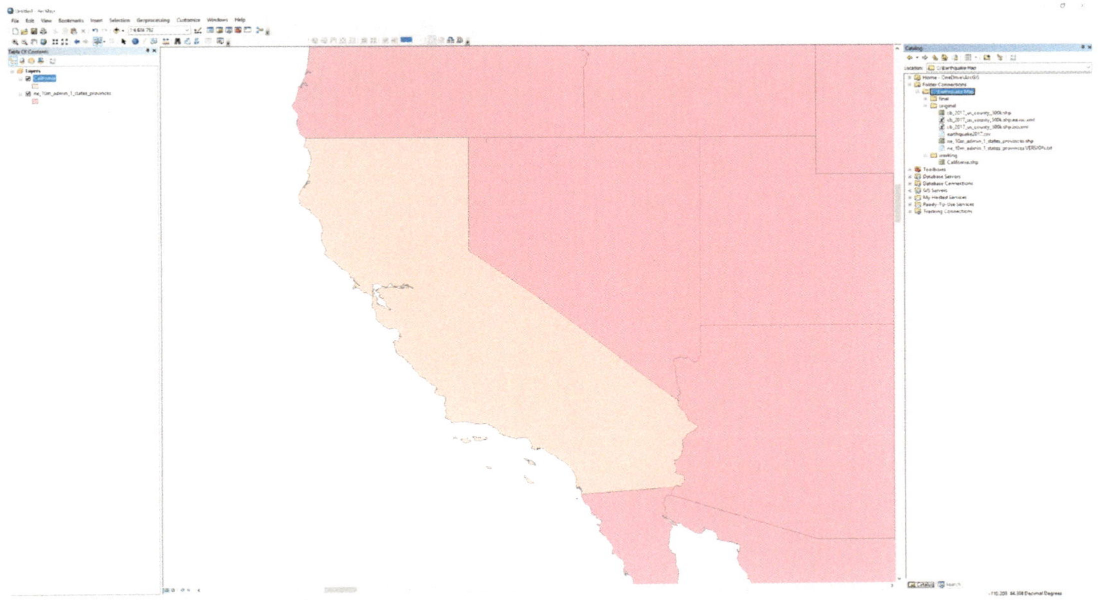

Figure 2.66: When you add new layers to ArcMap, the software randomly chooses colors.

Skill Drill: Adding the Humboldt County Boundary as a Basemap Layer

Using what you learned in the previous step, add the shapefile that contains the county boundaries to the map. You may get a Geographic Coordinate System Warning. Check the box that says, *Don't warn me again in this session*, and click *Close* (**Figure 2.67**).

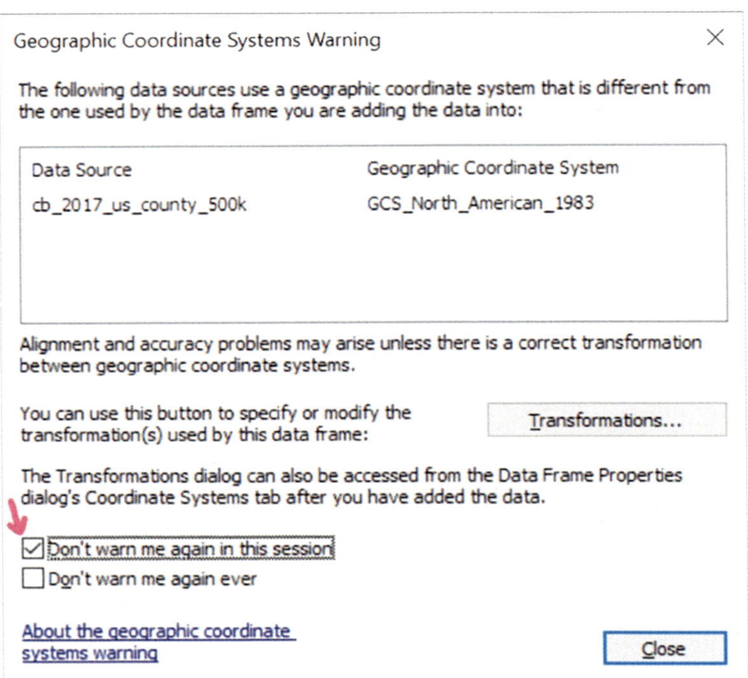

Figure 2.67: While it is OK to turn off this warning for one session. you should never turn it off permanently.

Use the direct selection tool to select Humboldt County. With Humboldt County selected, export the data as a new shapefile and save it to your *working* folder. Call the new shapefile Humboldt_County.shp. Add the Humboldt County shapefile to the map.

> ***Note: the California layer may also get selected. Don't worry about it. When you export, it only applies to the layer on which you right-click.***

When done, remove the layer that contains all the county boundaries for the United States. You should end up with three layers in your *Table of Contents*, one for Humboldt County, one for California, and one for the United States and Mexico. Now is an excellent time to save your work. In a previous tutorial, you learned how to set your map document properties to store relative paths. Do this now (**Figure 2.68**).

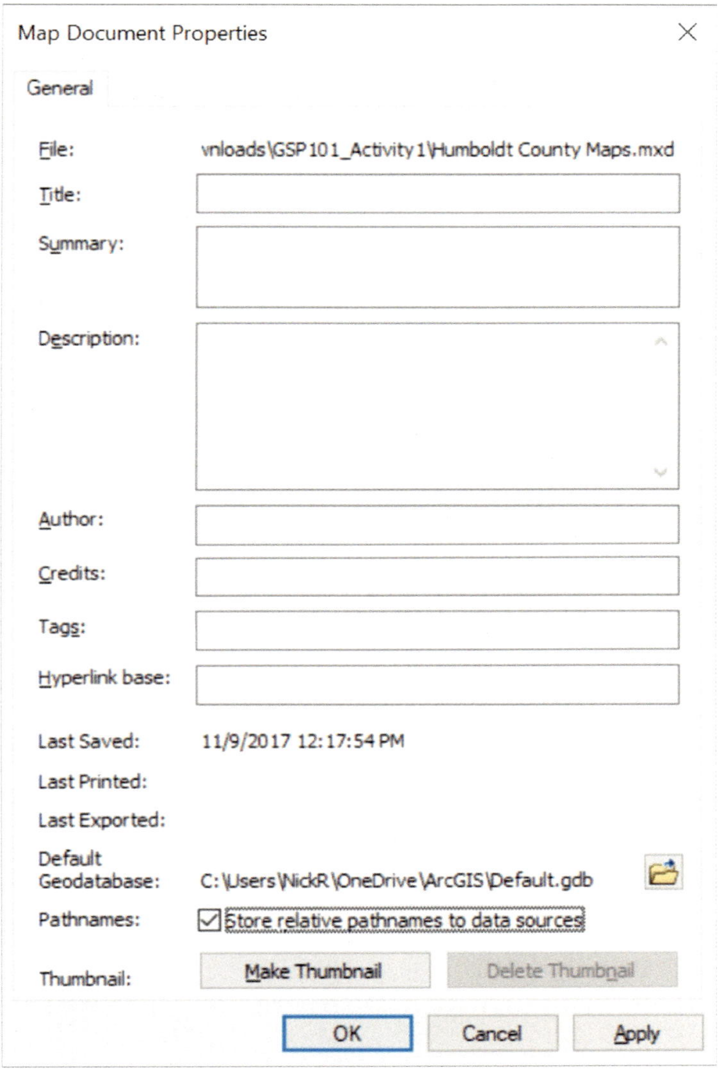

Figure 2.68: Always make sure to check the Store relative pathnames to data source option.

Then, save your map document to your workspace folder, *Earthquake_Map*. Call the map document *California Earthquakes* (**Figure 2.69**).

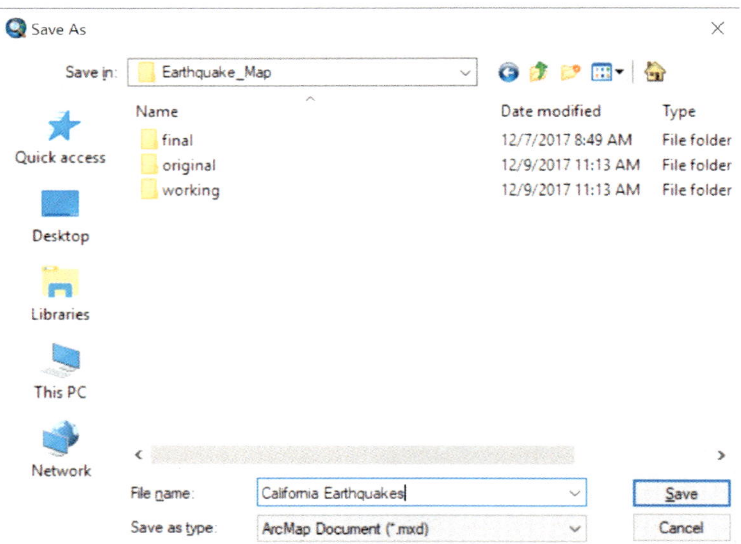

Figure 2.69: Generally, you should avoid spaces. However, empty spaces are allowed when naming map document files (.mxd)

Adding XY Data

As you learned in a previous step, a CSV is one of the plainest forms of geospatial data. As mentioned before, each column of data gets separated by a comma, and the first row in the table contains the field names. When adding XY data, it is the two columns, latitude and longitude, that are necessary to place this information on a map. The Longitude in decimal degrees is the X coordinate, and the Latitude in decimal degrees is the Y coordinate. You learn more about latitude and longitude in **Chapter 3**. ArcMap reads these numbers, the latitude, and longitude values in decimal degrees, to figure out where to place the points on the map. In ArcMap, from the File menu, select *Add XY Data* (**Figure 2.70**).

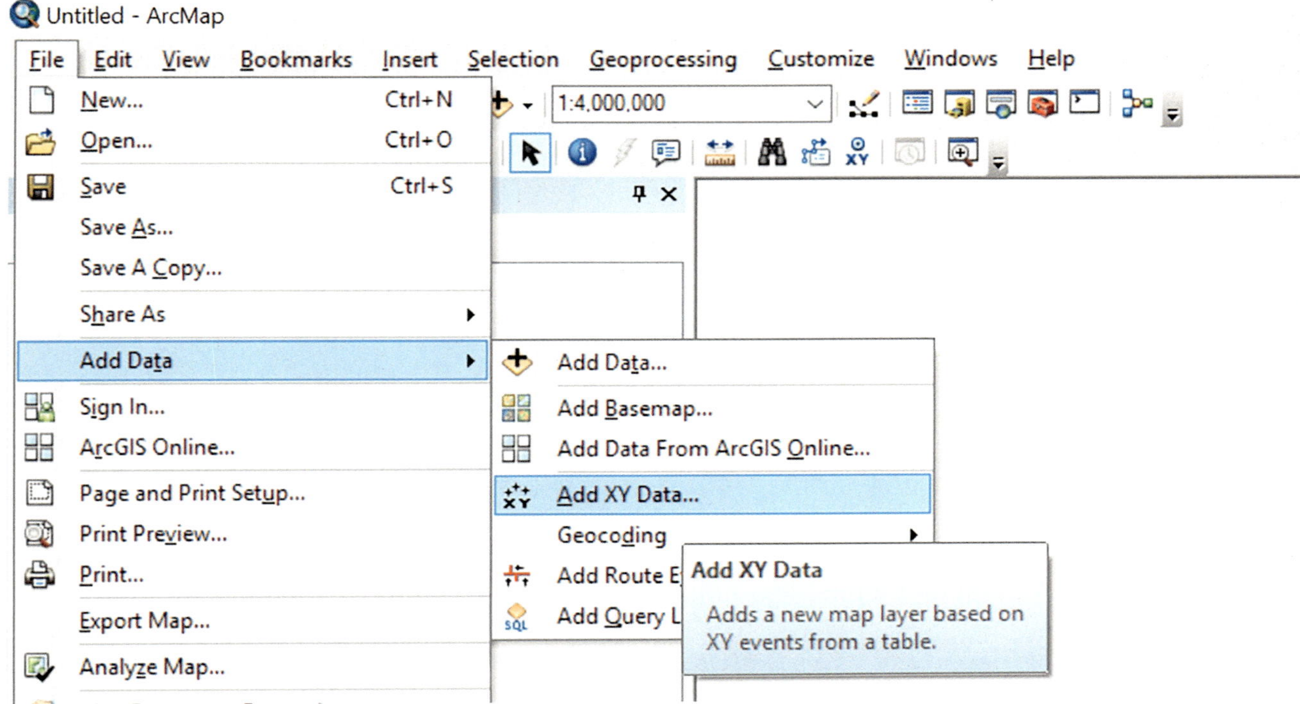

Figure 2.70: XY Data refers to data with coordinates, such as latitude and longitude.

When the *Add XY Data* window opens, click the yellow file folder icon and browse to the *original* folder (**Figure 2.71**).

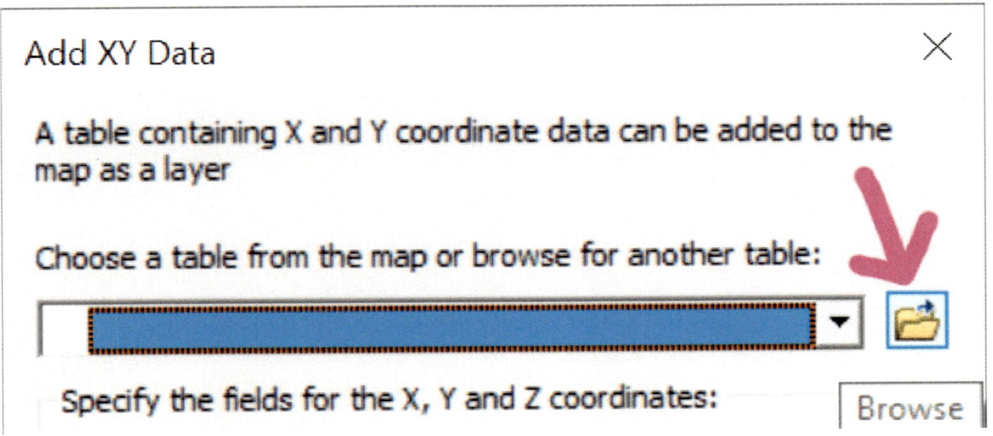

Figure 2.71: The yellow file folder icon allows you to browse to a specific folder.

Choose earthquake2017.csv and click Add (**Figure 2.72**).

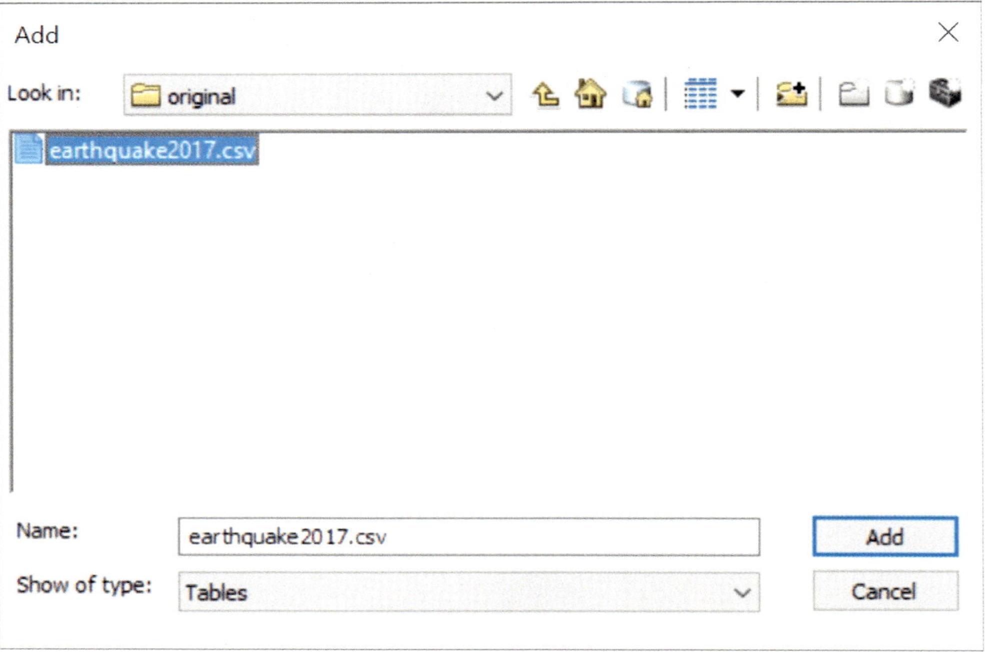

Figure 2.72: Be sure to save the CSV file to your *original* folder.

ArcMap reads the CSV table and tries to find the XY data automatically (**Figure 2.73**). In this instance, it assigns longitude as the X field and latitude as the Y field.

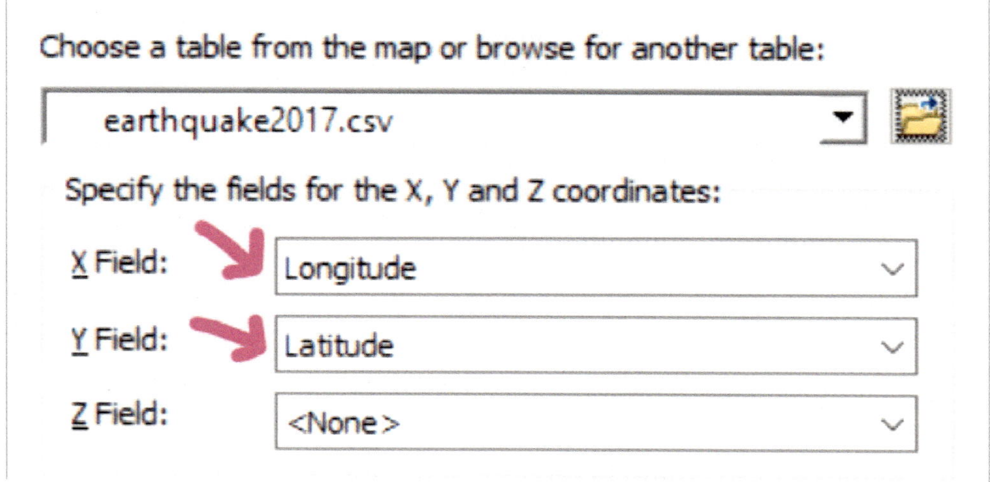

Figure 2.73: The most common XY data is latitude and longitude.

The next step is the most important one. Most readers make mistakes here. Many people assume that latitude and longitude coordinates are universal. **They are not**. There are many different spatial reference systems available when you make maps. Different spatial reference systems use *distinct values for latitude and longitude*. You learn more about spatial reference systems in **Chapter 4**. For now, understand that ArcMap does not know from which spatial reference system the latitude and longitude values come. Instead, it tries to guess. The mistake many make is that they allow ArcMap to guess incorrectly. In this instance, these latitude and longitude values come from a spatial reference system called *GCS WGS 1984*. Check to make sure that the Geographic Coordinate System is *GCS WGS 1984*. If not, then follow the next few steps to manually set the spatial reference information for the earthquake data (**Figure 2.74**). On the *Add XY Data* window, click the *Edit* button under *Coordinate System of Input Coordinates*.

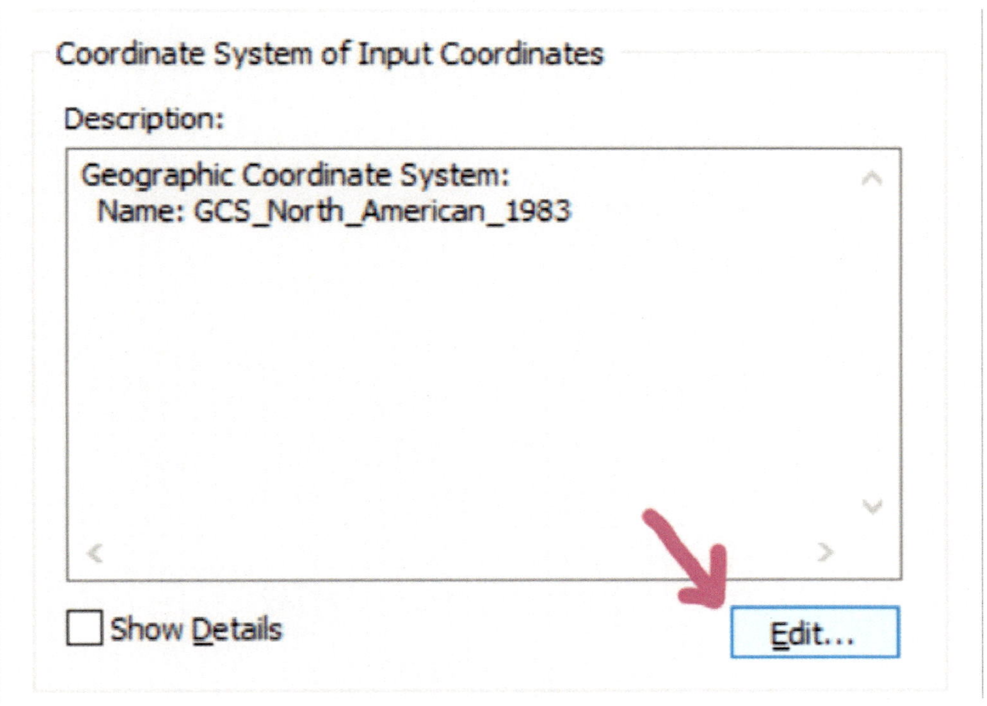

Figure 2.74: When adding XY data, always click the edit button. Never accept the default coordinate system.

The *Spatial Reference Properties* window opens up. Scroll to the top and expand the *Geographic Coordinate Systems* folder (**Figure 2.75**).

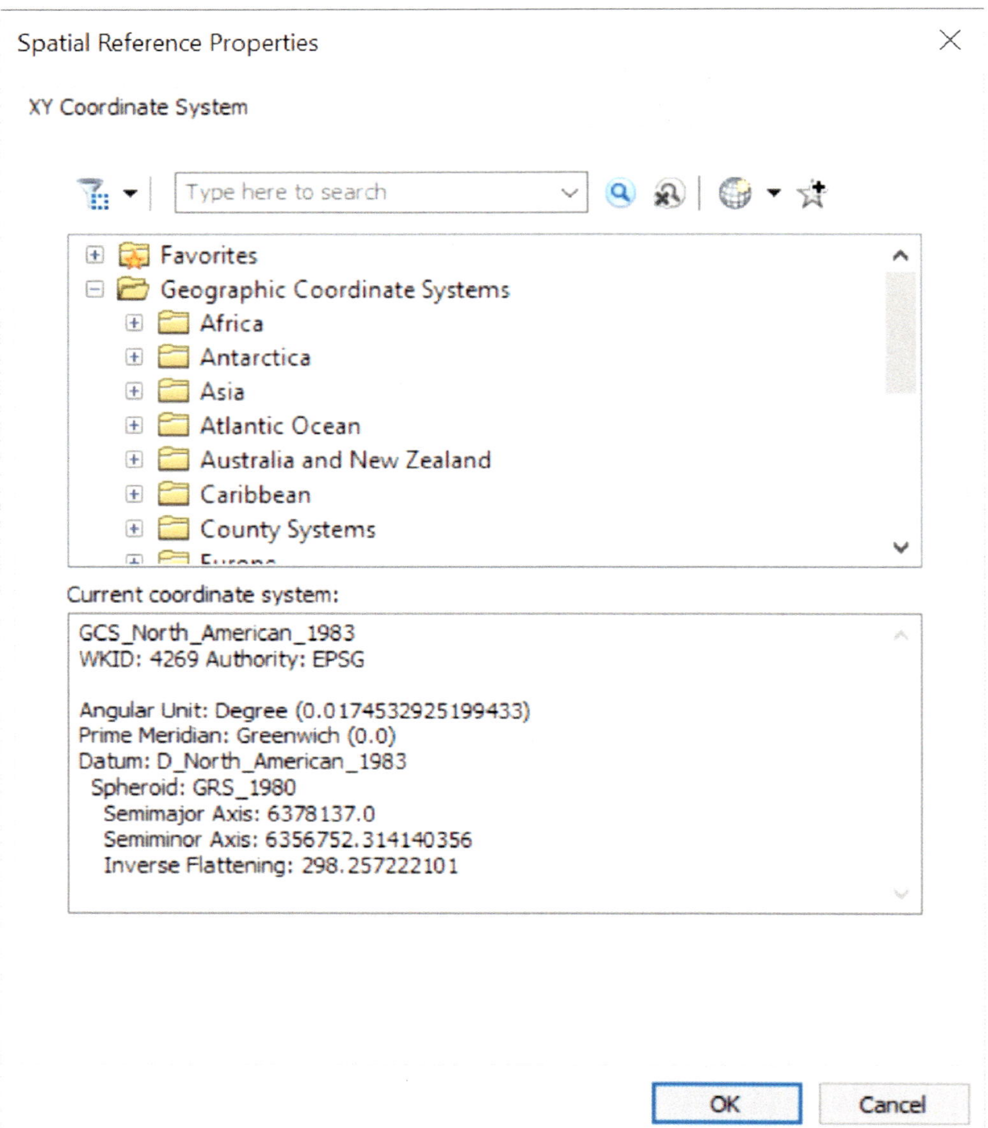

Figure 2.75: The *Spatial Reference Properties* window allows you to choose a spatial reference system from two main folders, *Geographic Coordinate Systems* and *Projected Coordinate Systems*.

Then scroll down until you see the *World* subfolder and expand it (**Figure 2.76**).

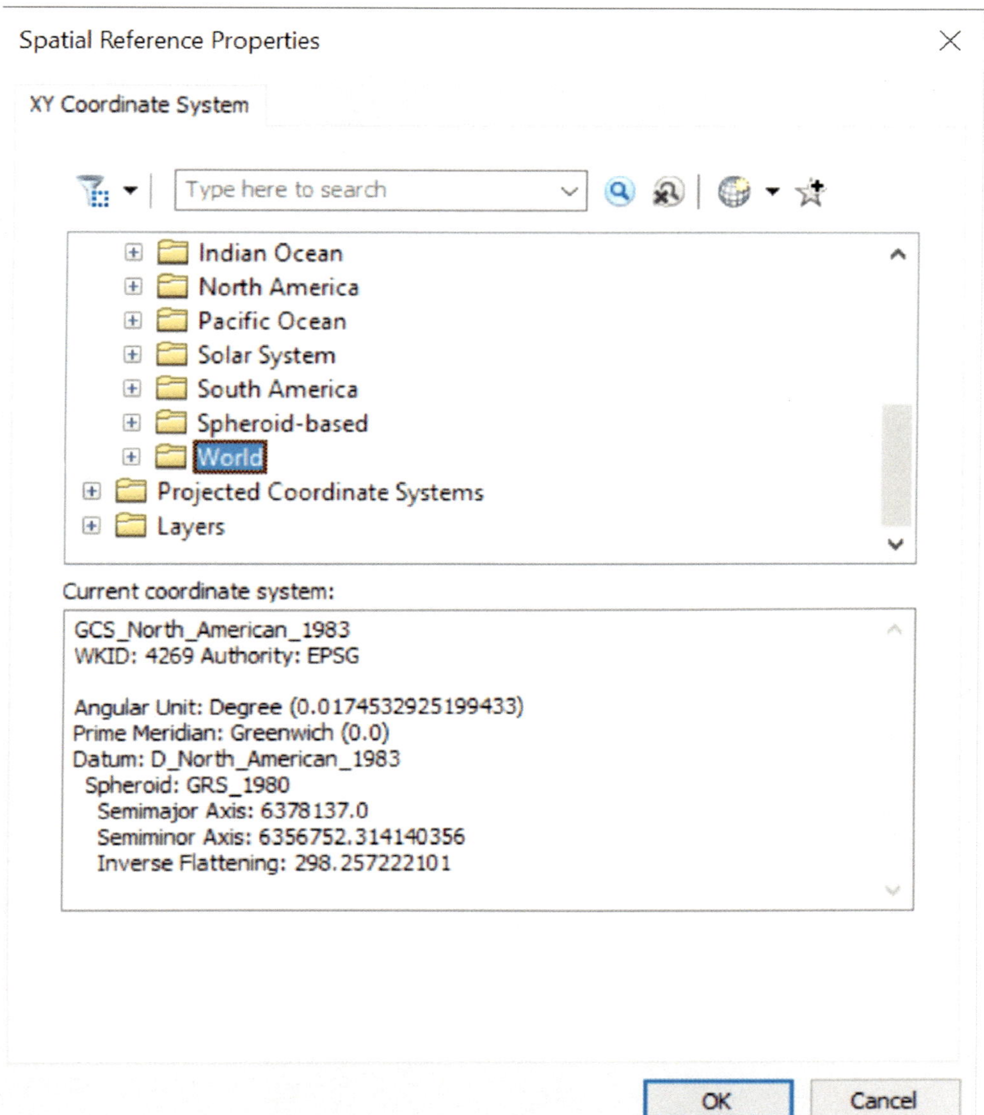

Figure 2.76: Be sure you are selecting the World folder from the *Geographic Coordinate Systems* folder.

Scroll down until you see *WGS 1984*. There are several versions. Select the simple one that only says *WGS 1984* (**Figure 2.77**). When you are ready, click *OK*.

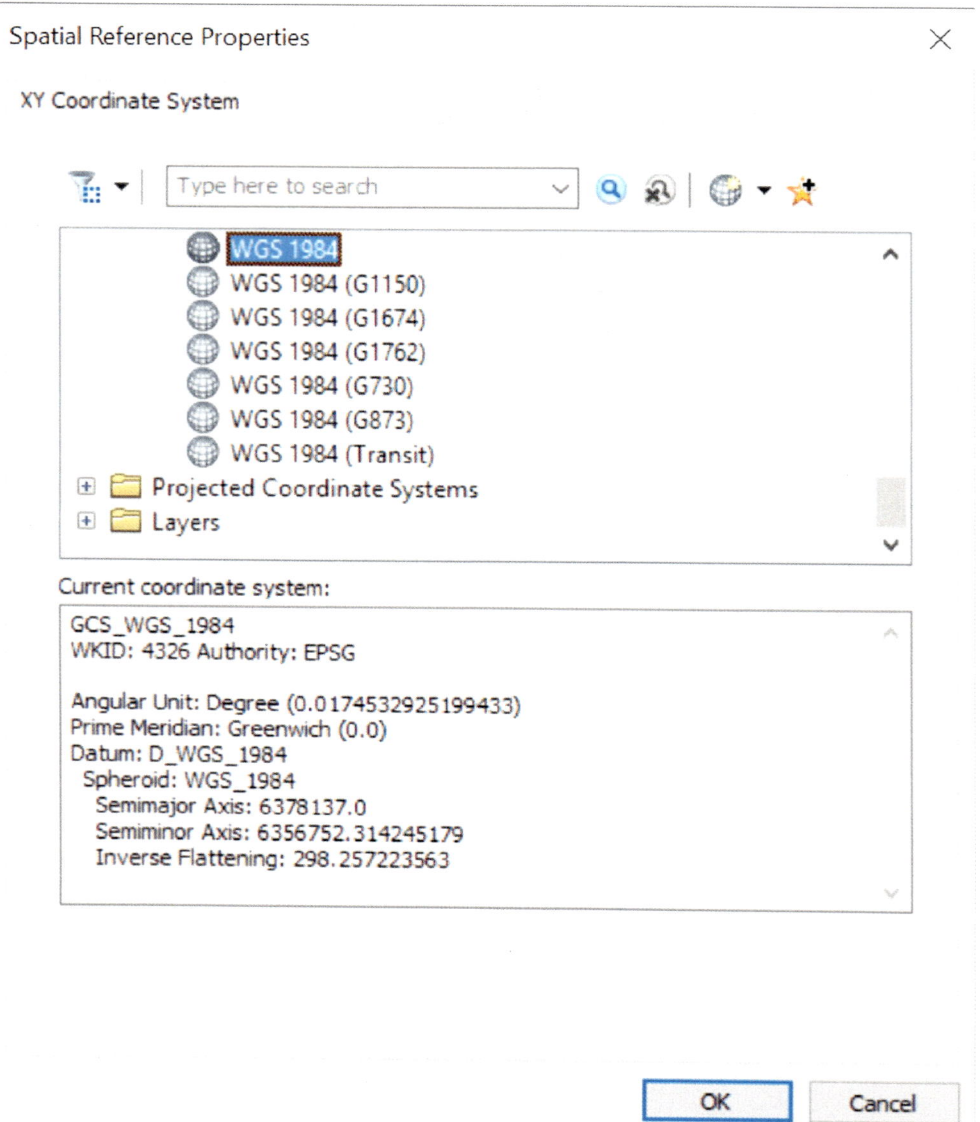

Figure 2.77: According to the NCEDC website, WGS 1984 is the correct spatial reference for the latitude and longitude values they provide.

Check to make sure everything matches the settings below (**Figure 2.78**). When you are ready, click *OK*.

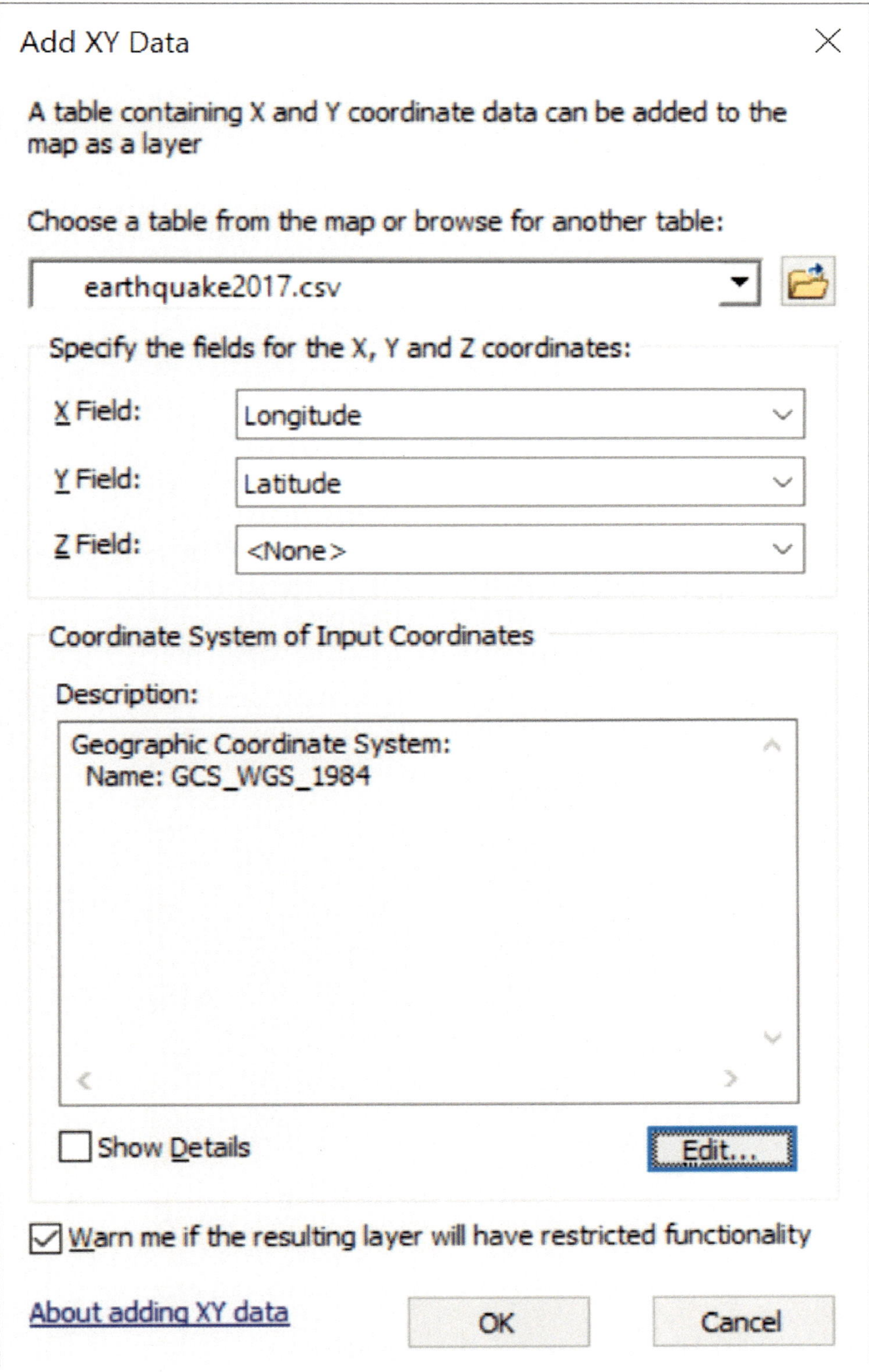

Figure 2.78: Your *Add XY Data* window should look like the image above.

You get an error message warning you that the table does not have an object id field (**Figure 2.79**). Take a moment to read the warning. When you add XY data, ArcMap creates a temporary representation of the data. It might look like a regular shapefile, but it does not have a database. To make it into a permanent shapefile with a database, you have to export the layer. Go ahead and click *OK* to close the warning.

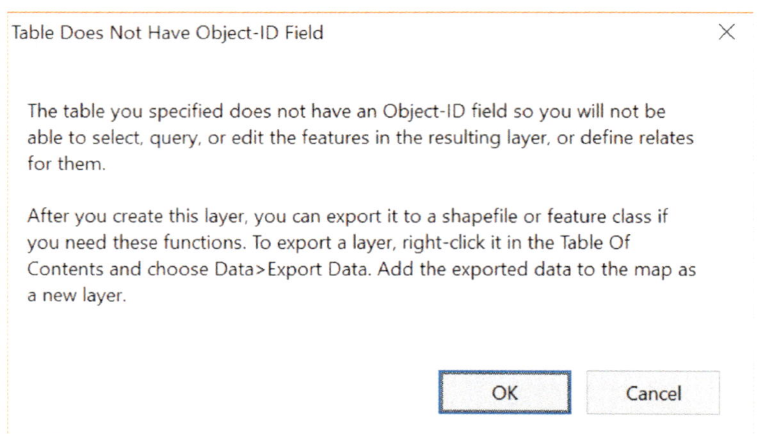

Figure 2.79: This message warns you that the table used to create this layer does not meet the database format requirements.

All the earthquakes above a magnitude of 3.0 get added to the map (**Figure 2.80**). In the *Table of Contents*, right click on earthquake2017.csv *events* and select *Zoom to Layer*.

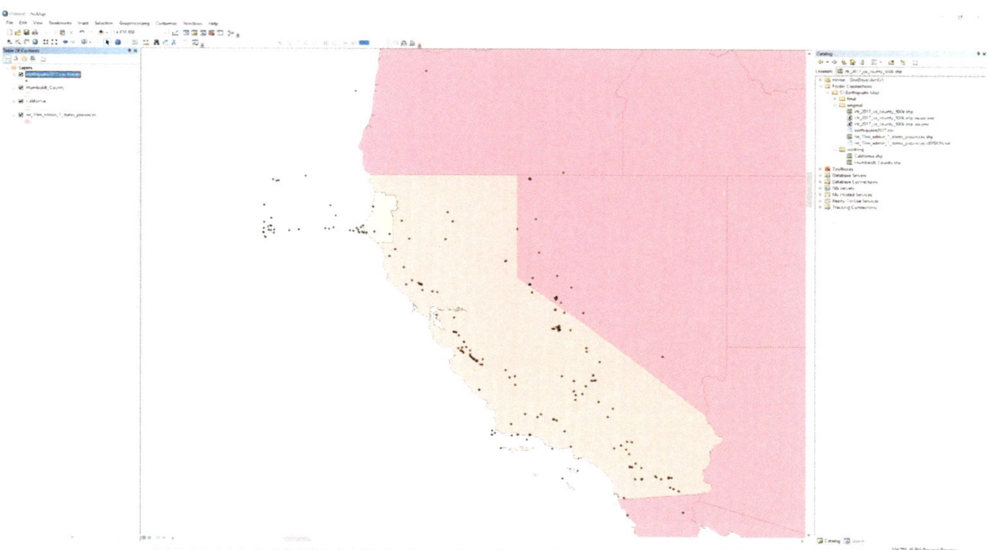

Figure 2.80: Initially, XY data gets added as a temporary events layer.

Make the earthquake layer a permanent shapefile by exporting it. The process is the same as when you exported the California selection and the Humboldt County selection.

However, in this instance, with nothing selected, the entire layer gets exported. In the *Table of Contents*, right-click on earthquake2017.csv *events* and select Data, then *Export Data*. Save the file to your *working* folder and name it CA_earthquakes_2017.shp. Add the exported data as a map layer. When done, right-click on the *events* layer and select *Remove*. You should now have four layers in the *Table of Contents*, the California earthquakes of 2017, Humboldt County, California, and the United States and Mexico. Take a moment to save your map document before continuing to the next step.

Changing the Map Projection of the Data Frame

The Earth is spherical, yet maps still managed to represent the Earth using a flat plane. Map projections make this possible. A **map projection** is the geometric transformation of the round earth onto a flat plane using mathematical equations. In **Chapter 3**, you learn more about map projections. What you need to understand now is that there are many different map projections. The one you choose changes the shape and appearance of features on the map. In this step, you change the map projection for the data frame to make the State of California appear closer to its actual size and shape on Earth. In ArcMap, zoom out until you see the entire continental United States (**Figure 2.81**). This view helps you to understand the difference in appearance between projections.

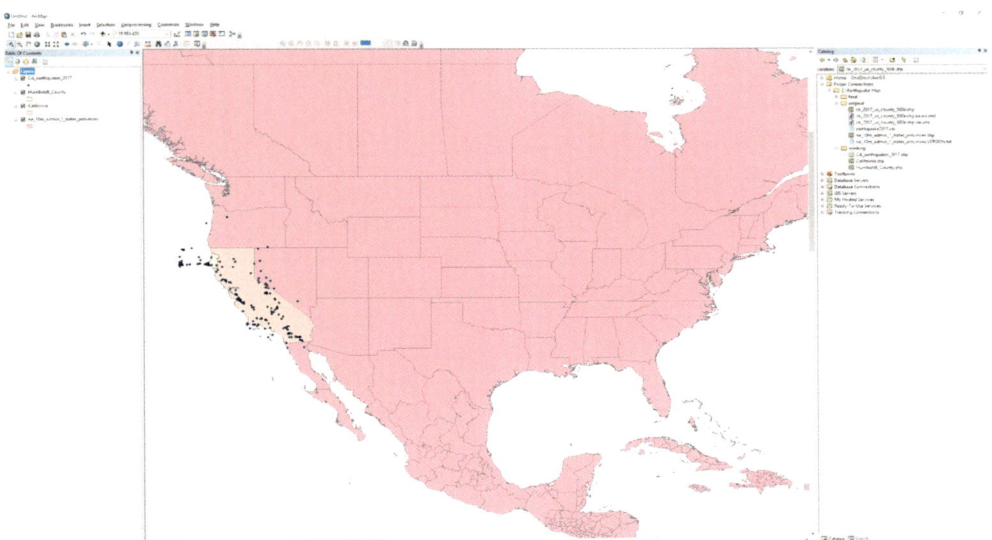

Figure 2.81: The United States has a slightly curved shape. When you see it represented with a flat top, you know it is not in the optimal projection.

Open the data frame properties. Recall from a previous tutorial, in the *Table of Contents* the word *Layers* appears above the four map layers. Esri made a lousy choice when naming this element in the *Table of Contents*. The name consistently leads to confusion. This item on the *Table of Contents* is *not* a layer. It represents the **data frame**.

Access the properties for the data frame by right-clicking on the word *Layers* in the *Table of Contents*. Select *Properties* from the contextual menu. When the Data Frame Properties window opens, click the *General* tab (**Figure 2.82**). Change the name of the data frame to something more descriptive, such as *California Earthquakes*.

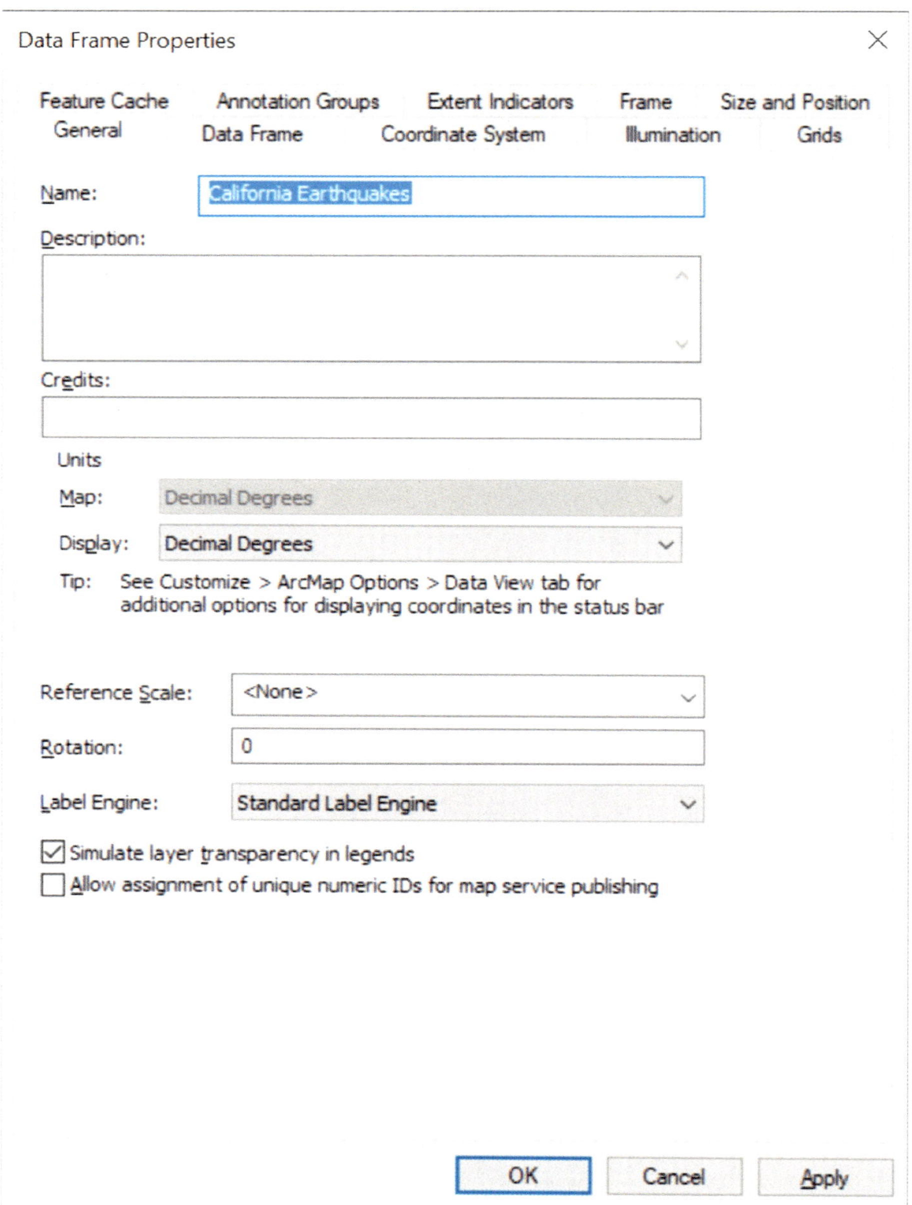

Figure 2.82: Changing the name of the data frame helps to add clarity to the *Table of Contents*.

Select the *Coordinate System* tab. Scroll down until you see the *Projected Coordinate Systems* folder. Expand the folder, then, expand the *Continental* subfolder. Locate the *North America* subfolder (**Figure 2.83**).

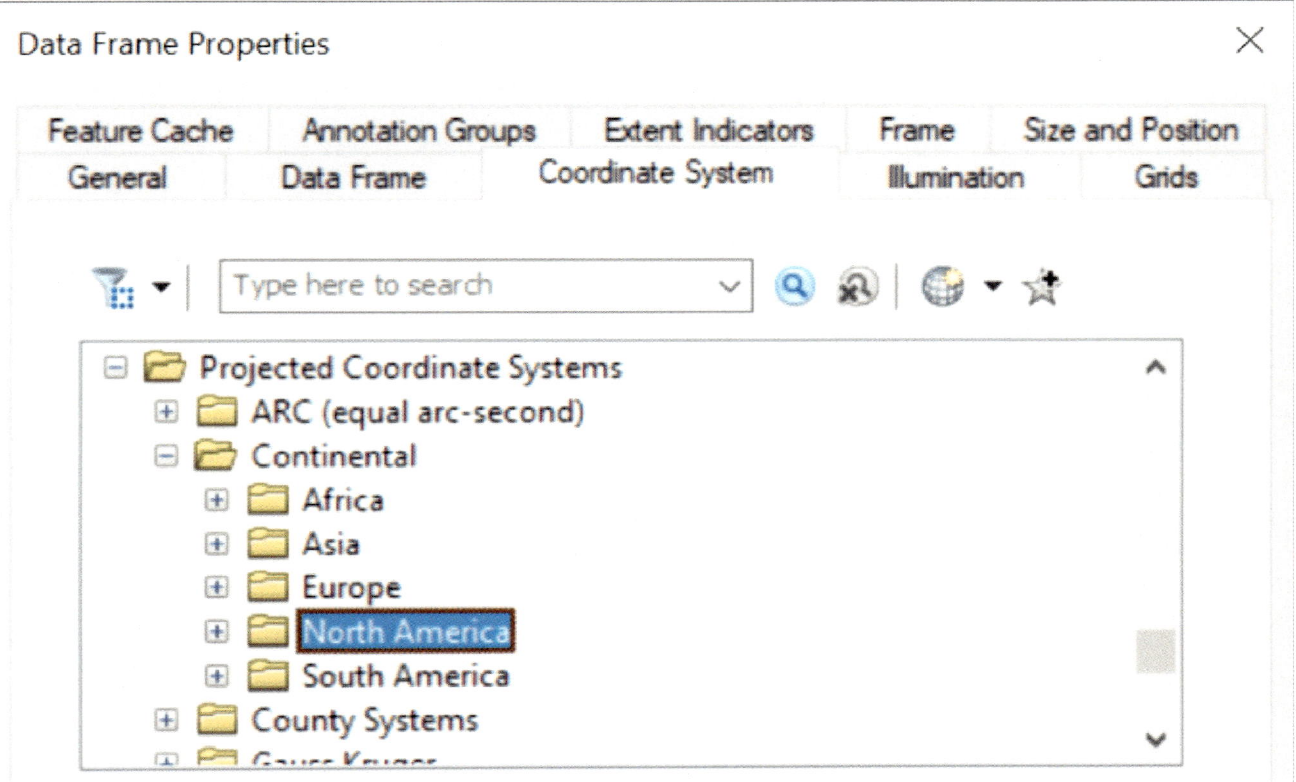

Figure 2.83: Be sure you are selecting the North America folder from the *Projected Coordinate Systems* folder.

Within the *North America* folder, select *North America Lambert Conformal Conic* (**Figure 2.84**) Then, click *OK*.

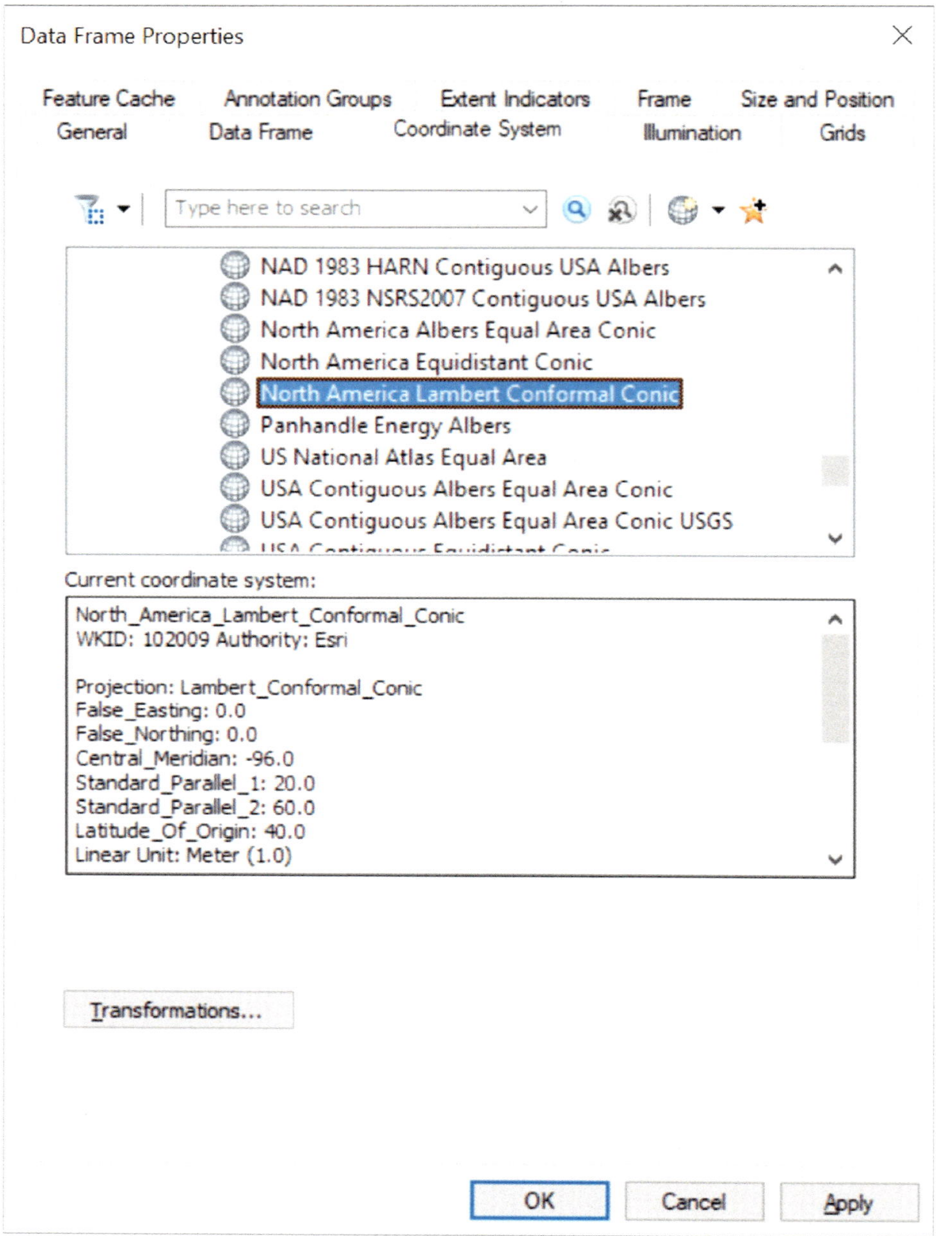

Figure 2.84: The North American Lambert Conformal Conic projection optimizes the shape and appearance of the North American Continent.

You may get a warning (**Figure 2.85**). ArcMap is telling you that the data frame coordinate system does not match some of the layers in the *Table of Contents*. Since we are only using it for display purposes, this is fine. Check the box that says, *Don't warn me again in this session*, and click *Yes* to close the warning.

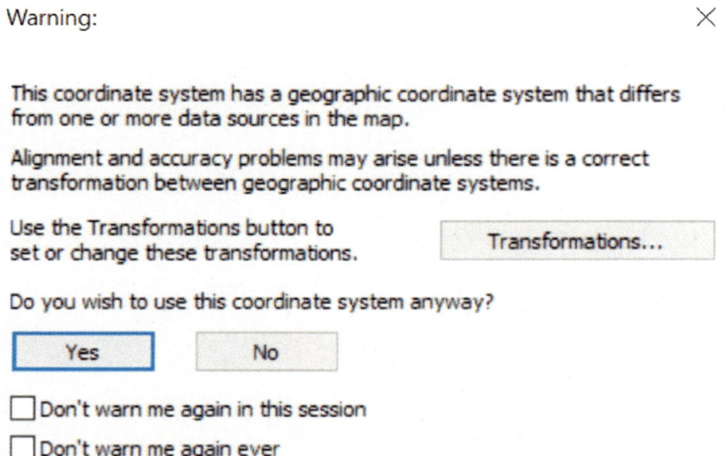

Figure 2.85: ArcMap is warning that not all of the layers have the same spatial reference system.

You should notice an immediate change to the size and shape of the United States (**Figure 2.86**). The Lambert Conformal Conic project works well to make the U.S. appear closer to the way it does on Earth. The size and shape of the individual states are also closer to reality.

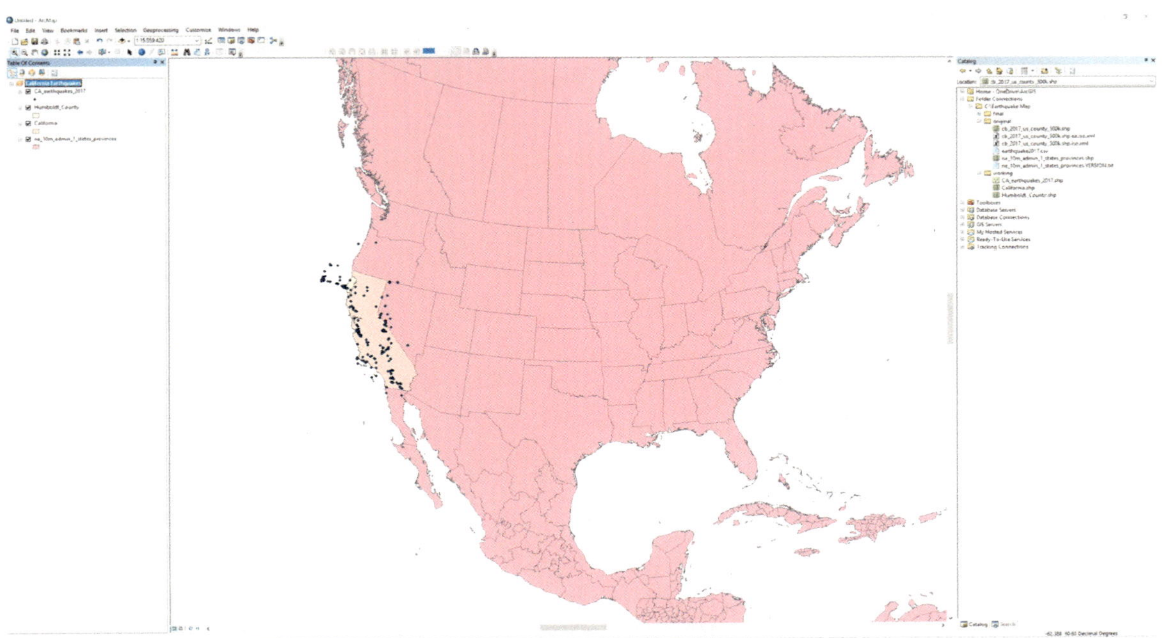

Figure 2.86: The North American Lambert Conformal Conic projection displays the correct shape of the United States.

Zoom back into your earthquake layer and save your map document before moving on to the next step.

Representing Earthquake Magnitude Using Graduated Symbols

At this point, your map displays the location of earthquake epicenters in California. You can also provide additional information to your map reader by also showing the relative magnitude of the earthquakes. As you learned in a previous tutorial, one of the most powerful features of GIS software is the connection between a database and the map. Each shapefile comes with a database table. In ArcMap, you refer to the database table as an **attribute table**. In the *Table of Contents*, right-click on the earthquake layer and open the attribute table. The first time you open the attribute table, it appears floating above the map.

Click and drag your attribute table towards the bottom of ArcMap. As you drag, position your cursor over the blue arrow that appears near the bottom. When you place your cursor over the blue arrow, the attribute table snaps to the bottom of ArcMap.

Take a moment to read through the attribute table. What type of information about earthquakes are provided by the NCEDC? Some of the information, such as the date and time, are easy to interpret. For other attributes, you may need to go back to the NCEDC website to learn more. In this step, you only need to understand the Magnitude attribute. The field named *Magnitude* records a series of numbers (**Figure 2.87**). These numbers determine the size of the point symbols on the map.

Table

CA_earthquakes_2017

FID	Shape *	DateTime	Latitude	Longitude	Depth	Magnitude	MagType	NbStations	Gap	Distance	RMS	Source	EventID
0	Point	2017/01/02 02:11:09.40	38.70833	-118.467	-2.15	3.91	ML	51	173	51	0.46	NCSN	72747035
1	Point	2017/01/02 13:05:08.31	38.3925	-118.92083	-2.19	3.68	ML	46	135	40	0.21	NCSN	72747170
2	Point	2017/01/02 18:01:16.49	38.39083	-118.90134	2.26	3.25	ML	26	135	38	0.16	NCSN	72747245
3	Point	2017/01/04 07:31:17.51	38.3785	-118.89283	2.88	3.04	ML	27	149	37	0.07	NCSN	72747675
4	Point	2017/01/05 19:18:55.89	38.42767	-118.88467	13.69	3.48	ML	26	139	39	0.12	NCSN	72748355
5	Point	2017/01/06 04:14:25.60	38.37717	-118.89917	3.57	3.31	ML	36	144	37	0.07	NCSN	72748665
6	Point	2017/01/06 16:36:15.66	38.42783	-118.883	15.09	3.06	Md	21	142	38	0.08	NCSN	72749155
7	Point	2017/01/09 13:14:36.40	38.43067	-118.8885	2.75	3.34	ML	26	175	39	0.16	NCSN	72750255
8	Point	2017/01/09 15:58:17.41	38.4005	-118.88184	14.84	3.04	ML	21	173	37	0.1	NCSN	72750405
9	Point	2017/01/10 16:23:55.74	40.30617	-124.069	32.82	3.12	ML	30	70	11	0.14	NCSN	72750720

1 (0 out of 214 Selected)

CA_earthquakes_2017

Figure 2.87: Notice that the first row in the original CSV table became the field names.

In the *Table of Contents*, right-click on the earthquake layer and select *Properties*. Navigate to the *Symbology* tab. On the left, choose *Quantities*, then *Graduated symbols*. Near the middle, use the drop-down menu to select *Magnitude* for the *Value* (**Figure 2.88**). Leave all other settings as default and click *OK*.

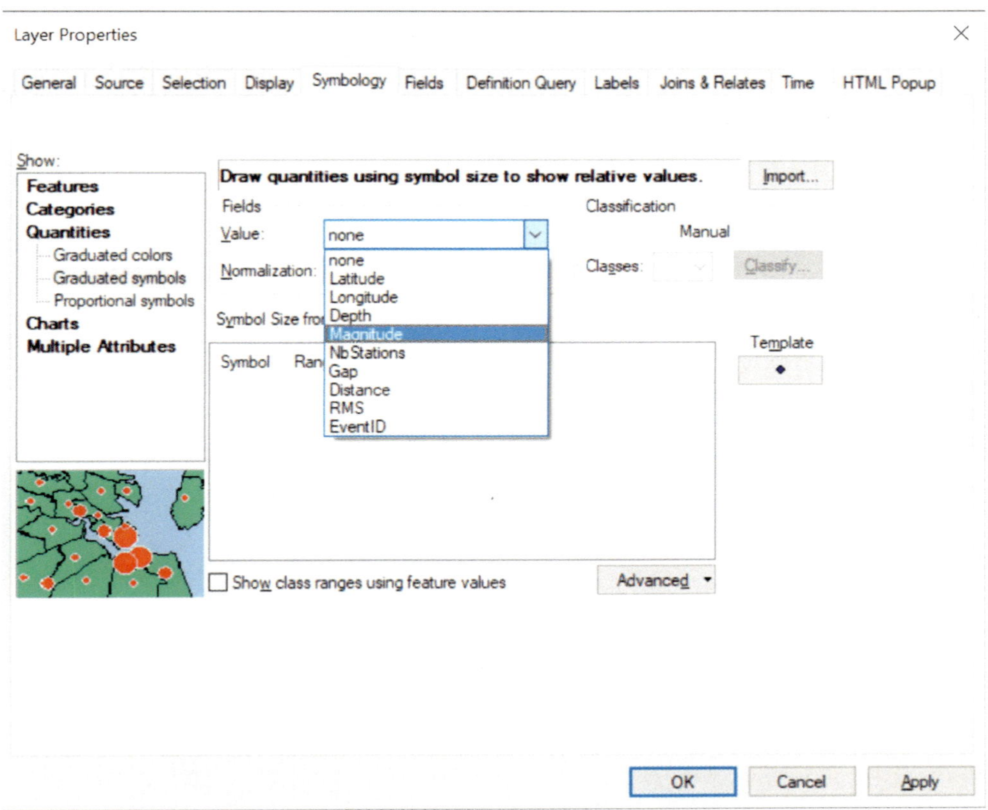

Figure 2.88: The drop-down menu provides a list of all the number-based attributes.

The earthquake data is now symbolized using different sized circles based on the magnitude value in the attribute table. Go ahead and close the attribute table.

You may also change the colors for the remaining layers as well. For this activity, I recommend choosing either a light color scheme or a dark color scheme for the basemap. Either one should have colors with low saturation, low value, or both. For example, a light color scheme might use light greys and pastels for the basemap features. A dark color scheme might use medium to dark greys instead. Your goal is to provide context without competing with your thematic data.

You work from the bottom up, starting with the states and provinces layer. It lays the foundation for the remainder of the map design. On the *Table of Contents*, click the colored rectangle under the states and provinces layer. When symbol selector opens, choose a something neutral for the Fill Color. Set the Outline Color and adjust the line weight (**Figure 2.89**).

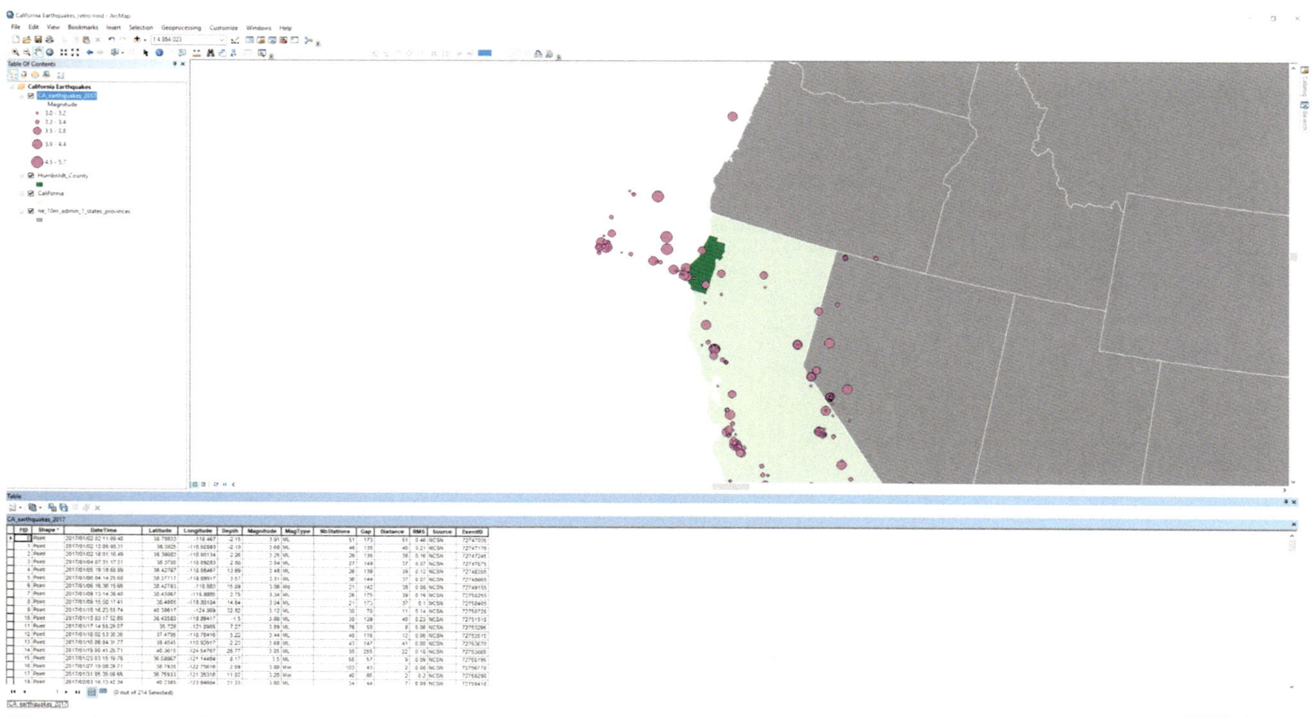

Figure 2.89: ArcMap chooses random colors. If you want to change your point symbol color, return to the *Symbology* tab.

Repeat these steps for the remaining layers, one at a time. Try to maintain your chosen color scheme throughout the process. If necessary, use the *Zoom In* tool on the *Layout* toolbar to get a closer look at the details.

Changing the Map Size and Position

When you first begin to design a map layout, you should always determine the size of the page on which you work. In this instance, create a poster-sized map of California earthquakes in 2017 that is **24 inches by 18 inches**. Include a map of California and an inset map of Humboldt County.

By default, ArcMap sets the page size to 8.5 inches by 11 inches. This setting is *rarely* acceptable in a professional setting. In this step, you change the page size. Then, you change the data frame size and position on the page. From the *Main* menu, select *View*, then *Layout view* (**Figure 2.90**).

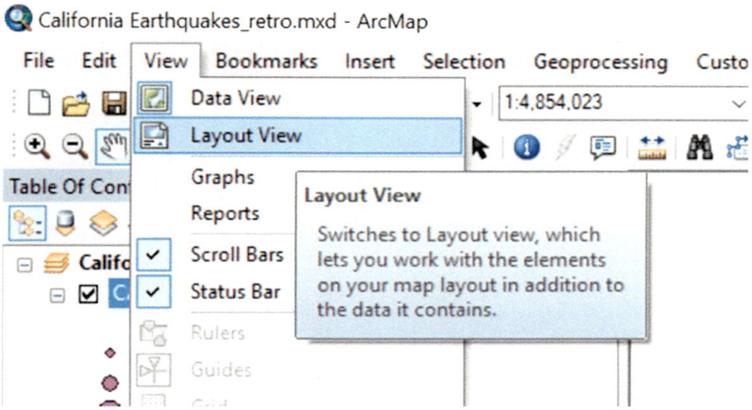

Figure 2.90: There are two ways to display the data frame in ArcMap, Data View, and *Layout View*.

The data frame window immediately changes to show the page size and the data frame position on the page. Across the top and along the left you see a ruler indicating the height and width of the page. Currently, the page size is 8.5 inches by 11 inches, which is the default setting in ArcMap. The *Layout* toolbar may also appear floating above the map. You may dock the toolbar on top (**Figure 2.91**). It is essential to understand each of the tools in the *Layout* toolbar. These tools are only active when ArcMap is in *Layout View*.

Figure 2.91: The layout tools only work while in *Layout View*.

On the left of the toolbar, you'll see two tools for zooming in and out of the map layout. The icon looks like a magnifying glass over a page. These tools are sometimes confused with the *Zoom In* and *Zoom Out* tool with which you previously worked. They work similarly. Instead of zooming in and out of the data, the zoom tools on the *Layout* toolbar zoom in and out of the *page layout*. Likewise, the *Pan* tool on the layout toolbar pans across the page layout and does not move the data. Next to the *Pan* tool is the *Zoom to Whole Page* tool. This tool is handy for quickly viewing the entire page layout. Take a moment to experiment with each of these tools before moving on to the next step.

Next, you change the page size. From the *Main* menu, select File, then Page and Print Setup (**Figure 2.92**).

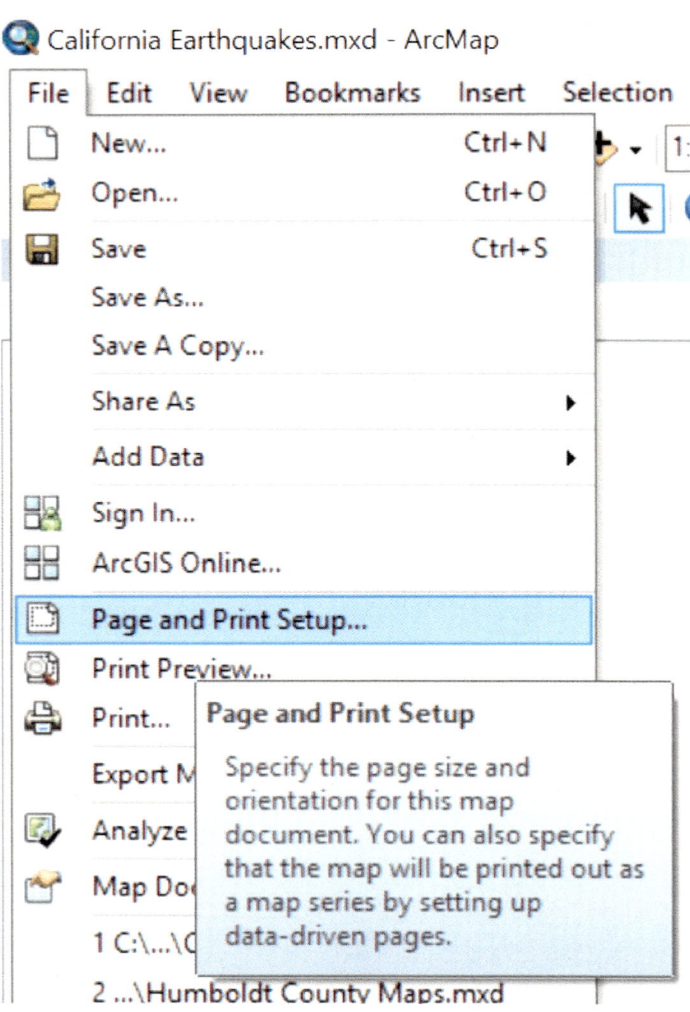

Figure 2.92: The page and print setup option allows you to change the size of the paper on which the data frame rests.

Uncheck the box next to Use Printer Paper Settings (**Figure 2.93**). For the Width, enter 24. For the Height, enter 18. When you are ready, click *OK*.

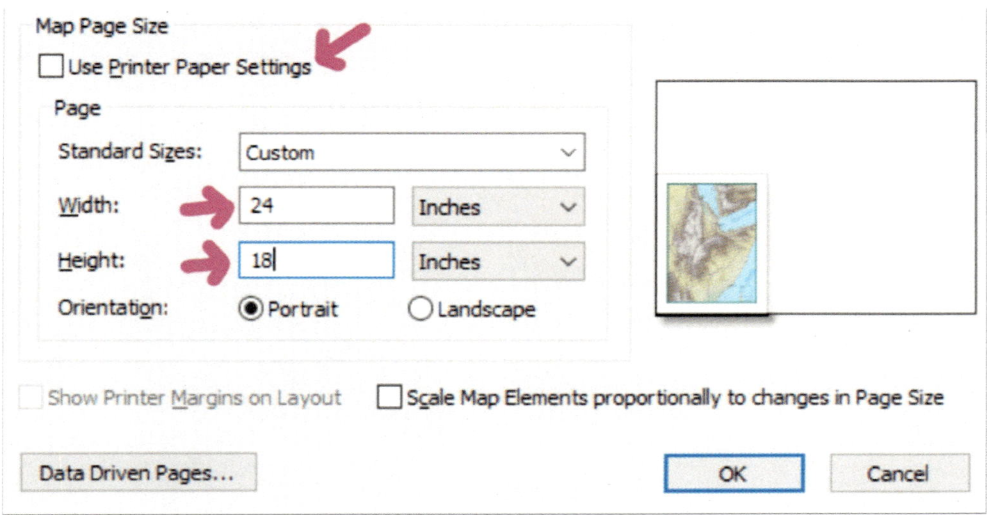

Figure 2.93: Check the lower left corner of the Page and Print Setup window to change these settings.

The page is now a large poster size of 24 inches by 18 inches. You can tell because the ruler along the top and the sides have changed to show the new dimensions. However, the data frame is still the default size (**Figure 2.94**).

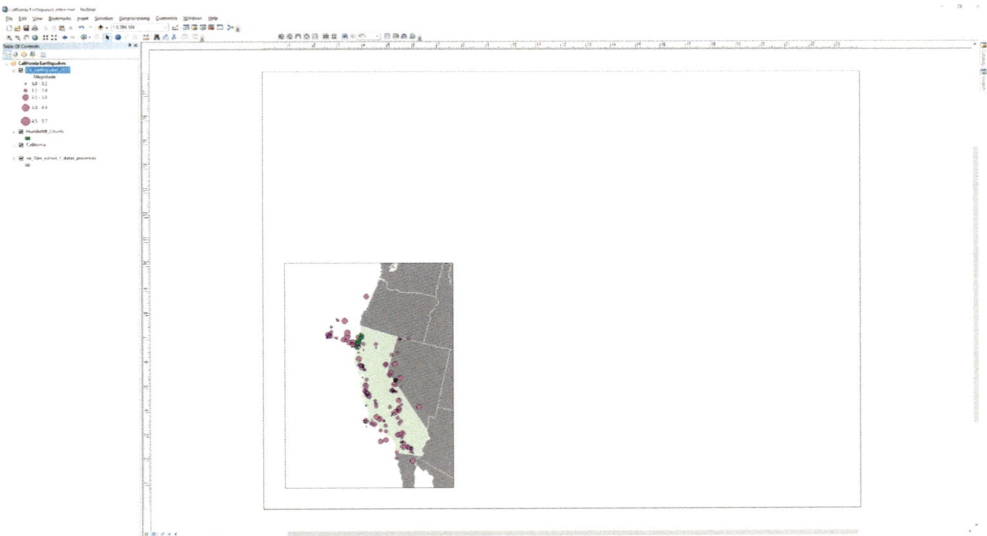

Figure 2.94: There are always two steps when setting up your map layout. Change the paper size and change the data frame size. Here you see an image of the new page size with original data frame size

To change the data frame size, open the data frame properties and navigate to the *Size and Position* tab (**Figure 2.95**). Change the width and the height values so that they match the page size. Under *Position*, change the X and the Y to zero.

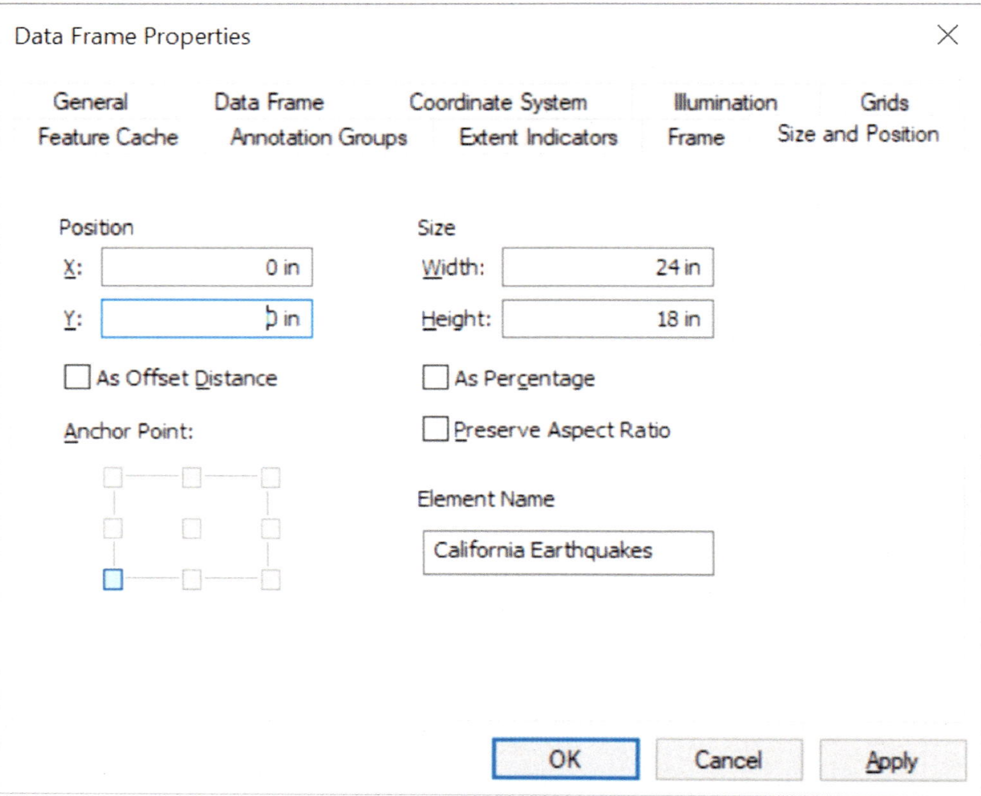

Figure 2.95: By default, the Position value anchors the data frame on the page relative to the lower left corner.

Navigate to the *Frame* tab (**Figure 2.96**). Change the Border to a thickness of 4.0 and change the border color to **dark navy**. Change the background color to **light cyan**. When you are ready, click *OK*. You may get another warning about the coordinate system. Click *Yes* to close the warning.

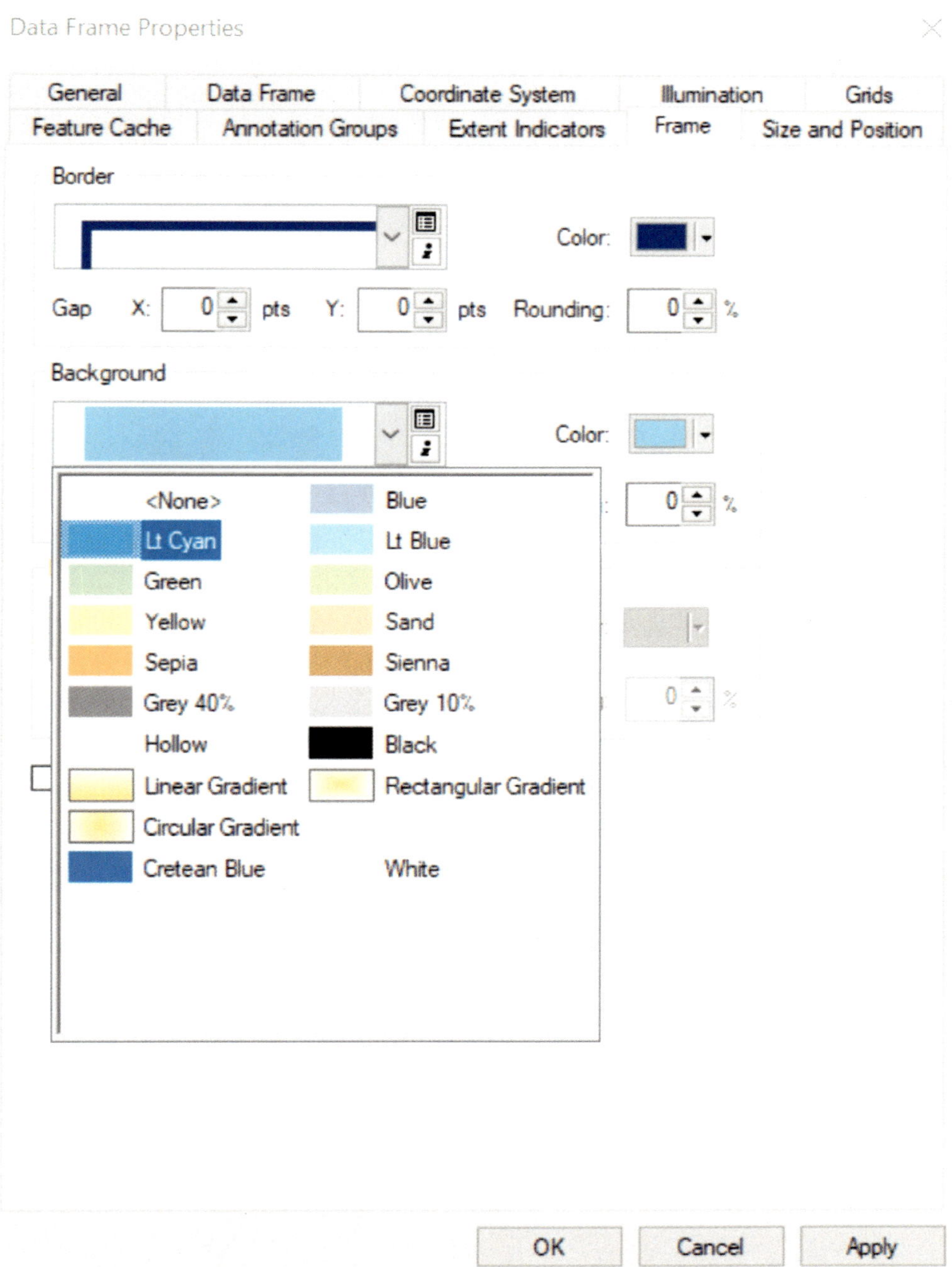

Figure 2.96: Do not confuse the Frame tab with the Data Frame tab.

On the *Table of Contents*, right-click on your earthquake layer and select *Zoom to Layer*. You should now have a poster size map of the earthquakes in California (**Figure 2.97**).

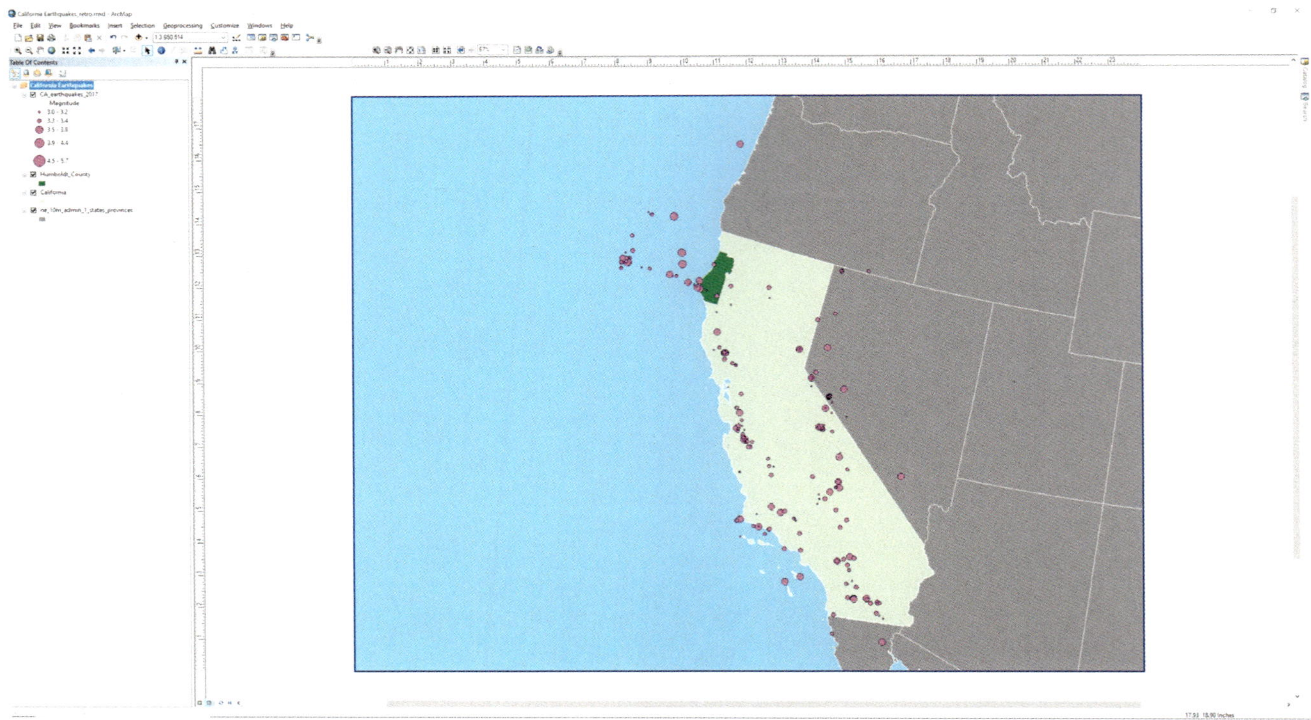

Figure 2.97: When you Zoom to Layer, ArcMap changes the map extent so that the entire layer gets centered on the page.

Creating an Inset Map

An **inset map** shows more detail for a specific region within the main map. In this step, you generate an inset map that shows more detail for the area around Humboldt County. In a previous tutorial, you learned how to create multiple data frames. Create a second data frame for the inset map. If you recall, you select *Insert* from the *Main* menu, then *Data Frame*. A new blank data frame appears in the *Table of Contents* (**Figure 2.98**). On the map, the new data frame appears near the center.

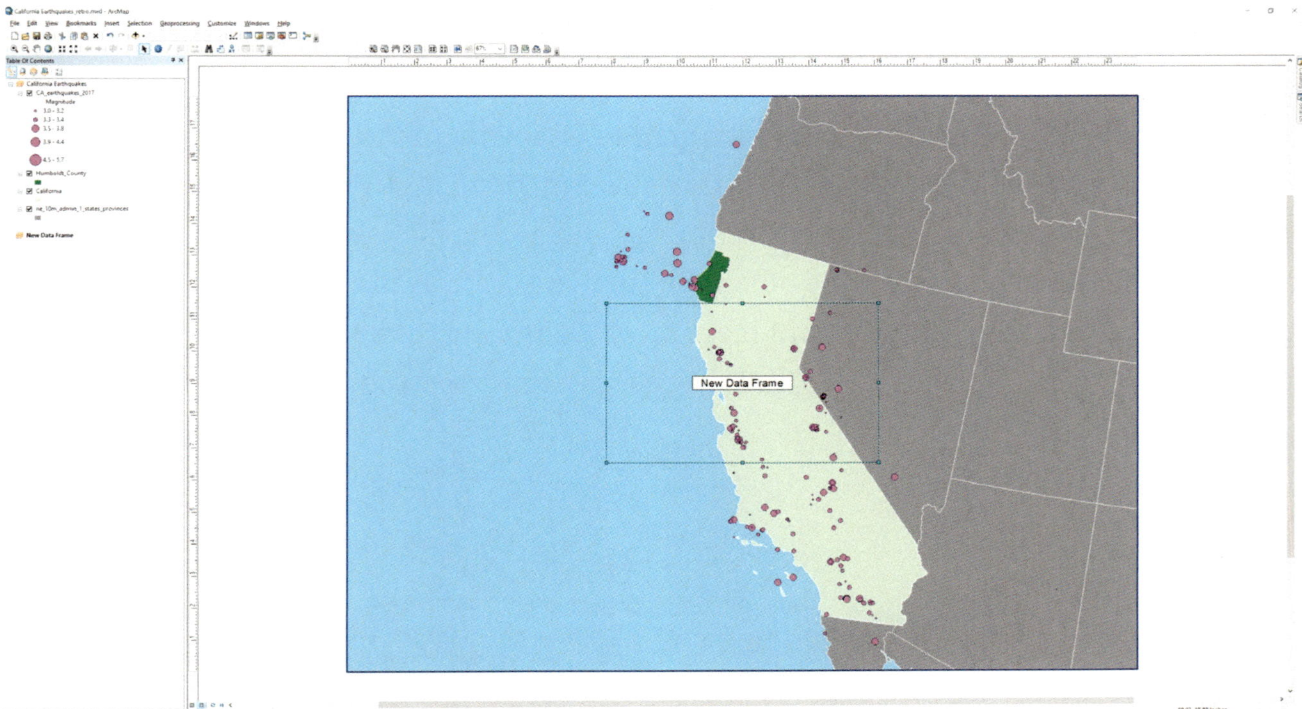

Figure 2.98: When you have multiple data frames, you need to arrange their position on the page.

Hold down the shift key and select each layer in the California Earthquakes data frame. Then, right-click and choose *Copy*. Then, right-click on the new data frame and select *Paste Layer(s)* (**Figure 2.99**). A copy of the layers appears in the new data frame.

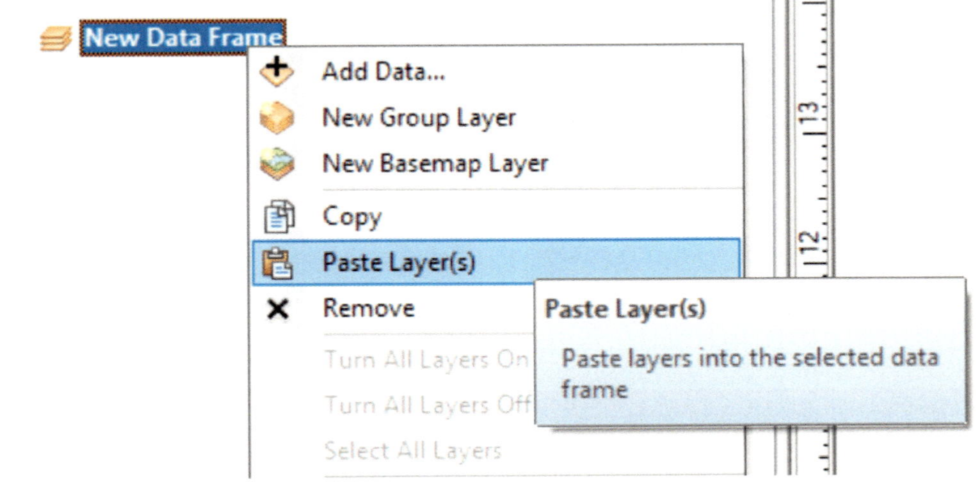

Figure 2.99: You can copy and paste multiple layers between data frames.

Using what you learned in previous steps, change the name of the new data frame to *Humboldt County Inset*. Using what you learned in a previous step, change the coordinate system to *North America Lambert Conformal Conic*. Change the size and position properties to set the width to **10 inches** and the height to **7 inches**. Anchor the new data frame to the lower left corner by setting the X and Y values to zero (**Figure 2.100**). Set the background color to blue and give it the same border as the main map. Zoom in close to Humboldt County. Be sure you can see all of the earthquakes off the coast (**Figure 2.100**). On the main map, move California to the right so that there is more space between the main map and the inset. It gives you room later when you want to place a scale bar.

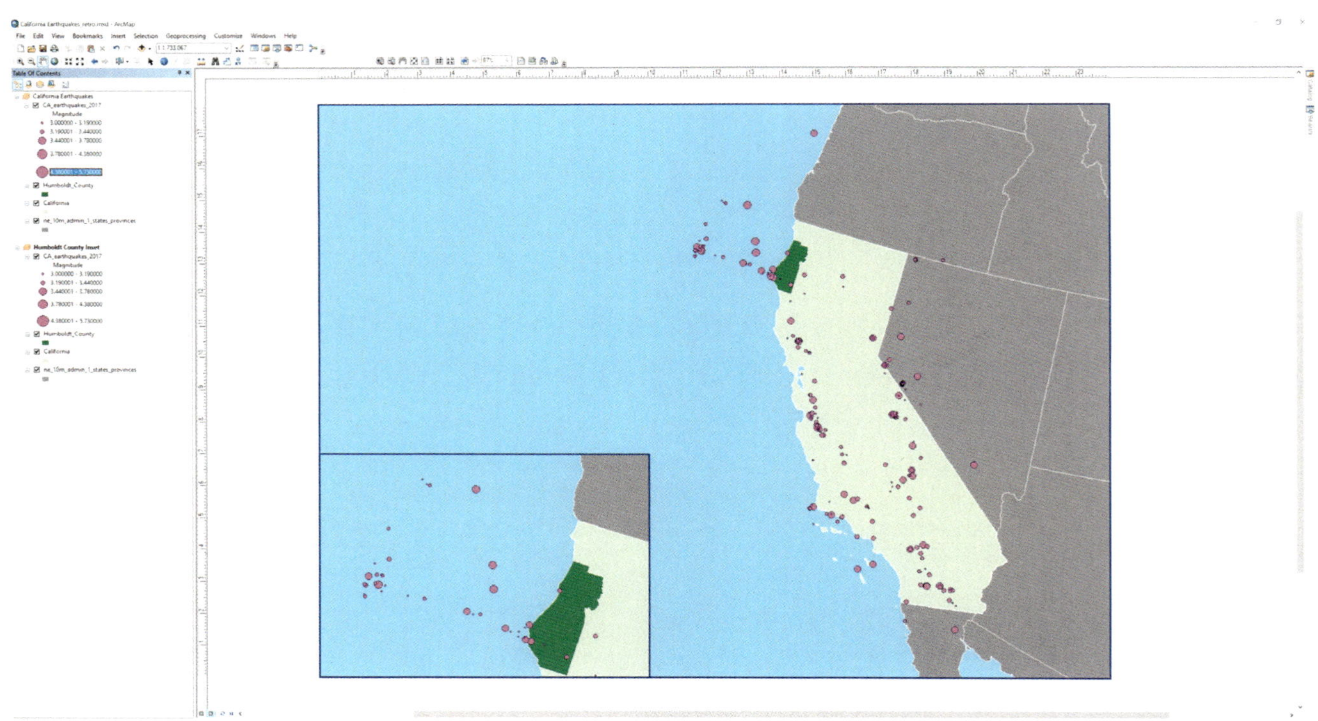

Figure 2.100: Resize your inset map so that it fits over the ocean in the lower left corner.

Take a moment to save your map document before moving on to the next step.

Inserting A Map Title

There are many common elements recognized by the map-reader. However, in this activity, you only need to add a few to the map. In the next few steps, you add a map title, a north arrow, a scale bar, and a legend. As a matter of courtesy, you also add an acknowledgment to Natural Earth, the Northern California Earthquake Data Center (NCEDC), and the United States Census Bureau.

When you have multiple data frames on a map, you need to add map elements to each one independently. Right-click on the data frame titled California Earthquakes and select *Activate*. The main map should now have a dotted outline. The dotted outline indicates which data frame is active (**Figure 2.101**).

Figure 2.101: The dotted outline indicates which data frame is active.

In a previous tutorial, you learned how to add toolbars to ArcMap. Right-click on the empty gray area and select *Draw* to add the *Draw* toolbar (**Figure 2.102**). When you first open the toolbar, it may appear floating above the map. You may dock the toolbar on top to keep it out of the way.

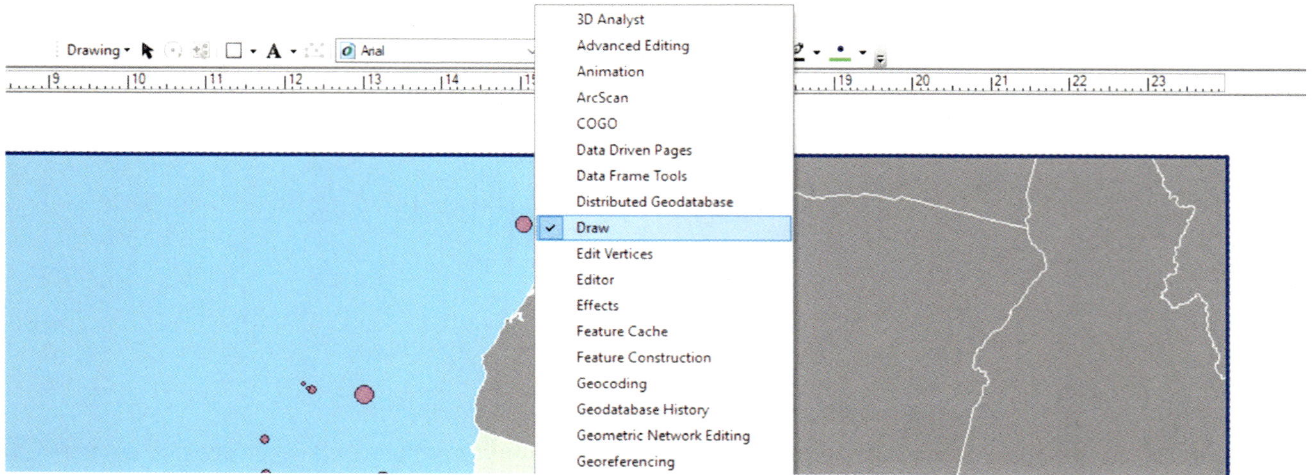

Figure 2.102: This image shows the *Draw* toolbar docked at the top. It also displays the contextual menu that appears when you click on the gray space at the top.

On the *Draw* toolbar, select the tool the looks like a letter **A**. Then, click somewhere near the top of the map to add a map title. Enter the title, *California Earthquakes, 2017*. When done, press the *Enter* key. Double-click the title to edit. The Properties window appears (**Figure 2.103**). Here, you can edit the text if needed. Click the button that says Change Symbol. On the Symbol Selector window (**Figure 2.104**), change the typeface to *Garamond*. Change the font size to 100. Click *OK* when you are ready. Then, click *OK* again.

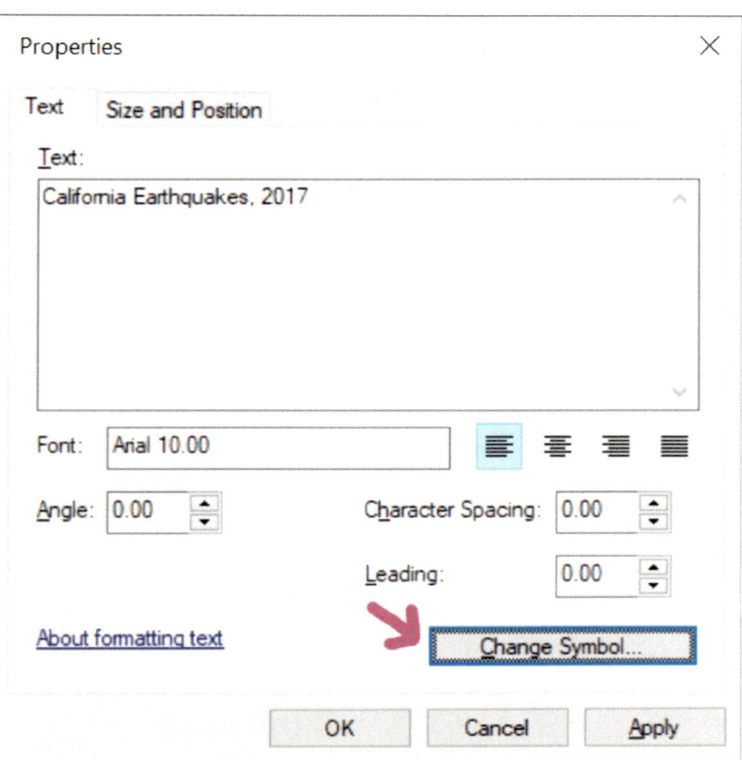

Figure 2.103: In ArcMap, sometimes you have to dig around for the right settings.

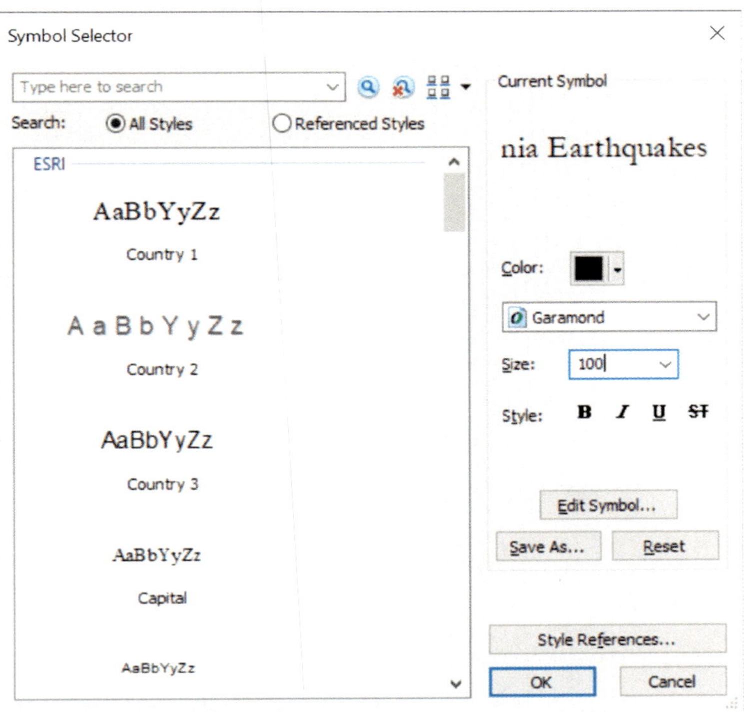

.Figure 2.104: You should try to limit the total number of typefaces on your map to two, one serif typeface, such as Garamond, and one sans serif typeface, such as Veranda.

You may need to move the title and adjust its position on the map. By default, the text does not have a background. The lack of background can make the text harder to read when there are map features underneath (**Figure 2.105**). Sometimes it is helpful to add a title bar across the top.

Figure 2.105: Avoid overlapping text and labels with other lines on the map, other map elements, or with each other. They should be easy to read. Consider moving such labels to an empty area or adding a background color to the text.

On the *Draw* toolbar, select the rectangle tool. Then click and drag to draw a rectangle across the top of the map. Any new feature you draw gets placed on top. Right-click on the title bar and select *Order*, then *Send Backward* (**Figure 2.106**). The title bar gets set under the map title.

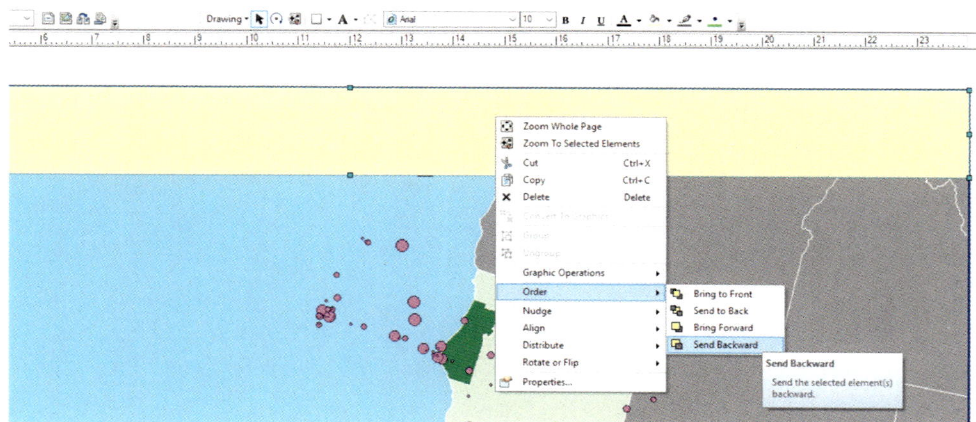

Figure 2.106: Any new feature you draw gets placed on top. In this instance, the title bar is covering the map title.

Just like the title text, the title bar also has properties you can edit. Right-click on the title bar and select *Properties*. Change the fill color to dark navy and the outline color to no color (**Figure 2.107**). On the size and position tab, change the width to 24 and the height to 1.75 (**Figure 2.108**). Change the X value to 0 and the Y value to 16.25. When you are ready, click *OK*.

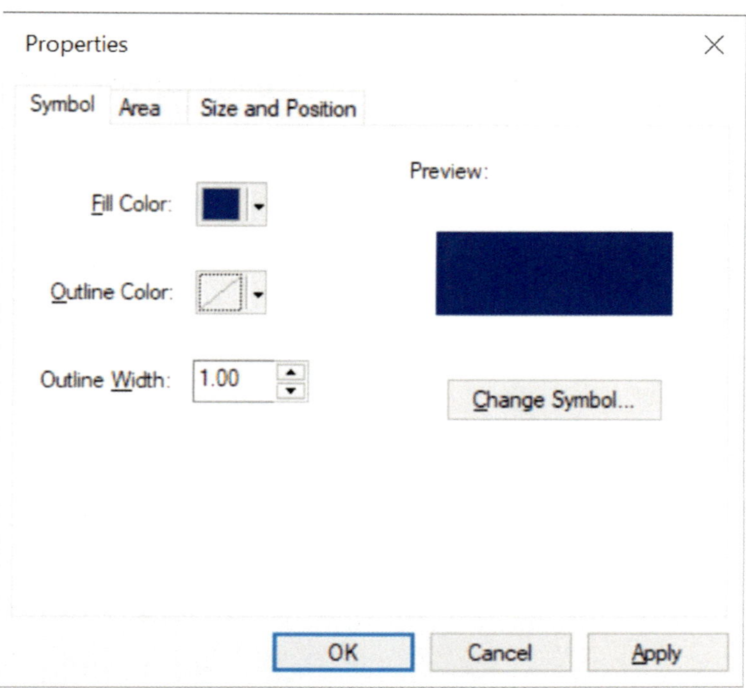

Figure 2.107: Most map elements have properties that you can edit.

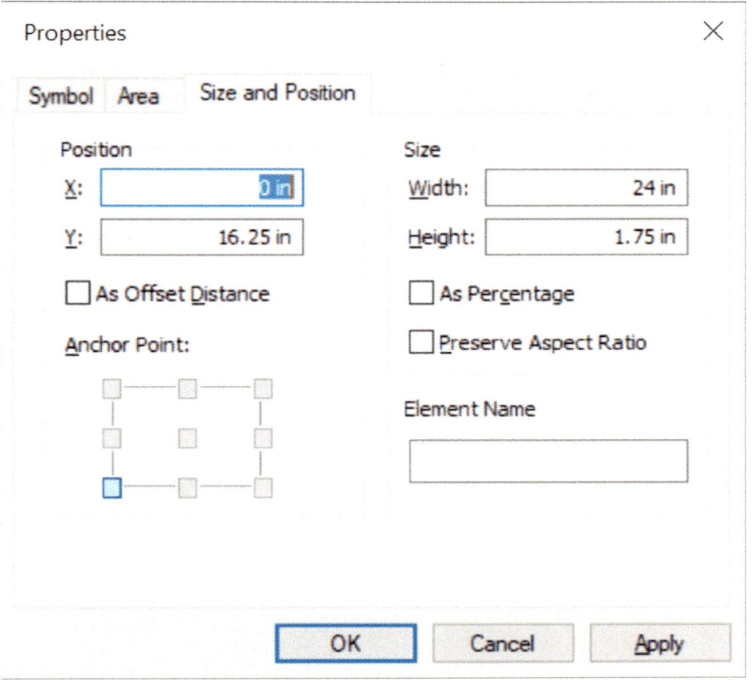

Figure 2.108: By default, the *Position* value anchors the title bar on the page relative to the lower left corner.

Go back and change the color of the title text to a 10% light gray, so that it contrasts with the color of the title background.

Inserting a Map Legend

A **map legend**, sometimes called a map key, helps to explain the symbology on the map. Not all maps need a map legend, as some simple features such as water bodies and roads may be self-explanatory. However, maps that contain thematic data, such as this earthquake map, need a map legend. From the *Main* menu, select *Insert*, then *Legend*. The Legend Wizard appears. By default, ArcMap tries to add all your layers to the map legend. In this instance, you only need the earthquake data. Click the button that has the double arrow pointing left to remove all the layers (**Figure 2.109**).

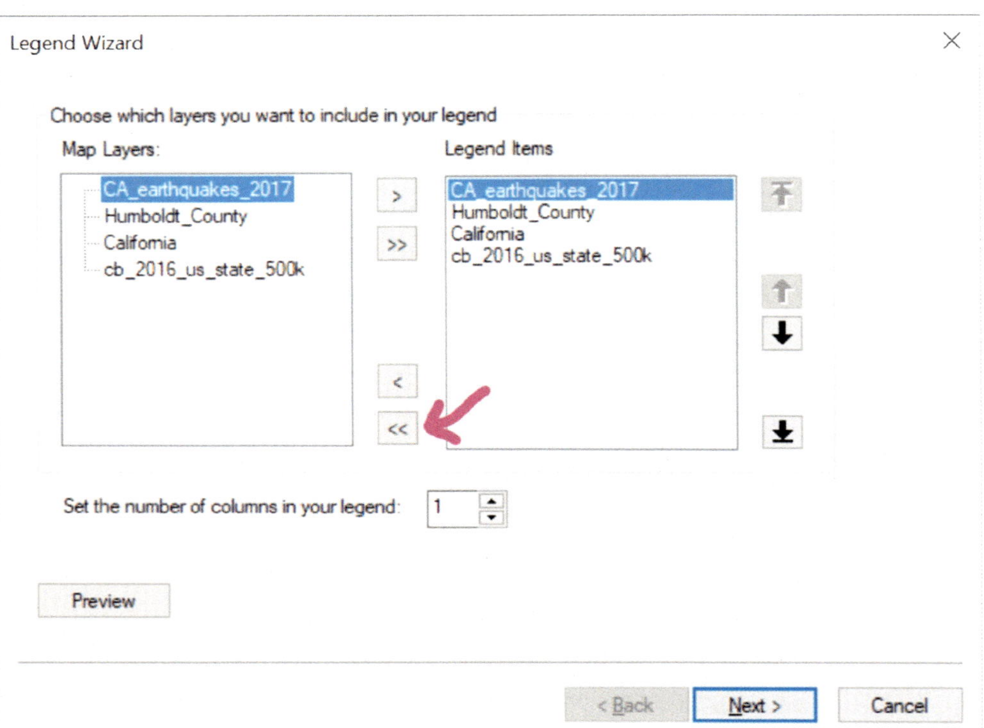

Figure 2.109: The legend wizard allows you to determine which layers to include.

Then, under Map Layers, select the earthquake layer and click the button with a single arrow pointing right (**Figure 2.110**). When you are ready, click *Next*. *Never* use the word "Legend," or a similar word, as a legend title. Your map readers already know that they are looking at a legend. Using this word is redundant, and it makes one look unprofessional. When choosing a legend title, utilize a descriptive title or leave it out altogether. In this instance, a descriptive title would be useful for explaining the meaning of the different sized circles. Change the title to *Earthquake Magnitude* (**Figure 2.111**) Change the font to *Garamond* to match the style of the map title. When you are ready, click *Next*.

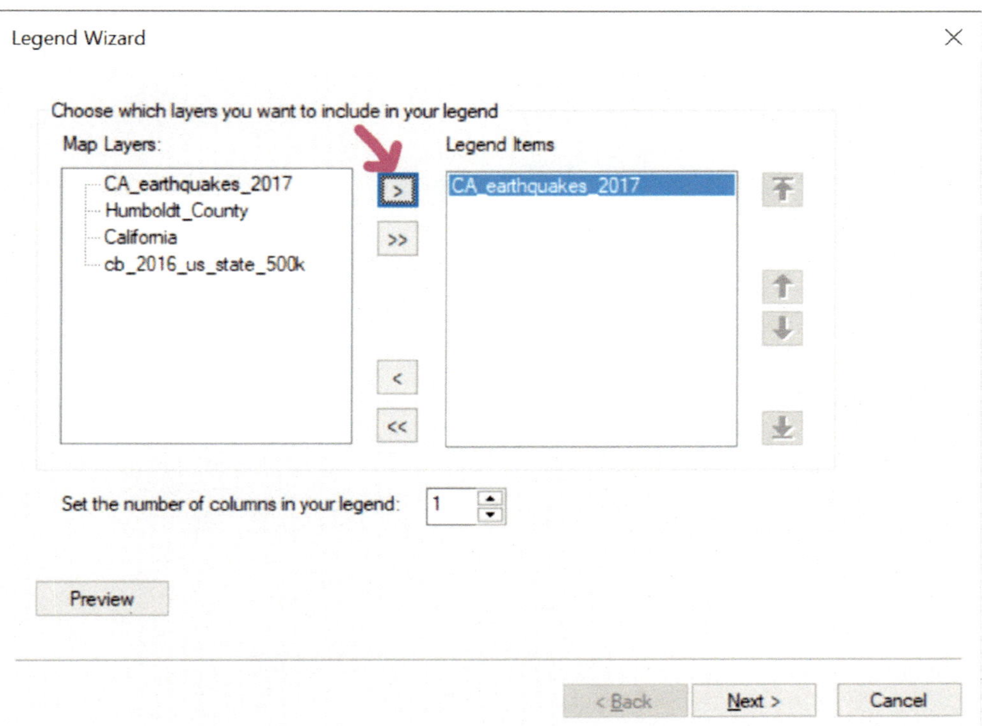

Figure 2.110: It is vital to make a conscious choice of which legend items to include. Never rely on the default settings.

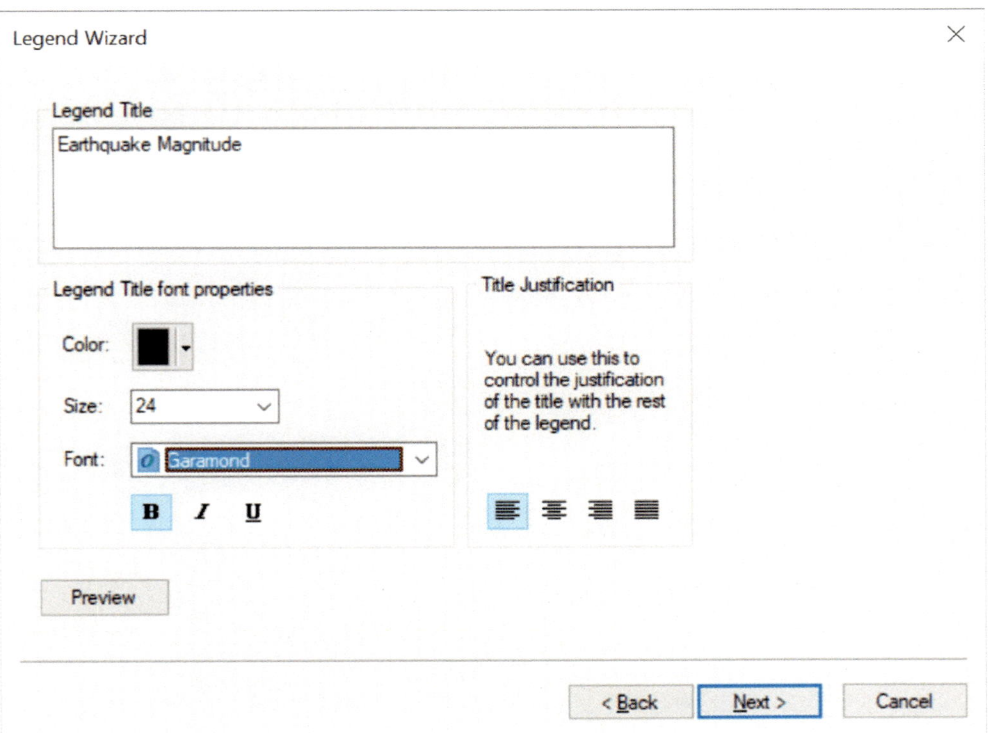

Figure 2.111: You do not always need a legend title. However, you should never accept the default.

Add a 1.0 point border, a gray 10% background, and a gray 50% drop shadow. Change the gap to 5 (**Figure 2.112**). When you are ready, click *Next*.

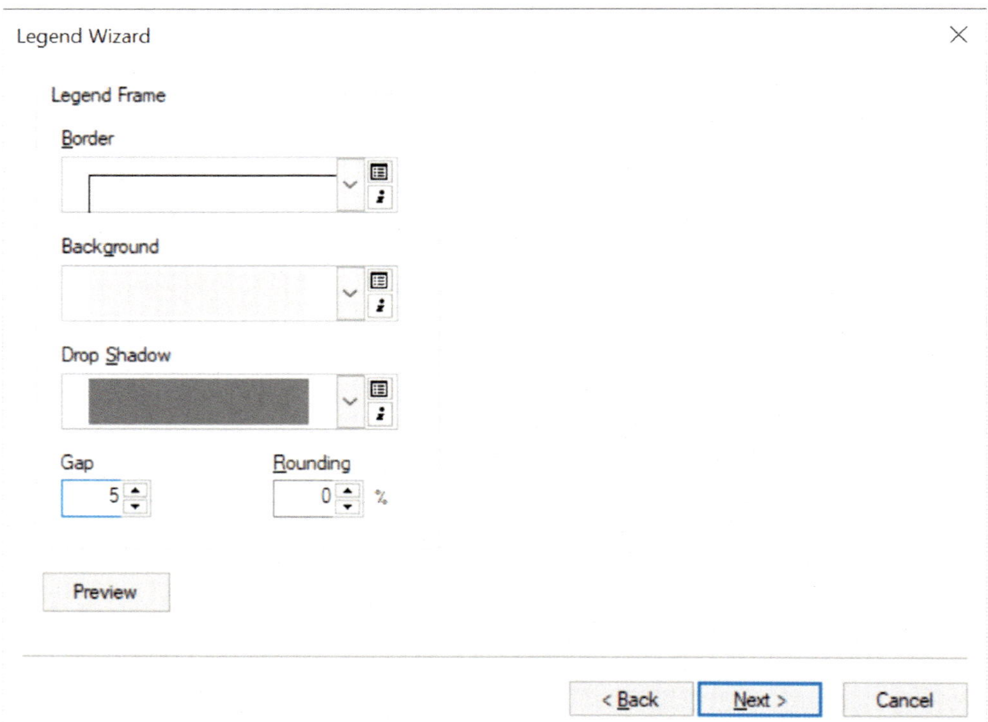

Figure 2.112: A legend background makes the legend items more comfortable to see.

The next window on the legend does not apply to our data (**Figure 2.113**). Go ahead and click *Next*.

The last window on the Legend Wizard allows you to change the space between the legend title and the legend items (**Figure 2.114**). In most cases, the default settings here work fine. When you are ready, click *Finish*.

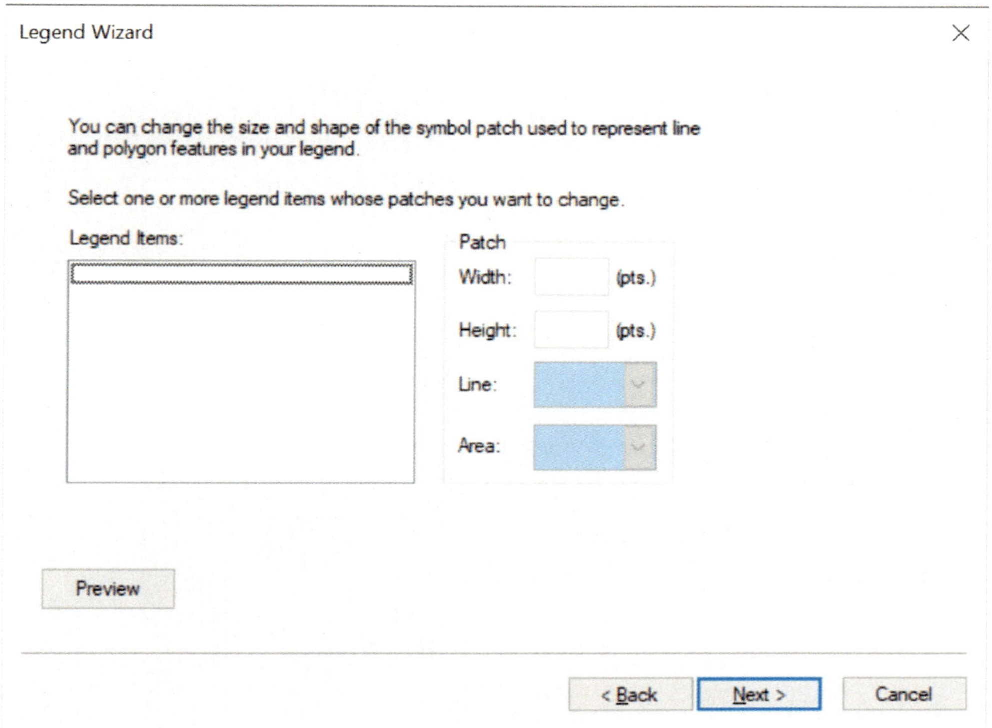

Legend Wizard

You can change the size and shape of the symbol patch used to represent line and polygon features in your legend.

Select one or more legend items whose patches you want to change.

Legend Items:

Patch

Width: (pts.)

Height: (pts.)

Line:

Area:

Preview

< Back Next > Cancel

Figure 2.113: You can change the size and shape of the symbol patch used to represent line and polygon features in your legend.

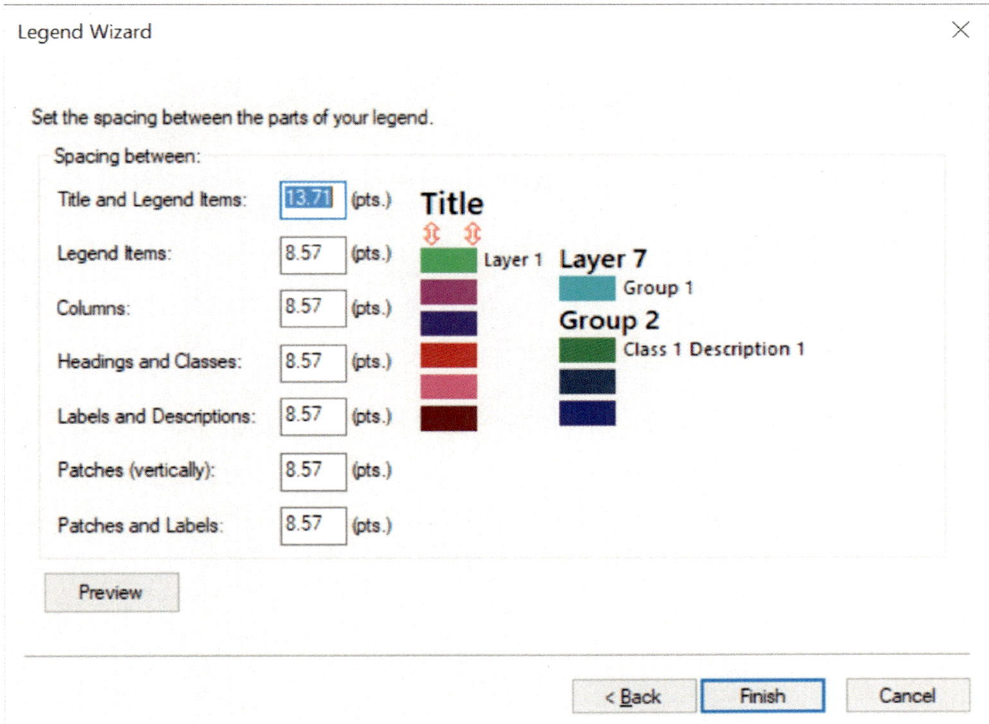

Figure 2.114: In this window, you can set the spacing between the parts of your legend.

Your legend appears near the center of the map. As a general rule, anything placed near the top of the map rises in the visual hierarchy. Secondary map elements, such as north arrows, scale bars, and legends should be placed somewhere in the lower half of the map if possible. In this instance, you may consider placing the legend in the Pacific Ocean, just to the right of the inset map. The legend still needs a little more work. The magnitude values have too many decimal places, and you should remove the layer name. Right-click on the legend and select *Properties*. On the Legend Properties window, navigate to the *Items* tab (**Figure 2.115**). On the lower left, click the button that says *Style*.

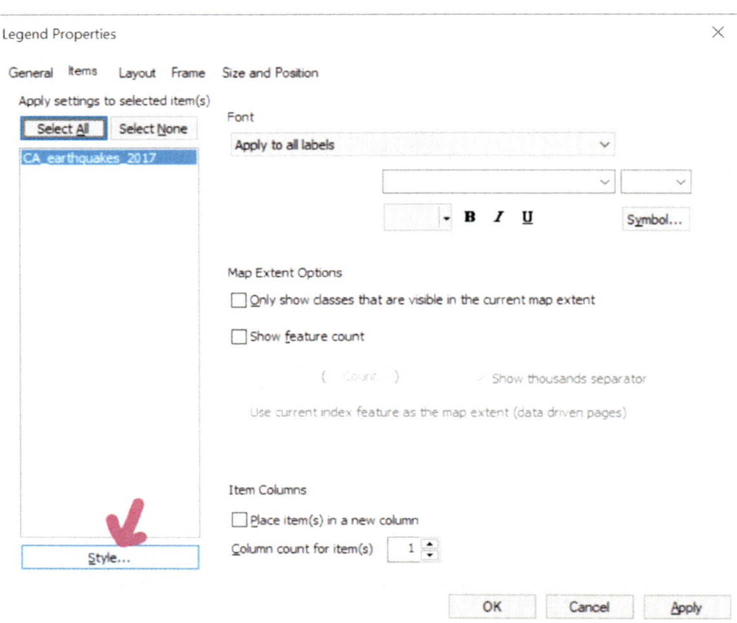

Figure 2.115: The items tab allows to you change settings on specific legend items.

There are many different legend styles from which to choose. However, the second style down on the left, *Horizontal Single Symbol Label Only*, removes the layer name from the legend (**Figure 2.116**). When you are ready, click *OK*, Then, click *OK* again.

The legend style updates, but the legend labels still have too many decimal places. To correct this issue, you must open the layer properties for the earthquake data in the *Table of Contents* under the California Earthquakes data frame. Then, navigate to the *Symbology* tab (**Figure 2.117**). Near the center of the *Symbology* tab, right-click on the word Label and select Format Label.

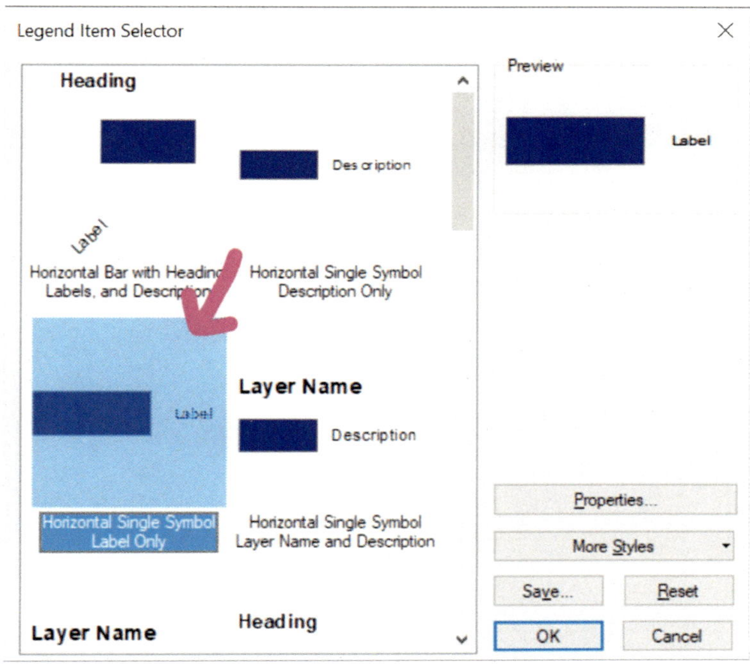

Figure 2.116: There are many legend styles from which to choose.

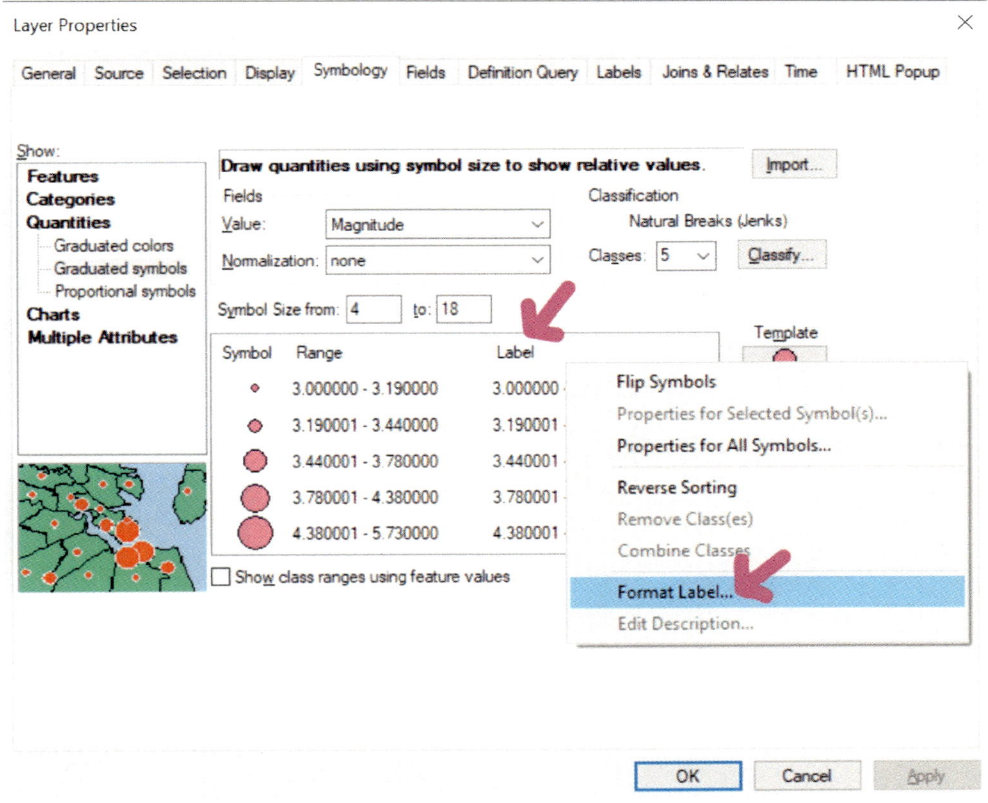

Figure 2.117: You can customize your legend labels on the Symbology tab.

When the Number Format window opens, enter 1 for the number of decimal places (**Figure 2.118**). When you are ready, click *OK*, then click *OK* again. You may need to adjust the size of the legend labels. Return to the *Items* tab on the Legend Properties and change the font size to 12 (**Figure 2.119**). Take a moment to save your map document before moving on to the next step.

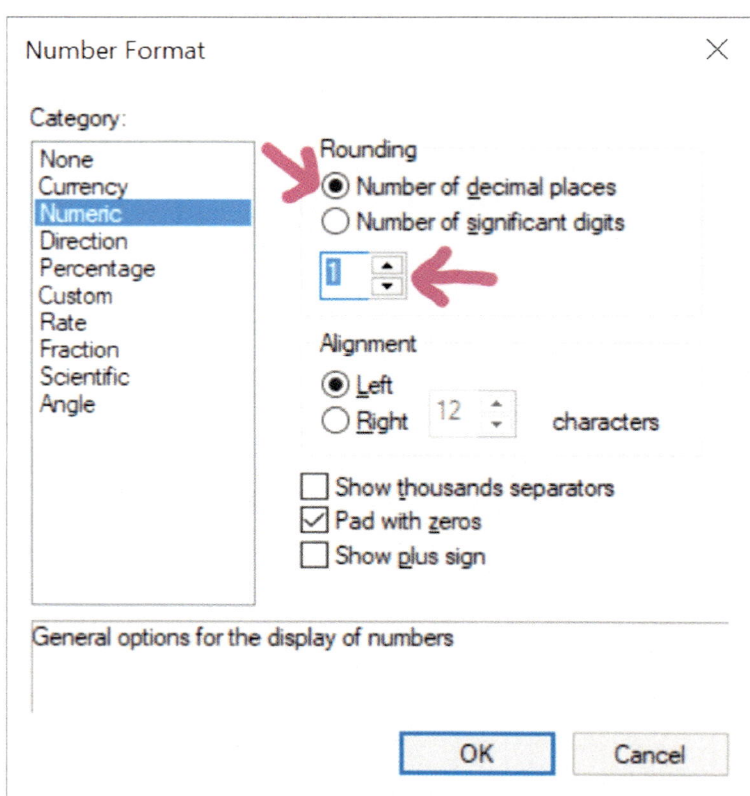

Figure 2.118: Never accept the default number of decimal places. Always choose the format that makes the most sense for your data.

Inserting a North Arrow, Scale Bar, and Acknowledgments

Cartographers use north arrows as a directional indicator. Not all maps need a north arrow, but including one is often expected. It is essential to include a directional indicator, such as a north arrow, when the top of the map is not due north.

From the *Main* menu select *Insert*, then select *North Arrow*. As you can see, there are many styles from which to choose (**Figure 2.120**). It is often tempting for the novice mapmaker to choose highly decorative styles. Picking gaudy north arrows is considered bad practice since it draws attention away from the map content. Always consider the map's purpose and audience when choosing a north arrow style. Scroll through the north arrow styles and pick a north arrow appropriate for an earthquake map. When you are ready, click *OK*.

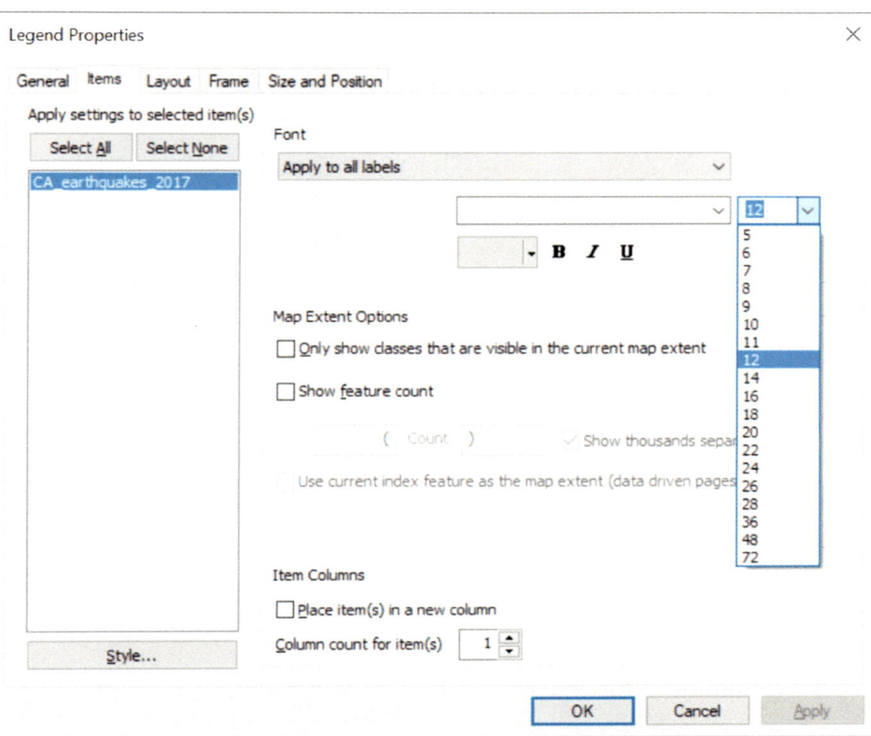

Figure 2.119: You can make adjustments to all labels using the options available on the *Items* tab.

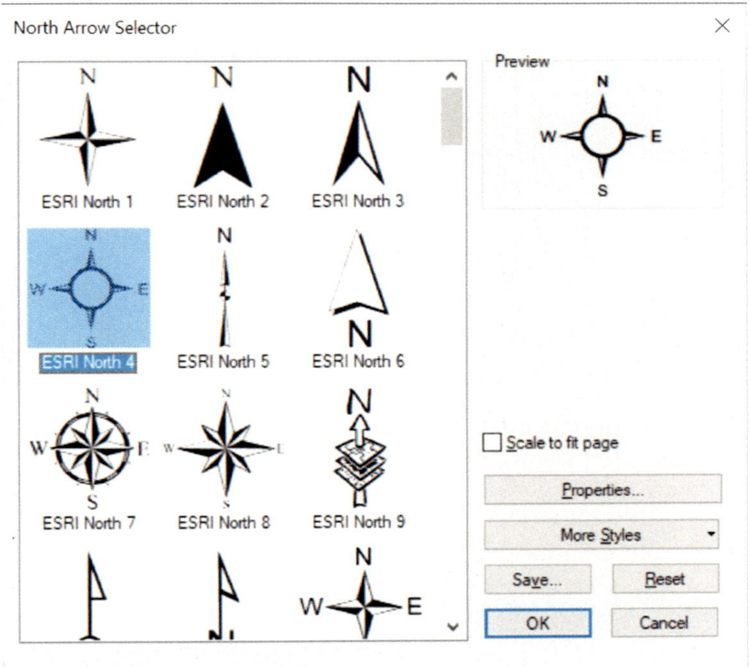

Figure 2.120: Aside from the symbol, there are other properties you can customize here.

Find an inconspicuous spot to place you north arrow. Resize the north arrow symbol as needed. Also include two scale bars, one for the main map, and one for the inset map. From the *Main* menu select *Insert*, then choose *Scale Bar*. On the Scale Bar Selector (**Figure 2.121**), choose a scale bar style. Then, click the button that says *Properties*.

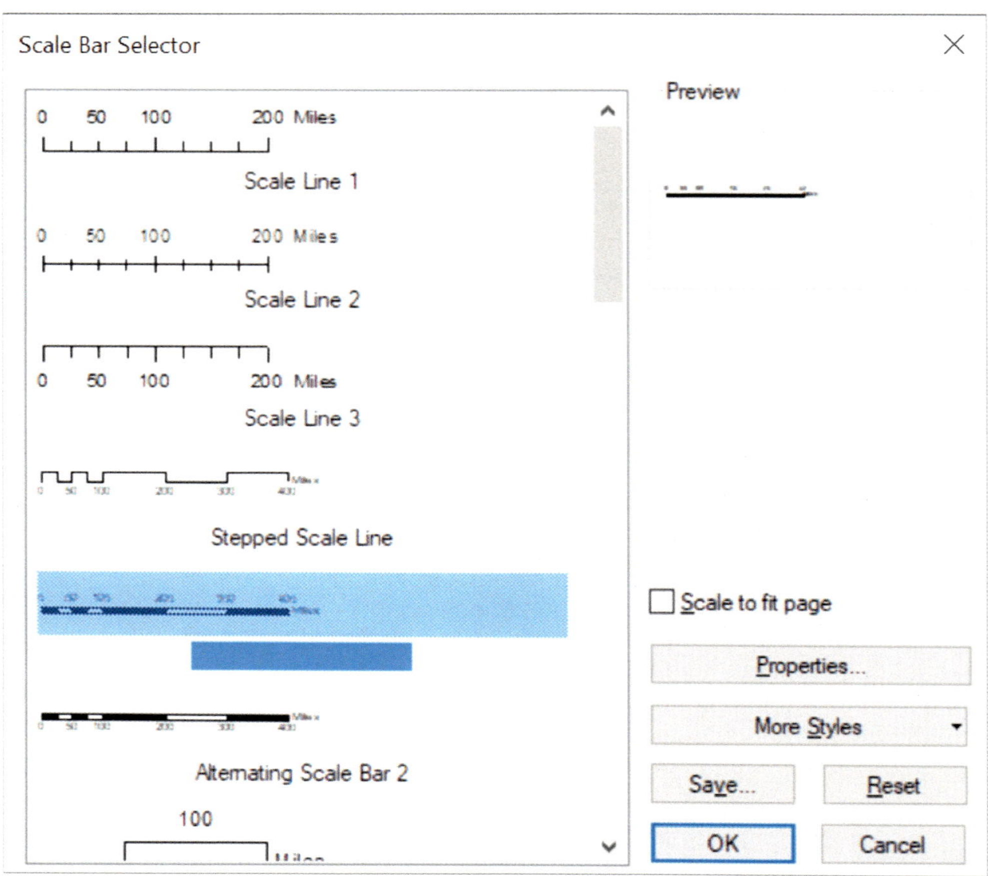

Figure 2.121: Like the north arrow properties, the Scale Bar Selector window provides many styles from which to choose.

There are many properties you can change on the scale bar including color, font, and units of measurement. In this instance, change the *Division Units* to *Kilometers* (**Figure 2.122**). You may leave the other settings default and click *OK*. Then, click *OK* again.

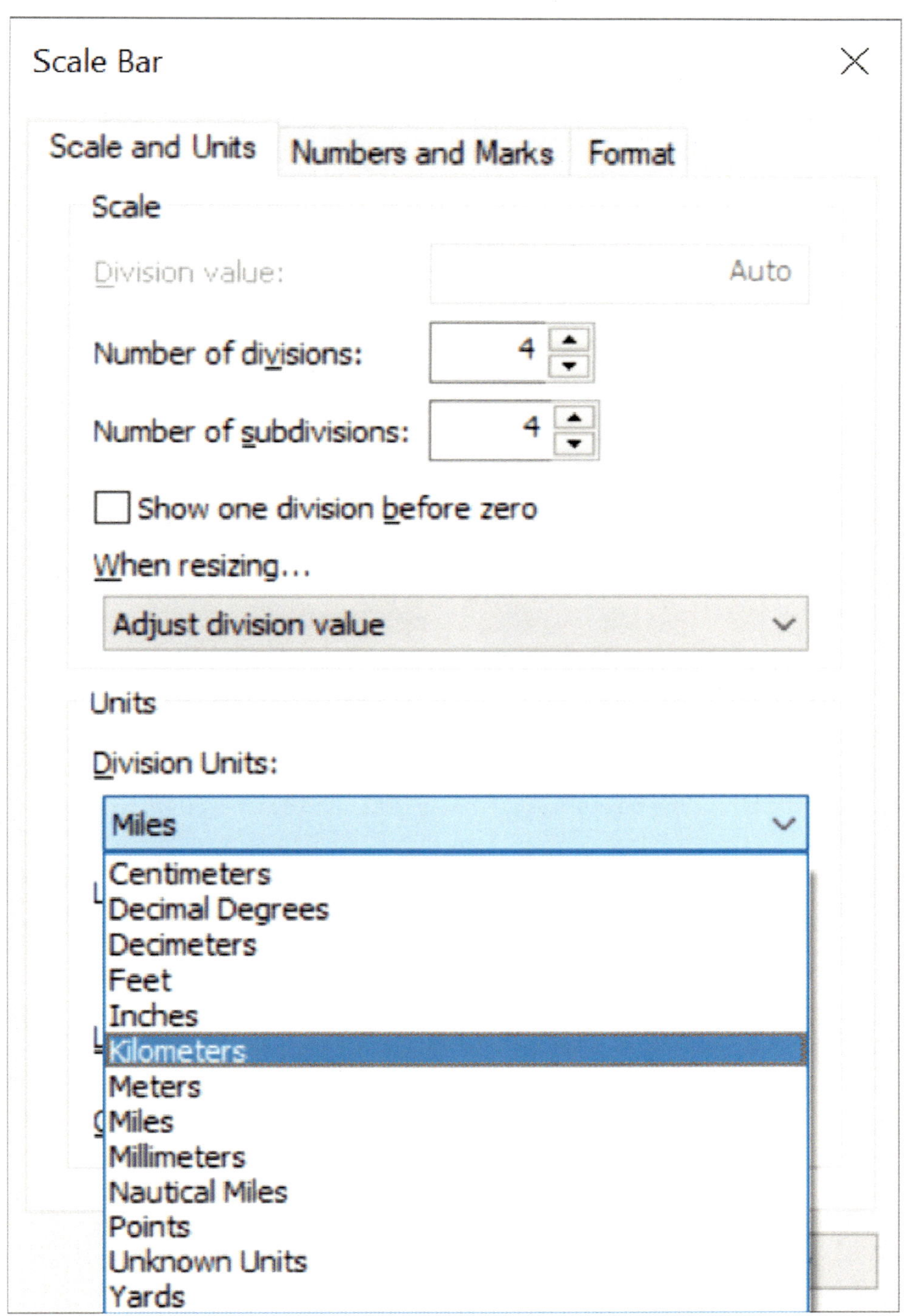

Figure 2.122: There are many options available to customize a scale bar. Take some time to experiment.

The scale bar appears in the middle of the map outlined by a blue bounding box. Move the scale bar to the lower right corner, under the legend. Use the handles on the bounding box to resize the scale bar (**Figure 2.123**). As you change the size, notice that the numbers change. Scale units should be in round numbers that the map user can efficiently use and understand. Resize the scale bar so that the number ends in an even **400 Kilometers**.

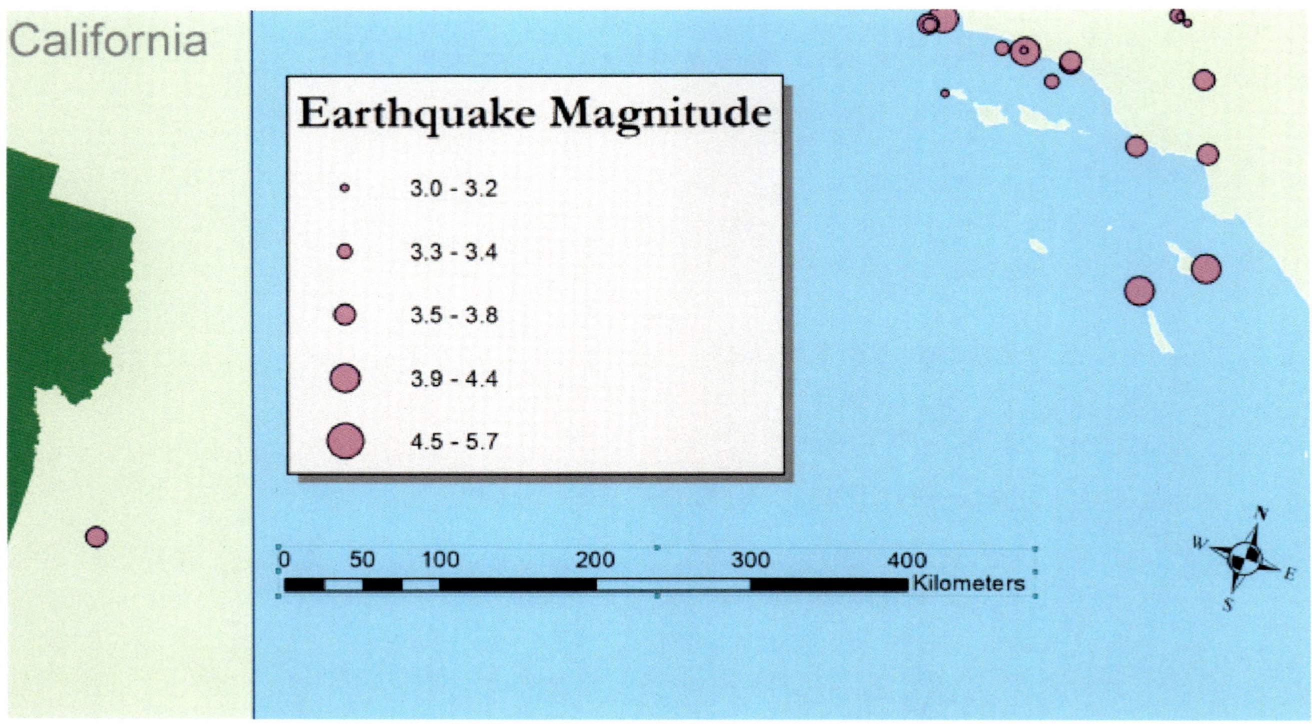

Figure 2.123: Use the bounding box to re-size the scale bar and set the length to an even number.

In the *Table of Contents*, right click on the Humboldt County Inset and select *Activate*. The inset map should now have a dotted outline. Repeat the steps to insert a scale bar. Use the same style as the main map and change the units to kilometers. Resize the scale bar so that it says **200 kilometers** and place it near the bottom of the inset map.

Caution! Be sure the inset data frame is activated so that the scale bar is accurate.

In the next step, you add your name and the year to the map. It is also a good idea to acknowledge your data sources. Using the text tool and 10-point font, add your name and the current year near an open space at the bottom of the map (**Figure 2.124**). Also include the following acknowledgment, "Earthquake data for this map were accessed through the Northern California Earthquake Data Center (NCEDC). State boundaries and country boundaries were obtained from Natural Earth. County boundaries were obtained from the United States Census Bureau." Change the font to *Ariel Narrow*. When done, take a moment to save your map document before moving on to the next step.

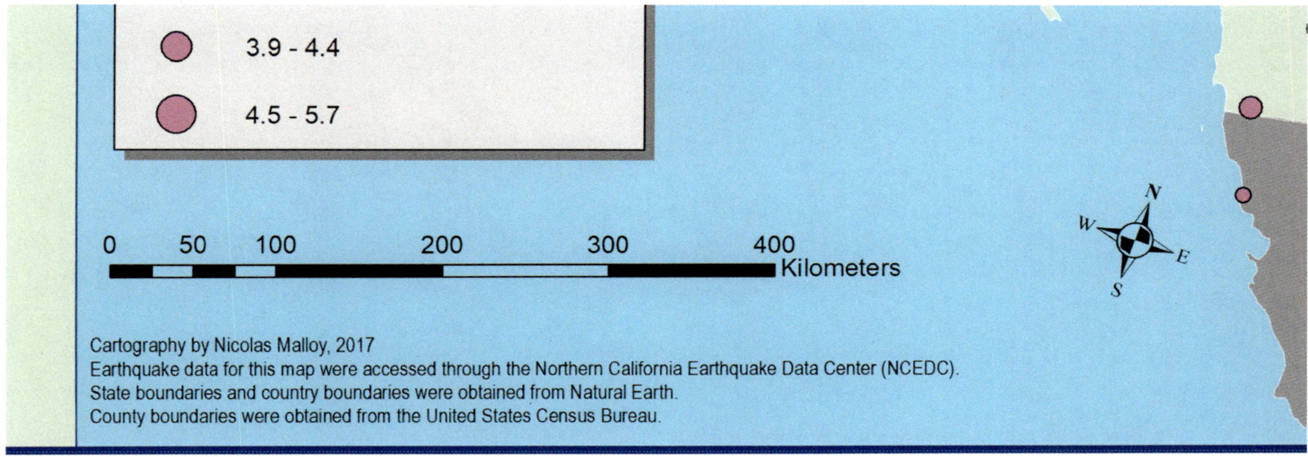

Figure 2.124: It is usually a good idea to acknowledge your data sources.

Exporting your map as a PDF file

Though it took much work to get to this point, the map is still relatively simple. There are many additional options to choose from regarding styles and colors. Before exporting your map, try to improve the look and feel of your map. You may want to experiment with moving the map elements to different positions. Include labels for Humboldt County and each of the States visible in both the main map and the inset map (**Figure 2.125**). Add a label for the Pacific Ocean as well. Use the best practices for typography discussed earlier in this chapter.

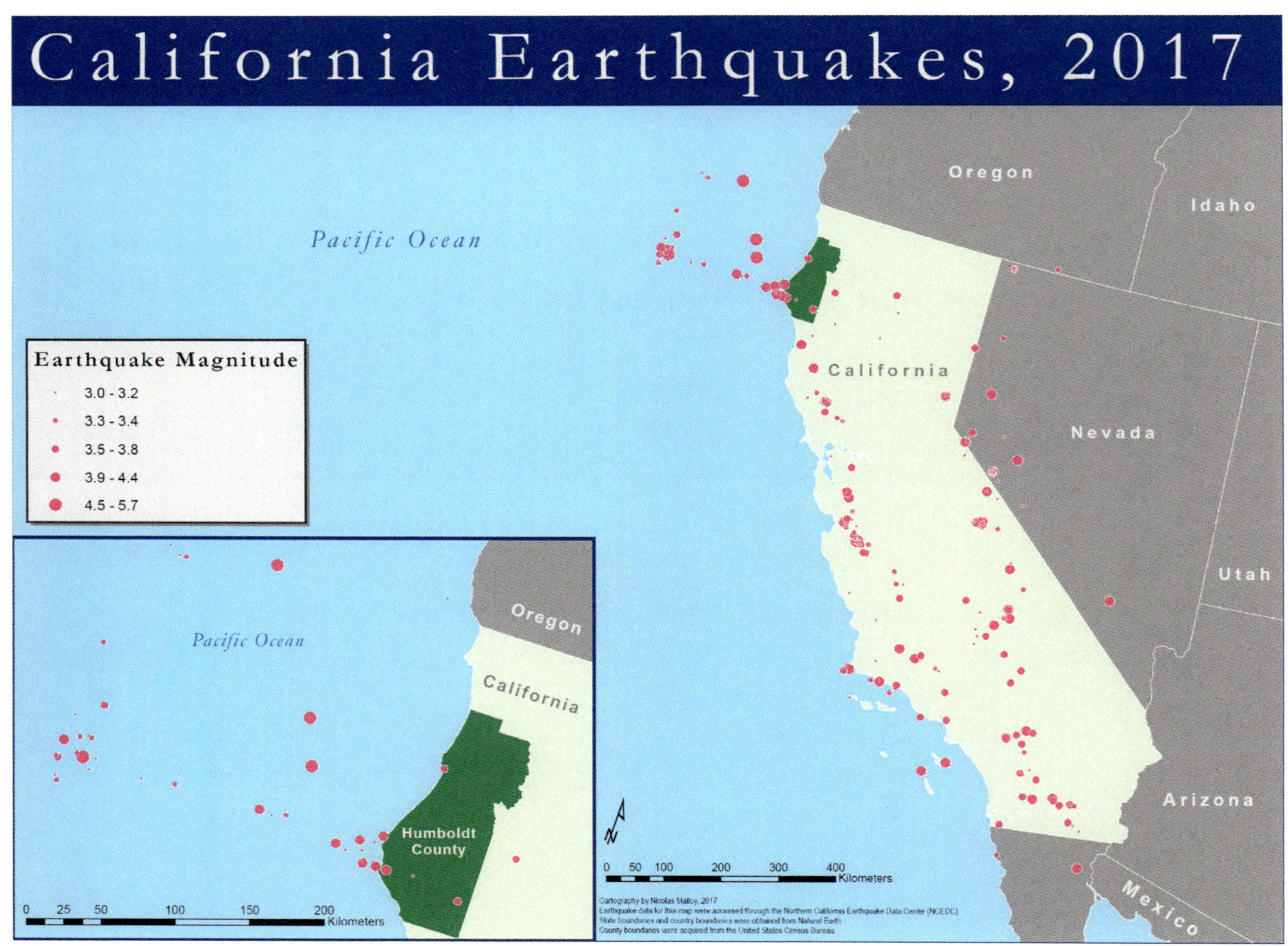

Figure 2.125: In this image, labels were added for the background States, Mexico, and the Pacific Ocean.

When done, you must export your map to a file format that can be seen and used by others. From the *Main* menu, select *File*, then *Export Map*. When the Export Map window opens, navigate to your *final* folder. Name the file *California Earthquakes 2017*.

Next to *Save as type*, choose *PDF* from the drop-down menu (**Figure 2.126**). Set the resolution to **300 dpi**.

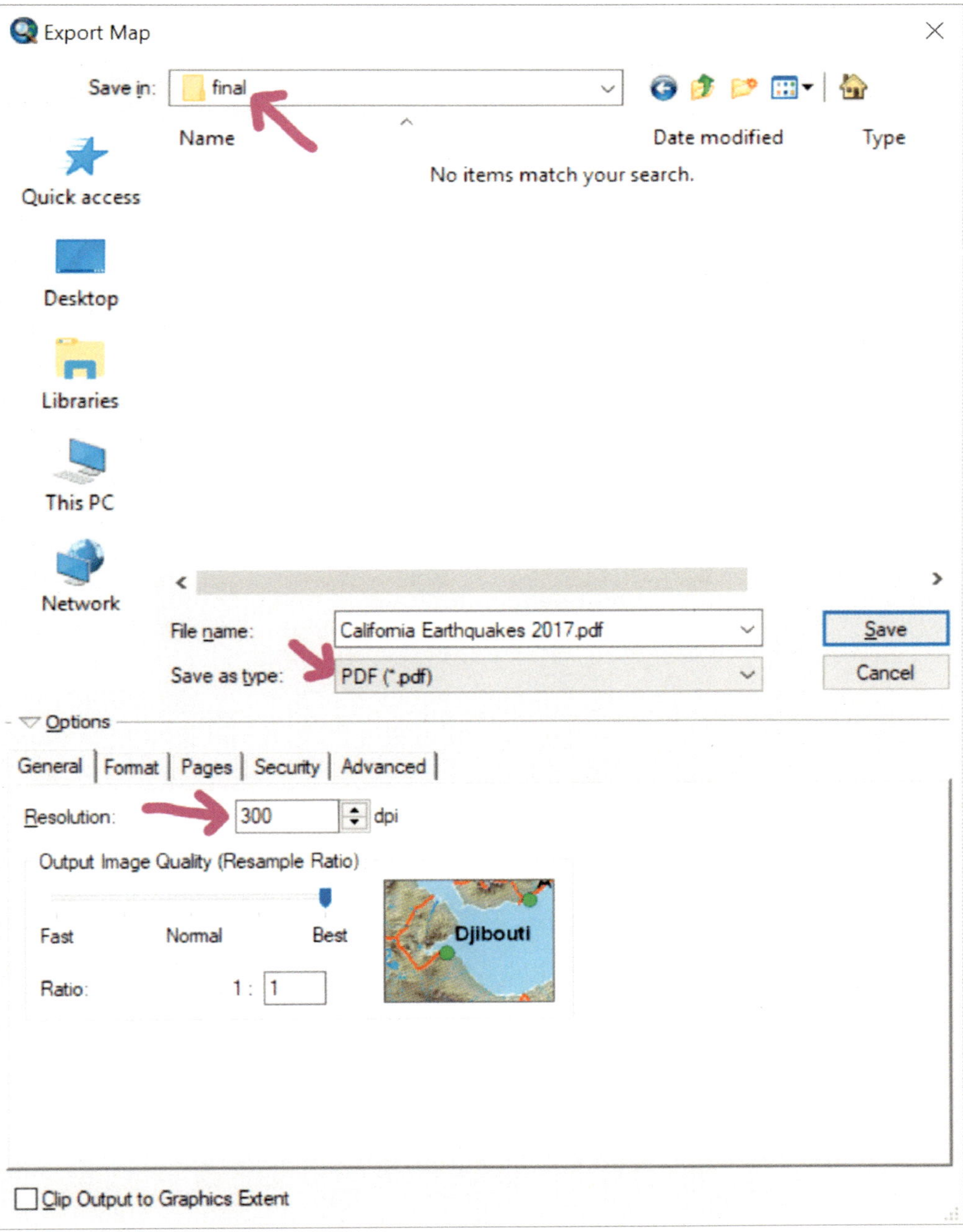

Figure 2.126: Always make sure the resolution is at least 300 dpi. Sometimes ArcMap tries to export at a lower resolution.

When you are ready, click *Save*. Then, in Microsoft Windows, open your *final* folder and check to make sure the PDF file is there. Open your PDF file and review the results. Check for any glaring errors and correct them if necessary.

Tutorial: Designing a Basemap

The goal of this activity is for students to learn how to design a basemap. In geospatial science, a **basemap** refers to a collection of data layers that are used to create a background for the map. The purpose of a basemap is to provide context for the primary data, both spatially and thematically. Typical basemap layers include roads, administrative boundaries, and aerial imagery. In this tutorial, you gather the data necessary to construct a basemap for the City of San Francisco. You then use what you learned in this chapter to design a basemap as the background for a thematic poster map.

ESTIMATED TIME TO COMPLETE THIS TUTORIAL: 6 HOURS

Learning Outcomes

Readers should be able to accomplish the following outcomes by the end of this tutorial:

» Summarize the steps for creating and organizing a project workspace folder structure
» Illustrate the ability to download data from a public source
» Demonstrate decompressing zipped files into a specific folder
» Practice changing map output size in ArcMap
» Apply symbology and color choices to map features
» Demonstrate proficiency in cartographic typography
» Insert essential map elements using ArcMap
» Export a map as a PDF file

Setting up Your Workspace

In a typical workflow, you work on geospatial data using a local hard drive. When done, you compress your data and back up your work to your cloud storage so that you can retrieve the files from anywhere. When referring to a **local hard drive**, it means you are working on data physically located on the computer in front of you.

In contrast, some computers also include networked drives. **Networked drives** link to cloud storage and save the data elsewhere. Examples include services like OneDrive or Google Drive. For this tutorial, use the **desktop** as your local hard drive location. You may also use an external USB drive if you plan to work in multiple places.

You must avoid using networked drives while you work. They increase the processing time and can cause technical glitches.

In this tutorial, you use a particular folder structure. Start by creating your **workspace folder** on the local hard drive. A **workspace** is a folder or series of folders that contain all of your project files. The top-level folder in your workspace should indicate the lab assignment or the project. Organize all of your work within the workspace folder. On your desktop, create a new folder and give it a descriptive name, such as *SF_Basemap*. Be sure there are *no spaces*. You may use underscores instead of spaces. Inside this folder, create the following three subfolders: original, working, and final. Having a standardized folder structure helps to keep a project organized, primarily when you are working with multiple partners. The folder structure you see here (**Figure 2.127**) is the standard used in each of the tutorials presented in this book.

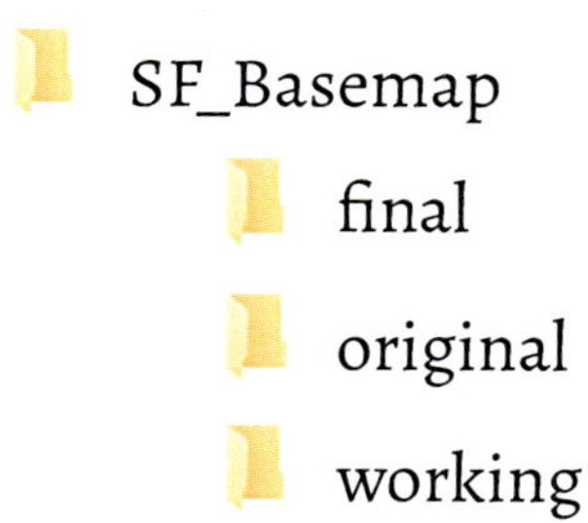

Figure 2.127: This diagram represents a basic folder structure used in this book.

As the name indicates, use the **original folder** for storing original, unaltered data. As you are working on a project, if for some reason your working version of the data gets lost or corrupted, you can go back to your *original* folder and find a fresh copy of the data. Use the **working folder** for data that you create or alter while working on your project. Use the **final folder** for storing any output you produce as a result of your work such as images, maps, tables, or reports. Setting up the standard folder structure for a project is good practice and a habit you want to develop.

Download Data from the DataSF Website

DataSF is a clearinghouse of public domain datasets available from the City and County of San Francisco (**Figure 2.128**). The data on this website is free to use under the ***Public Domain Dedication and License v1.0***[1]. Open the **Chrome** browser and navigate to the ***DataSF website***[2].

1 http://www.opendatacommons.org/licenses/pddl/1.0/

2 https://datasf.org/opendata/

If you do not have the Chrome browser, you can download it here[3].

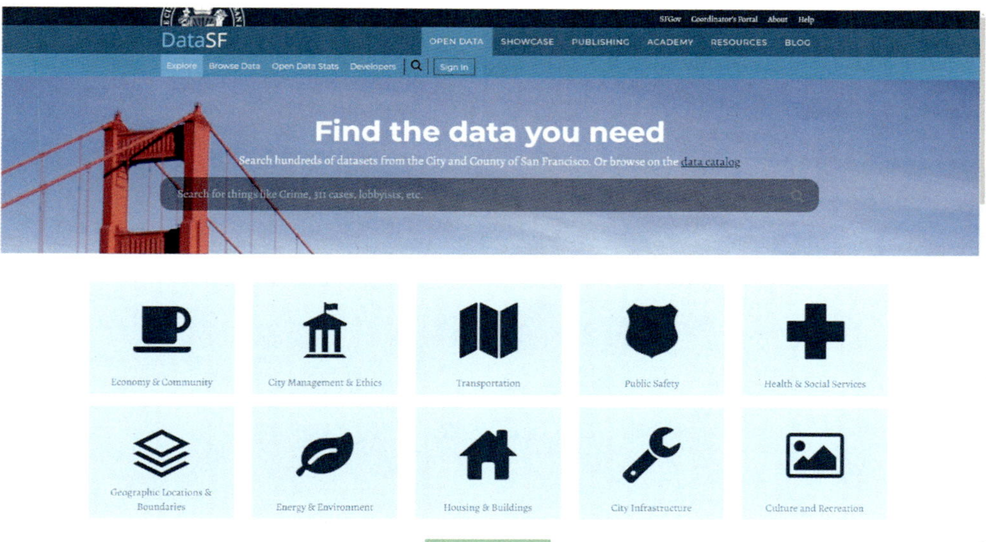

Figure 2.128: DataSF provides geospatial data organized under several themes.

On the search bar, type the keywords "SF Shoreline and Islands" and click the magnifying glass (**Figure 2.129**).

Figure 2.129: You can quickly search for data using keywords.

3 https://www.google.com/chrome/browser/desktop/index.html

When the results appear, click the link that says, *SF Shoreline and Islands*. When you do so, a map view appears with a visual display of the data over a Google Maps background (**Figure 2.130**). Near the right, click the *Export* tab. Then, right click on the *Shapefile* link, select *Save Link As*, then save the file to your *original* folder.

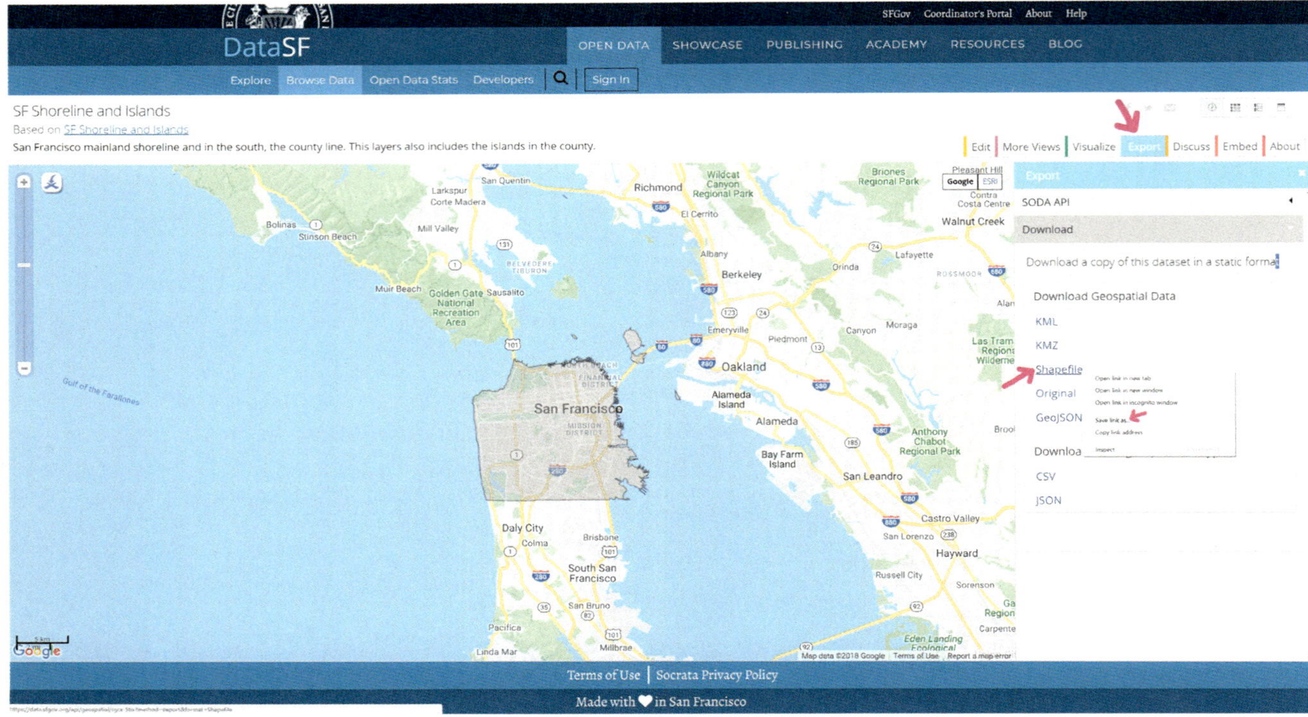

Figure 2.130: DataSF offers data in a variety of geospatial formats.

When done, click the *Browse Data* link near the top of the page. Repeat these steps using the following keywords:

- Recreation and Parks Properties
- SF Urban Tree Canopy
- Streets of San Francisco
- Muni Simple Routes
- Realtor Neighborhoods

You should also search for buildings using the keyword *Building Footprints*. The steps for downloading this dataset are a little different. It takes you to an external link where you download the file in geodatabase format. A **geodatabase** is a container file that can hold many types of GIS data layers. It is a way to organize related datasets. Click the *Building Footprints* link. On the Building Footprints page, right-click the button that says ZIP, select *Save Link As*, then save the file to your *original* folder.

Be sure to download each file to your original folder. Leave all the files compressed for now. You decompress the files in a later step.

Refreshing a Folder in the Catalog Tree

Locate ArcMap on your computer and launch the software. If you are using Microsoft Windows 10, click the windows button and type ArcMap to find the desktop application. Launch the ArcMap software. When you first launch ArcMap, a window appears that gives you the option to either open a blank map or to open any of your recent map documents. Choose to open a blank map document.

In this step, you start by working with the *Catalog Window*. The catalog window is where you manage your geospatial data in ArcMap. Start by clicking on the Connect to Folder button (**Figure 2.131**). Navigate to your workspace folder, *SF_Basemap*, on your local hard drive.

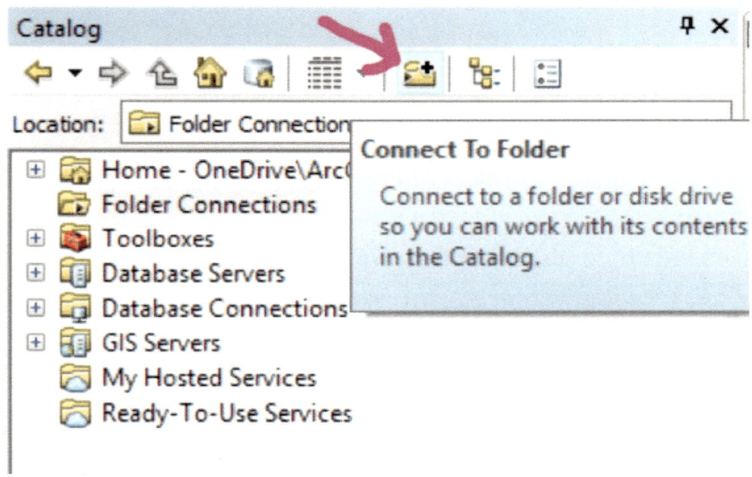

Figure 2.131: You connect to a folder so that the contents appear in the *Catalog Window*.

In this step, it is essential to select your primary workspace folder, *SF_Basemap*, to add it to the *Catalog Window* (**Figure 2.132**). When ready, click *OK*.

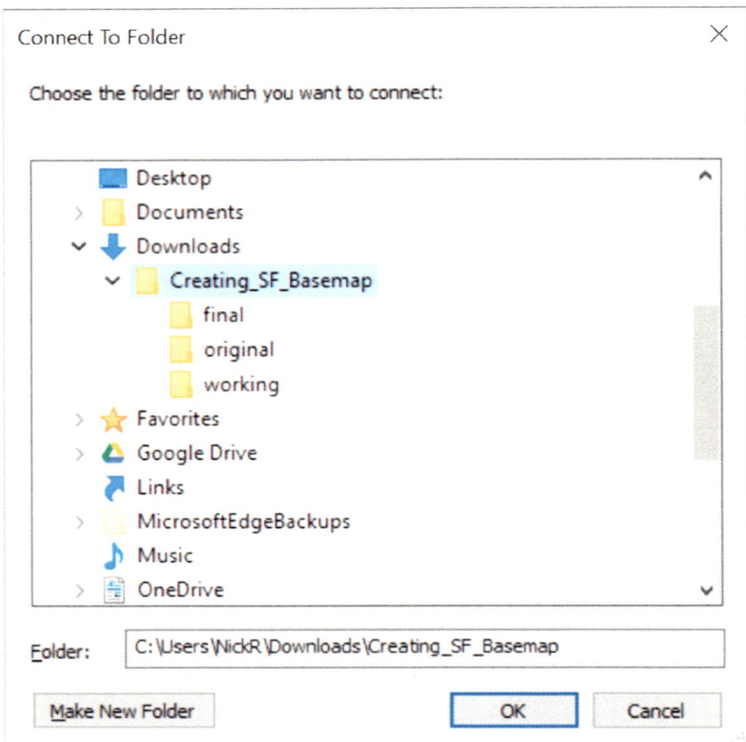

Figure 2.132: Connect to Folder adds the folder to the *Catalog Window*.

Once you add your workspace folder to the *Catalog Window*, expand the folder by clicking on the plus sign. You should see your three subfolders inside. In previous tutorials, you decompressed the files *before* opening ArcMap. In this instance, you **did not** decompress the files for reasons that should be clear in the next step. Expand the *original* folder to view the contents. Assuming that you did not decompress the files, nothing appears under the *original* folder in the *Catalog Tree*. The folder seems empty because ArcMap cannot make use of compressed zip files (**Figure 2.133**).

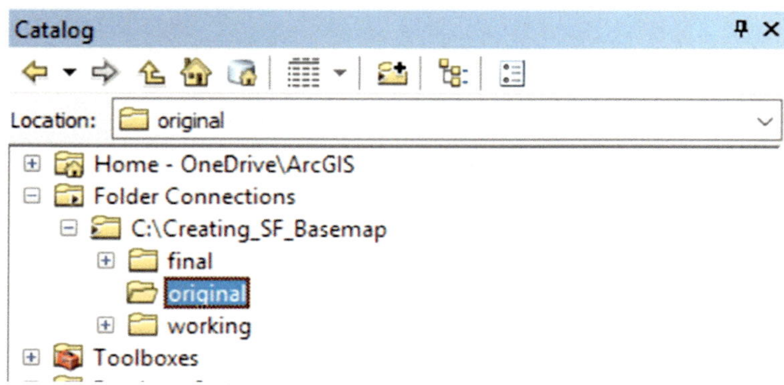

Figure 2.133: ArcMap does not display compressed zip files in the *Catalog Tree*.

In a previous tutorial, you learned how to decompress a file using 7zip. In Microsoft Windows, navigate to your *original* folder. You should see the zip file for the *SF Shoreline and Islands*. Right-click on the file, select *7zip*, then Extract Here. Be sure to delete the zip file when you are done decompressing it. You won't need it anymore. Eliminating the zip file saves space and helps to avoid confusion later.

When you first connect to a folder, the *Catalog Window* correctly lists the items within a folder. However, if you later make changes to the items in the folder outside of ArcMap, the contents will not automatically update to display the alterations. You must **refresh** the folder to update the *Catalog Tree*. Right-click on the *original* folder in the *Catalog Tree* and select *Refresh* (**Figure 2.134**).

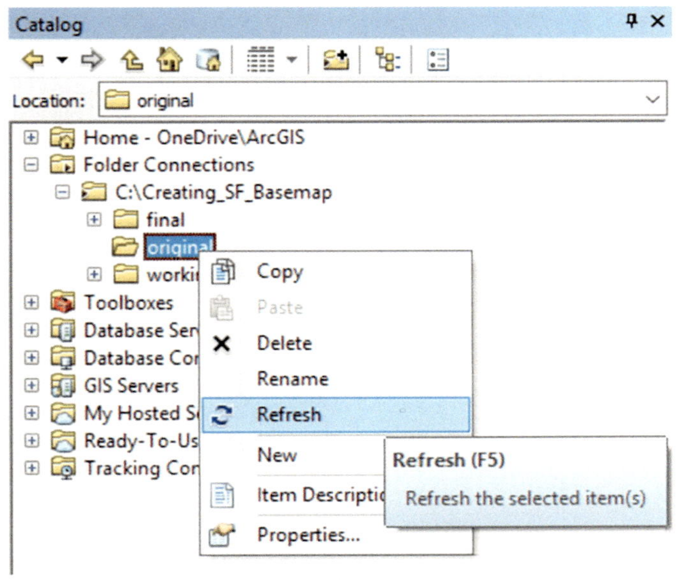

Figure 2.134: The *Catalog Window* in ArcMap does not always keep track of changes made in Microsoft Windows.

Expand the *original* folder in the *Catalog Tree*. You should now see the shapefile for SF Shoreline and Islands listed (**Figure 2.135**).

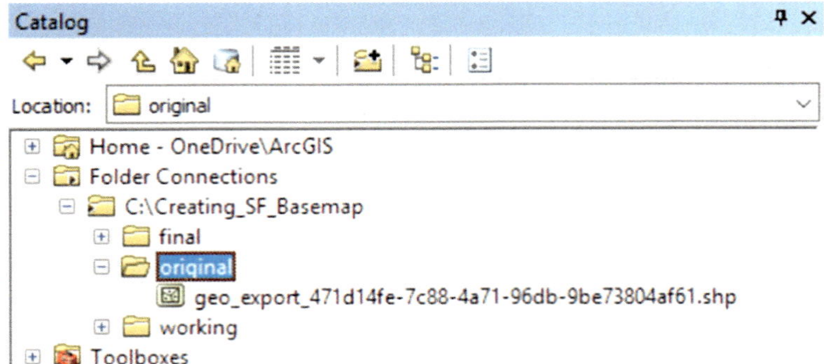

Figure 2.135: The default naming convention used by DataSF is not user-friendly.

The default naming convention used by DataSF begins with the words "geo_export" followed by a series of numbers and letters. Regrettably, the file name is not very user-friendly. All of the other files downloaded also have a similar name. To help avoid confusion, you rename each of the files immediately after decompressing them. In a previous tutorial, you learned how to rename files using the *Catalog Window*. Rename the shapefile and give it a more human-friendly name such as *shoreline* (**Figure 2.136**)

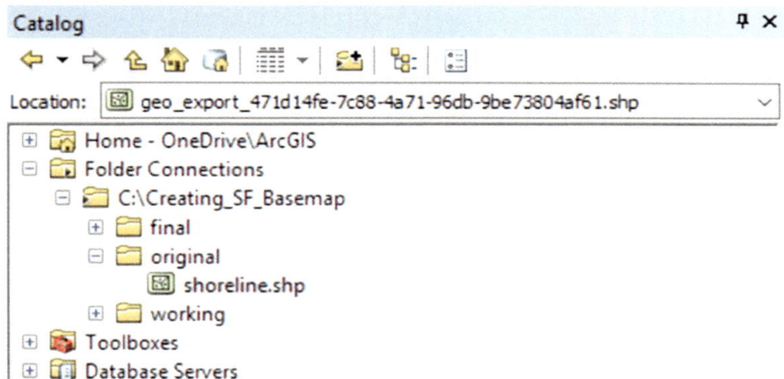

Figure 2.136: Always use the *Catalog Window* to rename shapefiles.

Repeat these steps for the remaining files. Avoid decompressing more than one file at a time. Each time you decompress a file, refresh the *original* folder and rename it. Be sure there are no spaces in the name. When done, remember to delete the compressed zip files from the *original* folder. You do not need them anymore. Removing them saves space and helps to prevent confusion later.

With the building footprints, you need to decompress it *twice* to get to the data. After refreshing the *original* folder in the *Catalog Tree*, you should see the geodatabase, represented by a silver cylinder (**Figure 2.137**). Click the plus sign to expand the geodatabase. Then, click the plus sign next to the feature dataset that starts with G84. Rename the first feature class *buildings*.

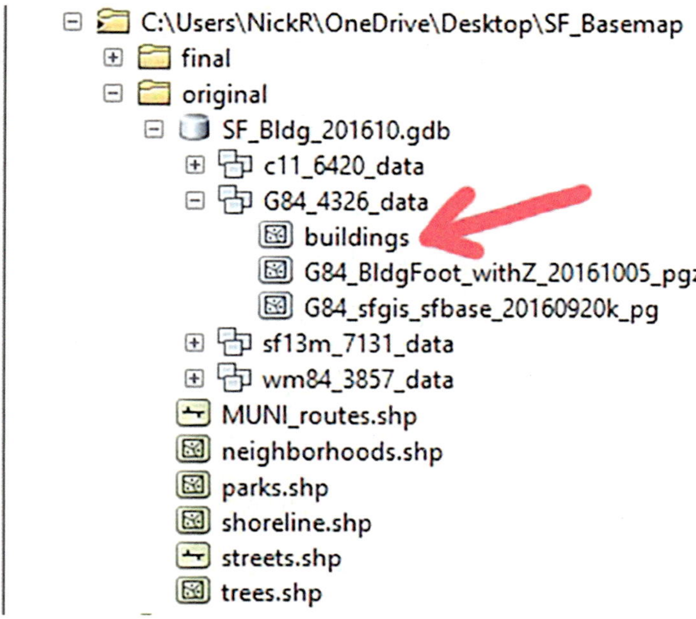

Figure 2.137: As you can see, a geodatabase can store many layers within.

After renaming all the files, add each one to the map document (**Figure 2.138**). You may ignore any geographic coordinate system warnings for now.

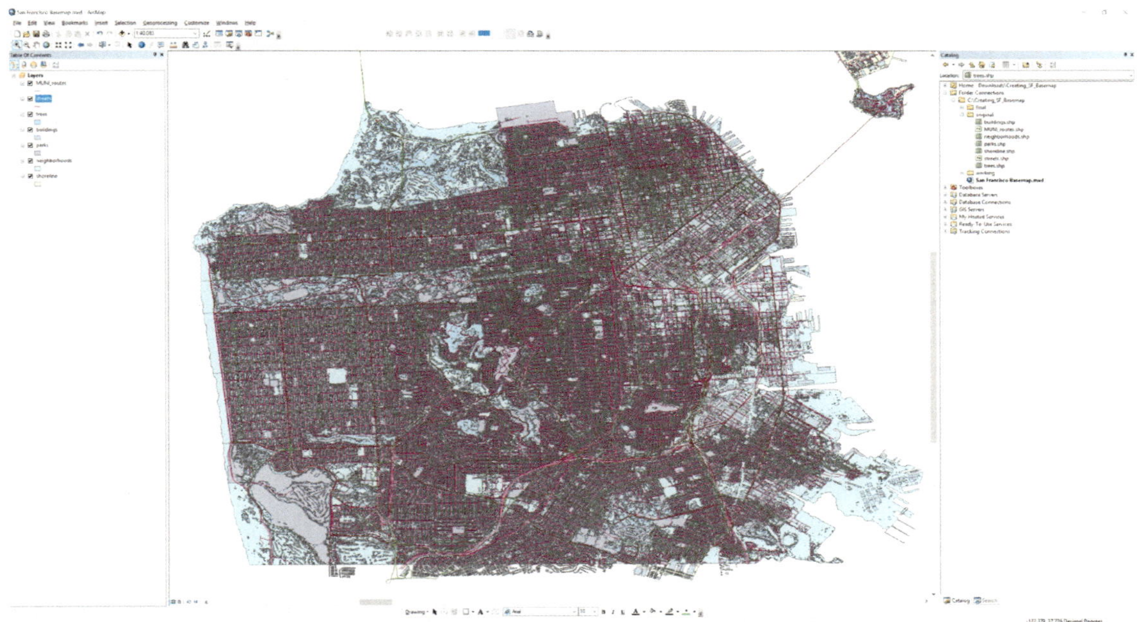

Figure 2.138: ArcMap assigns random colors when adding data.

In a previous tutorial, you learned how to set your map document properties to store relative paths. Do this now (**Figure 2.139**). Then, save your map document to your workspace folder, *SF_Basemap*. Call the map document *San Francisco Basemap*.

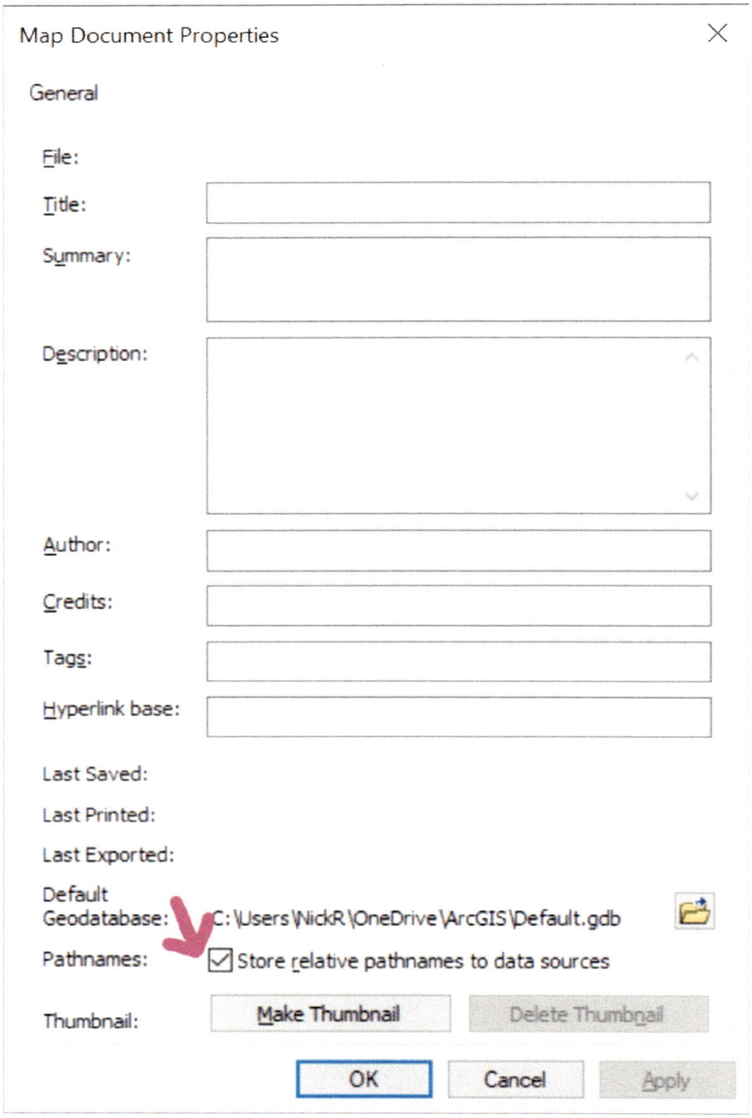

Figure 2.139: Always make sure to check the Store relative pathnames to data source option.

Changing the Map Projection of the Data Frame

The Earth is spherical, yet maps still managed to represent the Earth using a flat plane. Map projections make this possible. A **map projection** is the geometric transformation of the round earth onto a flat plane using mathematical equations. In **Chapter 3**, you learn more about map projections. What you need to understand now is that there are many different map projections. The one you choose changes the shape and appearance of features on the map. In this step, you change the map projection for the data frame to make the City of San Francisco appear closer to its actual size and shape on Earth.

Open the data frame properties. If you recall from a previous activity, in the *Table of Contents*, the word *Layers* appears above the map layers. If you remember from previous tutorials, this item on the *Table of Contents* is not a layer. It represents the **data frame**. Access the properties for the data frame by right-clicking on the word Layers in the *Table of Contents*. Select, *Properties* from the contextual menu. When the Data Frame Properties window opens, click the *General* tab (**Figure 2.140**). Change the name of the data frame to something more descriptive, such as *San Francisco Basemap*.

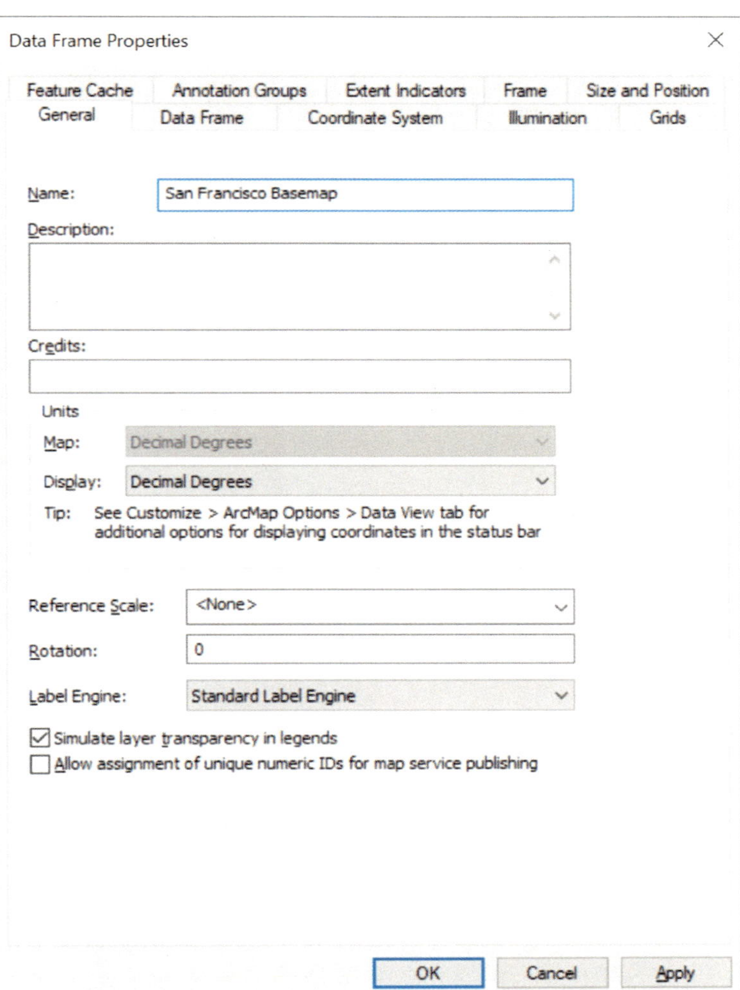

Figure 2.140: Changing the name of the data frame helps to add clarity to the *Table of Contents*.

Select the Coordinate System tab (**Figure 2.141**). Scroll down until you see the *Projected Coordinate Systems* folder. Expand the folder, then, expand the *UTM* folder. Locate and open the folder that says *NAD 1983*. Select *NAD 1983 UTM Zone 10N* from the list and click *OK*.

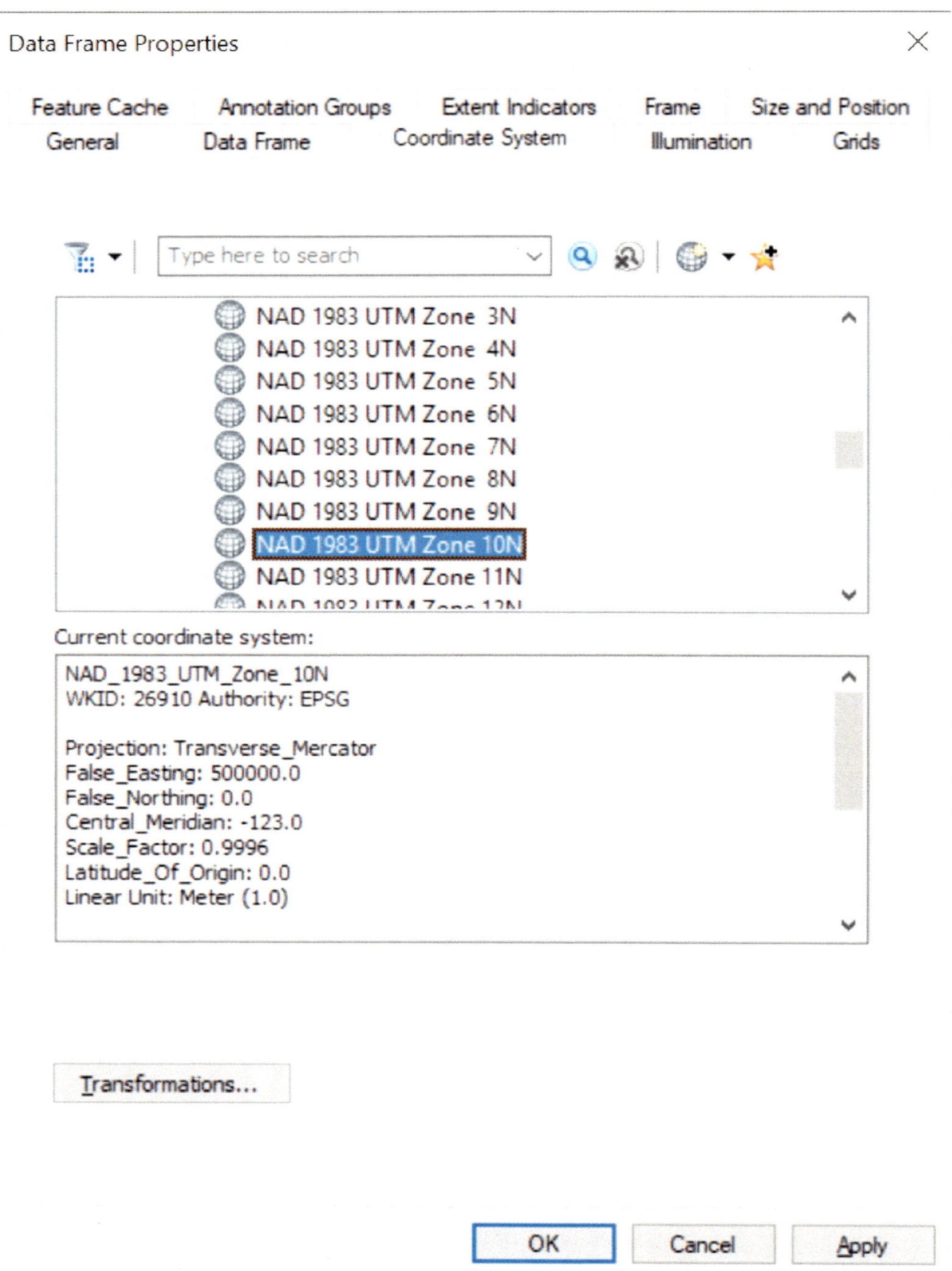

Figure 2.141: This spatial reference system optimizes the size and shape of Northern California.

You may get a warning (**Figure 2.142**). ArcMap is telling you that the data frame coordinate system does not match some of the layers in the *Table of Contents*. Since we are only using it for display purposes, this is fine. Check the box that says, *Don't warn me again this session*. Click *Yes* to close the warning.

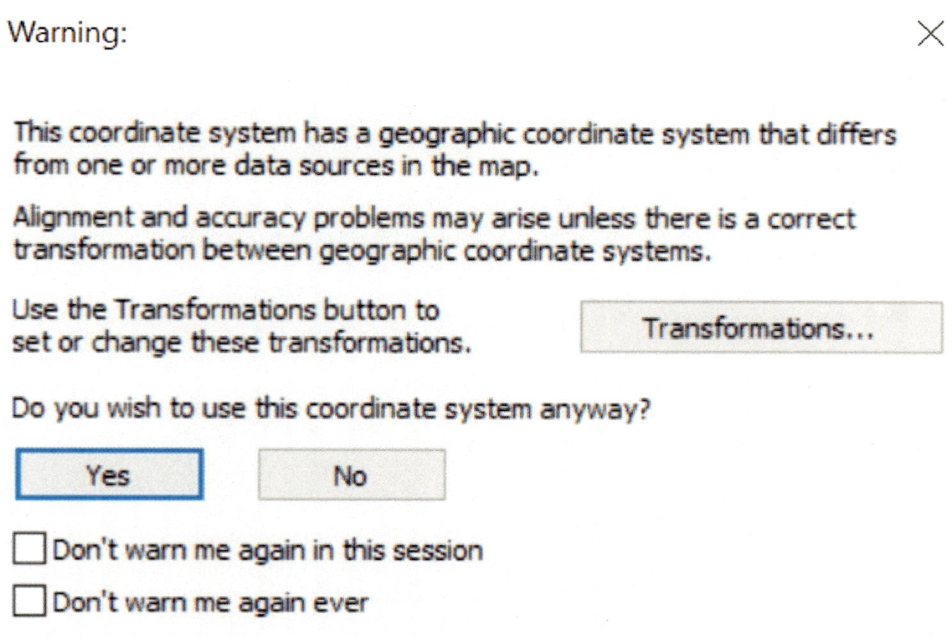

Figure 2.142: This warning appears when you have layers with different Geographic Coordinate Systems loaded in the *Table of Contents*.

At such a large scale, you may not notice a dramatic change to the shape of the city. Nevertheless, optimizing the appearance of the spatial reference system displayed on the data frame is an excellent habit to develop for cartographic design. Take a moment to save your map document before moving on to the next step.

Changing the Map Size and Position

Before starting work on the cartographic design, always prepare your map layout ahead of time. Start by determining the size of the page on which you be working. In this activity, you are making a poster-sized map that displays your choice of one out of two themes. You can choose the historic fire damage from the 1906 earthquake as the theme, or you can pick sea level rise and flooding. The poster size is **32 inches** by **18 inches**. Include a map of the City of San Francisco. Descriptive text and other map elements surround the main map on the poster.

By default, ArcMap sets the page size to 8.5 inches by 11 inches. This size is rarely acceptable in a professional setting. In this step, you change the page size. Then, you change the data frame size and position on the page. From the main menu, select *View*, then *Layout View* (**Figure 2.143**).

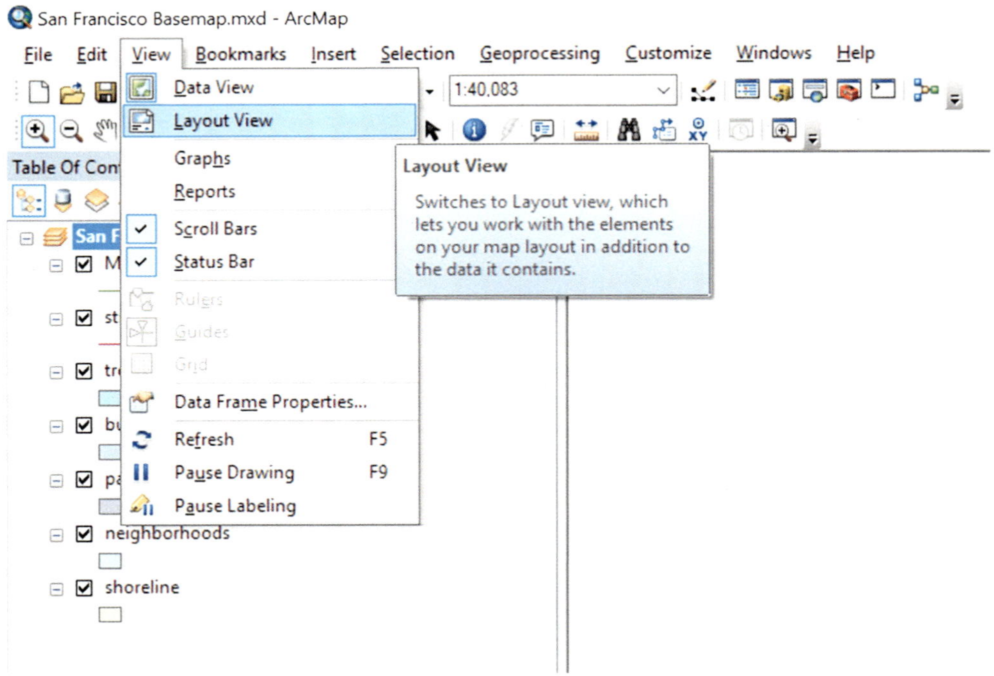

Figure 2.143: There are two ways to display the data frame in ArcMap, *Data View*, and *Layout View*.

The data frame window immediately changes to show the page size and the data frame position on the page (**Figure 2.144**). Across the top and along the left you see a ruler indicating the height and width of the page. Currently, the page size is 8.5 inches by 11 inches, which is the default setting in ArcMap. The *Layout* toolbar may also appear floating above the map. You may dock the toolbar on top.

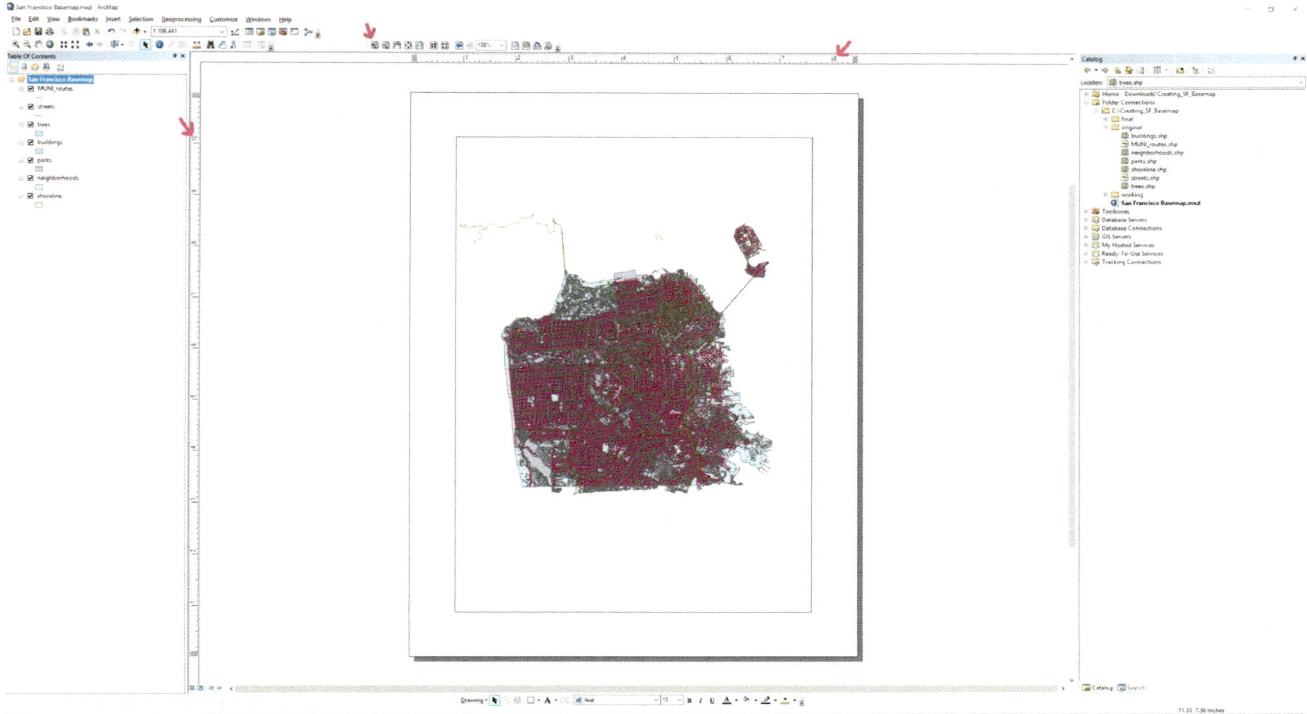

Figure 2.144: As you can see, ArcMap defaults to a layout for an 8.5 by 11 sheet of paper.

It is essential to understand each of the tools in the *Layout* toolbar (**Figure 2.145**). These tools are only active when ArcMap is in *Layout View*.

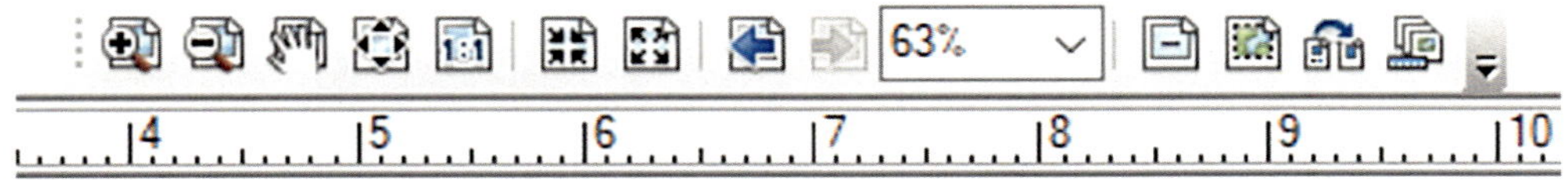

Figure 2.145: The layout tools only work while in *Layout View*.

On the left of the toolbar, you see two tools for zooming in and out of the map layout. The icon looks like a magnifying glass over a page. These tools are sometimes confused with the *Zoom In* and *Zoom Out* tool with which you previously worked. They work similarly. Instead of zooming in and out of the data, the zoom tools on the *Layout* toolbar zoom in and out of the page layout. Likewise, the *Pan* tool on the layout toolbar pans across the page layout and does not move the data. Next to the *Pan* tool is the *Zoom to Whole Page* tool. This tool is handy for quickly viewing the entire page layout. Take a moment to experiment with each of these tools before moving on to the next step.

In a previous tutorial, you learned how to change the page size. From the main menu, select *File*, then *Page and Print Setup*. Uncheck the box next to *Use Printer Paper Settings* (**Figure 2.146**). For the width, enter 32. For the height, enter 18. When you are ready, click *OK*.

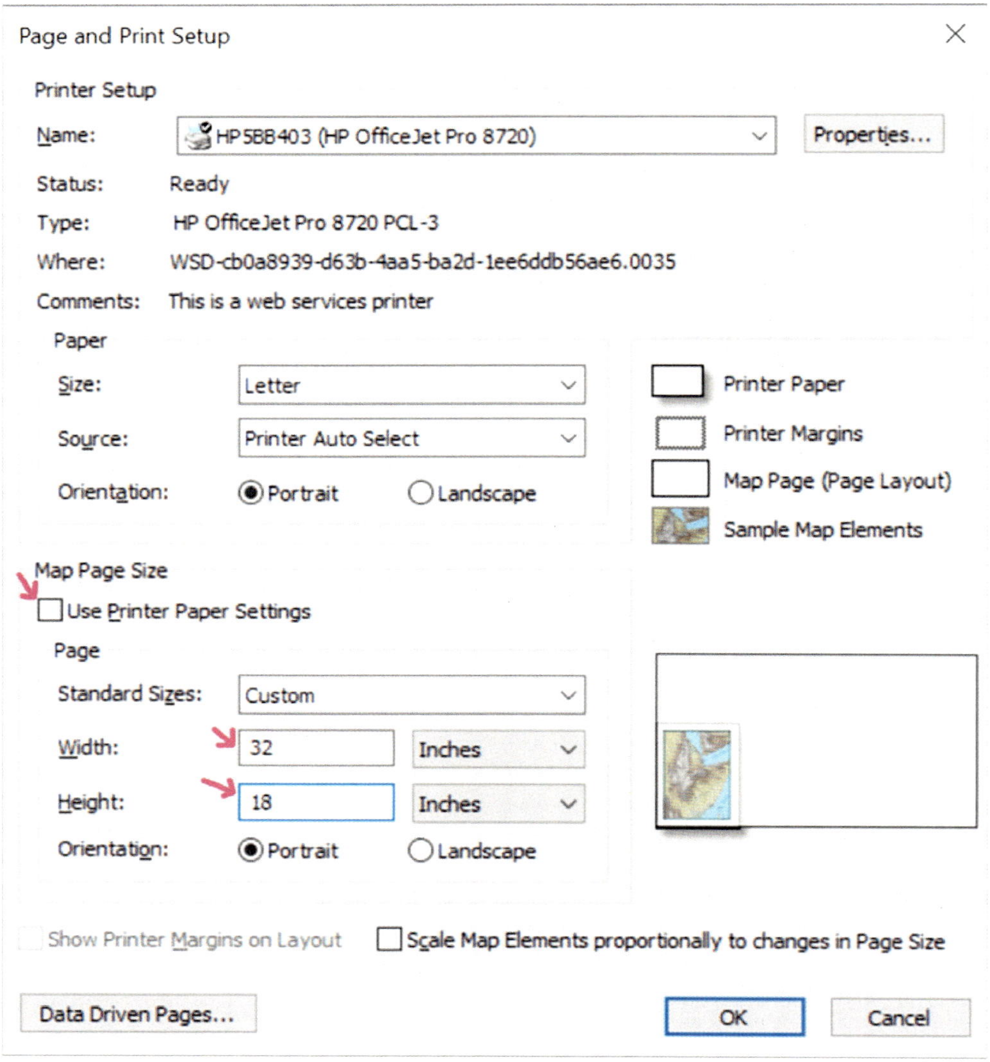

Figure 2.146: Check the lower left side of the *Page and Print Setup* window to change these settings.

The page is now a large poster size of 32 inches by 18 inches. You can tell because the ruler along the top and the sides have changed to show the new dimensions (**Figure 2.147**). However, the data frame is still the default size.

Figure 2.147: There are always two steps when setting up your map layout. Change the paper size and change the data frame size.

In a previous tutorial, you learned how to change the data frame size and position. Unlike that last tutorial, you don't match the size of the paper. Instead, the data frame has a different dimension and location on the page. To change the data frame size, open the data frame properties and navigate to the *Size and Position* tab (**Figure 2.148**). Change the width value to **19** and the height value to **13**. Change the *Anchor Point* so that it is in the center position. Then, enter **10.5** for the X value. Enter **9** for the Y value. When done, click *OK*.

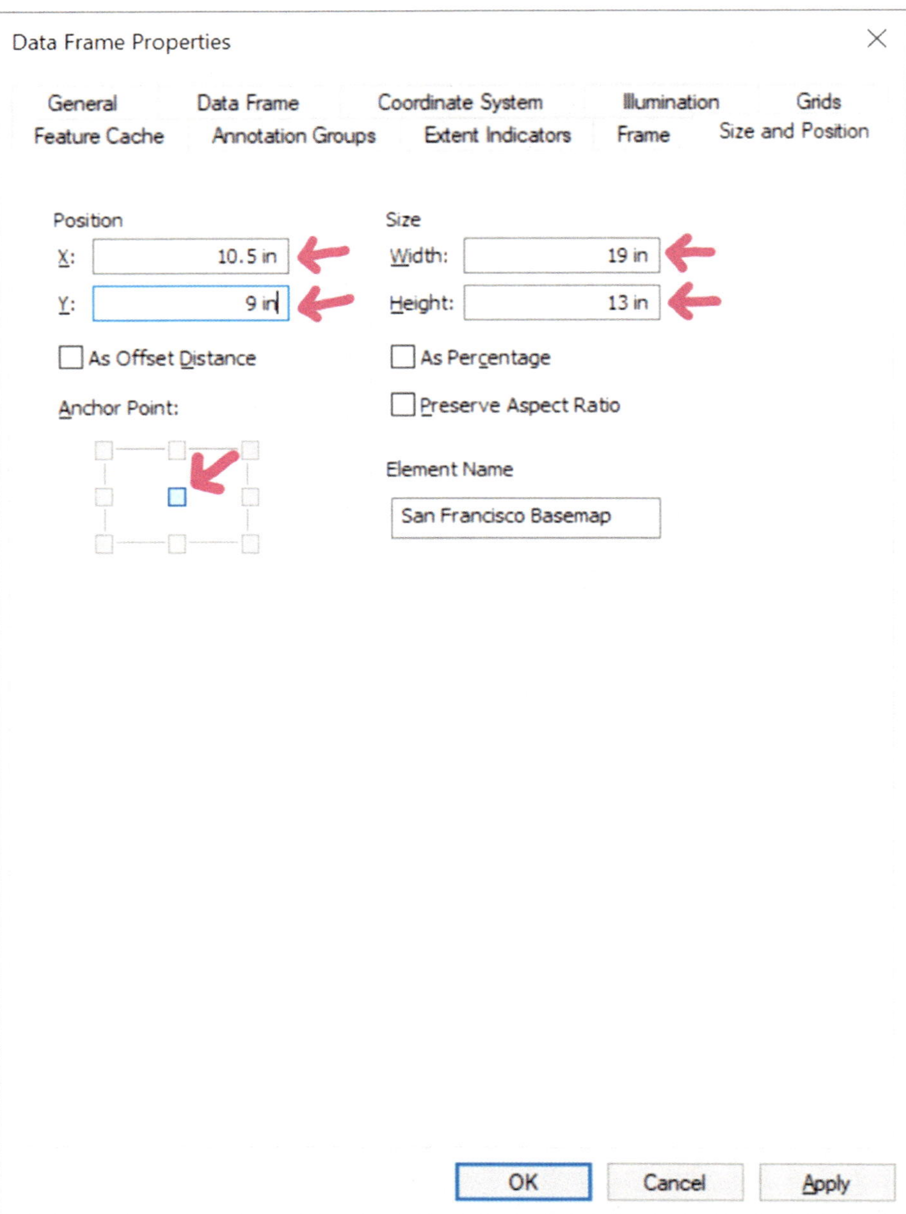

Figure 2.148: By default, the Position value anchors the data frame on the page relative to the lower left corner. In this instance, you anchor the dataframe relative to the center of the page.

When done, you should see your dataframe positioned on the left side of the paper (**Figure 2.149**).

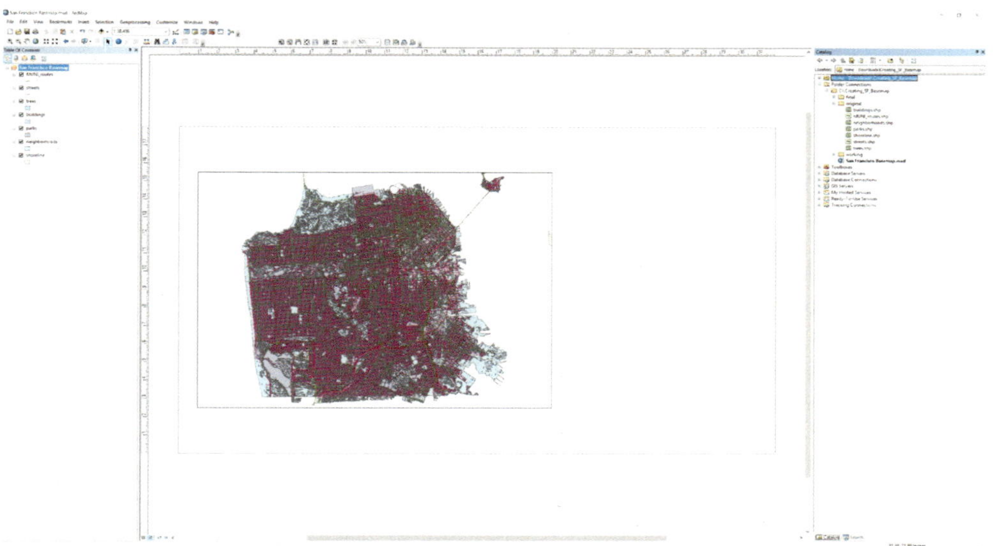

Figure 2.149: Here the data frame and the paper have different dimensions.

Preparing the Layout

Before delving into the cartographic design of the San Francisco Basemap, it is essential to understand its relationship and placement relative to the other elements on the poster. In a typical workflow, many cartographers create one or more rough mockups on scratch paper as a first step in the design process (**Figure 2.150**). Though it may seem like an extra step, this process can save you time. This tutorial provides a specific layout for you, so you do not need to make sketches. However, it is essential to know that this element in the workflow can be useful.

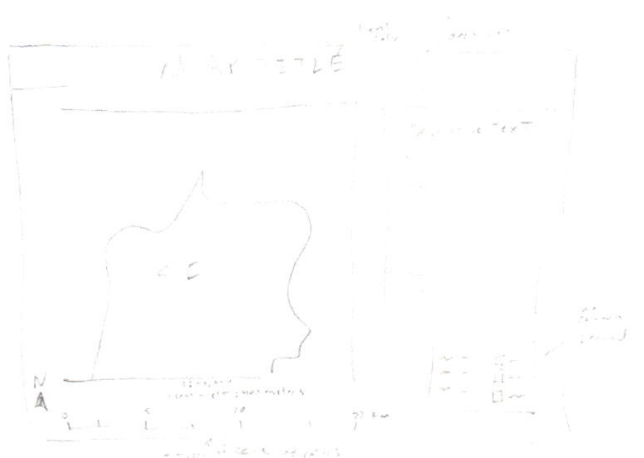

Figure 2.150: A rough sketch of the poster layout.

The next step in the layout process is positioning the features on the map and setting the map scale. In ArcMap, uncheck all of the layers in the *Table of Contents* except for the shoreline.

Unchecking the layers makes ArcMap run faster since it does not have to draw so many features.

To see the poster design clearly and improve efficiency, maximize the ArcMap program on your monitor. Minimize the *Catalog Window* by clicking on the pushpin icon. On the *Layout* toolbar, click the *Zoom to Whole Page* button. The layout should now take up most of your screen (**Figure 2.151**). In the next few steps, you use tools from the *Standard* toolbar, the *Tools* toolbar, and the *Layout* toolbar to set the map extent and position the City of San Francisco within the data frame.

Figure 2.151: Maximizing your view of the layout improves efficiency.

Using the *Zoom In* and *Zoom Out* on the *Tools* toolbar, zoom into the City of San Francisco so that it is visible within the data frame. Do not worry about any of the islands for now.

A representative fraction (RF) is the ratio between the map distance and the ground distance and determines the scale for the map. You learn more about representative fractions and map scale in **Chapter 4**. For now, enter 1:40,000 for the RF on the *Standard* toolbar (**Figure 2.152**).

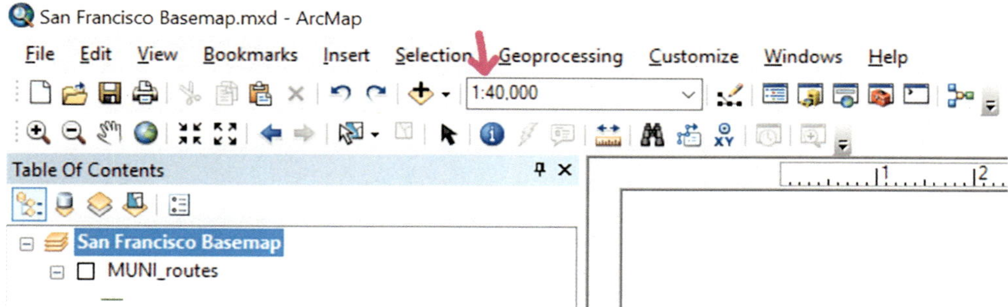

Figure 2.152: A representative fraction (RF) is the ratio between the map distance and the ground distance.

Use the *Zoom In* tool on the *Layout* toolbar and zoom into the space on the page just under the data frame (**Figure 2.153**). Be careful not to use the wrong zoom tool, or you change the map scale. On the *Tools* toolbar, use the *Pan* tool to move the shoreline layer to the bottom-center of the data frame. The goal is to try to cover up where the data ends. You want to give the impression that the land continues south beyond the map.

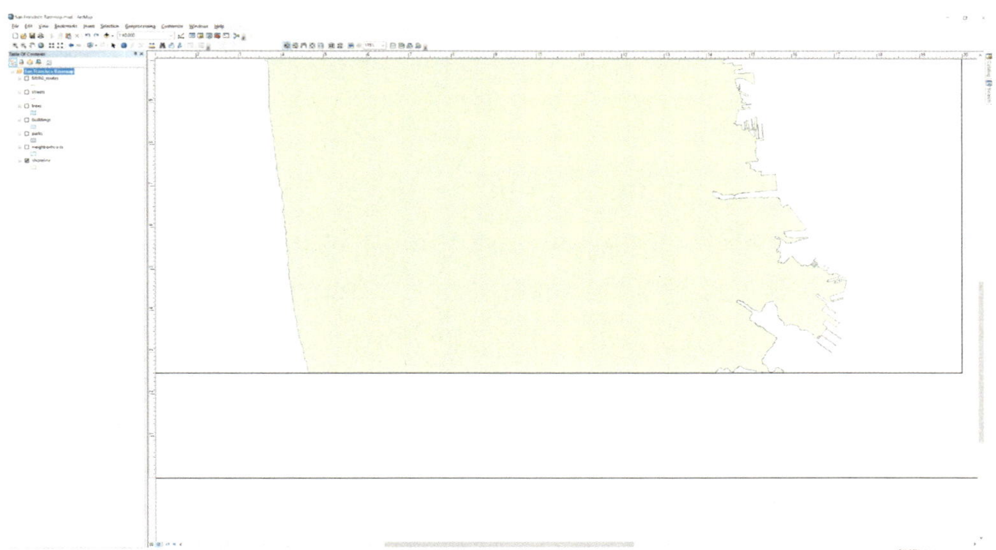

Figure 2.153: The bottom edge of the shoreline layer is hidden from view.

When done, use the *Zoom to Whole Page* tool to view the entire poster. Create a spatial bookmark. A **bookmark** is an easy way to save the map scale and position information if you accidentally move the map while you work. From the *Main* menu across the top, select *Bookmarks*, then choose *Create Bookmark*. Name the bookmark *City of San Francisco* (**Figure 2.154**).

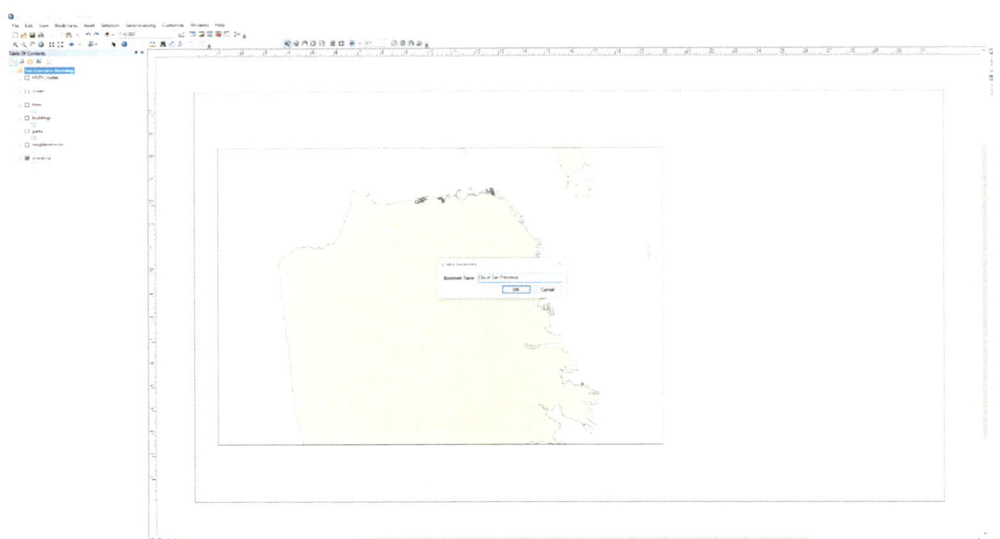

Figure 2.154: A bookmark saves the map scale and position information.

Before moving on to the next step, try out the bookmark by intentionally moving the position of the shoreline layer within the data frame. Then, use the bookmark to return to your saved position.

In this chapter, you learned about borders and neatlines. A **map border** is usually a rectangular box that delineates a map and all elements within. A **neatline** is a type of border, which delineates the geographic extent of the map content. It is acceptable practice to have map elements reside between the border and the neatline. In this instance, the default border around the dataframe serves as a *neatline* because it *delineates the geographic extent* of the map. In a previous tutorial, you learned how to change the background and border color of the data frame. Open the data frame properties and navigate to the *Frame* tab. Change the Border to a thickness of 4.0 and choose a border color. Change the background color to your choice of blue.

From the *Main* menu choose *Insert*, then *Neatline*. When the Neatline Window opens, click the radio button next to *Place inside margins* (**Figure 2.155**). Change the Gap to zero. For the background choose a very light and neutral color such as Grey 10%. When ready, click *OK*.

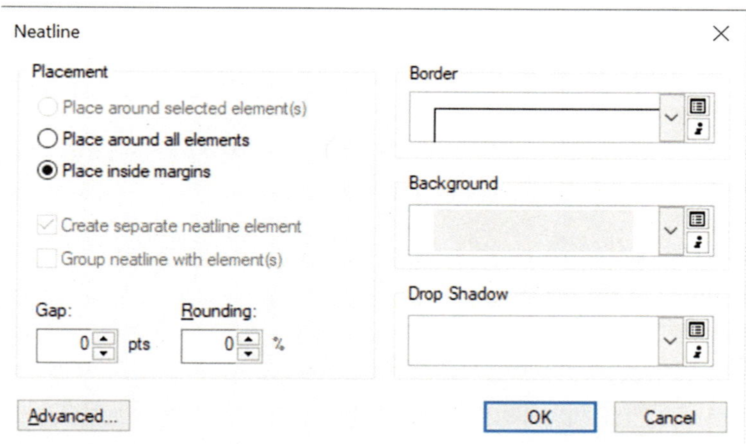

Figure 2.155: You can use the *Neatline* window to create a background for the poster.

As you can see, the element you just placed on the poster is technically not a neatline because it delineates more than just the geographic extent (**Figure 2.156**). In this step, you used the tool to create a background color on the page of the poster.

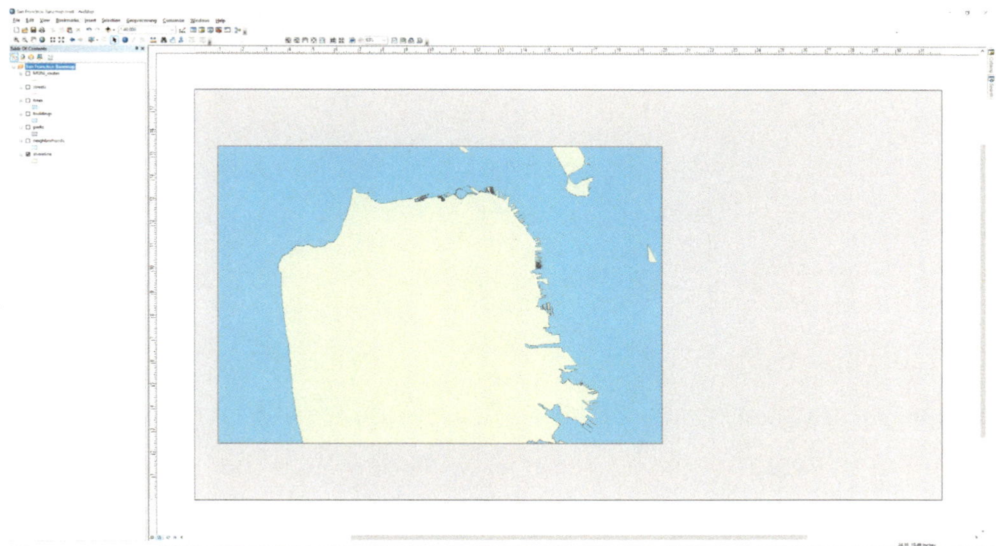

Figure 2.156: The background colors help to define the boundaries between the map and the poster page.

Adjusting Line Weight and Color

In this chapter, you learned about **visual variables**, the graphic representation of information using visualization techniques that clarify symbols and features from one another. Among the visual variables to consider for this activity are hue, saturation, and value, the three dimensions that characterize color. When choosing a basemap color pallet, keep the map purpose and audience in mind. The goal of a basemap is to provide context for the primary data or theme. In this instance, the basemap serves as the background for your choice of one out of two themes. You can choose the historic fire damage from the 1906 earthquake as the theme, or you can pick sea level rise and flooding. Whether it is fire damage or flooding, you want your theme communicated to the map reader clearly and effortlessly.

For this activity, I recommend choosing either a light color scheme or a dark color scheme for the basemap. Either one should have colors with low saturation, low value, or both. For example, a light color scheme might use light greys and pastels for the basemap features. A dark color scheme might use medium to dark greys instead. Your goal is to provide context *without* competing with your thematic data.

Start by arranging the layers in the *Table of Contents* using the following order from top to bottom:

» neighborhoods

» muni routes

» streets

» trees

» buildings

» parks

» shoreline

Work from the bottom up, starting with the shoreline layer. It lays the foundation for the remainder of the map design. On the *Table of Contents*, click the colored rectangle under the shoreline layer. When the symbol selector opens, choose something neutral for the Fill Color (**Figure 2.157**). Set the Outline Color to No Color. For most of the polygon layers, you want to remove the outlines.

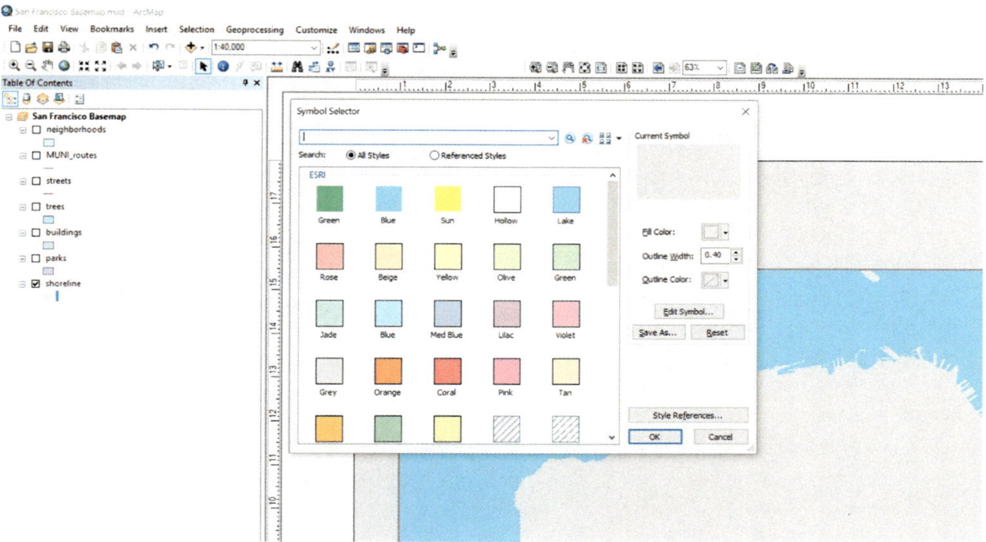

Figure 2.157: You can make quick edits to feature symbols and colors directly from the *Table of Contents*.

Repeat these steps for the parks, buildings, and trees, one at a time. Try to maintain your chosen color scheme throughout the process (**Figure 2.158**). You need to remove the outlines for each of these layers. If necessary, use the *Zoom In* tool on the *Layout* toolbar to get a closer look at the details.

Figure 2.158: The shoreline, parks, buildings, and trees make up the foundation of the basemap. This example uses a light color scheme.

For linear features, such as the streets, you need to adjust both the color and the line weight. Turn on the streets layer. In the *Table of Contents*, click the symbol for the streets. By default, ArcMap sets the line width to 1.00. For this map, 1.00 is a little too heavy. Change the width to 0.75. Change the color to something that matches your chosen color scheme (**Figure 2.159**).

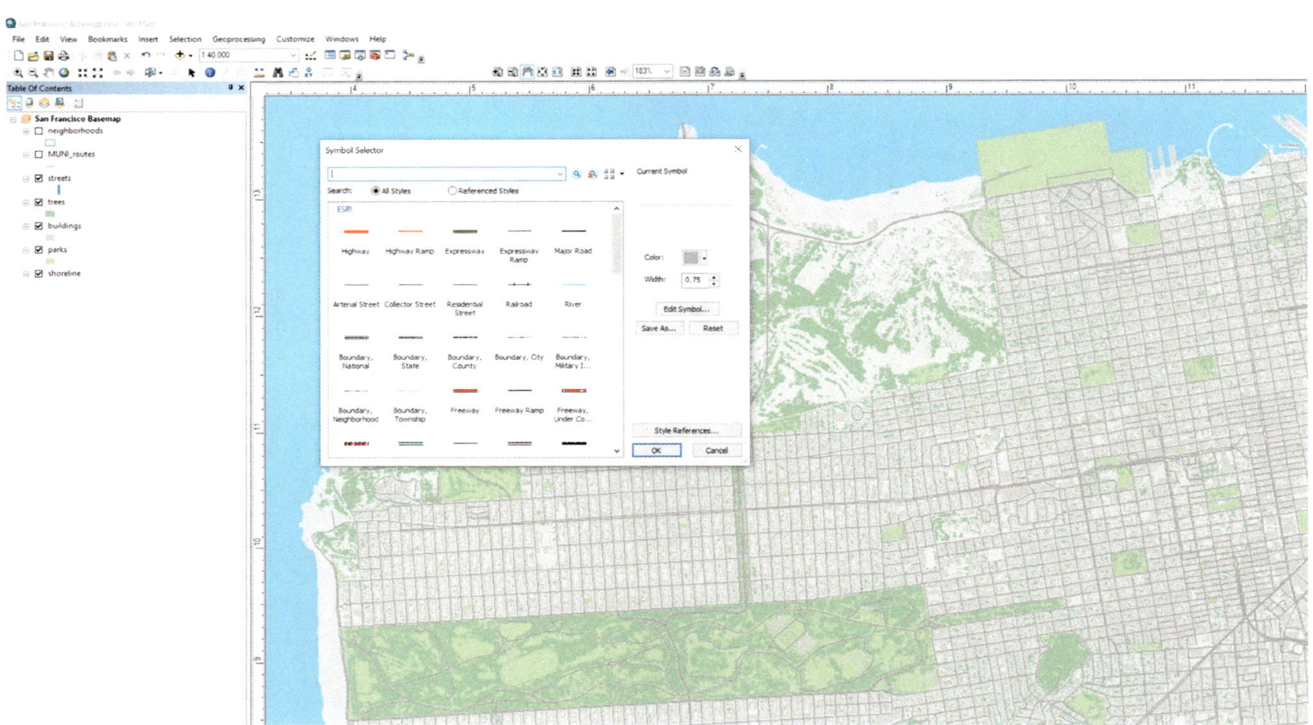

Figure 2.159: The streets, MUNI routes, and neighborhoods serve as accents on the basemap.

Repeat these steps for the MUNI routes. Because the MUNI routes are on top of the streets, you may want to choose a different width and color (**Figure 2.160**). The goal is for the map reader to distinguish between the two linear features while keeping both features relatively low in the visual hierarchy.

You may be wondering why the neighborhoods layer is on top of the stack in the *Table of Contents*. For this map, the neighborhoods layer serves as an accent. Turn on the neighborhoods layer. Because the neighborhoods layer is on top, you also need to make the layer transparent. Open the neighborhood properties and navigate to the *Display* tab (**Figure 2.161**). Change the transparency to 90%.

Remove the *Fill Color*. Change the *Outline Width* to 5.00. Choose an Outline Color that works with your chosen color scheme. It should be different enough so that these lines on the map are confused with the road layers.

Figure 2.160: This example uses a slightly thinner line weight and a lighter color for the MUNI routes.

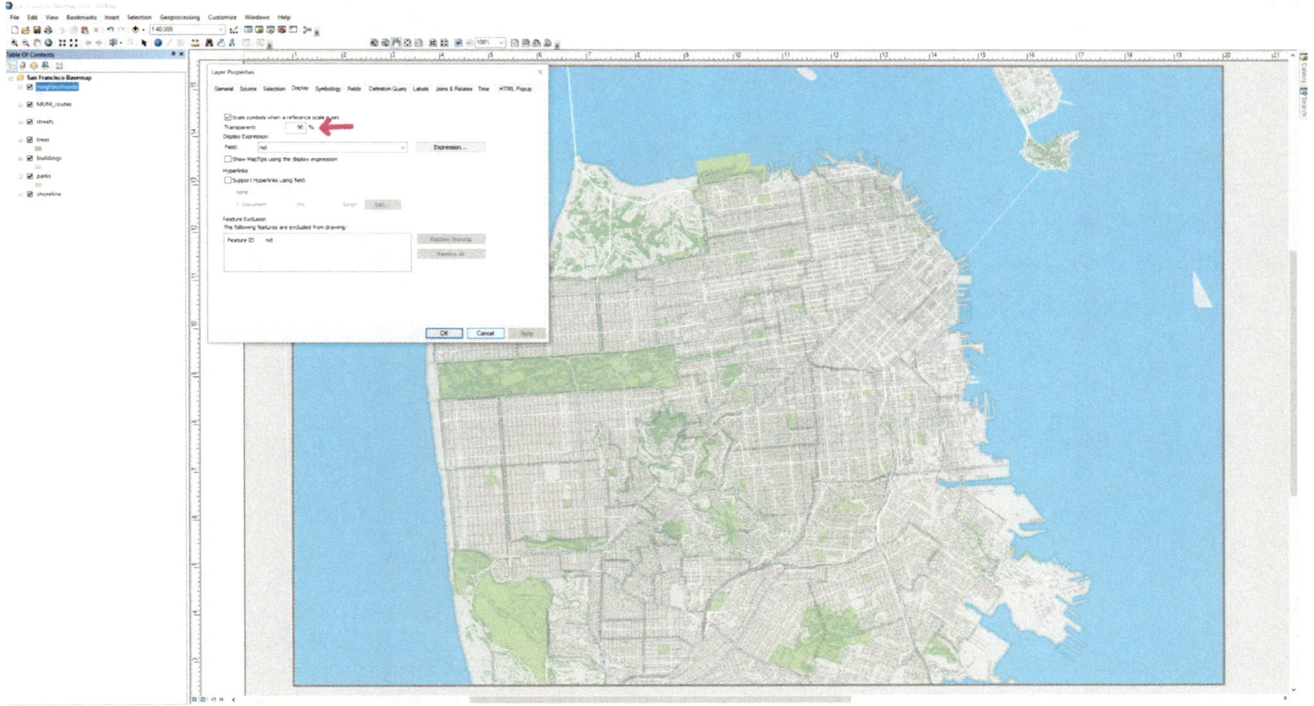

Figure 2.161: The higher the number, the more transparent the layer.

Cartographic Typography

In this chapter, you learned that **typography** is the art and technique of type placement, arrangement, and design for aesthetics, communication, and readability of written language. A skilled cartographer views the typography on a map as both a means to communicate language and as a visual variable. Typography is useful in establishing an effective visual hierarchy to reduce initial confusion and enhance communication. On this map, you practice working with type and label placement by choosing ten neighborhoods in the City of San Francisco. The purpose of these labels is to identify points of interest as well as provide a geographic frame of reference to the map reader.

In ArcMap, open the properties for the neighborhoods layer and navigate to the *Labels* tab. Near the top left corner of the *Layer Properties* window, check the box that says *Label features in this layer* (**Figure 2.162**). Use the drop-down menu next to the Label Field and choose *nbrhood* from the list of fields. When done, click *OK*.

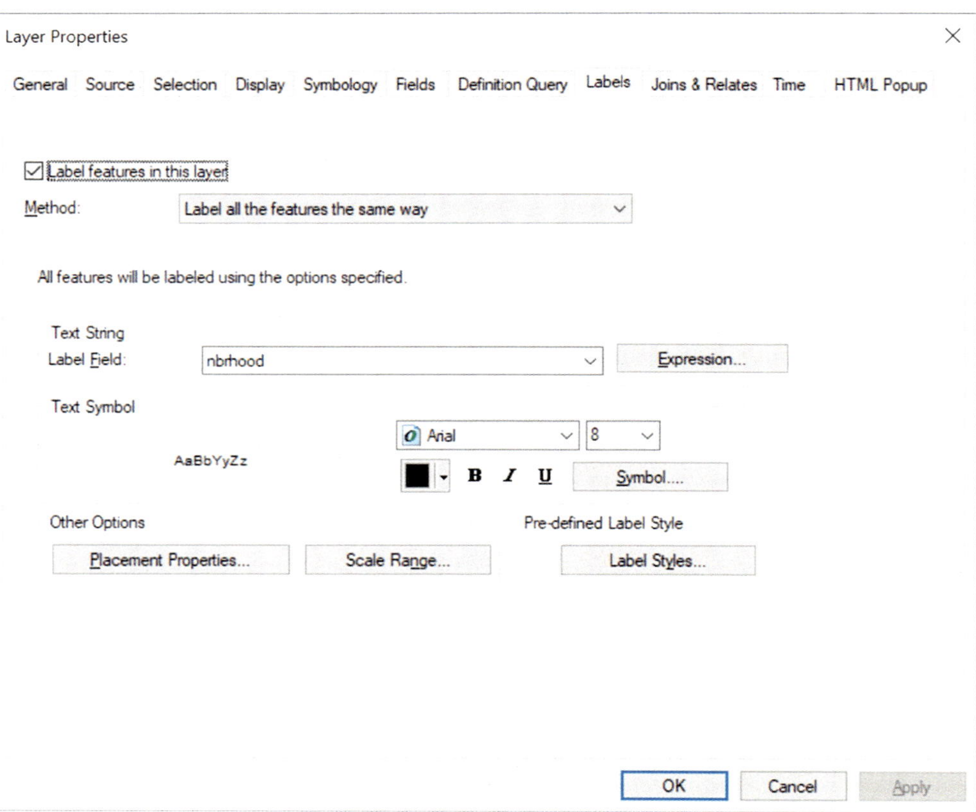

Figure 2.162: The Labels tab provides many options for working with text.

ArcMap dynamically places labels on the map and does a passable job for many purposes (**Figure 2.163**). For advanced cartography, this only provides a place to start. Most projects require precise adjustments to label placement. For this map, these labels serve as a reference for you to choose ten place names. You eventually turn off the automatic labels.

Figure 2.163: Skilled cartographers use the automatic label placement only as a starting point.

In a previous tutorial, you learned how to activate toolbars. Activate the *Draw* toolbar and dock it at the top of the ArcMap window. Choose your first area of interest and click the *Text* tool on the *Draw* toolbar. The icon looks like the letter **A**. Add the label over the region of interest (**Figure 2.164**). If necessary, use the *Zoom In* tool on the *Layout* toolbar. Don't worry about the appearance and exact location for now. You refine the labels later. Repeat these steps for the remainder of your chosen areas of interest. Be careful with your spelling. When done, open the neighborhoods layer properties and turn off the automatic labels by unchecking the box in the Labels tab.

Turn off the large datasets such as buildings and trees to speed up the drawing time in ArcMap. Be sure to turn them back on when finished with label placement.

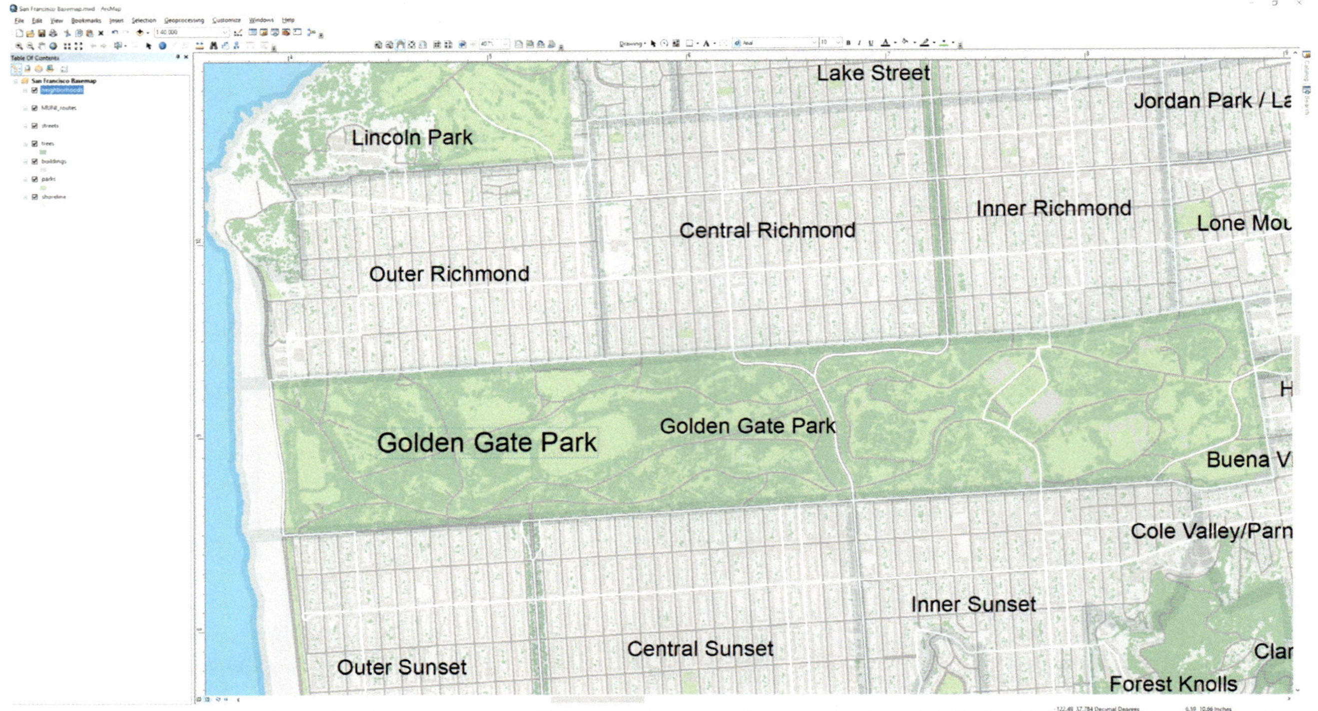

Figure 2.164: By default, ArcMap uses the Arial typeface for labels.

In this chapter, you learned about typeface and font. A **typeface** encompasses style and symbology across an entire alphabet including letters, numbers, and punctuation. A **font** is a subset of a typeface. It describes a specific size and style of a typeface. Arial, Times, and Courier are all typefaces. Arial italic 10-point, Times regular 8-point, Courier bold 12-point, are all fonts because they describe a style (italic, regular, bold) and size (10, 8, and 12-point). You also discovered that professional type designers classify typefaces into many distinct categories, often based on historical periods and geographic origin. This tutorial uses three broad classifications, *serif*, *sans serif*, and *display*. Sans means without in French. **Sans serif typefaces** do not have any serifs at the end of stems, while **serif typefaces** do. **Display typefaces** are a catchall category of highly stylized typefaces used to catch the reader's attention.

Cartographic convention suggests that you should limit the total number of typefaces on map. Typically, this means one serif and one sans serif typeface. In this instance, you may also use a display typeface for the poster title to convey a personality, elicit an emotion, or to catch the map-reader's attention. Just be sure to never use display type for main map content.

You can edit individual labels by selecting the label, then by right-clicking on it to open the properties window (**Figure 1.65**).

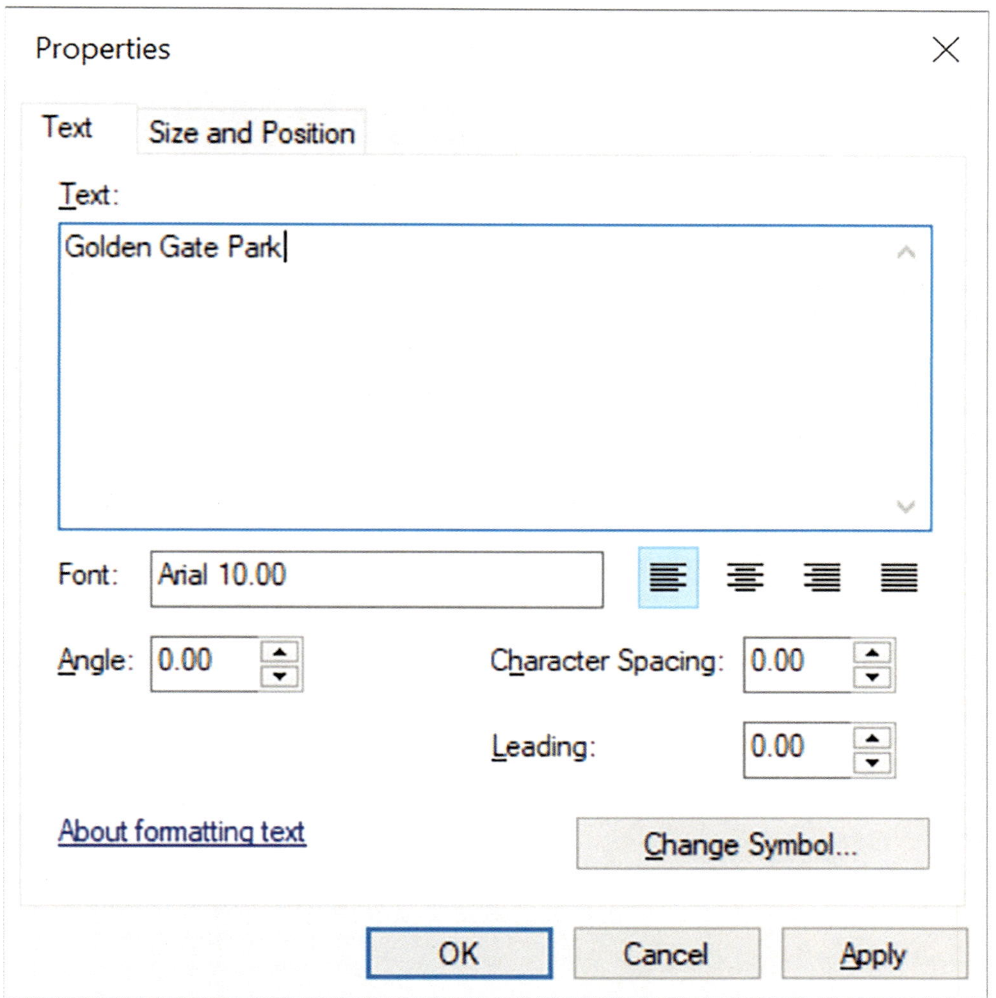

Figure 1.65: The label properties window provides many options for working with text.

Click the Change Symbol button for more options. When *Symbol Selector* window opens, choose a sans serif typeface such as Ariel or Veranda. Choose a font size, and a color (**Figure 2.166**). When ready click *OK*, then *Apply* to preview your changes.

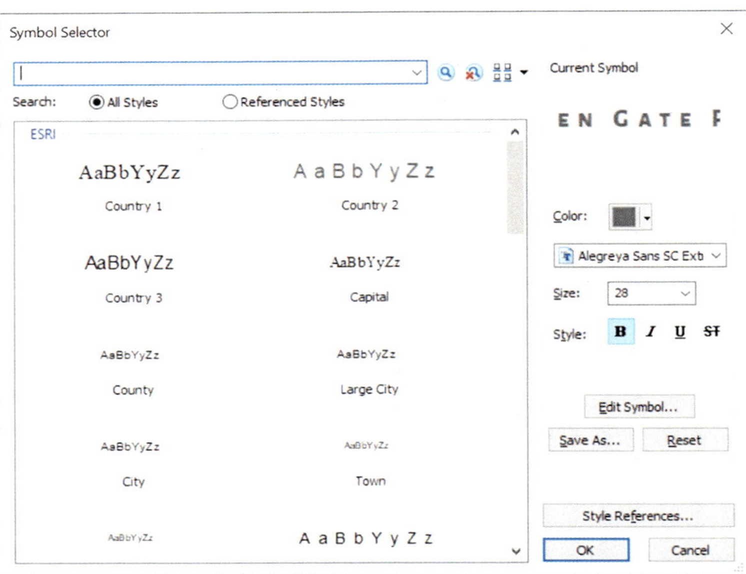

Figure 2.166: This example uses Alegreya Sans SC as the choice for a sans serif typeface, available from *Google Fonts*[4].

If necessary, Change the Angle and Character Spacing (**Figure 2.167**). When ready, click *OK*.

A **halo** is a drop-shadow effect placed under the text. Including a halo behind a label improves legibility by breaking up lines that cross through the text, such as the streets and MUNI routes. You add a halo effect by opening the label properties, clicking the *Change Symbol* button, then clicking the *Edit Symbol* button. When the *Editor* window appears, navigate to the *Mask* tab. Click the radio button next to *Halo*. The halo is white and sized at 2.0 by default. You can change the color by clicking on the *Symbol* button (**Figure 2.168**).

4 https://fonts.google.com/

Figure 2.167: The angle and character spacing follow the shape of the Golden Gate Park.

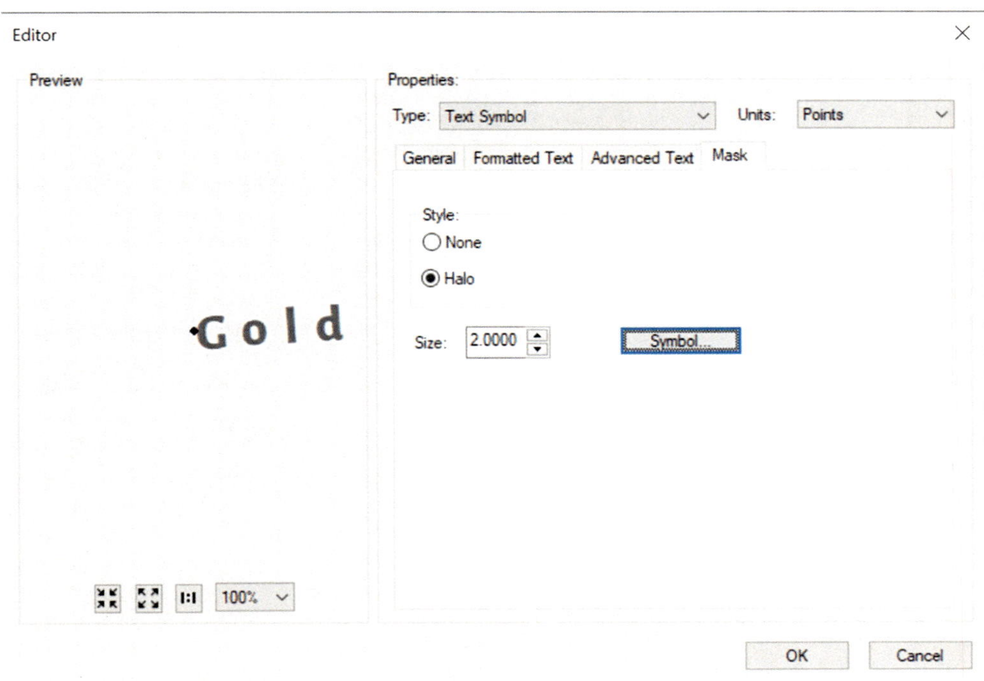

Figure 2.168: When choosing a halo color, pick one with high contrast or one that blends in with the map background.

The best strategy for picking halo color and weight is to determine if you need high contrast or if you need to blend in with the background. If you have a busy background, a contrasting halo is often the best choice. If your background has a mostly uniform color with only a few linear features, choosing the same color as the background is a subtle improvement (**Figure 2.169**).

Figure 2.169: In this example, the halo matches the green background color of the park.

Skill Drill: Practicing Cartographic Typography

Using what you learned in the previous step, place, and style the remaining labels for the areas of interest around the City of San Francisco (**Figure 2.170**). Use the *Layout* toolbar to zoom in and out of the poster page as you work to get a sense of how your typographic choices are working. Be sure to re-save the map document often so that you do not lose your work. As the last step, add a label for the Pacific Ocean and the San Francisco Bay. As you learned in this chapter, cartographic convention suggests that labels for water features should be italicized and use a *serif* typeface, such as Garamond or Times New Roman.

Skill Drill: Choose a Map Theme

In a previous step, you downloaded data from the ***DataSF website***[5]. In this step, you can choose the historic fire damage from the 1906 earthquake as the poster theme, or you can pick sea level rise and flooding.

5 https://datasf.org/opendata/

Return to and enter one of the following keywords in the search bar:

» Areas Damaged by Fire Following the 1906 Earthquake
» 108" Inundation Vulnerability Zone Line

Download the data to your *original* folder and decompress the file. Refresh your *original* folder and rename the shapefile. When done, add the data to the map document. Edit the color and symbology to distinguish your chosen theme from the basemap (**Figure 2.171**).

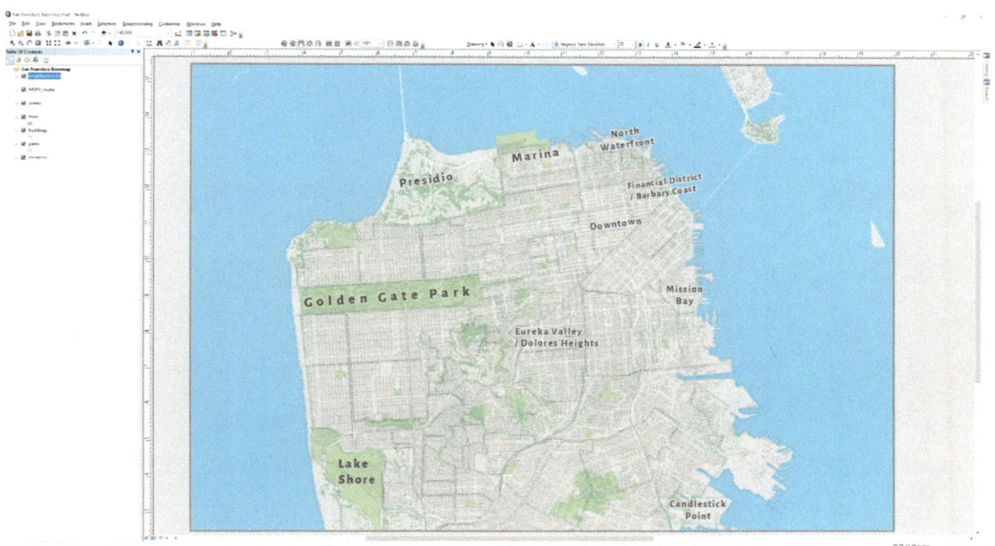

Figure 2.170: The map labels several points of interest throughout the City of San Francisco.

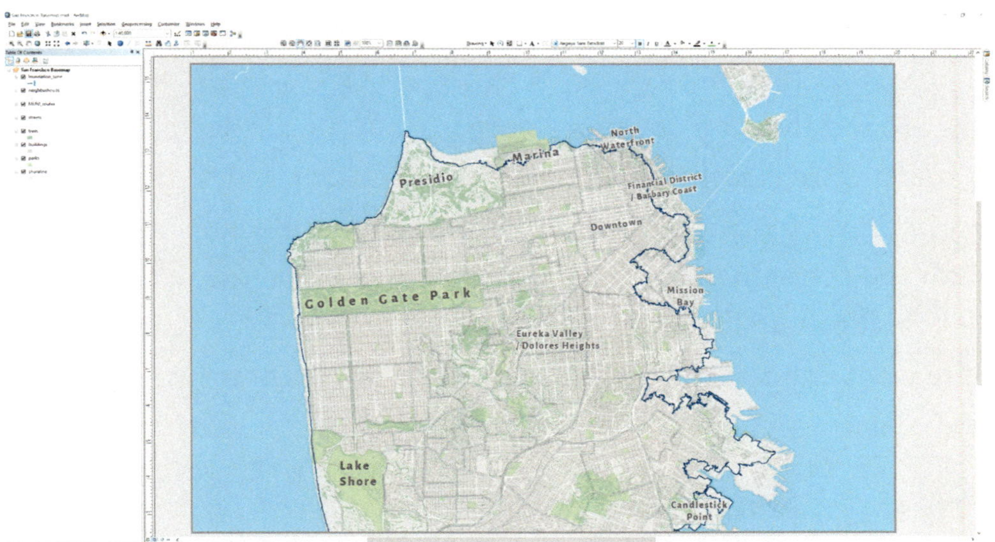

Figure 2.171: In this example, the 108" Inundation Vulnerability Zone Line appears in dark blue.

Skill Drill: Finalizing the Poster

In a previous tutorial, you learned how to create a map poster by inserting map elements and labels. Using what you learned, finalize the poster layout. Incorporate a title and a title bar for the poster above the map. You may use a decorative typeface if appropriate. To the right of the map, use the *Rectangle Text* tool on the *Draw* toolbar to provide a block of descriptive text (**Figure 2.172 and Figure 2.173**).

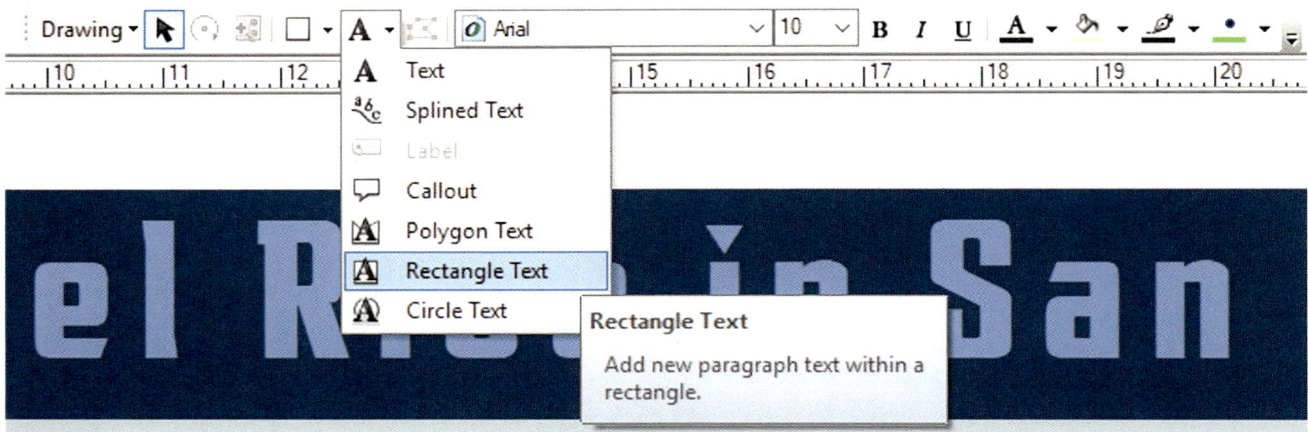

Figure 2.172: You can access additional text tools using the drop-down menu.

You may search the internet for appropriate text. If your theme is sea level rise, the ***metadata on the DataSF website***[6] might also be helpful. If your theme is fire damage from the 1906 earthquake, the ***USGS has a webpage***[7] that might interest you.

You may use the text from either source. Be sure to give the appropriate source credit.

In the margins around the map, include a north arrow, a scale bar, a representative fraction (RF), and a word statement (**Figure 2.174**). You can find the last two on the *Insert* menu under *Scale Text*. Your map legend does not need a title but should include all the layers **except for the shoreline**. For each layer, open the *Properties* window and navigate to the *General* tab. Clean up the labels for the layers by removing any underscores. The layer names are what appear in the map legend, and you want to avoid underscores on the final map.

6 http://onesanfrancisco.org/node/148

7 https://earthquake.usgs.gov/earthquakes/events/1906calif/18april/

Areas Damaged by Fire Following the 1906 Earthquake
Pacific Ocean
San Francisco Bay
Presidio
Marina
North Waterfront
Financial District / Barbary Coast
Downtown
Golden Gate Park
Mission Bay
Eureka Valley / Dolores Heights
Lake Shore
Candlestick Point
The California earthquake of April 18, 1906 ranks as one of the most significant earthquakes of all time. Today, its importance comes more from the wealth of scientific knowledge derived from it than from its sheer size. Rupturing the northernmost 296 miles (477 kilometers) of the San Andreas fault from northwest of San Juan Bautista to the triple junction at Cape Mendocino, the earthquake confounded contemporary geologists with its large, horizontal displacements and great rupture length. Indeed, the significance of the fault and recognition of its large cumulative offset would not be fully appreciated until the advent of plate tectonics more than half a century later. Analysis of the 1906 displacements and strain in the surrounding crust led Reid (1910) to formulate his elastic-rebound theory of the earthquake source, which remains today the principal model of the earthquake cycle.
Excerpted from Ellsworth, 1990.
Fire Damage
San Francisco Neighborhoods
MUNI Routes
Streets
Trees
Buildings
Parks
1:40,000
1 centimeter = 400 meters
0 2.5 5 10 15 20 Kilometers

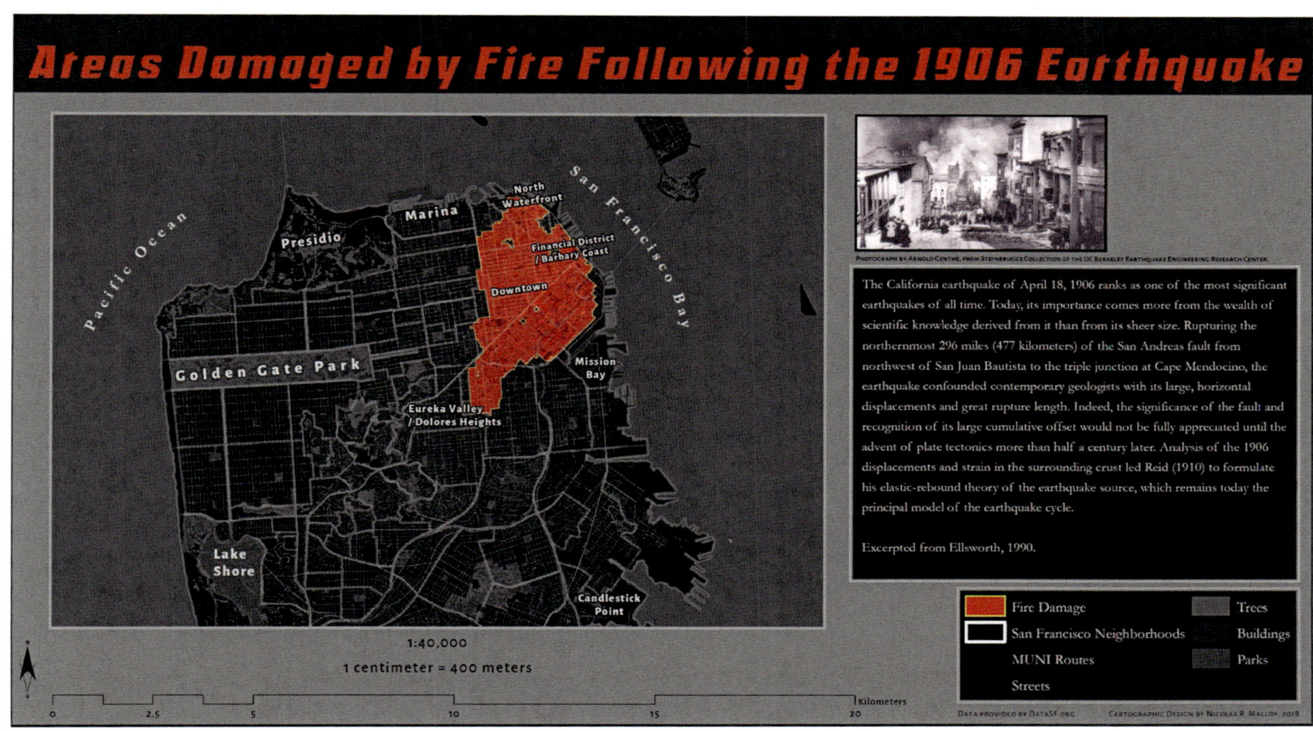

Figure 2.173: In this example, the theme of the poster is areas damaged by fire following the 1906 earthquake.

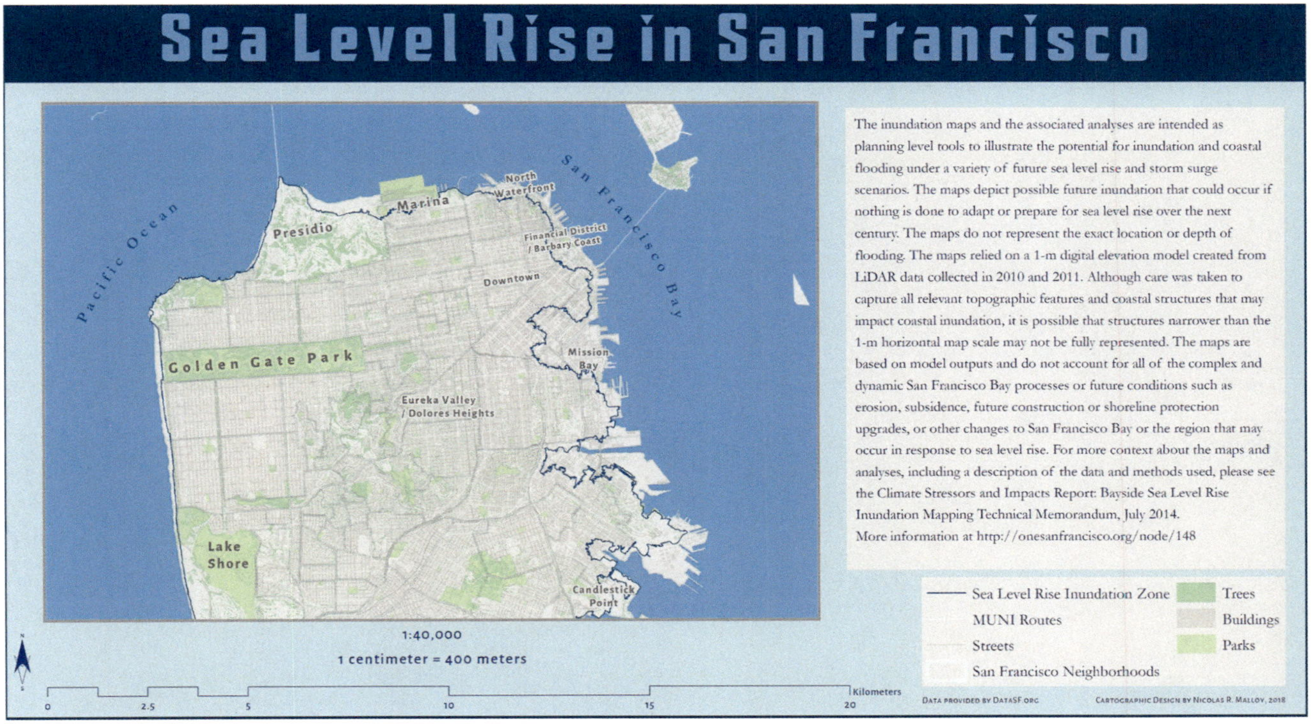

Sea Level Rise in San Francisco
Pacific Ocean
San Francisco Bay
Presidio
Marina
North Waterfront
Financial District / Barbary Coast
Downtown
Golden Gate Park
Mission Bay
Eureka Valley / Dolores Heights
Lake Shore
Candlestick Point
The inundation maps and the associated analyses are intended as planning level tools to illustrate the potential for inundation and coastal flooding under a variety of future sea level rise and storm surge scenarios. The maps depict possible future inundation that could occur if nothing is done to adapt or prepare for sea level rise over the next century. The maps do not represent the exact location or depth of flooding. The maps relied on a 1-m digital elevation model created from LiDAR data collected in 2010 and 2011. Although care was taken to capture all relevant topographic features and coastal structures that may impact coastal inundation, it is possible that structures narrower than the 1-m horizontal map scale may not be fully represented. The maps are based on model outputs and do not account for all of the complex and dynamic San Francisco Bay processes or future conditions such as erosion, subsidence, future construction or shoreline protection upgrades, or other changes to San Francisco Bay or the region that may occur in response to sea level rise. For more context about the maps and analyses, including a description of the data and methods used, please see the Climate Stressors and Impacts Report: Bayside Sea Level Rise Inundation Mapping Technical Memorandum, July 2014.
More information at http://onesanfrancisco.org/node/148
Sea Level Rise Inundation Zone
MUNI Routes
Streets
San Francisco Neighborhoods
Trees
Buildings
Parks
1:40,000
1 centimeter = 400 meters
0 2.5 5 10 15 20 Kilometers
Data provided by DataSF.org
Cartographic Design by Nicolas R. Malloy, 2018

Figure 2.174: In this example, the theme of the poster is Sea Level Rise in San Francisco.

Add an acknowledgment to DataSF.org at the bottom of the poster. Also, include your name and the year. When done creating your map poster, export the map as a PDF file with a resolution of at least **300 dpi**. Save it to your final folder.

Warning! Due to the large size of the datasets on this map, ArcMap could take up to 20 to 30 minutes to export.

When exporting, be patient! When working with large datasets, the ArcGIS software can take a long time to run tools and export maps. Some operations can take ten minutes. Others can take more than an hour. The amount of time depends on the memory and processing power of the computer. Unfortunately, ArcGIS software is also prone to crashing. When it is working on a task, try to avoid clicking anywhere on the software, as this could trigger a crash. Sometimes it is best to wait.

Principal Terms

ascender
ascender line
attribute table
baseline
basemap
bold font style
bowl
cap height
cap line
cartographic conventions
cartography
Catalog Tree
Catalog Window
cognitive expectations
color space
counter
descender
descender line
descriptive text
display typeface
field
field names
figure-ground phenomenon
final folder
graphic scale
graticule
HSV color space
hue
inset map
italic font style
kerning
label placement
leading
letterform
local hard drive
map border
map legend
map projection
meridians
navigation map
neatline
networked drives
north arrow
orientation
original folder
parallels
point size
qualitative classification
quantitative classification
record
reference map
regular font style
representative fraction (RF)
sans serif typefaces
saturation
serif
serif typefaces
shape
size
source note
spacing
stem
stress
Table of Contents
thematic map
tracking
typeface
typography
value

verbal scale
visual balance
visual center
visual hierarchy
visual variables
working folder
workspace
workspace folder
x-height

Chapter 3: Geodesy and Transformation

Overview

Since ancient Greece, mathematicians and philosophers have speculated about the size and shape of Earth. The Greek mathematician Pythagoras was one of the first to advocate the idea of Earth as a sphere. Since then, there have been many estimates on the circumference of Earth by those that followed. **Chapter 3** presents the discipline at the root of geospatial science, geodesy. Geodesy is a branch of applied mathematics. It is the science of measuring and representing the size and shape of Earth, the exact position of points on the planet, and the study of Earth's gravitational and magnetic fields as they change over time.

Learning Outcomes

Readers should be able to accomplish the following outcomes by the end of this chapter:

- Discuss the science and history of geodesy
- Explain the process of modeling Earth as a mathematically defined surface
- Describe the elements and purpose of a geodetic datum
- Recognize commonly used datums for North America
- Express an understanding of a geographic coordinate system
- Demonstrate an understanding of geometric transformations and map projections.
- Recognize frequently used map projections and their uses

Measuring Earth

Geodesy is a branch of applied mathematics. It is the science of measuring and representing the size and shape of Earth, the exact position of points on Earth, and the study of Earth's gravitational and magnetic fields as they change over time. Since ancient Greece, mathematicians and philosophers have speculated about the size and shape of Earth. The Greek mathematician Pythagoras was one of the first to advocate the idea of a spherical earth. Since then, there have been many estimates on the circumference of Earth by those that followed. Around 250 B.C., the ancient Greek philosopher Eratosthenes made a remarkably accurate estimate for his time based on the equation:

$$C = (360^\circ \div \Theta) \times (s)$$

In this equation, the arc s is the distance between two points on a sphere, exactly north, and south from each other. One divides 360° by theta Θ, an angle created by extending an imaginary line from each point to the center of Earth. Then, one multiplies this value by s (**Figure 3.01**).

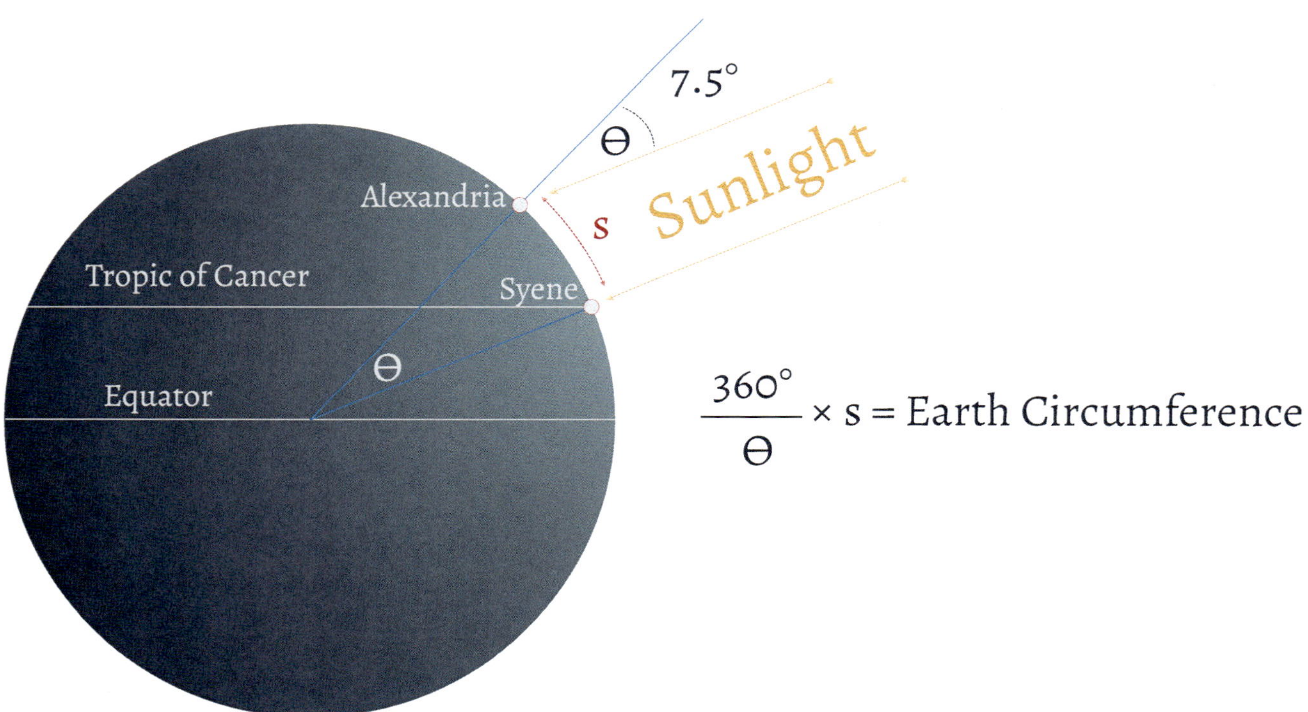

Figure 3.01: A simplified diagram of Eratosthenes' method

Eratosthenes observed that in the town of Syene, on the day of the summer solstice, the midday sun shone to the bottom of a well, casting no shadow. This phenomenon led him to conclude that the sun's rays were perpendicular to the surface of the well[1] on the day of the summer solstice. Eratosthenes also measured the angle of the shadow created from an obelisk in the city of Alexandria on the same day in a later year. This measurement provided Eratosthenes with the angle *theta* Θ. The length between Alexandria and Syene provided him with *distance s*. Eratosthenes estimated that the circumference of Earth was approximately 250,000 stadia. There is some variation on how long each stadion was, but his estimate is equal to about **25,000 miles**. The currently accepted value for the circumference of Earth is **24,901 miles**. As one can see, Eratosthenes was very close! Watch the video on Eratosthenes by Derek Owens to learn more (**Figure 3.02**).

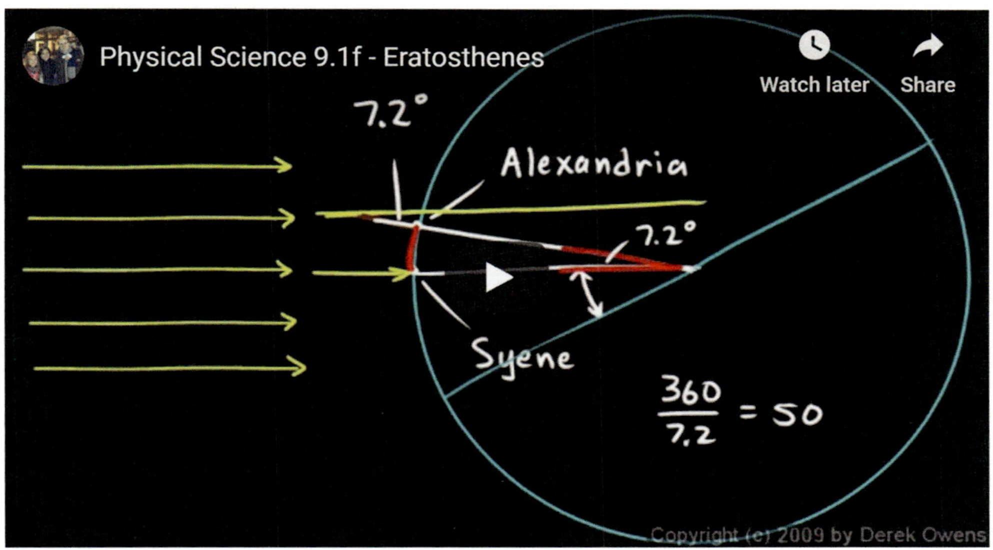

Figure 3.02: Physical Science 9.1f-Eratosthenes by David Owens describes the methods used by Eratosthenes. URL: *https://youtu.be/O6KOSvYHAmA*

The Geoid

In the 17th century, Sir Isaac Newton developed the theory of gravity. Newton theorized that the rotation of Earth would introduce outward centrifugal forces which would counteract the pull of gravity. These forces would cause Earth to take the shape of an oblate ellipsoid, meaning it is a bit wider around the equator than around the poles. In 1735, the French Academy of Sciences sponsored scientific expeditions to measure portions of Earth close to the equator and near the Arctic, which proved Newton's theory.

1 The sun cast no shadow because the rays were parallel to the sides of the inner wall of the well.

The need to measure elevation resulted in the development of additional models of Earth. **Elevation** is a relative measurement equal to the height above mean sea level. Because elevation is relative to mean sea level, changes in mean sea level result in changes of elevation. Accurate vertical measurements, such as elevation, require a different model of Earth called a geoid. A **geoid** is a surface model of Earth based on gravity. No landmasses exist on the geoid surface model and mean sea level covers the entire planet. Places where gravity is stronger or weaker change mean sea level, creating an undulating shape. Watch the video on the Gravity field and steady-state Ocean Circulation Explorer (GOCE) and the Geoid by the European Space Agency (ESA) to learn more (**Figure 3.03**).

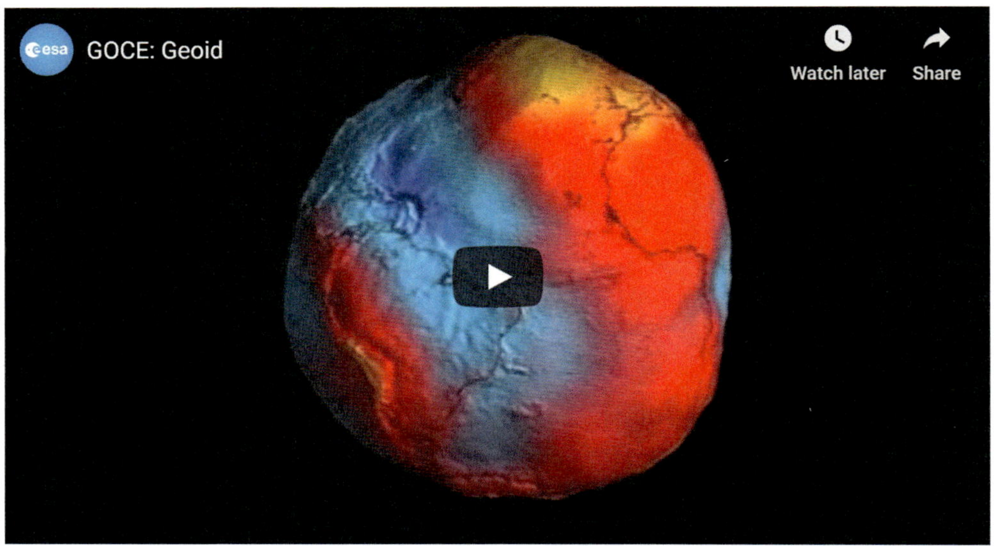

Figure 3.03: This video provides a detailed explanation of the concepts related to the geoid. URL: *https://youtu.be/qu-075pe5GY*

The Reference Ellipsoid

The irregular, non-geometrical shape of the geoid is challenging to work with directly. For mapping purposes, one transforms the geoid into a smooth mathematically defined surface model called the **reference ellipsoid** (**Figure 3.04**). Because the geoid and the reference ellipsoid are not an exact match, there are *multiple* reference ellipsoids used today, each optimized for different regions of Earth.

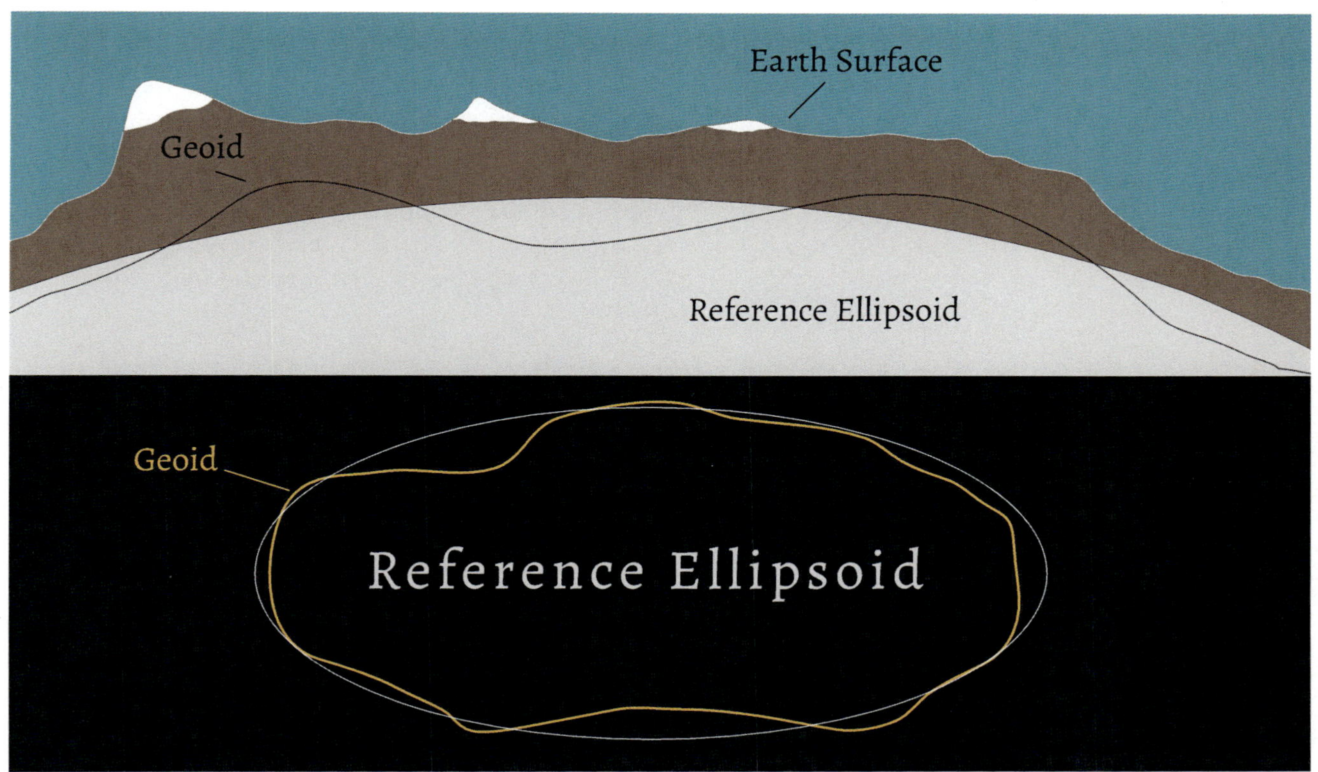

Figure 3.04: The brown area represents the surface of the Earth. The dotted line represents the *geoid* model of Earth. The smooth grey surface represents the *reference ellipsoid* model of Earth. The lower section of this graph depicts the geoid-ellipsoid surfaces overlaid on top of each other.

Geodetic Datums

Reference ellipsoids, along with a network of carefully surveyed control points, establish a geodetic datum. **Geodetic datums** are standard reference points or reference ellipsoids that serve as a base for calculating positions on and above the surface of Earth. One surveys a control point at the **origin** of a geodetic datum, where the geoid and the reference ellipsoid perfectly align. All other survey points reference back to the *origin*. As the distance increases away from the origin, the surfaces between the geoid model and the reference ellipsoid model become separated, with the reference ellipsoid passing above and below the geoid. As a result, geodesists develop different geodetic datums for the best fit in different parts of the world. Some datums use the earth's center of mass as the origin. In this instance, satellite measurements calculate the position of the origin. For accuracy, geodesists calculate a position on Earth using two types of geodetic datums, horizontal geodetic datums, and vertical geodetic datums. **Horizontal geodetic datums** form the basis for computing horizontal locations: north, south, east, and west. Vertical geodetic datums form the basis for computing elevation. Watch *What are Geodetic Datums?*, by the COMET Program/MetEd to learn more (**Figure 3.05**).

Figure 3.05: The video *What are Geodetic Datums?*, by the COMET Program/MetEd explains the concepts behind datums. URL: *https://youtu.be/kXTHaMY3cVk*

North American Datum of 1927

The United States Standard Datum, with an origin in Meades Ranch, Kansas, was officially established in 1901. This datum was an extension of the New England Datum 1879, based on the **Clark 1866 ellipsoid**. In 1913, Canada and Mexico expanded the United States standard datum, adding their control points to the network. It took many years to correct all control points uniformly across North America. This datum is now known as the **North American Datum 1927 (NAD27)**. Watch *How Were Geodetic Datums Established?*, by the COMET Program/MetEd to learn more (**Figure 3.06**).

North American Datum of 1983

In 1970, the National Geodetic Survey (NGS) was formed as part of the National Oceanic and Atmospheric Administration (NOAA). Using modern survey techniques, the NGS spent the next dozen years establishing the **North American Datum 1983 (NAD83)**, which was meant to replace the NAD27 Datum. The reference ellipsoid for the North American Datum 1983 is the **Geodetic Reference System of 1980 (GRS 80)** ellipsoid.

The North American Datum of 1983 is a horizontal **geocentric datum** that references the origin of the ellipsoid at Earth's center of mass. Watch *What Is the Status of Today's Geodetic Datums?*, by the COMET Program/MetEd to learn more (**Figure 3.07**).

Figure 3.06: The video *How Were Geodetic Datums Established?*, by the COMET Program/MetEd describes the history of geodetic datums. URL: *https://youtu.be/-oUFqg1Lw1U*

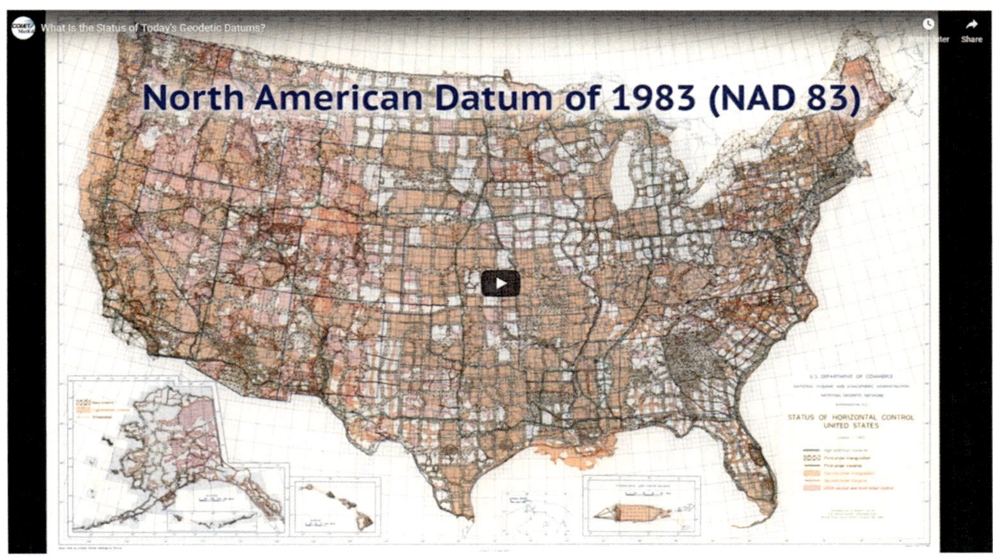

Figure 3.07: This video, *What Is the Status of Today's Geodetic Datums?*, by the COMET Program/MetEd, provides an overview of modern datums. URL: *https://youtu.be/ZFe_rZccGOQ*

World Geodetic System 1984 (WGS84)

In 1984, the U. S. Department of Defense established the World Geodetic System 1984 (WGS84), a geocentric datum and coordinate system. The **World Geodetic System of 1984** uses a reference ellipsoid of the same name, **WGS84 ellipsoid**, which, like NAD83, is calculated from the earth's center of mass. Most latitude and longitude values one acquires from the internet via public sources nearly always use the WGS84 as the datum. WGS 84 is also the reference system for the Global Positioning System (GPS). **Chapter 5** covers Global Positioning Systems in detail. Watch *What's Next for Geodetic Datums?*, by the COMET Program/MetEd to learn more (**Figure 3.08**).

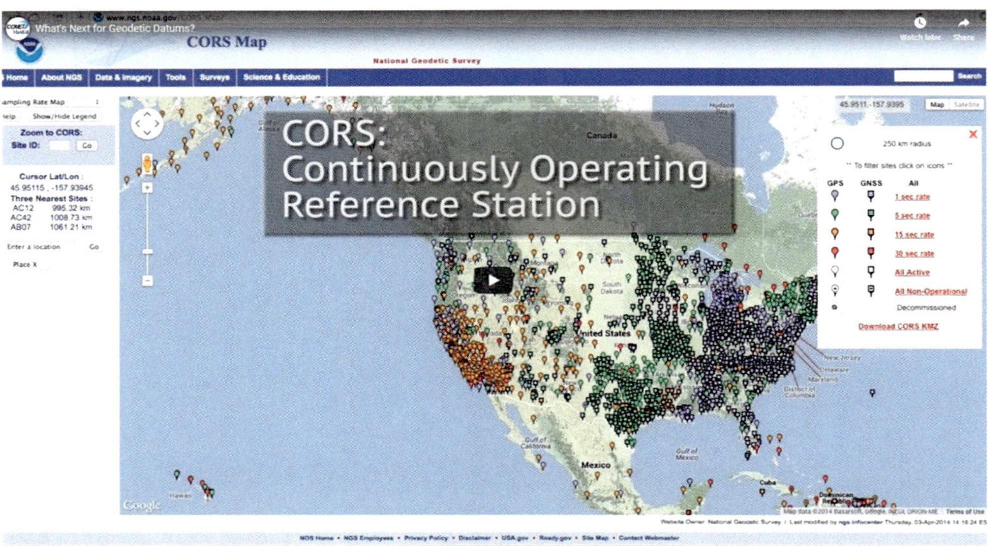

Figure 3.08: The video, *What's Next for Geodetic Datums?*, by the COMET Program/MetEd discusses the future of geodetic datums. URL: *https://youtu.be/w69xc_U1Rao*

Parallels and Latitude

Around 150 BC, Ptolemy began to experiment with creating a grid of lines, called a graticule, to determine a position on Earth. A **graticule** is a network of parallels and meridians, draped over a horizontal datum. **Parallels** are east-west lines that run parallel to the equator and are equally spaced above and below the equator. One measures the distance from one parallel to the next using degrees of **latitude**. Measurements start at the **equator**, which is 0° latitude (**Figure 3.09**). From there, latitude continues north and south until it reaches 90° at each pole.

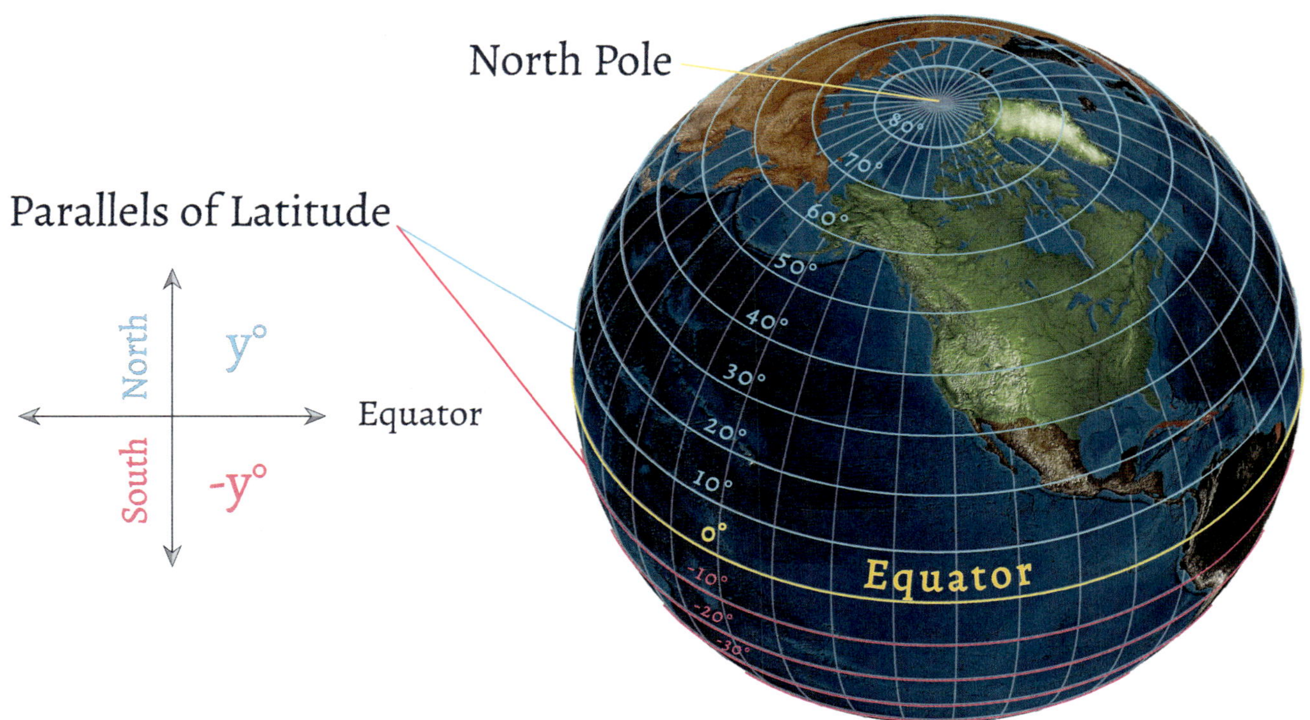

Figure 3.09: The blue lines represent north latitude while the pink lines represent south latitude. One expresses south latitude as a negative number.

Meridians and Longitude

Meridians are a series of lines that run north-south and pass through the poles. The name comes from the Latin word for noon. One measures the distance from one meridian to the next using degrees of **longitude**. Historically the origin of longitude was an arbitrary location, usually set by a cartographer's nation of origin. In 1884, the International Meridian Conference defined the international standard using the British Prime Meridian. The **prime meridian** is 0° longitude, defined by the north-south axis of a telescope at the Royal Observatory in Greenwich, England. Longitude is measured east-west along the equator, starting at the prime meridian, until it reaches 180° in both directions. (**Figure 3.10**).

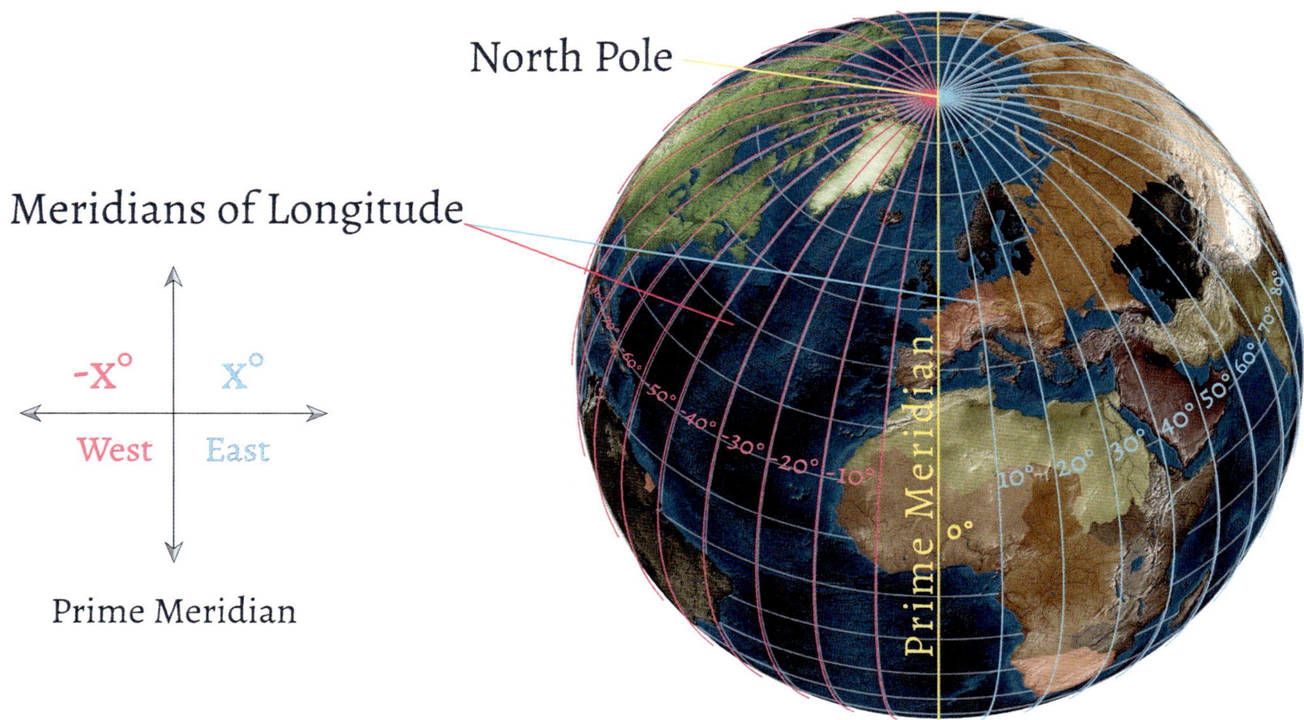

Figure 3.10: The blue lines represent east longitude while the pink lines represent west longitude. One expresses west longitude as a negative number.

A **great circle**, made from the circumference of a sphere, divides Earth into equal halves (**Figure 3.11**). The equator and every opposing pair of meridians form a great circle. However, a great circle can follow any direction, as long as it divides Earth into equal halves. Great circles are frequently used in navigation since they represent the shortest distance from one point on Earth's surface to another.

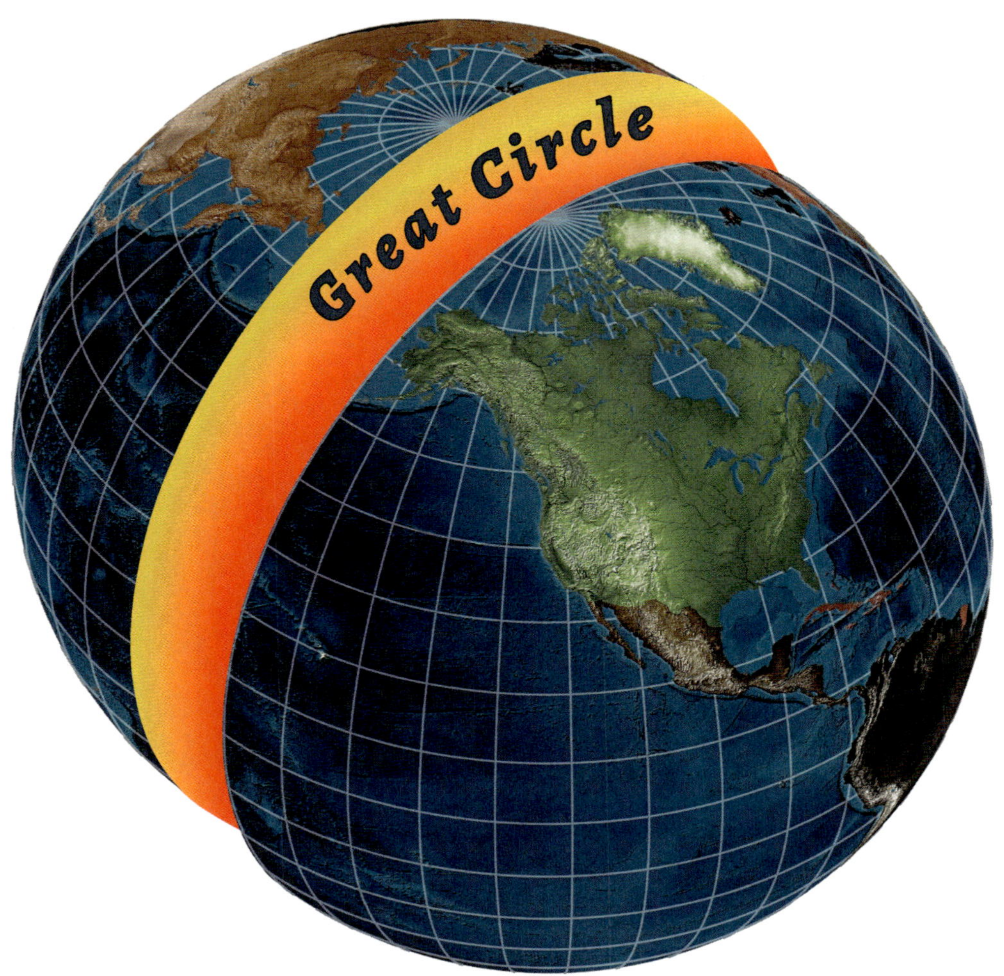

Figure 3.11: A great circle splits a sphere into equal halves. As a result, it represents the shortest distance from one point to another on a sphere.

Geographic Coordinate System

Latitude and longitude are measured using a sexagesimal, or base 60, system. Similar to units of time, one divides degrees of latitude and longitude into fractions of 60, so each degree consists of 60 minutes, and each minute consists of 60 seconds. The horizontal datum, the prime meridian, and the angular units of latitude and longitude, together form a **geographic coordinate system (GCS)**. **Chapter 4** covers geographic coordinate systems in more detail. Watch *Longitude and Latitude Song* by Tom Glazer & Dottie Evans to learn more (**Figure 3.12**).

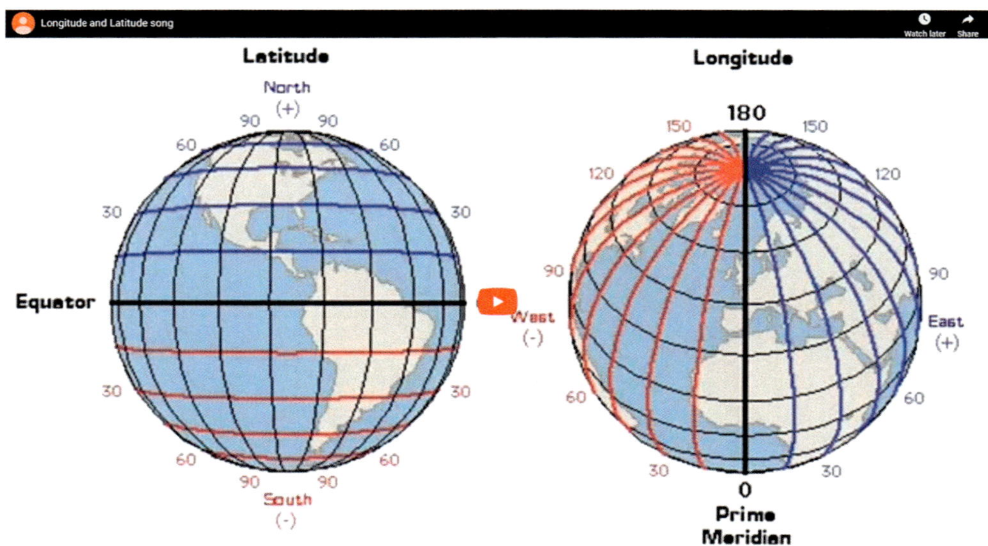

Figure 3.12: *Longitude and Latitude Song* by Tom Glazer & Dottie Evans explains how one measures latitude and longitude using a catchy tune. Enjoy! URL: *https://youtu.be/MjDqhLUzCpE*

Geometric Transformations

Locations on Earth are nearly always represented using a flat rectangular map. Most people never consider how this is possible. To illustrate this point, imagine cutting apart a tennis ball and trying to rearrange the parts into a flat rectangle. It would not take long to realize that it is physically impossible. Like a tennis ball, Earth is also spherical, yet maps still managed to represent Earth using a flat plane. Map projections make this possible. A **map projection** is the geometric transformation of the round earth onto a flat plane using mathematical equations. One cannot perform this transformation without a high degree of distortion.

Map projections change the shape and appearance of geospatial data, so every mapmaker ought to have a rudimentary understanding of them. Watch *Why all world maps are wrong* by Vox to learn more (**Figure 3.13**).

Figure 3.13: The video *Why all world maps are wrong* by Vox explains the concepts related to map projections in detail. URL: *https://youtu.be/kIID5FDi2JQ?list=PLSfHj8toBl19xfQnz3EcC23PGQ8FRoywj*

For more about map projections, visit ***Jason Davies' website***[2] to interact with different map projections and discover how they distort the geometry of Earth.

The Developable Surfaces

Map projections begin with a **reference ellipsoid** or an **authalic sphere**. The geometry of the authalic sphere makes it easier with which to work. For maps of global scale, it has little impact on accuracy. An ellipse is more mathematically sophisticated but offers a higher degree of accuracy when needed. Every latitude and longitude coordinate on earth is first plotted, point by point, onto a reference ellipsoid or authalic sphere. The result is known as a **generating globe**. The coordinate points are plotted point by point from the generating globe onto a developable surface using mathematical equations to create a map projection. A **developable surface** is any surface that either already exists as a 2-dimensional flat plane or a surface that can be cut and laid flat to become a 2-dimensional plane without distorting the original surface beyond altering its continuity. For example, one can cut a straight line along the side of a cone. While this breaks the continuity of the surface, one can lay it flat. One may also cut a cylinder and

2 https://www.jasondavies.com/maps/transition/

lay it flat in the same way. Watch *Map Projections of Earth*, a clip from the PBS show *Life by the Numbers* to learn more (**Figure 3.13**).

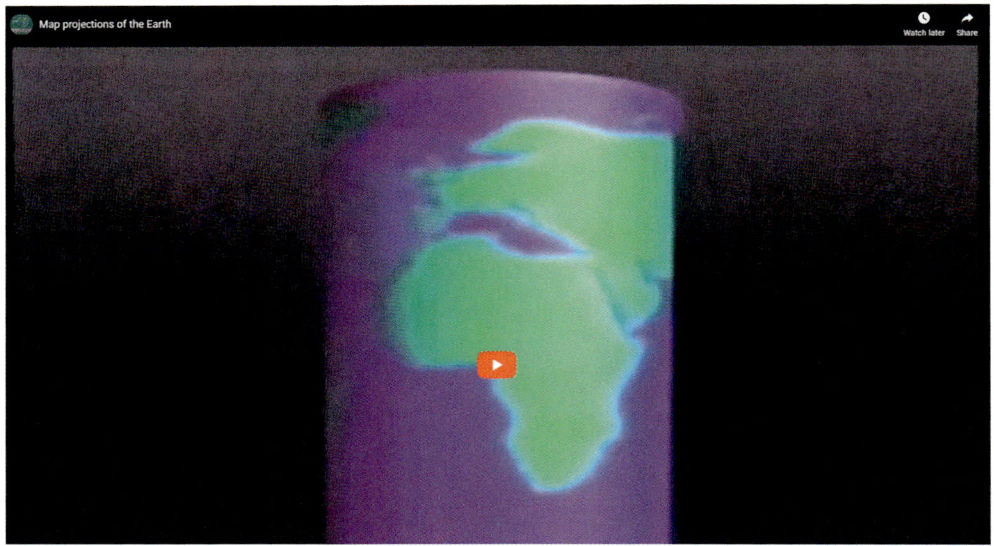

Figure 3.14: The video, *Map Projections of Earth*, a clip from the PBS show *Life by the Numbers* provides an overview of the concepts related to map projections. URL: *https://youtu.be/X4wgFSHZXBg*

This chapter examines three basic types of developable surfaces:

» Cones
» Cylinders
» Planes

A **conic projection surface** looks like a cone stuck on the earth, and works well for countries that have an east-west orientation, such as the United States (**Figure 3.15**). A **cylindrical projection surface** can be used to project the entire globe and works well for a narrow band around the center of the cylinder (**Figure 3.16**). A **planar projection surface** is well suited to circular regions, such as the North and South Poles. Cartographers also often call this surface **azimuthal** (**Figure 3.17**).

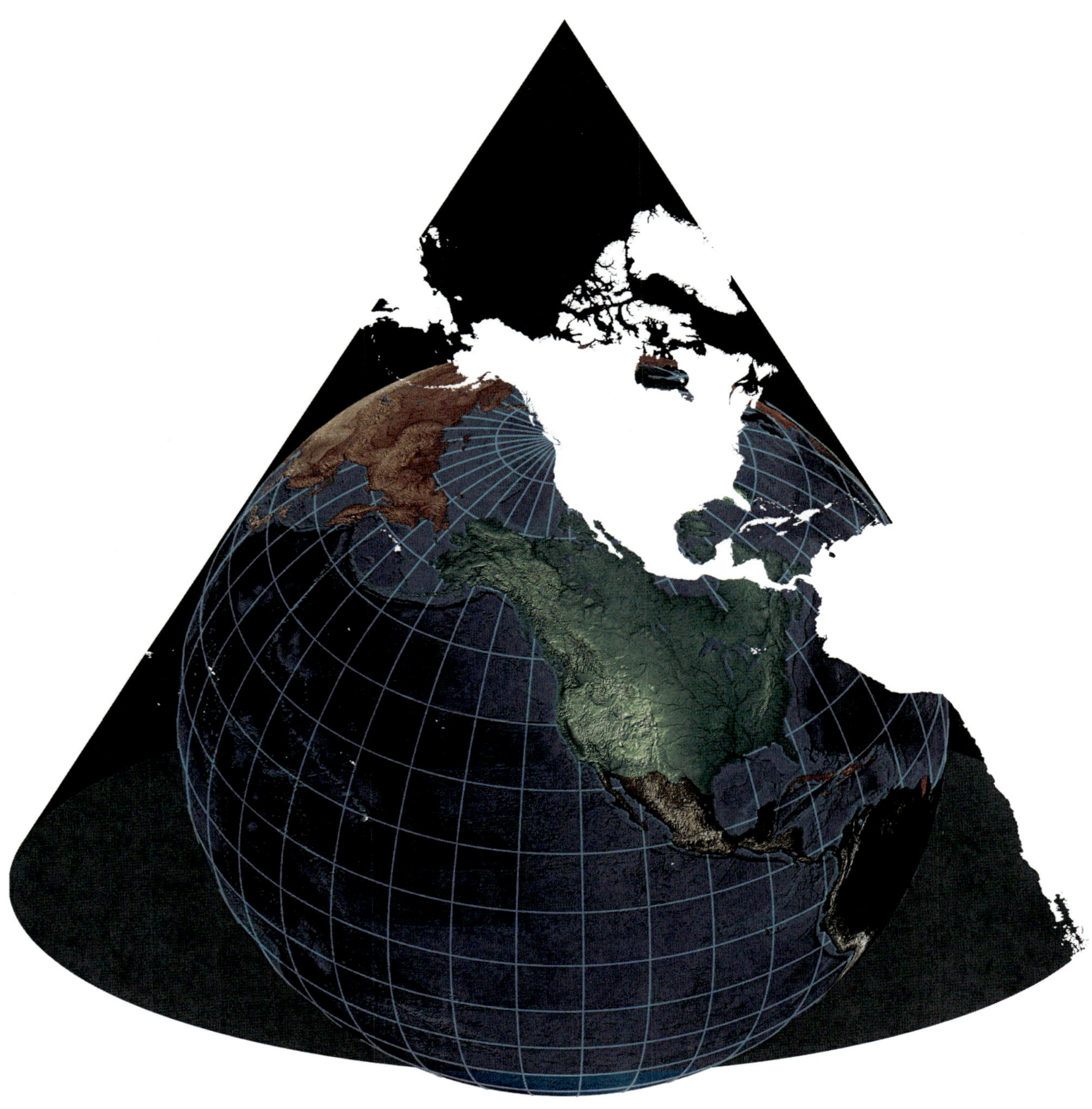

Figure 3.15: A conic map projection works well to preserve shape and area along a narrow east-west region. Distortion is most significant near the poles.

Figure 3.16: A cylindrical map projection works well for maintaining shape and direction. Areas become distorted away from the center.

Aspect and Case

The map projection **aspect** refers to the orientation of the developable surface relative to the generating globe (**Figure 3.18**). Each developable surface has three possible aspects relative to the generating globe: normal, transverse, and oblique. A **normal aspect**, sometimes called a **polar aspect**, is centered on the earth's imaginary axis that intersects the poles of the generating globe. In a cylindrical projection, this means the cylinder wraps around the equator.

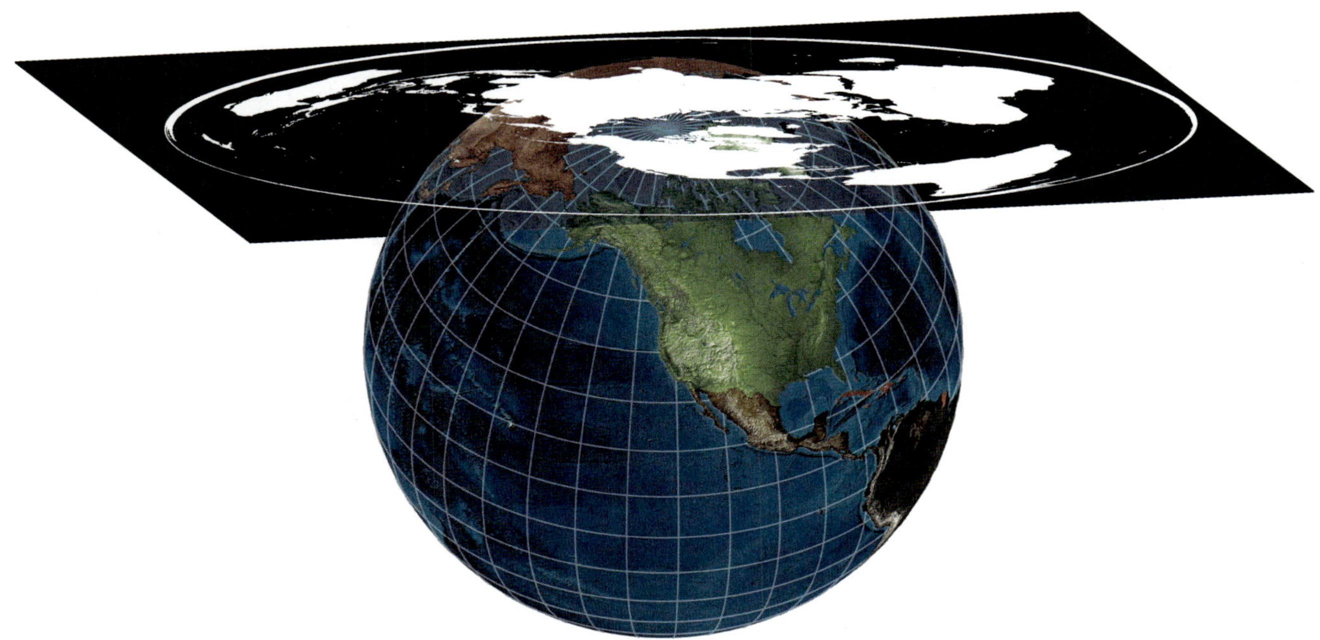

Figure 3.17: Planar map projections work well for circular regions, like the North Pole. Many also have specialized applications related to distance and direction.

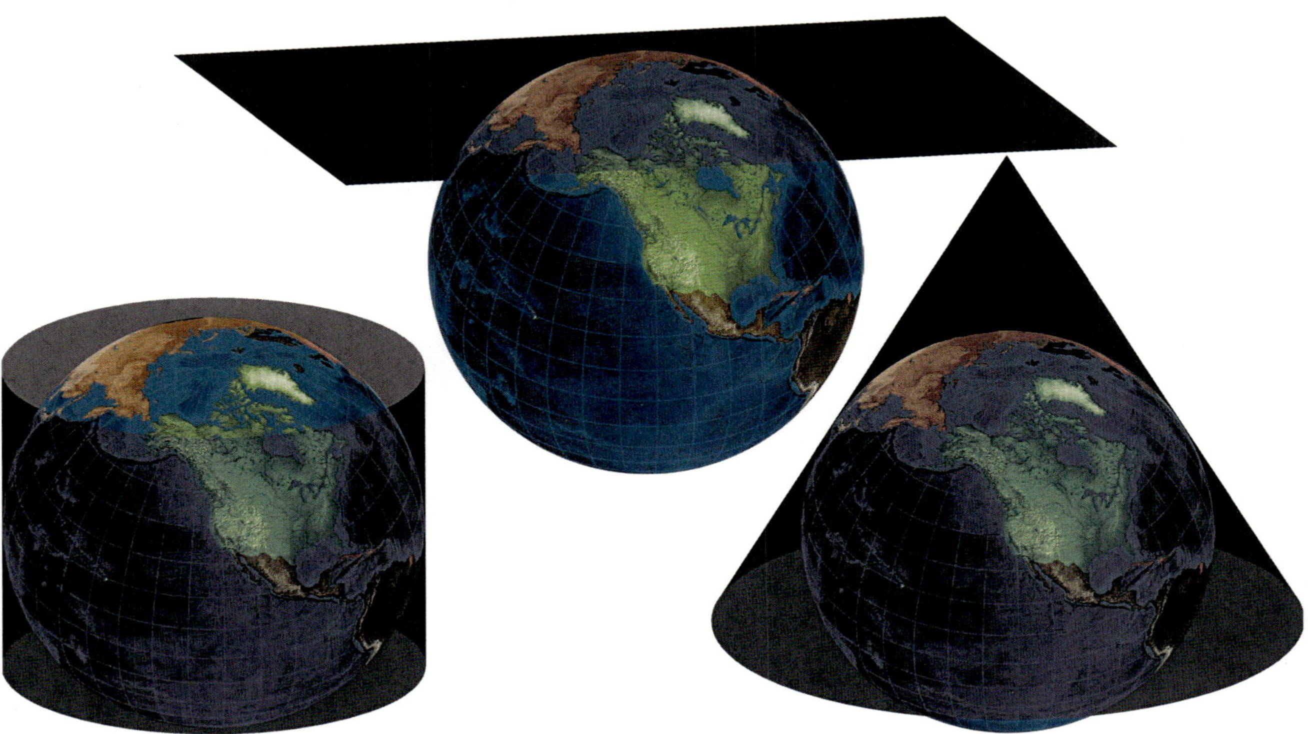

Figure 3.18: The normal aspect is used most commonly with large east-west extents, such as for a map of Asia. With planar projections, cartographers frequently use it for maps of the North and South Pole.

A **transverse aspect**, sometimes called an **equatorial aspect**, is centered on an imaginary axis that intersects the equator of the generating globe (**Figure 3.19**). In a cylindrical projection, this means the cylinder wraps around a pair of meridians. A transverse aspect is used most commonly with long north-south extents, such as for a map of Chile.

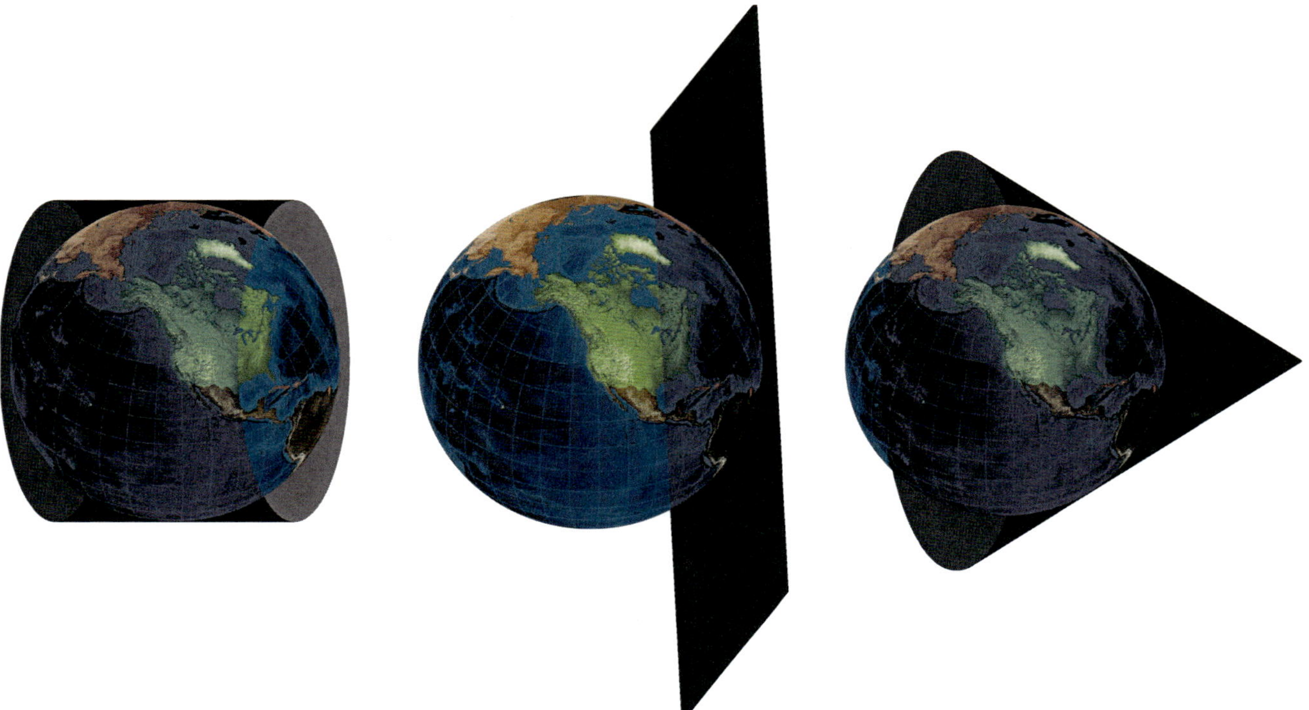

Figure 3.19: With cylindrical projections, cartographers use the transverse aspect for large north-south regions.

An **oblique aspect** refers to a developable surface oriented in any other direction other than normal and transverse (**Figure 3.20**). The map projection **case** refers to how the developable surface intersects or touches the generating globe.

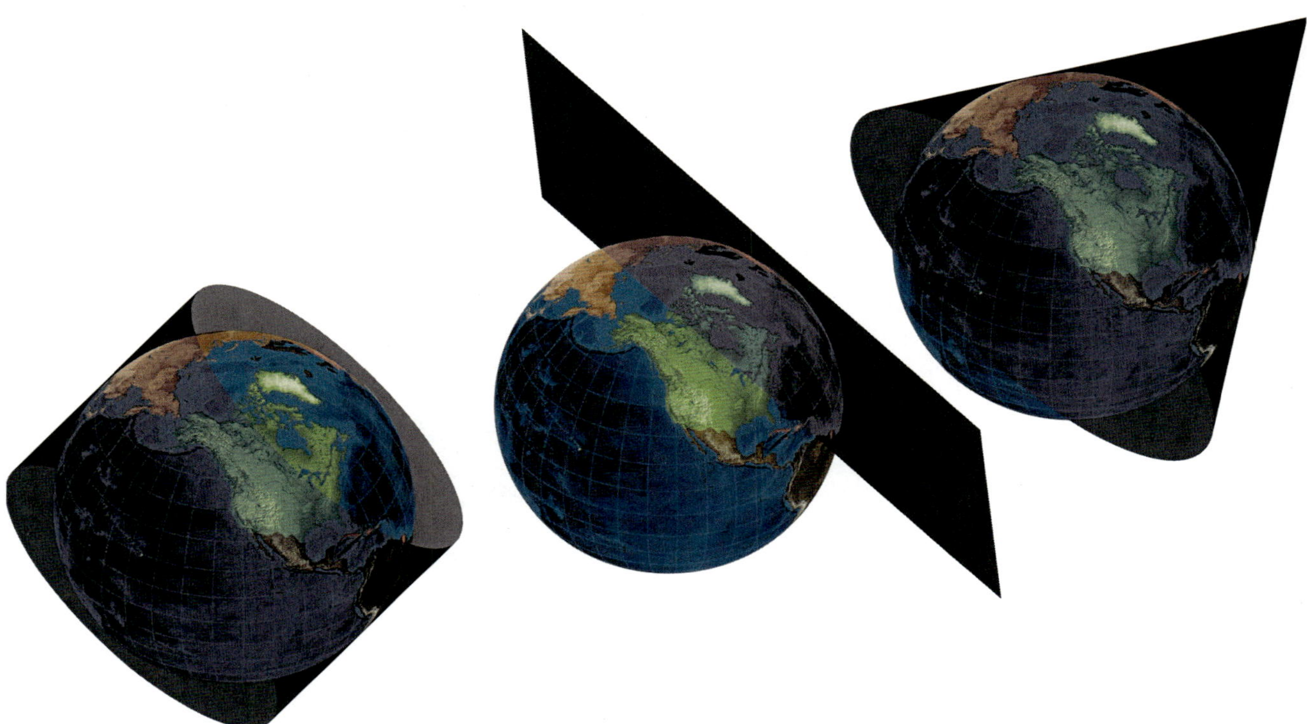

Figure 3.20: Cartographers often use an oblique aspect for specific purposes or locations, such as a navigation map for air travel.

Each developable surface has two possible cases, tangent, and secant. The place where the developable surface touches or intersects the generating globe is called a **line of tangency**. If the surface touches a generating globe along a single line of tangency, it is considered to have a **tangent case** (**Figure 3.21**).

In a cylindrical projection, this line of tangency will always be a great circle. Any time a line of tangency coincides with a parallel or a meridian it is called either a **standard parallel** or a **standard meridian**. Planar projections, also called azimuthal projections, are very different from cylindrical and conic in that they do not have standard parallels or standard meridians. In the tangent case, the planar surface touches the generating globe at a single point called a **standard point**.

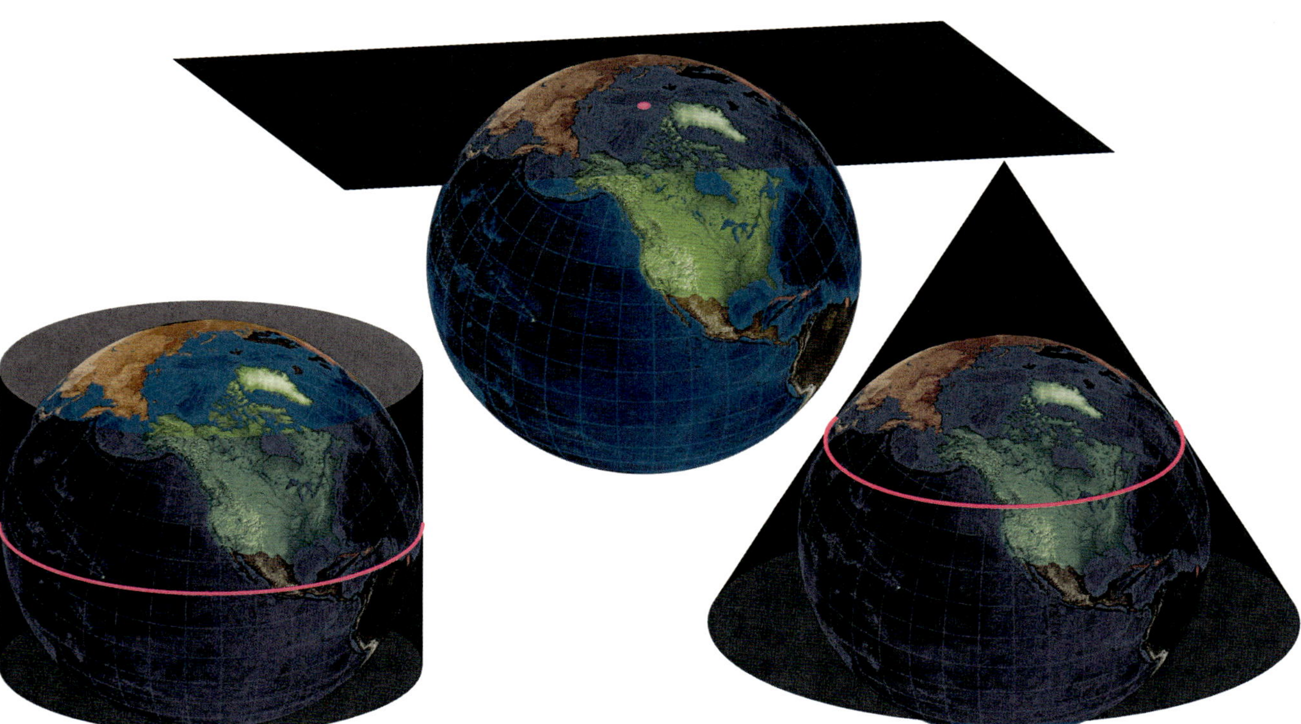

Figure 3.21: The pink lines represent places where the developable surface touches the generating globe. In the planar projection, a pink dot represents the point of contact.

A **secant case** occurs when a developable surface that cuts through the generating globe creating two lines of tangency (**Figure 3.22**). In planar projections, the planar surface cuts through the generating globe creating a circle called a **standard circle**.

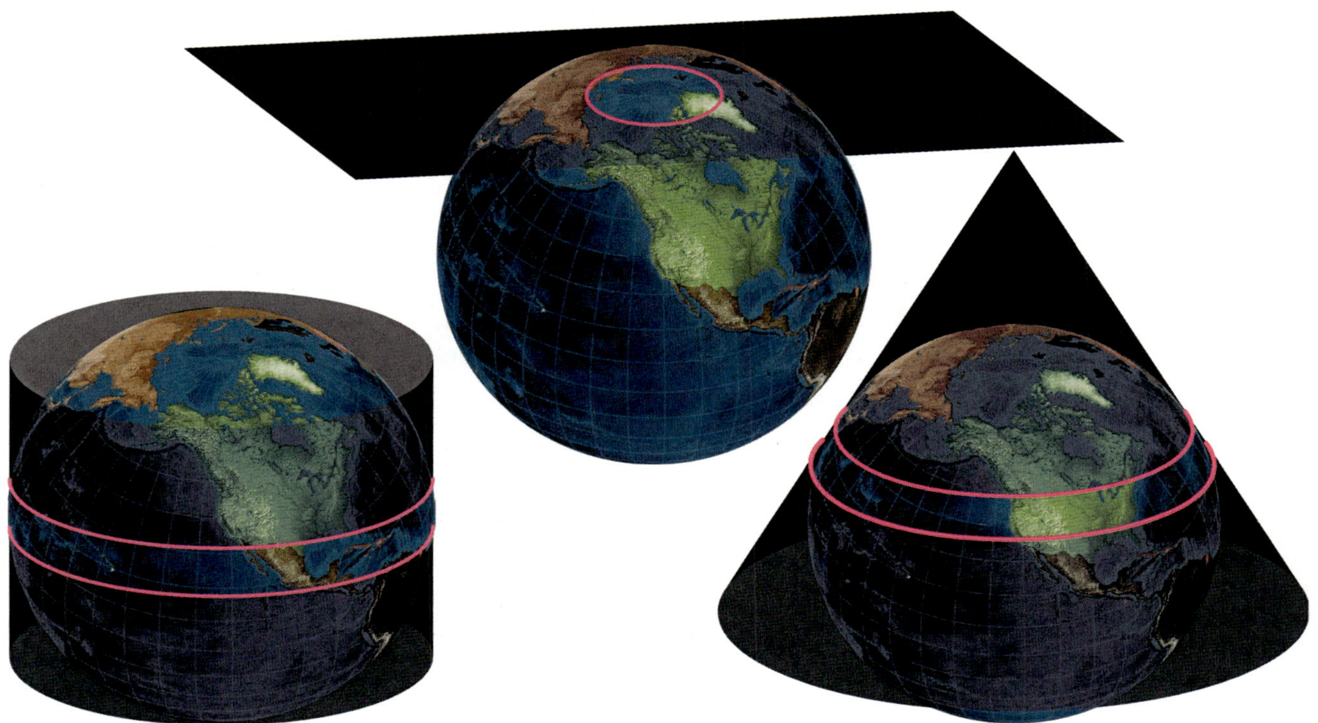

Figure 3.22: The pink lines represent places where the developable surface intersects the generating globe.

Preserved Properties

The number of possible map projections is infinite, and there are hundreds published over the years. To make sense of it all, cartographers use the developable surface, the aspect, and the preserved properties to classify map projections. **Preserved properties** are geometric characteristics maintained with a high degree of accuracy. As mentioned earlier, it is not possible to project Earth onto a flat rectangular map without a high degree of geometric distortion. Distortion occurs in one or more of the following properties:

» Area

» Shape

» Distance

» Direction

» Continuity

Area

In terms of preserved properties, **area** refers to the size of the continents or region of interest on the map as compared to the equivalent size on the globe. Cartographers use equal-area map projections when the map's purpose relates to the spatial extent or quantitative attributes such as population density. Globally, **equal-area projections** maintain the area at the expense of angles and shapes. However, some of these projections may also preserve shape at a local scale. For example, the Albers Equal-Area projection centered on the continental United States does well for preserving the shape of the country but does poorly for places farther away (**Figure 3.23**).

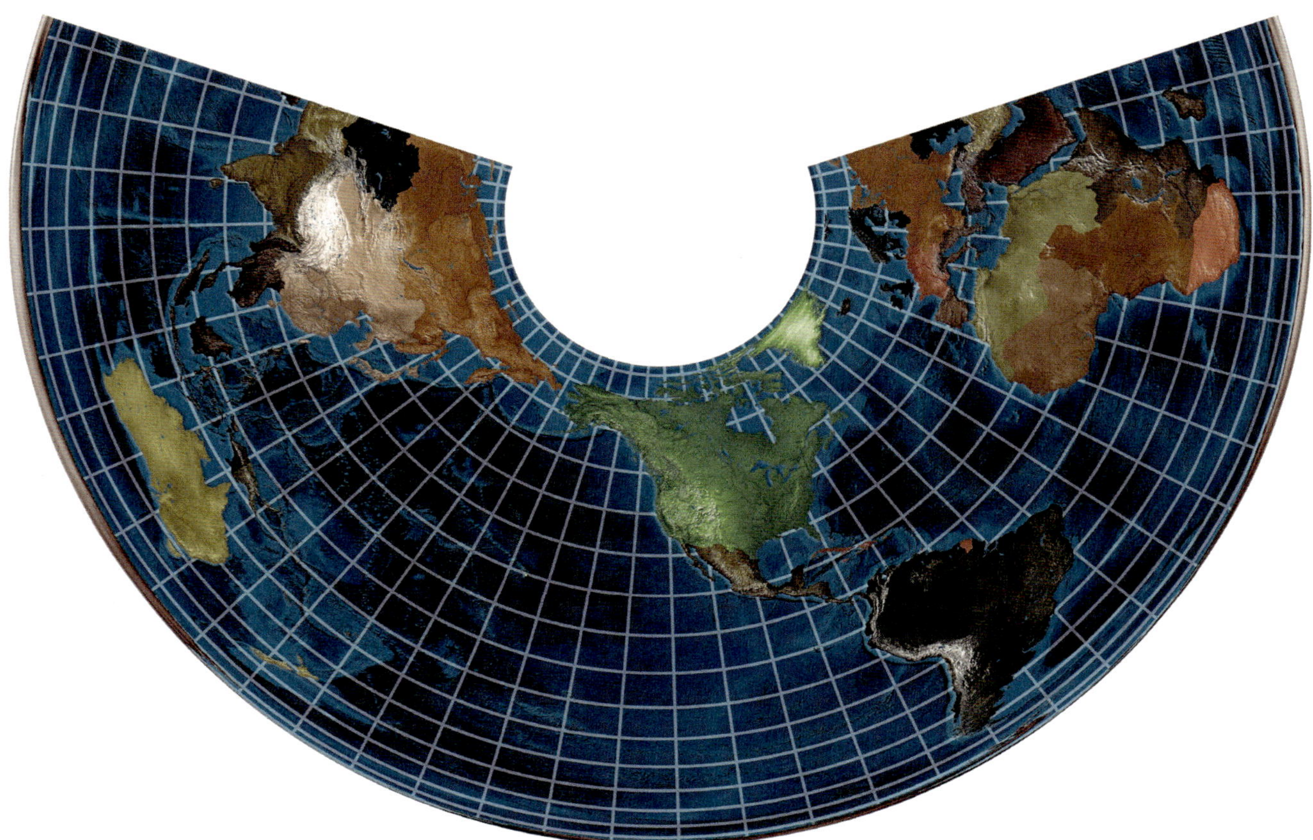

Figure 3.23: This map uses the Albers Equal-Area Projection. While areas are maintained all over the map, shape deteriorates near the poles.

Shape

Regarding preserved properties, **shape** refers to how the continents or region of interest conform to the related shapes on the globe. For this reason, cartographers often call map projections that preserve the shape of continents **conformal projections**. Conformal projections preserve shape and angle locally, often at the expense of size (**Figure 3.24**).

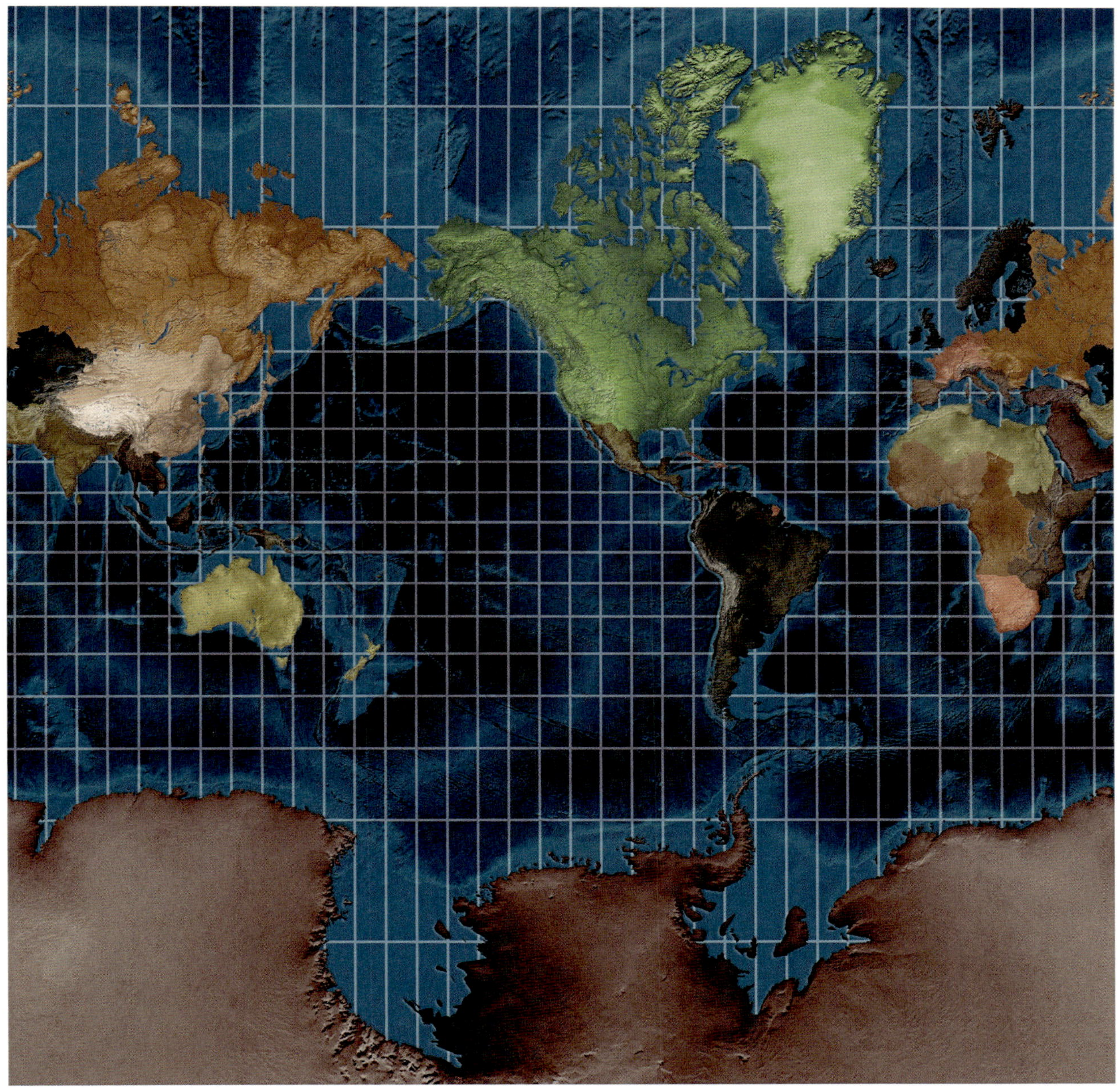

Figure 3.24: Navigators around the world used the Mercator projection for ocean voyages due to its simplicity and ease of use. Gerardus Mercator never intended this projection for use as a representation of Earth.

Distance

As it pertains to preserved properties, **distance** represents the accuracy of the map scale. **Map scale** describes the relationship between the distance represented on a map as compared to the equivalent distance on Earth. For example, one inch on a map may represent one mile on Earth. The relationship between the two defines the map scale. Chapter 4 covers the details related to map scale. Globally, it is challenging to preserve distance at a consistent scale throughout the map. However, some map projections specialize in preserving distances along one or more lines. Cartographers call maps that preserve distance, **equidistant projections** (**Figure 3.25**).

Direction

When speaking of preserved properties, direction refers to the accuracy and consistency of a bearing as compared to the same course on Earth. Cartographers often use the term **azimuth**, which is a direction, measured in degrees from 0° to 360°, clockwise from true north. They use **azimuthal projections** to preserve direction (**Figure 3.25**). Chapter 5 discusses the concepts related to direction in additional detail.

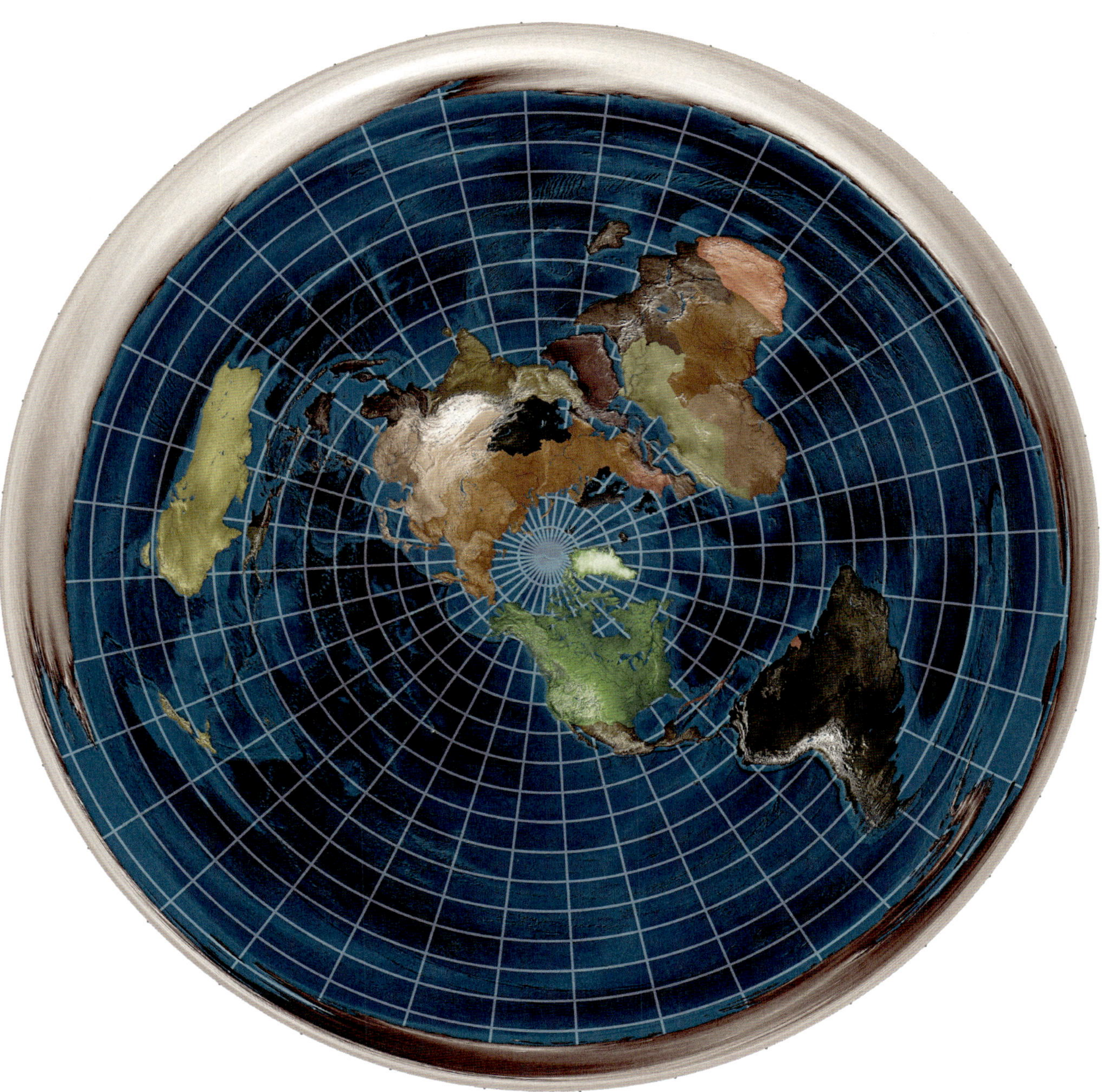

Figure 3.25: This world map uses an azimuthal equidistant projection to preserve direction and scale along any line originating from the standard point. In this instance, the standard point centers on the North Pole. However, cartographers may center this projection anywhere on the globe.

Continuity

Continuity is the preservation of proximity between features on a map. Some projections break continuity in favor of a high degree of accuracy over small regions of Earth (**Figure 3.26**).

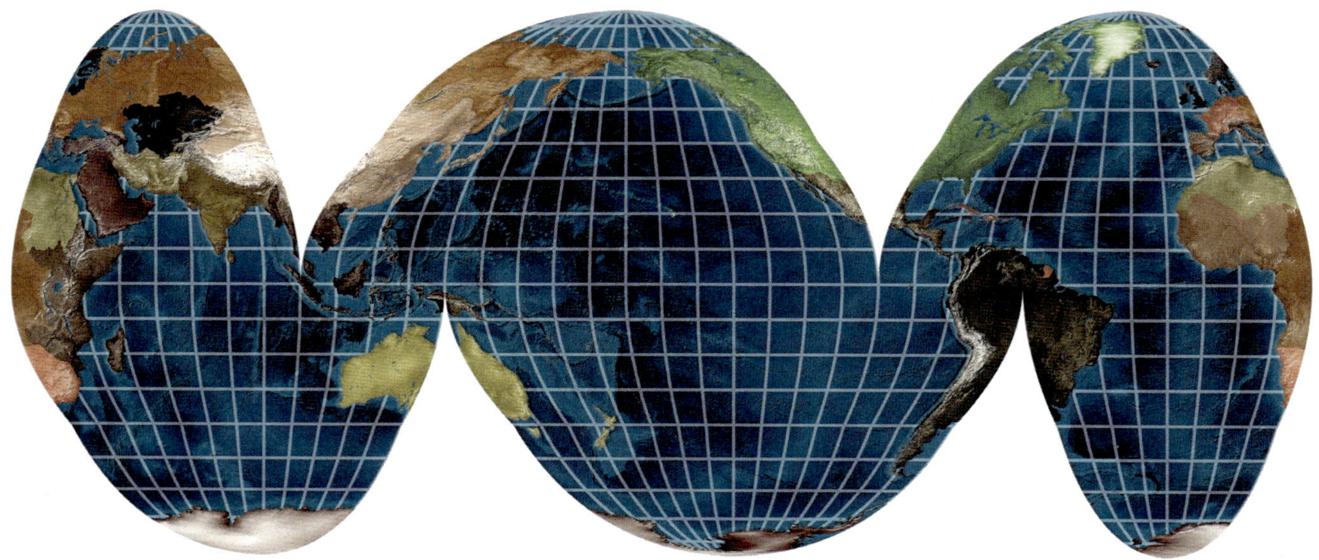

Figure 3.26: The Goode Homolosine projection sacrifices continuity to depict the world oceans with a high degree of accuracy.

There are no bad or good projections except for the one shown in **figure 3.26**. However, some have more practical applications than others. Cartographers carefully select projections to minimize distortion in the area mapped. At a local scale, it is possible to preserve multiple properties, but globally, it is difficult to do. The key to choosing a projection is to determine what sort of distortion needs minimization for the map's purpose. Some projections are specifically designed to preserve one or more of these properties, and also reduce interruptions in parts of the map significant to its purpose. Watch *What Does Earth Look Like?*, by Vsauce to learn more (**Figure 3.27**).

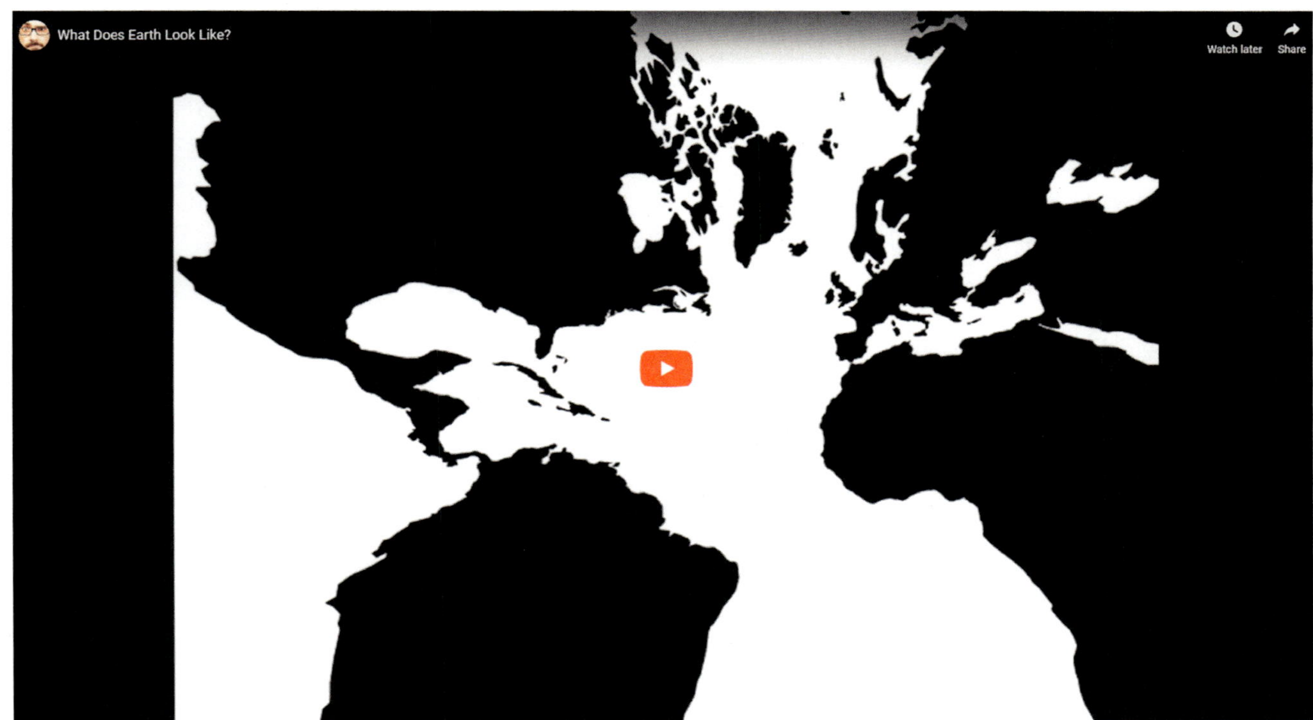

Figure 3.27: The video, *What Does Earth Look Like?*, by Vsauce discusses the concepts related to map projections in detail. URL: *https://youtu.be/2lR7s1Y6Zig*

Cylindrical Projections

The Mercator Projection

The Flemish cartographer Gerardus Mercator designed the most famous cylindrical map projection in 1569. The **Mercator projection** is a normal aspect tangent case cylindrical conformal projection developed for navigation (**Figure 3.28**). The meridians are equally spaced across the map while the spacing between parallels increases uniformly towards the poles. This method produces a map in which all **rhumb lines**, lines of constant compass direction, are straight. Due to extreme distortion at the poles, the Mercator projection is cut off at 80° north and 80° south. Navigators around the world once used the Mercator projection for ocean voyages due to its simplicity and ease of use. Any straight line they drew on the map could be used to set a compass bearing. As long as they followed that bearing, they would arrive at their destination. Rhumb lines, however, *do not* represent the shortest distance from one point to another on a sphere. A *great circle*, not a rhumb line, represents the shortest distance between two points on a sphere[3]. As a result, modern-day navigation uses great circles rather than rhumb lines, and for a period, use of the Mercator projection fell by the wayside.

The Mercator projection experienced a resurgence in the United States and around the world when it was frequently adopted as a wall map representing Earth in schools and public buildings. Gerardus Mercator never intended this projection for use as a representation of Earth. This utilization of the Mercator projection was a poor choice because it greatly exaggerates the size landmasses near the North and South Poles.

Since the average person did not have cartographic training, the Mercator projection left a false impression about the size of American and European countries when compared to countries around the equator. This perception and the use of the Mercator projection in classrooms led to an intense debate raised by the German historian and journalist Arno Peters. As a result of this debate, by the late 1970s, the Mercator projection again saw a decline in use.

3 The exception is when a great circle and a rhumb line coincide, as is the case with all great circle meridian pairs and the equator.

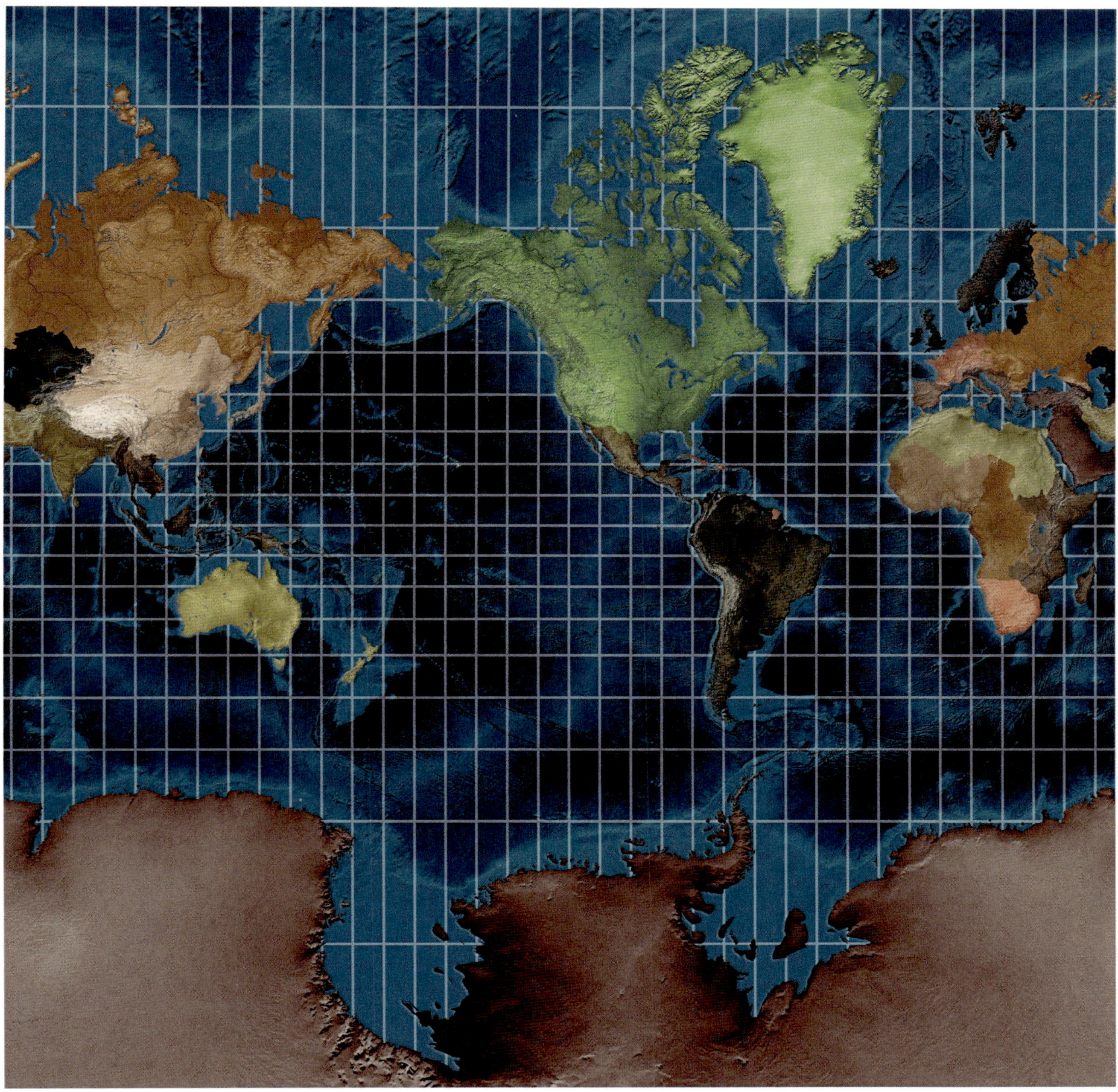

Figure 3.28: Navigators around the world used the Mercator projection for ocean voyages due to its simplicity and ease of use. Gerardus Mercator never intended this projection for use as a representation of Earth.

Today the Mercator is used with more frequency than ever before. The uniform square shape of the Mercator projection makes it highly suitable for the digital map tile system used by Google Maps and other online web maps. Though it is mostly the same projection, it is now commonly referred to as the **Web Mercator** (**Figure 3.29**).

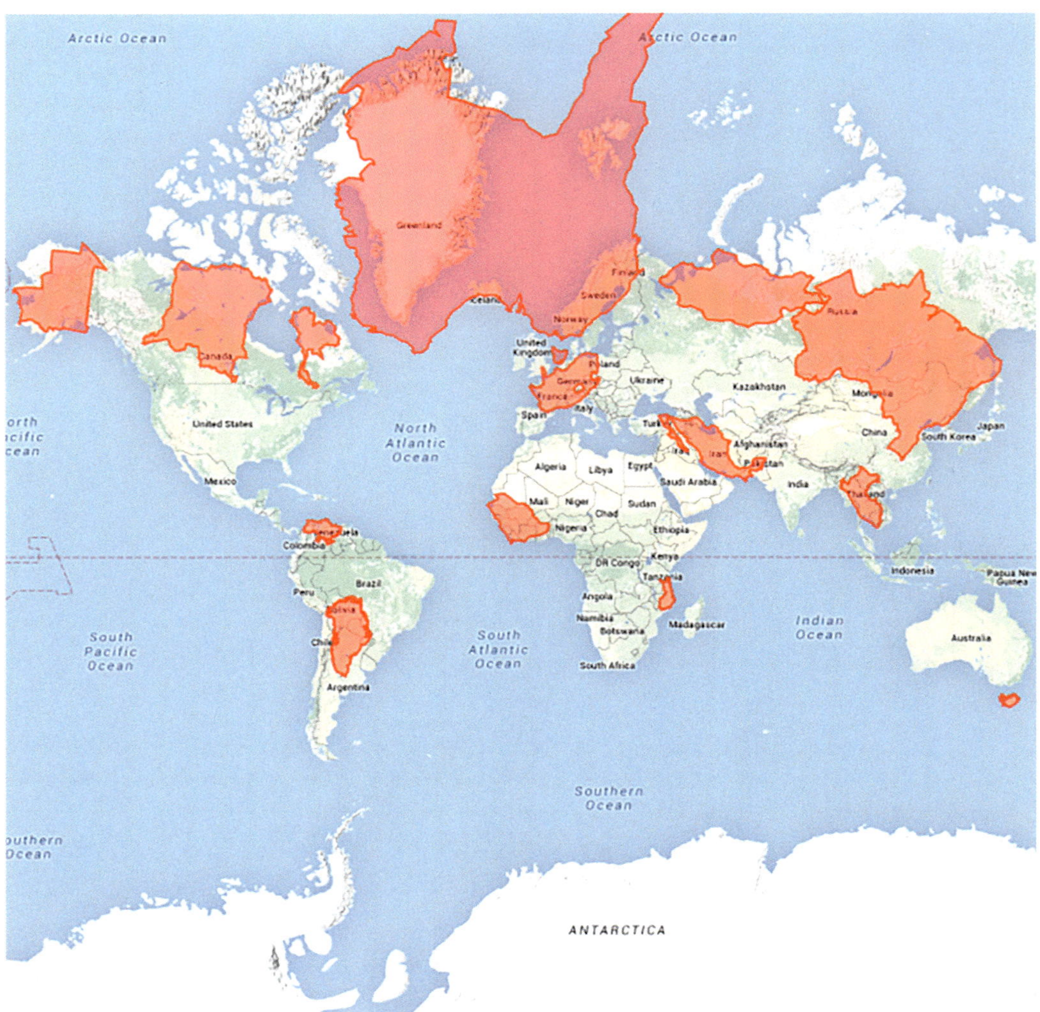

Figure 3.29: To observe how the Mercator map projection can distort areas on a map, play the Mercator puzzle on Google maps. URL: *https://gmaps-samples.googlecode.com/svn/trunk/poly/puzzledrag.html*

Gall's Orthographic

In the late 1800s, a Scottish clergyman and astronomer named James Gall published equations for the **Gall's Orthographic projection**, a normal aspect secant case cylindrical equal-area projection with standard parallels at 45° north and 45° south. This projection was never considered to be of much use. Though it maintained the area of continents, it significantly distorted the shape (**Figure 3.30**).

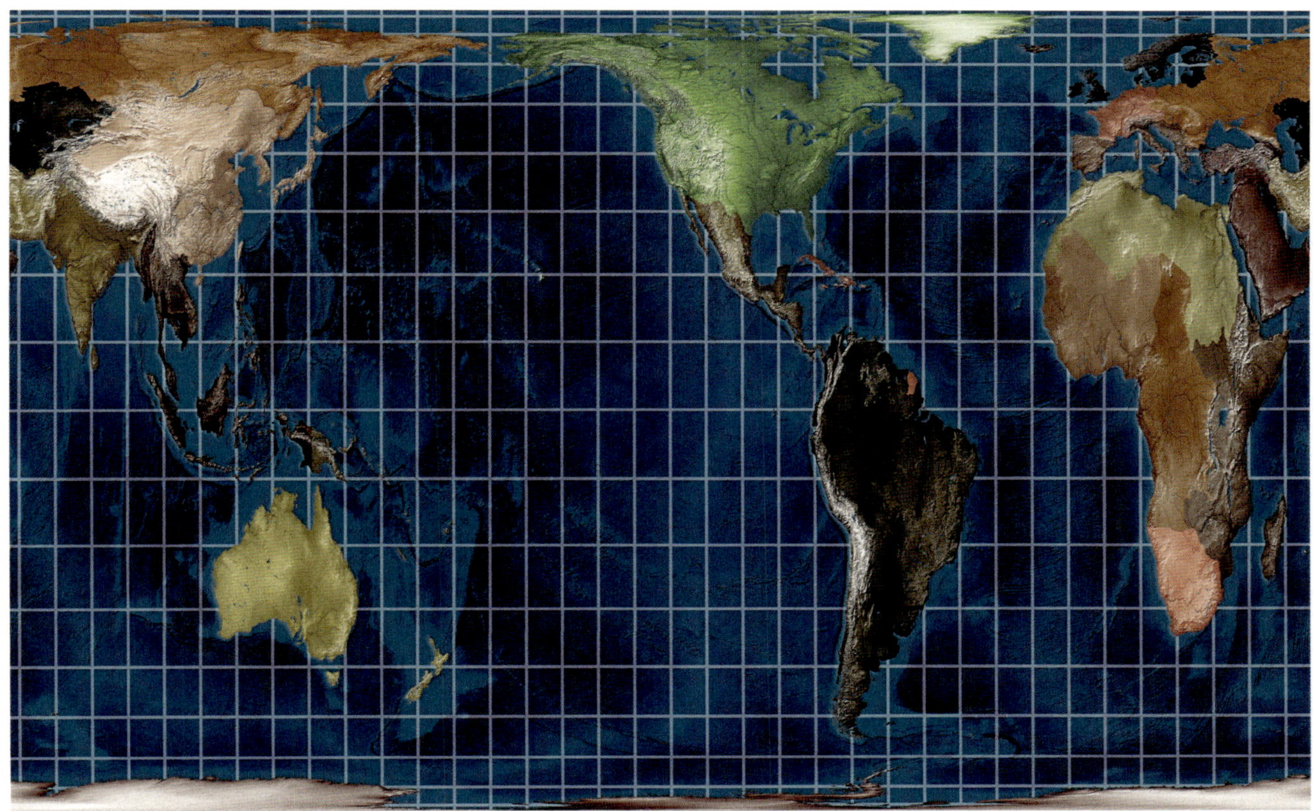

Figure 3.30: Arno Peters promoted this projection based on social equality. However, cartographers agree that the Gall-Orthographic projection is a poor choice for representation of Earth due to the extreme distortion of the shape of continents.

In 1973, Arno Peters presented the same projection published by James Gall as his new invention. He renamed it for himself calling it the Peters projection. Arno Peters was not a cartographer. He was a historian and journalist who presented the idea that his projection was the only non-racist world map. He asserted that the Mercator projection created a racial bias by exaggerating the size of European and North American countries. This assertion generated intense debate, especially among the cartographically literate. Arno Peters swayed public opinion because he presented the map as a political correctness issue when racial tensions, at least in the United States, were very high. Many agreed that people equate size with importance and that the Mercator projection created a bias regarding social equality (**Figure 3.31**).

Figure 3.31: Watch the clip from the TV show West Wing. In this clip, the Cartographers for Social Equality are lobbying the White House to make it mandatory for every school in America to teach geography using the Gall-Peters Projection instead of the Mercator. This clip highlights some of the arguments made by Arno Peters. *URL: https://youtu.be/vVX-PrBRtTY*

Cartographers countered that the Mercator projection was intended for navigation and not as a representation of the world and therefore was not racist. As a representation of Earth, what many now refer to as the Gall-Peters projection, was also a poor choice due to the extreme distortion of the shape of continents. A prominent cartographer named Arthur Robinson described the Gall-Peters projection as "somewhat reminiscent of wet, ragged long winter underwear hung out to dry on the Arctic Circle." There were, and are, better choices available for equal-area projections, including the widely celebrated **Equal Earth projection** created in 2018 by Bojan Šavrič, Tom Patterson, and Bernhard Jenny (**Figure 3.32**).

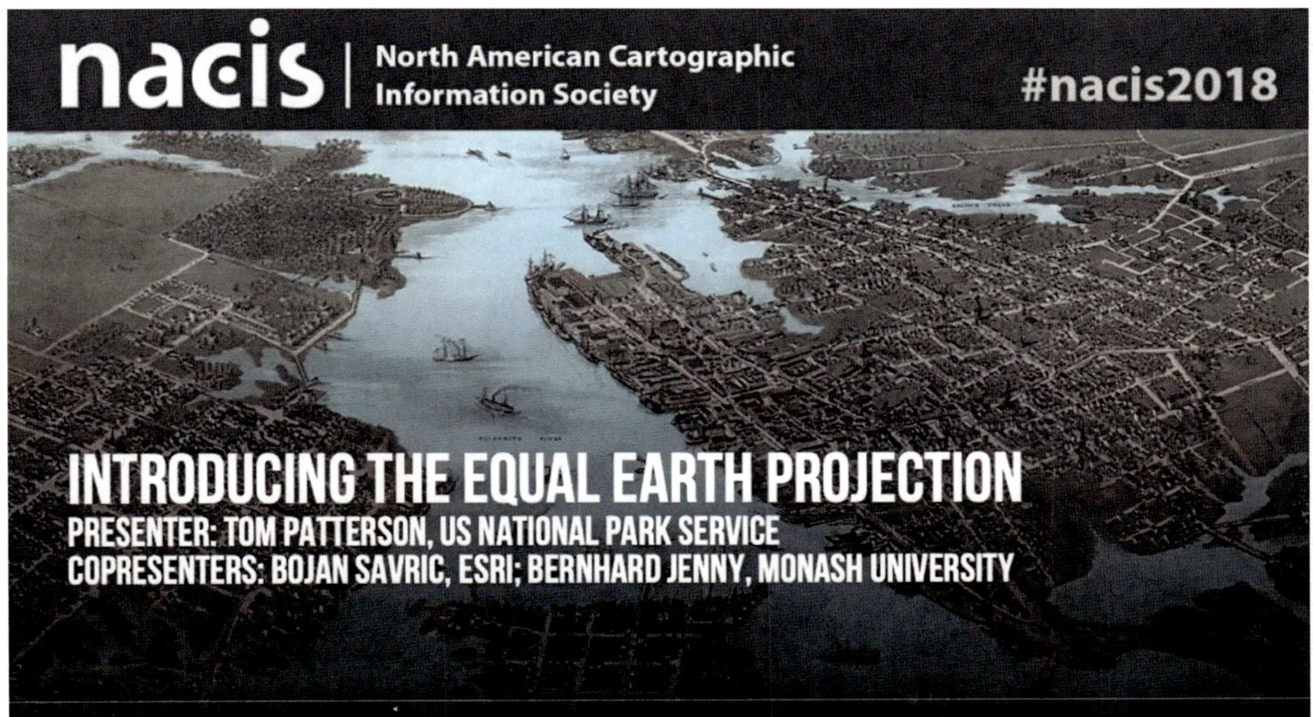

Figure 3.32: Watch this presentation published by the North American Cartographic Information Society. Presenter: Tom Patterson, US National Park Service. Copresenters: Bojan Šavrič, Esri; Bernhard Jenny, Monash University. URL: *https://youtu.be/m5Te3JZfPDM*

Transverse Mercator

In 1772, the mathematician and astronomer Johann Heinrich Lambert published *Beyträge zum Gebrauche der Mathematik und deren Anwendung*, translated and republished by Waldo R. Tobler in 2011 under the title *Notes and Comments on the Composition of Terrestrial and Celestial Maps*. In this publication, Lambert presented several map projections widely used today. One of these was the **Transverse Mercator**, a transverse aspect cylindrical projection, used with both tangent and secant case (**Figure 3.33**). Lambert rotated the Mercator by 90° producing lines of tangency that coincided with a meridian and its antipodal meridian. An **antipodal meridian**, or antimeridian, is any meridian 180° (opposite) of any other meridian on a sphere. While no longer used for navigation, this projection created a map with minimal distortion of area, distance, and shape along any meridian. The preserved properties made the Transverse Mercator particularly well suited for highly accurate maps depicting regions or countries with a north-south orientation. The ability to create consistent, narrow strips around the globe makes this projection well suited for a map series depicting different areas with the same projection. For this reason, the Transverse Mercator is one of the most widely used map projections in standardized spatial reference systems. Chapter 4 discusses the use of the Transverse Mercator in global and regional spatial reference systems.

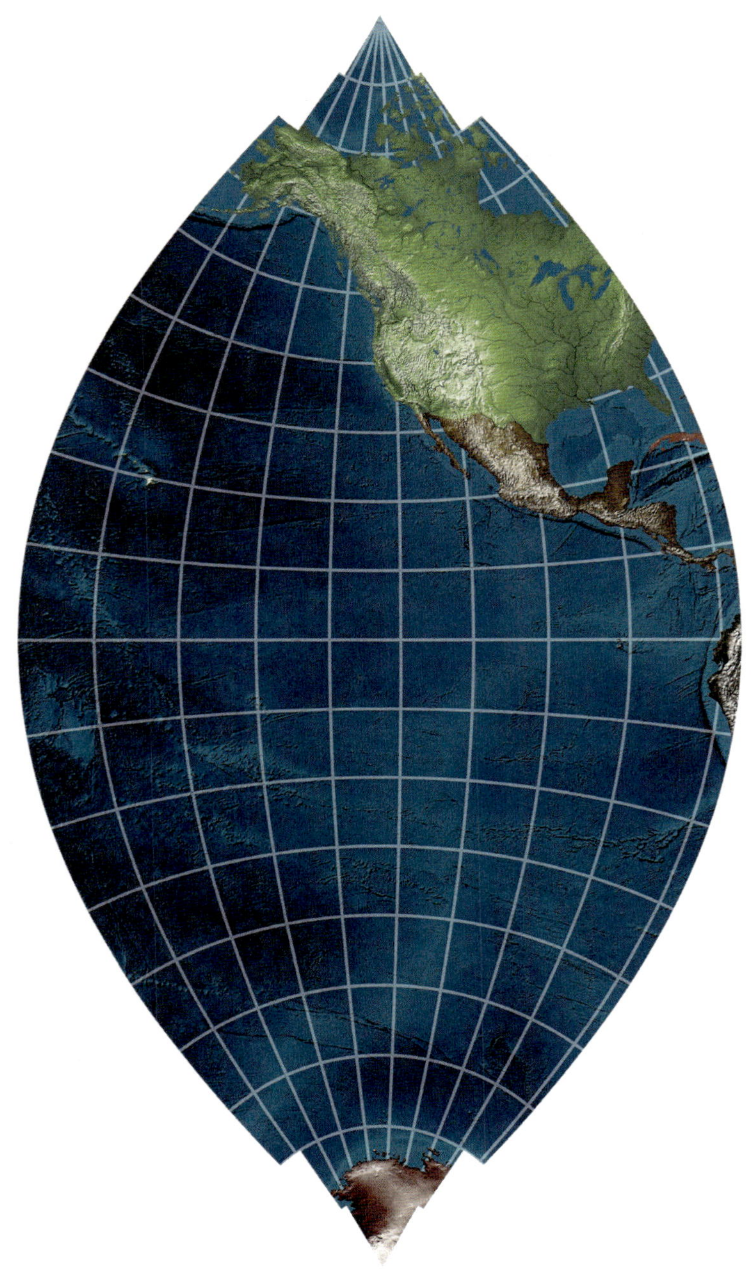

Figure 3.33: The Transverse Mercator is one of the most widely used map projections due to its use in many standardized spatial reference systems.

Conic Projections

Unlike cylindrical projections, conic projections are not well suited for world maps. However, conic projections with a polar aspect are superior for maps of mid-latitude regions with an east-west orientation, such as Europe or the United States, as they preserve distance along the standard parallels.

Lambert Conformal Conic

Published by Johann Heinrich Lambert in 1772, the **Lambert Conformal Conic projection** is a normal aspect secant case conic conformal projection. As the name suggests, this projection preserves shape using two standard parallels. Parallels appear as curved arcs, closer together near the center of the map. Meridians are equally spaced, and cross parallels at right angles, preserving shape. The Lambert Conformal Conic projection is well suited for mid-latitude regions or countries with an east-west orientation. For this reason, cartographers often use it for the coterminous United States (**Figure 3.34**).

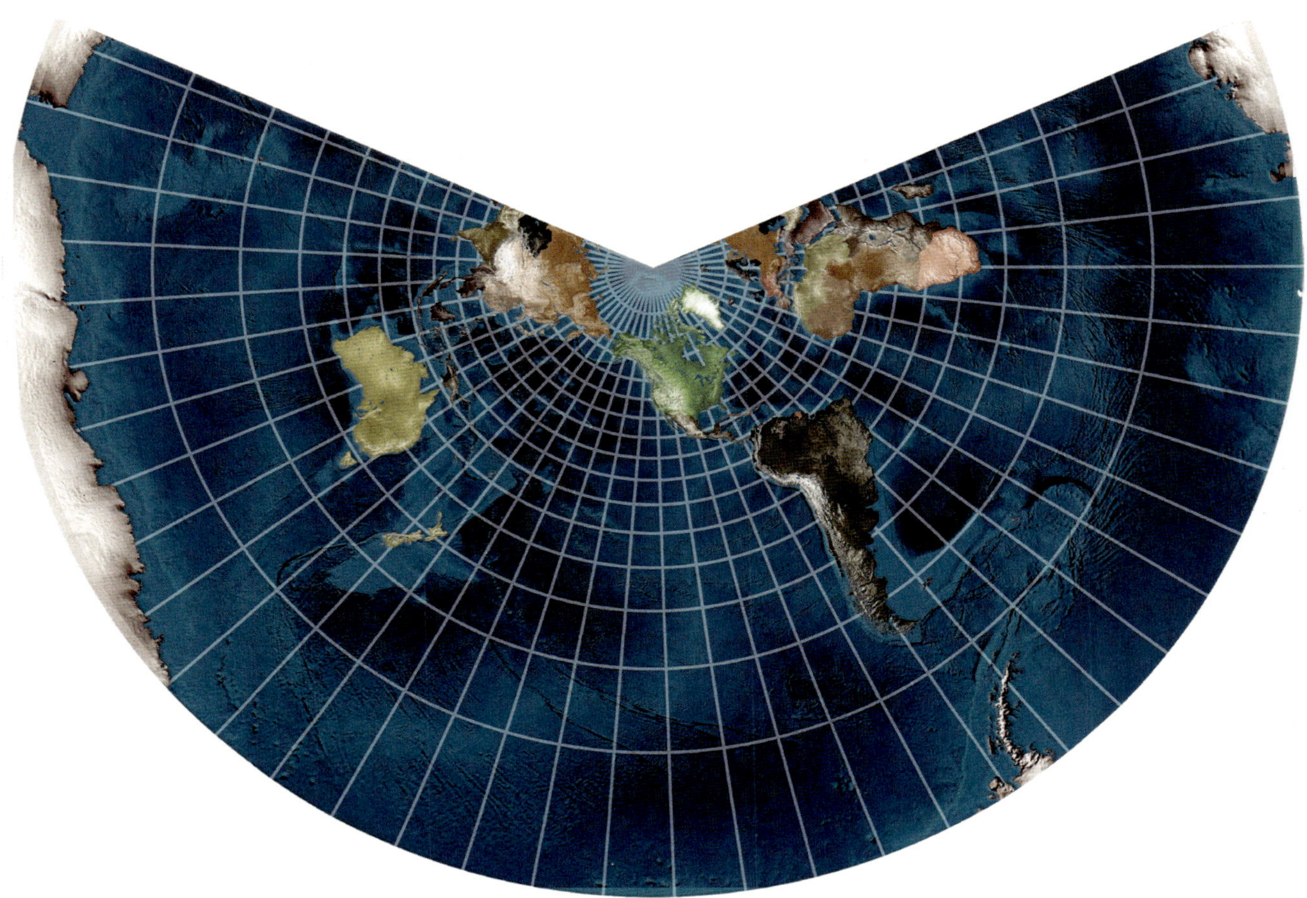

Figure 3.34: The Lambert Conformal Conic projection works well for preserving distance and direction near or between standard parallels, which also makes it useful for navigation over localized regions.

Albers Equal Area Conic

Published by Heinrich Christian Albers in 1805, the **Albers Equal Area Conic projection** is another normal aspect conic projection, and one of the most frequently used equal-area projections for the coterminous United States (**Figure 3.35**). Similar in composition and appearance to Lambert Conformal Conic, parallels are laid out as a series of concentric arcs and meridians are equally spaced and cross parallels at right angles.

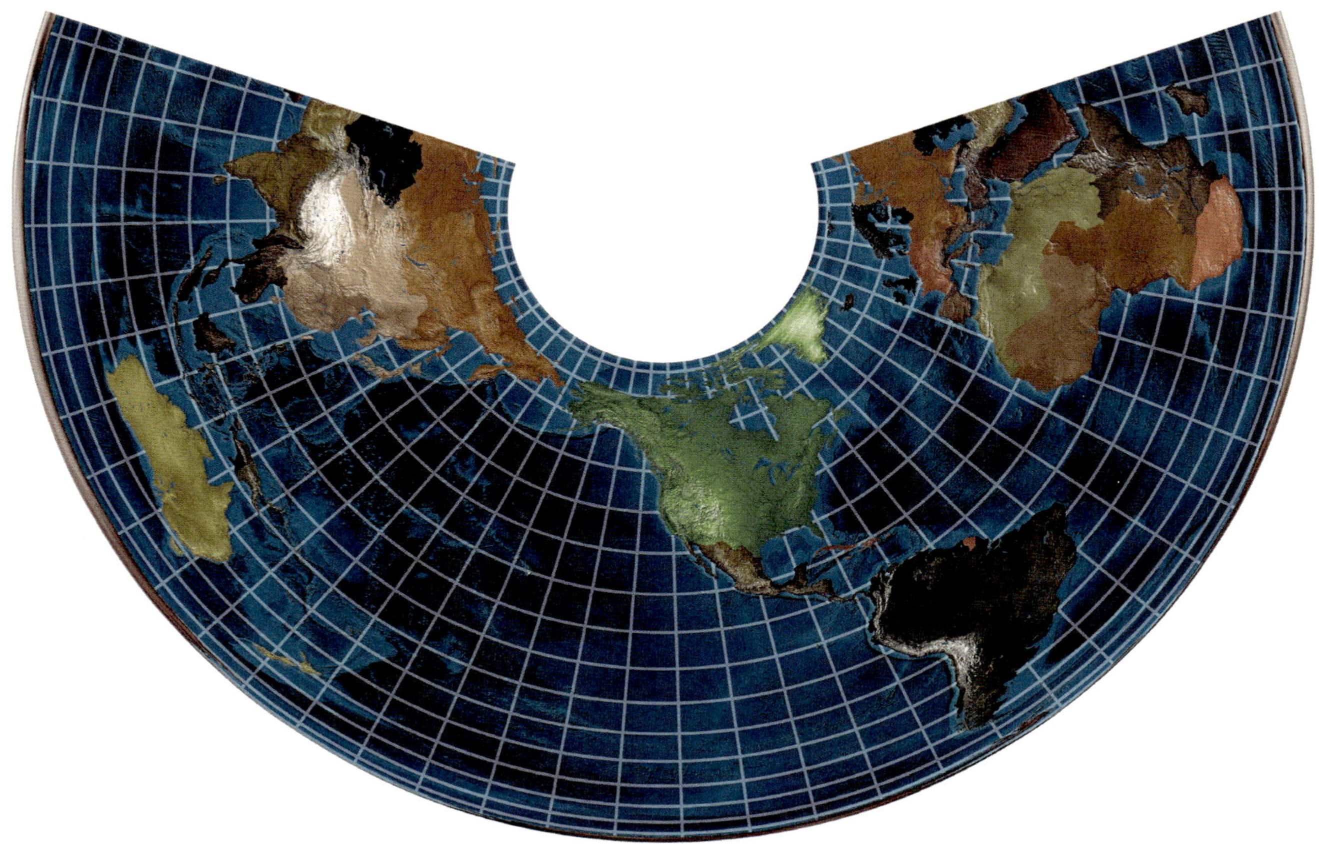

Figure 3.35: The Albers Equal-Area Conic projection is particularly useful for comparing statistical data from one region to another, such as U.S. Census data, where preserving relative area is essential.

Planar Projections

Planar projections are very useful for mapping regions with a circular extent, such as the North or South pole, where there is no dominant orientation in any one direction. Cartographers often use the following five planar projections:

» Orthographic
» Stereographic
» Gnomonic
» Lambert Azimuthal Equal-Area
» Azimuthal Equidistant

True-perspective Projections

The first three on the list of planar projections are known as **true-perspective projections**. These projections have origins dating back over 2000 years. A **perspective** is a common point or direction from which the geometric transformation occurs. To visualize this, imagine a globe made of glass with all of the continents painted on the surface. Now imagine shining a light from inside or through the globe onto a flat projection screen (**Figures 3.36, 3.38, and 3.40**). The light would cast shadows of the continents on the screen, changing larger or smaller depending on where the light was. Conceptually, the light source in this example is the perspective.

Continuing the conceptual light source and glass globe analogy, imagine one shines a light toward the globe while standing at a distance of six meters or more[4]. In this scenario, the light rays shining through the globe would be parallel (**Figure 3.36**). A similar concept forms the basis for the orthographic projection. The perspective is at an infinite distance from the generating globe.

Figure 3.36: The perspective for the orthographic projection is conceptually at an infinite distance from the generating globe. The result is a map that looks similar to Earth as seen from a great distance. An image of the Orthographic projection.

4 Light rays become parallel at a distance of about six meters.

The result is a spherical map that looks similar to how Earth would look if viewed from a great distance (**Figure 3.37**).

Figure 3.37: The orthographic projection is limited in that it can only display one hemisphere at a time, and is neither conformal nor equal area. It is best used for showing the spherical shape of Earth.

Imagine a tangent planar surface touching the glass globe at one point. The point of perspective, our conceptual light source for the stereographic projection, sits opposite of the point of tangency for the planar surface. So, for example, if the point of tangency were at the North Pole, the point of perspective would be at the South Pole (**Figure 3.38**).

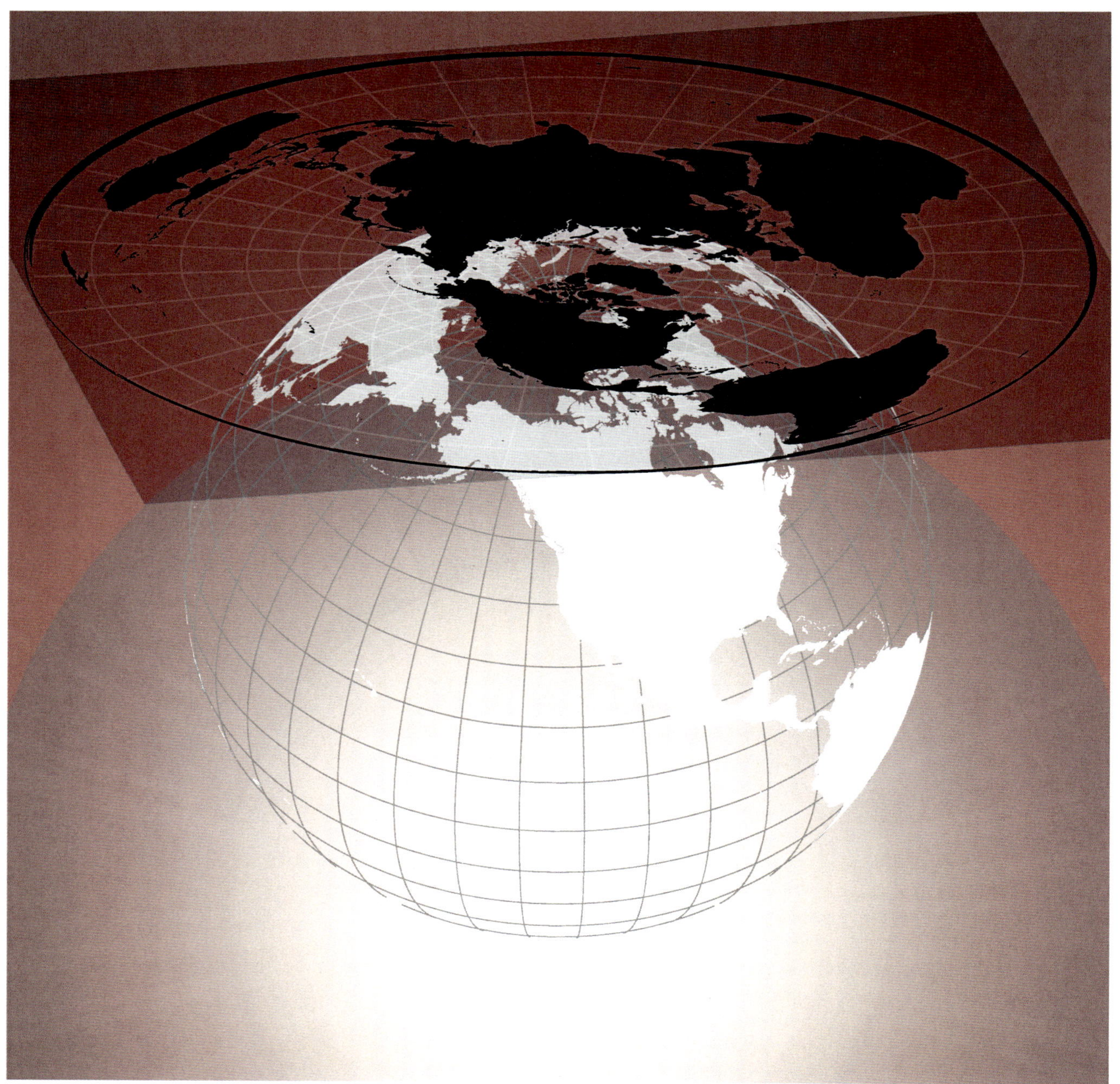

Figure 3.38: In this example, the light source sits at the South Pole while the plane touches the globe at the North Pole.

The stereographic projection is conformal and commonly used for maps of polar areas and star maps (**Figure 3.39**).

Figure 3.39: Cartographers may center the stereographic projection anywhere on the globe. In this image, the projection centers on Northern California.

For the **gnomonic projection**, the point of perspective is located precisely in the center of the generating globe (**Figure 3.40**).

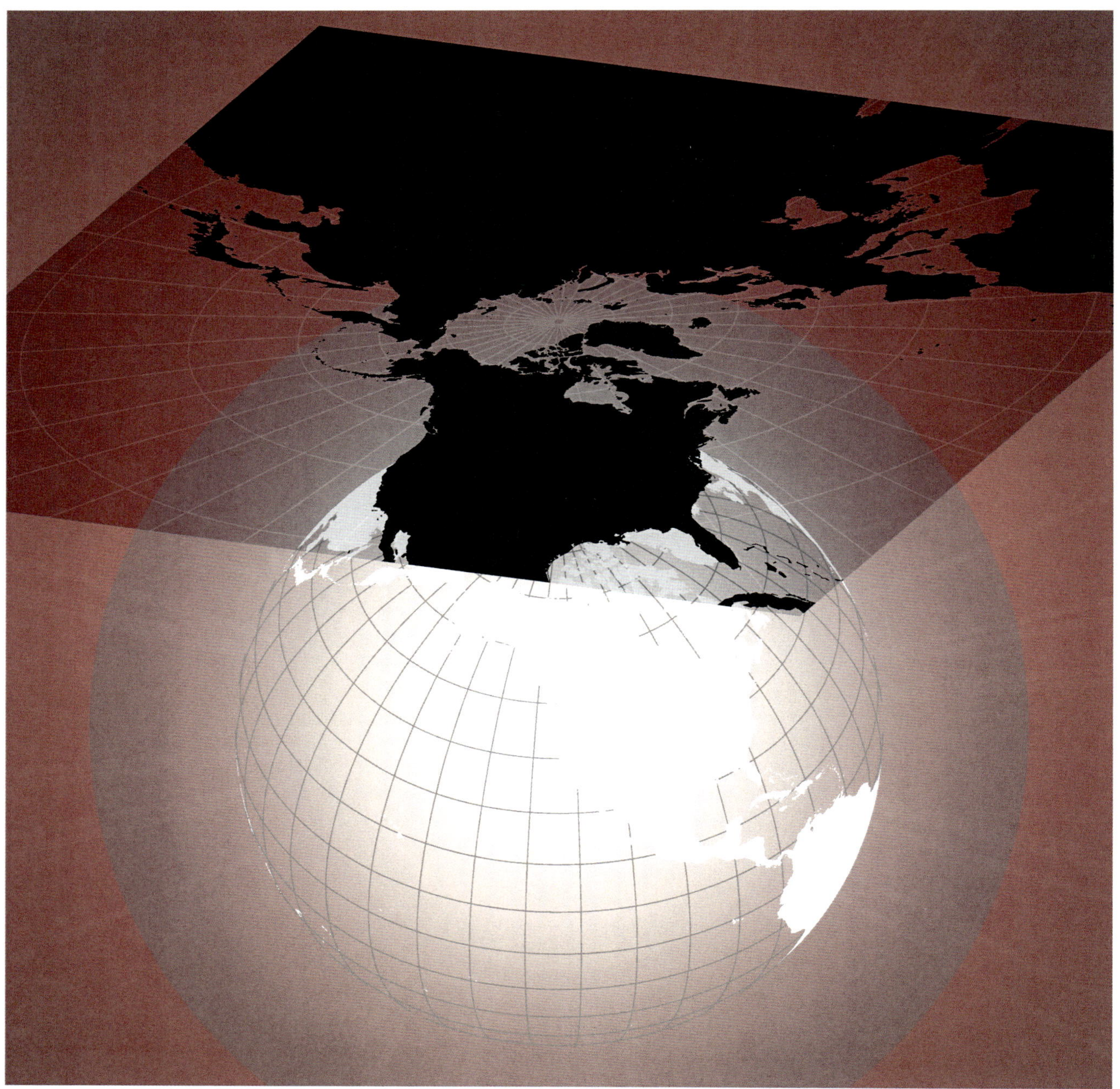

Figure 3.40: Because the point of perspective is from the center of the globe, the gnomonic projection cannot show a full hemisphere.

The gnomonic projection is neither conformal nor equal area, and scale is highly distorted (**Figure 3.41**).

Figure 3.41: One useful feature of the gnomonic projection is that all straight lines are great circles, which makes it useful for plotting the shortest path from one point to another.

Lambert Azimuthal Equal-Area

The last two planar projections are not true-perspective projections. Also published by Johann Heinrich Lambert in 1772, the **Lambert Azimuthal Equal-Area projection** is a tangent case planar projection. This projection preserves area and the direction from the center of the map (**Figure 3.42**).

Figure 3.42: The Lambert Azimuthal Equal-Area projection works well to show areas with a circular extent or to show different equal-area hemispherical perspectives of Earth using all three aspects.

Azimuthal Equidistant Projection

The Azimuthal Equidistant projection is a tangent case planar projection commonly used on maps centered on the poles (**Figure 3.43**). It has a useful feature in that all distances measured from the point of tangency are true to scale and distance.

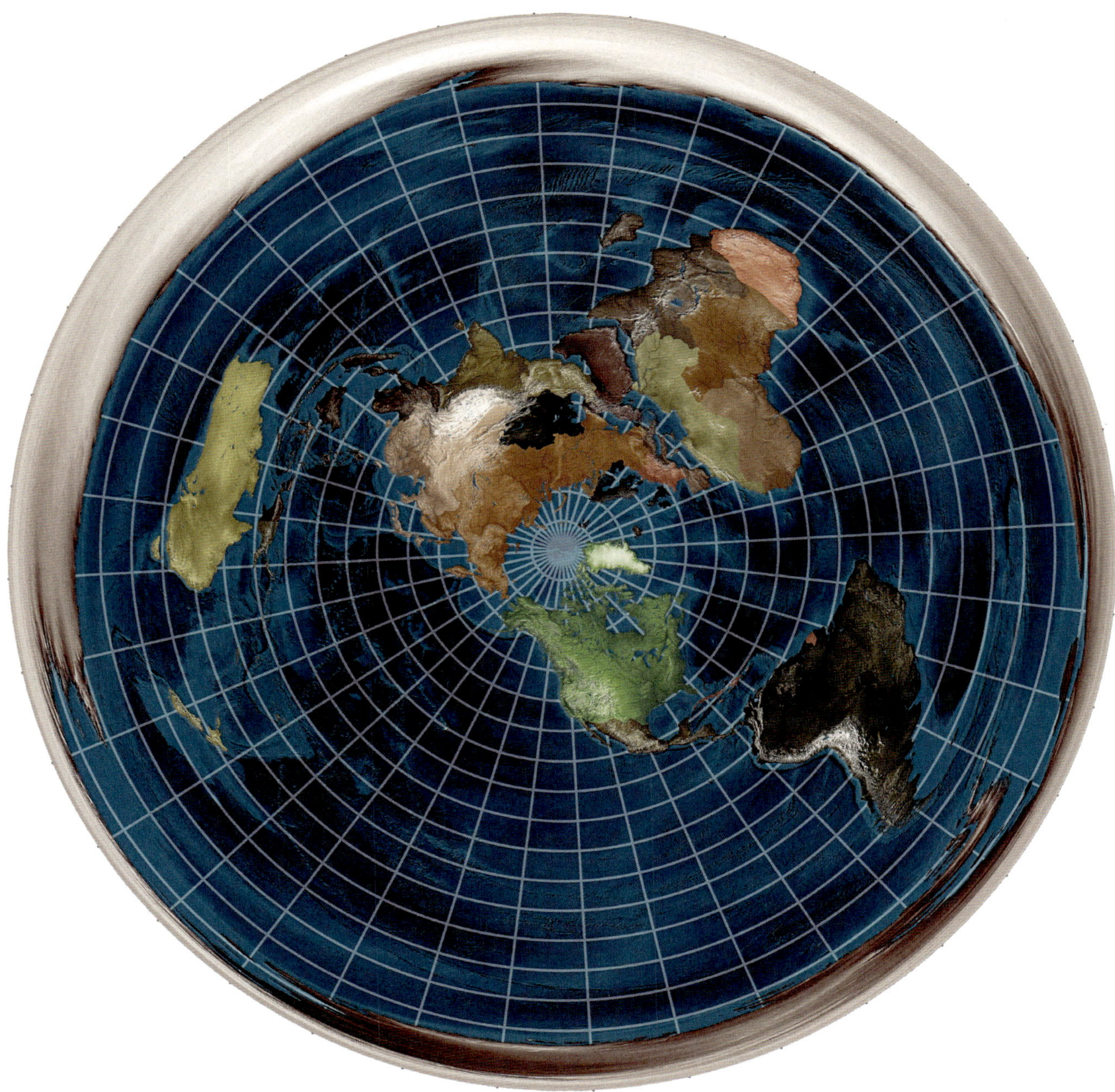

Figure 3.43: Cartographers often use the Azimuthal Equidistant projection in world maps for aviation where the map centers on a specific airport. In this example, it centers on the North Pole.

Compromise Projections

There are some projections, called **compromise projections**, which do not try to preserve any of the map projection properties completely but rather may preserve a little of several. Many of these projections are instead designed to appear aesthetically pleasing, balancing shape and area. These are best for world maps where one needs to see the general shape, size, and spatial relationship of the continents. One such projection is the **Robinson projection**, published in 1963 by the American cartographer Arthur H. Robinson (**Figure 3.44**). The Robinson projection draws all parallels as straight lines, and all meridians except the central meridian are curved.

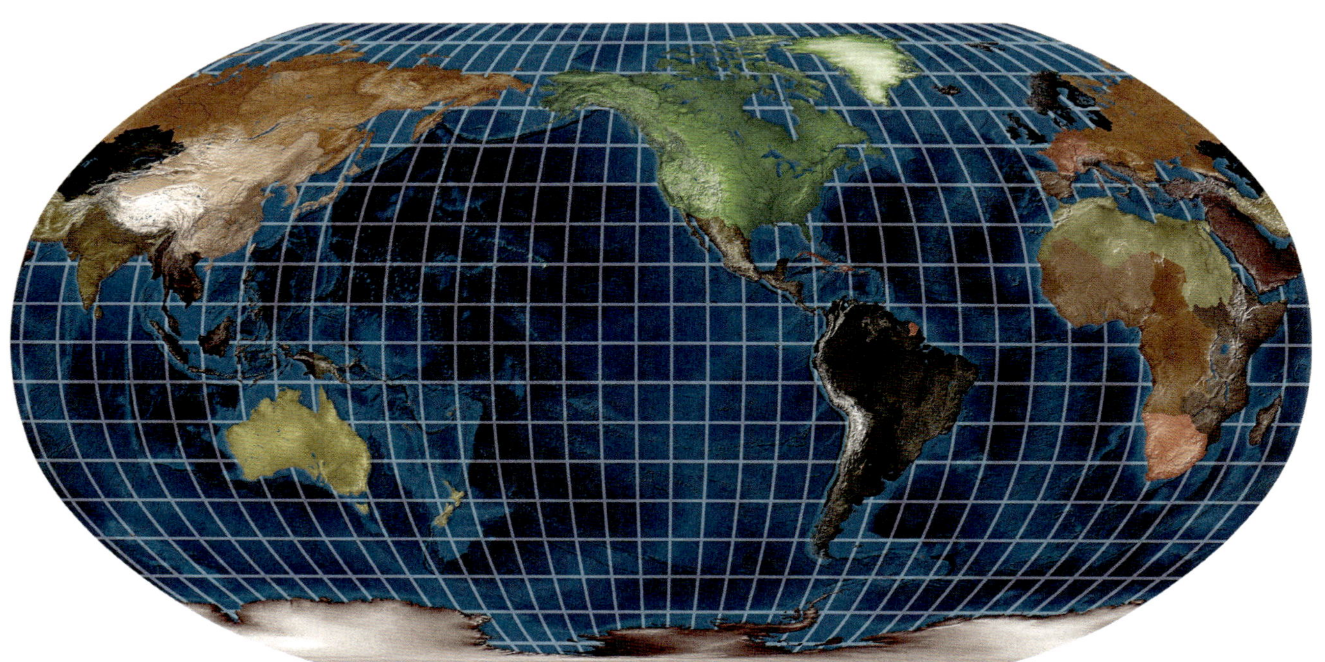

Figure 3.44: The Robinson is a popular projection for world maps and used for many years by the National Geographic Society.

The Robinson projection does not use any of the standard developable surfaces. Instead, this type of projection is known as a **pseudocylindrical projection**. Conceptually, a pseudocylindrical projection is comprised of many cylindrical surfaces of varying sizes (**Figure 3.45**).

Figure 3.45: The Robinson projection has a partial similarity to cylindrical projections, but is more mathematically sophisticated.

Several other compromise projections are featured in this popular comic by xkcd: ***What your favorite map projection says about you***[5].

5 https://xkcd.com/977/

Tutorial: Learning About Projections Using ArcGIS

The goal of this activity is to learn about standard map projections by exploring their properties using ArcGIS. In this activity, you create and organize a workspace folder using a standardized folder structure. You then download and decompress the data from the Natural Earth website. After delving into several map projections, you pick your map projection and customize it based on a location you choose. You create a map of world subregions using your personalized map projection.

ESTIMATED TIME TO COMPLETE THIS TUTORIAL: 3 HOURS

Learning Outcomes

Readers should be able to accomplish the following outcomes by the end of this tutorial:

» Summarize the steps for creating and organizing a project workspace folder structure

» Illustrate the ability to download data from a public source

» Demonstrate decompressing zipped files into a specific folder

» Describe the difference between standard map projections

» Illustrate how to create a customized map projection

» Practice changing map output size in ArcMap

» Apply symbology and color choices to map features

» Insert essential map elements using ArcMap

» Export a map as a PDF file

Setting up Your Workspace

In a typical workflow, you work on geospatial data using a local hard drive. When done, you compress your data and back up your work to your cloud storage so that you can retrieve the files from anywhere. When referring to a **local hard drive**, it means you are working on data physically located on the computer in front of you.

In contrast, some computers also include networked drives. **Networked drives** link to cloud storage and save the data elsewhere. Examples include services like OneDrive or Google Drive. For this tutorial, use the **desktop** as your local hard drive location. You may also use an external USB drive if you plan to work in multiple places.

You must avoid using networked drives while you work. They increase the processing time and can cause technical glitches.

In this tutorial, you use a particular folder structure. Start by creating your **workspace folder** on the local hard drive. A **workspace** is a folder or series of folders that contain all of your project files. The top-level folder in your workspace should indicate the lab assignment or the project. Organize all of your work within the workspace folder. On your desktop, create a new folder and give it a descriptive name, such as *World_Map*. Be sure there are *no spaces*. You may use underscores instead of spaces. Inside this folder, create the following three subfolders: original, working, and final. Having a standardized folder structure helps to keep a project organized, primarily when you are working with multiple partners. The folder structure you see here (**Figure 3.46**) is the standard used in each of the tutorials presented in this book.

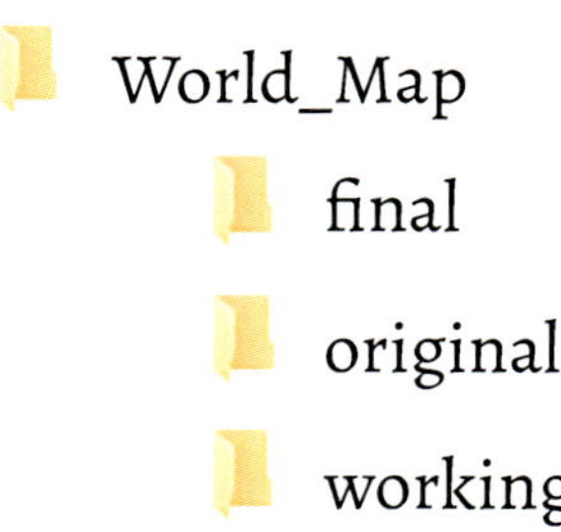

Figure 3.46: This diagram represents a basic folder structure used in this book.

As the name indicates, use the **original folder** for storing original, unaltered data. As you are working on a project, if for some reason your working version of the data gets lost or corrupted, you can go back to your *original* folder and find a fresh copy of the data. Use the **working folder** for data that you create or alter while working on your project.

Use the **final folder** for storing any output you produce as a result of your work, such as images, maps, tables, or reports. Setting up the standard folder structure for a project is good practice and a habit you want to develop.

Downloading Data from Natural Earth

Natural Earth is a website created to provide vector and raster data to meet the needs of cartographers using a variety of software applications. The data on this website is free to use without restrictions.

Open the **Chrome** browser and navigate to the ***Natural Earth website***[1]. Click the *Downloads* link. Under Small scale data, click the *Cultural* link (**Figure 3.47**).

If you do not have the Chrome browser, you can download it here[2]**.**

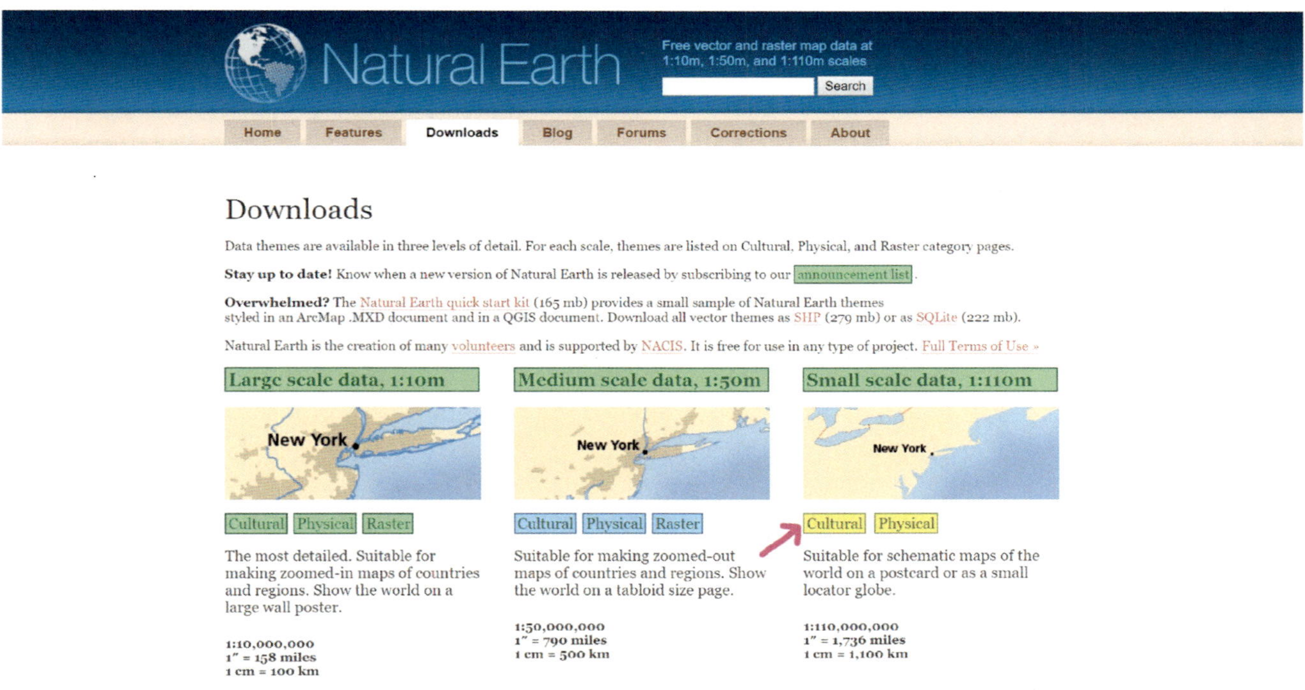

Figure 3.47: Natural Earth provides physical and cultural data in large, medium, and small scale. Double-click or tap twice to view the image in a larger size.

1 http://www.naturalearthdata.com/

2 https://www.google.com/chrome/browser/desktop/index.html

Under *Admin 0-Countries*, right-click *Download countries* (**Figure 3.48**). Select, *Save Link As*, then navigate to your *original folder* and save.

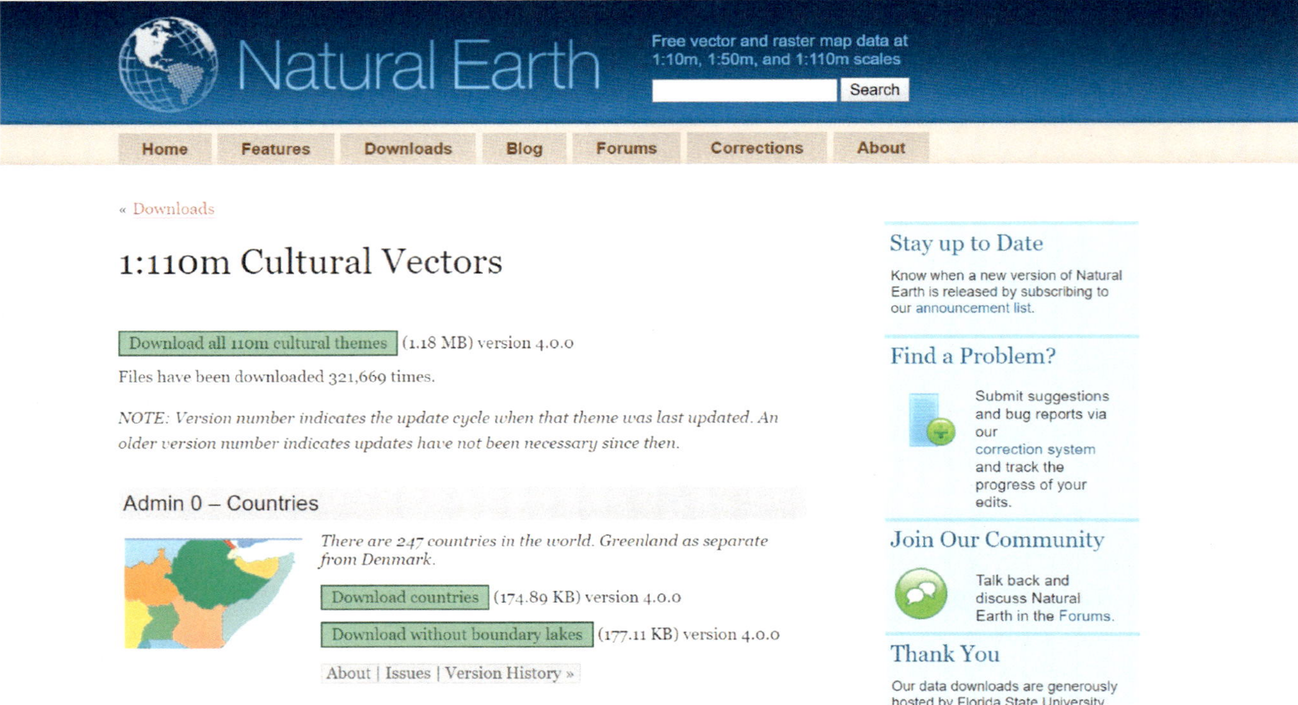

Figure 3.48: The image here shows the Natural Earth Cultural Vectors download page for small-scale data.

When you are ready, hit the back button and return to the main downloads page. Click *Physical* under *Small scale data*. On the Physical Vectors page, scroll down until you see *Graticules* (**Figure 3.49**). Right-click on the button that says *Download 10*. Select, *Save Link As*, then navigate to your *original folder* and save.

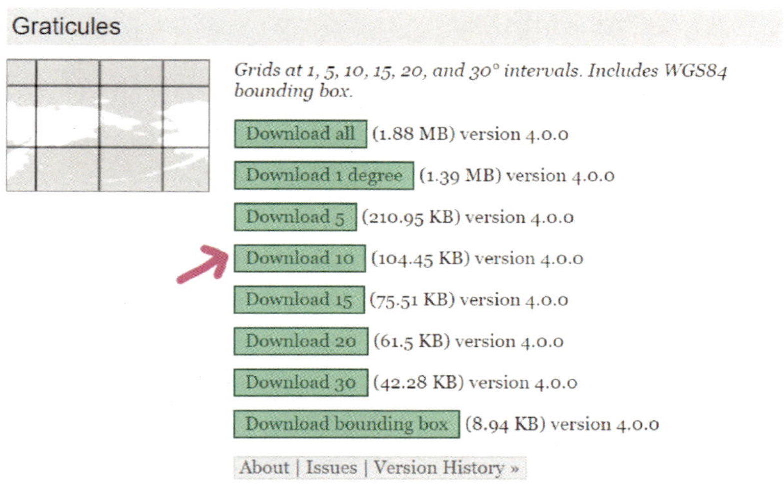

Figure 3.49: The image here shows the Natural Earth Physical Vectors download page for small-scale data. Graticules is near the bottom.

In a previous activity, you learned how to decompress a file using 7zip. In Microsoft Windows, navigate to your *original folder*. You should see the two zip files, one for countries and one for graticules (**Figure 3.50**). For each zip file, right-click, select 7zip, then *Extract Here*.

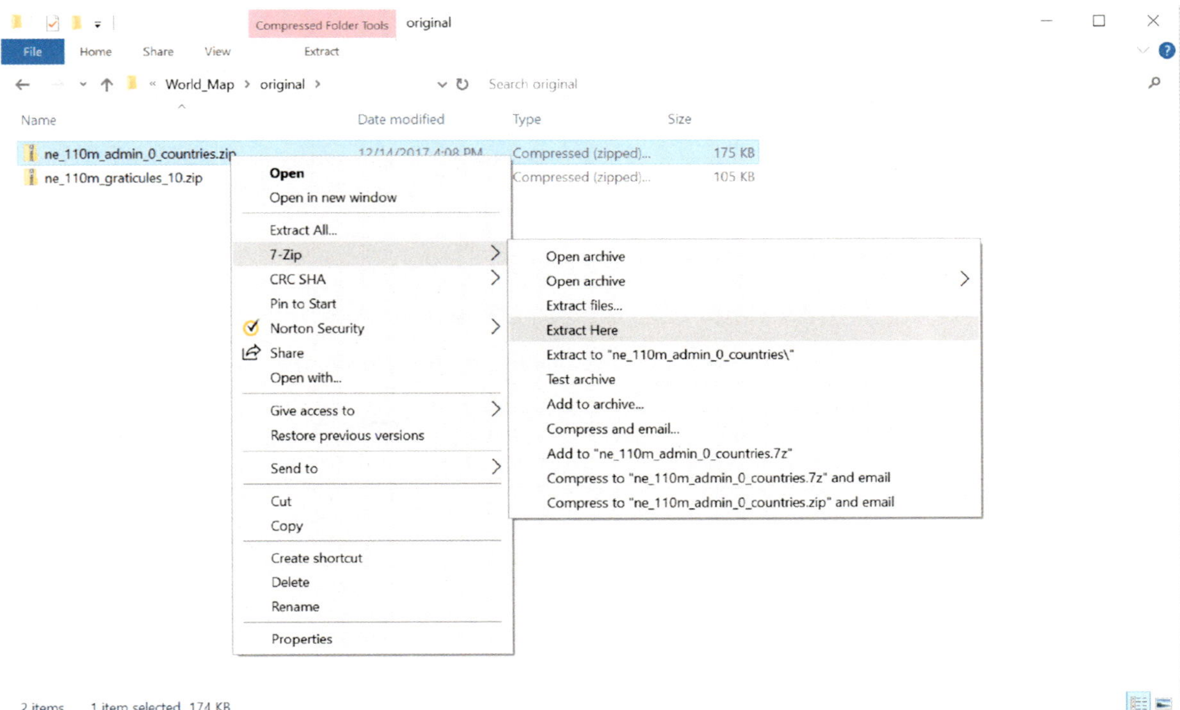

Figure 3.50: You can access the 7zip software by right-clicking on a compressed file.

Be sure to delete the zip files when you are done decompressing them (**Figure 3.51**). You won't need them anymore. Eliminating them saves space and helps to avoid confusion later.

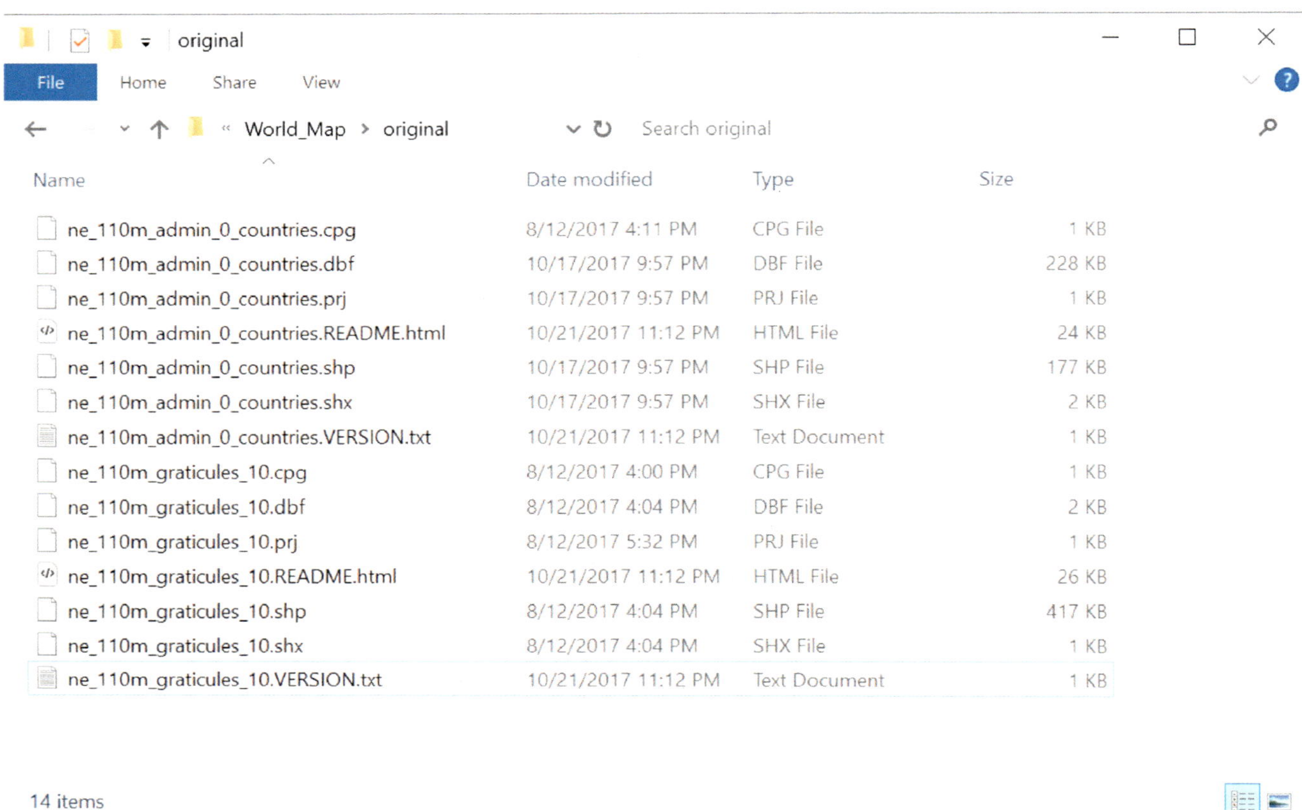

Figure 3.51: As you can see, many parts make a shapefile. It must use these files together to work correctly. NEVER delete any of the component files. Pay attention when deleting unwanted zip files.

Adding Data to ArcMap

Locate ArcMap on your computer and launch the software. When you first launch ArcMap, a window appears that gives you the option to either open a blank map or to open any of your recent map documents (**Figure 3.52**). Choose to open a blank map document.

The ArcMap user interface has three main windows that you use regularly, the **Table of Contents**, the **data frame**, and the **Catalog Window** (**Figure 3.53**). The *Table of Contents* displays a list of data frames and map layers loaded into the map documents. The data frame, sometimes spelled dataframe, represents the layers as a map and defines the map extent. The Catalog Window displays a hierarchical view of folder connections and data as a **Catalog Tree**.

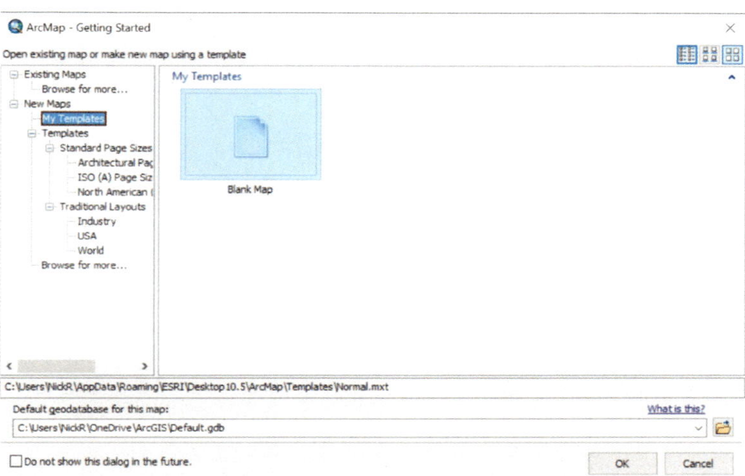

Figure 3.52: The ArcMap Getting Started window provides several options.

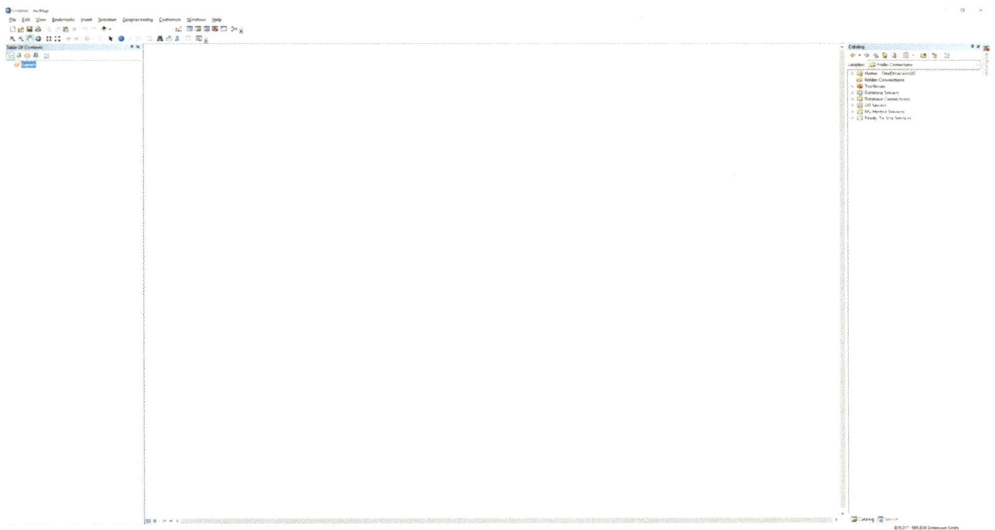

Figure 3.53: From left to right, the three main windows are the Table of Contents, the data frame, and the Catalog Window.

The icon for the catalog window looks like a yellow file cabinet (**Figure 3.54**). If your catalog window is missing, you can find the icon on the toolbar across the top. Click to open it back up.

Figure 3.54: Click the yellow file folder icon if your Catalog Window is missing.

In this step, you start by working with the Catalog Window. The catalog window is where you manage your geospatial data in ArcMap. Start by clicking on the *Connect to Folder* button (**Figure 3.55**). Navigate to your workspace folder, *World_Map*, on your local hard drive.

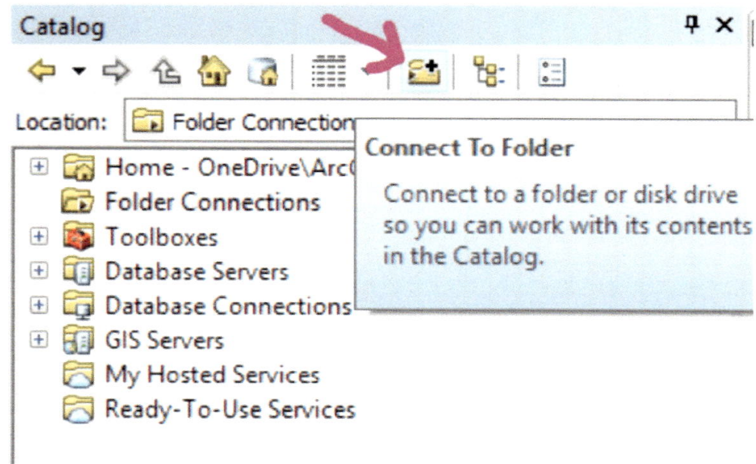

Figure 3.55: You connect to a folder so that the contents appear in the Catalog Window.

In this step, it is essential to select your primary workspace folder, *World_Map*, to add it to the Catalog Window (**Figure 3.56**). When ready, click *OK*.

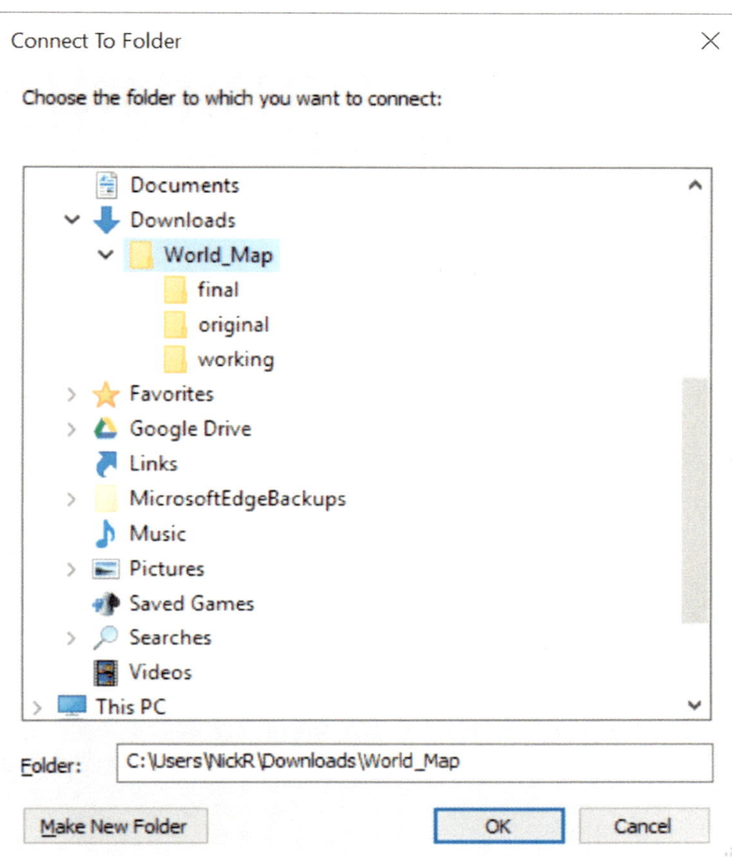

Figure 3.56: Connect to Folder adds the folder to the Catalog Window.

Once you add your workspace folder to the Catalog Window, expand the folder by clicking on the plus sign. You should see your three subfolders inside. Expand the original folder to view the contents. This display of folders and files within the Catalog Window is sometimes called the **Catalog Tree**.

There are several ways to add data to ArcMap. The most direct method is to drag and drop files from the Catalog Window. In this step, you add the countries to the map document. From the Catalog Window, select the countries shapefile from the *original folder*. In this example, a green square represents the shapefile called ne_110_admin_0_countries.shp. Click and drag the file over the data frame, the interior window pane in ArcMap, then release. The countries layer appears in the data frame window (**Figure 3.57**).

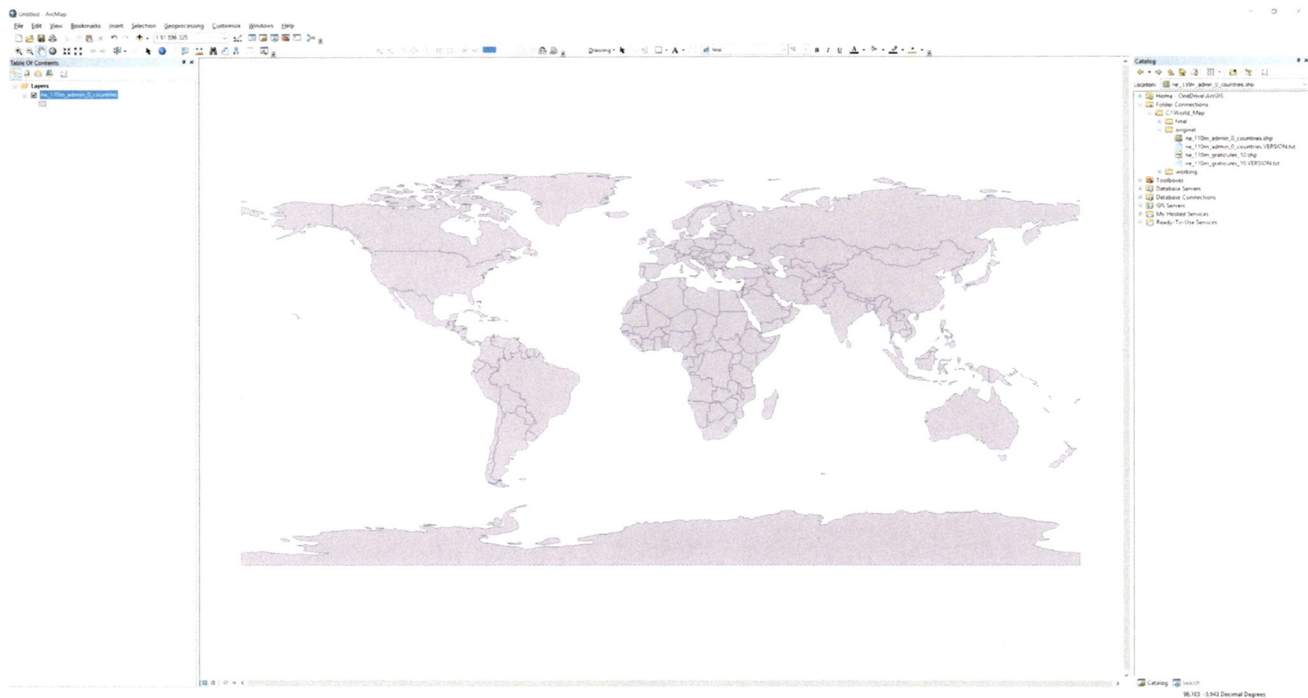

Figure 3.57: The countries layer is an example of a vector data model representing a polygon feature.

Repeat these steps to add the graticules shapefile. You may get a Geographic Coordinate Systems Warning. Ignore the warning for now. Check the box next to *Don't warn me again this session* and click *Close* (**Figure 3.58**). When you first add the graticules to the map, they appear above the continents. In the Table of Contents, drag the graticule layer under the countries layer, so ArcMap draws them underneath (**Figure 3.59**). Now is an excellent time to save your work.

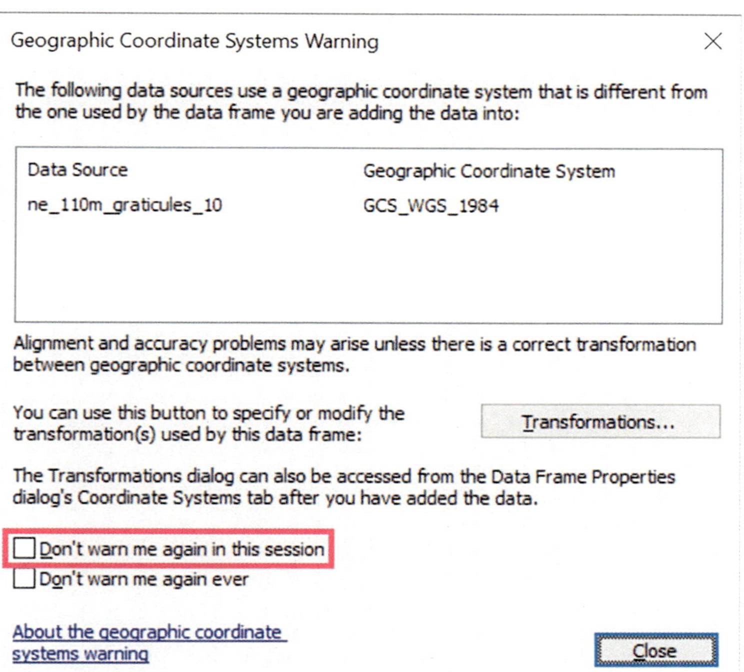

Figure 3.58: While it is OK to turn off this warning for one session. you should never turn it off permanently.

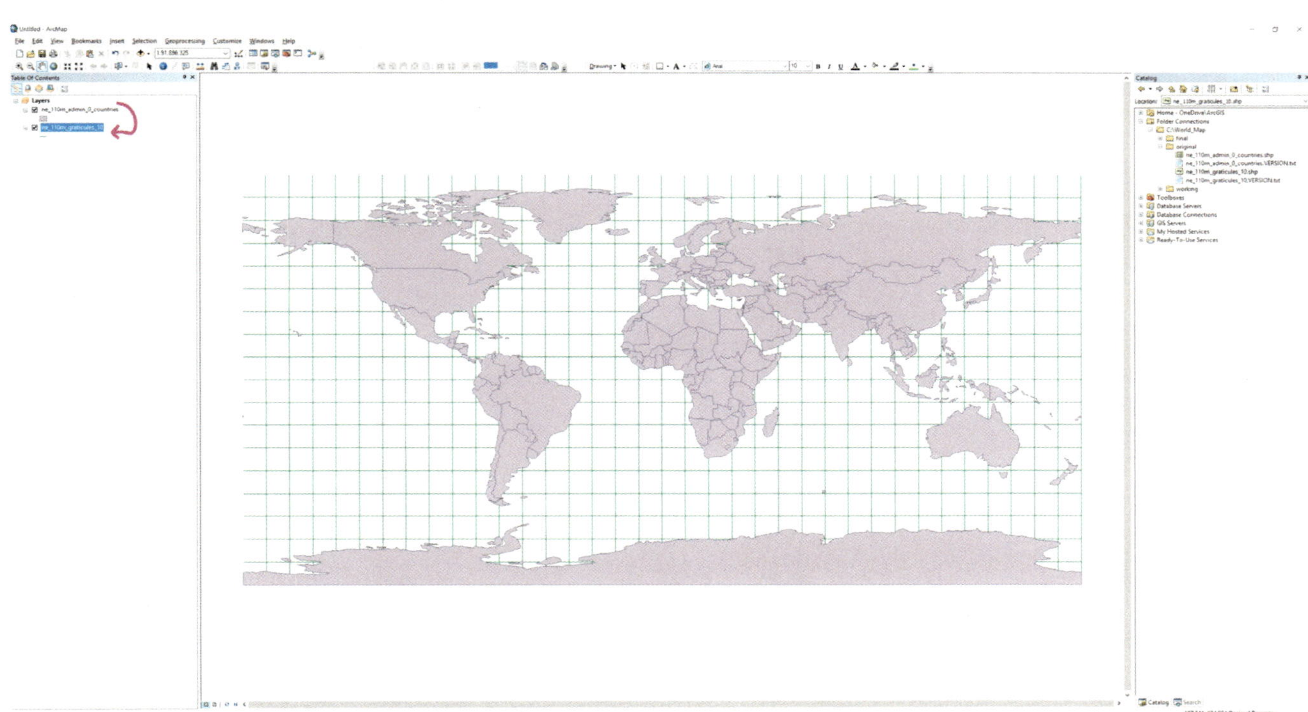

Figure 3.59: When the Table of Contents is set to List by Drawing Order, the order listed determines the order drawn on the map.

In a previous activity, you learned how to set your map document properties to store relative paths (**Figure 3.60**). Do this now. Then, save your map document to your workspace folder, World_Map. Call the map document, "Global Subregions."

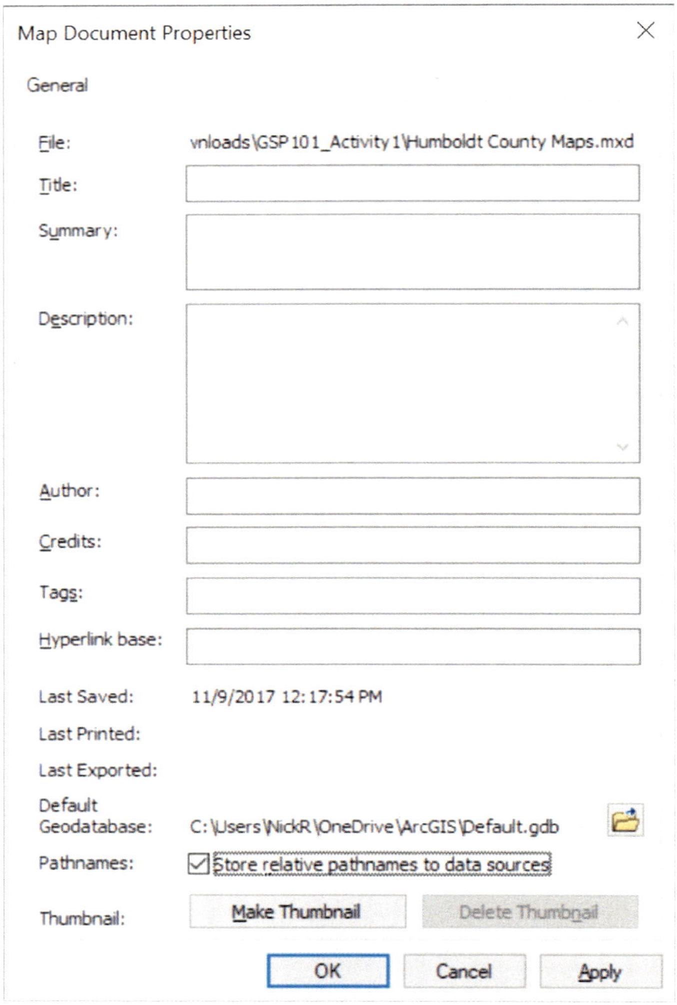

Figure 3.60: The store relative pathnames to data sources setting allows ArcMap to remember the location of the data sources relative to its current position.

Symbolizing the Map by Subregions

In a previous activity, you learned how to rename the data frame and change the background color. Open the data frame properties and navigate to the General tab. Change the name to *World Map*. Then, on the *Frame* tab, change the background color to Gray 10%. When you are ready, click *OK*. You may get a repeat of the coordinate system warning. Ignoring it during this activity is OK. Check the box next to *Don't warn me again this session* and click *Yes*, to close the warning (**Figure 3.61**).

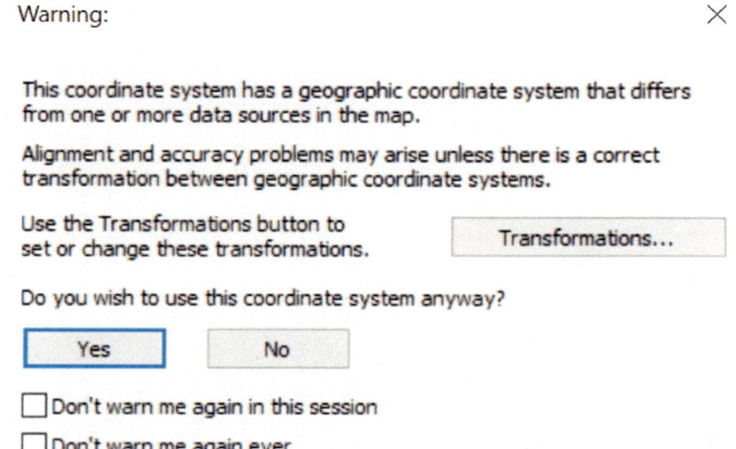

Figure 3.61: This warning appears when you have layers with different Geographic Coordinate Systems loaded in the Table of Contents.

In the Table of Contents, right-click on the countries layer to open the attribute table (**Figure 3.62**). Take a moment to read through the table to find out what kind of data is available about the various countries around the world. When done, you may close the attribute table.

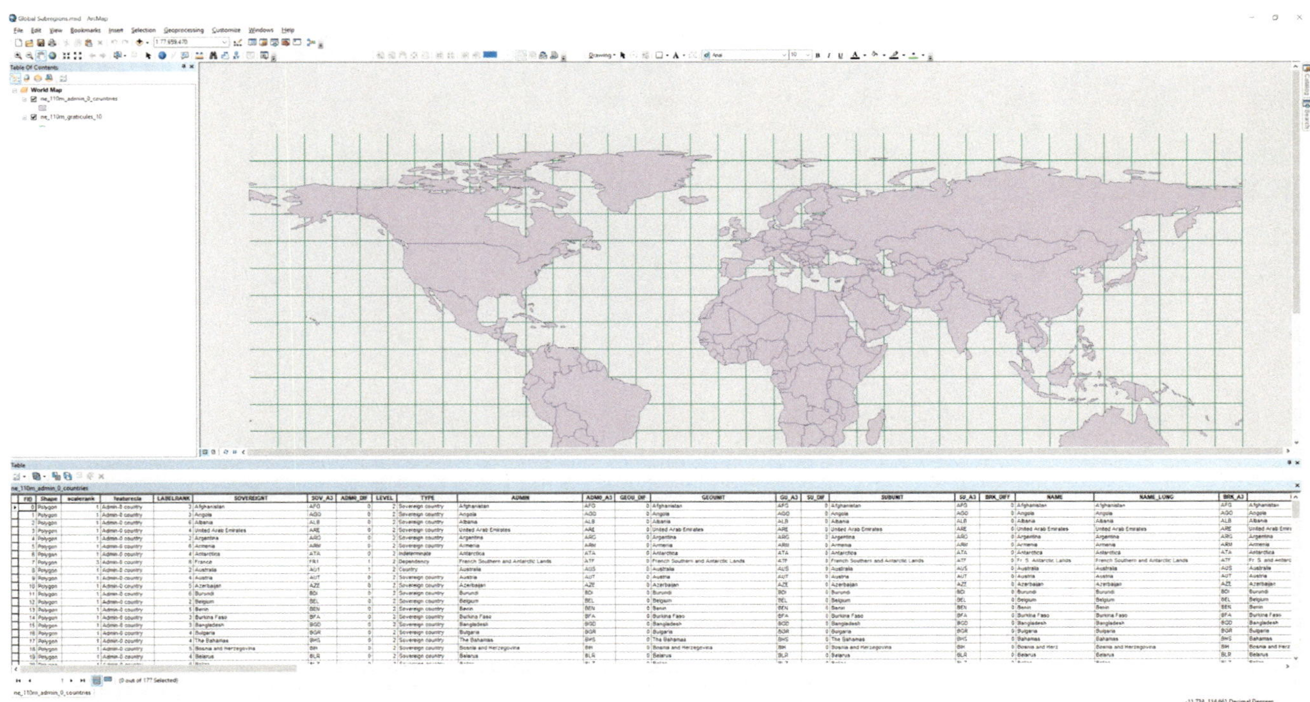

Figure 3.62: The attribute table contains useful information about countries around the world, including population and economic statistics.

In the Table of Contents, right-click on the countries layer again to open the properties. Navigate to the *Symbology* tab. On the left, choose *Categories* and select *Unique values*. Uncheck the box next to *all other values*. Under the *Value Field*, use the drop-down menu to select *SUBREGION* from the list of attributes. Then, click the button that says *Add all values*. Right-click on the color ramp and uncheck *Graphic View* (**Figure 3.63**). Choose *Dark Glazes* from the list of choices. When you are ready, click *OK*.

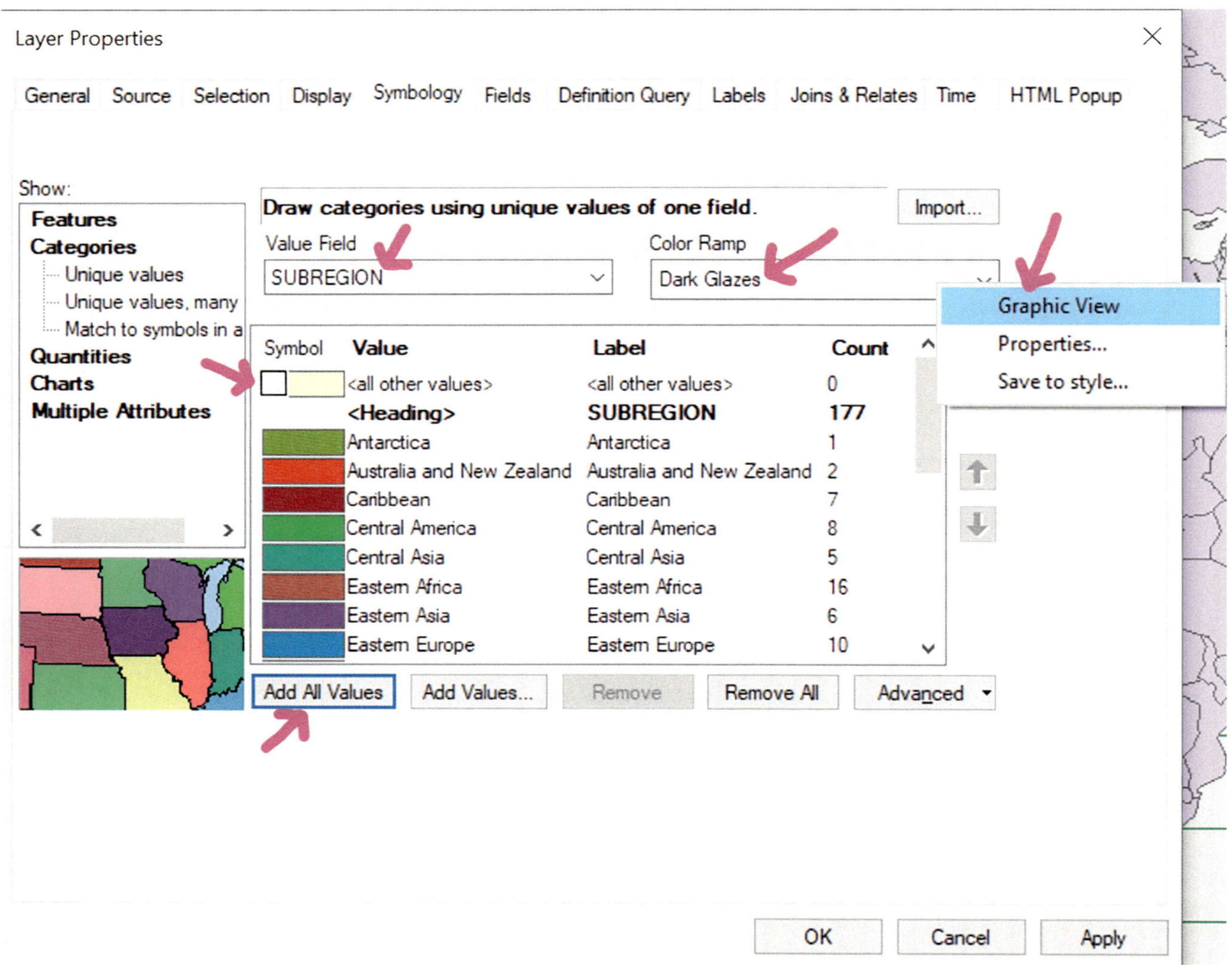

Figure 3.63: The arrows point to each step described above.

Another way to change colors is to click the symbol in the Table of Contents. In the Table of Contents, click the line symbol for the graticules (**Figure 3.64**). When the symbol selector appears, change the color to Gray 70% and the line width to 0.25.

Figure 3.64: You can edit symbols directly in the Table of Contents, though this limits your options.

The countries should be color-coded by subregions in the graticules should lie underneath (**Figure 3.65**)

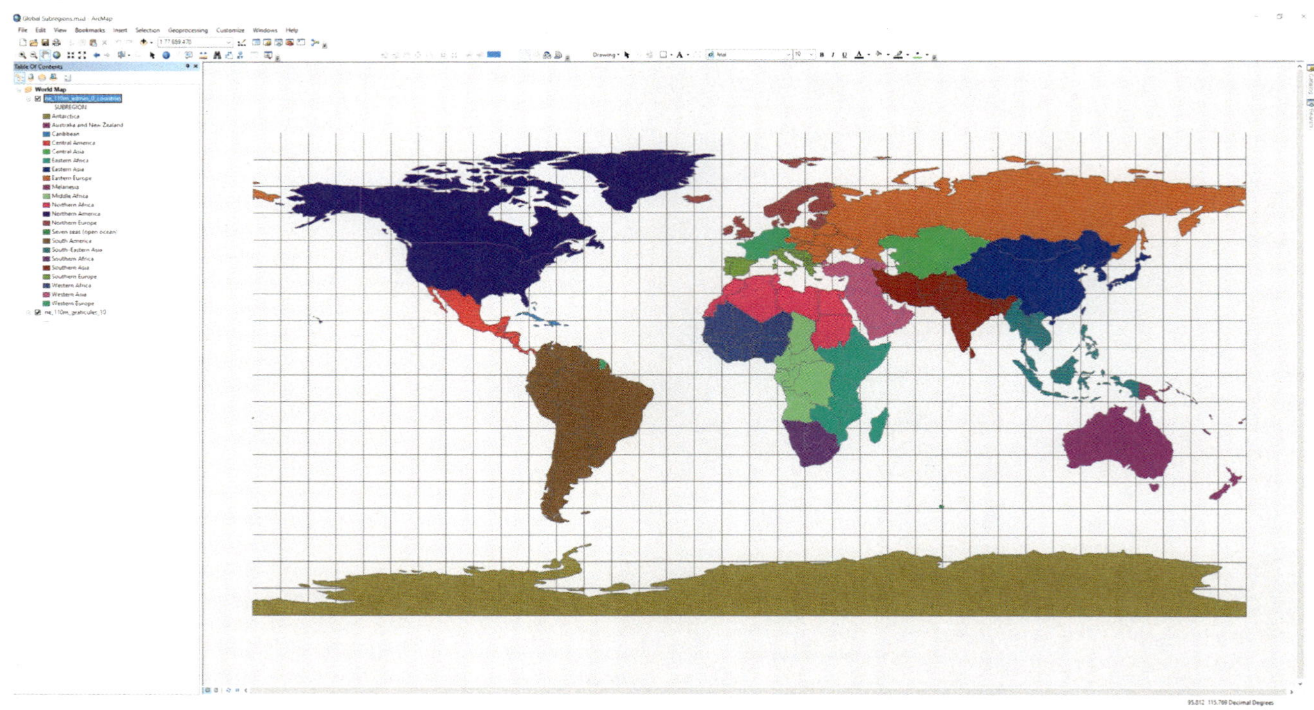

Figure 3.65: ArcMap uses the color ramp to assign colors to the subregions randomly. Your results may vary.

Exploring Map Projections

In this chapter, you learned that Earth is spherical. However, maps still managed to represent Earth using a flat plane. A flat representation is made possible by using map projections. A **map projection** is a geometric transformation of the round Earth onto a flat plane using mathematical equations. The process of transforming a spherical object onto a flat plane distorts area, shape, distance, direction, and continuity. These are known as the five **preserved properties** because some map projections can maintain one or more of these characteristics with a high degree of accuracy. However, no map projection can retain all five of these qualities at the same time.

In ArcMap, the data frame window displays the map using **project-on-the-fly**. Even though the layers in the Table of Contents might use different projections and coordinate systems internally, ArcMap tries to line them up on the screen using the coordinate system defined in the data frame properties. What this means is that you can change the projection in the *data frame* to alter the appearance of the map onscreen without modifying the original data files.

Cylindrical Projections

Cylindrical projections transform the globe onto a flat plane using a cylinder as the **developable surface**. In this step, you use ArcMap to explore one of the most commonly used cylindrical projections, the **Mercator projection**.

Open the data frame properties. Navigate to the *Coordinate System* tab (**Figure 3.66**). Expand the *Projected Coordinate Systems* folder.

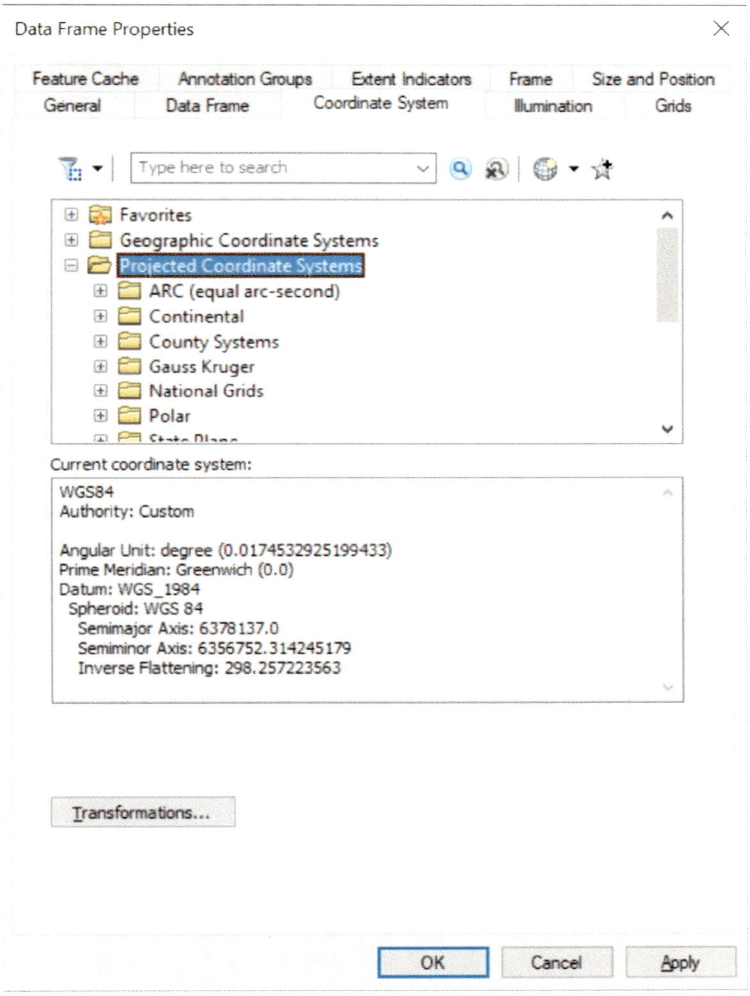

Figure 3.66: The Coordinate Systems tab in the data frame properties allows you to choose a spatial reference system from two main folders, Geographic Coordinate Systems and Projected Coordinate Systems.

Scroll down until you see the *World* folder. Expand the *World* folder and select *Mercator (world)*. When you are ready, click *OK*. Click *Yes*, when the warning appears. On the *Tools* toolbar, click the *Full Extent* button, which looks like a globe (**Figure 3.67**).

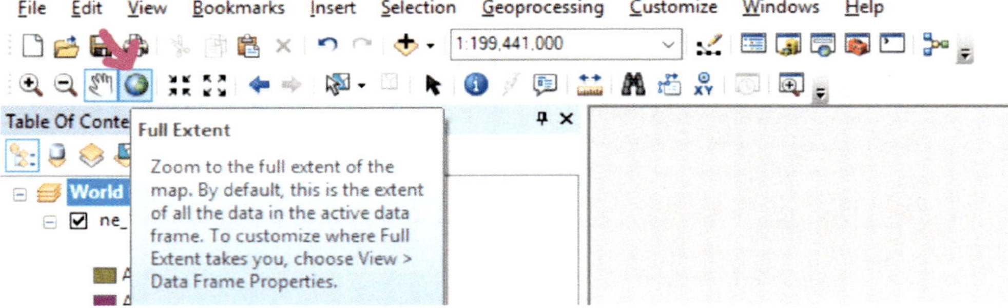

Figure 3.67: The *Full Extent* button zooms to the extent of all the data on the map.

Take a moment to review the text earlier in this chapter regarding the Mercator projection. When done, continue to the next step.

By default, the Mercator projection centers on 0 degrees longitude (**Figure 3.68**). However, in ArcMap, you can customize map projections to a certain extent. What you are allowed to do varies by map projection. For the Mercator projection, it is possible to center the map on any longitude you wish. In this step, you align the map on the opposite side of the globe, 180 degrees longitude.

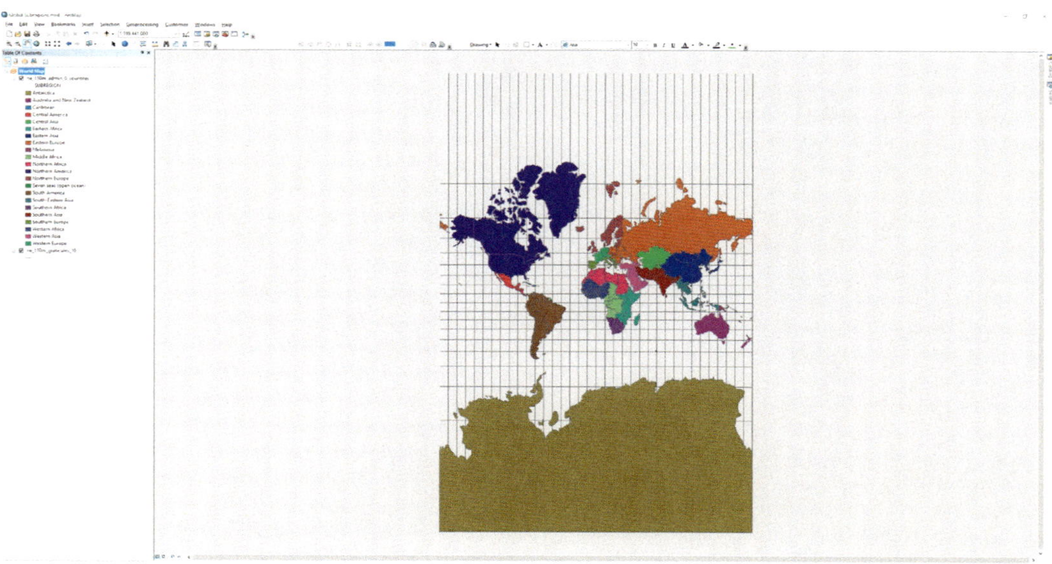

Figure 3.68: The Mercator projection maintains the shape, or conformity, of the continents, but the sizes get altered dramatically.

Open the properties for the data frame and navigate to the coordinate system tab (**Figure 3.69**). Double-click on the *Mercator (world)* projection. When the *Projected Coordinate System Properties* window appears, change the name to *World Mercator 180*.

In the *Value* field for the **central meridian**, enter 180. When done, click *OK*. Then, click *OK* again to apply the changes to the data frame. Click *Yes* to close the warning.

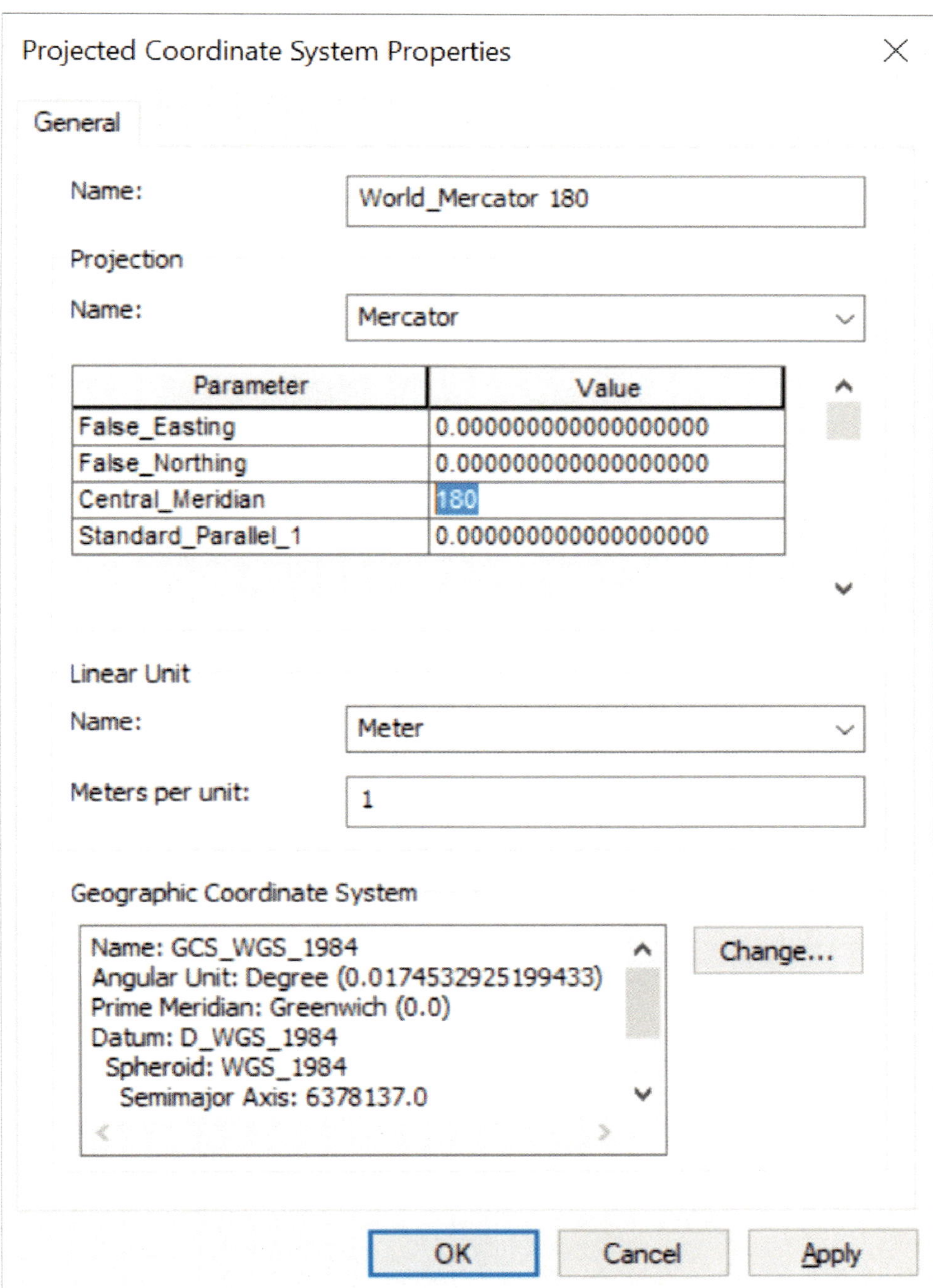

Figure 3.69: Changing the central meridian is the most common parameter you can change. Some projections allow you to edit additional features.

You should immediately notice that the map rotates so that the Pacific Ocean is in the center (**Figure 3.70**). You have just created your first customized projection!

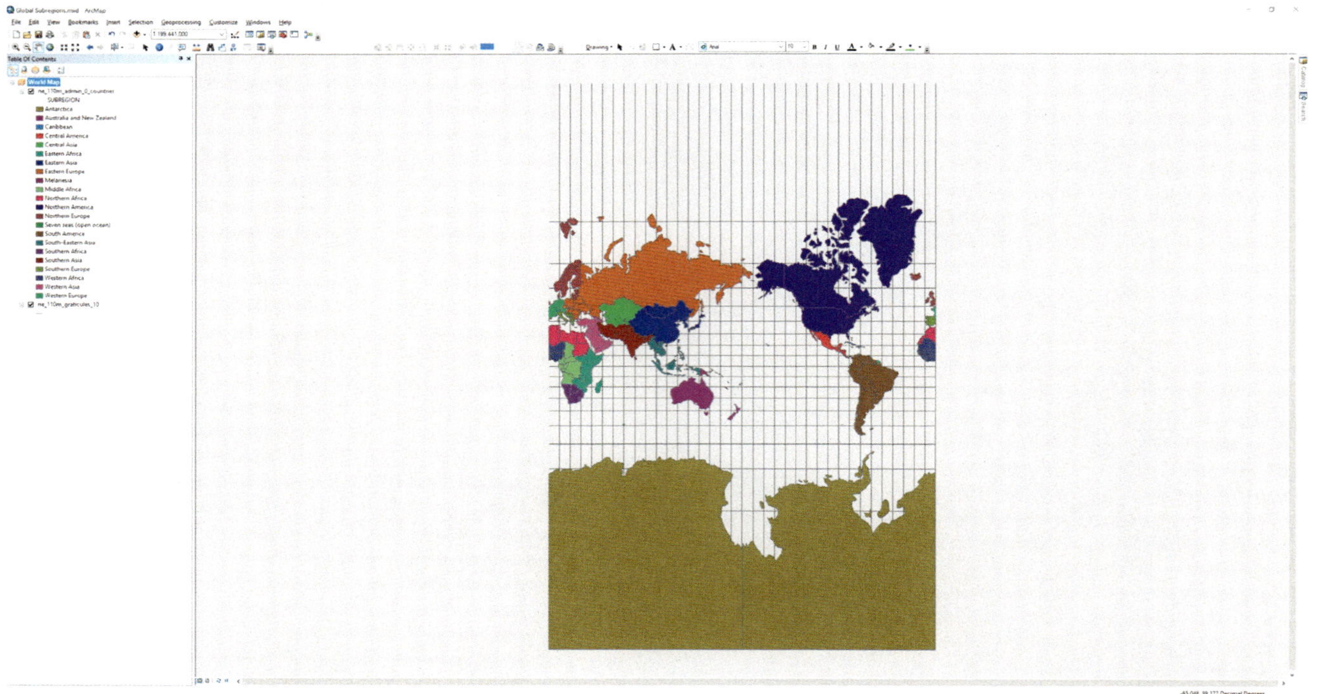

Figure 3.70: This image shows the Mercator projection centered at 180 degrees longitude.

Another commonly used cylindrical projection is the **Transverse Mercator**. Open the properties for the world map data frame (**Figure 3.71**). Navigate to the *Coordinate System* tab. Expand the *Projected Coordinate Systems* folder. Scroll down until you see the *UTM* folder. Expand the *UTM* folder and scroll down until you see the *NAD 1983* folder. Expand the folder and choose *NAD 1983 UTM Zone 10N*.

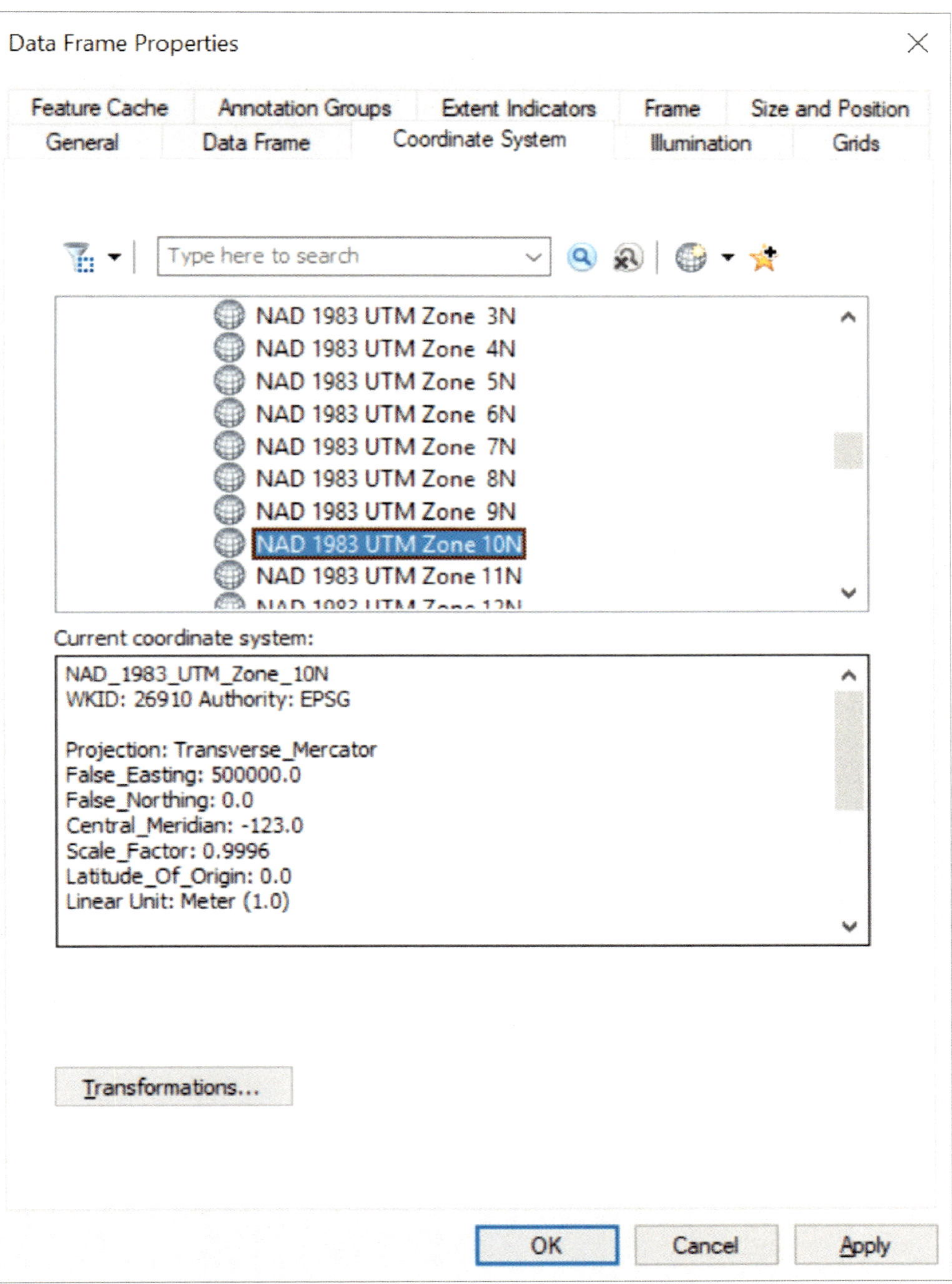

Figure 3.71: UTM stands for Universal Transverse Mercator.

When you are ready, click *OK*. Click *Yes*, when the warning appears. On the *Tools* toolbar, click the *Full Extent* button. By default, Zone 10 centers in Northern California (**Figure 3.72**).

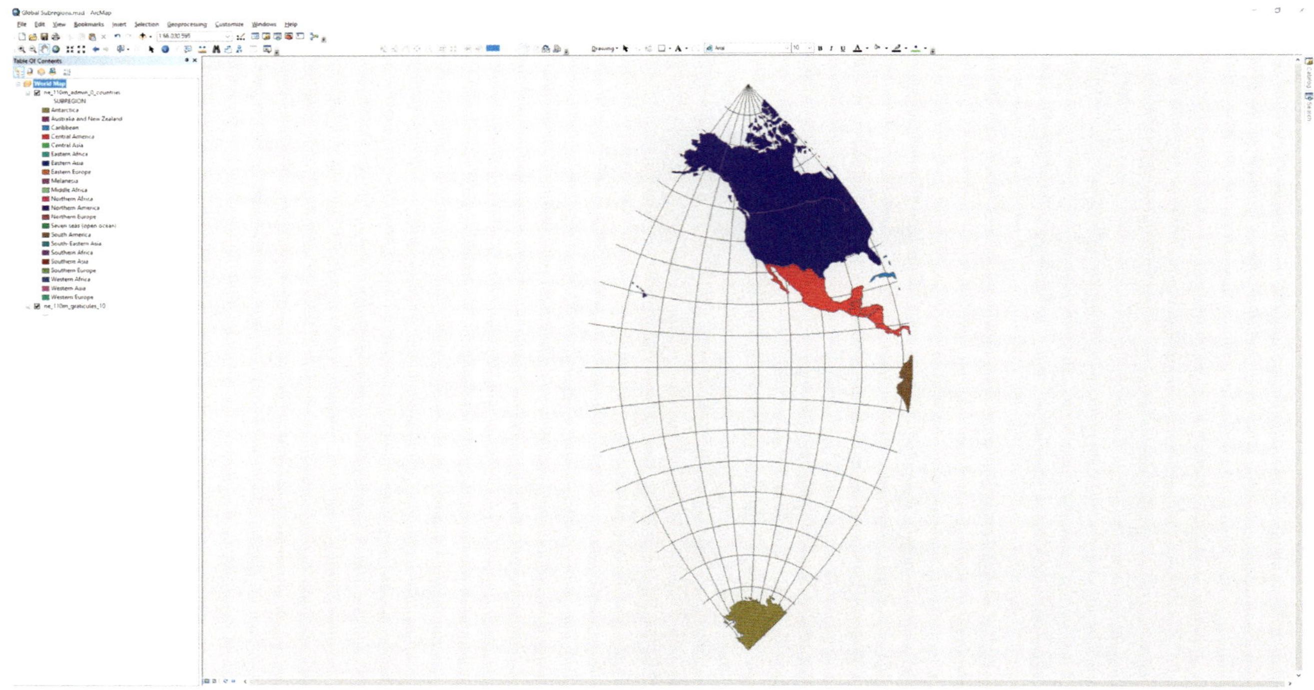

Figure 3.72: The Transverse Mercator projection has a high degree of accuracy along the central meridian.

Take a moment to review the text earlier in this chapter regarding the Transverse Mercator projection. When done, continue to the next step.

Normally, when you want to use the Transverse Mercator for a different region, you would choose the appropriate zone. However, it is possible to center the Transverse Mercator to a location manually. Using what you learned, create a customized Transverse Mercator projection. Change the central meridian to 57 (**Figure 3.73**). Name the customized projection *Transverse Mercator 57*.

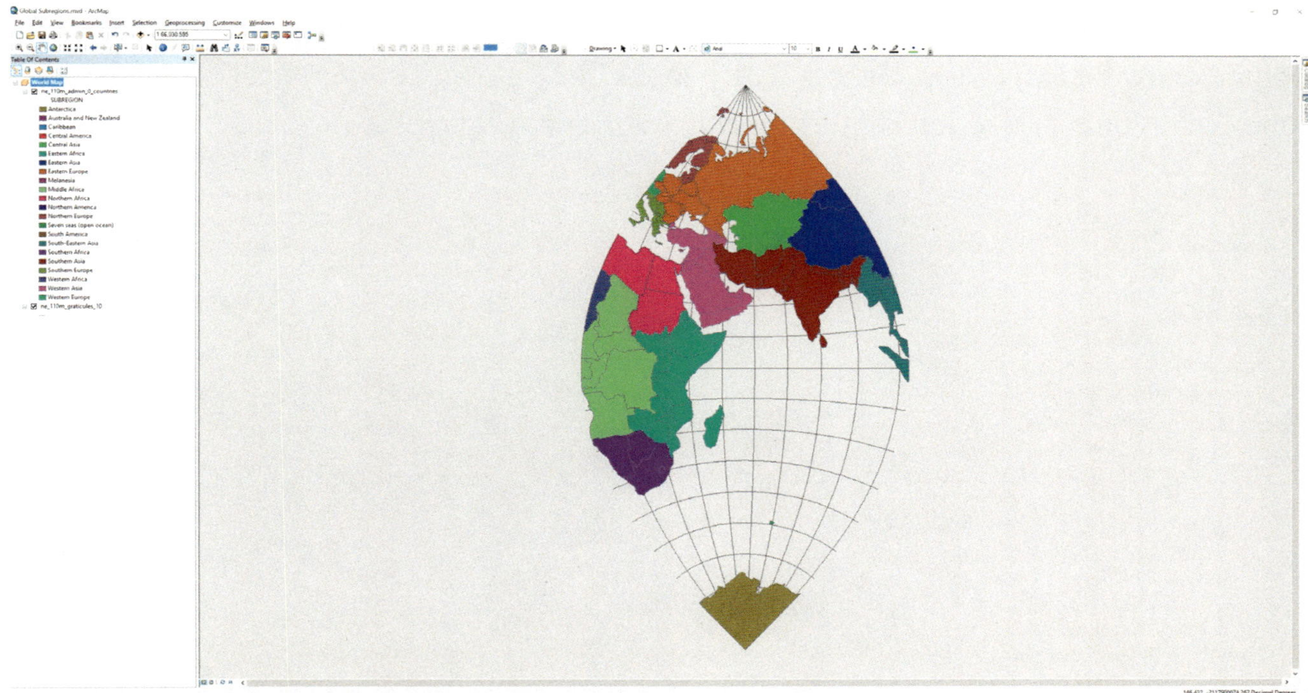

Figure 3.73: The Transverse Mercator now centers on the opposite side of the globe.

Conical Projections

Conical projections transform the globe onto a flat plane using a cone as the **developable surface**. In this step, you use ArcMap to explore one of the most commonly used conical projections, **Lambert Conformal Conic**. Open the properties for the world map data frame. Navigate to the *Coordinate System* tab. Expand the *Projected Coordinate Systems* folder. Scroll down until you see the Continental folder. Expand the Continental folder, then expand the North America folder. Choose North America Lambert Conformal Conic. When you are ready, click *OK*. Click *Yes*, when the warning appears. On the *Tools* toolbar, click the *Full Extent* button, which looks like a globe. This version of the Lambert Conformal Conic projection centers on North America (**Figure 3.74**).

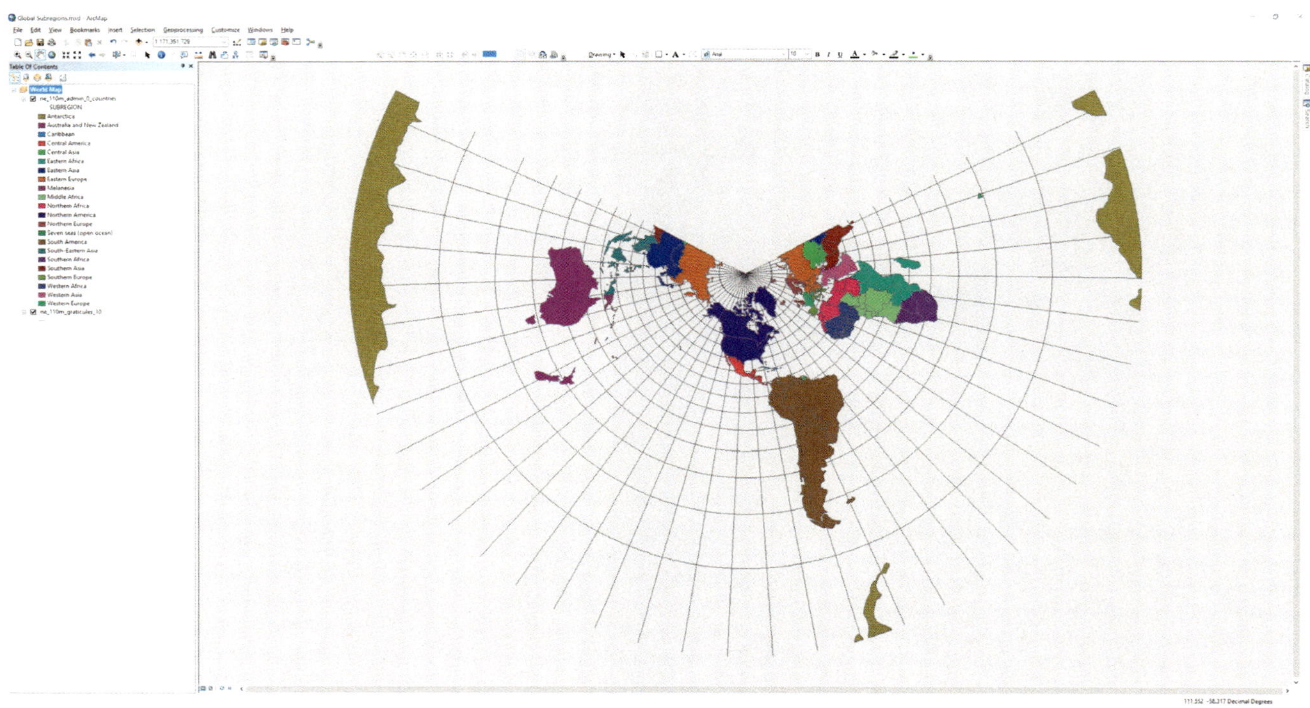

Figure 3.74: The Lambert Conformal Conic maintains the shape (conformity) of the continents, but only at mid-latitudes.

Take a moment to review the text earlier in this chapter regarding the Lambert Conformal Conic projection. When done, continue to the next step.

Skill Drill: Create a Custom Map Projection

Using what you learned earlier, create a customized Lambert Conformal Conic projection centered on Japan (**Figure 3.75**). Change the central meridian to 140. Name the customized projection *Japan Lambert Conformal Conic 140*.

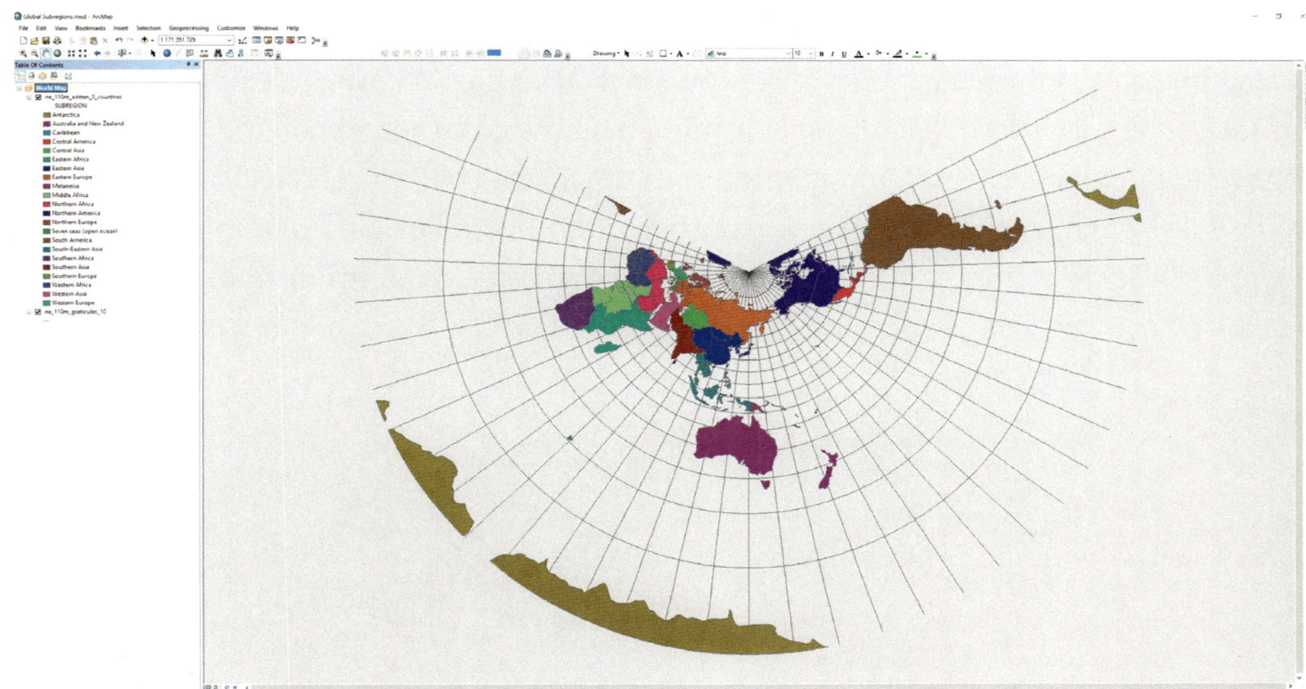

Figure 3.75: The Lambert Conformal Conic maintains the shape, or conformity, of the continents, but only at mid-latitudes. Here it is centered on Japan.

Planar Projections

Planar projections transform the globe using a plane as the **developable surface**. In this step, you use ArcMap to explore one of the most commonly used planar projections, the **Stereographic projection**. Open the properties for the world map data frame. Navigate to the *Coordinate System* tab. Expand the *Projected Coordinate Systems* folder. Scroll down until you see the *Polar* folder. Expand the *Polar* folder. Choose, *North Pole Stereographic*. When you are ready, click *OK*. Click *Yes*, when the warning appears. On the *Tools* toolbar, click the *Full Extent* button. This projection centers on the North Pole (**Figure 3.76**). If you wanted a map centered on the South Pole, you could choose the South Pole Stereographic. However, customizing the North Pole Stereographic provides an opportunity to change a different projection parameter, the **latitude of origin**.

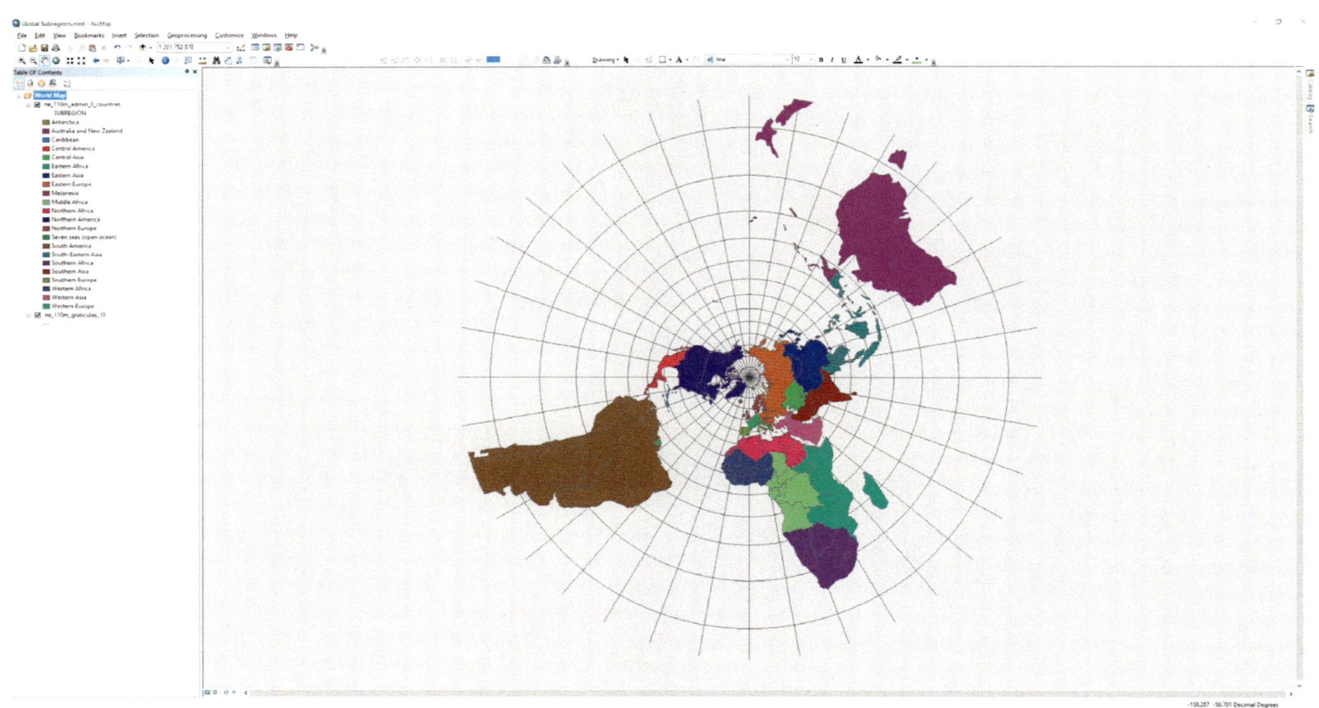

Figure 3.76: The Stereographic projection is conformal and commonly used for maps of polar areas and star maps.

Take a moment to review the text earlier in this chapter regarding the Stereographic projection. When done, continue to the next step.

Open the *Projected Coordinate System Properties* for the North Pole Stereographic projection (**Figure 3.77**). Change the Latitude of Origin from 90 to -90. This step flips the map so that it centers on the south pole. Name the customized projection *North Pole Stereographic Flipped*.

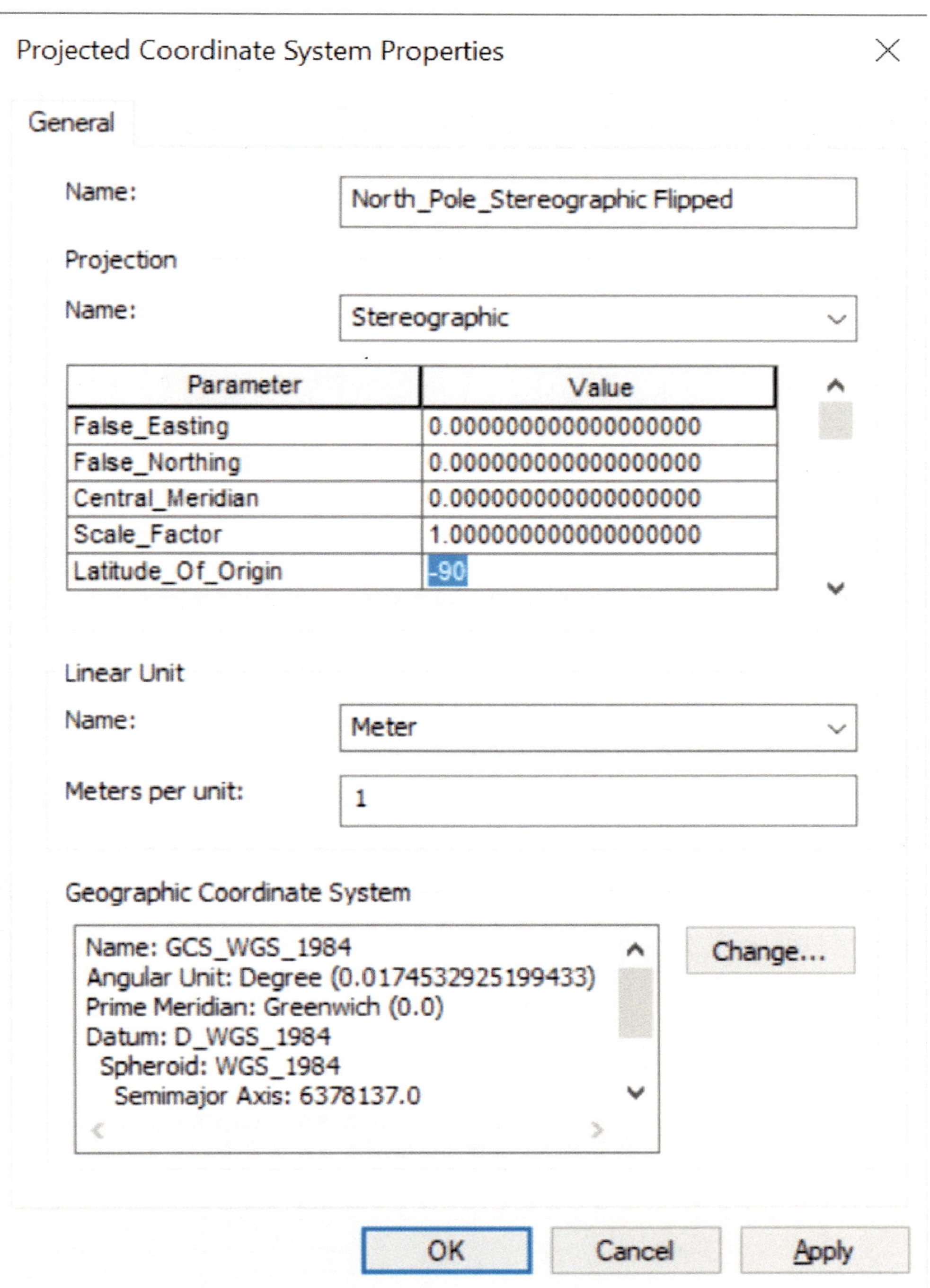

Figure 3.77: Instead of changing the central meridian as before, you change the latitude of origin.

When you are ready, click *OK*. On the *Tools* toolbar, click the *Full Extent* button. You should see the South Pole (**Figure 3.78**).

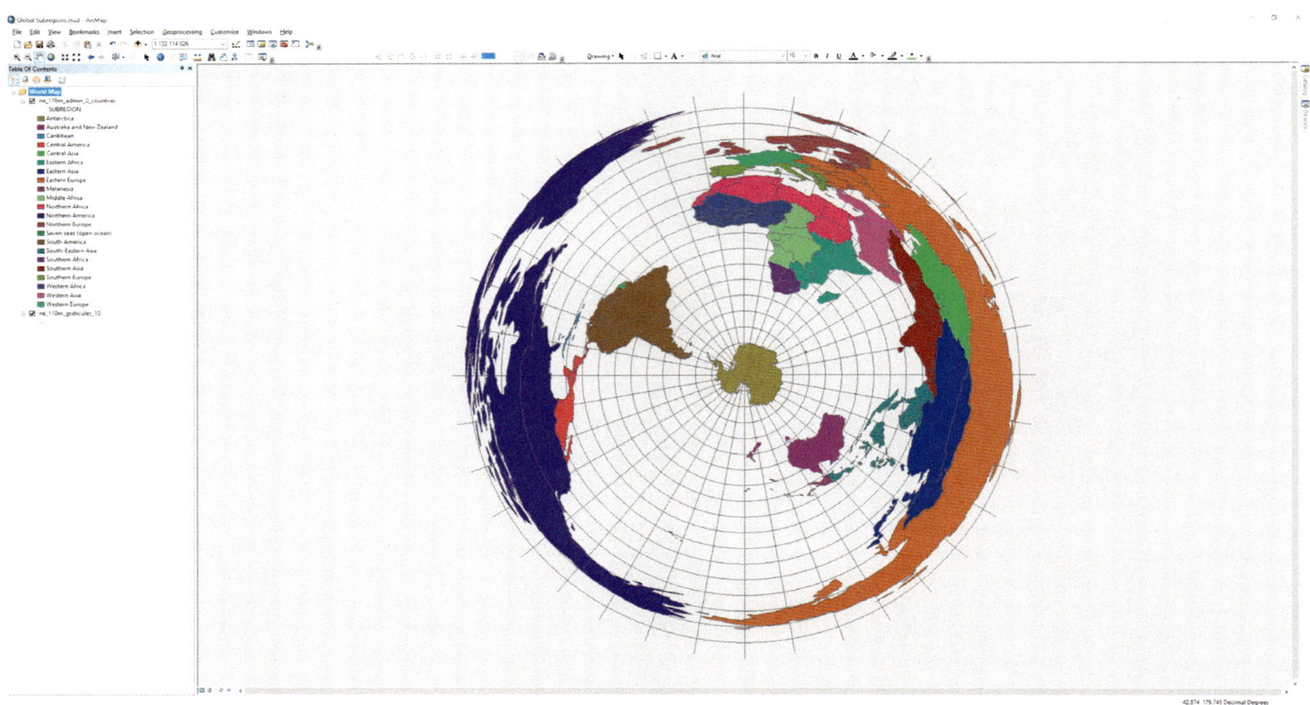

Figure 3.78: Flipping the North Pole Stereographic creates the same projection as the South Pole Stereographic.

Another commonly used planar projection is the **Azimuthal Equidistant projection**. Open the properties for the world map data frame. Navigate to the *Coordinate System* tab. Expand the *Projected Coordinate Systems* folder. Scroll down until you see the *Polar* folder. Expand the *Polar* folder. Choose, *South Pole Azimuthal Equidistant*. When you are ready, click *OK*. Click *Yes*, when the warning appears. On the *Tools* toolbar, click the *Full Extent* button. Though it is a planar projection and also centered on the south pole, it appears entirely different than the stereographic projection from the previous step (**Figure 3.79**).

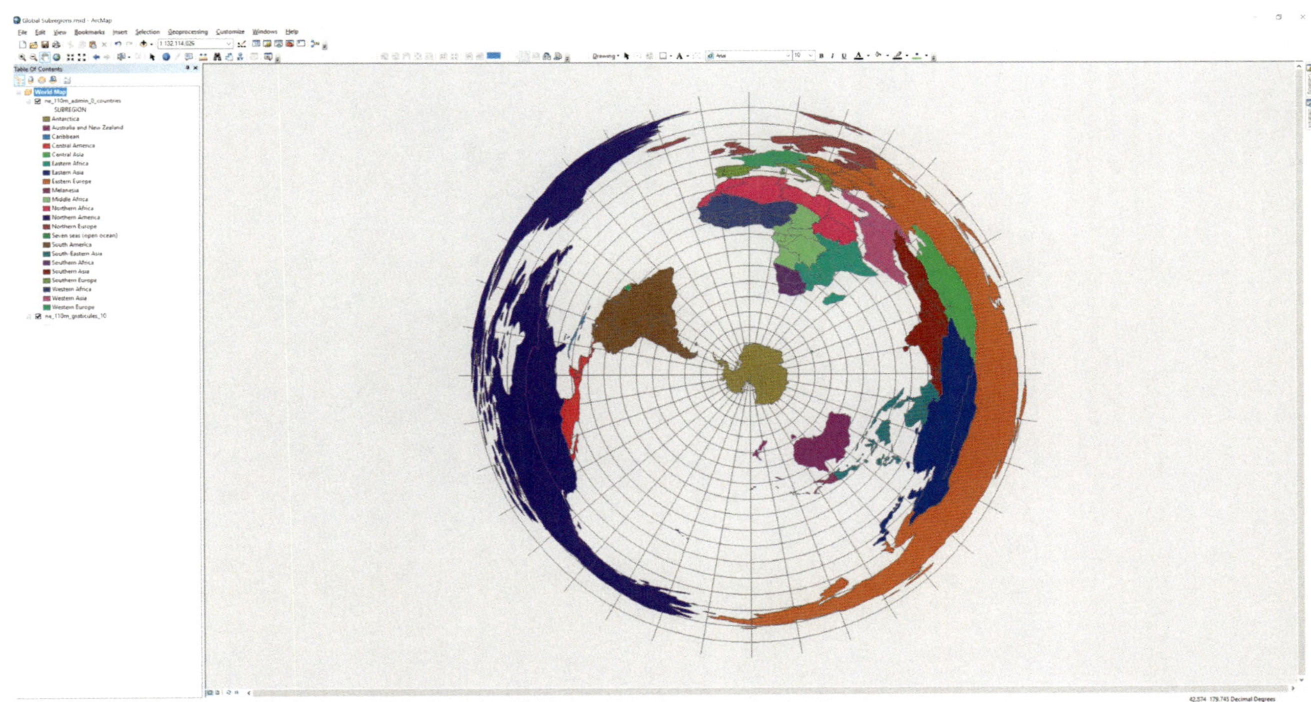

Figure 3.79: Cartographers frequently use the Azimuthal Equidistant projection used in world maps for aviation where the map centers on a specific airport.

Take a moment to review the text earlier in this chapter regarding the Azimuthal Equidistant projection. When done, continue to the next step.

Compromise Distortion Projections

Compromise distortion projections do not try to preserve any of the map projection properties. Instead, they are designed to appear aesthetically pleasing, balancing shape, and area. Using what you have learned, change the data frame projection to *Robinson (world)*. Locate it in the *World* folder under *Projected Coordinate Systems*. Change the Robinson projection so that it is centered on Humboldt County using a central meridian of -124 (**Figure 3.80**). Name the customized projection *Robinson Humboldt*.

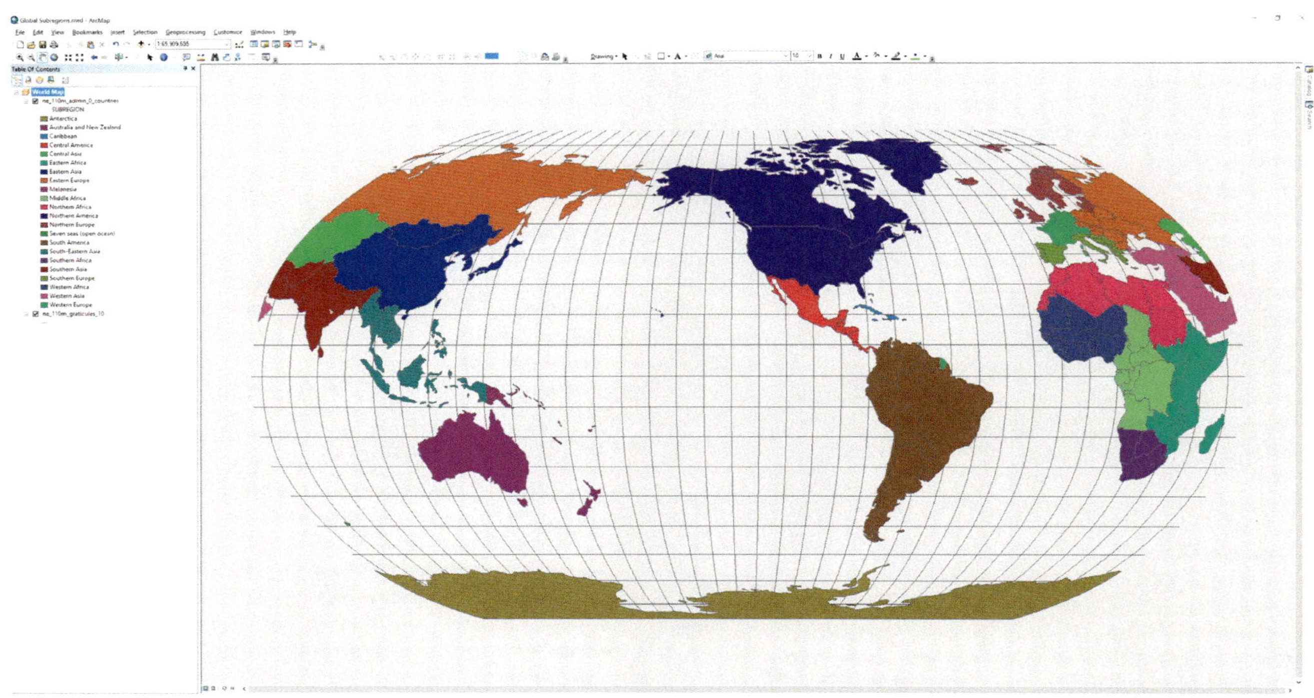

Figure 3.80: The Robinson projection does not use any of the standard developable surfaces. Instead, this type of projection is known as a pseudocylindrical projection.

Take a moment to review the text earlier in this chapter regarding compromise distortion projections. When done, continue to the next step.

Skill Drill: Choose Your Projection

Take some time to explore additional projections **not covered** in this chapter. You should not use the following projections in this step:

» Albers Equal Area Conic
» Azimuthal Equidistant
» Equal Earth
» Gall's Orthographic
» Gnomonic
» Lambert Azimuthal Equal Area
» Lambert Conformal Conic
» Mercator
» Orthographic
» Robinson
» Stereographic
» Transverse Mercator

Choose a projection that you find interesting. After choosing a projection, search the internet for more information about the projection to determine the type based on developable surface, aspect, case, and preserved properties. If possible, customize the projection so that it centers on a country or region you would like to visit someday. You may need to change the central meridian, the latitude of origin, or both[3].

3 Some projections do not allow for these types of customizations.

Using the skills you acquired in previous chapters, create a map poster with a size of 16 by 9 inches (**Figure 3.81**). Include a map title with a title bar using the name of the map projection. Add some descriptive text within a shaded text box below the map explaining what you know about the projection and any changes you made. Add a map legend. The legend should only display the subregions. Include a descriptive legend title. Be sure there are no underscores in the legend labels or title. Also, add your name and the year near the bottom of the poster. Include an acknowledgment for Natural Earth that says, "Data for this map was obtained from Natural Earth." You do not need a scale bar nor a north arrow for this map, so leave them out. When done creating your map poster, export the map as a PDF file with a resolution of at least 300 dpi. Save it to your *final* folder.

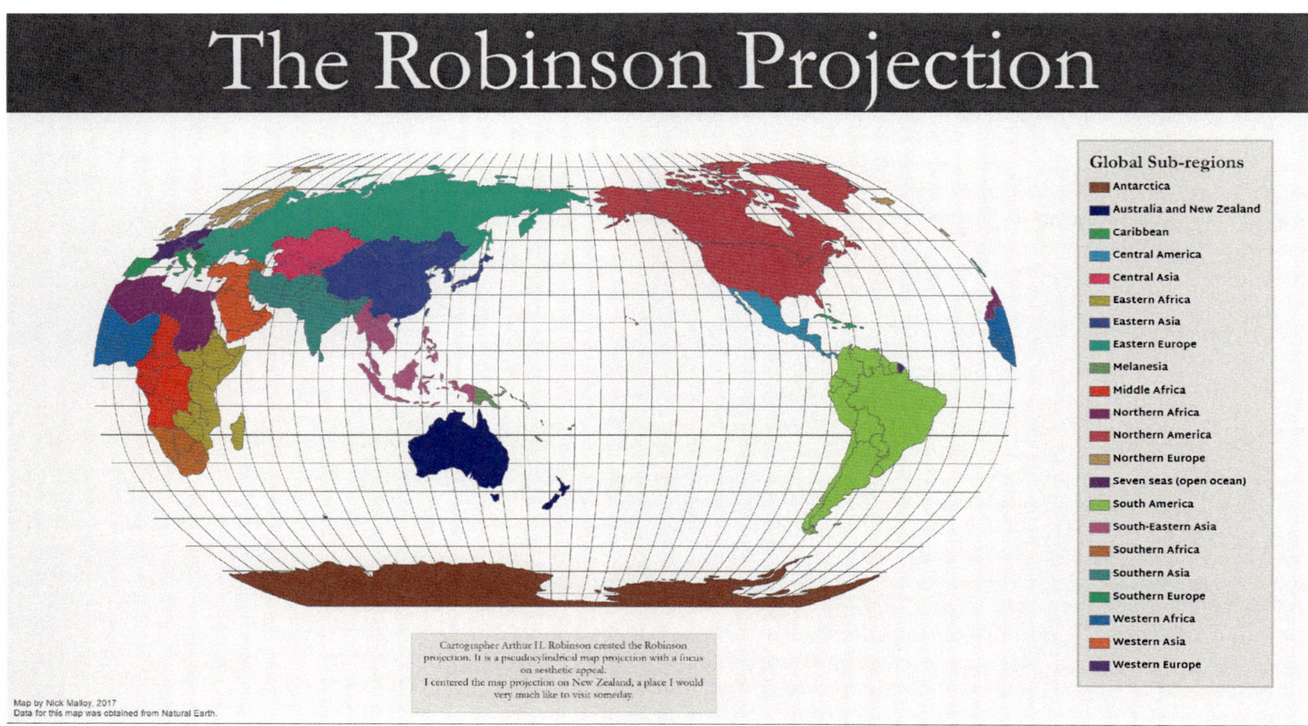

Figure 3.81: This example uses the Robinson projection, by Arthur H. Robinson. You should choose a *different* projection.

Tutorial: Working with Projections

The goal of this activity is to learn how to manage map projections using ArcGIS. In this activity, you create and organize a workspace folder using a standardized folder structure. You then download and decompress the data from several different data sources. After inspecting the spatial reference properties, you use the *Project* tool to create a collection of data in a specific map projection.

Estimated time to complete this tutorial: 2 hours

Learning Outcomes

Readers should be able to accomplish the following outcomes by the end of this tutorial:

» Summarize the steps for creating and organizing a project workspace folder structure

» Illustrate the ability to download data from a public source

» Demonstrate decompressing zipped files into a specific folder

» Practice looking up properties for both a data frame and a shapefile in ArcGIS

» Describe the difference between the data frame projection and a dataset's projection

» Illustrate how to use the *Project* tool

» Practice compressing a project level folder with contents intact

Setting up Your Workspace

In a typical workflow, you work on geospatial data using a local hard drive. When done, you compress your data and back up your work to your cloud storage so that you can retrieve the files from anywhere. When referring to a **local hard drive**, it means you are working on data physically located on the computer in front of you.

In contrast, some computers also include networked drives. **Networked drives** link to cloud storage and save the data elsewhere. Examples include services like OneDrive or Google Drive. For this tutorial, use the **desktop** as your local hard drive location. You may also use an external USB drive if you plan to work in multiple places.

You must avoid using networked drives while you work. They increase the processing time and can cause technical glitches.

In this tutorial, you use a particular folder structure. Start by creating your **workspace folder** on the local hard drive. A **workspace** is a folder or series of folders that contain all of your project files. The top-level folder in your workspace should indicate the lab assignment or the project. Organize all of your work within the workspace folder. On your desktop, create a new folder and give it a descriptive name, such as *Map_Projections*. Be sure there are *no spaces*. You may use underscores instead of spaces. Inside this folder, create the following three subfolders: original, working, and final. Having a standardized folder structure helps to keep a project organized, primarily when you are working with multiple partners. The folder structure you see here (**Figure 3.82**) is the standard used in each of the tutorials presented in this book.

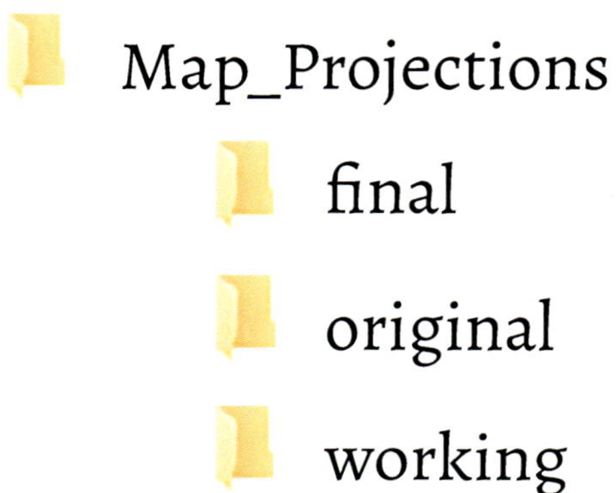

Figure 3.82: This diagram represents a basic folder structure used in this book.

As the name indicates, use the **original folder** for storing original, unaltered data. As you are working on a project, if for some reason your working version of the data gets lost or corrupted, you can go back to your *original* folder and find a fresh copy of the data. Use the **working folder** for data that you create or alter while working on your project. Use the **final folder** for storing any output you produce as a result of your work, such as images, maps, tables, or reports. Setting up the standard folder structure for a project is good practice and a habit you want to develop.

Downloading Data from Natural Earth

Natural Earth is a website created to provide vector and raster data to meet the needs of cartographers using a variety of software applications. The data on this website is free to use without restrictions.

Open the **Chrome** browser and navigate to the ***Natural Earth website***[1]. Click the Downloads link. Under Large-scale data, click the Cultural link (**Figure 3.83**).

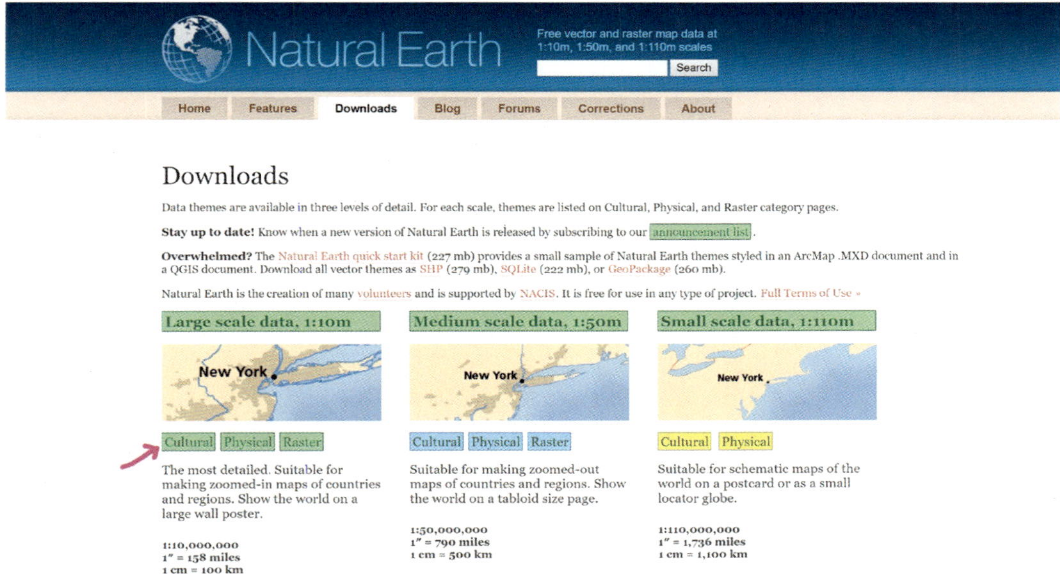

Figure 3.83: Natural Earth provides physical and cultural data in large, medium, and small scale.

Under Admin 1-States, Provinces, right-click *Download states and provinces* (**Figure 3.84**) Select, *Save Link As*, then navigate to your *original* folder and save.

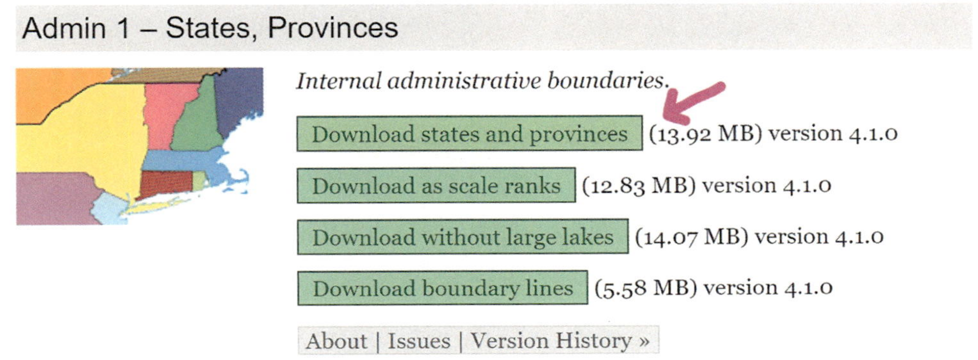

Figure 3.84: The image here shows the Natural Earth Cultural Vectors download page for small-scale data.

---

1 http://www.naturalearthdata.com/

Skill Drill: Download Data From Natural Earth

Repeat these steps for the Populated Places (**Figure 3.85**) and the Roads datasets (**Figure 3.86**). In a previous activity, you learned how to decompress a file using 7zip. In Microsoft Windows, navigate to your *Original* folder. You should see the three zip files, one for states and provinces, one for populated places, and one for roads. For each zip file, right-click, select 7zip, then *Extract Here*. When done, be sure to delete the zip files. You do not need them anymore. Removing them saves space and prevents confusion later.

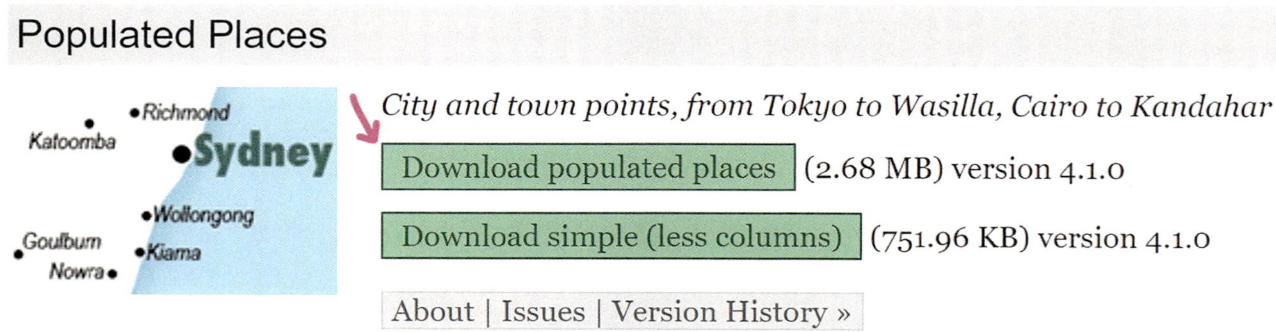

Figure 3.85: Natural Earth provides data on populated places as point features.

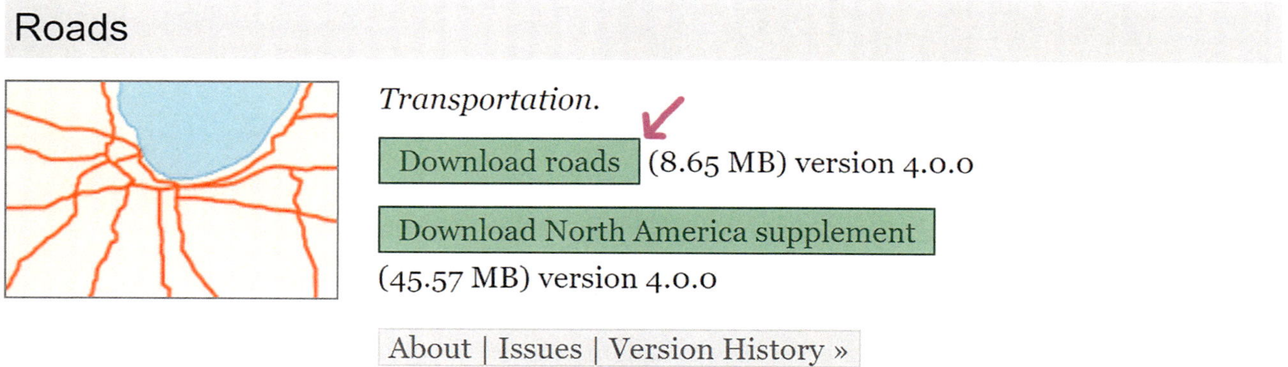

Figure 3.86: Natural Earth provides data on transportation as linear features.

Adding Data to ArcMap

Locate ArcMap on your computer and launch the software. When you first launch ArcMap, a window appears that gives you the option to either open a blank map or to open any of your recent map documents. Choose to open a blank map document. The ArcMap user interface has three main windows that you use regularly, the **Table of Contents**, the **data frame**, and the **Catalog Window** (**Figure 3.87**). The *Table of Contents* displays a list of data frames and map layers loaded into the map documents. The data frame, sometimes spelled dataframe, represents the layers as a map and defines the map extent. The Catalog Window displays a hierarchical view of folder connections and data as a **Catalog Tree**.

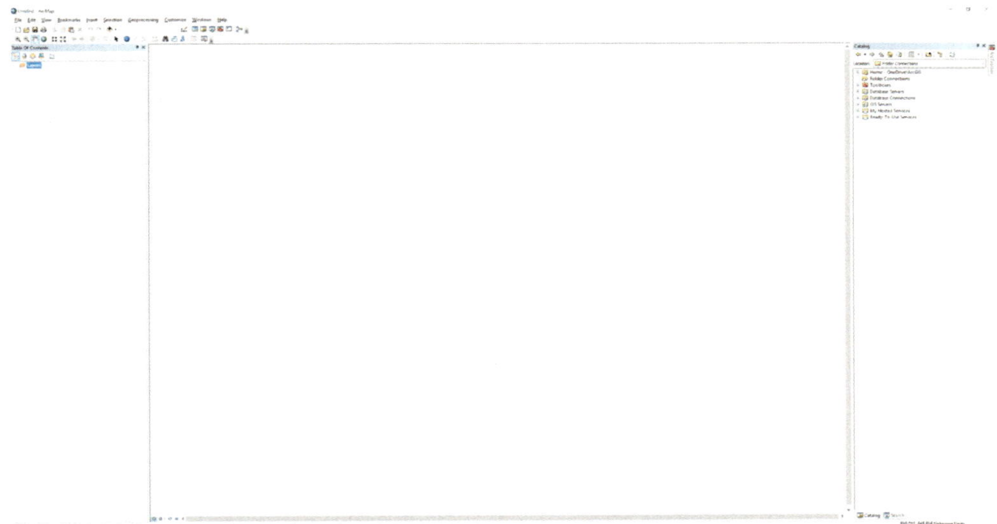

Figure 3.87: From left to right, the three main windows are the Table of Contents, the data frame, and the Catalog Window.

The icon for the catalog window looks like a yellow file cabinet (**Figure 3.88**). If your catalog window is missing, you can find the icon on the toolbar across the top. Click to open it back up.

Figure 3.88: Click the yellow file folder icon if your Catalog Window is missing.

In this step, you start by working with the Catalog Window. The catalog window is where you manage your geospatial data in ArcMap. Start by clicking on the *Connect to Folder* button (**Figure 3.89**). Navigate to your workspace folder, *Map_Projections*, on your local hard drive.

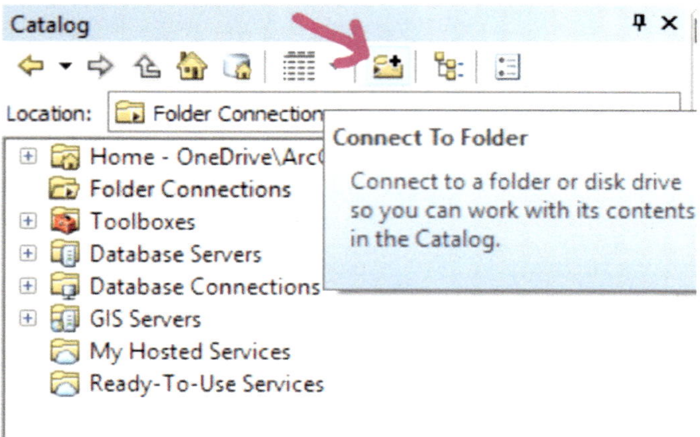

Figure 3.89: You connect to a folder so that the contents appear in the Catalog Window.

In this step, it is essential to select your primary workspace folder, *Map_Projections*, to add it to the Catalog Window (**Figure 3.90**). When ready, click *OK*.

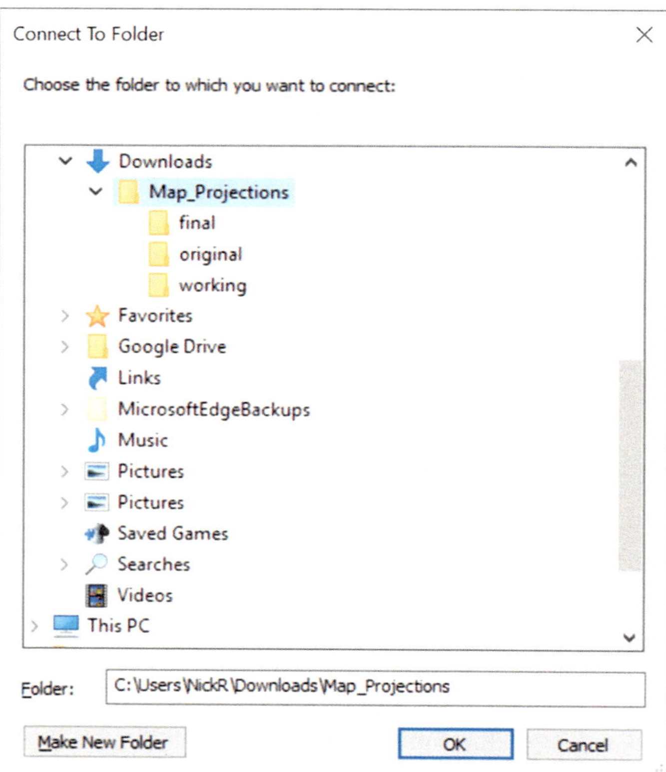

Figure 3.90: *Connect to Folder* adds the folder to the Catalog Window.

Once you add your workspace folder to the Catalog Window, expand the folder by clicking on the plus sign. You should see your three subfolders inside. Expand the *original* folder to view the contents. This display of folders and files within the Catalog Window is sometimes called the **Catalog Tree** (**Figure 3.91**).

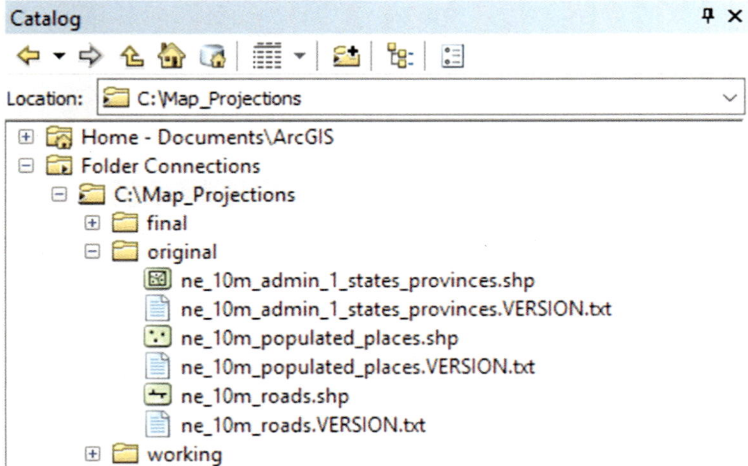

Figure 3.91: The contents of the *original* folder are displayed and organized in the Catalog Window. This display of folders and files is sometimes called the Catalog Tree.

In a previous activity, you learned how to add data to the map by dragging and dropping shapefiles from the Catalog Window to the data frame. Add the shapefile for the states and provinces data (**Figure 3.92**).

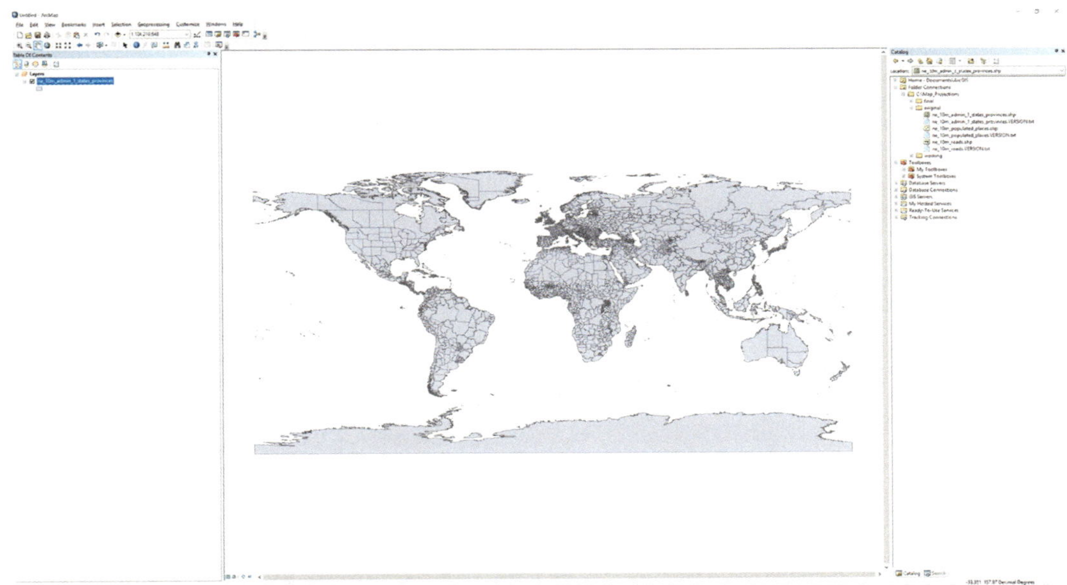

Figure 3.92: The states and provinces layer is an example of a vector data model representing a polygon feature.

Skill Drill: Adding Data and Saving the Map Document

Repeat these steps to add the populated places and the roads shapefiles. In a previous activity, you learned how to set your map document properties to store relative paths (**Figure 3.93**). Do this now. Then, save your map document to your workspace folder, *Map_Projections*. Call the map document "Working with Projections."

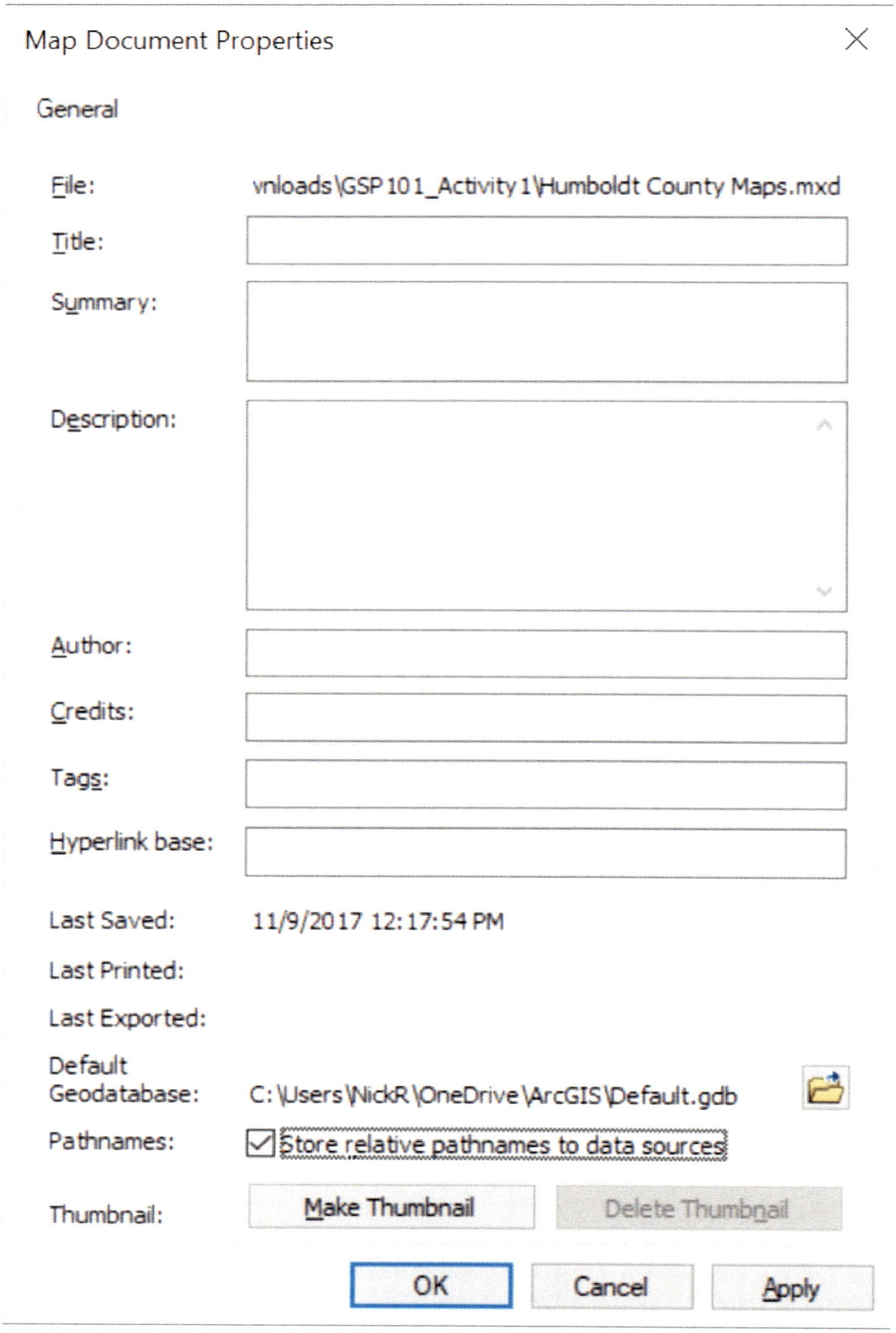

Map Document Properties

General

File: vnloads\GSP 101_Activity 1\Humboldt County Maps.mxd

Title:

Summary:

Description:

Author:

Credits:

Tags:

Hyperlink base:

Last Saved: 11/9/2017 12:17:54 PM

Last Printed:

Last Exported:

Default Geodatabase: C:\Users\NickR\OneDrive\ArcGIS\Default.gdb

Pathnames: ☑ Store relative pathnames to data sources

Thumbnail: Make Thumbnail Delete Thumbnail

OK Cancel Apply

Figure 3.93: Always make sure to check the Store relative pathnames to data source option.

Checking the Spatial Reference System

In this chapter, you learned that a flat representation of the spherical Earth is made possible by using map projections. A **map projection** is a geometric transformation of the round Earth onto a flat plane using mathematical equations. In ArcGIS, the data frame window displays the map using **project-on-the-fly**. Even though the layers in the Table of Contents might use different projections and coordinate systems internally, ArcGIS tries to line them up on the screen using the coordinate system defined in the data frame properties. What this means is that you can change the projection in the data frame to alter the appearance of the map onscreen without modifying the original data files.

When using data for display purposes, project-on-the-fly is not a problem. However, in Chapter 7, you learn how to conduct a geospatial analysis. When doing so, it is critical for each dataset used in the analysis to use the same spatial reference system to avoid introducing unwanted spatial errors. A **spatial reference system** defines a geographic location and includes the map datum, the **coordinate reference system**, and if relevant to the data, the map projection. Chapter 4 discusses spatial reference systems in more detail. For now, focus on checking the spatial reference system properties in ArcMap.

In the Table of Contents, the word *Layers* appears above the three map layers. I believe Esri chose very poorly when naming this element in the Table of Contents. The name consistently leads to confusion. This item on the Table of Contents is not a layer. It represents the **data frame**. Access the properties for the data frame by right-clicking on the word *Layers* in the Table of Contents. Select, Properties from the contextual menu (**Figure 3.94**).

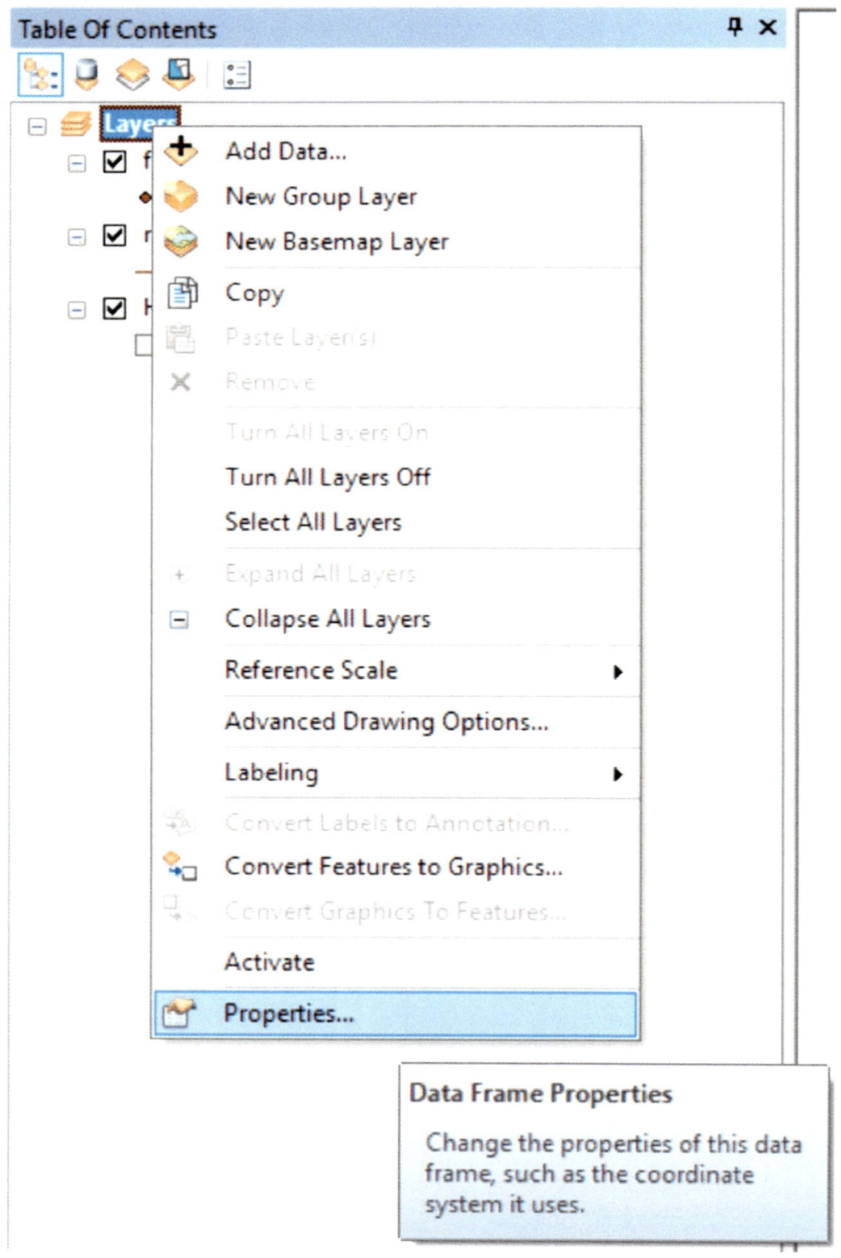

Figure 3.94: You can access contextual menus in the Table of Contents by right-clicking on layers and data frames.

When the *Data Frame Properties* window opens, click the *General* tab. Change the name of the data frame to something more descriptive, such as *Global Cities*. Click *Apply*. The name updates in the Table of Contents. Navigate to the *Coordinate System* tab and review the contents under *Current coordinate system*. Here you see information about the spatial reference system of the data frame (**Figure 3.95**). It starts with GCS WGS 1984. In this example, the datum used for this coordinate system is WGS 1984.

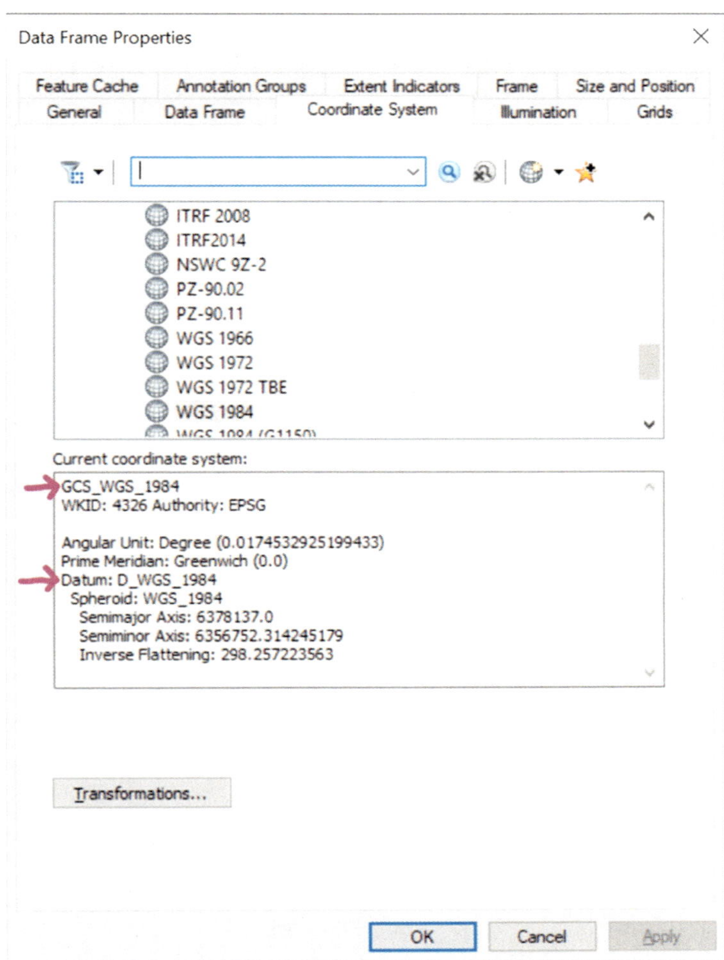

Figure 3.95: The Coordinate Systems tab in the data frame properties allows you to choose a spatial reference system from two main folders, Geographic Coordinate Systems and *Projected Coordinate Systems*.

Don't worry if you don't understand what this means. Chapter 4 covers coordinate systems in detail. For now, make a note that if you see the abbreviation **GCS**, it stands for **Geographic Coordinate System**. A **geographic coordinate system (GCS)** is a coordinate-based spatial reference system using latitude and longitude. An important fact to remember is that a GCS is also a *datum-based* coordinate system. This format makes it useful for defining a location on the surface of the earth. However, it is more complicated to use for measuring distances and areas because it uses angles in degrees as the primary units of measure. You may notice that there is no information about projections. Currently, the data frame only uses a geographic coordinate system and a datum[2].

2 You may be wondering why does the data frame display the map as two-dimensional (2D) instead of spherical (3D)? This issue is another potential problem with ArcGIS's project-on-the-fly behavior.

ArcGIS uses a **display projection** for showing geographic data on the screen. According to ***Esri's GIS Dictionary***[3], a display projection is a pseudo-Plate Carree projection used by ArcGIS to present data that is in a geographic coordinate system. The ArcGIS software maps the angular values of the geographic coordinate system to the display screen just as with the values from a projected coordinate system. Unfortunately, this **project-on-the-fly** behavior can lead you to mistakenly believe you are working with projected data when you are not. Better and less confusing behavior would be to display geographic data as a 3D model of Earth.

Close the data frame properties window. By default, the data frame adopts the spatial reference of the first layer added to the Table of Contents. If this is true, you can infer that the states and provinces layer has the same spatial reference as the data frame, GCS WGS 1984.

Open the *Layer Properties* for the states and provinces by right-clicking on the layer in the Table of Contents and selecting *Properties* from the contextual menu. Navigate to the *Source* tab (**Figure 3.96**).

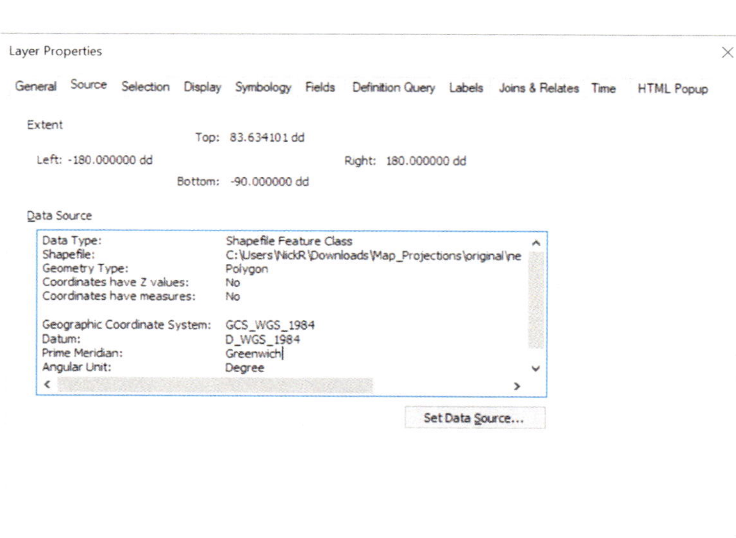

Figure 3.96: The *Source* tab provides information on the data, including file type, file location, and spatial reference information.

As you can see, the states and provinces layer also has the Geographic Coordinate system GCS WGS 1984. Close the *Layer Properties* window.

3 https://support.esri.com/en/other-resources/gis-dictionary

Skill Drill: Checking Spatial Reference Systems

Repeat these steps and check the spatial reference systems for the populated places and the roads. Make a mental note about whether they are all the same or if they are different from the states and provinces and the data frame.

Using the Project Tool

In a previous tutorial, you changed the spatial reference system of the data frame to view the appearance of different projections. In this instance, the ArcGIS software used project-on-the-fly to alter the display on the screen. However, the original data remained unchanged. In most cases, the spatial reference system of geospatial data is immutable. In computer science, an **immutable** object or property is one whose state cannot be modified after creation. What this means in this example is that the spatial reference systems for the shapefiles representing populated places, roads, and states and provinces cannot be changed.

Earlier, you learned that geographic coordinate systems (GCS) are not practical to use for measuring distances and areas because they use angles in degrees as the primary units of measure. For analysis, using a projected coordinate system (PCS) is necessary to improve both ease of use and for improving accuracy. Also mentioned earlier, when conducting a geospatial analysis, it is critical for each dataset used in the analysis to use the same spatial reference system to avoid introducing unwanted spatial errors.

You might be wondering, how can I use projected data if the spatial reference system is immutable? The answer is to create a new dataset with the desired spatial reference system and projection.

ArcGIS uses a tool called **Project** to create a new copy of a dataset with the spatial reference system you define. The *Project* tool takes the information from the original dataset and performs a mathematical transformation to convert one spatial reference system to another. It then uses the converted spatial information to create a copy of the original data. Once copied, the spatial reference system of that copy also becomes immutable.

In ArcMap, click the toolbox button located on the *Standard* menu. The icon looks like a red toolbox. The ArcToolbox may appear floating above the map. Dock it to the side by dragging it to the left in a similar manner in which you docked the attribute table in previous activities. Expand the *Data Management Tools* (**Figure 3.97**). Next, expand the *Projections and Transformations* toolbox. Locate the *Project* tool. Double-click the tool to open it.

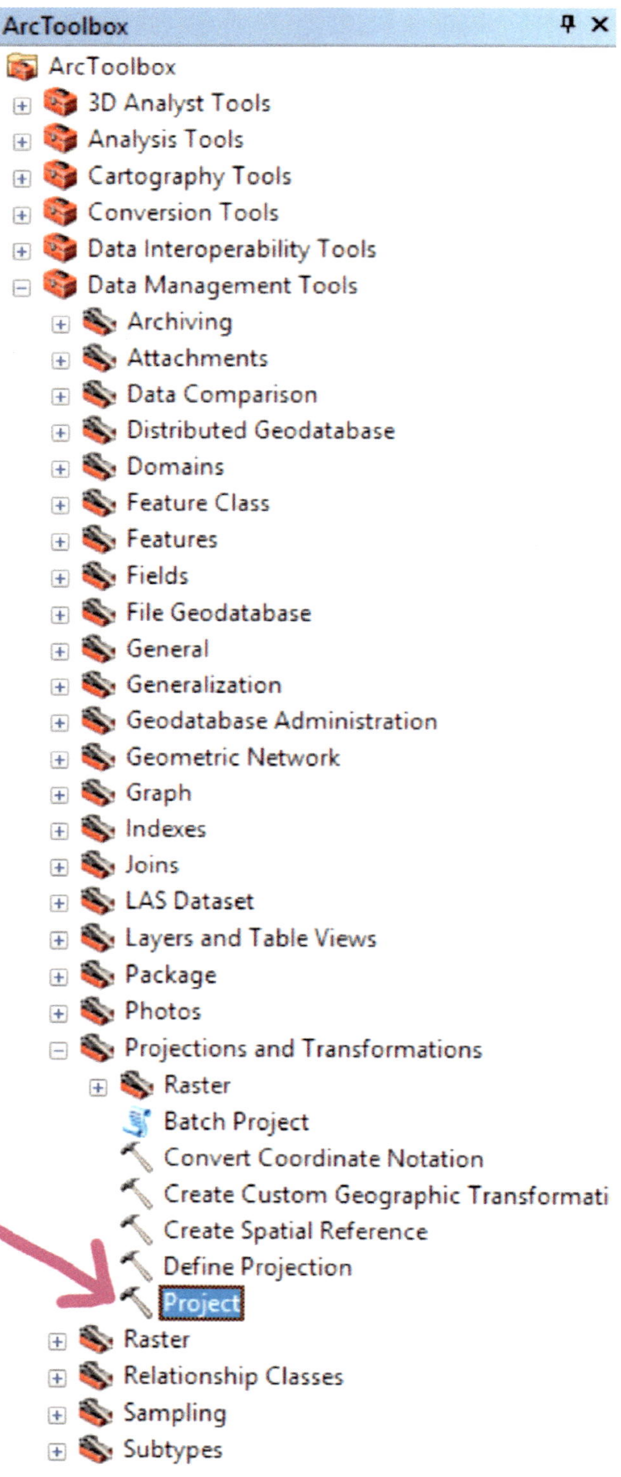

Figure 3.97: The *Project* tool is used for vector data.

For the *Input Dataset or Feature Class*, use the drop-down menu to select the layer representing states and provinces. For the *Output Dataset or Feature*, Class click the yellow file folder icon and browse to your *working* folder. Name the file States_Provinces_NAD83_UTM10N.shp.

For the *Output Coordinate System*, click the button on the right. When the *Spatial References Properties* window opens, click the plus sign next to *Projected Coordinate Systems* to expand the folder. Scroll down until you see the *UTM* folder. Expand it, then open the *NAD 1983* folder. Select NAD 1983 UTM Zone 10N and *click* OK. You may leave all other default settings and *click* OK (**Figure 3.98**).

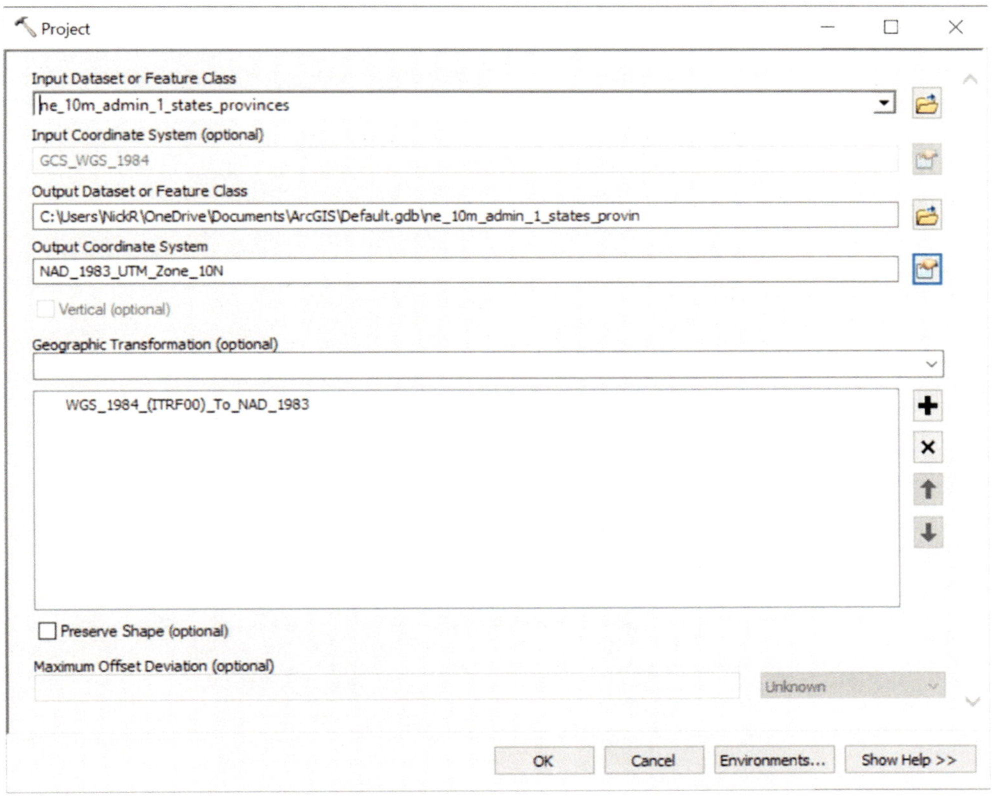

Figure 3.98: Check to make sure your settings match those in this image.

When the *Project* tool completes, ArcMap may not automatically add the new layer to the map. In the Catalog Window, open the *working* folder and locate the new shapefile. Add the data to the map. You may get a warning. ArcMap is telling you that the data frame coordinate system does not match the layer you are about to add to the Table of Contents. Since ArcGIS uses project-on-the-fly, you might not realize it when data you add has a different spatial reference system. In this instance, the data frame uses GCS

WGS 1984, and the new states and provinces layer uses NAD 1983 UTM Zone 10N. Check the box that says, Don't warn me again in this session, and click Close.

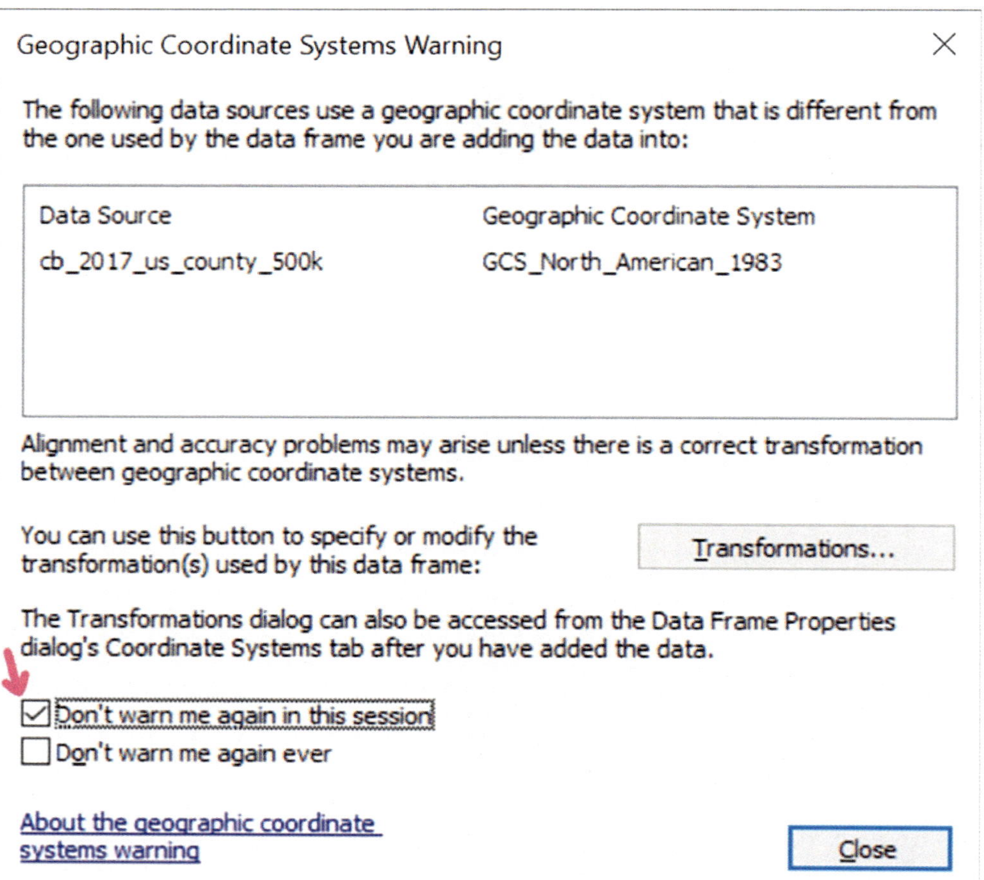

Figure 3.99: While it is OK to turn off this warning for one session, you should never turn it off permanently.

Since you do not need it anymore, remove the original states and provinces layer from the Table of Contents. Eliminating unnecessary layers helps to keep the Table of Contents organized and helps to prevent confusion.

Zoom to the new states and provinces layer. You may notice that the layer with the UTM spatial reference system does have the same extent as the original data. You learn more about the UTM system in Chapter 4. For now, it is only necessary to know that the UTM system uses the **Transverse Mercator projection**. In this instance, the Transverse Mercator projection centers on Humboldt County, California (**Figure 3.100**).

Take a moment to review the text earlier in this chapter regarding the Transverse Mercator projection. When done, continue to the next step.

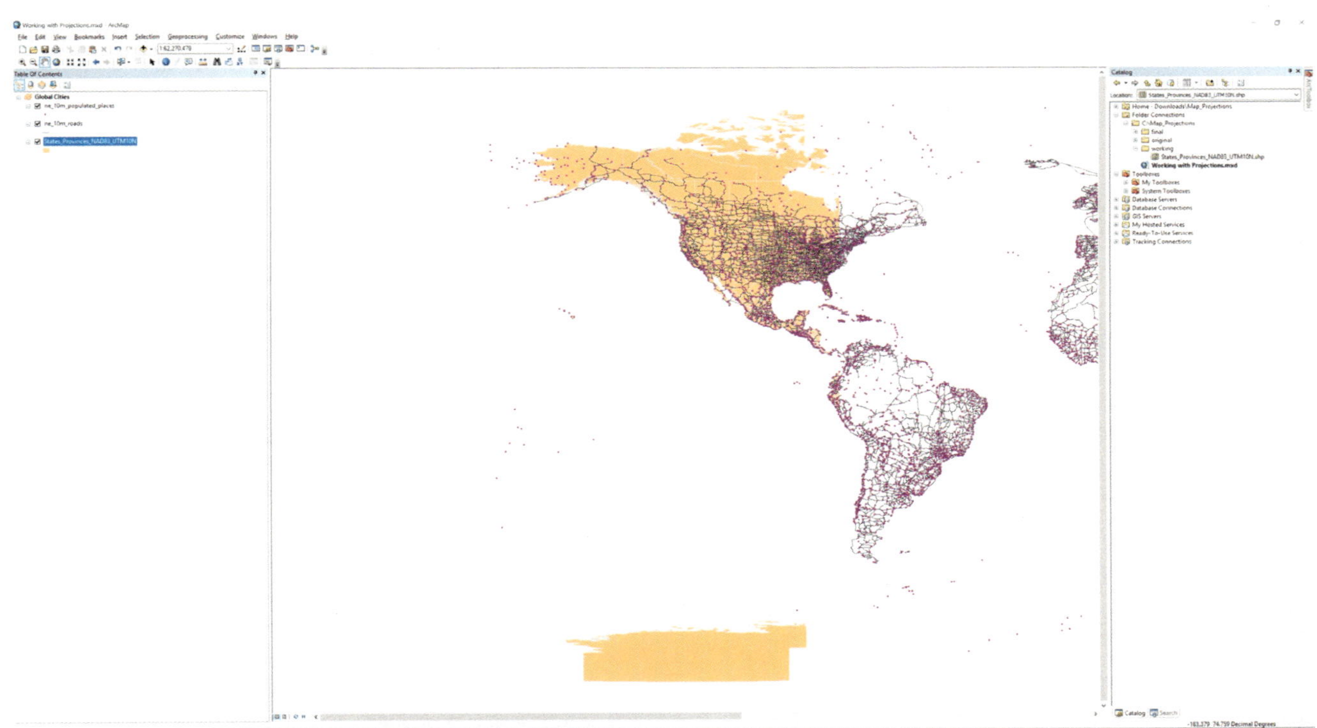

Figure 3.100: Though the new dataset uses the Transverse Mercator Projection, the data frame still uses the default display projection for geographic data.

Skill Drill: Using the Project Tool

Use the *Project* tool to create new datasets for the roads and the populated places layers that use NAD 1983 UTM Zone 10N. Be sure to save the output to your *working* folder. Add the new data to the Table of Contents and remove the original data layers. Also, change the spatial reference system of the data frame to match that of the layers (**Figure 3.101**). When done, be sure to save your map document.

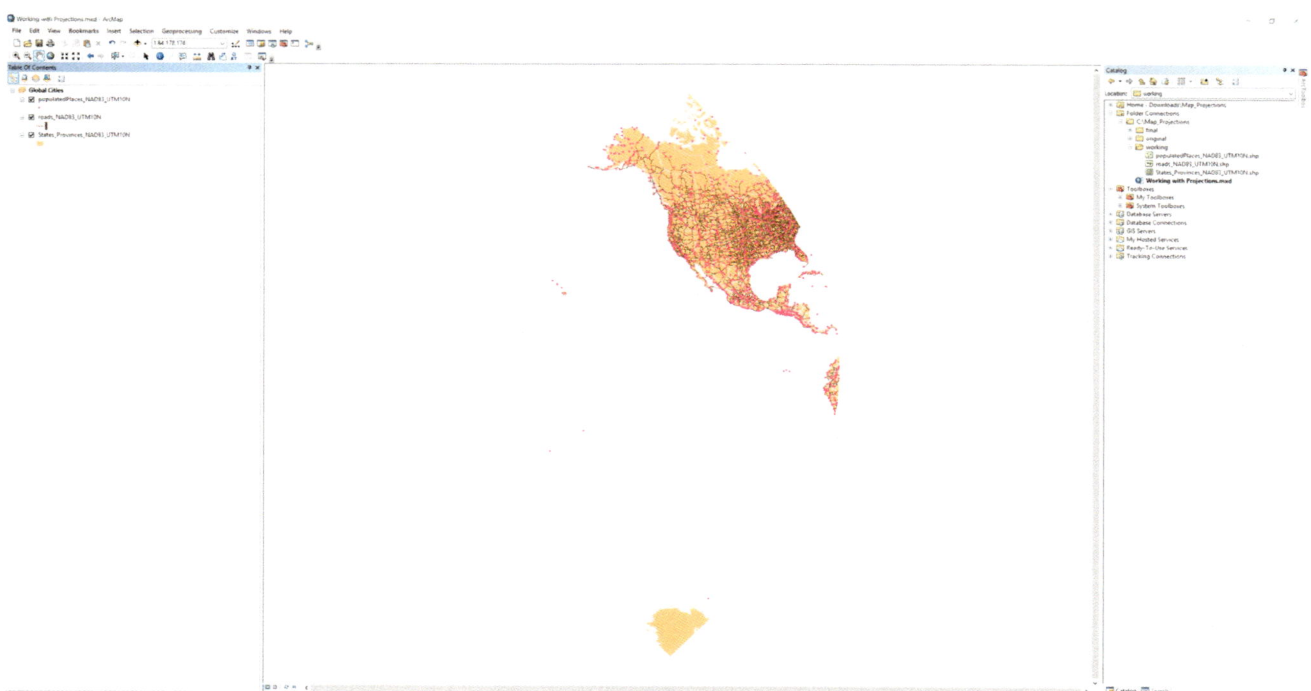

Figure 3.101: All of the layers and the data frame are now in the UTM spatial reference system.

Skill Drill: Compressing the Workspace Folder as a 7z file

Sometimes, when working with geospatial data, it is necessary to move the data from one location to another. Compressing your workspace folder with the sub-folder structure intact is a safe way to do so. The zip file format is a universal file compression type. However, other file compression types, such as 7z, sometimes work better. After saving your work, close your map document. Compress your *Map_Projections* folder as a 7z file. When done, you should see your workspace folder compressed as a 7z file (**Figure 3.102**).

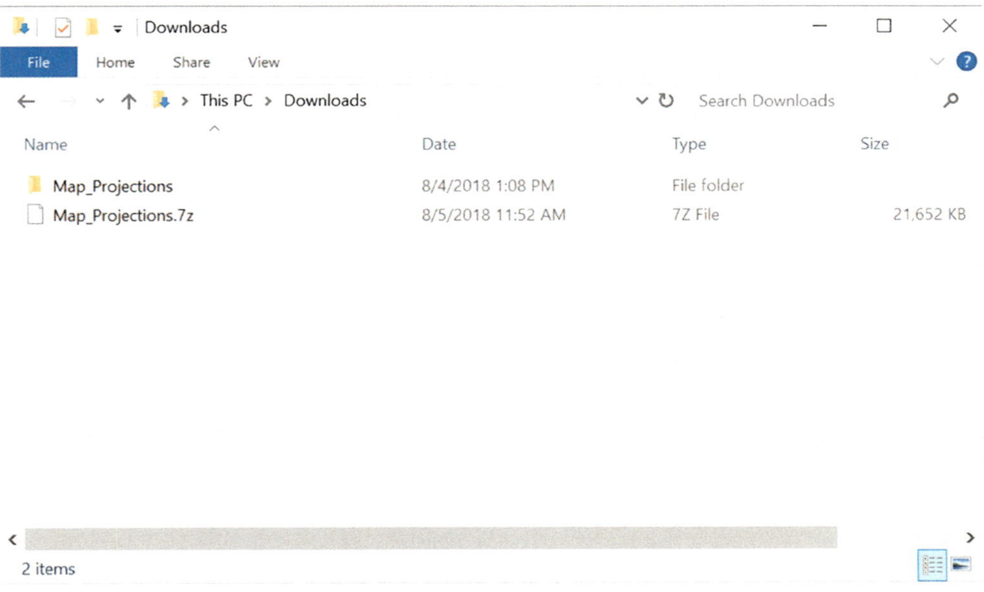

Figure 3.102: Check the file type to be sure it is a 7z file.

Principal Terms

Albers Equal Area Conic projection
antipodal meridian
aspect
authalic sphere
azimuth
Azimuthal Equidistant projection
azimuthal projections
case
Catalog Tree
Catalog Window
Clark 1866 ellipsoid
compromise distortion projections
conformal projections
conical projections
conic projection surface
continuity
coordinate reference system
cylindrical projections
cylindrical projection surface
data frame
datum
developable surface
display projection
elevation
equal-area projections
Equal Earth projection
equator
equatorial aspect
equidistant projections
final folder
Gall's Orthographic projection
generating globe
geocentric datum
geodesy
geodetic datum
Geodetic Reference System of 1980 (GRS 80)
geographic coordinate system (GCS)
geoid
gnomonic projection
graticule
great circle
horizontal geodetic datums
immutable
Lambert Azimuthal Equal-Area projection
Lambert Conformal Conic projection
latitude
line of tangency
local hard drive
longitude
map projection
map scale
Mercator projection
meridians
networked drives
normal aspect
North American Datum 1927 (NAD27)
North American Datum 1983 (NAD83)
oblique aspect
origin
original folder
parallels
perspective
planar projections
planar projection surface
polar aspect
preserved properties
area
distance
shape

prime meridian
project-on-the-fly
Project tool
pseudocylindrical projection
reference ellipsoid
rhumb lines
Robinson projection
secant case
spatial reference system
standard circle
standard meridian
standard parallel
standard point
Stereographic projection
Table of Contents
tangent case
transverse aspect
Transverse Mercator projection
true-perspective projections
vertical geodetic datums
WGS84 ellipsoid
working folder
workspace
World Geodetic System of 1984 (WGS84)

Chapter 4: Map Scale and Spatial Reference Systems

Most people have the idea that coordinate systems are static, unchanging definitions of where they are. One can log on to google maps and look up their latitude and longitude coordinates and feel confident that these numbers have a universal meaning that does not change. In reality, the numbers one sees on google maps are just one of many versions of latitude and longitude coordinates that can define their location. Determining a position on earth in a way that is meaningful to others is a difficult challenge. In part, the difficulty is due to the variations in map projections and datums used across the world, which can change longitude and latitude coordinates in different ways. It may seem like a small detail, yet boundary definitions and positional information can have significant legal, political, and military consequences. Chapter 4 presents how distance and location are defined and communicated using map scale and spatial reference systems.

Learning Outcomes

Readers should be able to accomplish the following outcomes by the end of this chapter:

- Explain the difference between map scale versus scope
- Express three ways of representing map scale
- Recognize large-scale and small-scale maps
- List common distance equivalents
- Perform scale conversions
- Determine an unknown scale with a variety of references
- Describe early land partitioning systems
- Identify the differences between geographic and projected spatial reference systems
- Demonstrate correct notation for a variety of spatial reference system

MAP SCALE

Most people are familiar with the term large-scale and small-scale as a measure of **scope**. For most, describing something as grand, large, or complex is synonymous with large-scale in terms of scope. Similarly, when describing something as minor, small, or simple, it is synonymous with small-scale. When it comes to maps, scale refers primarily to measurement and extent, which many find counterintuitive and confusing. **Map scale** describes the relationship between distance on Earth and the way that a map represents the same distance. It is the ratio of map distance over ground distance (**Figure 4.01**). In this book, **ground distance** refers to the distance on Earth. For example, one might come across a map where one inch on the map is equivalent to one mile on earth. One mile represents the ground distance. The relationship between the map distance and the ground distance defines the map scale.

$$\text{Scale} = \frac{\text{Map Distance}}{\text{Ground Distance}}$$

Figure 4.01: In this definition of scale, ground distance represents the distance on Earth.

There are three common ways to express scale:

» Representative fraction
» Verbal scale
» Graphic Scale

Watch *Understanding Map Scale*, by Marcos Luna to learn more (**Figure 4.02**).

Figure 4.02: This video by Marcos Luna discusses map scale in additional detail. URL: *https://youtu.be/n7na5SP3pAU*

Representative Fraction

A **representative fraction (RF)** is merely the relationship described above, map distance over ground distance, expressed as a ratio or fraction. An RF is written using the notation 1/x or 1:x, where 1 represents the distance on the map and x represents the distance on the ground. Because the RF is a ratio, it is independent of any unit of measurement (**Figure 4.03**). For example, an RF of 1:24,000 means that any unit measured on the map is 24,000 times that same unit on the ground. Therefore, one inch on this map would represent 24,000 inches on the ground. One centimeter on this map would represent 24,000 centimeters on the ground.

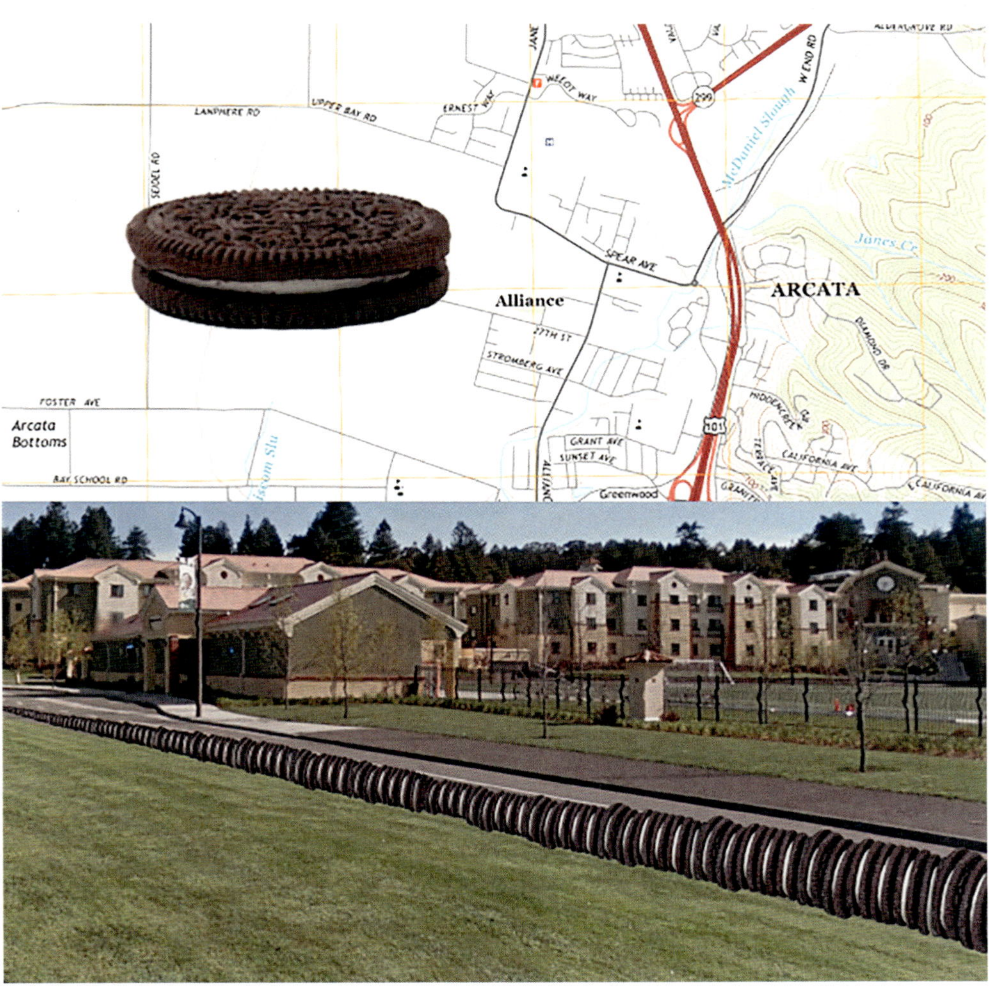

Figure 4.03: One can substitute any unit, and it will always be right. So, if one wanted to use a novel unit of measurement such as an Oreo cookie, one Oreo cookie on the map would represent the length of 24,000 Oreo cookies on the ground.

Verbal Scale

A **verbal scale** uses a word statement to express scale. Verbal scales have mixed units because the statement will refer to the map using one type of unit while referring to the distance on the ground using another type of unit. For example, one may refer to a map that has an RF of 1:63,360 using the verbal scale, *one inch to one mile*. The map distance uses inches while the ground distance uses miles. Thus the units of measurement are mixed. A verbal scale is a user-friendly format that easily allows the map-reader to understand the relationship between the map and the real world.

> ***It is acceptable for verbal scales to represent a close approximation to the actual map scale since the map reader likely uses another means for measurement or calculation.***

When converting an RF to a verbal scale, it is okay to round off to the nearest whole number. For example, one may refer to a map that has an RF of 1:16,000 is *one inch to 133.333 feet*. This phrase is not a user-friendly verbal scale, so it would be perfectly acceptable to say *one inch to 130 feet*. For this reason, cartographers seldom use verbal scales as the primary map scale indicator in reference maps or navigation maps where accurate distance calculation is essential.

Graphic Scales

A **graphic scale** is a symbol representing ground distance. One of the most frequently used graphic scale symbols is a **scale bar**[1], typically a line or bar divided into sections using tick marks or colored regions (**Figure 4.04**). As with other map elements, there are usually many choices when it comes to selecting the style and appearance of the scale bar. Although it may be tempting to choose a more visually complex version, simpler is generally better, both to reduce visual weight and improve clarity and legibility.

Each division on a scale bar represents a unit of measure on the ground typically in feet, meters, miles, or kilometers. One reads scale bars from left to right, starting at zero and ending with the total distance represented by the scale bar. Sometimes scale bars come with an **extension scale**, which is an addition to the left of the zero on the scale bar. Typically, extension scales are the length of one division and are themselves divided into fractions of units for more precise measurements. When an extension scale is present, it is essential to remember to add the length of the extension to the total distance when using the entire scale bar to measure a feature on the map. Forgetting to add the

1 Other terms for scale bar include bar scale and line scale.

extension scale to the total length is a common mistake. Scale bars have the advantage of maintaining the correct proportion when enlarging or reducing map size. If one alters the size of the map in any way, the verbal scale and the RF will be incorrect as they are dependent on the original size of the map. For this reason, the scale bar is the one most commonly used on printed maps.

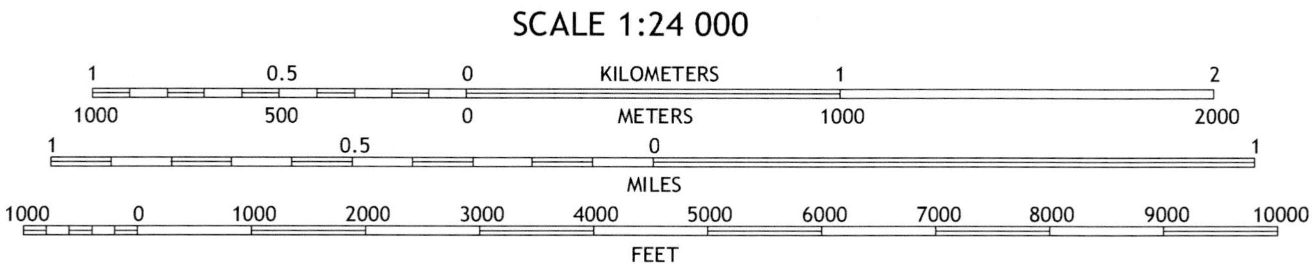

Figure 4.04: In this image, each of the three scale bars includes an extension to the left of the zero marks.

Large Scale and Small Scale

Since one expresses the map scale as a ratio or fraction of map distance over ground distance, when using the terms **large-scale** or **small-scale**, it typically means the opposite of the everyday usage of these terms. Smaller fractions have a higher number in the denominator, and bigger fractions have a lower number in the denominator. For example, ½ of a pie is larger than ¼ of the pie even though the denominator 4 is a higher number than the denominator 2.

For example, a map with an RF of 1:100,000,000 is smaller in scale than a map with an RF of 1:1,000. This concept may seem simple, but it is often a source of confusion. A map with an RF of 1:100,000,000 would probably show the entire world at once. The scope is large, so it would seem intuitive to call it a large-scale map. However, the fraction is very small, and thus, it is a *small-scale* map. The terms large-scale and small-scale are somewhat arbitrary, though useful for making broad categories in maps.

A large scale map is typically something like a city or county map. Large scale maps allow the mapmaker to include many features, whereas small scale maps get crowded very quickly. Think about how the level of detail changes on a Google map as one zooms out from large to small scale, from the neighborhood level to the city, state, country, and the world. Generally speaking, anything larger than 1:250,000 refers to a **large-scale map**. Cartographers refer to maps between 1:250,000 and 1:1,000,000 as a **medium-scale map**. Anything smaller than 1:1,000,000 refers to a **small-scale map**.

Converting Scale

Sometimes it may be necessary to convert scale expression from one form to another. For example, one can convert an RF of 1:24,000 into a verbal statement by changing the units of the denominator to feet. The verbal scale would read, *one inch to 2,000 feet*. It is possible to verbally express an RF of 1:24,000 literally by stating, "one inch on the map is equivalent to 24,000 inches on the ground." This phrase would be a true statement. However, verbal statements use mixed units for clarity. In this case, one would convert 24,000 inches into feet by dividing by 12. To create a scale bar, sometimes it is easiest to start from the verbal scale. In this example, 1 inch on the map is equivalent to 2000 feet on the ground. One can draw a scale bar by creating a series of 1-inch divisions and labeling each division in increments of 2000 feet.

Here are some common unit conversions frequently used for converting map scale:

- Inches × 2.54 = Centimeters
- Centimeters × 0.3937007874 = Inches
- Feet × 0.3048 = Meters
- Meters × 3.280839895 = Feet
- Yards × 0.9144 = Meters
- Meters × 1.093613298 = Yards
- Miles × 63,360.0 = Inches
- Miles × 5,280.0 = Feet
- Kilometers × 0.6213711922 = Miles

Determining Scale

One might come across a map or an aerial photo with an unknown scale. In this situation, one uses recognizable features of known size or refer to another map that includes the same extent and has a known scale. For example, suppose someone acquired a historical photograph of an aerial image depicting the Humboldt State University campus and needed to determine the scale of the image (**Figure 4.05**). Since the Redwood Bowl football field would be visible on the image, one can use the length of the field to determine scale.

Figure 4.05: This image depicts Humboldt State University in the 1970s. Source: Humboldt State University Library.

The football field represents 100 yards on the ground and thus used as the denominator of the scale ratio. Suppose measuring the football field on the image using a ruler resulted in a length of one inch. This number would provide one with the numerator.

Immediately, one could determine the scale verbally, “One inch to 100 yards.” Likewise, one could draw a graphic scale using one-inch increments to create a scale bar. One could determine the RF by converting yards to inches and simplifying the fraction (**Figure 4.06**). Baseball diamonds, tennis courts, and other sports facilities are generally all a specific size and thus used in this manner.

$$\text{Photograph RF} = \frac{\text{Map Distance}}{\text{Ground Distance}} = \frac{1 \text{ inch}}{100 \text{ yards}} = \frac{1 \cancel{\text{inch}}}{3600 \cancel{\text{inches}}} = \frac{1}{3600} \text{ or } 1{:}3600$$

Figure 4.06: Remember that an RF always has a numerator of 1, so one needs to simplify the fraction after measuring.

Determining Scale using a Reference Map

In most cases, one may not have a sports field handy and will have to determine scale using a known source. For example, suppose one had an aerial photo of Eureka, California. To determine the scale of the photo, one could use a map of Eureka with a known scale. Measuring the distance between two points on the photo, such as two street intersections, would give one the numerator of the scale ratio. Measuring the same two points on the map and using the map scale to determine the distance on the ground would give one the denominator. Then as in the previous example, one could use conversions to generate an RF, a verbal scale, and a scale bar.

Determining Scale using Geographic Coordinates

There may be a time when one needs to determine scale using geographic coordinates such as latitude and longitude. To determine the numerator of the scale ratio, one would measure the distance between two lines of longitude or two lines of latitude on the map. Next, one would have to convert latitude or longitude degrees into other units of measure such as miles or kilometers to determine the denominator of the scale ratio. To accomplish this, one has to know the length of a degree of longitude or the length of a degree of latitude. Because a geographic reference system uses an ellipsoid base, distances between longitude and latitude are not consistent. They depend upon the location and the reference ellipsoid used.

On the WGS 84 ellipsoid, the difference between the length of 1° of latitude ranges from 68.703 statute miles near the equator to 69.407 at the poles. Be aware that even using these precise numbers has an inherent amount of uncertainty and error since the reference ellipsoid is a generalization derived from a geoid model of the earth. Recall that lines of latitude are parallel, so for most purposes, it is safe to assume this distance is consistent at any longitude and that 1° of latitude is equal to approximately 69 miles. Latitude and longitude use a **sexagesimal numeral system**, a base 60 system commonly used to measure time and degrees. Every 1 degree divides into 60 minutes (60'), and every 1 minute divides into 60 seconds (60"). Since one assumes 1° of latitude is equal to 69 miles, one can divide 69 miles by 60 minutes to find out how many miles there are in 1 minute. Similarly, one can find out how many miles there are in one second (**Figure 4.07**).

1° of latitude = 69 miles

1° of latitude = 60 minutes

$$\frac{\text{69 miles}}{\text{60 minutes}} = \frac{\text{1.15 miles}}{\text{1 minute}}$$

1 minute of latitude = 1.15 miles

1° of latitude = 3600 seconds

$$\frac{\text{69 miles}}{\text{3600 seconds}} = \frac{\text{0.02 miles}}{\text{1 second}}$$

1 second of latitude = 0.02 miles

Figure 4.07: 1° of latitude is equal to 69 miles.

Converting Longitude to Ground Distance

The distance between 1° of longitude strongly relates to location. Recall that meridians of longitude converge at the poles. At the equator, the distance between 1° of longitude is approximately 69 miles. At the poles, the distance between degrees of longitude is 0 miles because all lines of longitude converge at the poles. Everything in between has a constant rate of change depending upon the degree of latitude. Therefore, one must calculate the distance from 1° of longitude to the next. The distance in miles between 1° of longitude is equal to the cosine of the corresponding latitude multiplied by 69 miles (**Figure 4.08**). Once on has map distance over the ground distance, one can use the conversion method described above to generate an RF, a verbal scale, or a scale bar.

$$1^\circ \text{ of longitude} = \cos(x^\circ \text{ latitude}) \times 69 \text{ miles}$$

Figure 4.08: One can find the length in miles of any degree of longitude by using this formula.

Scale Factor

Cartographers refer to a map scale measured locally as an **actual scale**. An actual scale will only be accurate along one line or near one point on the map, very close to the place measured. The actual scale often differs from the overall scale of the map, called the principal scale. One uses the **principal scale,** based on the scale of the generating globe, to construct the map projection. The relationship between the actual scale and the principal scale is called a **scale factor (SF)**. Defined as the actual scale divided by the principal scale (**Figure 4.09**), it provides a means of evaluating uncertainty and error related to size and distance.

$$\text{Scale Factor (SF)} = \frac{\text{Actual Scale}}{\text{Principal Scale}}$$

Figure 4.09: Scale Factor (SF) can serve as an indicator of accuracy and distortion throughout the map.

One obtains the scale factor by dividing the denominator of the principal scale by the denominator of the actual scale. For example, suppose one had a map with an RF of 1:24,000. One then measures a small area in the upper right corner and determines that the scale at this location is 1:20,998. The actual scale is 1:20,998, and the principal scale is 1:24,000. The scale factor at this location would be 24,000 ÷ 20,998, or 1.143 (**Figure 4.10**).

$$SF = \frac{\dfrac{1}{20{,}998}\ \text{(Actual Scale)}}{\dfrac{1}{24{,}000}\ \text{(Principal Scale)}}$$

$$SF = \frac{24{,}000\ \text{(Denominator of the Principal Scale)}}{20{,}998\ \text{(Denominator of the Actual Scale)}}$$

$$SF = 1.143\ \text{(Actual scale is 14.3\% larger than the principal scale)}$$

Figure 4.10: Though the math may seem counterintuitive, when one divides by fractions, the denominators and numerators are flipped.

A scale factor of 1 means that the actual scale and the principal scale are the same. A scale factor of less than one indicates that the scale measured locally is smaller in scale than the overall map scale. A scale factor of greater than one means that the scale measured locally is larger in scale than the overall map scale. Knowing the range of the scale factor throughout the map is a good indicator of error and uncertainty related to size and distance.

Spatial Reference Systems

A **spatial reference system** defines a geographic location and may include the map datum, the coordinate reference system, and the map projection. Be aware that many commonly use the term coordinate system synonymously for the term spatial reference system. This use of terminology is not entirely correct. A **coordinate reference system (CRS)** is more specific in that it refers to a *coordinate-based* reference system to define location. Some spatial reference systems, such as *land partitioning systems*, do not have coordinates and use other means of determining location.

Land Partitioning Systems

A **land partitioning system** is a spatial reference system used to divide the property into units such as sections, tracks, parcels, or lots. This book classifies these systems into two broad categories, regular and irregular. **Regular land partitioning systems** use standardized geometries and consistent divisions to define boundaries. The U.S. Public Land Survey System is the most frequently used regular land partitioning system in the United States. **Irregular land partitioning systems** often use natural features, sometimes in conjunction with standardized geometries, to define boundaries. Frequently used irregular land partitioning systems include the system of metes and bounds, the Seigneurial system, and the league and labor system.

Metes and Bounds

An early land partitioning system used in Europe, and brought by the English to the original United States colonies was called metes and bounds. **Metes and bounds** is an irregular land partitioning system that defines property boundaries using descriptions of geographical or anthropogenic features and landmarks. For example, one may describe a property boundary defined by metes and bounds as "A parcel of land beginning with a corner at the intersection of Hodgson Ave. and Main St., thence south to Martin Slough, thence east to the corner of Madrone St. and Darcy St., Thence north to the apple tree by the stone wall, thence west back to the point of beginning."

Metes and bounds partitions often follow natural features and leave a distinctive, but random, pattern on the landscape, somewhat like a crazy quilt (**Figure 4.11**). While following natural features might seem like a logical way to divide up the land, many markers used in this system, such as the apple tree in the example above, may move or disappear, making clear delineation of boundaries difficult and prone to error. Metes and bounds can be found in most of the East Coast states and also parts of Texas. This randomness is visible from the air in property bounding features like trees and fences, and also in the shape of administrative regions, such as counties.

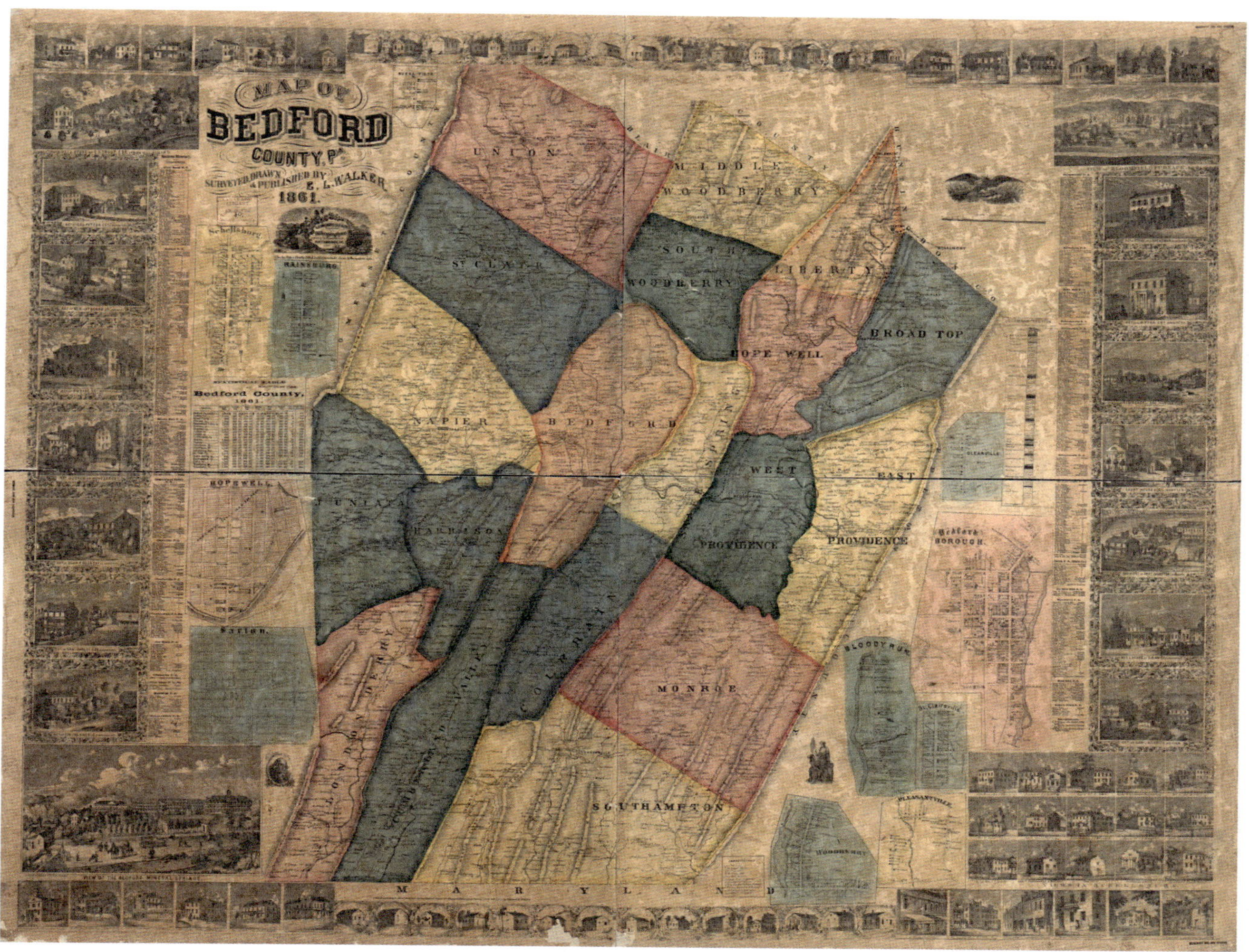

Figure 4.11: The pattern of metes and bounds on the landscape often follows natural features, and may appear random. Source: Library of Congress.

Seigneurial System

Another irregular system used in early surveying was the **Seigneurial System**, a system of land distribution established in New France in 1627. This system was designed for river access, dividing the land into **seigneuries**, long narrow strips frequently called **long lots** or **ribbon farms**, each with access to the river (**Figure 4.12**). In North America, settlers primarily used these seigneuries in New France, as well as parts of the early colonies and territories of the United States (**Figure 4.13**).

League and Labor

The **league and labor system** was similar to the Seigneurial System. A league and a labor was the allotment given to settlers in some Spanish territories. A **league** was 4,400 acres of grazing land, and a **labor** was 170 acres of farmland, typically in a floodplain. Settlers were often granted both for farming and ranching. While these can be harder to spot on the landscape, there are still some areas in Texas, California, and New Mexico, where one can still see these early land divisions (**Figure 4.14**).

Figure 4.12: Ribbon farms leave a distinctive pattern on the landscape, as seen in this image of the Mississippi River. Source: USGS.

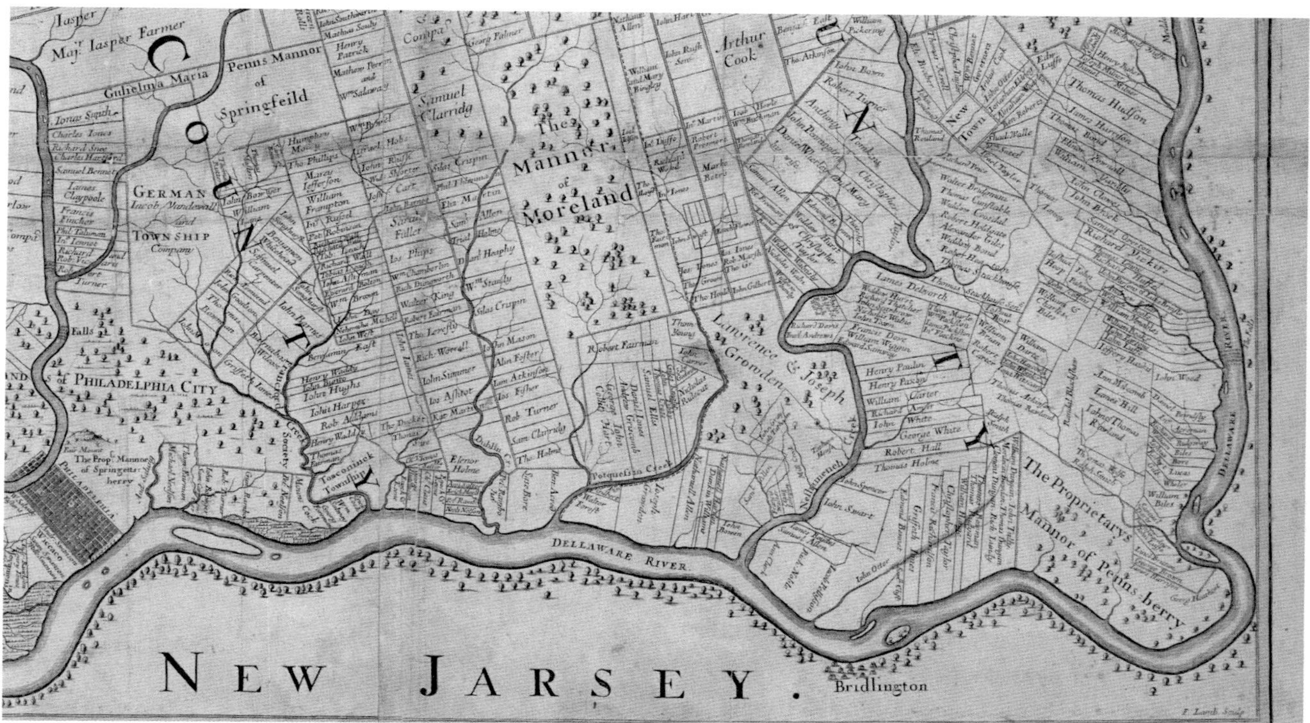

Figure 4.13: This parcel map of the Province of Pennsylvania in 1681 shows long lots along the Delaware River and other nearby tributaries. Source: Library of Congress.

Figure 4.14: This image shows the pattern of land division created by labors in New Mexico. Source: USGS.

U. S. Public Land Survey System

People used grid systems throughout the world in ancient and modern civilizations. As the US began to acquire territory, surveyors determined that a more regular system of land partitioning was needed. The land ordinance of 1785 established the **U. S. Public Land Survey System (PLSS)**. The PLSS is a systematic way of subdividing and describing the land in the United States (**Figure 4.15**). The U.S. Public Land Survey System is comprised of square grids, divided into townships and ranges. As a result, the PLSS is sometimes called the **Township and Range System**. Land surveyors would select an initial point at the intersection of a parallel and a meridian. The starting parallel was called the **baseline**, and the starting meridian was named the **principal meridian**. The first PLSS survey was measured in Ohio in the northwest corner of the state at what is now called the *1st Principal Meridian*. Several methods were tested, but once this system was refined, it was used to survey new territories added to the United States.

Surveyors gave each new principal meridian and baseline a name. They sometimes used a sequential number, such as the *2nd Principal Meridian* or the *3rd Principal Meridian*. In other cases, surveys gave the principal meridian and baseline a geographic name. For example, the westernmost principal meridian and baseline in the conterminous United States are called the *Humboldt Meridian and Baseline* (**Figure 4.16**).

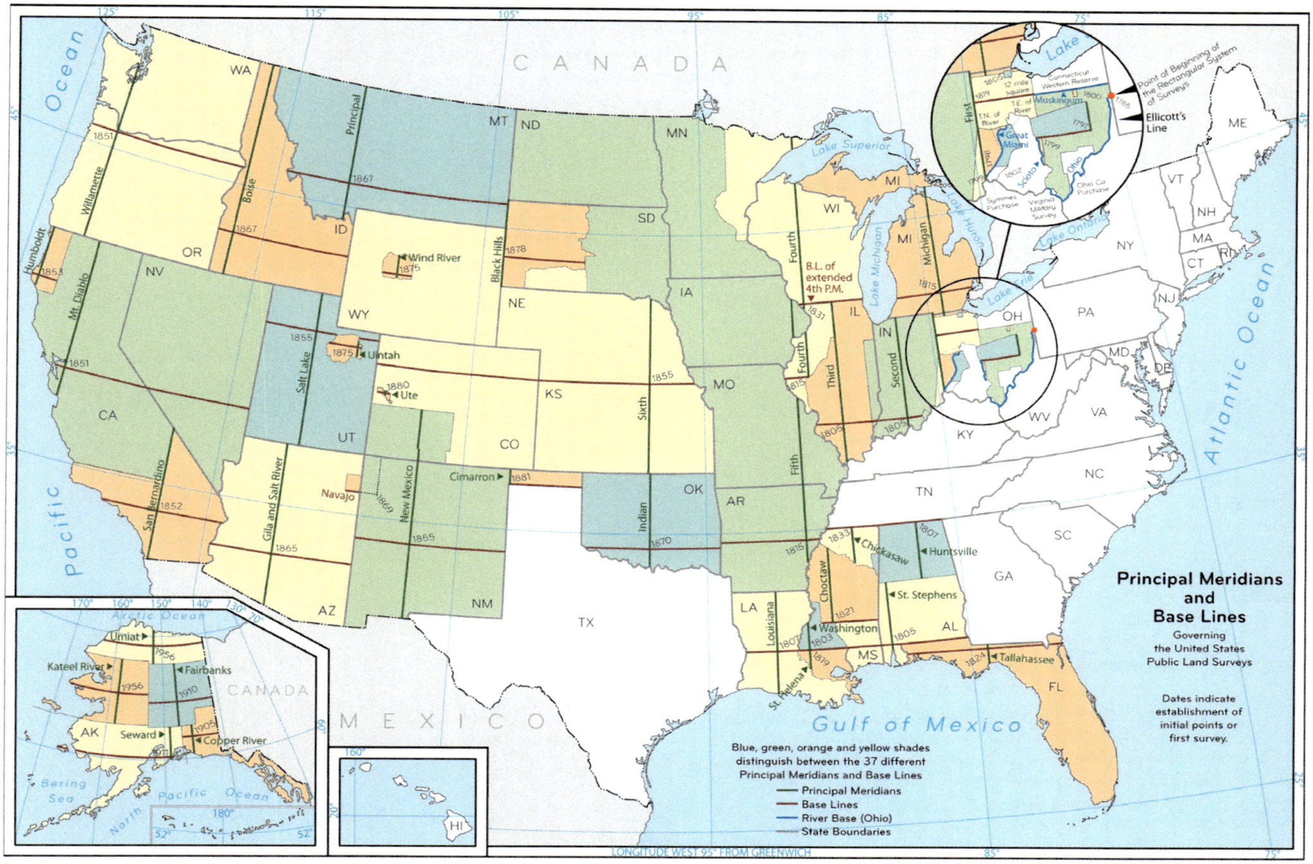

Figure 4.15: This image shows a map titled *Principal Meridians and Base Lines Governing the United States Public Land Surveys*, published in 2012 by the U.S. Bureau of Land Management.

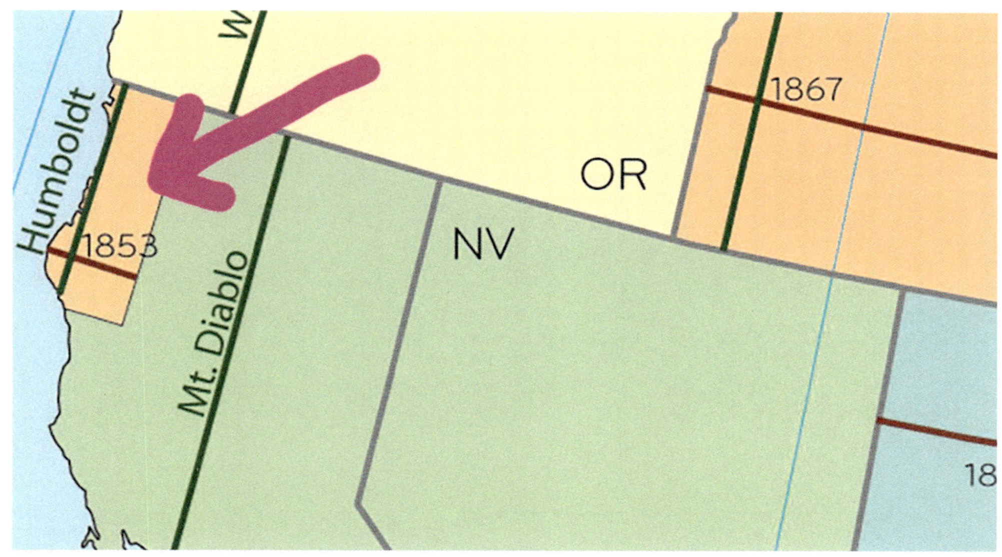

Figure 4.16: This image shows a close up of the Principal Meridians and Base Lines Governing the United States Public Land Surveys, published in 2012 by the U.S. Bureau of Land Management.

Township and Range

Once surveyors established the principal meridian and baseline, they measured a grid of 6-mile quadrilaterals called **townships**, outward north, south, east, and west. The rows in the grid of 6-mile quadrilaterals, separated by **township lines**, are also called **townships**. The columns in the grid of 6-mile quadrilaterals, separated by **range lines**, are called **ranges**. Surveyors established range lines every 6 miles east and west of the principal meridian. Surveyors established township lines every 6 miles north and south of the baseline, each following a parallel. A notation based on the row and column or township and range defines the location of each township (**Figure 4.17**). For example, T1N, R1E identifies the township located one row north of the baseline and one column east of a given principal meridian.

Township Sections

Each 6-square-mile township is further divided into 36 sections of one square mile each. A zigzag method of numbering the sections begins with section 1 at the upper right corner of the township and ending with section 36 the lower right corner of the township (**Figure 4.18 and 4.19**). One partitions smaller units of land through a series of fractional divisions of ½ or ¼ sections.

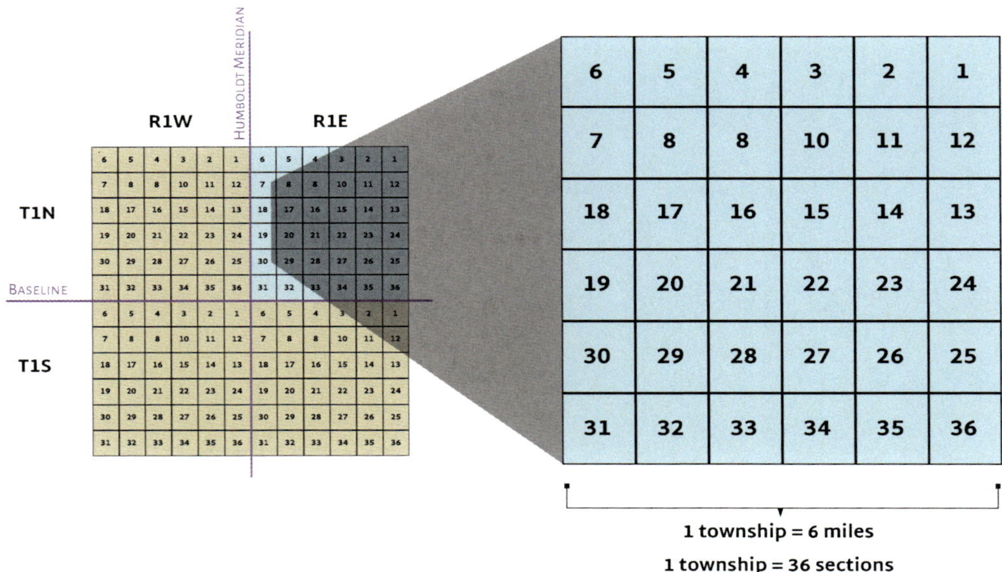

Figure 4.17: In this example, the township and range shown in blue can be described as T1N R1E Humboldt Meridian and Baseline.

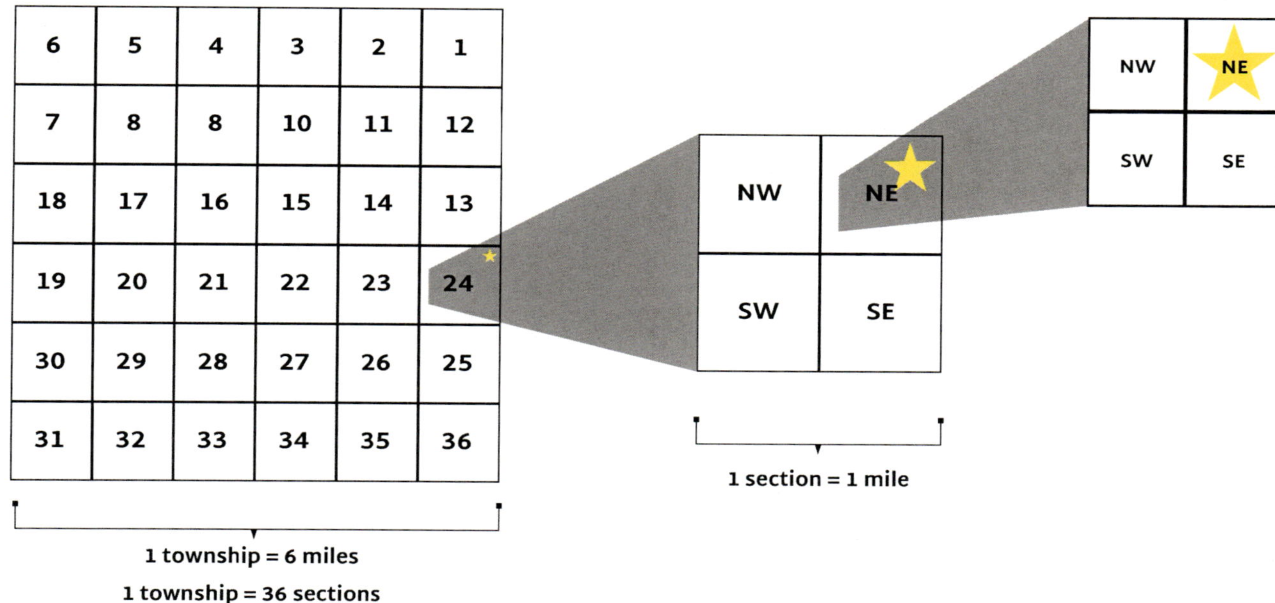

Figure 4.18: In this example, one can locate the gold star at the NE 1/4 of the NE 1/4 of Section 24.

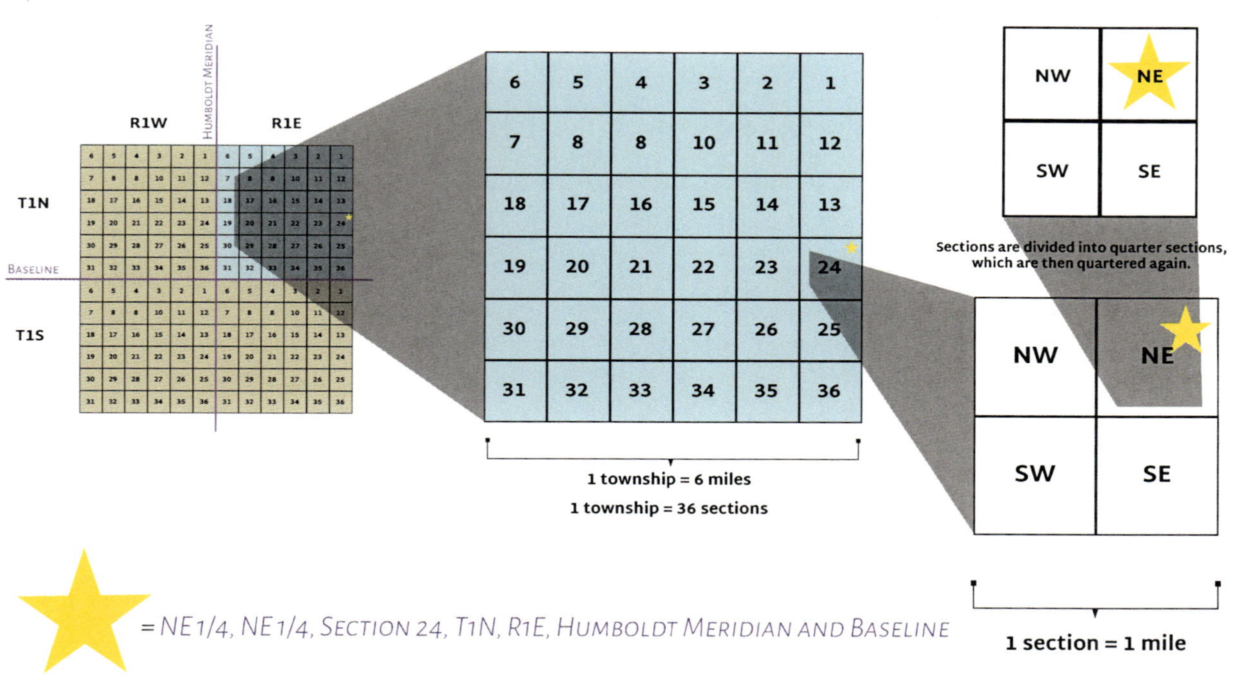

Figure 4.19: In this example, one can locate the gold star at the NE 1/4 of the NE 1/4 of Section 24, of Township 1 North, Range 1 East, of the Humboldt Meridian and Baseline. The image shows the correct notation format next to the gold star.

Correction Lines

Even though the PLSS was an effort to establish a regular system of land partitioning, it was far from perfect, and many irregularities still occurred. As discussed previously, the distance between each meridian of longitude continually changes as one travels north or south. Because range lines were following meridians, the distance between them became shorter the farther north they traveled. The result was townships that were not perfect 6 miles squares. To remedy this problem surveyors established **correction lines** at every fourth township by establishing new range lines east and west of the principal meridian. Other sources of irregularities included geographic obstacles such as rivers, lakes, or mountains, as well as human error.

Geographic Coordinate Systems (GCS)

A **geographic coordinate system (GCS)** is a coordinate-based spatial reference system using latitude and longitude. An important fact to remember is that a geographic coordinate system is ellipsoid-based, which makes it useful for defining a location on the surface of the earth. However, a geographic coordinate system makes measuring distances and areas complicated. Geographic coordinates are traditionally written using degrees minutes seconds (DMS) notation: dd° mm' ss" N or S, ddd° mm' ss" E or W. Using this notation, dd° is the number of whole degrees, mm' is the number of minutes, and ss" is the number of seconds. The letter N for locations north of the equator or the letter S for locations south of the equator follow the degrees minutes and seconds of latitude. The letter W for locations west of the Prime Meridian or the letter E for locations east of the Prime Meridian follow the degrees minutes and seconds of longitude (**Figure 4.20**).

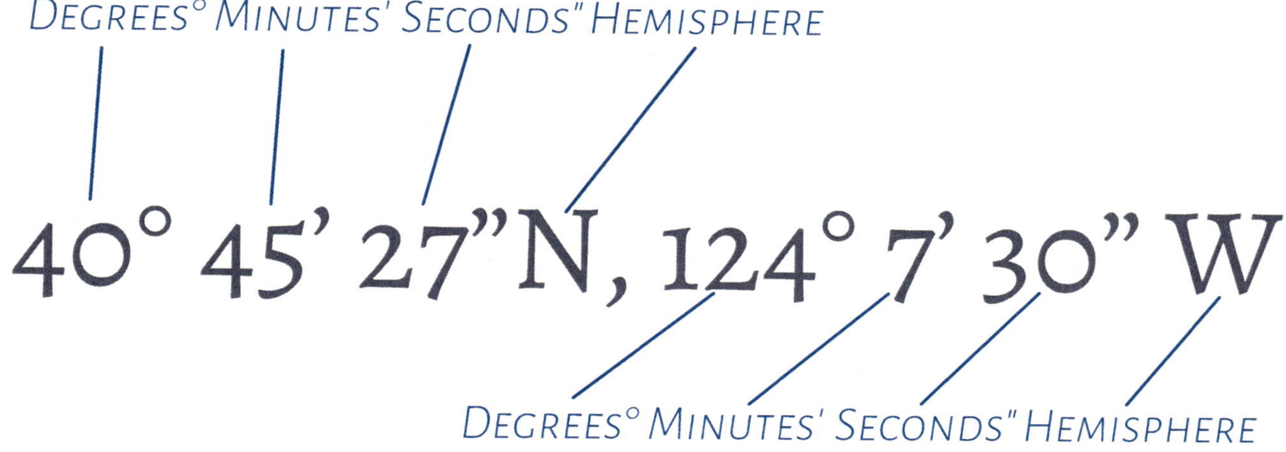

Figure 4.20: The correct notation starts with latitude, followed by longitude.

Converting to Decimal Degrees

Sometimes, it is necessary to convert the traditional notation into an alternative decimal format called **decimal degrees (DD)**. Using decimal degrees makes it easier for GIS software to manage coordinates because it can treat latitude and longitude coordinates as decimal-based numbers. Chapter 7 discusses how a database stores numbers in additional detail.

$$\text{Decimal Degrees} = dd^{\circ} + (mm' \div 60) + (ss'' \div 3600)$$

Figure 4.21: To convert latitude and longitude into decimal degrees, one can use the formula shown here.

Recall from the discussion of latitude and longitude in Chapter 3 that a negative sign indicates south latitude or west longitude. Using the formula above, the geographic coordinates in decimal degrees for values shown in Figure 4.20 is **40.7575, -124.125**.

Measuring Distance Using Geographic Coordinates

Sometimes, one must determine geographic coordinates on a map in a place that does not line up with any specific latitude or longitude marking. One does this by measuring the proportional distances between latitude and longitude indicators on a map. For example, the USGS Arcata North 7.5-minute series quadrangle indicates latitude and longitude with a series of black tick marks just inside the neatline (**Figure 4.22**). The USGS sets each of these tick marks 2.5 minutes apart. One can use these marks to calculate coordinates for any point on the map. Suppose one wanted to find the geographic coordinates for the center of the Arcata Plaza in Arcata, California. The Arcata Plaza does not land exactly along one of the tick marks on the map. Since the exact distance between each tick mark is a known quantity, this poses no problem. To find the geographic coordinates for the intersection, one measures the proportional distance across to the next tick mark between north and south for latitude and between east and west for longitude. One could then use these proportions to calculate how many minutes and seconds to add or subtract from the nearest tick mark, depending on the direction (**Figure 4.23**).

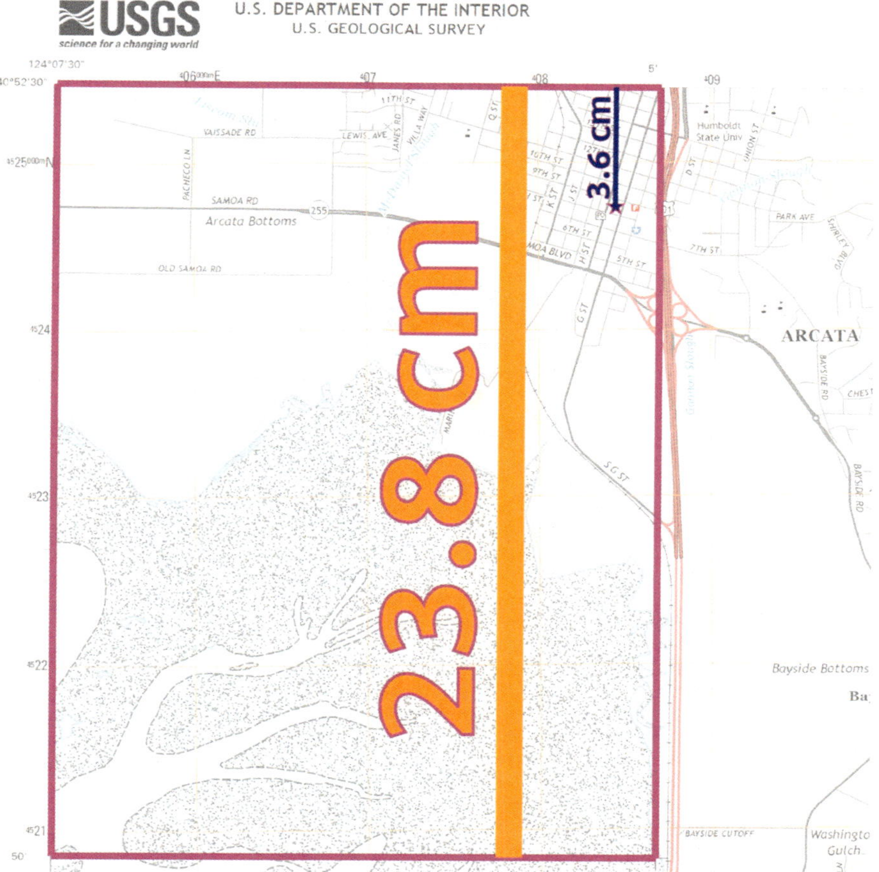

Figure 4.22: The next tick mark below the top left corner represents a line of latitude, 40° 50' 00" N. Here, the distance on the map between the lines of latitude is 23.8 centimeters. The distance to the Arcata Plaza is 3.6 centimeters.

Total Distance betwen ticks of latitude = 23.8 cm

Distance to Arcata Plaza = 3.6 cm

$$\frac{3.6}{23.8} = 0.15$$

The Arcata Plaza is 15% of the total distance between the tick marks.

Figure 4.23: Using the proportional distance is an effective way to estimate geographic coordinates.

Projected Coordinate Systems (PCS)

The **Cartesian coordinate system**, named for the French mathematician René Descartes, is a planar, two-dimensional grid coordinate system based on two axes, the x-axis, and the y-axis. Once one transforms the globe into a plane, a Cartesian coordinate system defines locations on the map. This type of spatial reference system is commonly called a **projected coordinate system (PCS)**. For large-scale maps, projected coordinate systems measure distances and areas with a high degree of precision and accuracy. The x-axis represents positions in east-west directions while the y-axis represents positions in north-south directions. When converting latitude and longitude into Cartesian coordinates, the grid is divided into four quadrants using the equator and the prime meridian as the x and y-axes (**Figure 4.24**). Cartographers assign negative numbers to west and south latitude and longitude coordinates using decimal degrees. However, many projected coordinate systems do not use latitude and longitude and take measures to avoid negative numbers.

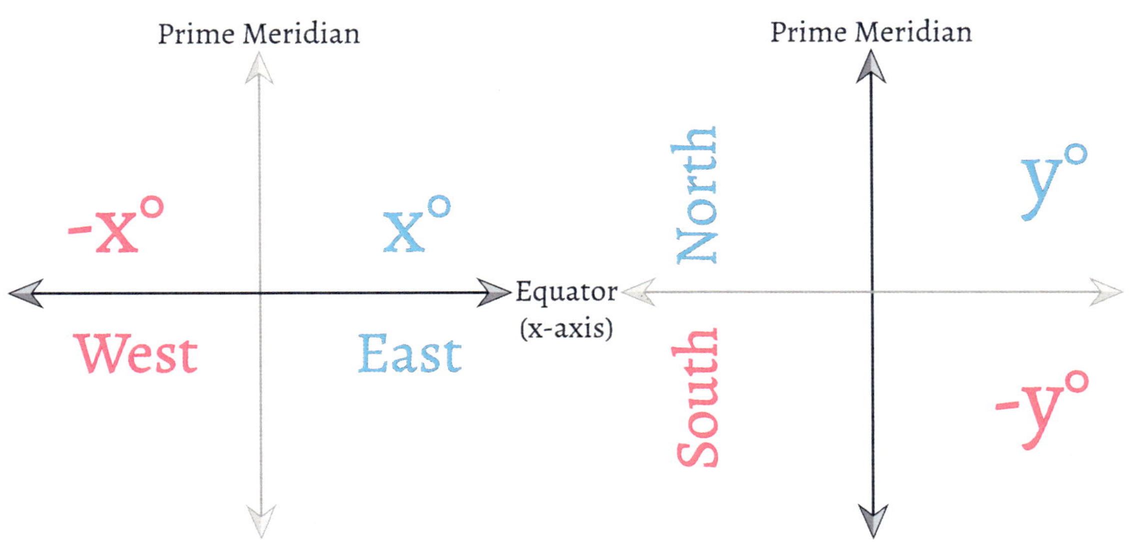

Figure 4.24: Values are either positive or negative based on the quadrant.

All projected coordinate systems originate from a specific geographic coordinate system and datum.

The geodetic datum used as the basis for any projected coordinate system influences the position of coordinates. It is a fact that many readers find confusing. Always be aware of which datum on which the projected coordinate system is based, making sure it is the optimal datum for the region of interest.

Universal Transverse Mercator

The **Universal Transverse Mercator (UTM) system** is an international grid coordinate system based on the Transverse Mercator projection (**Figure 4.25**). It consists of a series of 60 north-south strips of the earth called gores. Created by rotating the Transverse Mercator projection so that it centers on each gore, each is 6° longitude across. The gores are cut in half at the equator creating a series of 120 sections called **UTM zones**.

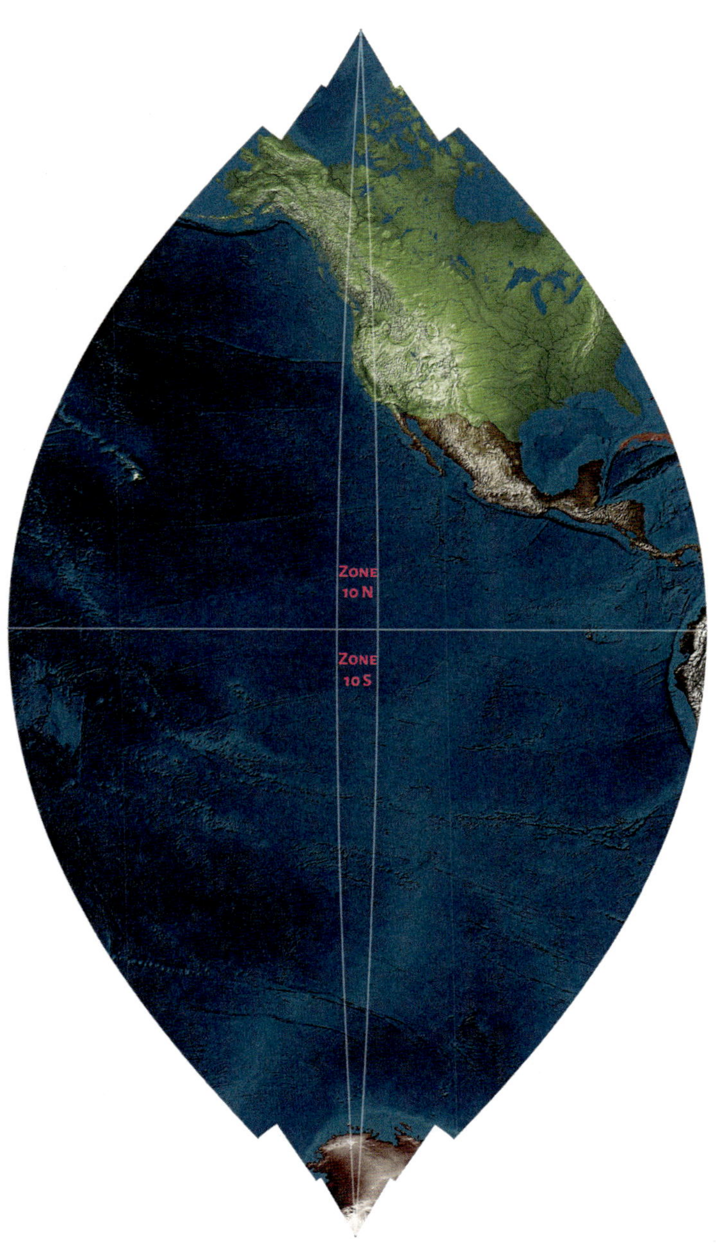

Figure 4.25: Here, one can see UTM Zone 10, split between north and south by the equator. Zone 10 covers most of California except for Los Angles and other Southern California counties, which lay in Zone 11.

The UTM system extends 84° latitude north and 80° latitude south. Each UTM zone includes a number and a letter for identification. Zone 1 starts at 180° west. The numbers increase moving from west to east. The last zone ends with the number 60, at 180° east, centered over the eastern end of New Zealand (**Figure 4.27**). The letter N for the northern hemisphere or the letter S for the southern hemisphere follows the zone number. Each UTM zone has Cartesian grid coordinates utterly independent from the others.

UTM Eastings and Northings

The UTM system ensures that all coordinates in the system are positive numbers using a system of eastings and northings. One may imagine this as forcing all coordinates into the first zone of the Cartesian coordinate system. The UTM system accomplishes this by designating an arbitrary point of origin for the x-axis and y-axis, to the southwest of each zone (**Figure 4.26**).

An **easting** is the x-axis coordinate measured in meters east of the origin. The **central meridian**, a line of tangency for the Transverse Mercator projection, of each zone has a **false easting** of 500,000 meters. A false easting is simply the value given to grid coordinate systems on the x-axis so that all x-coordinates in the system are positive numbers. Eastings are often a source of confusion for many readers. They invariably want to know where the origin is, and the answer is always the same, "it does not matter." The origin of every zone in the northern hemisphere is west of the zone, and at the equator. Every zone centers on 500,000 m east. The coordinates everywhere else in the zone are relative to that origin.

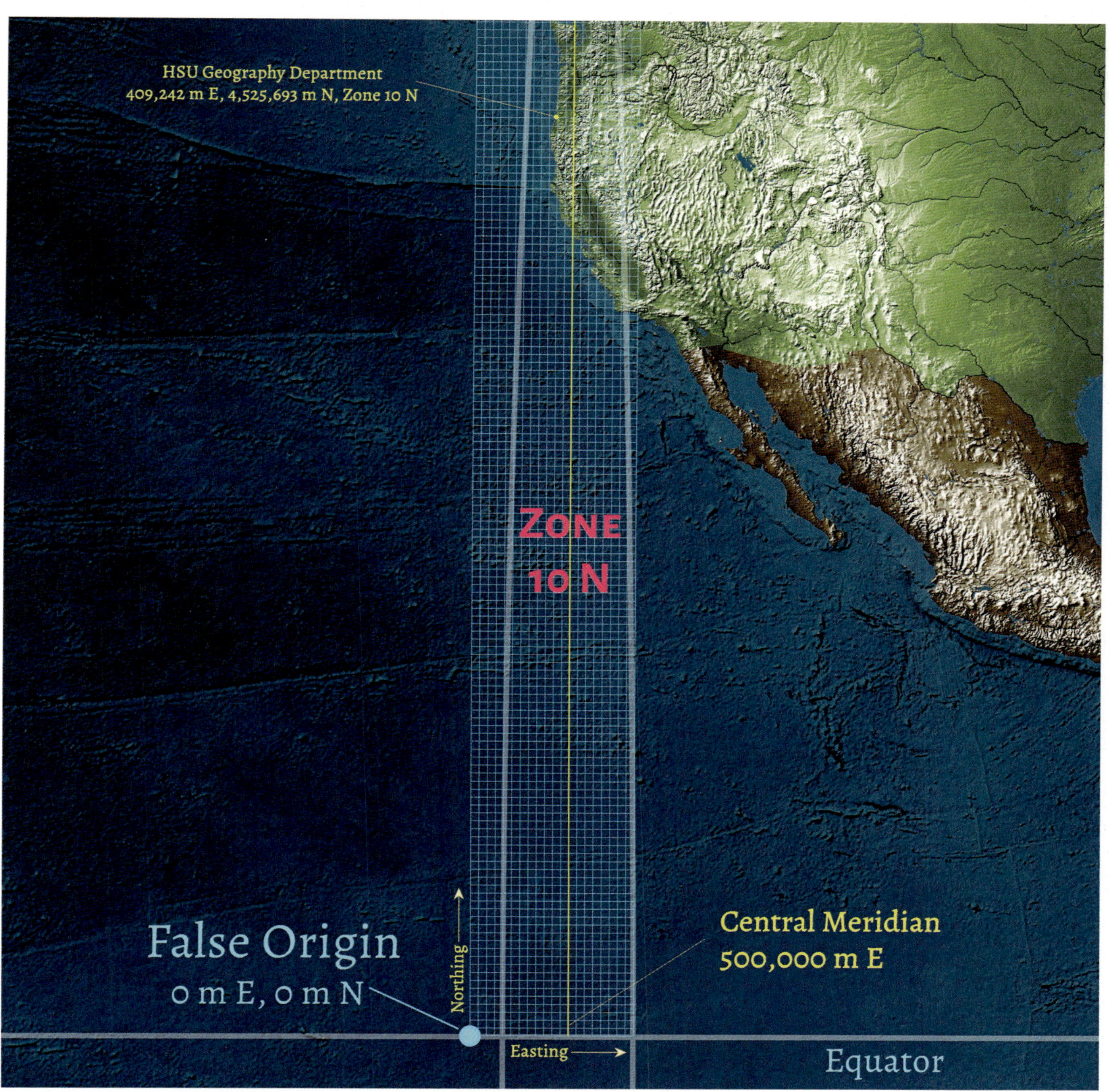

Figure 4.26: The UTM system uses coordinates made up of eastings and northings, measured in meters. The false origin is used to keep the numbers positive.

The y-axis has a value called a **northing**. In the northern hemisphere, this represents the number of meters north of the equator, which has a value of 0 meters N. In the southern hemisphere, the UTM system assigns the equator a **false northing** of 10,000,000 meters, which corresponds to 80° S (**Figure 4.27**). Like the false easting, a false northing for the southern hemisphere ensures that all the coordinates in the system are positive numbers.

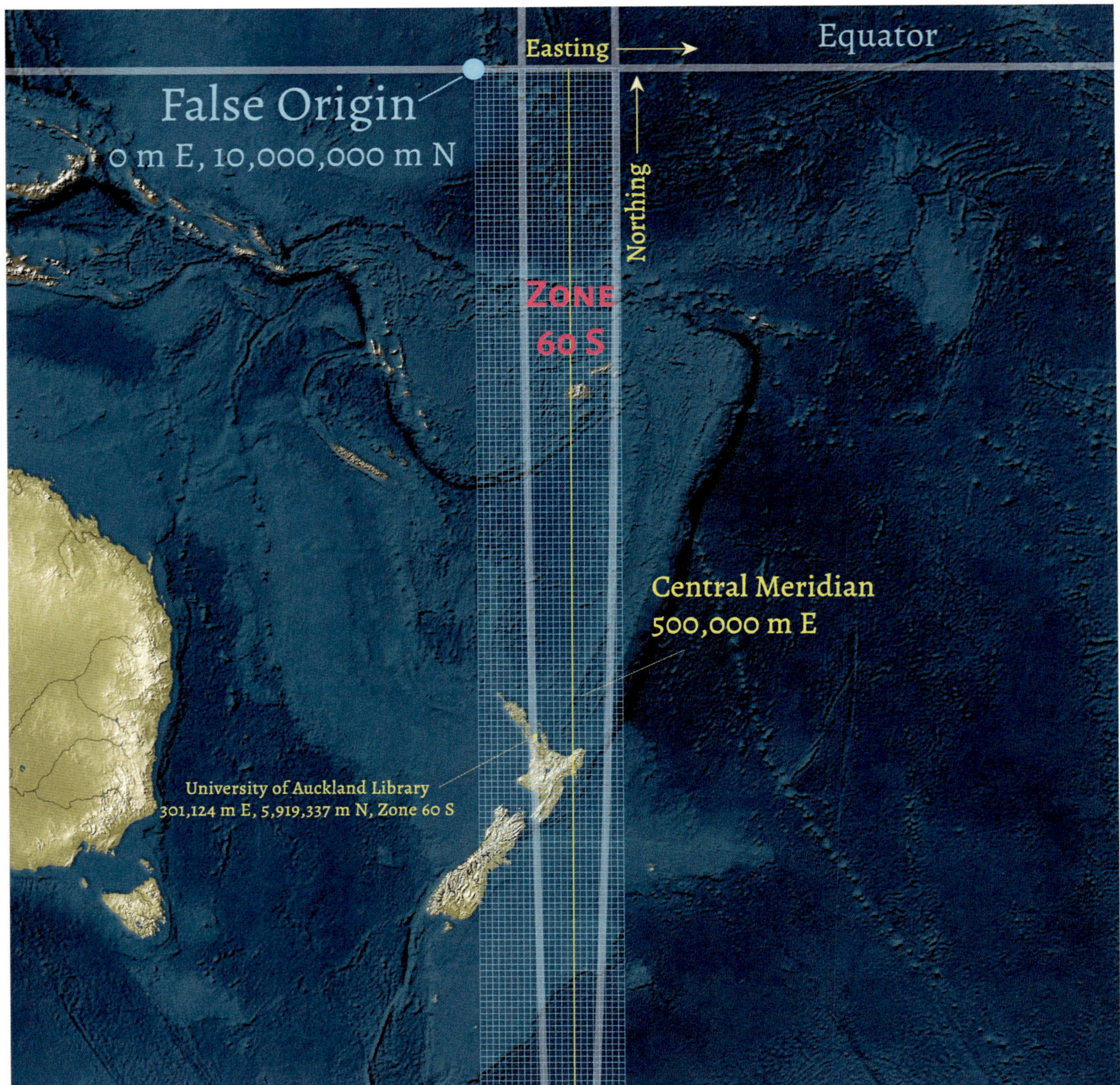

Figure 4.27: Here in the southern hemisphere, the equator has a false northing value of 10,000,000 meters N.

Coordinates in the UTM system are written using the notation x m E, y m N followed by zone number and hemisphere (**Figure 4.28**). For example, the approximate location of the Geography Department office at Humboldt State University in UTM coordinates would be written using the notation: 409,242 m E, 4,525,693 m N, Zone 10 North. The UTM system is highly accurate, having a scale factor that ranges 1 +/- 0.0004 within each zone. Due to its accuracy, global extent, and consistent structure, the Universal Transverse Mercator system is the most useful spatial reference system in the world.

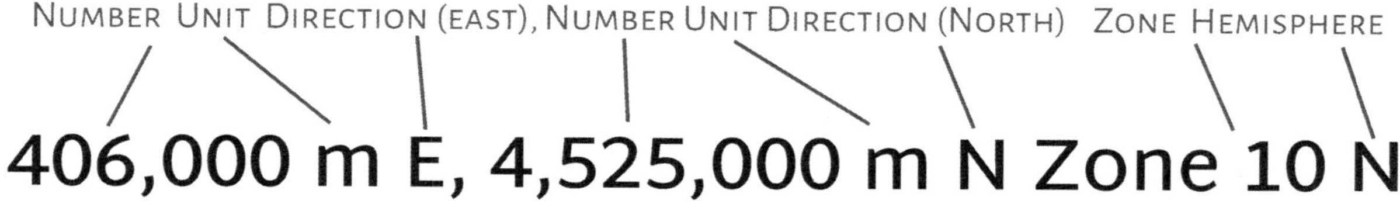

Figure 4.28: The correct notation for UTM starts with easting in meters, northing in meters, the zone, and the hemisphere.

One can determine their UTM coordinates on a map using the same method described in the example above for determining geographic coordinates. USGS 7.5-minute series quadrangles indicate UTM coordinates with a series of blue tick marks just outside the neatline. Each UTM tick mark is **1000 meters** apart. Just as with geographic coordinates, one measures the proportional distance across to the same intersection, north-south, and east-west. One could then use these proportions to calculate how many meters to add or subtract from the nearest tick mark, depending on the direction to the intersection. Numbers increase moving north and to the east and decrease moving south and to the west.

To save time, instead of calculating coordinates manually, one can also use a tool called a roamer. A **roamer** is usually a set of corner-rulers printed on a transparent sheet of plastic. If one does not have a roamer, one can construct a roamer using the scale bar (**Figure 4.29**).

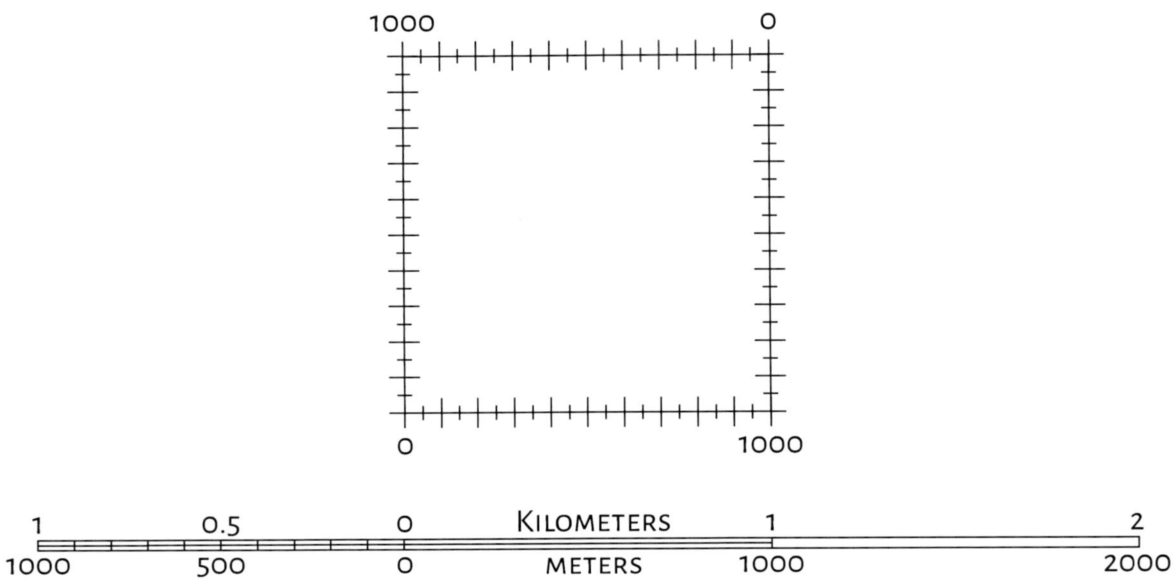

Figure 4.29: Construct a roamer using the scale bar and a piece of paper to draw the lines and ticks.

Each ruler on the UTM roamer is marked in meters specific to a standard map scales, such as 1:24,000 or 1:62,500. UTM coordinates are then obtained by placing the roamer on the map and using the ruler on it to determine how many meters to add or subtract from the closest tick marks to one's location of interest (**Figure 4.30**).

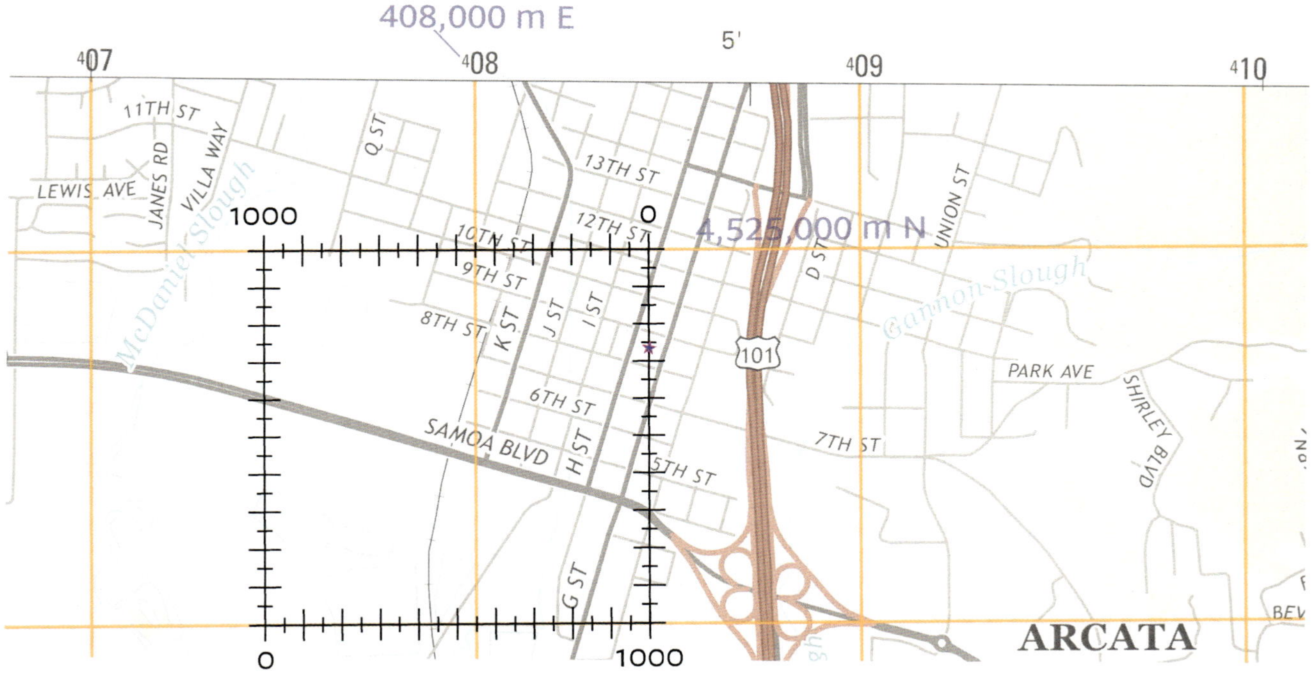

Figure 4.30: The roamer is aligned with the sides of the grid and passes through the point of interest. Each tick mark on the roamer is 50 meters.

If a roamer is not available, one should be able to determine one's UTM coordinates manually.

State Plane Coordinate System

Many countries establish local projected coordinate systems for highly accurate measurements over a small region. The United States uses a projected coordinate system called the **State Plane Coordinate (SPC) system**. Like the UTM, the State Plane Coordinate System is a grid coordinate system that uses eastings and northings so that all coordinates are positive numbers. Unfortunately, the State Plane Coordinate System does not have a consistent structure nationwide. Thus, it is only useful for localized regions within a specific state and does not work well across state lines. The most recent version, the **State Plane Coordinate System of 1983 (SPC 83)**, uses the North American Datum 1983 (NAD83).

Projections

Each state uses one of two map projections. States with a predominantly north-south extent use the Transverse Mercator projection. States with a predominantly east-west extent use the Lambert Conformal Conic projection. However, even the use of projections is not always consistent with the directional orientation of the state. For example, due to the east-west orientation of SPC Zones that subdivide the state, California uses the Lambert Conformal Conic projection even though the state itself has a north-south extent.

SPC83 Zones

Each zone in the United States also has a unique shape, with boundaries following state and county lines. Nearly every state divides into a variable number of zones. Unlike UTM zones, which follow lines of longitude, State Plane zones typically follow county boundaries, for administrative simplicity. Most states have anywhere between 2 to 6 zones and use cardinal directions to designate zone names. Names such as North, South, East, West, or Central identify the zones. States with more than three zones tend to use numbers instead. For example, California uses a series of Roman numerals from I to VI. Texas sometimes labels theirs North, North Central, Central, South Central and South, and other times they use numbers. Also, some states, like Montana, have recently combined all of their zones into a single region covering the entire state and no longer have zone designations at all. Zones in the SPC83 system also have a unique set of standardized codes, called the Federal Information Processing Standard (FIPS) codes. These codes are used as an alternative to zone names or numbers to help computer systems manage large datasets.

Like the UTM system, eastings and northings in SPC83 ensure that all coordinates are positive numbers. However, there are some significant differences. Measurements used for eastings and northings in the State Plane Coordinate System are inconsistent across the nation. Initially, the State Plane Coordinate System measured eastings and northings in feet. The State Plane Coordinate System of 1983 converted to the metric system during its creation. However, most maps published by the United States Federal and State governments still use feet when indicating tick mark increments, even if it is in the SPC83. Watch *Two Right Feet? U.S. Survey Feet vs. International Survey Feet* by the COMET Program/MetEd to learn more (**Figure 4.31**).

Figure 4.31: This video discusses the State Plane System in additional detail. URL: *https://youtu.be/crK1hp-WbKs*

Another inconsistency of measurement is the way the State Plane Coordinate System defines false eastings and northings. The false easting assigned to the center of each zone varies by state based on the map projection. Zones using the Lambert Conformal Conic projection have a central meridian with a larger false easting than zones using the Transverse Mercator projection because they are bigger east to west. Specific values vary by state. Every zone in the SPC83 system has a different northing value, with an origin somewhere south of the zone coinciding with a parallel of latitude.

Coordinates in the State Plane Coordinate System are written using the notation x ft E, y ft N or x m E, y m N if using the metric system. The state name follows this notation. Then, the zone name or zone number using Roman numerals if it is California or nothing for states without zones (**Figure 4.32**) completes the notation. For example, the approximate location of the Geography Department office at Humboldt State University in SPC coordinates would be written using the notation: 5,987,361 ft E, 2,209,680 ft N, California I.

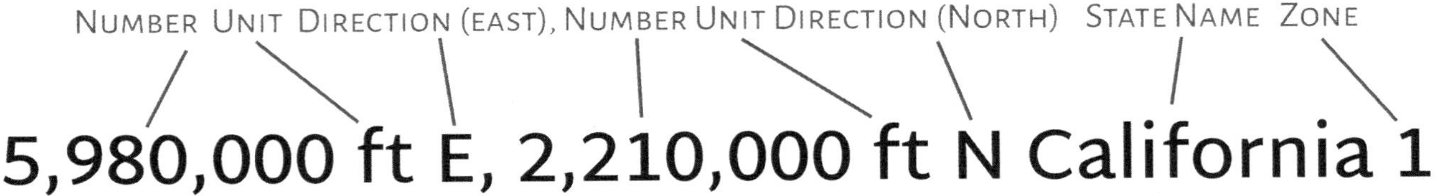

Figure 4.32: The correct notation for SPC 83 starts with easting in feet or meters, northing in feet or meters, state name, and zone.

The method for determining SPC coordinates on a map is nearly identical to the UTM system. USGS Arcata North 7.5-minute series quadrangle indicates SPC coordinates using a series of black tick marks just outside the neatline. Each SPC tick mark is **10,000 feet** apart. As with UTM, one uses proportional distances or a roamer to determine SPC coordinates on the map (**Figure 4.33**).

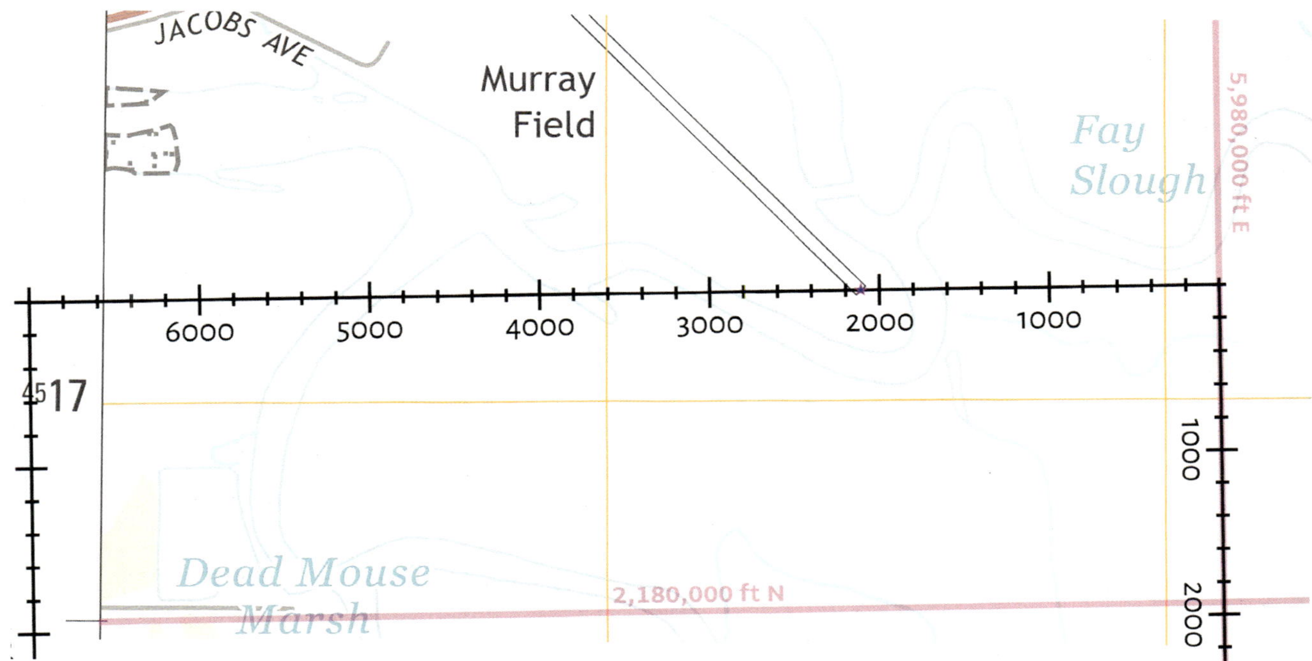

Figure 4.33: The roamer is aligned with the sides of the grid and passes through the point of interest. Each tick mark on the roamer is 200 feet.

Due to inconsistencies between each state and between each zone, using the State Plane Coordinate System can sometimes be problematic, if not a little frustrating, especially if the area of interest crosses county or state lines into another zone. However, if the area of interest lays within a single county, the State Plane Coordinate System is highly accurate with a scale factor that ranges 1 +/- 0.0001.

TUTORIAL: WORKING WITH SCALE

In this chapter, you learned that map scale describes the relationship between distance on the ground and the way the map represents that distance. In this book, the ratio of map distance over ground distance defines scale (**Figure 4.34**).

Figure 4.34: Scale is the relationship between a distance measured on the map and the corresponding distance in the real world.

There are three common ways to express scale:

» Representative fraction

» Verbal scale

» Scale bar

In this activity, you learn how to determine scale and distance using the ***Arcata South USGS Topographic Quadrangle***[1] and the ***Arcata North USGS Topographic Quadrangle***[2].

Download the PDF files using the links above.

ESTIMATED TIME TO COMPLETE THIS TUTORIAL: 2 HOURS

Learning Outcomes

Readers should be able to accomplish the following outcomes by the end of this tutorial:

» Demonstrate competence in calculating and converting map scale

» Explain the relationship between distance, elevation, and slope

» Measure distances using a graphic scale

» Determine distance using a representative fraction

» Determine an unknown scale using a reference map

» Formulate a word statement to communicate map scale

1 https://wp.me/a6siVq-1L9

2 https://wp.me/a6siVq-2Q

Measuring Distance Using a Graphic Scale

In this chapter, you learned that a **scale bar** is a graphic symbol representing ground distance. You also learned that scale bars come with an **extension scale**, which is an addition to the left of the zero on the scale bar. Scale bars have the advantage of maintaining the correct proportion when enlarging or reducing map size. In this example, two points of interest, point A and point B are measured by drawing a line between them (**Figure 4.35**).

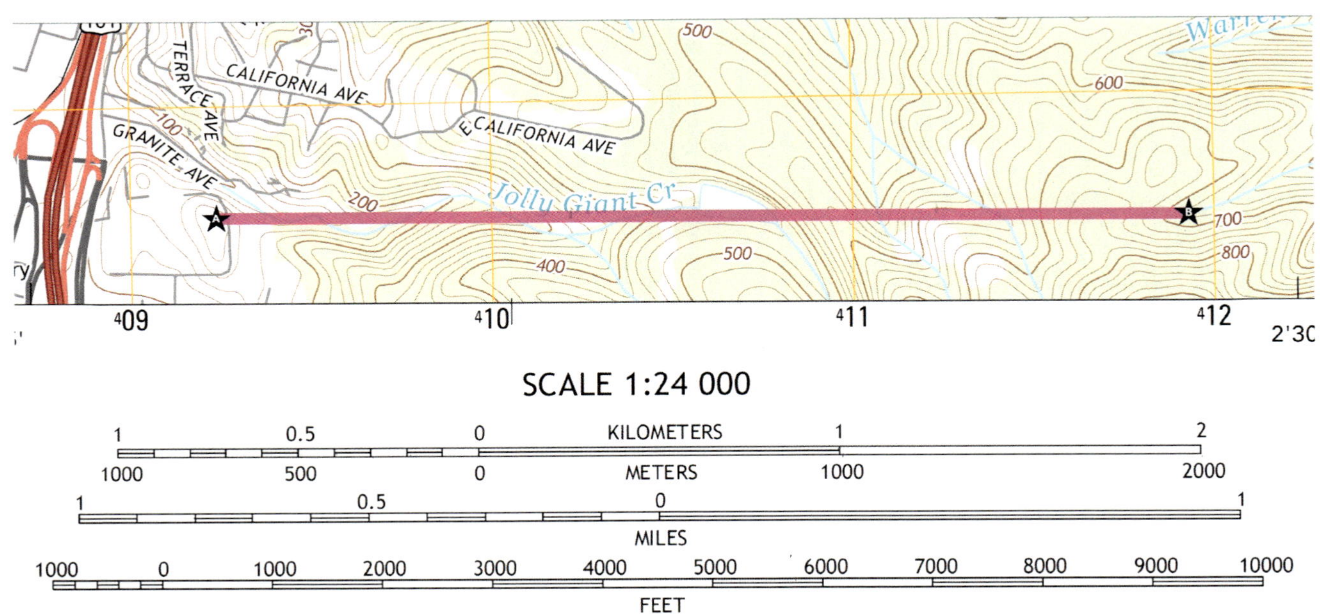

Figure 4.35: The stars labeled A and B are the points of interest. This measurement does not incorporate in changes in elevation.

On a separate piece of paper, write down the answers to the following questions:

1. What is the approximate distance in *kilometers* between point A and B (**Figure 4.36**)?
2. What is the approximate distance in *meters* between point A and B (**Figure 4.36**)?

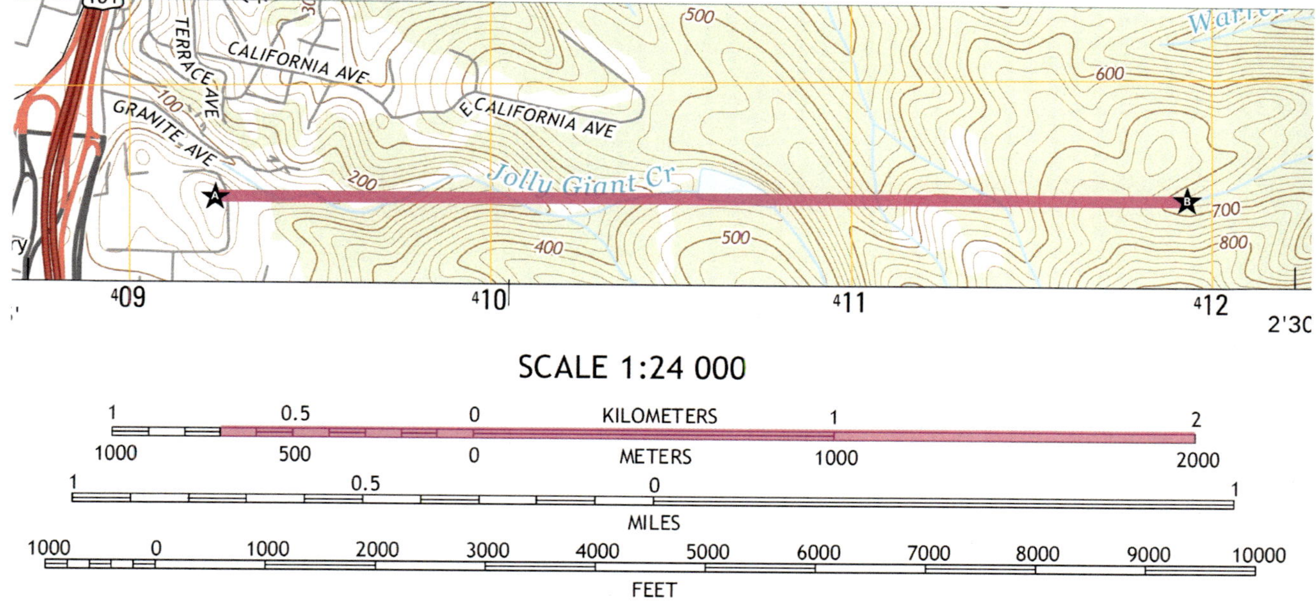

Figure 4.36: The purple line over the Kilometers/Meters scale bar is the approximate distance between point A and B. This measurement does not incorporate in changes in elevation.

3. What is the approximate distance in miles between point A and B (**Figure 4.37**)?

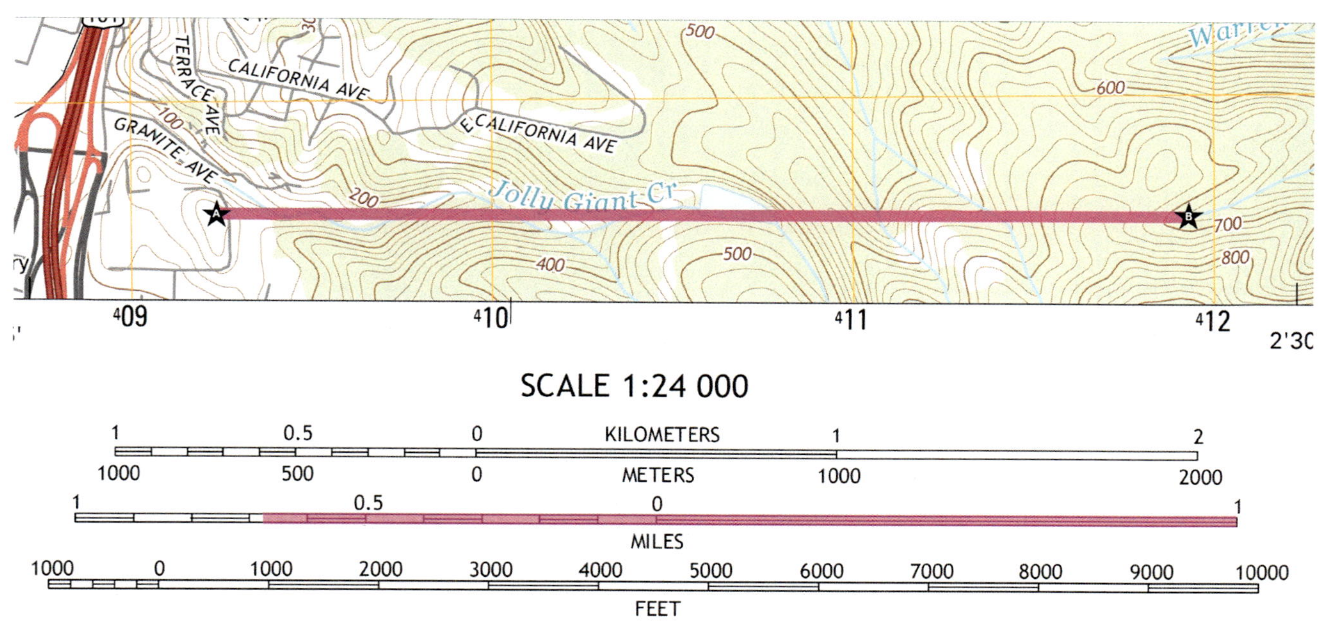

Figure 4.37: The purple line over the Miles scale bar is the approximate distance between point A and B. This measurement does not incorporate in changes in elevation.

4. What is the approximate distance in feet between point A and B (**Figure 4.38**)?

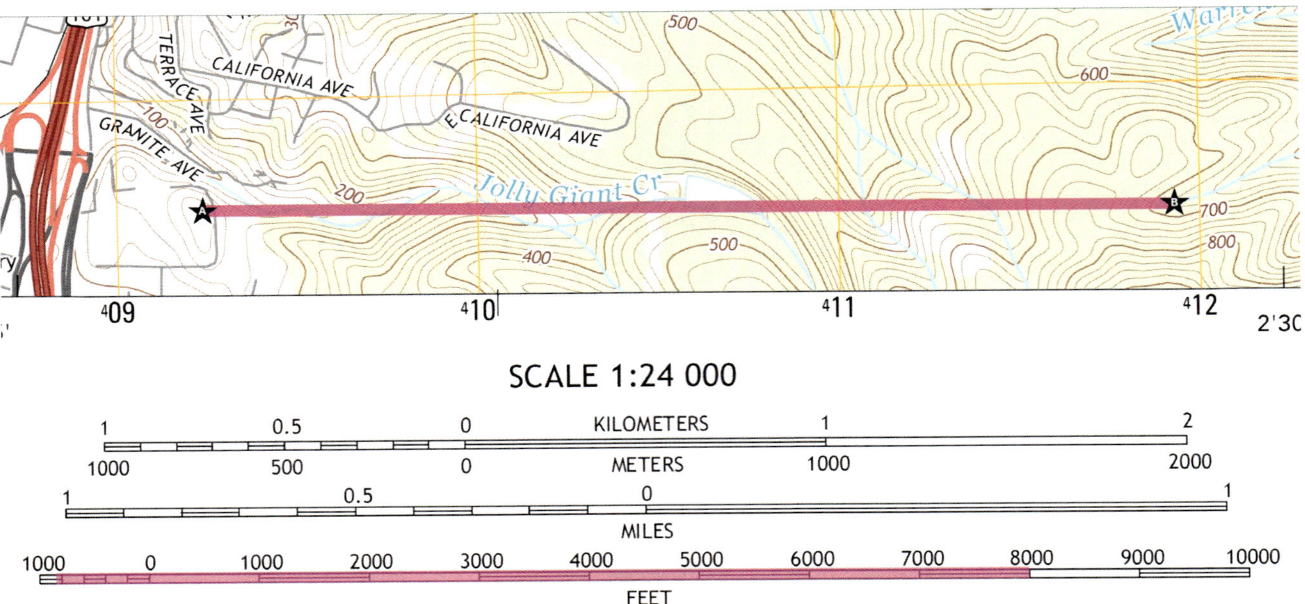

Figure 4.38: The purple line over the Feet scale bar is the approximate distance between point A and B. This measurement does not incorporate in changes in elevation.

Calculating Slope When Determining Distance

The previous example showed a line drawn between point A and point B on the Arcata North USGS Topographic Quadrangle. You were asked to estimate the distance between these two points using a graphic scale bar. However, the distances measured do not factor changes in elevation. In this step, you use the Pythagorean Theorem to calculate the slope distance. To review concepts related to the Pythagorean Theorem, watch the following video by Khan Academy (**Figure 4.39**).

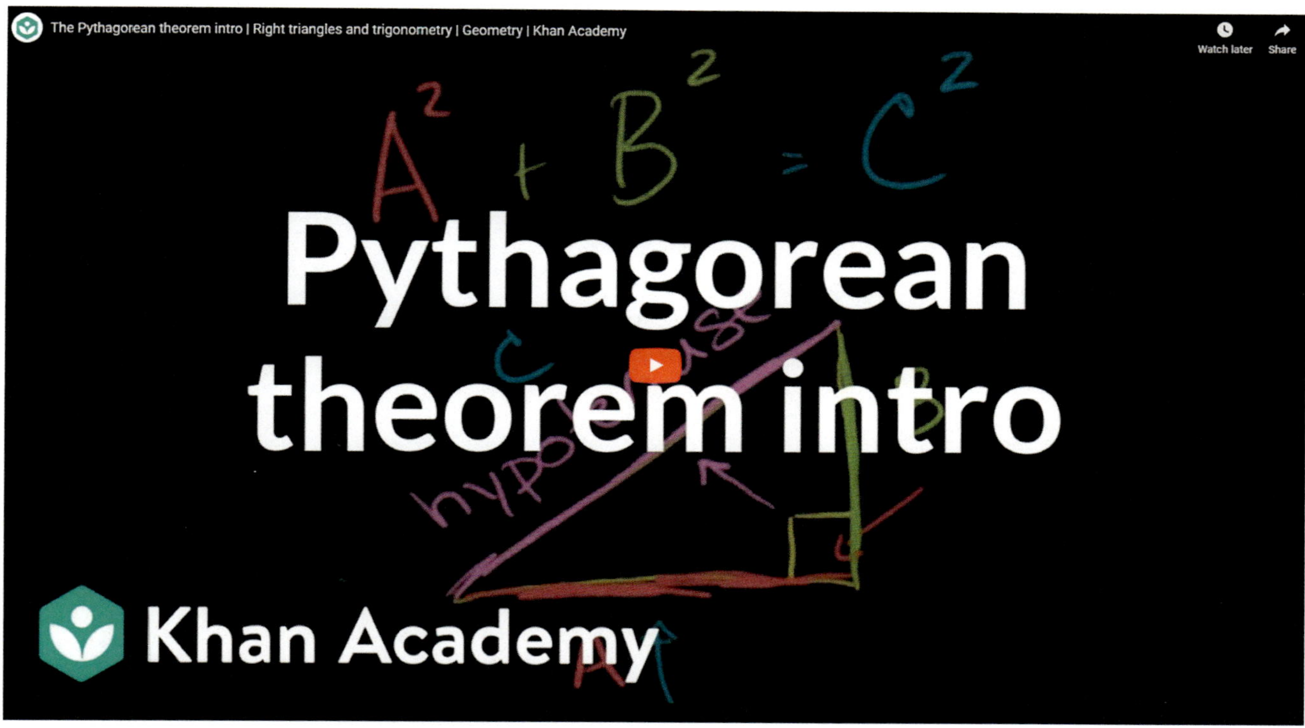

Figure 4.39: This video by Khan Academy discusses the Pythagorean theorem in additional detail. URL: *https://youtu.be/AA6RfgP-AHU*

Let us assume that Point A on the map is at an elevation of 90 feet above sea level. Let us also assume the point B on the map is 680 feet above sea level. Last, for our purposes here, assume that there are a gradual rise and constant slope between the two points on the map (**Figure 4.40**).

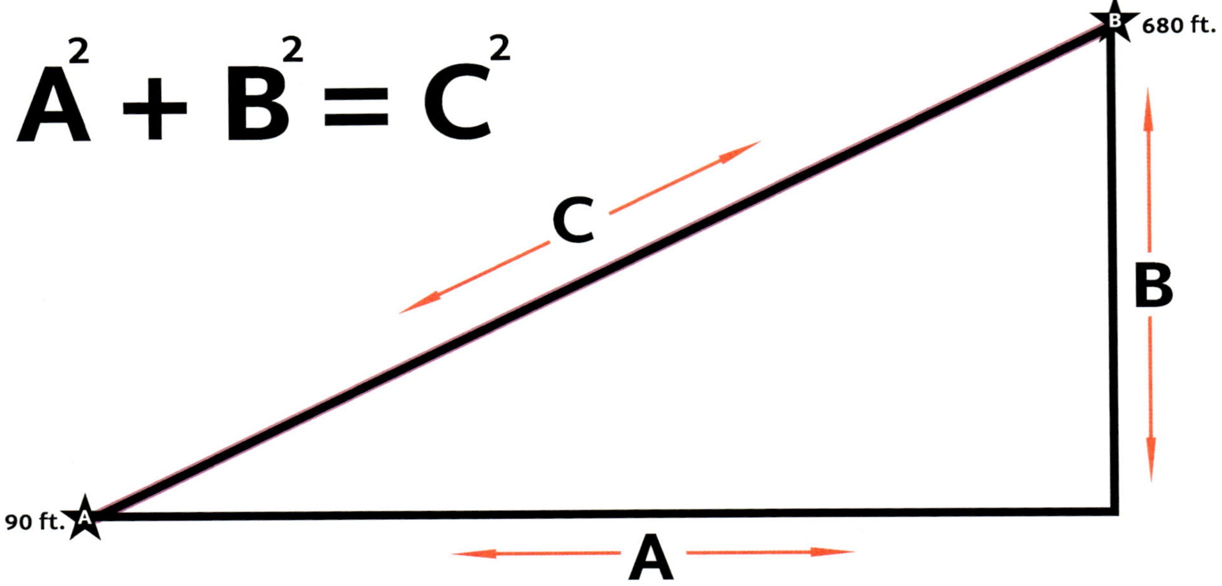

Figure 4.40: Recall that the Pythagorean Theorem measures the hypotenuse of a triangle, where C is the distance along the slope.

On a separate piece of paper, write down the answers to the following question:

5. Using your answer from Question 4 and the Pythagorean Theorem, what is the approximate distance in feet between points A and B on the map given the changes in elevation (**Figure 4.41**)?

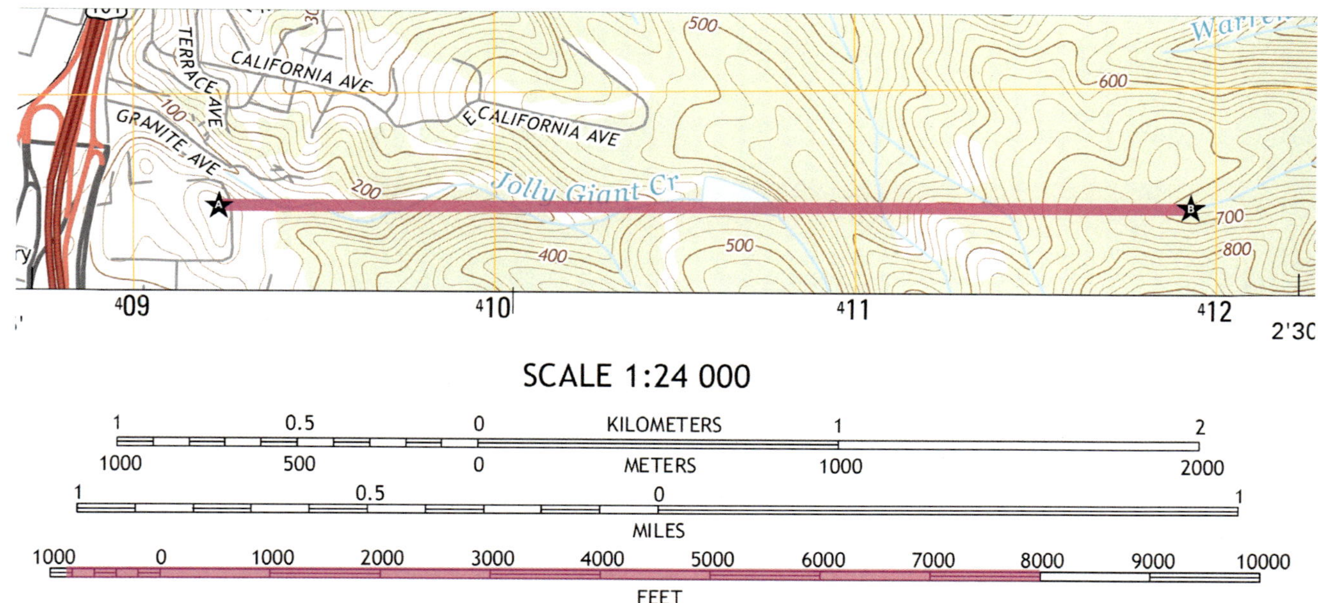

Figure 4.41: The purple line over the Feet scale bar is the approximate distance between point A and B.

Converting Map Distance to Ground Distance Using a Representative Fraction (RF)

As you learned in this chapter, a **representative fraction (RF)** is the ratio between the distance on the map and the distance on the ground. An RF is written using the notation 1/x or 1:x, where 1 represents the distance on the map and **x** represents the distance on the ground. Because the RF is a *ratio*, it is independent of any units of measurement.

On a separate piece of paper, write down the answers to the following question:

6. What is the RF of the Arcata North USGS Topographic Quadrangle (**Figure 4.42**)?

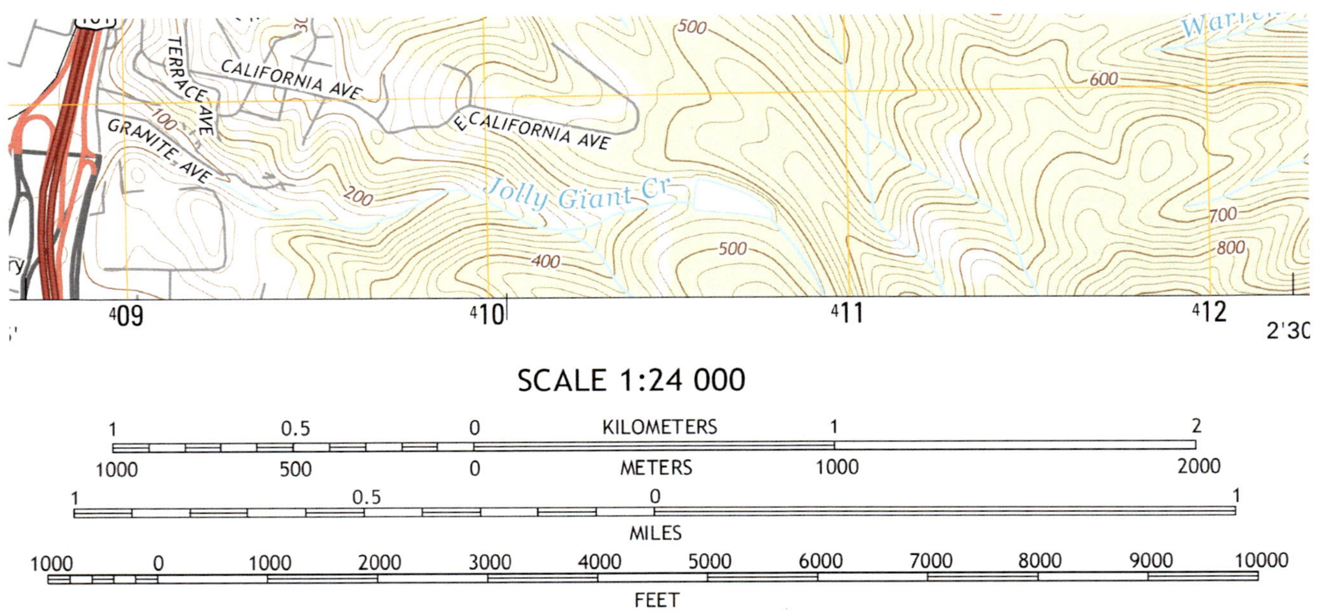

Figure 4.42: A USGS Topographic Quadrangle displays the RF near the bottom of the map, just outside the neatline.

Calculating the Ground Distance

You can use a representative fraction to estimate distances between two points on the map. For example, suppose you had a map with an RF of **1:48,000**. On the map, you marked two points. Using a ruler, you measured that the distance on the map between the two points was **6 inches**. You could use this information to find out the distance between the two points in miles. An RF of 1:48,000 means that 1 inch on the map is equal to 48,000 inches on the earth (ground inches). Therefore, 6 inches on the map is equal to 288,000 inches on the earth (6 x 48000 = 288000). There are 63,360 inches in 1 mile. To find out how many miles there are between the two points, convert inches to miles by dividing 288000 by 63360. The answer is that there are approximately 4.5 miles between the two points (**Figure 4.43**).

RF 1:48000 **1 inch on the map = 48000 inches on the ground** **1 mile = 63360 inches**

$$6 \text{ map inches} \times \frac{48000 \text{ ground inches}}{1 \text{ map inch}} = 288000 \text{ ground inches}$$

$$288000 \text{ inches} \times \frac{1 \text{ mile}}{63360 \text{ inches}} = 4.5 \text{ miles}$$

Figure 4.43: Though this example uses inches, it is essential to remember that an RF has no units. It is the ratio between map distance and ground distance.

On a separate piece of paper, write down the answers to the following question:

7. Using a ruler, you found that the distance between two points on the Arcata North USGS Topographic Quadrangle was **13.2 inches**. Given the representative fraction printed on the map (**Figure 4.44**), how many *miles* are there between the two points?

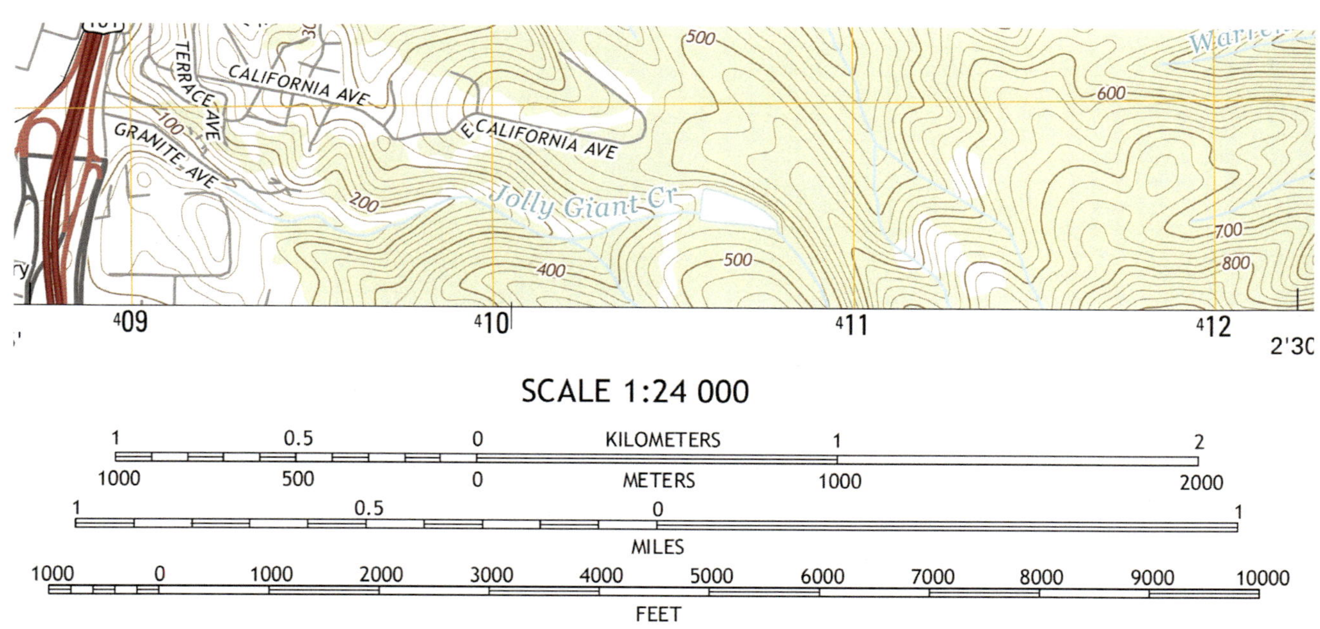

Figure 4.44: A USGS Topographic Quadrangle displays the RF near the bottom of the map, just outside the neatline.

Determining Scale Using a Representative Fraction

It is also possible to determine an unknown scale by using the representative fraction from a source with a known scale, such as a map. For example, suppose you acquired an old aerial photograph but were unsure of the scale (**Figure 4.45**).

Figure 4.45: The oxidation pond in this image provides an excellent feature to reference. Source: USGS.

You could use the scale from a known source, such as a map, to determine the scale of the photo (**Figure 4.46**).

U.S. DEPARTMENT OF THE INTERIOR
U.S. GEOLOGICAL SURVEY

Figure 4.46: The Arcata South USGS Topographic Quadrangle contains the same feature as the photograph.

Map scale describes the relationship between distance on Earth and the way that a map represents the same distance. It is the ratio of map distance over ground distance (**Figure 4.47**).

Figure 4.47: In this definition of scale, ground distance represents the distance on Earth.

To determine the scale of the old photograph, you need to know how the distance on the photo relates to the distance on the ground (**Figure 4.48**).

Figure 4.48: The definition of scale for the photograph is photo distance over ground distance.

To start, measure the distance of a prominent feature on the photo that you could also identify on a map. In this instance, the let us assume that the width of the divide in the oxidation pond, when measured on the photograph, is **six inches** (**Figure 4.49**).

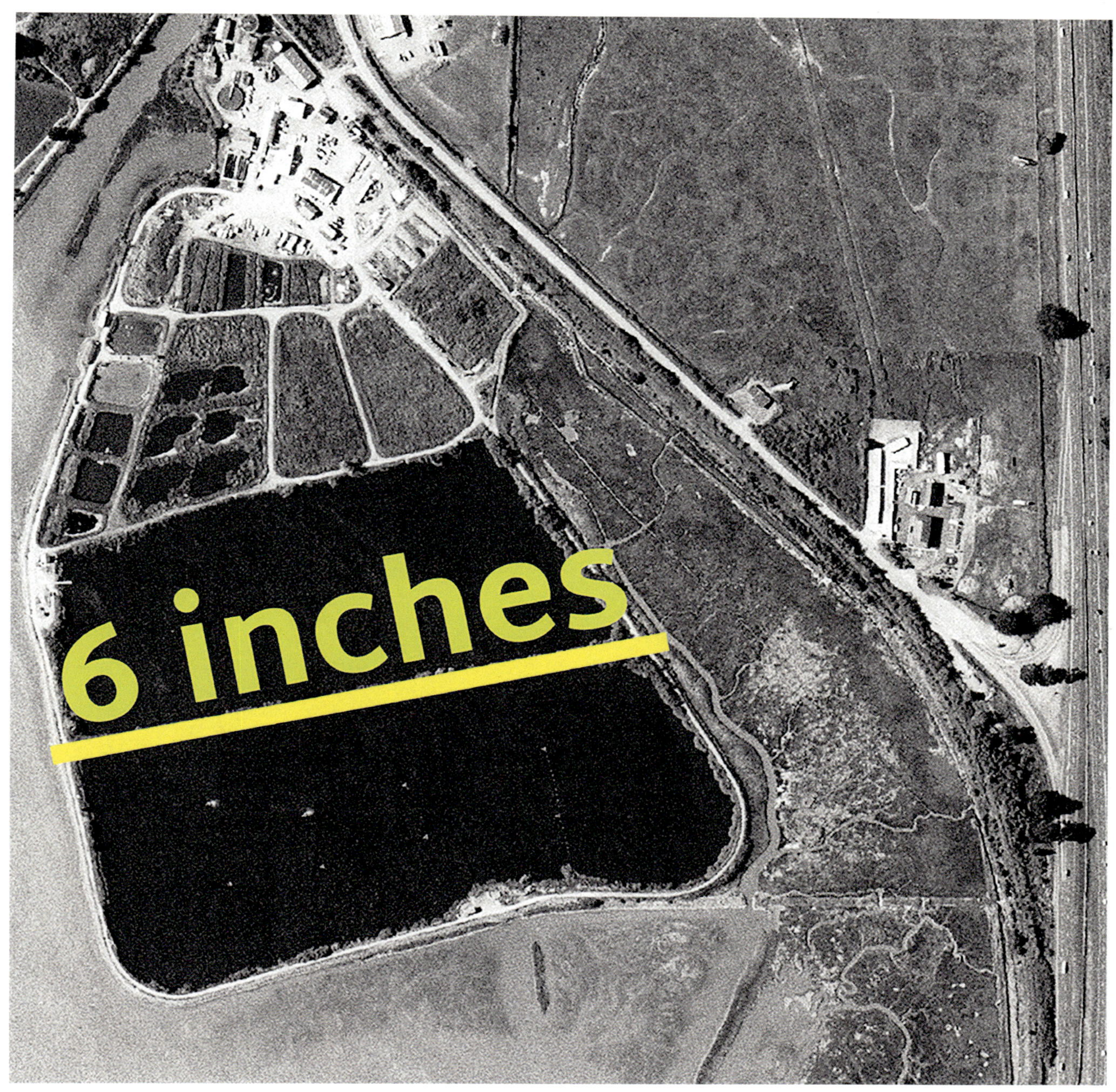

Figure 4.49: The oxidation pond is six inches on the old photograph.

By measuring the distance on the photograph, you obtain half of the ratio for determining scale, the *photo distance*. Next, determine how far this distance represents on earth, which we call *ground distance*. Let us assume that the same distance, when measured on the Arcata South USGS Topographic Quadrangle is **one inch** (**Figure 4.50**).

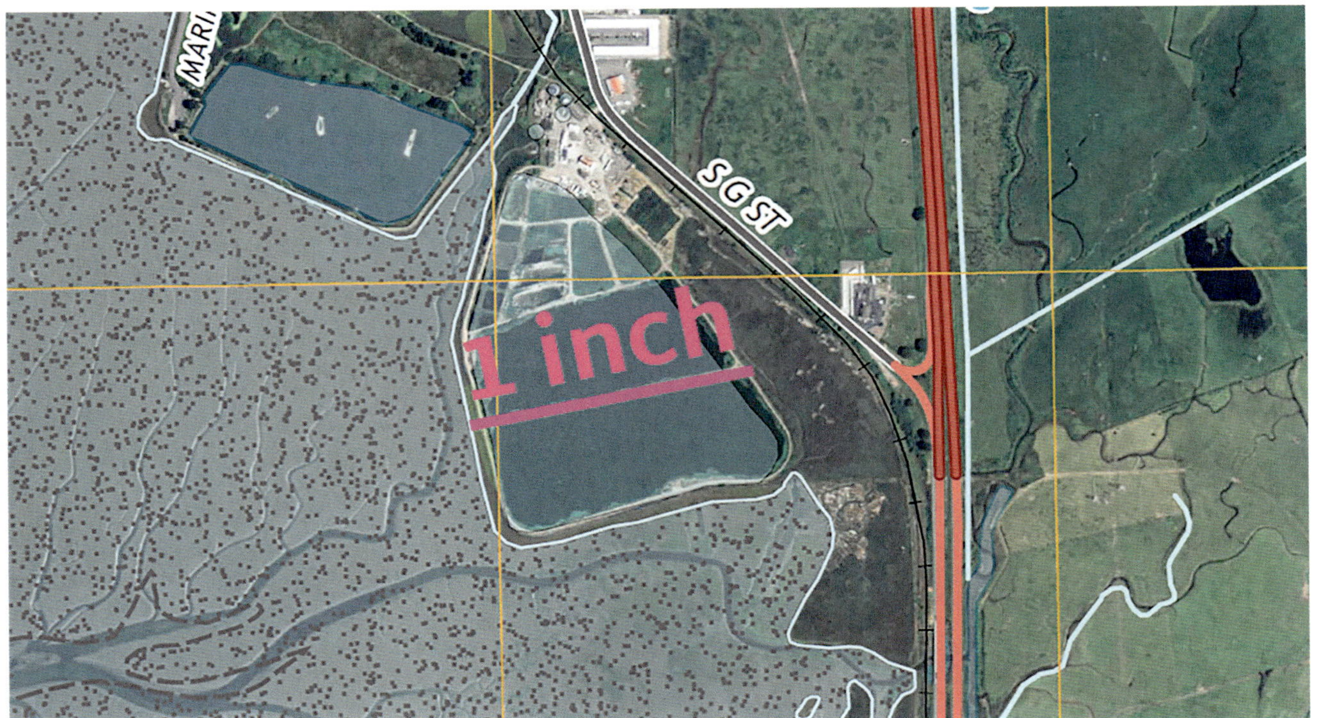

Figure 4.50: The oxidation pond is one inch on the map.

Skill Drill: Converting Map Distance to Ground Distance Using a Representative Fraction (RF)

In the previous step, you learned how to convert map distance to ground distance using a Representative Fractions (RF). The RF for the Arcata South USGS Topographic Quadrangle is 1:24000.

On a separate piece of paper, write down the answers to the following questions:

8. What is the *ground distance* of the width of the oxidation pond, given that the distance on the map was 1 inch and the map RF is 1:24,000?
9. Based on your answer to question 8 and the definition of scale, what is the scale of the photo? Be sure to reduce the fraction so that you have the number 1 in the numerator and can represent the ratio as 1:x.
10. What would be the best way to express the scale of the old photograph as a *word statement*?
11. Is the old photograph larger or smaller in scale than the Arcata South USGS Topographic Quadrangle?

Tutorial: Working with Spatial Reference Systems

In this chapter, you learned that a spatial reference system defines a geographic location. Some spatial references use coordinates, and others do not. In this activity, you will learn how to identify a position on a map using the correct notation for several spatial reference systems, including the following:

» U.S. public land survey system (PLSS)
» Geographic Coordinate Systems (GCS)
» Universal Transverse Mercator (UTM) system
» State Plane Coordinate (SPC) system

ESTIMATED TIME TO COMPLETE THIS TUTORIAL: 3 HOURS

Learning Outcomes

Readers should be able to accomplish the following outcomes by the end of this tutorial:

» Use the correct notation for spatial reference systems
» Determine distances on a spherical Earth using a GCS
» Define a position using a GCS
» Define a position using PLSS
» Define a position using SPC
» Define a position using UTM

Geographic Coordinate Systems

As you learned in a previous chapter, parallels and meridians are measured in degrees using latitude and longitude. Latitude and longitude use a **sexagesimal numeral system**, a base 60 system commonly used to measure time and degrees. Every 1 degree is divided into 60 minutes (60') and every 1 minute is divided into 60 seconds (60") (**Figure 4.51**).

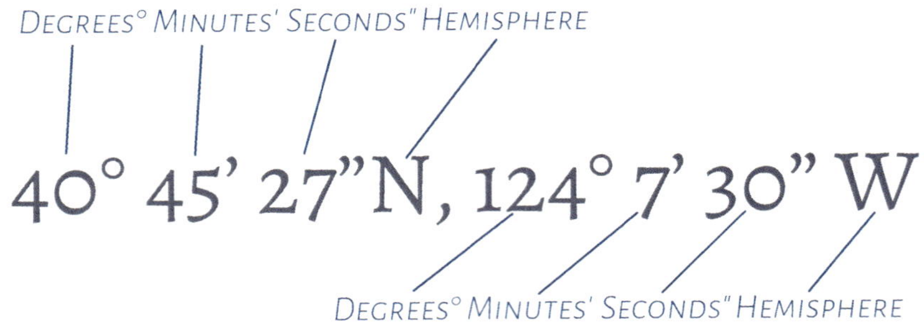

Figure 4.51: The correct notation starts with latitude, followed by longitude.

Subtracting Degrees Minutes and Seconds

Sometimes, it is difficult to determine how latitude and longitude translate to distance on the ground. With latitude, it is a little less complicated because the distance between lines of latitude is relatively constant, with just a minor variation between the equator and the poles (**Figure 4.52**). For our purposes here, let us assume that 1° of latitude is equal to approximately 69 miles.

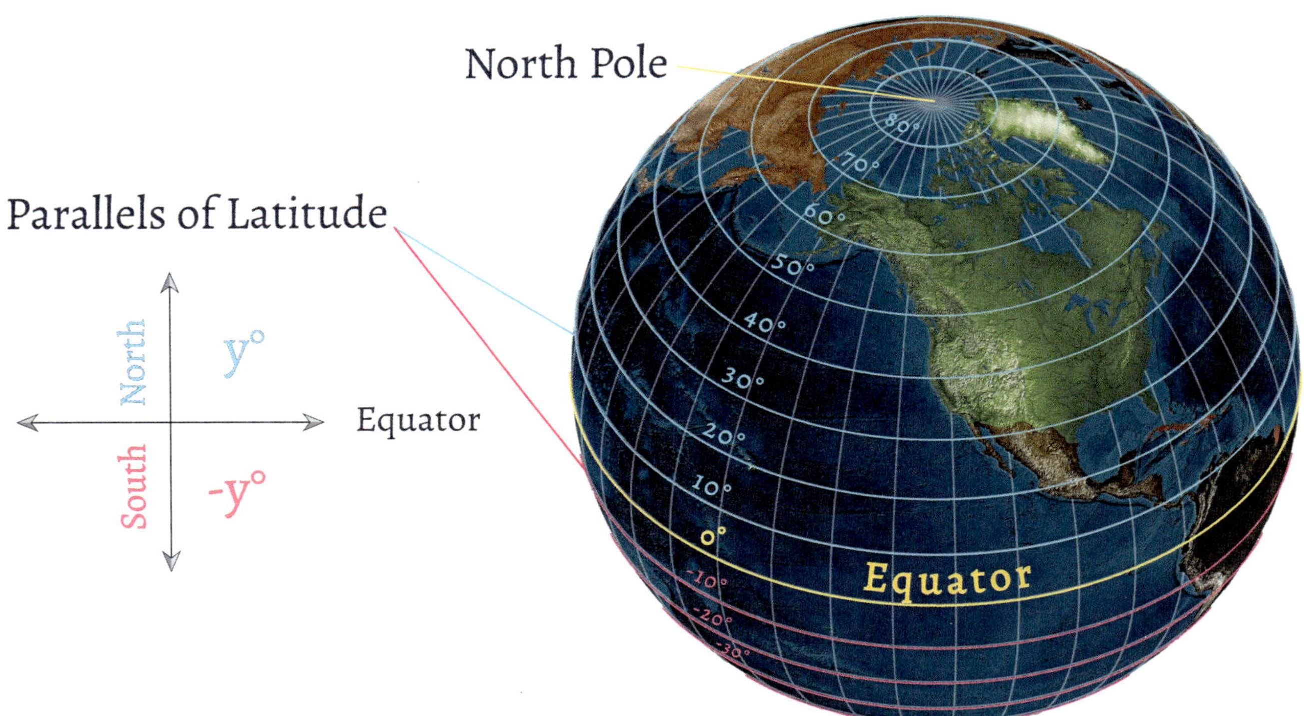

Figure 4.52: On the WGS 84 ellipsoid, the difference between the length of 1° of latitude ranges from 68.703 statute miles near the equator to 69.407 at the poles.

You can calculate North-South distances based on latitude coordinates. For example, suppose you were sailing a ship from Humboldt Bay, California to Glacier Bay, Alaska and wanted to know how many miles north you were traveling. Let us assume that Glacier Bay is located at 58° 29' 14" N latitude and that Humboldt Bay is located at 40° 45' 27" N latitude. Start by finding the difference in latitude (**Figure 4.53**). Remember, when using a sexagesimal numeral system it is very similar to adding and subtracting time. On a piece of paper, practice subtracting 40° 45' 27" from 58° 29' 14". Keep in mind that you borrow 60 from the next unit up the line. For example, you can't subtract 27 seconds from 14 seconds. To fix this, you borrow 60 seconds from 29 minutes. 29 minutes becomes 28 minutes and 14 seconds becomes 74 seconds. 74 seconds minus 27 seconds is 47 seconds. Use the same method to subtract 45 minutes from 28 minutes. 58° 29' 14" minus 40° 45' 27" is a difference of 17° 43' 47" in latitude.

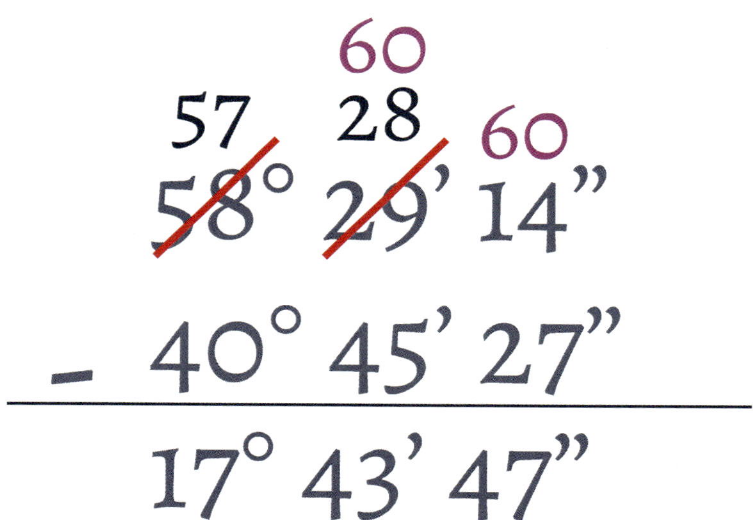

Figure 4.53: Every 1 degree is divided into 60 minutes (60') and every 1 minute is divided into 60 seconds (60").

Now it's time for you to try. On a separate piece of paper, write down the answer to the following question:

1. Suppose you wanted to know the difference in latitude between two points, C and D, on the Arcata North USGS Topographic Quadrangle. Point D is located at 40° 57' 30" N latitude, and location C is located at 40° 55' 00" N latitude (**Figure 4.54**). What is the difference in latitude?

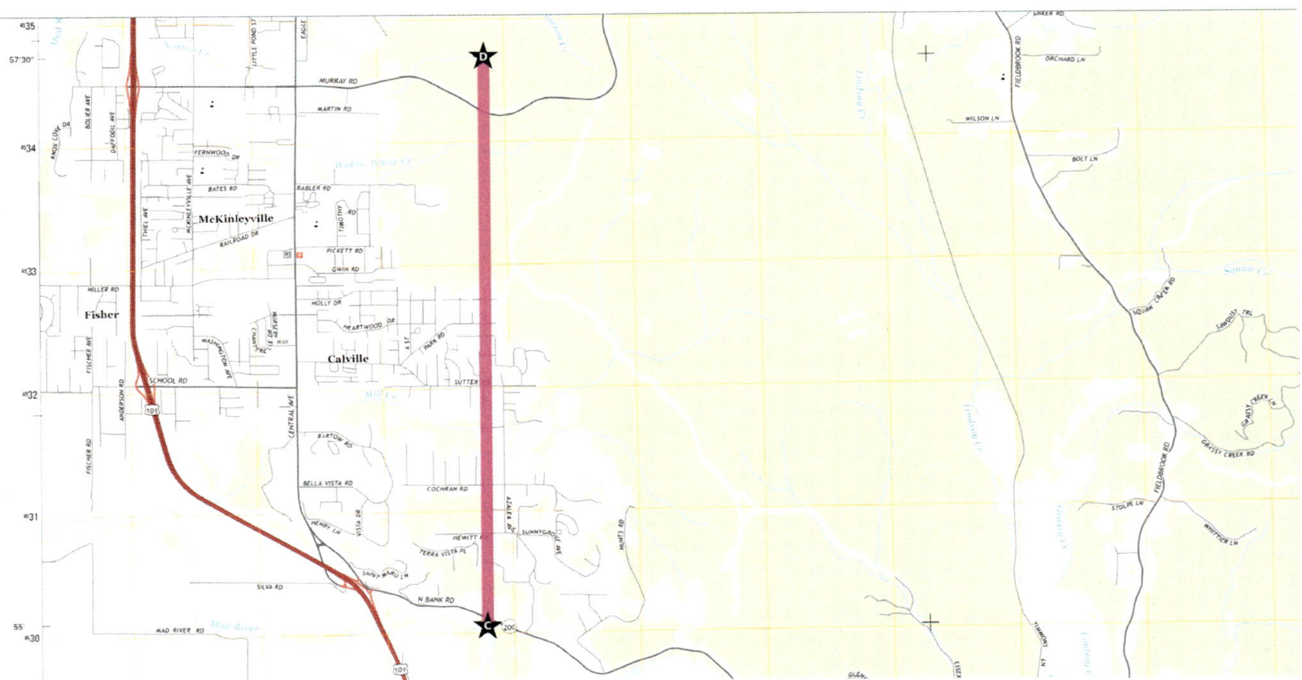

Figure 4.54: The stars labeled C and D are the points of interest.

Converting Latitude to Miles

In the previous step, you learned how to subtract degrees minutes and seconds to find the difference in latitude from one point to another. In this step, you will convert the difference in latitude to miles. For example, you found that the difference in latitude between Humboldt Bay and Glacier Bay is 17° 43' 47". For our purposes here, let us assume that 1° of latitude = 69 miles. Therefore, 17° is equal to 1173 miles (17 x 69 = 1173). Use the same relationship to determine how many miles there are in 1 minute of latitude. 1° of latitude is equal to 69 miles, and 1° of latitude is also equal to 60 minutes. By dividing 69 by 60, you know that there are 1.15 miles per minute of latitude. Using this information, you can estimate that there are 49.45 miles in 43 minutes of latitude (1.15 x 43 = 49.45). Repeat this process to find out how many miles there are in 1 second of latitude (**Figure 4.55**). 1° of latitude is equal 69 miles, and 1° of latitude is also equal to 3600 seconds. By dividing 69 by 3600, you know that there are 0.02 miles per second of latitude. Using this information, you can estimate that there are 0.94 miles in 47 seconds of latitude (0.02 x 47 = 0.94). The total miles add up to 1223.39 (1173 + 49.45 + 0.94 = 1223.39).

1° of latitude = 69 miles   1° of latitude = 60 minutes   1° of latitude = 3600 seconds

$$\frac{69 \text{ miles}}{1 \text{ degree}} \qquad 17 \text{ degrees} \times \frac{69 \text{ miles}}{1 \text{ degree}} = 1173 \text{ miles}$$

$$\frac{69 \text{ miles}}{60 \text{ minutes}} = \frac{1.15 \text{ miles}}{1 \text{ minute}} \qquad \frac{1.15 \text{ miles}}{1 \text{ minute}} \times 43 \text{ minutes} = 49.45 \text{ miles}$$

$$\frac{69 \text{ miles}}{3600 \text{ seconds}} = \frac{0.02 \text{ miles}}{1 \text{ second}} \qquad \frac{0.02 \text{ miles}}{1 \text{ second}} \times 47 \text{ seconds} = 0.94 \text{ miles}$$

17° 43' 47" latitude = 1173 miles + 49.45 miles + 0.94 miles = 1223.39 miles

Figure 4.55: Through a series of conversions and multiplication, you get a distance of 1223.39 miles.

On a separate piece of paper, write down the answer to the following question:

2. Based on your answer to question 1, how many miles are there between point C and point D on the Arcata North USGS Topographic Quadrangle?

Converting Longitude to Miles

As you learned in this chapter, the distance between 1° of longitude is more strongly related to location than latitude. Calculating distances using longitude can be much more complicated. Lines of longitude move closer together as you near the poles (**Figure 4.56**). Therefore, the distance between one degree of longitude is continuously changing as you move north or south.

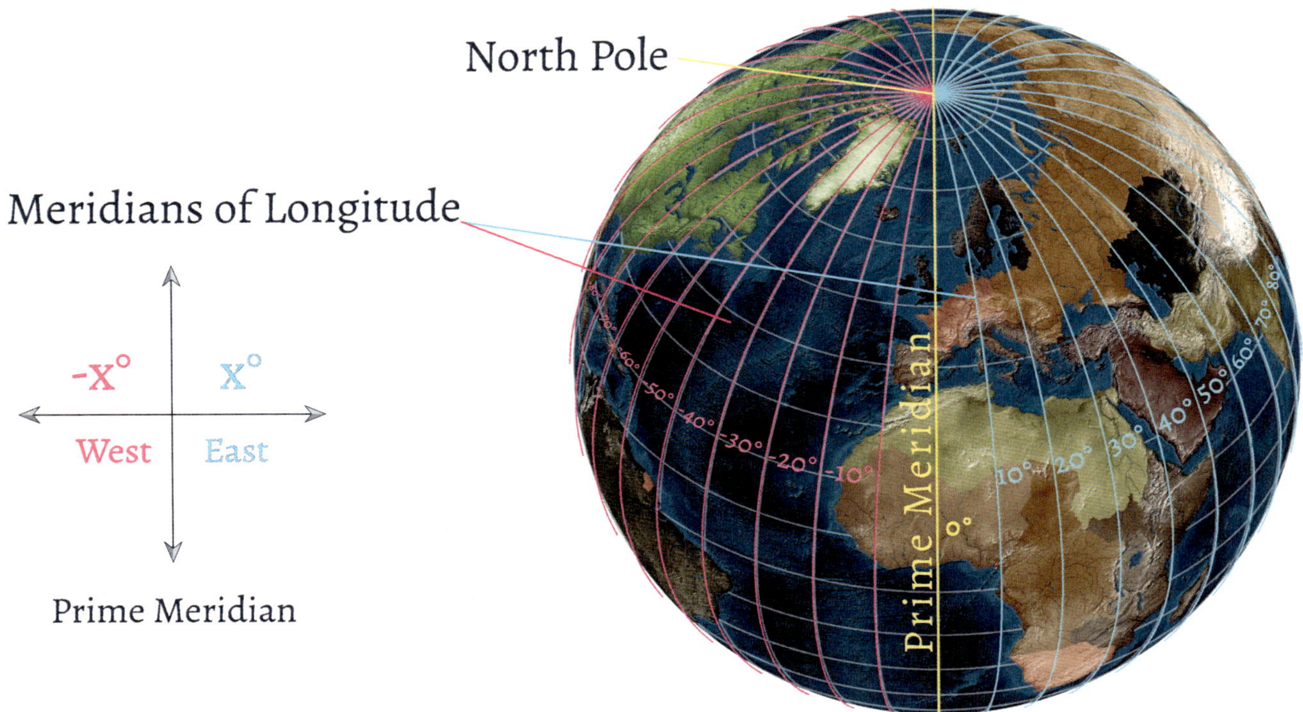

Figure 4.56: At the equator, the distance between 1° of longitude is about 69 miles. At the poles, the distance between degrees of longitude is 0 miles because all lines of longitude converge at the poles.

Luckily, there is a simple formula for calculating the distance between lines of longitude (**Figure 4.56**).

$$1^\circ \text{ of longitude} = \cos(x^\circ \text{ latitude}) \times 69 \text{ miles}$$

Figure 4.56: You can find the approximate length in miles of any degree of longitude by using this formula.

For example, the latitude at Glacier Bay is approximately 58.5°. The distance between 1° of longitude at Glacier Bay is about 36 miles (cos(58.5) x 69 = 36.05).

On a separate piece of paper, write down the answer to the following questions:

3. Assuming that Humboldt County is located at approximately 41° N latitude. What is the distance in miles between 1° longitude in Humboldt County? You may round to the nearest mile.
4. In Humboldt County, point E is located at 124° 7' 30" W longitude, and point F is at 124° 2' 30" W longitude. Based on your answer to question 3 and using what you learned when you answered question 2, what is the approximate distance from east to west in miles between points E and F?

Defining a Location on a Map Using Geographic Coordinates

The *Arcata South USGS Topographic Quadrangle*[1] displays the latitude and longitude at each of the four corners of the neatline (**Figure 4.57**).

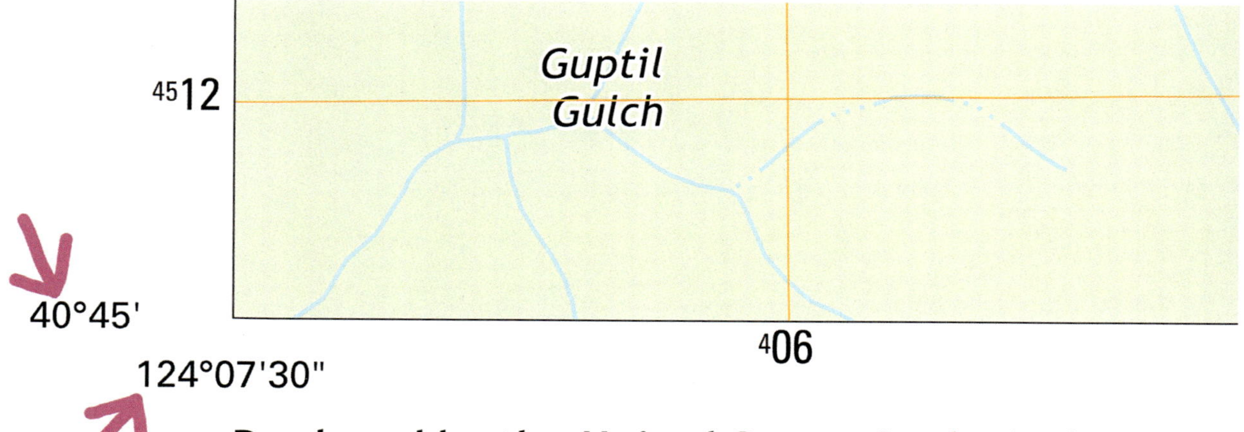

Figure 4.57: The bottom of the neatline runs along 40° 45' 00" N latitude, and the left side of the neatline runs thru 124° 07' 30" W longitude.

As you learned in this chapter, longitude in the western hemisphere increases as you travel west and decreases as you move east. Likewise, longitude increases as you journey north and decreases as you move south. The map indicates every 2.5 minutes of either

1 https://wp.me/a6siVq-1L9

latitude or longitude using a black tick mark along the inside of the neatline. It also abbreviates the geographic coordinates in between the corners (**Figure 4.58**).

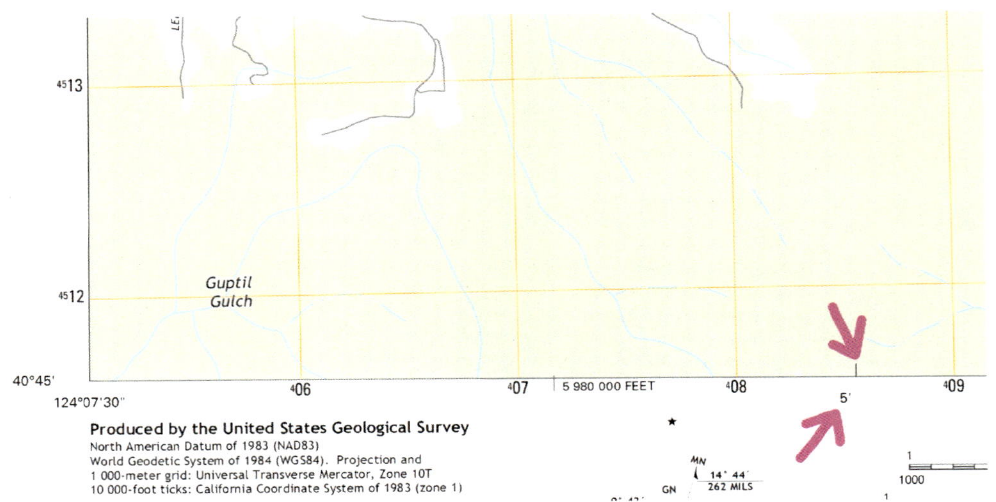

Figure 4.58: The black tick on the inside of the neatline indicates a line of longitude. Though abbreviated, the longitude is 124° 5' 00".

Sometimes you may need to estimate the geographic coordinates for a particular location on a map, in between the tick marks. You can do this by measuring the proportional distance from your point of interest to the nearest tick mark. For example, suppose you wanted to determine the geographic coordinates for the Arcata Plaza on the Arcata South USGS Topographic Quadrangle (**Figure 4.59**).

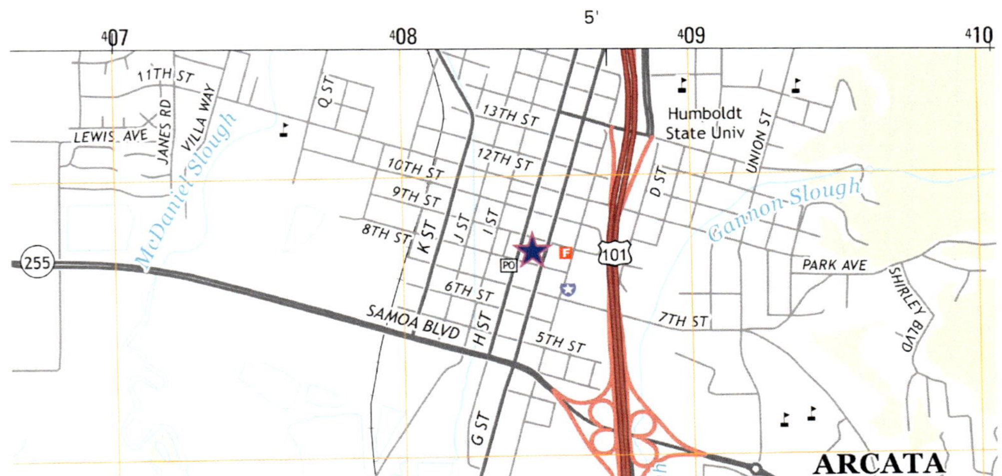

Figure 4.59: The blue and magenta start marks the location of the Arcata Plaza.

Start by drawing a grid around your point of interest (**Figure 4.60**). For geographic coordinates, use the black tick marks on the inside of the neatline as a guide. Because each tick mark for both latitude and longitude are 2.5 minutes apart, you may expect the grid to be square shaped. As you learned in this chapter, lines of longitude are closer together than lines of latitude in Humboldt County. As a result, the grid you draw between tick marks appears rectangular with a north-south (portrait) orientation.

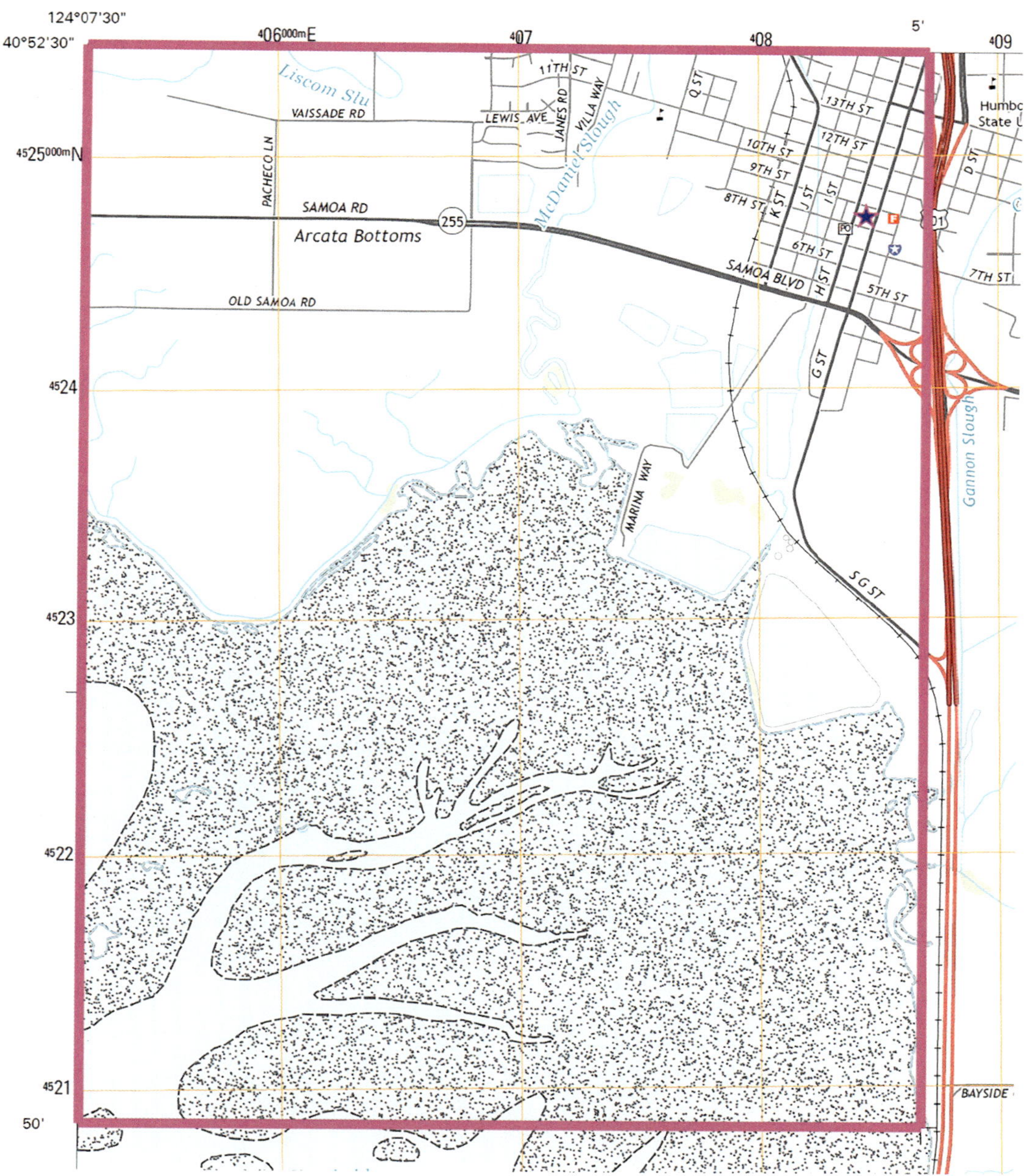

Figure 4.60: The purple lines delineate the boundaries of the geographic coordinates marked on the map. Each tick mark is 2.5 minutes apart.

To estimate the latitude, measure the distance from the nearest tick to the Arcata Plaza. The unit of measure does not matter, though some units, such as centimeters are easier to work with mathematically. In this example, the distance from the neatline, 40° 52' 30"N latitude, to the Arcata Plaza is 3.6 centimeters.

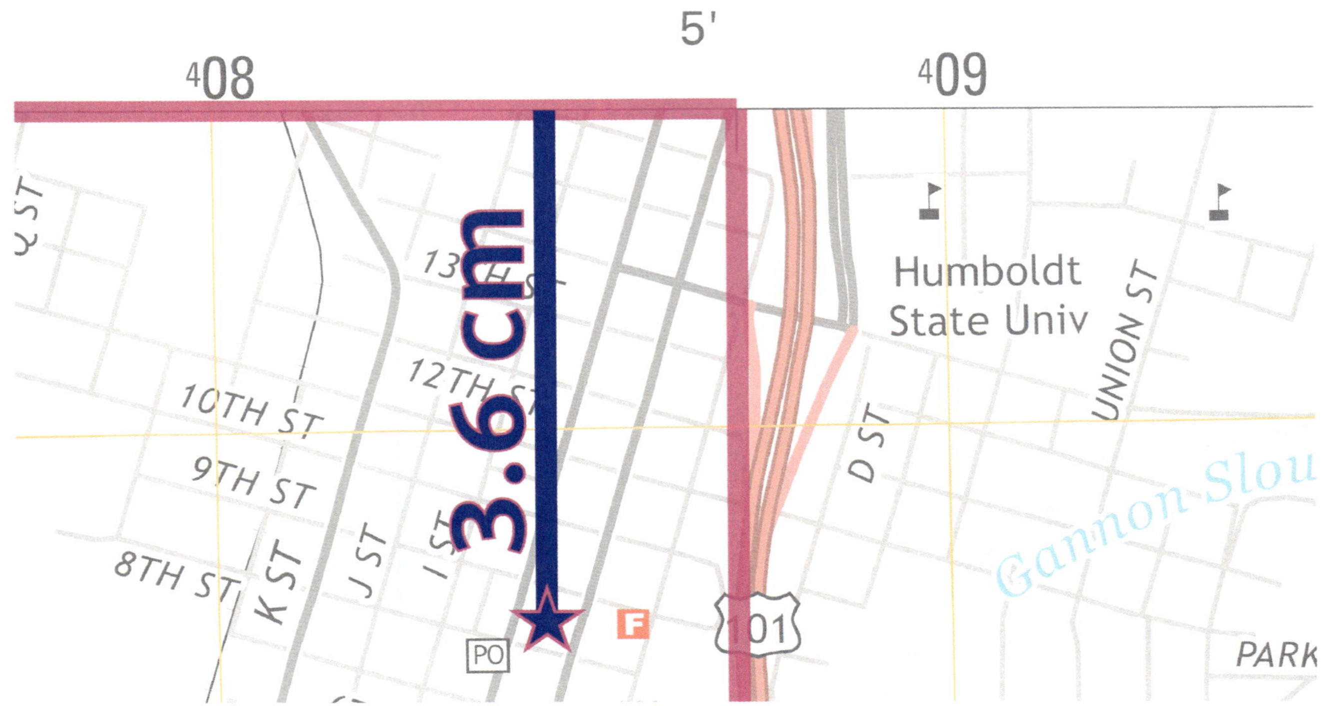

Figure 4.61: The top neatline boundary represents a line of latitude, 40° 52' 30" N. Here, the distance on the map from the line of latitude to the Arcata Plaza is 3.6 centimeters.

Next, measure the distance across, from one line of latitude to the next on the map (**Figure 4.62**). In this example, the distance between the distance from the neatline, 40° 52' 30" N latitude, to the next tick down, 40° 50' 00" N latitude, is 23.8 centimeters.

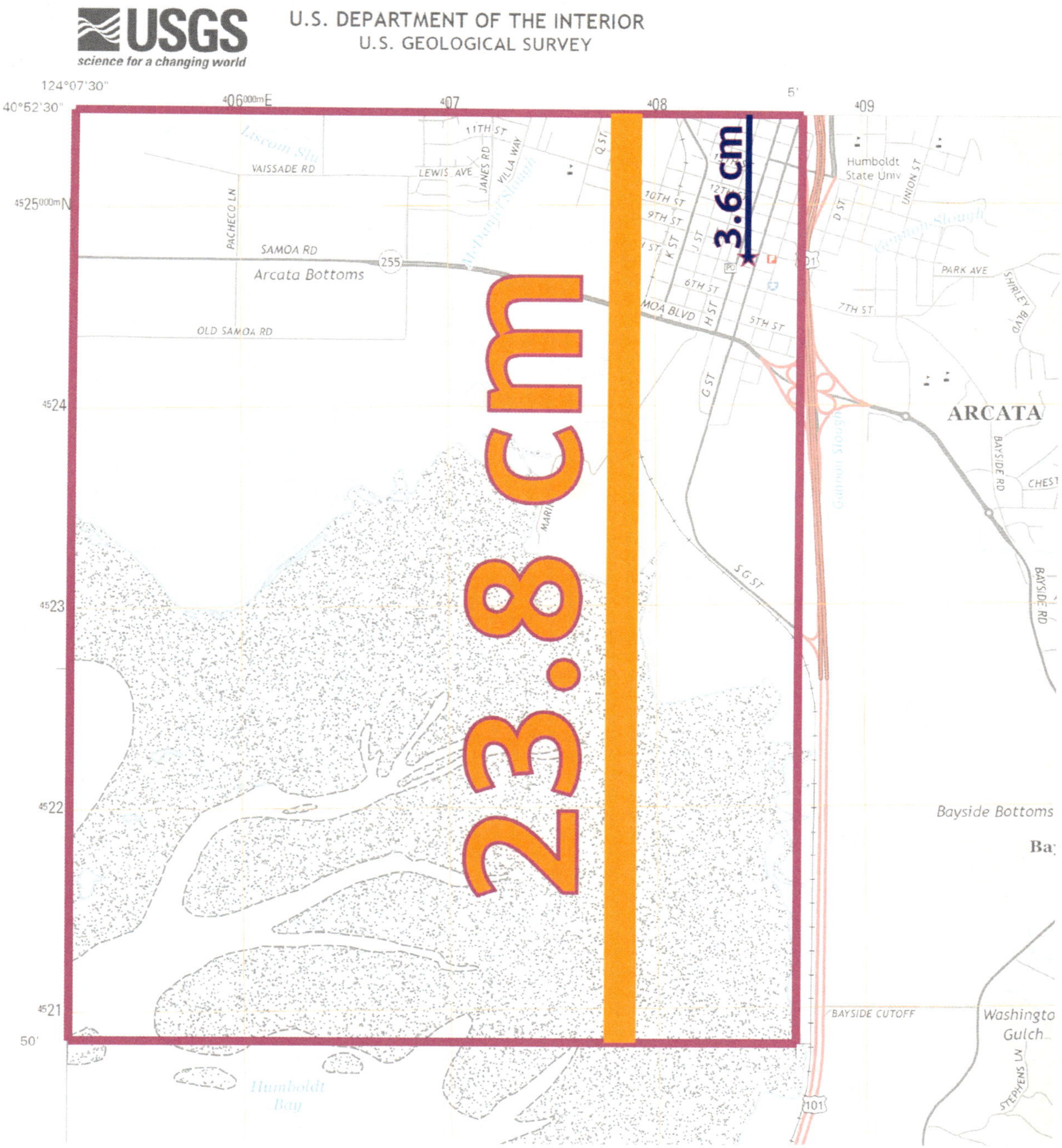

U.S. DEPARTMENT OF THE INTERIOR
U.S. GEOLOGICAL SURVEY

Figure 4.62: The next tick mark below the top left corner represents a line of latitude, 40° 50' 00" N. Here, the distance on the map between the lines of latitude is 23.8 centimeters.

With the distance from the top of the map to the Arcata Plaza and the total distance between tick marks measured, you can find the proportional distance. 3.6 centimeter divided by 23.8 centimeters is approximately 0.15 (**Figure 4.63**). The distance from the top of the map to the Arcata Plaza is about 15% of the total distance between tick marks.

Total Distance betwen ticks of latitude = 23.8 cm

Distance to Arcata Plaza = 3.6 cm

$$\frac{3.6}{23.8} = 0.15$$

The Arcata Plaza is 15% of the total distance between the tick marks.

Figure 4.63: Using the proportional distance is an effective way to estimate geographic coordinates.

The next step is to determine what this percentage means regarding degrees, minutes, and seconds. You know that the distance between the latitudinal tick marks is 2.5 minutes.

On a separate piece of paper, write down the answer to the following questions:

5. What is 15% of 2.5 minutes? Convert your answer to seconds.
6. Given that latitude decreases as you move south, use your answer from question 5 to determine the latitude of the Arcata Plaza in degrees, minutes, and seconds. Round to the nearest second. Be sure to use the correct notation for latitude.

You can also determine longitude using the same method. Measure the distance to the nearest line of longitude (**Figure 4.64**). Then, measure the total distance between longitudinal tick marks and find the proportions. Let us assume that the entire distance between longitudinal tick marks on the map is 18.1 cm. Let us also understand that the distance from the tick mark east of the Arcata Plaza is 1.3 cm.

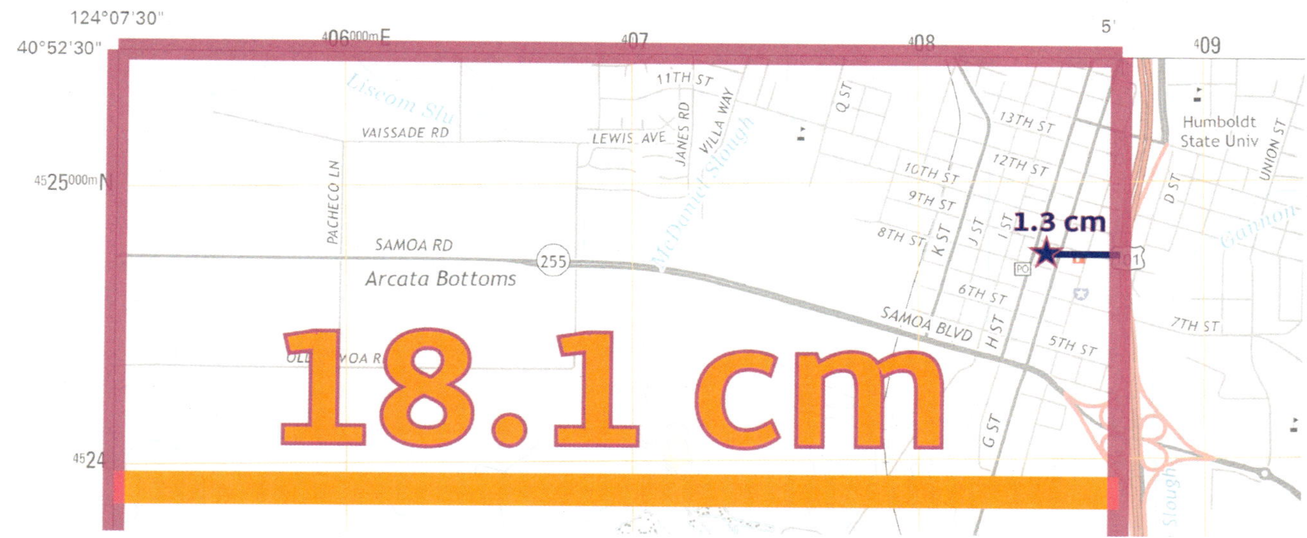

Figure 4.64: The next tick mark east the top left corner represents a line of longitude, with a marking abbreviated as 5'. Here the distance on the map between the lines of longitude is 18.1 centimeters, and the distance between the Arcata Plaza to the nearest tick is 1.3 centimeters.

On a separate piece of paper, write down the answer to the following questions:

7. The longitude at the top left corner is 124° 07' 30" W. The longitudinal tick mark east of the top left corner has a marking abbreviated as 5'. What is the longitude at this tick mark in degrees minutes and seconds? Be sure to use the correct notation for longitude
8. Let us assume that the total distance between longitudinal tick marks is 18.1 centimeters, and the distance to the Arcata Plaza is 1.3 centimeters. What percentage of the total is the distance from the Arcata Plaza to the nearest longitudinal tick mark? Round to the nearest percent.
9. Given that the distance between the latitudinal tick marks is 2.5 minutes use your answer from question 8 to determine how many seconds are there from the nearest longitudinal tick mark to the Arcata Plaza?
10. Given that longitude increases as you move west, use your answer from question 9 to determine the longitude of the Arcata Plaza in degrees, minutes, and seconds. Round to the nearest second. Be sure to use the correct notation for longitude.

Defining a Location Using the U.S. Public Land Survey System

As you learned in this chapter, the land ordinance of 1785 established the **U. S. public land survey system (PLSS)** (**Figure 4.65**). The U.S. Public Land Survey System consists of a grid of sections referred to as township and ranges. Land surveyors would select an initial point at the intersection of a parallel and a meridian. The starting parallel is the baseline, and the starting meridian is the principal meridian. For example, the westernmost principal meridian and baseline in the conterminous United States are called the *Humboldt Meridian and Baseline* (**Figure 4.66**).

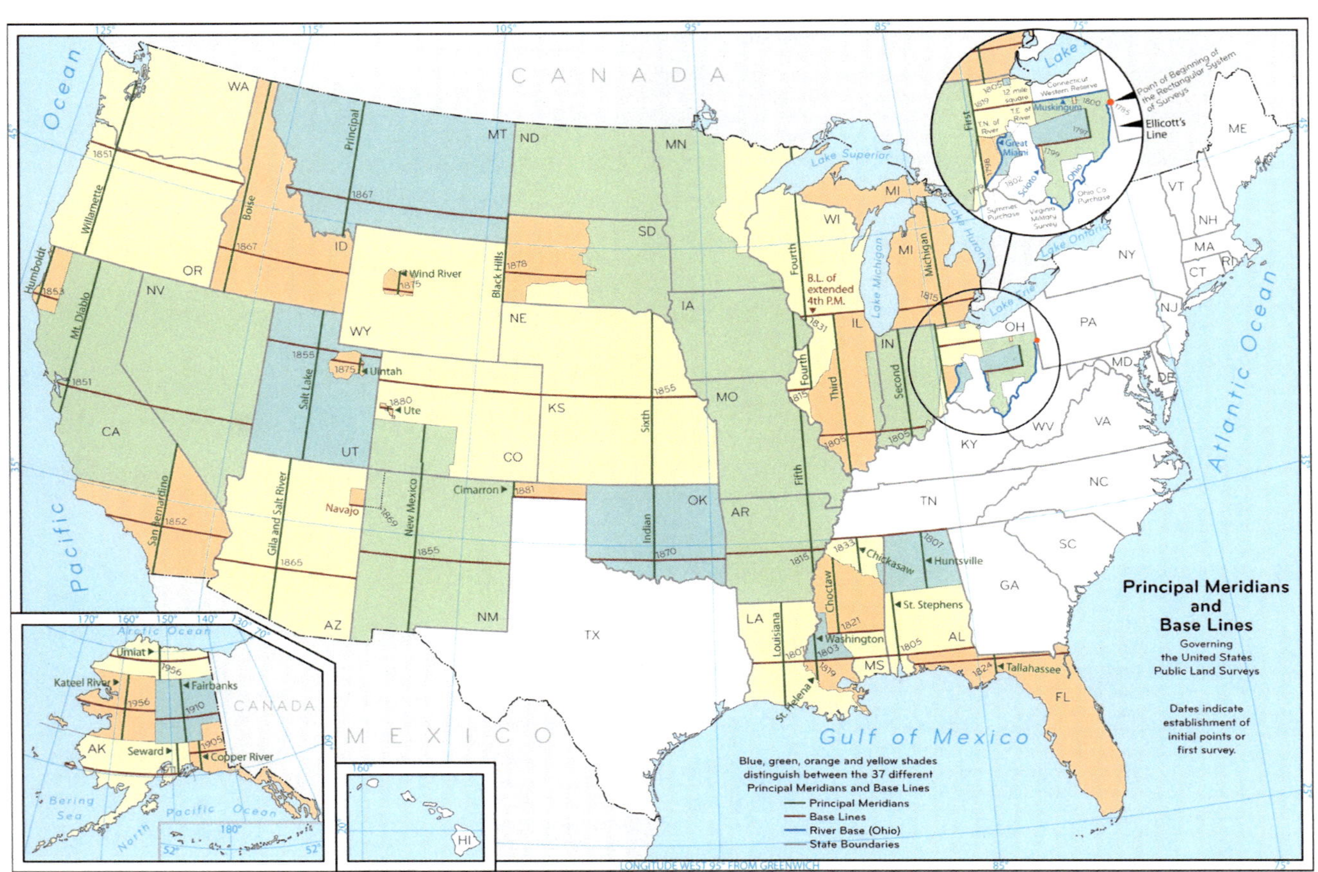

Figure 4.65: The map here shows the principal meridians and baselines governing the United States Public Land Surveys, published in 2012 by the U.S. Bureau of Land Management.

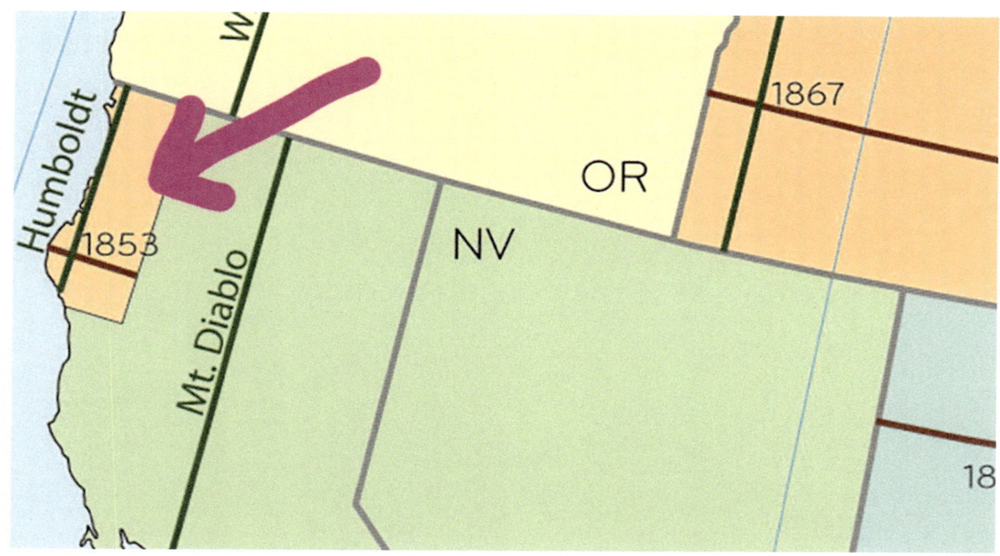

Figure 4.66: The westernmost principal meridian and baseline in the conterminous United States are called the Humboldt Meridian and Baseline.

Once surveyors established the principal meridian and baseline, they measured a grid of 6-mile quadrilaterals called townships and ranges. Each 6-square-mile township further divides into 36, one-square-mile sections (**Figure 4.67**).

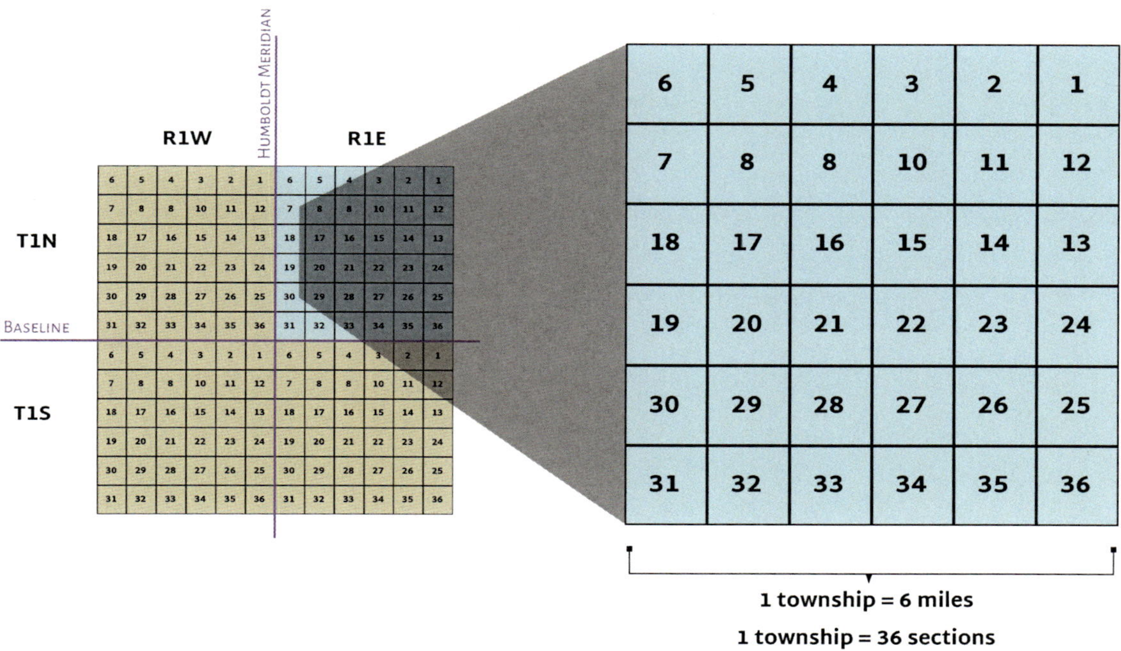

Figure 4.67: In this example, the township and range shown in blue can be described as T1N R1E Humboldt Meridian and Baseline.

A zigzag method of numbering the sections begins with section 1 at the upper right corner of the township and ending with section 36 the lower right corner of the township. A series of fractional divisions of ½ or ¼ section are used to partition smaller units of land (**Figure 4.68**). The correct notation for the PLSS system begins with the smallest fractional division and moves up to the most extensive fractional division, each separated by commas. The section number, township, and range follow the fractional divisions. The notation ends with the name of the meridian and baseline.

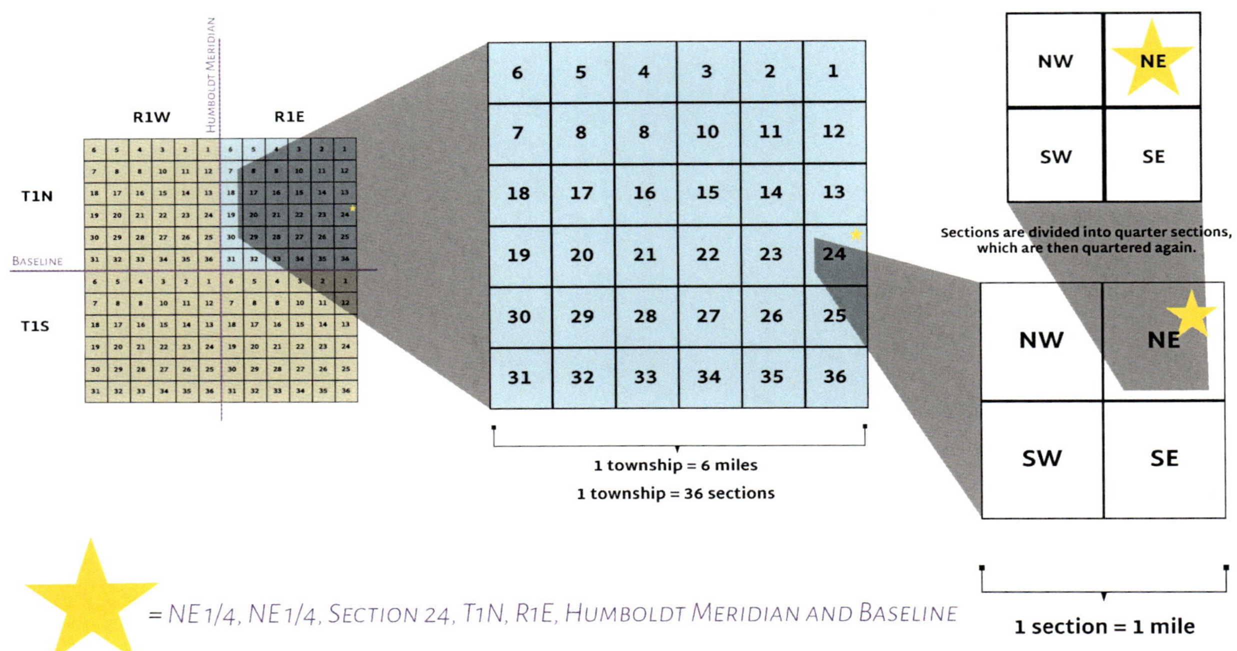

Figure 4.68: In this example, the gold star is located at the NE 1/4 of the NE 1/4 of Section 24, of Township 1 North, Range 1 East, of the Humboldt Meridian and Baseline. The correct notation format is shown in the image next to the gold star.

For example, suppose you wanted to define the location of the soccer field at Humboldt State University using the U.S. Public Land Survey System. Based on the site, you already know that it uses the Humboldt Meridian and Baseline. According to the map, the township is six rows north of the baseline. The range is one row east of the principal meridian (**Figure 4.69**).

The numbers on the map indicate the sections within T6N, R1E. The HSU soccer field is located at Section 28, T6N, R1E, Humboldt Meridian, and Baseline. To be more specific with the U.S. PLSS, you need to divide the section into quarters (**Figure 4.70**).

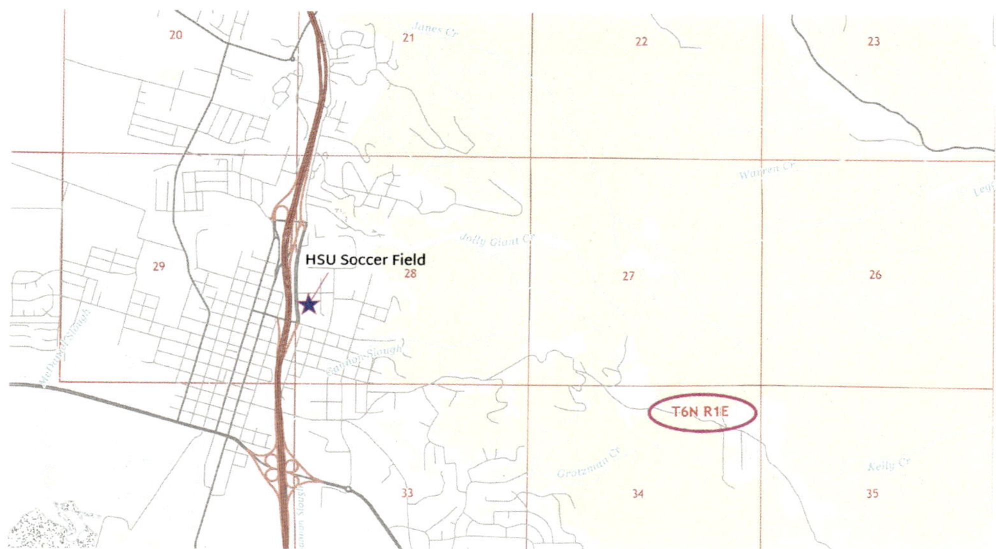

Figure 4.69: The star indicates the location of the HSU soccer field, Section 28, T6N, R1E, Humboldt Meridian, and Baseline.

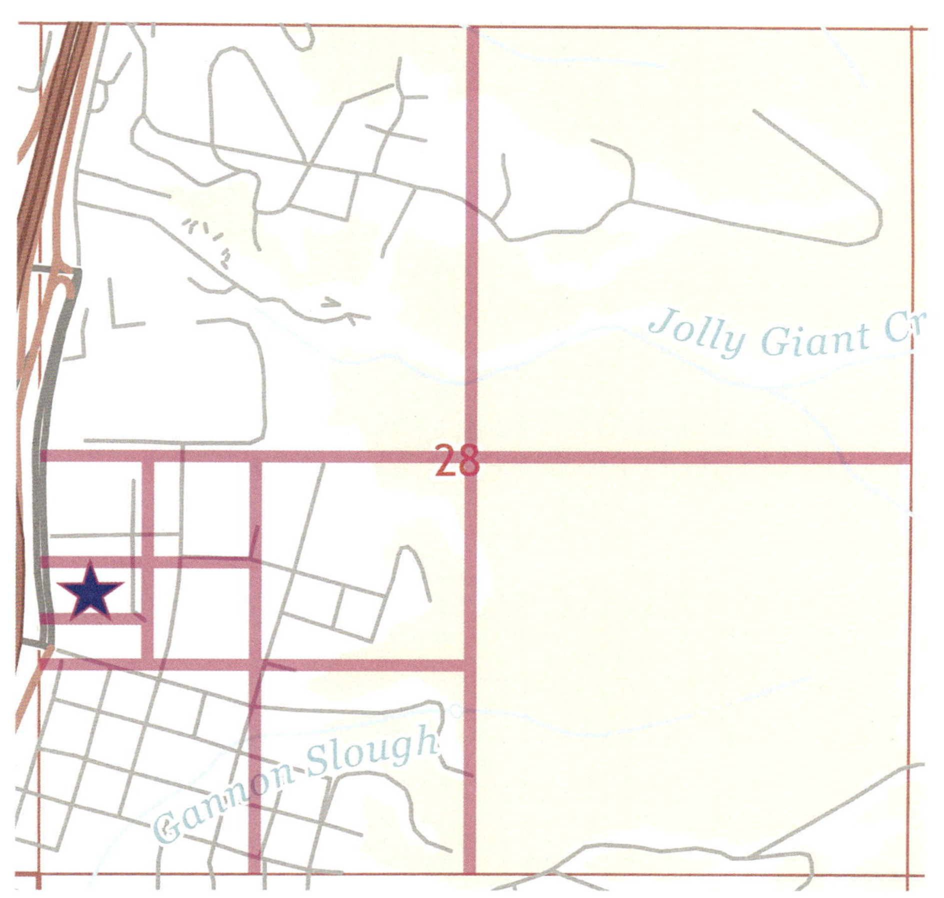

Figure 4.70: The star indicates the location of the HSU soccer field, N ½, SW ¼, NW ¼, SW ¼, Section 28, T6N, R1E, Humboldt Meridian, and Baseline.

Continue to divide the sections into new quarters and halves until you are satisfied with the level of detail. The proper notation for the U.S. PLSS begins with the smallest level of detail. In this example, the Humboldt State University soccer field is located at the N ½, SW ¼, NW ¼, SW ¼, Section 28, T6N, R1E, Humboldt Meridian, and Baseline.

On a separate piece of paper, write down the answer to the following questions:

11. What is the location of the Arcata Plaza using U.S. PLSS (**Figure 4.71**)? Choose the level of detail that is the most practical with this system. Be sure to use the correct notation for PLSS.

Figure 4.71: The star indicates the location of the Arcata Plaza.

Defining a Location Using the State Plane Coordinate System of 1983

As you learned in this chapter, many countries have established a local plane coordinate systems (PCS) that are highly accurate over a small region. The United States uses a PCS called the State Plane Coordinate (SPC) System. The most recent version of the State Plane Coordinate System of 1983 (SPC 83) uses the North American Datum 1983 (NAD83). Nearly every state is divided into a variable number of zones. Humboldt County is located in California Zone 1. SPC 83 also uses a system of false eastings and northings measured in either feet or meters. The Arcata South USGS Topographic Quadrangle displays the SPC coordinates using a series of black tick marks that extend outside of the neatline. Unlike with the geographic coordinates, the neatline does not coincide with easting and northing ticks. When you draw the lines for the SPC grid, you will see that they do not align with the sides of the map or other grids. Each SPC tick mark is ten thousand feet apart. At least one tick mark on each side of the neatline will display the full easting or northing coordinate (**Figure 4.72**).

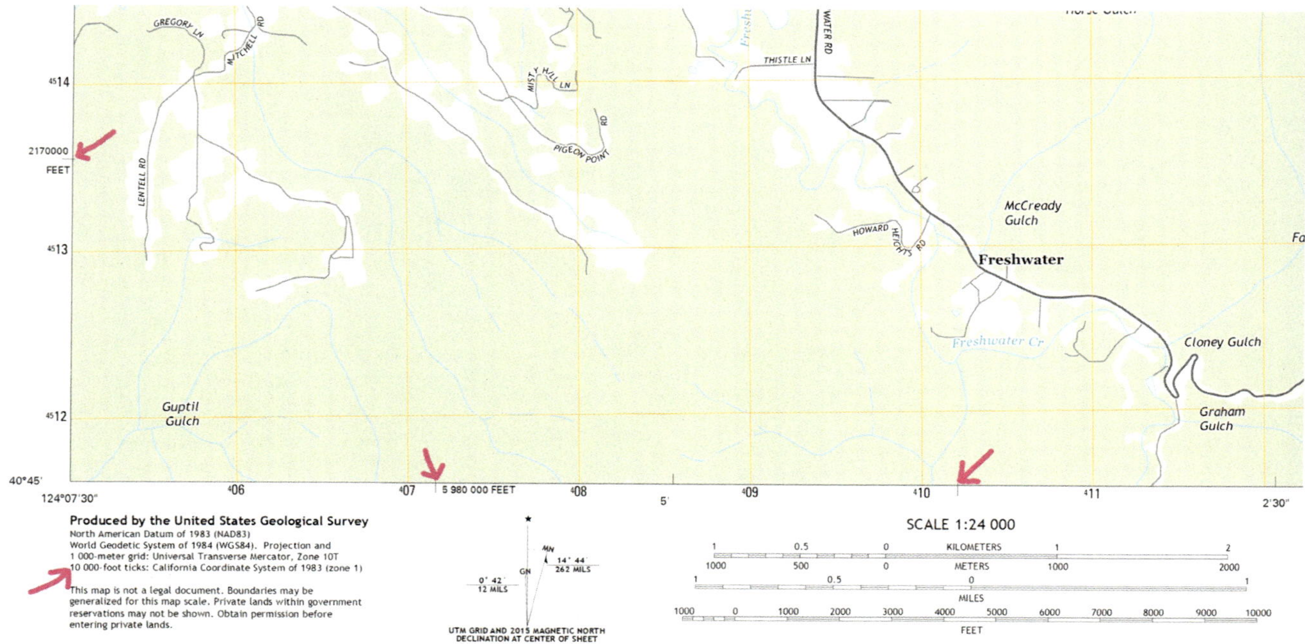

Figure 4.72: On the Arcata South USGS Topographic Quadrangle, the SPC ticks are 10,000 feet apart. The state and zone are printed on the lower left corner of the map.

Coordinates in the SPC system are written using the notation number ft E, number ft N or number m E, number m N if one uses the metric system. The state name follows this notation, and if applicable, the zone (**Figure 4.73**).

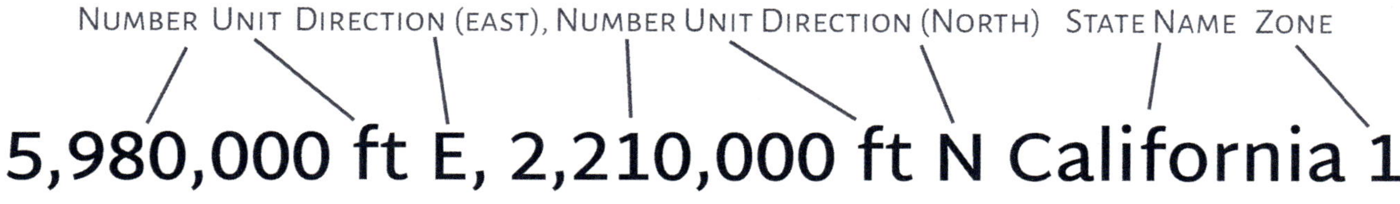

Figure 4.73: The correct notation for SPC 83 starts with easting in feet or meters, northing in feet or meters, state name, and zone.

For example, suppose you wanted to determine the SPC coordinates for the east end of the runway at Murray Field. Start by drawing a grid around your point of interest. Use the tick marks on either side of the map to help maintain the proper direction for the gridlines (**Figure 4.75**).

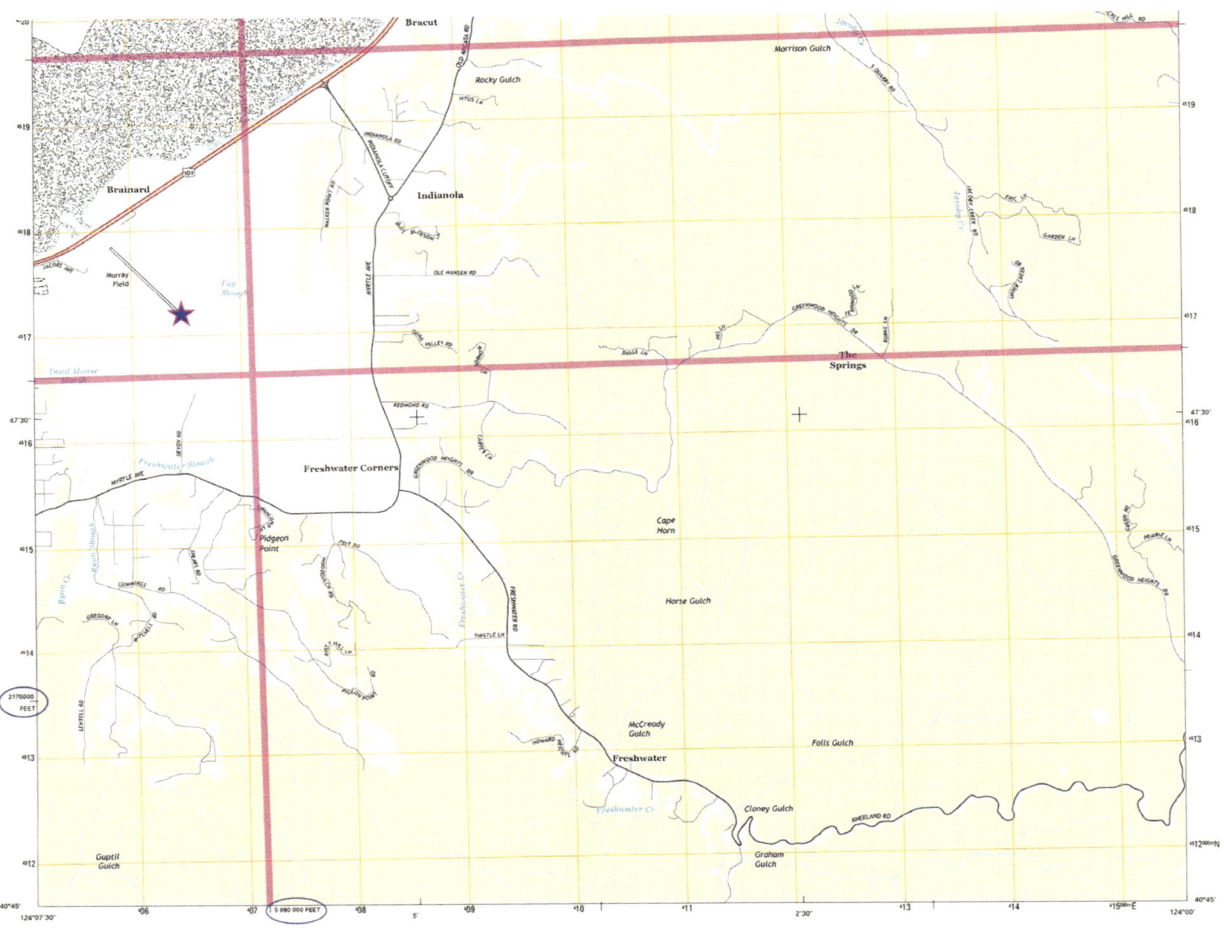

Figure 4.74: Notice how the gridlines do not line up with the edges of the map or with any other grids. Each planar coordinate system has a unique Grid North.

Before measuring, determine the easting and northing values of the nearest tick marks. In this example, the gridline to the right of the Murray Field runway coincides with the tick labeled 5,980,000 feet (**Figure 4.75**). The gridline below the Murray Field runway is one tick above the one marked 2,170,000. Thus, you must add ten thousand feet. The gridline below the Murray Field runway is at 2,180,000 feet N.

Figure 4.75: Each SPC tick mark is ten thousand feet apart.

Once the gridlines are drawn on the map, measure the distance in feet from the edge of the nearest gridline to the point of interest. You can use the scale bar, or you can make a portable scale called a roamer (**Figure 4.76**).

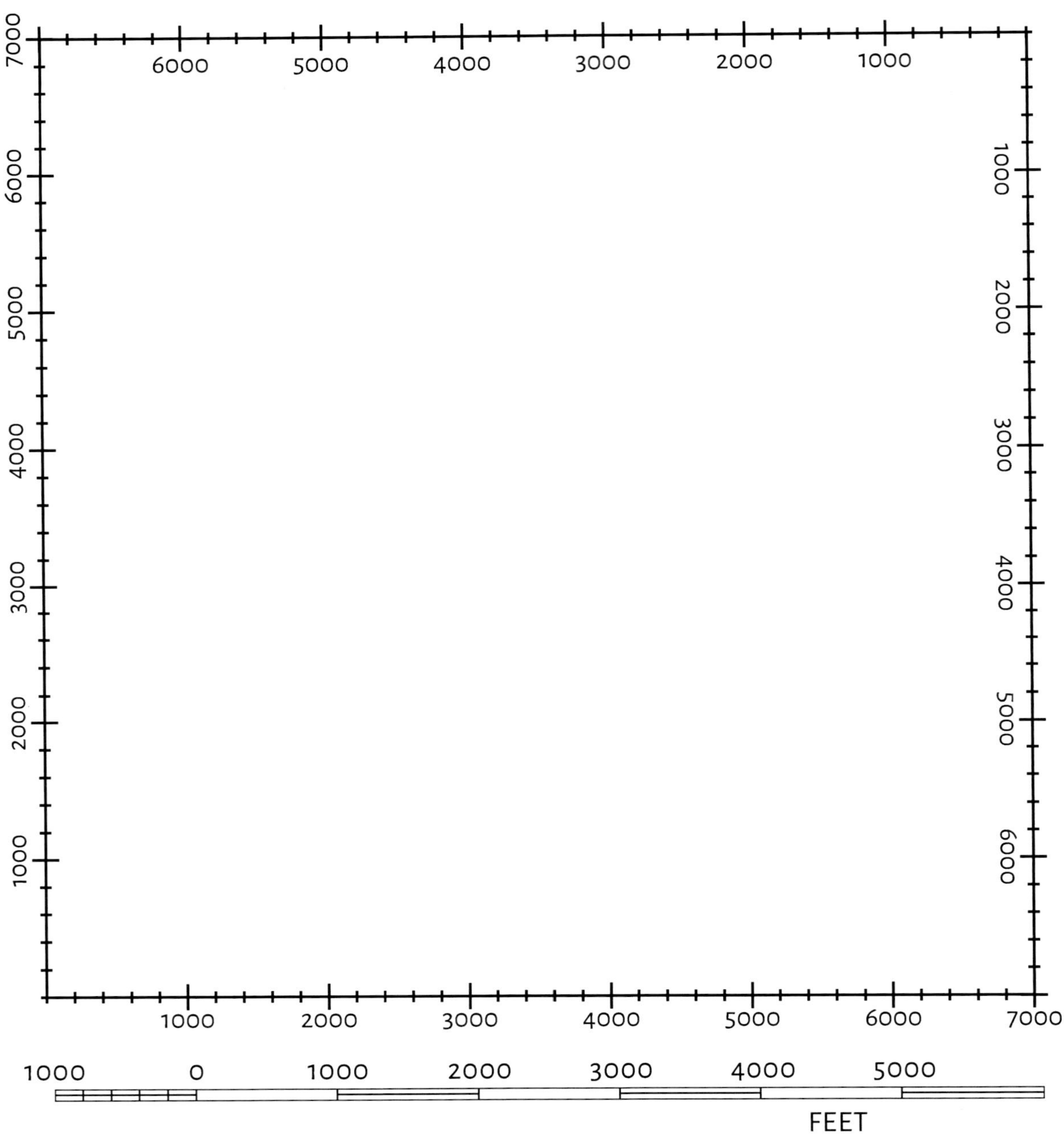

Figure 4.76: Construct a roamer using the scale bar and a piece of paper to draw the lines and ticks.

Start with determining the easting by measuring from the gridline to the right of the Murray Field runway. Easting increases as you travel east and decreases as you move west. Because your point of interest is to the left, you must subtract the distance in feet from 5,980,000 ft E. Likewise, when you measure north from the gridline, you must add the number in feet to 2,180,000 ft N.

On a separate piece of paper, write down the answer to the following questions:

12. Approximately how many feet west of the gridline marked 5, 980,000 ft E is the east end of Murray Field runway (**Figure 4.77**)?
13. Approximately how many feet north of the gridline marked 2,180,000 ft N is the east end of Murray Field runway (**Figure 4.77**)?
14. What are the SPC coordinates for the east end of the Murray Field runway? Be sure to use the correct notation.

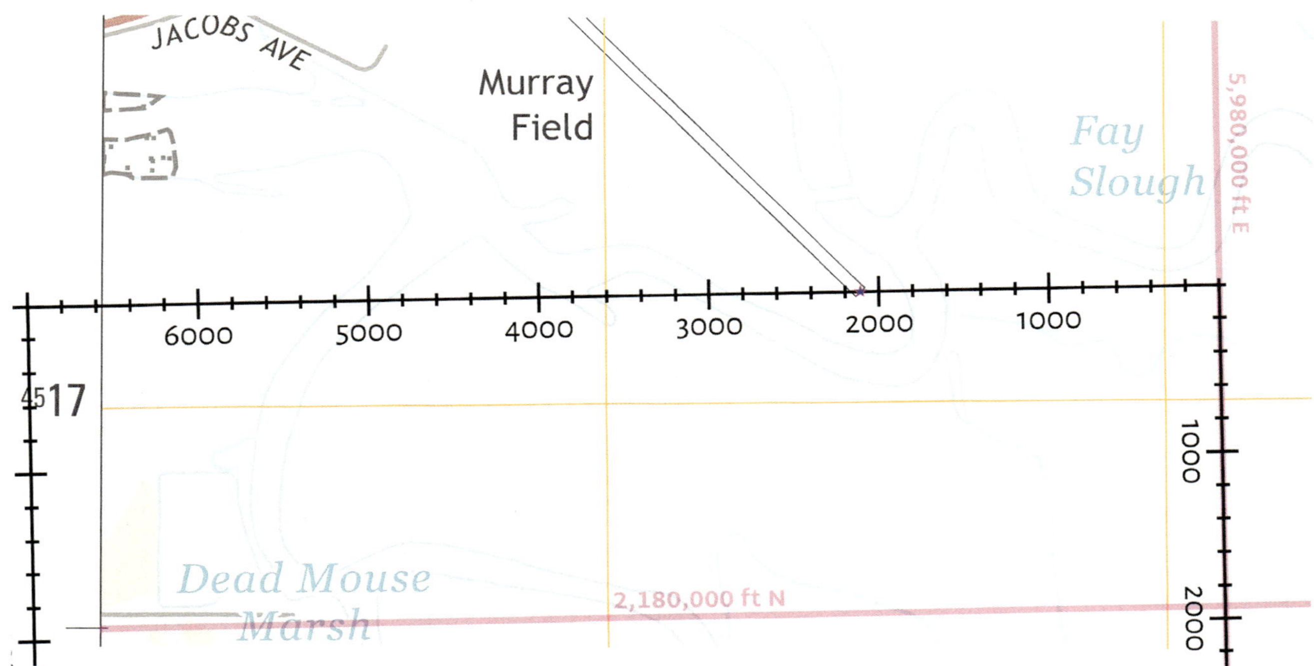

Figure 4.77: The roamer is aligned with the sides of the grid and passes through the point of interest. Each tick mark on the roamer is 200 feet.

Defining a Location Using the Universal Transverse Mercator (UTM) System

As you learned in this chapter, the **Universal Transverse Mercator (UTM) System** is an international grid coordinate system based on the Transverse Mercator projection. It consists of a series of 60 north-south strips of the earth called gores. Created by rotating the Transverse Mercator projection so that it is centered on each gore, each is 6° longitude wide. The gores are cut in half at the equator creating a series of 120 sections called **UTM zones**. The UTM system extends 84° latitude north and 80° latitude south. Each UTM zone is given a number and a letter. The UTM system also uses a system of false eastings and northings measured in meters. The Arcata South USGS Topographic Quadrangle displays the UTM coordinates using a series of orange gridlines. On a paper map, blue tick marks are sometimes used. You may notice that the UTM lines do not align with the sides of the map or with other grids. The full notation for the UTM coordinates is printed at least once on each side of the map. The remainder of the gridlines has abbreviated numbers. You must add three zeros to the end of the abbreviated numbers to get the full value (**Figure 4.78**).

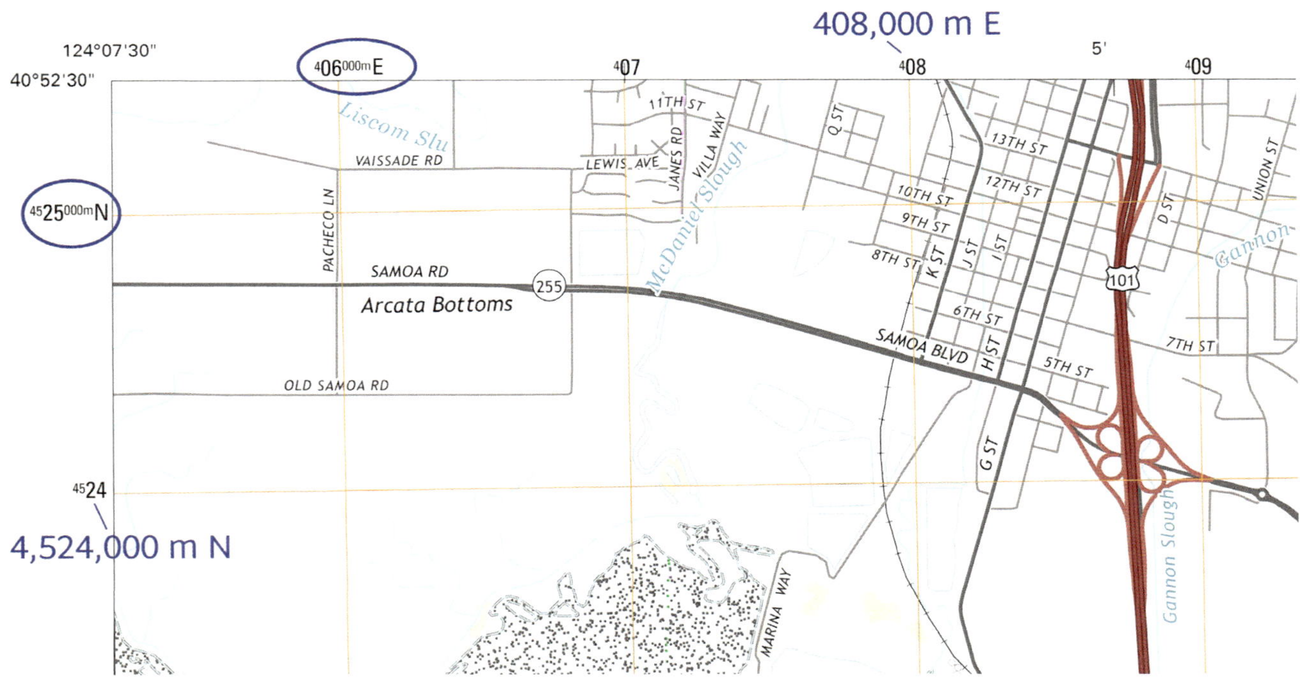

Figure 4.78: On the Arcata South USGS Topographic Quadrangle, the UTM gridlines are 1,000 meters apart. The zone is printed on the lower left corner of the map.

For example, a UTM easting abbreviated as 408 reads as 408,000 m E. A UTM northing abbreviated as 4524 reads as 4,524,000 m N. Coordinates in the UTM system are written using the notation number m E, number m N. Easting and northings, followed by the zone and the hemisphere (**Figure 4.79**).

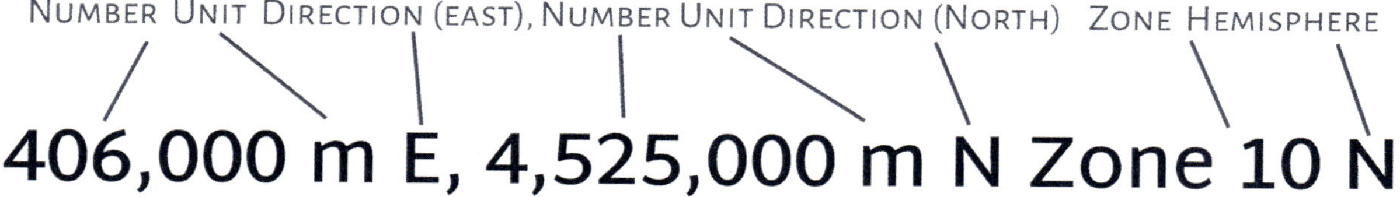

Figure 4.79: The correct notation for UTM starts with easting in meters, northing in meters, the zone, and the hemisphere.

For example, suppose you wanted to determine the UTM coordinates for the Arcata Plaza. If you are working with a paper map, you might need to start by drawing a grid around your point of interest using tick marks. On the digital version of the Arcata South USGS Topographic Quadrangle, the gridlines are already drawn on the map. Next, make a note of the closest easting and northing values. In this example, the closest easting is the gridline to the left of the Arcata Plaza at 408,000 m E. The closest northing is the gridline above the Arcata Plaza at 4,525,000 m N (**Figure 4.80**).

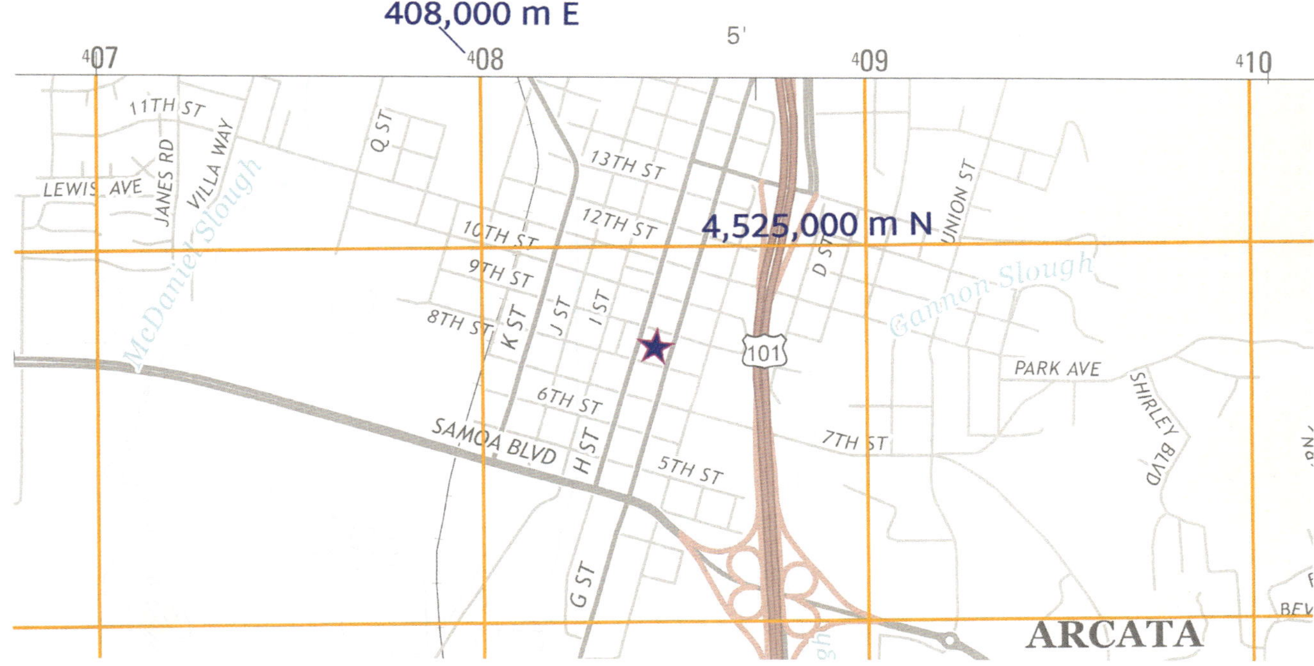

Figure 4.80: The star indicates the location of the Arcata Plaza. The orange gridlines indicate UTM eastings and northings.

Measure the distance in feet from the edge of the nearest gridline to the point of interest. You can use the scale bar, or you can make a portable scale called a roamer (**Figure 4.81**).

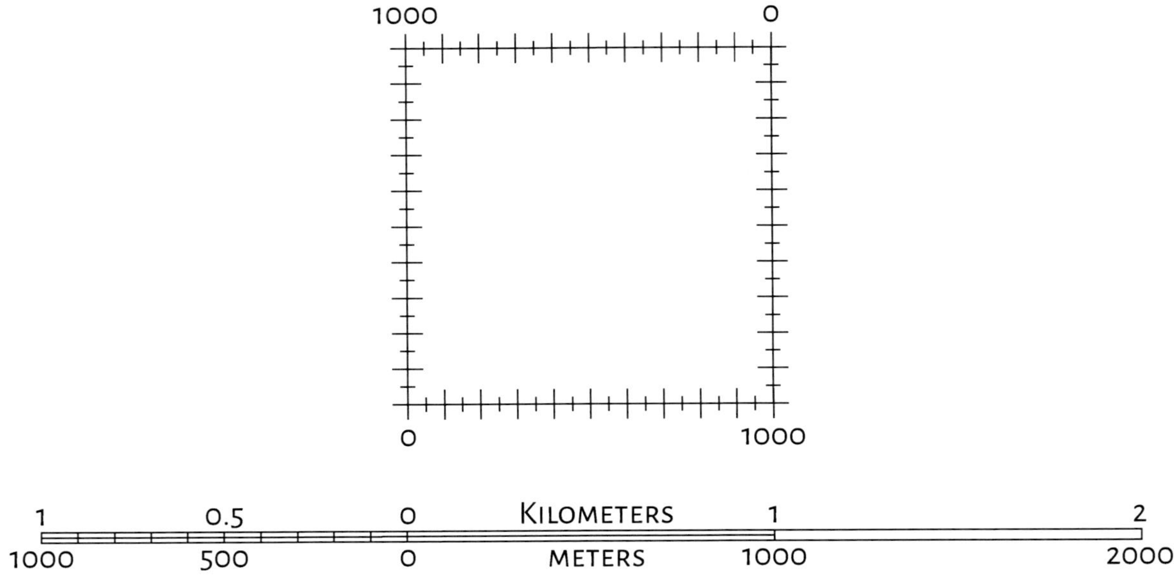

Figure 4.81: Construct a roamer using the scale bar and a piece of paper to draw the lines and ticks.

Start with determining the easting by measuring from the gridline to the left of the Arcata Plaza. Easting increases as you travel east and decreases as you move west. Because your point of interest is to the right, you must add the distance in meters to 408,000 m E. Likewise when you measure south from the gridline you must subtract the distance in meters from 4,525,000 m N.

On a separate piece of paper, write down the answer to the following questions:

15. Approximately how many meters east of the gridline marked 408,000 m E is the Arcata Plaza (**Figure 4.82**)?
16. Approximately how many meters south of the gridline marked 4,525,000 m N is the Arcata Plaza (**Figure 4.82**)?
17. What are the UTM coordinates the Arcata Plaza? Be sure to use the correct notation.

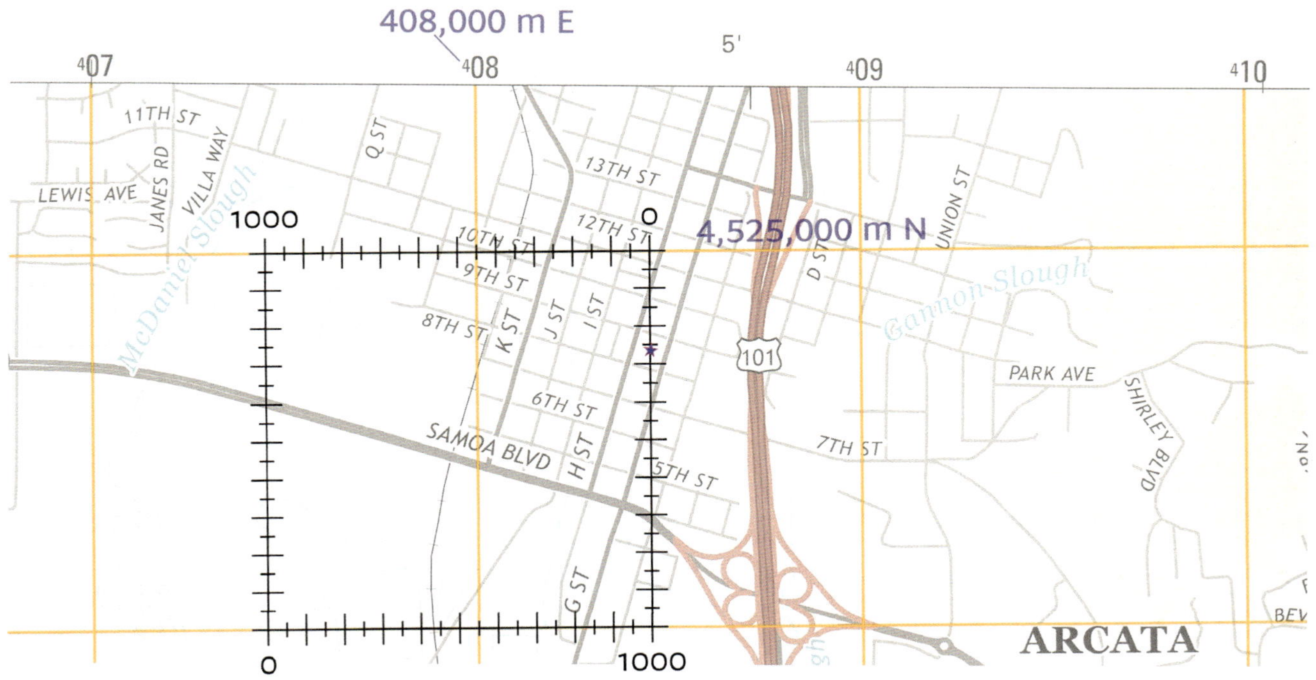

Figure 4.82: The roamer is aligned with the sides of the grid and passes through the point of interest. Each tick mark on the roamer is 50 meters.

Principal Terms

actual scale
baseline
Cartesian coordinate system
central meridian
coordinate reference system (CRS)
correction lines
decimal degrees (DD)
easting
extension scale
false easting
false northing
geographic coordinate system (GCS)
graphic scale
ground distance
irregular land partitioning system
labor
land partitioning system
large-scale
large-scale map
league
league and labor system
long lots
map scale
medium-scale map
metes and bounds
northing
principal meridian
principal scale
projected coordinate system (PCS)
range lines
ranges
regular land partitioning system
representative fraction (RF)
ribbon farms
roamer
scale bar
scale factor (SF)
scope
Seigneurial System
seigneuries
sexagesimal numeral system
small-scale
small-scale map
spatial reference system
State Plane Coordinate (SPC) system
State Plane Coordinate System of 1983 (SPC 83)
Township and Range System
township lines
townships
U. S. public land survey system (PLSS)
U. S. Public Land Survey System (PLSS)
Universal Transverse Mercator (UTM) system
Universal Transverse Mercator (UTM) System
UTM zones
verbal scale

Chapter 5: Mobile Mapping Fundamentals

Most readers of this book may expect some level of mobile mapping, geospatial fieldwork involving the measurement of a position, an elevation, a perimeter, or an area to define positional information on Earth. One aspect of mobile mapping is a land survey, the direct application of geodesy, linking mathematical models of Earth to physical reality through precise field measurements. While most readers may not expect to achieve a surveyor's level of precision, this book discusses some of the underlying concepts related to land surveying, and to a broader extent, mobile mapping. Chapter 5 presents a series of methods and equipment for mapping data in the field. This chapter differs from others due to the hands-on nature of field collection that is difficult to translate into a digital textbook. The activities included in this chapter have far less focus on software and incorporate some outdoor activities that readers will have to perform.

Learning Outcomes

Readers should be able to accomplish the following outcomes by the end of this chapter:

» Demonstrate competence in the ways of defining direction on a map
» Determine magnetic declination values for a specific region
» Perform conversions between different directional systems
» Describe the components of an orienteering compass
» Explain field methods for position and measure using triangulation, trilateration, and traverse
» Discuss the principles behind the Global Positioning System (GPS)
» Identify sources of GPS error and uncertainty
» List ways of augmenting GPS accuracy
» Summarize the limitation of GPS technology
» Complete a data collection plan for a project

Direction Systems

All directional systems measure direction relative to a directional baseline. A **directional baseline** is a line of reference between a location on Earth and a standard reference point. In most direction systems, the directional baseline refers to the north. There are three types of north commonly used in geospatial science:

» True north

» Magnetic north

» Grid north

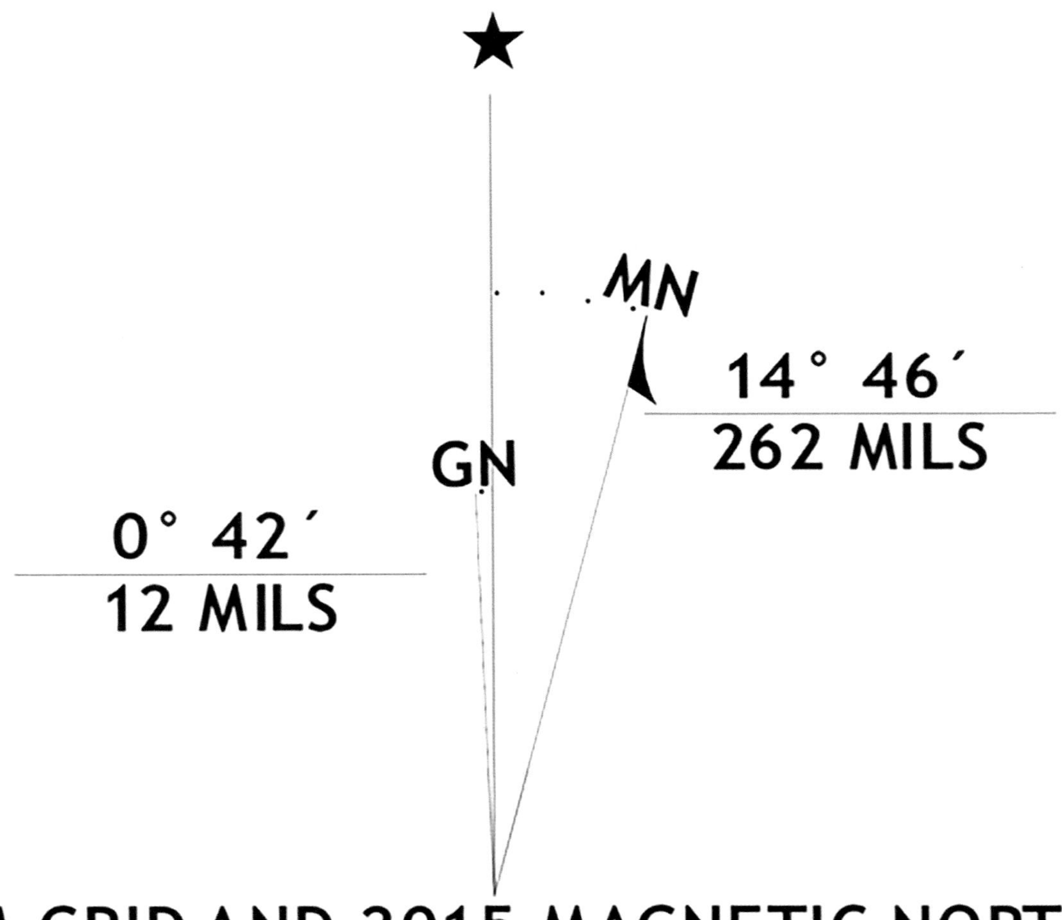

Figure 5.01: This image shows a declination diagram that includes true north, grid north, and magnetic north.

True North

True north, sometimes referred to as **geographic north**, is the northern end on the axis of Earth's rotation[1]. Because true north is a relatively permanent and stable position on Earth, most directional systems are relative to a true north reference line, which is any meridian on earth (**Figure 5.02**). True north makes directional calculations and measurement accurate and straightforward. However, finding true north is not always straightforward. To find true north another type of north is used, called magnetic north.

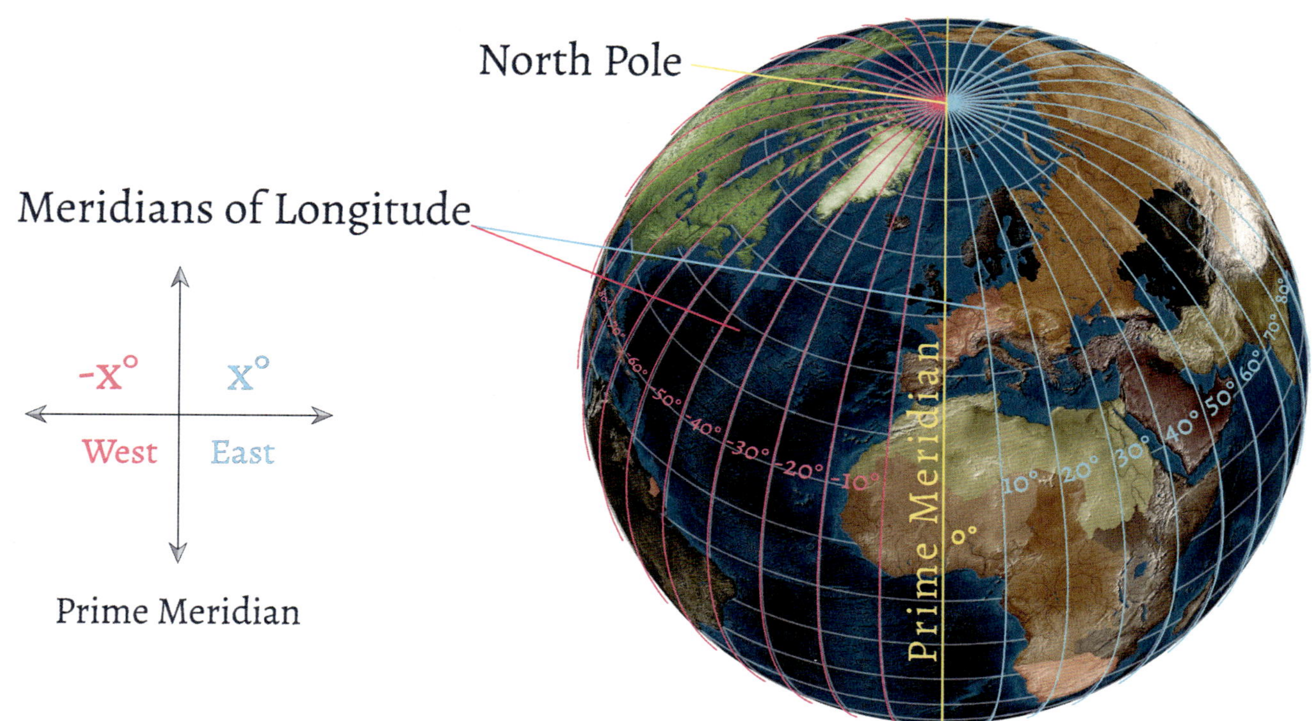

Figure 5.02: Each meridian is a true north reference line.

Magnetic North

Magnetic north is the northern pole of Earth's geomagnetic field, which wanders around the north pole. Most readers that have used a compass are familiar with the concept of magnetic north. However, one might be surprised to find that magnetic north rarely coincides with the location of true north.

1 Many also refer to this location as the North Pole

Geomagnetism is not a fully understood phenomenon but is measurable, and to some degree, predictable. Barring any local magnetic disturbance, the magnetized end of a compass needle will always point towards the north magnetic pole. However, the location of this pole is not stable and changes position over time (**Figure 5.03**).

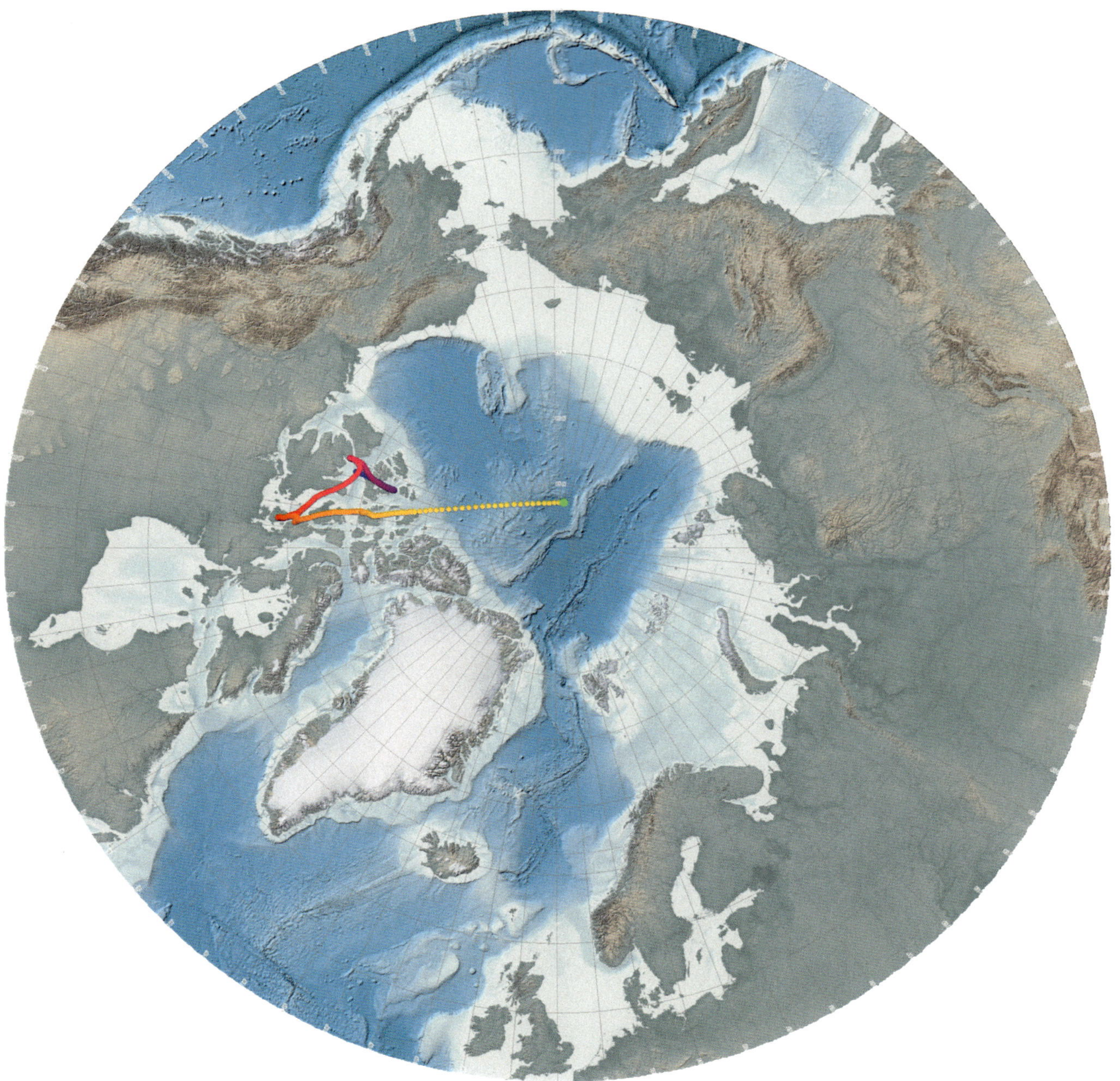

Figure 5.03: This map shows the historical track of the magnetic north pole 1590–2020 (predicted). The green dot indicates the current (2015) position of the north magnetic pole. Source: NOAA Declination Map Viewer.

The rate of change is not always predictable, and since the year 2000, the rate of movement of the magnetic pole has increased significantly, with rapid movement in the past few years. Some scientists believe that this is an indication of an imminent pole reversal, which has happened several times before. However, the pole-reversal process is not instantaneous, sometimes taking thousands of years, so readers should have plenty of warning before their compass needles point south! Watch Geomagnetic innovations 2015 CO-LABS winner, foundational technology by COLABS Colorado to learn more (**Figure 5.04**).

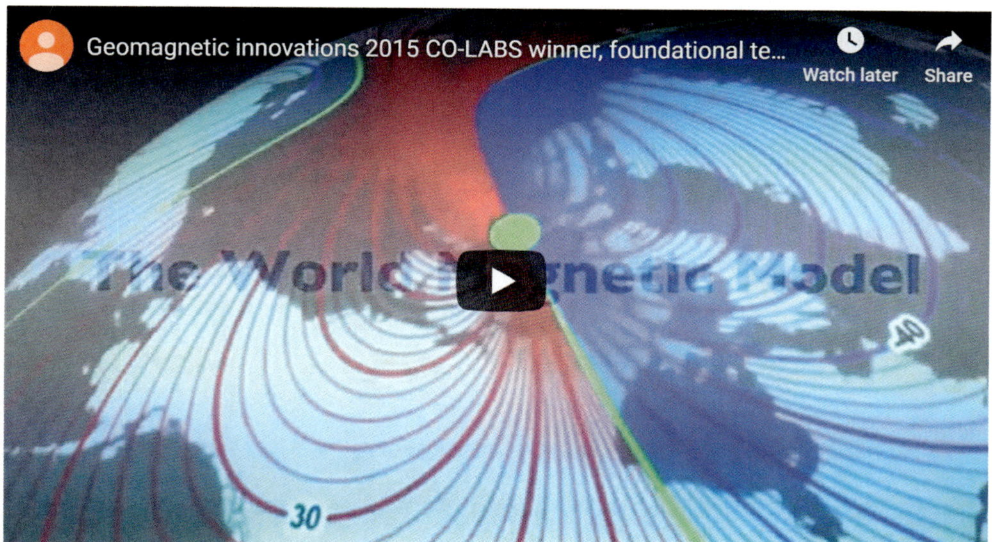

Figure 5.04: This video discusses the concepts related to magnetic declination. URL: *https://youtu.be/9xBoGMJ9NkA*

Magnetic Declination

When using a compass, one compensates for the migration of magnetic north by setting the magnetic declination. **Magnetic declination** is the angular deviation between true north and magnetic north. Magnetic declination changes annually, so it is essential to know the current declination and to make sure to set one's compass declination correctly. One determines the local magnetic declination by using an **isogonic map** (**Figure 5.05**). An isogonic map shows a series of **isogonic lines**, lines indicating the magnetic declination along those lines as well as the rate of change over time. On an isogonic map, the line where magnetic north and true north are the same, 0° declination, is called an **agonic line**.

Figure 5.05: This map shows an isogonic map of the continental United States as of 2019. Red lines on the map indicate positive declination (east). Blue lines on the map indicate negative declination (west). The green line running through the Midwest is the agonic line, which has 0° declination. Source: NOAA Declination Map Viewer.

One may also determine local magnetic declination using a USGS Quadrangle map for their region. The USGS Quadrangles have the declination printed near the bottom of the map in the form of a **declination diagram**. If too much time has passed from the printed date, the declination on either the USGS quadrangle or the isogonic map will no longer be accurate. Luckily, NOAA provides an online calculator for determining one's local magnetic declination (**Figure 5.06**).

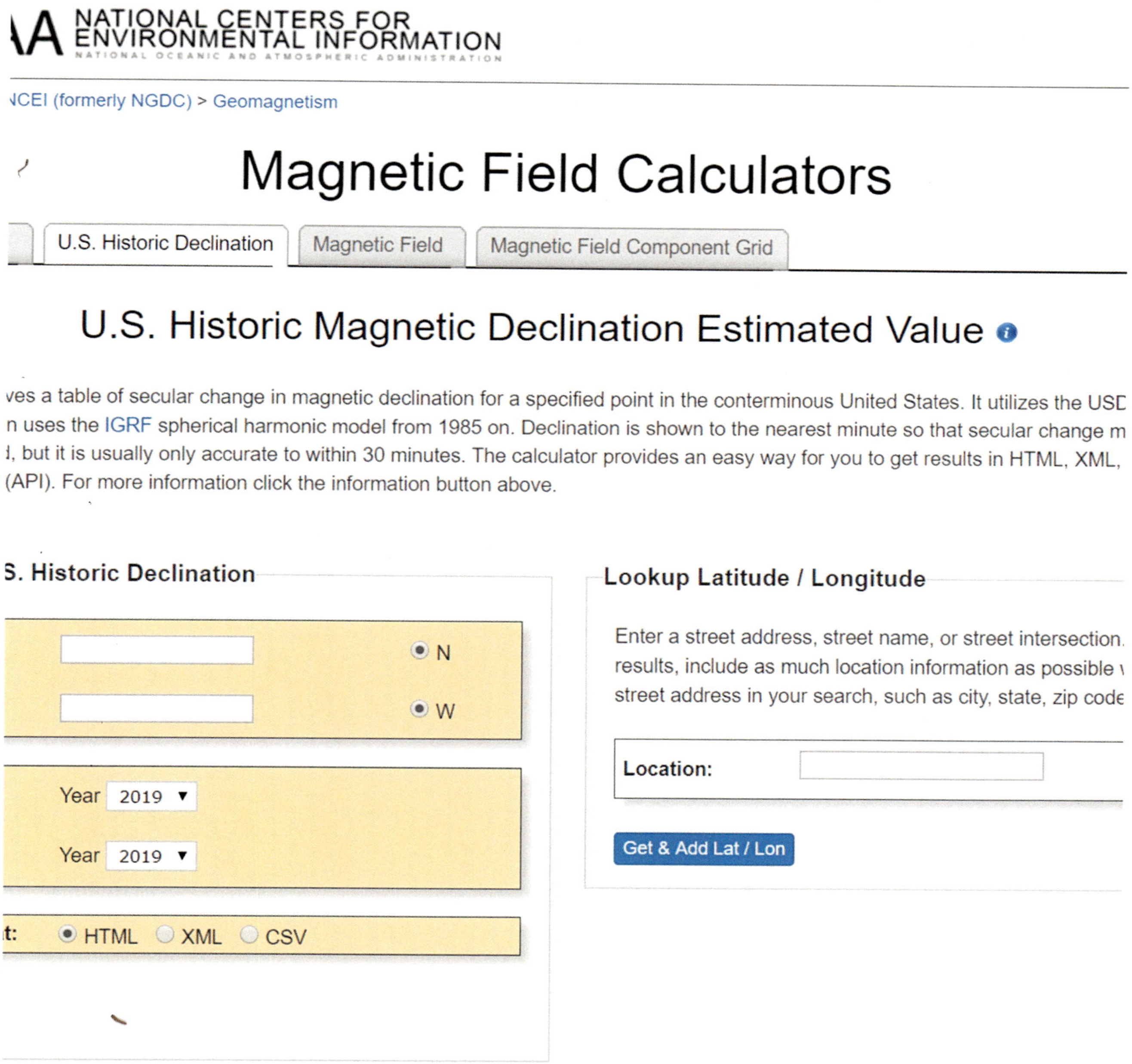

Figure 5.06: One can look up their local declination using the NOAA Magnetic Field Calculator.
URL: *https://www.ngdc.noaa.gov/geomag/calculators/magcalc.shtml#declination*

Grid North

The third type of north is called grid north. Grid north is the northerly direction along the grid lines of a coordinate system on a map. Coordinate system grids do not always follow true north. For example, the Universal Transverse Mercator (UTM) is made up of a regular grid of vertical and horizontal lines. The vertical line at the center of the UTM zone is a meridian and points to true north. While all meridians converge towards true north, the vertical lines in the UTM grid do not. Farther away from the center the vertical lines lead away from true north (**Figure 5.07**).

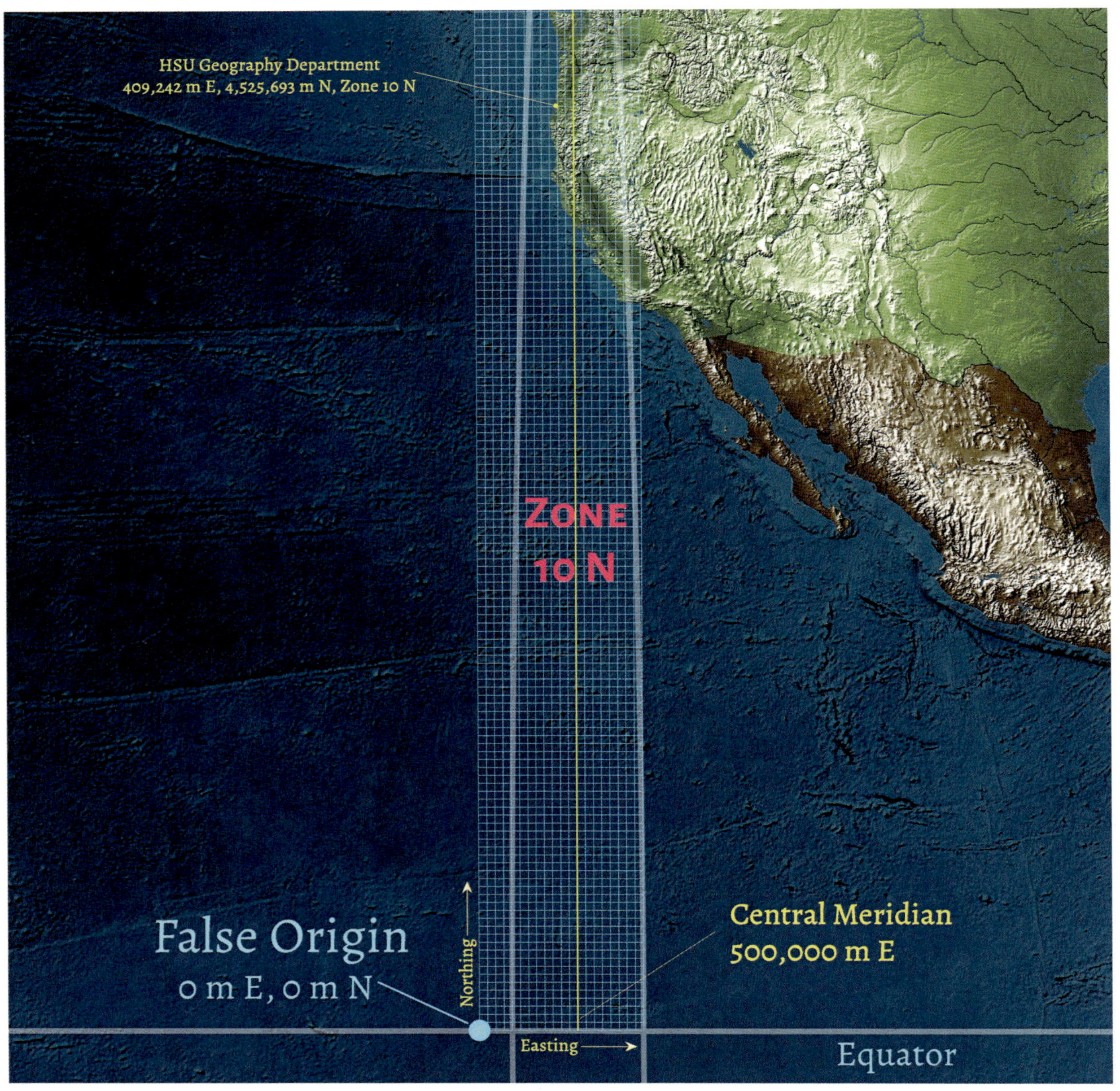

Figure 5.07: The central meridian points to true north. However, the remaining grid lines do not.

Many maps are too large in scale to show the entire UTM zone. As a result, the grid lines will appear to be slanted when compared to the sides of the map (**Figure 5.08**). The angular difference between these north-south gridlines and true north is called grid declination. One finds the grid declination on the declination diagram of a USGS quadrangle (**Figure 5.01 and Figure 5.08**).

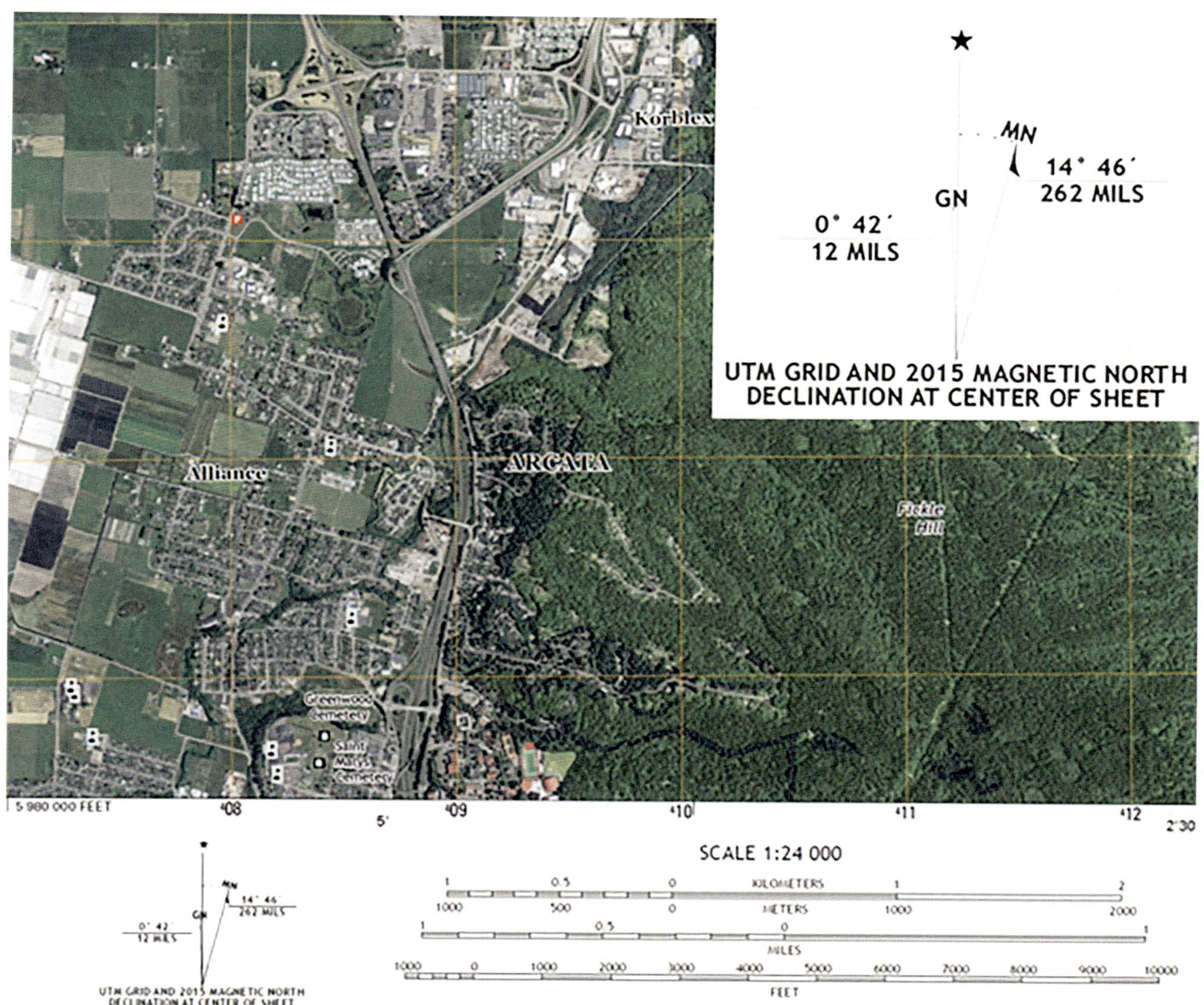

Figure 5.08: This image shows a UTM grid.

INDICATING DIRECTION

As discussed earlier, all directional systems measure direction relative to a directional baseline. A **directional baseline** is a line of reference between a location on Earth and a standard reference point. In most direction systems, the directional baseline refers to **true north**, which is the northern end of the axis of Earth's rotation. Azimuth and bearing are examples of directional systems that use true north as a directional baseline.

Azimuth

An **azimuth** indicates direction by using true north as a baseline and measures clockwise from 0° to 360°. Azimuths are easy to use because one number defines each direction. Treating direction as a single number also makes it simple to use for calculations and in GIS software. Only a single number expressed in degrees, indicates an azimuth. For example, if one faces true north and then rotates their body 45 degrees clockwise, their azimuth would be written as 45° (**Figure 5.09**).

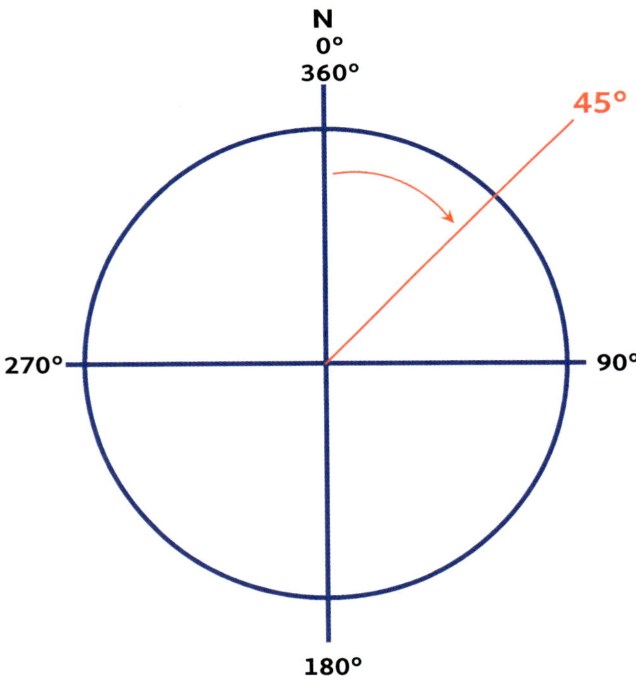

Figure 5.09: An azimuth uses a single number, which makes it simple to use for calculations and in GIS software.

When determining a **back azimuth**, the opposite direction, add 180° if the azimuth is less than 180°. Subtract 180° if the azimuth is higher than 180°. To calculate the back azimuth in this example, add 180 to 45 degrees to get 225° (**Figure 5.10**).

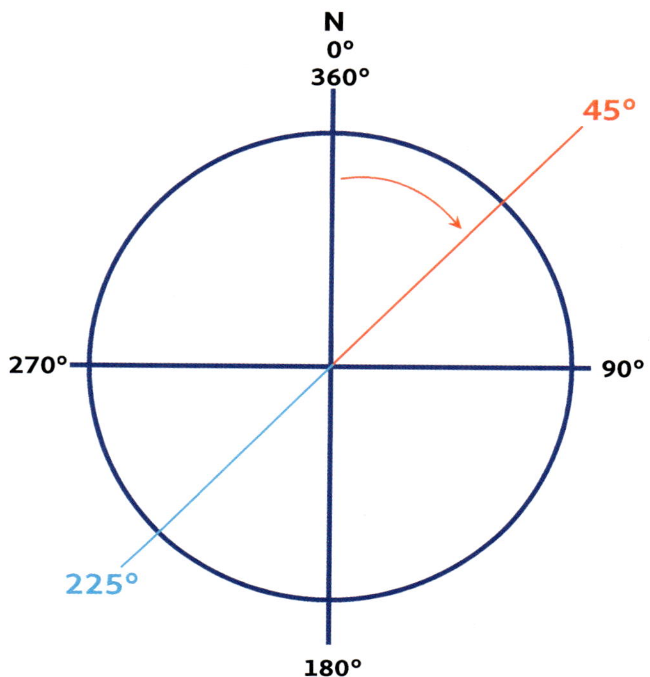

Figure 5.10: Land surveyors sometimes use a back azimuth.

Bearings

A **bearing** indicates direction by using the angular difference away from a north or south baseline, ranging from 0° to 90°. Determining one's bearing begins with a north or south baseline, whichever is closer to one's direction. One then measures the angle east or west from the baseline. One writes the bearing using the following notation: N or S (baseline) degrees° E or W (orientation). For example, if once faces true north, and then rotates their body 45 degrees clockwise, the bearing is written as N45°E (**Figure 5.11**).

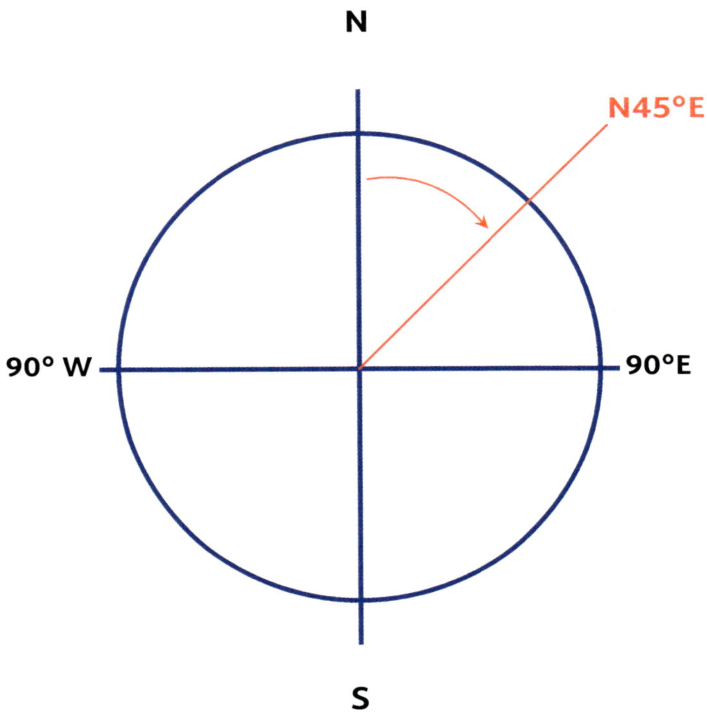

Figure 5.11: Bearings are not as computer friendly but still used in navigation.

A bearing notation is never greater than 90°, as once one rotates more than 90° from the baseline, one is now closer to the other baseline and would start the notation there. For example, if one faces north and rotates 91° west, one would be closer to facing south, so the bearing would be S89°W.

Sometimes one also needs to find the opposite direction from which one faces. A **back bearing** is the opposite direction from a bearing and is readily determined by merely changing the letters. In this example, one writes the back bearings as S45°W (**Figure 5.12**).

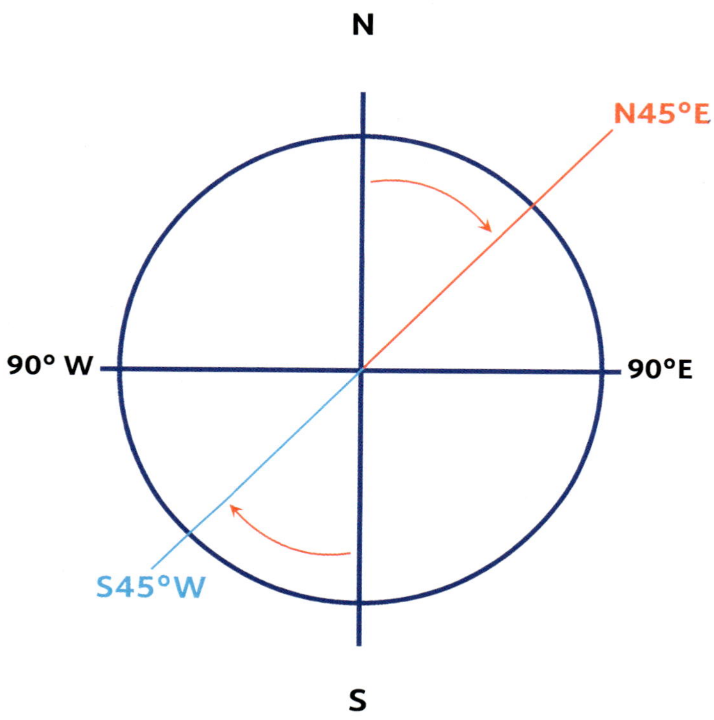

Figure 5.12: A back bearing is used to find the opposite direction one faces.

Orienteering and Ranging

When collecting data in the field, a useful tool to have is a compass. While there are many different types of compasses available on the market, the compass that this book discusses is the orienteering compass.

The orienteering compass has the following components (**Figure 5.13**):

» Rotating magnetic needle inside a liquid filled capsule

» Orienteering arrow

» Bezel with azimuth graduations and cardinal directions

» Index pointer

» Clear baseplate with meridian lines

» Declination scale and adjusting key

» Sighting mirror

» Targeting sight

» Lanyard

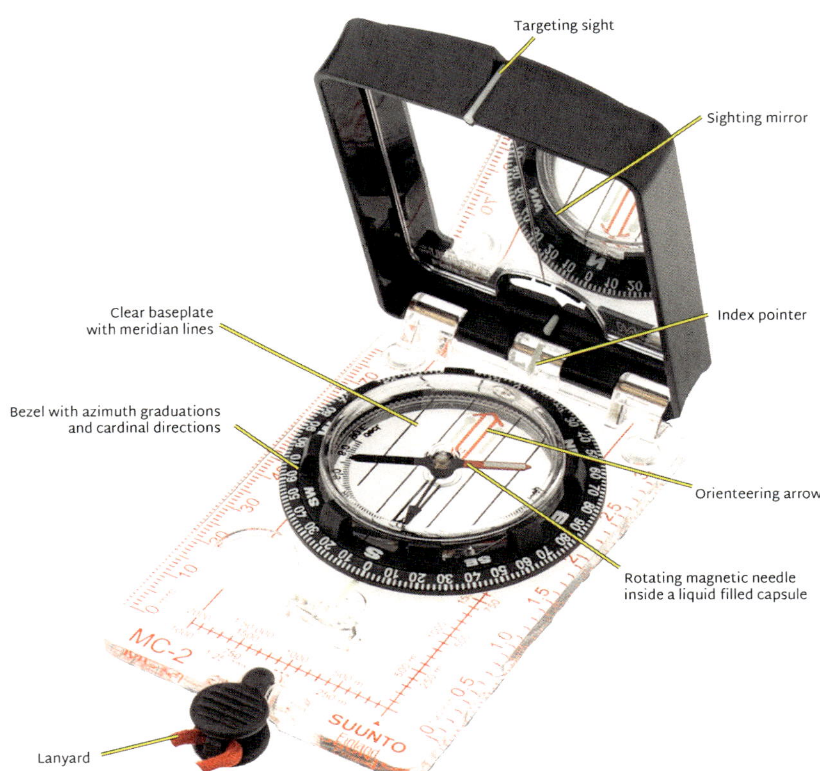

Figure 5.13: This image shows an overview of the orienteering compass and its components.

When held in a level position, the **rotating magnetic needle** floats freely inside the liquid filled capsule (**Figure 5.14**). Compass manufacturers distinguish the portion of the needle that points to magnetic north using red paint, a different shape, or other markings. One should refer to the compass manual for specifics if one cannot identify which end is magnetic north.

Figure 5.14: This image shows the magnetic needle and orienteering arrow.

The **orienteering arrow** generally appears as an arrow outline that rotates with the **bezel** with azimuth graduations and meridian lines. The north end of the orienteering arrow sometimes painted red, can be rotated to line up with the northern end of the magnetic needle. The azimuth can then be read just under the **index pointer**. When determining the azimuth of a distant object, the **lanyard** is used to steady the compass and provide a consistent distance from the compass to the eyes of the user.

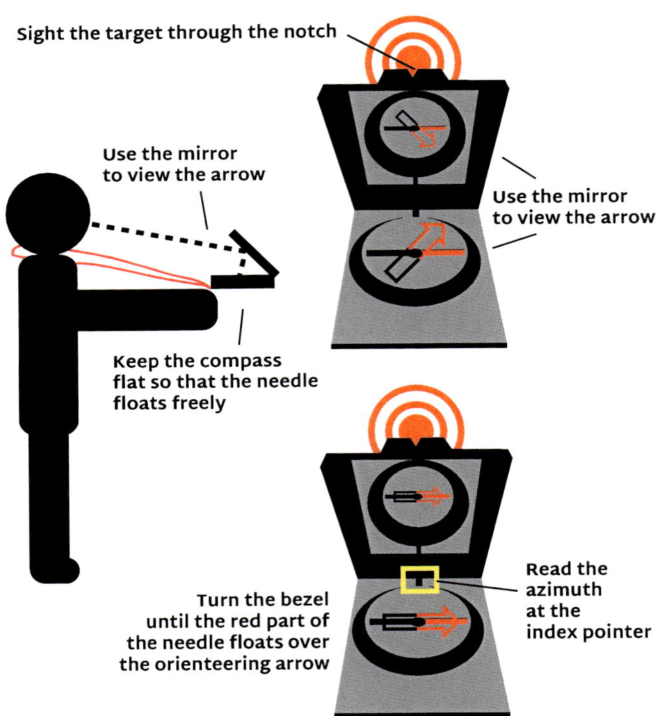

Figure 5.15: The targeting sight lines up with the distant object, and one uses the sighting mirror when rotating the bezel and reading the azimuth.

On a USGS topographic quadrangle, azimuth is determined by aligning the edge of the orienteering compass along the line between two points of interest, then rotating the meridian lines so that they line up with meridians on the map. The **adjusting key** rotates the declination scale to compensate for local magnetic declination. Watch *Navigation Toolbox: Compass bearings* by James van Oppen to learn more (**Figure 5.16**).

Navigating by Compass

Navigating to a place using a compass is sometimes more difficult than plotting the location on a map. In most cases, travel in a constant direction is not possible due to terrain or local obstructions. In this case, one must plot the route as a series of azimuths leading to one's destination. Then one uses the compass to navigate each section of the route. One rotates the bezel on the compass to locate the first azimuth under the index pointer. Then one holds the compass steady so that the needle floats freely as one's body rotates until the north end of the magnetic needle lines up with the north end of the orienteering arrow. One now faces the azimuth indicated under the index pointer. Many readers make mistakes on the next step. When following the route, it is important not to stare at the compass. Instead, one uses the compass to sight a distant landmark on the horizon along the azimuth line, such as a distant tree or mountain peak. Then one keeps their eyes facing up towards the landmark as one travels along the correct course. Once one reaches this landmark, one stops and repeats the steps for the next leg of the journey.

Figure 5.16: This video discusses the concepts related to compass bearings. URL: *https://youtu.be/dovinoVk6xA*

The triple-legged walk

Accurately navigating by compass takes practice. The **triple-legged walk** is an exercise one does to increase one's accuracy. In an open area, one places a nickel on the ground between one's feet. One set the compass to an arbitrary direction between 0° and 120° azimuth, holding the compass steady so that the needle floats freely. One rotates their body until the north end of the magnetic needle lines up with the north end of the orienteering arrow. One now faces the azimuth indicated under the index pointer. One uses the compass targeting notch to sight on a distant landmark or object on the horizon. Keeping one's eyes up and on the landmark walk 40 steps. Once one walks 40 steps, one adds 120° to the azimuth on the compass. So for example, if one sets the first azimuth to 20°, one now sets the next azimuth to 140° and walks an additional 40 steps. After walking the second leg, one adds another 120° to the azimuth and walks 40 steps. So in the previous example, the original azimuth was 20°, the second leg was 140°, and the third leg was 260°. Once one completes these steps, they should now stand over the nickel. Chances are one will not end up exactly where one started. With practice, one can end up closer and closer to the nickel. For a more significant challenge, increase the number of paces or add more legs to the course. Watch *Woodsmanship 101–Navigation 1* by Backcountry Hunters & Anglers to learn more (**Figure 5.17**).

Figure 5.17: This video covers navigation by compass. URL: *https://youtu.be/4yli86dgFDw*

Resection

Out in the field, one uses a USGS topographic quadrangle and a compass to find one's position using a method called **resection**. This technique involves plotting lines that cross one's position. Observing one's surroundings, one locates two or three landmarks that they can identify on the map. Using the compass, one acquires the azimuth of each landmark relative their position. One calculates a back azimuth[2] for each landmark and sets this value on the compass using the bezel and the index pointer. One orients the compass on the map along the back azimuth utilizing the meridian lines. Once the meridian lines on the compass align with the meridians on the map, one draws a line from their landmark back towards their position using the compass as a straight edge. One repeats this process for each landmark, locating one's position where all of the lines cross. Watch *Tip Of The Week–"Resection"–How To Determine Your Location With A Map And Compass (E9)* by IntenseAngler to learn more (**Figure 5.18**).

2 Sometimes this term is referred to as a back-sight

Figure 5.18: This video describes how to determine one's location using resection. URL: *https://youtu.be/Bf3_SHEebNQ*

Intersection

One may also locate the position of a distant landmark or object using a method called intersection. The intersection method uses a series of foresight measurements, or azimuths, taken from two or more positions. Using the compass, one can acquire the azimuth to the target at one's current location. Moving to a second and third position, one obtains additional azimuths to the destination. Once one records each azimuth, one plots the lines on a map the same way used for the resection method. The place where all the lines cross is the position of the landmark or object. Watch Intersection and Resection by RangerKSchool to learn more (**Figure 5.19**).

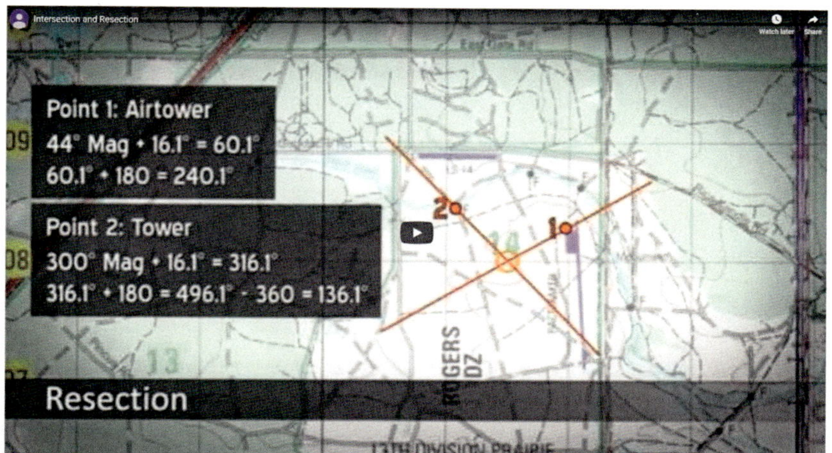

Figure 5.19: This video discusses intersection and resection. URL: *https://youtu.be/8Cotfsb2sLA*

Positioning and Measure

At some point, it is likely that one's geospatial fieldwork involves some form of measurement related to a position, an elevation, a perimeter, or an area. When taking field measurements, it is useful to have a **control point**, a highly accurate starting point, as a reference. A control point is a previously surveyed position, such as a USGS survey mark. A **survey mark** is a small engraved metal disc used to mark vertical and horizontal control points. The *National Geodetic Survey (NGS)*[3] maintains the National Spatial Reference System (NSRS), which has records on hundreds of thousands of control points across the United States. Watch *Geospatial Infrastructure: Informing Adaptation to Sea Level Rise* and *Precision and Accuracy in Geodetic Surveying* by the COMET Program/MetEd to learn more (**Figure 5.20 and Figure 5.21**).

Theodolite and Total Station

There are several standard tools used for positioning and measure. Earlier, this chapter discussed the methods of determining position using a compass. While it is possible to use a compass to delineate a boundary or establish a position, other tools are more suited to this task. A **transit** is a land-surveying instrument that measures vertical and horizontal angles using a tripod, telescope, and bubble level. A **theodolite** is a modern version of the transit that includes electronic components. Watch *Theodolite 1–Intro & Setup* by OTENBuildingCourses to learn more(**Figure 5.22**).

Figure 5.20: This video discusses surveying as it relates to sea level rise. *URL: https://youtu.be/HsRNiPitTzo*

3 http://www.ngs.noaa.gov/datasheets/

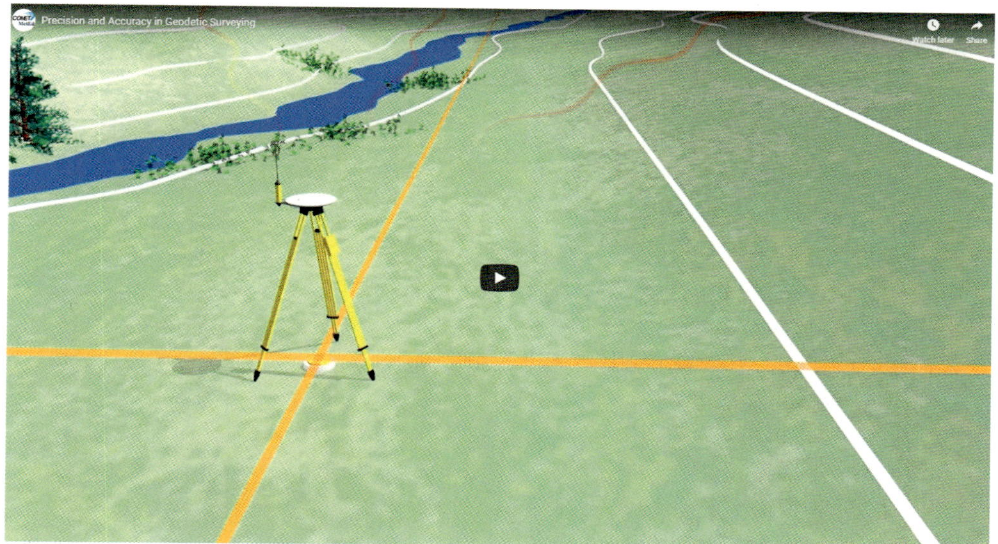

Figure 5.21: This video discusses precision and accuracy in geodetic surveying. URL: *https://youtu.be/ApKw5qWqYF8*

Figure 5.22: This video demonstrates how to set up a theodolite. URL: *https://youtu.be/QUX9_1fRnlo*

Figure 5.23: In this image, a total station used to create a stem map on the Humboldt State University L.W. Schatz Demonstration Tree Farm. Upper Image from left to right: Dr. Aaron Hohl, Nicolas R. Malloy, Marcio Pagano Aragona.

Figure 5.24: In this image, students practice using a total station during a fieldwork activity organized by Nicolas R. Malloy. From left to right: Pamela Aparecida Melo, Raiza Tinoco Borges, Felipe Ribeiro De Toledo Camargo.

A more expensive, but highly useful piece of equipment, is called a total station (**Figure 5.23 and Figure 5.24**). The **total station** works like a theodolite, but also includes **electronic distance measurement (EDM)** capabilities. EDM devices emit electromagnetic waves at the target and measure the distance using the return signal.

Traverse

A simple way of measuring out from a control point is called a traverse. A **traverse** starts at a known position, such as a **control point**, and measures angles and distance to another point. One completes the traverse when the angle and distance to each additional point are measured. A **closed traverse** ends at the starting point. Any other traverse that does not stop at the starting point is an **open traverse**. Once one calculates the azimuths in a traverse, one uses the angles and distance from the control point or origin, to calculate grid coordinates (x,y), by using the following equation (**Figure 5.25**).

$$X_b = X_a + (ab \times \sin\theta)$$

$$Y_b = Y_a + (ab \times \cos\theta)$$

Figure 5.25: In this equation, the control point is the letter a, the transit point is the letter b, and the azimuth is the symbol theta θ. X and Y represent grid coordinates. The distance between point a and point b, multiplied by the sine or cosine of θ, provides the X and Y coordinates of point a and point b respectively.

Watch Working with Azimuths Part 1 and Part 2 by Suorsafam to learn more (**Figure 5.26 and Figure 5.27**).

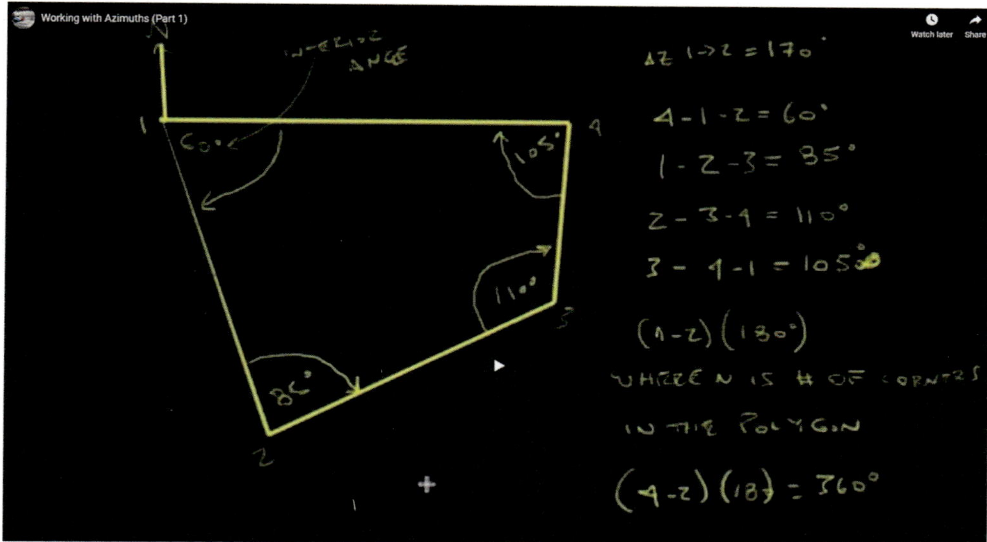

Figure 5.26: This video reviews the concepts related to measuring a traverse. URL: *https://youtu.be/HmGHYwUovn8*

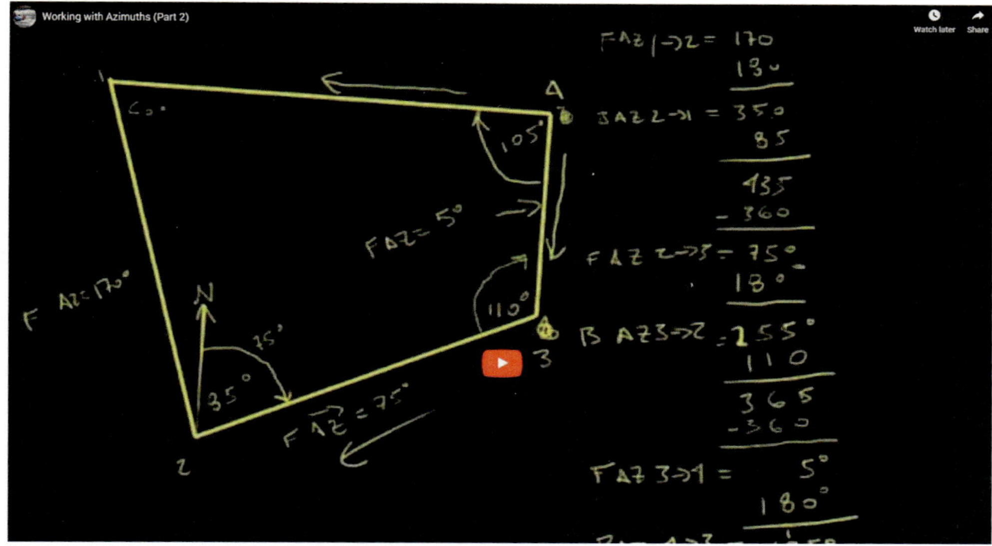

Figure 5.27: This video reviews the concepts related to measuring a traverse. URL: *https://youtu.be/WSK7ZEifT-w*

Triangulation versus Trilateration

Another standard survey method used to extend horizontal control points is triangulation. **Triangulation** is a method of calculating distances using the geometry of triangles, where the length and angle of one or more sides are known. In this case, the known length called a base is the distance between the control point and a second point. Distance to a third or any number of additional points can be determined using the length of the base and the angle from the base to the third point. Any number of other triangles can be calculated using the trigonometric law of sines. **Trilateration** is another survey method that uses the geometry of triangles. This method applies only measured distances from point to point and the trigonometric law of cosines to find interior angles. One determines a location using the intersection of the distances from these points. Watch *What is triangulation?* By Dr. Chris Tisdell to learn more (**Figure 5.28**).

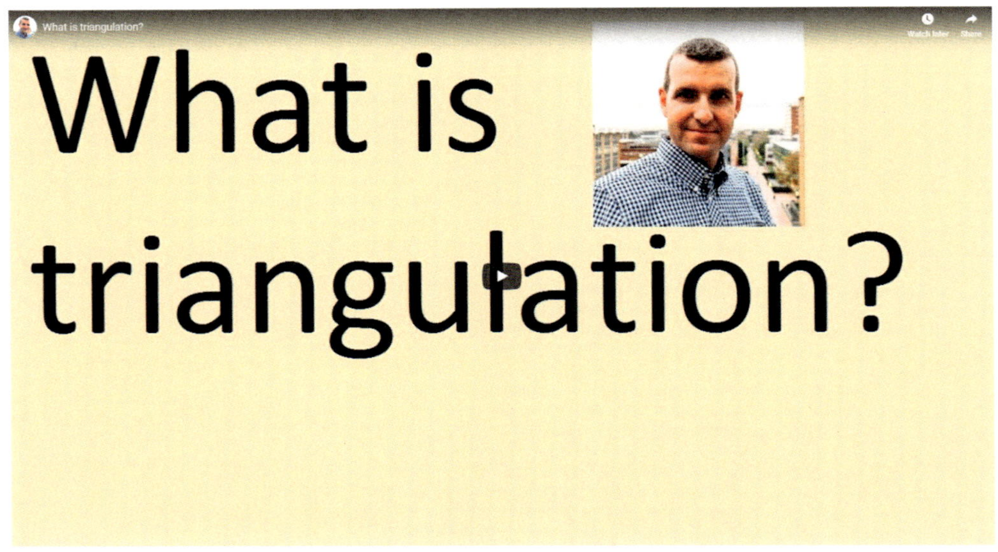

Figure 5.28: In this video, Dr. Chris Tisdell explains triangulation. URL: *https://youtu.be/Nv_oiLPJOVo*

Global Navigation Satellite Systems

Global navigation satellite systems (GNSS) is the generic term for worldwide radio navigation systems using satellites to provide geographic positioning. The United States Department of Defense (DOD) established the first operational GNSS in 1994 called the **Navigation System with Time and Ranging (NAVSTAR) Global Positioning System**. Watch Global Positioning System: "On Target with GPS" circa 2008 NASA–US Air Force by Jeff Quitney to learn more (**Figure 5.29**).

Figure 5.29: This video explains some details related to the Global Positioning System. URL: *https://youtu.be/hOSnE-XRGFE*

This system, now commonly called the **Global Positioning System (GPS)**, is comprised of three segments:

» Space segment
» Control segment
» User segment

The **space segment** initially consisted of a core constellation of 24 GPS satellites. Arranged into six equally spaced orbital planes surrounding Earth and positioned approximately 20,200 km above Earth's surface, each satellite makes two revolutions around Earth in a day. Initially, up to four satellites occupied each orbital plane for a total of 24 core satellites in the system. Since then the number of core satellites has increased to 27. Also, there are some spare satellites used to maintain coverage when servicing core satellites or when they become decommissioned. Though not considered part of the core satellite constellation, additional satellites increase GPS performance. As of June 24, 2015, *www.gps.gov* reported 31 operational satellites in orbit with three more to launch in 2016. The **control segment** is made up of a master control station, monitoring stations, and antennas located around the world. Monitoring stations track GPS satellites, ground antennas transmit and receive signals to and from satellites, and the master control station utilizes this information to ensure the health and accuracy of the GPS constellation.

Any devices, such as GPS receivers, or applications that use GPS technology, make up the **user segment** (**Figure 5.30**). There are many types of GPS receivers ranging from less than a hundred dollars to thousands of dollars. The term recreation grade refers to lower-cost receivers. These types of units are suitable for automobile navigation and outdoor recreational activities such as **geocaching**, a kind of treasure hunt using geographic coordinates. **Mapping-grade GPS** units are ideal for the collection of geospatial data and can offer up to 1-meter accuracy. These units are capable of recording features such as waypoints, tracks (lines), and polygons. These features can be downloaded to a computer and used with GIS software. Survey grade GPS units can cost thousands of dollars but are capable of accuracy within a centimeter. **Survey-grade GPS** units are suitable for engineering and construction applications.

Satellite Ranging and Space Trilateration

The global positioning system calculates a position on Earth using satellite ranging and space trilateration. **Satellite ranging** is the method of finding the distance from a satellite to a location on earth. The exact location of each satellite must be known to calculate these distances. The ground control stations, along with the regular orbital track of the satellites, make this possible. Each satellite broadcasts a radio signal traveling at the speed of light. The time it takes the radio signal to go from a satellite to a GPS receiver is used to measure distance. Measuring the speed of light requires precise atomic clocks. Each GPS satellite is equipped with these atomic clocks, each synchronized with other satellites in what is known as **GPS time**.

Unfortunately, GPS receivers must also help compute the amount of time it takes for the satellite signals to reach the ground. It would be cost prohibitive to equip each GPS receiver with an atomic clock. The satellites broadcast a signal called a pseudorandom noise (PRN) code to correct clock error in the GPS receivers. A **pseudorandom noise code** is a repeating radio signal that appears to be randomly distributed noise. GPS receivers internally track the timing of the repeating pattern in the code. When the GPS receiver receives the pseudorandom noise code from a satellite, it compares the difference in the pattern to measure how long it took the signal to reach the receiver.

Figure 5.30: In this image, students practice using GPS receivers during a fieldwork activity organized by Nicolas R. Malloy. Top, from left to right: Cintia Farias De Souza, Felipe Ribeiro De Toledo Camargo, Clara Dos Santos Baptista, Raphael Garcia Da Silva Luiz Pereira. Middle: Nicolas R. Malloy. Bottom, from left to right: Cintia Farias De Souza, Marcio Pagano Aragona.

By multiplying the time difference in the pseudorandom noise code pattern by the speed of light, a GPS receiver can calculate the distance to the satellite. **Space trilateration** is a method of determining a position on Earth using the distances from satellites. The GPS satellite constellation functions as a control network, similar to survey control points. Ranging from three satellites is enough to fix a horizontal position using trilateration. A fourth satellite is required to determine elevation and to correct for receiver clock errors as well. Watch *How GPS works? Trilateration explained* by Tobiasz Karoń to learn more (**Figure 5.31**).

Figure 5.31: This video discusses the concepts related to space trilateration. URL: *https://youtu.be/4O3ZVHVFhes*

GPS Uncertainty and Error

Collecting geospatial data using GPS units requires an understanding of the sources of uncertainty and error and the limitations of the global positioning system. The United States Department of Defense (DOD) initially developed NAVSTAR for military applications. As a defensive measure, they included intentional degradation and error in the system called **selective availability (SA)**. At any time, the DOD could introduce this error into the system when necessary for military defense. In the early years of GPS, the DOD had selective availability turned on continuously. In May 2000, Pres. Bill Clinton directed the U.S. government to discontinue the use of selective availability. This policy was enacted to make the use of GPS more reliable for research and commercial applications. Currently, a new series of satellites without the selective availability feature called GPS III are in development.

With selective availability no longer an issue, that leaves one with a series of errors called user equivalent range errors. **User equivalent range errors (UERE)** are sources of error in satellite ranging that contribute to the **total error budget**, the cumulative sum of range errors from each source. Watch *GPS Receiver Sources of Error–SixtySec* by ExploreGate to learn more (**Figure 5.32**).

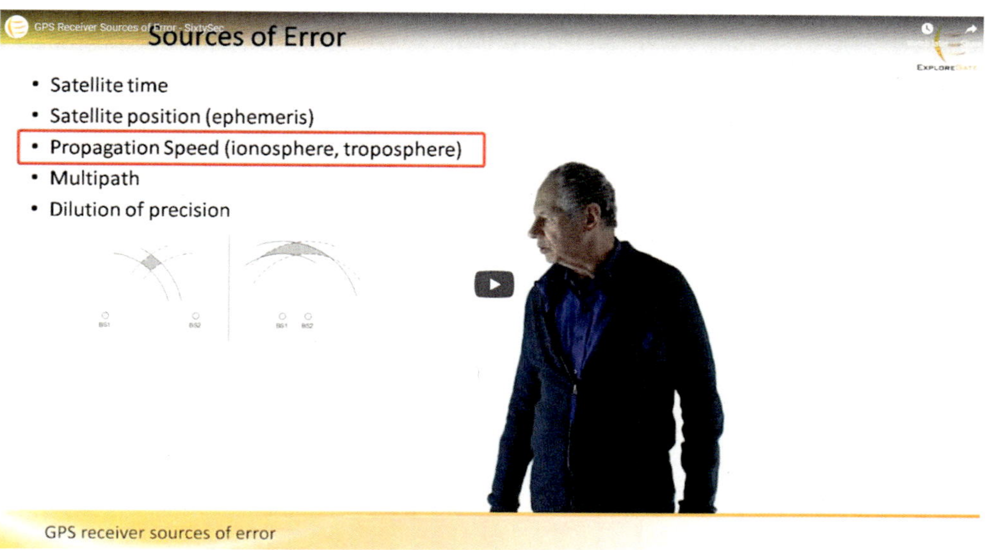

Figure 5.32: This video covers the different sources of GPS error. URL: *https://youtu.be/BaucPwdn4YU*

The sources of GPS error include the following:

- Satellite clock
- Orbital
- Atmospheric
- Receiver clock
- Signal Masking
- Multipath

Clock Error

The atomic clocks onboard GPS satellites have to be extremely accurate, but can still experience a small amount of error over time. The master ground control station accounts for **clock error** by broadcasting corrections to satellites using standard GPS time, which are then transmitted down to GPS receivers.

Ephemeris Bias

The non-spherical nature of Earth's gravitational pull can cause satellite orbits to shift over time. This shift in orbit causes an error called **ephemeris bias**. As with the atomic clocks, the master ground control station tracks the exact location of each satellite and broadcasts the corrections. Each satellite, in turn, transmits corrections down to GPS receivers.

Atmospheric Error

When GPS signals enter Earth's atmosphere, the signals can experience a slight deflection and delay. The amount of **atmospheric error** can vary by location, season, and even the time of day. Monitoring stations transmit corrections to the GPS satellites, which are then passed on to GPS receivers. Since atmospheric conditions can change quickly, one cannot always correct this source of error in real time. It leaves a certain amount of error unaccounted. Signals from satellites close to the horizon pass through more of the atmosphere, increasing this type of error.

Receiver Clock Error

Receiver clock error relates to the GPS receiver and local environmental conditions. GPS receivers are equipped with inexpensive clocks that are less stable than the atomic GPS satellite clocks. The error introduced by the receiver clocks is minimized using the pseudorandom noise code.

Signal Masking

Signal masking is a source of error introduced by the local environmental conditions such as canopy cover, tall buildings, or topography. These obstructions can block the signal from one or more satellites reducing GPS accuracy.

Multipath Error

Sometimes the GPS radio signal will bounce off these local obstructions before reaching the antenna causing a **multipath error**. When this happens, the GPS receiver acquires the same signal from multiple sources or paths. There is no systematic way to correct this type of error. Sometimes the best method is to move away from obstructions to acquire a signal and then use other equipment such as a compass, measuring tape, or an EDM device to measure the distance and direction to the point of interest.

Geometric Dilution of Precision

The arrangement of satellites in the sky changes hour by hour. The specific positions of satellites overhead can cause a different type of error called **geometric dilution of precision (GDOP)**. Satellites high in the sky and close together create more significant areas of uncertainty. The best arrangement to minimize the geometric dilution of precision is to have a satellite overhead with three other satellites evenly spaced and close to the horizon. One expresses the geometric dilution of precision numerically as a multiplier of user equivalent range errors:

GDOP	Description	Application
1	Perfect, does not magnify UERE	Military grade applications
2 -3	Excellent	Civilian survey grade applications
4–6	Good	Vehicle navigation, environmental resource management
7 +	Poor	Not recommended for positioning. Rough estimates only

GPS Augmentation Systems

Though the NAVSTAR control segment includes measures to reduce GPS error, there are additional sources of GPS augmentation available. **GPS augmentation** is the use of external information to improve the accuracy of the GPS satellite signal. These augmentation systems include the following:

- Differential GPS
- Wide Area Augmentation System
- Inertial Navigation Systems

Differential GPS

An augmentation system called **differential GPS (DGPS)** uses a base station to continually monitor its position and broadcast corrections to mobile receivers, called **rovers**, in the local area. The base station is located at a carefully surveyed control point. Because this receiver is stationary, it knows its location precisely and can determine the amount of satellite range error present in each GPS satellite signal. These corrections, called a **differential correction**, are broadcast locally to GPS receivers in real time. GPS receivers equipped with specialized antennas can take advantage of **real-time differential GPS (RTDGPS) corrections**. Less expensive GPS receivers without real-time differential correction capabilities can still take advantage of differential GPS augmentation using a system called **post-mission differential GPS (PMDGPS)**. To do this, one uploads data from the GPS receiver to a free service such as the ***online positioning user service (OPUS)***[4] provided and maintained by the National Geodetic Survey (NGS). Post-mission differential GPS has the advantage of being inexpensive and usable with any GPS receiver. Watch *How does DGPS Work?* By CSC Admin to learn more (**Figure 5.33**).

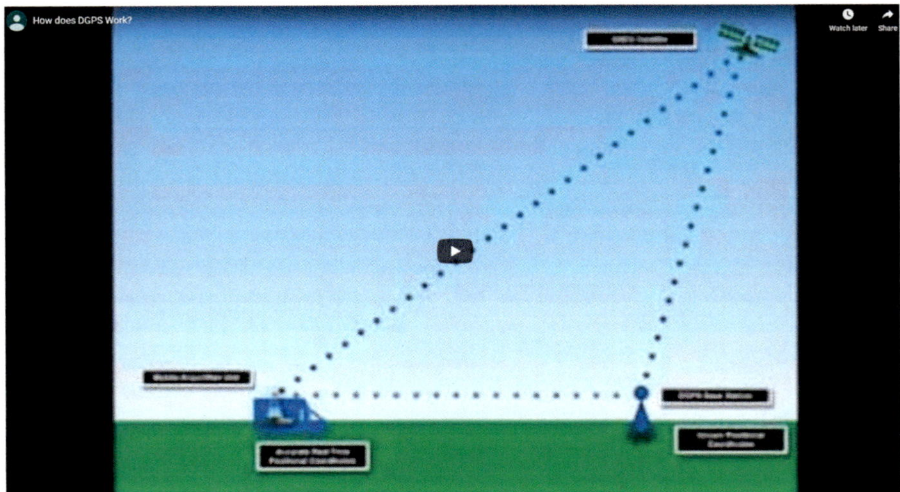

Figure 5.33: This video provides additional information on DGPS. URL: ***https://youtu.be/tKX7e7kJ9wc***

Wide Area Augmentation System

The federal aviation administration (FAA) implemented the **wide area augmentation system (WAAS)** for civilian aviation navigation. The wide area augmentation system uses a series of wide area reference station (WRS) sites located at precisely surveyed control points. Information collected by each of the WRS sites is forwarded to the **WAAS master station (WMS)** and then uploaded to geostationary communication satellites. **Geostationary satellites** orbit Earth at the same rate as Earth's rotation to maintain a constant position above Earth. These satellites broadcast the augmentation messages across the nation, available to any WAAS enabled GPS receiver. WAAS is also known as Wide Area Differential GSP (WADGPS). Watch *Wide Area Augmentation System (WAAS)* by TetraTechAMT to learn more (**Figure 5.34**)

4 http://www.ngs.noaa.gov/OPUS/

Figure 5.34: This video describes the wide area augmentation system. URL: *https://youtu.be/a2l2ftZ_CAY*

Inertial Navigation Systems

One uses Inertial navigation systems (INS) whenever GPS signals are unavailable such as underground or deep in the ocean. **Inertial navigation systems (INS)** use motion and acceleration to track the position of a GPS receiver continuously. INS capable receivers are equipped with motion sensors, such as accelerometers, and computers that calculate changes in velocity. The receivers must first be initialized using the GPS signal to provide its position and current velocity to provide accurate location information. Once the GPS receiver loses a signal, the INS detects changes in geographic positions and accurately records this information on the receiver. This feature is particularly useful when working in areas prone to signal masking and multipath error.

Redundancy Protocols

As GPS receivers are becoming more accurate and less expensive, it is essential to remember that all technology can eventually fail. Not recognizing this fact leads to tragic consequences. When conducting fieldwork, never rely solely on GPS units. Always have a map, a compass, and a cell phone or other means of communication.

Tutorial: Indicating Direction Using Azimuth and Bearing

As discussed in this chapter, all directional systems measure direction relative to a directional baseline. A directional baseline is a line of reference between a location on the earth and a standard reference point. In most direction systems, the directional baseline refers to true north, which is the northern end of the axis of Earth's rotation. Azimuth and bearing are examples of directional systems that use true north as a directional baseline. In this activity, you learn how to indicate direction utilizing both azimuth and bearing and practice how to convert from one directional system to another.

ESTIMATED TIME TO COMPLETE THIS TUTORIAL: 1 HOUR

Learning Outcomes

Readers should be able to accomplish the following outcomes by the end of this tutorial:

» Explain the difference between azimuth and bearing
» Indicate a direction using an azimuth
» Indicate a direction using a bearing
» Convert directional notation between azimuth and bearing

Azimuth

An **azimuth** indicates direction by using true north as a directional baseline and measuring clockwise from 0° to 360°. Azimuths are easy to use because one number defines each direction. Treating direction as a single number also makes it simple to use for calculations and in GIS software. Only a single number, written in degrees, indicates an azimuth. For example, if you were facing true north, and then rotate your body 45 degrees clockwise, your azimuth would be written as 45° (**Figure 5.35**).

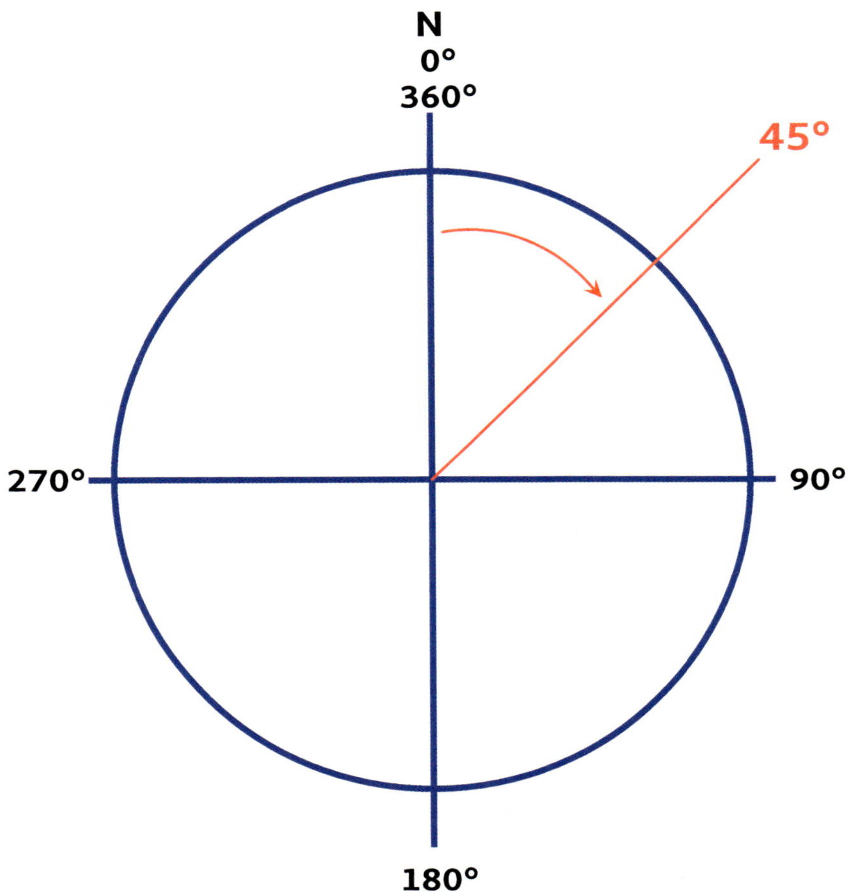

Figure 5.35: An azimuth uses a single number, which makes it simple to use for calculations and in GIS software.

When determining a **back azimuth**, the opposite direction, add 180° if the azimuth is less than 180°. Subtract 180° if the azimuth is greater than 180°. To calculate the back azimuth in this example, add 180 to 45 degrees to get 225° (**Figure 5.36**).

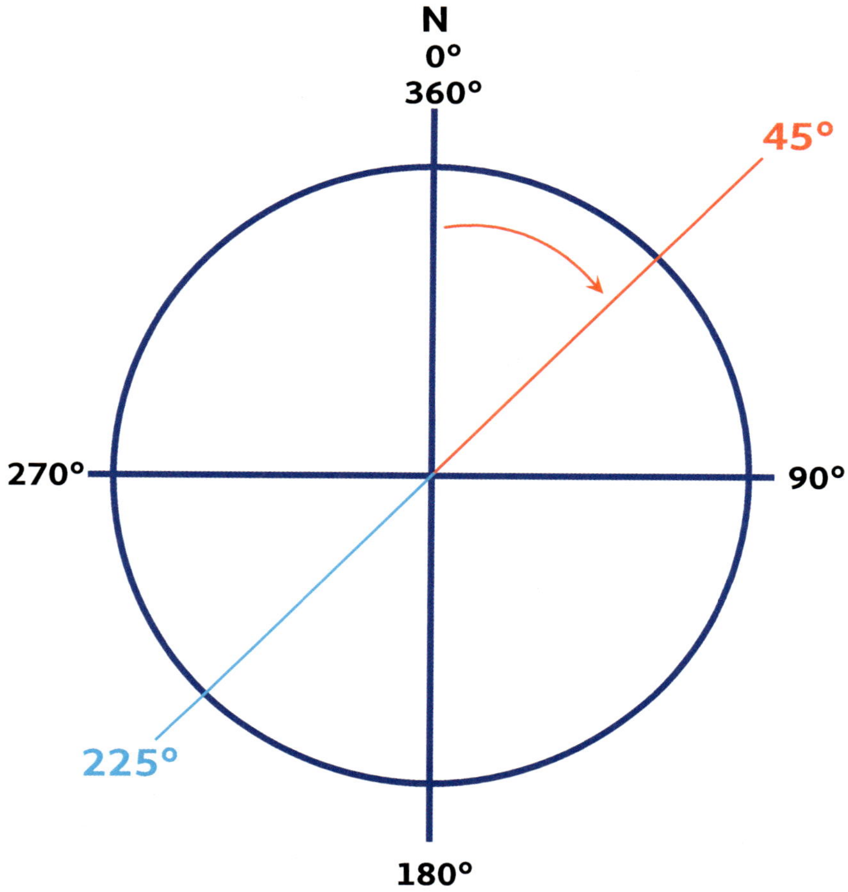

Figure 5.36: Surveyors sometimes use a back azimuth.

On a separate piece of paper, write down the answers to the following questions:

1. If you were facing true north and turned your body 45 degrees counter-clockwise, what would be your azimuth?
2. What would be the back azimuth for question 1?
3. If you were facing south and turned your body 20 degrees clockwise, what would be your azimuth?
4. What would be the back azimuth for question 3?

Bearing

A bearing indicates direction by using the angular difference away from a north or south baseline, ranging from 0° to 90°. Determining your bearing begins with a north or south baseline, whichever is closer to your direction. You then measure the angle east or west from the baseline. You write the bearing using the following notation: N or S (baseline) degrees° E or W (orientation) For example, if you were facing true north, and then rotate your body 45 degrees clockwise, your bearing would be written as N45°E (**Figure 5.37**)

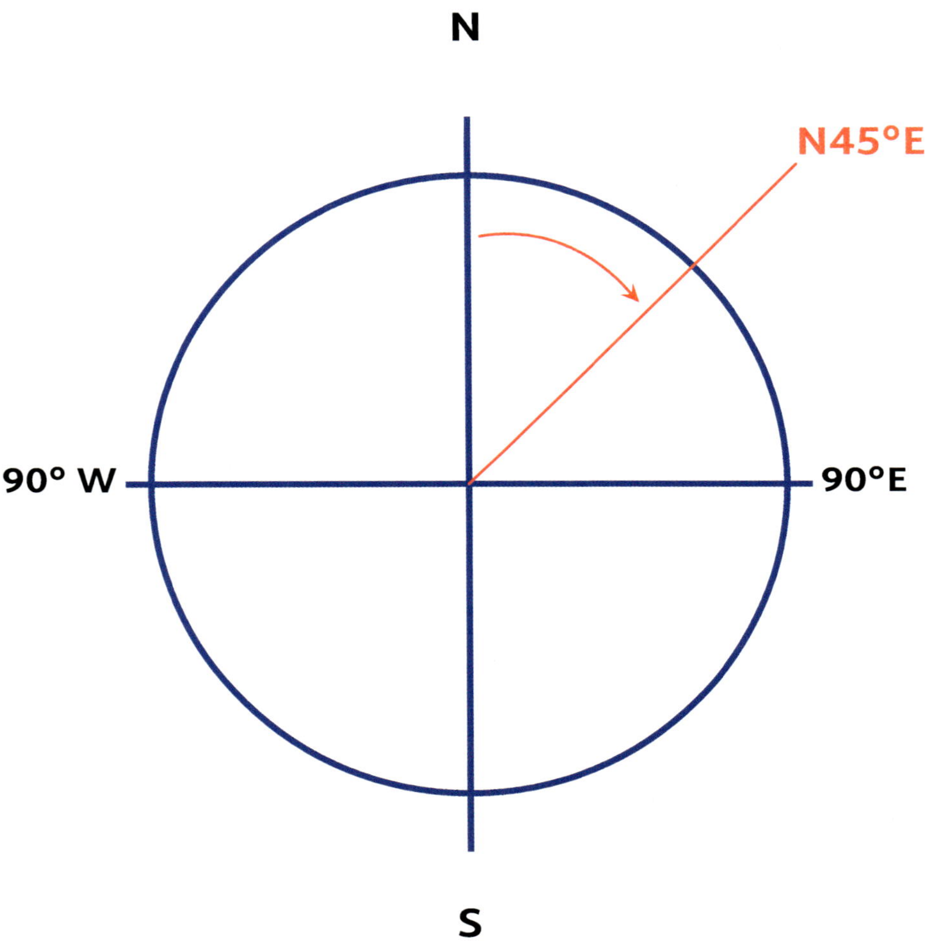

Figure 5.37: Bearings are not as computer friendly but still used in navigation.

Sometimes you also need to find the opposite direction from which you face. A back bearing is the opposite direction from a bearing and is readily determined by merely changing the letters. The back bearings would be written as S45°W (**Figure 5.38**).

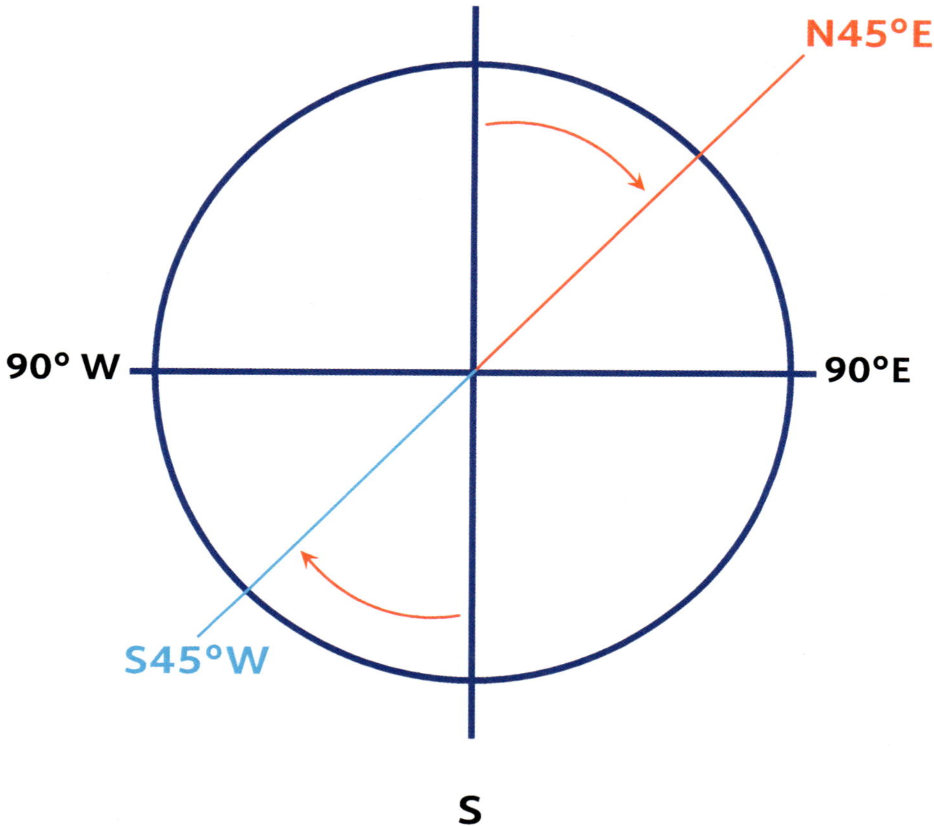

Figure 5.38: A back bearing is used to find the opposite direction you are facing.

On a separate piece of paper, write down the answers to the following questions:

5. If you were facing true north and turned your body 45 degrees counter-clockwise, what would be your bearing?
6. What would be the back bearing for question 5?
7. If you were facing south and turned your body 20 degrees clockwise, what would be your bearing?
8. What would be the back bearing for question 7?
9. If you had an azimuth of 175°, how would you write the same direction using a bearing?
10. If you had a bearing of N25°W, how would you write the same direction using an azimuth?

Tutorial: Geocaching Basics

Geocaching is a great way to learn how to operate a GPS receiver while engaging in a casual activity with friends. While it is possible to use a smartphone during this activity, you should take the time to learn how to operate the GPS receiver.

Estimated time to complete this tutorial: 2 hours

Learning Outcomes

Readers should be able to accomplish the following outcomes by the end of this tutorial:

- » Apply concepts learned related to Global Positioning Systems
- » Practice navigation using a GPS receiver
- » Demonstrate the ability to enter and locate a waypoint using a GPS receiver
- » Document their geocache experience by writing a summary

Learning About Geocaching

1. Before heading out into the field, watch each the videos below to learn about geocaching (**Figures 5.39, 5.40, 5.41, 5.42, 5.43, 5.44, 5.45, and 5.46**).
2. Create an account on the ***geocaching website***[1] and log in.
3. Choose five geocache points and write down the geographic coordinates.

Figure 5.39: What is Geocaching?URL: ***https://youtu.be/1YTqitVK-Ts***

1 https://www.geocaching.com/

Figure 5.40: Finding a Geocache. URL: *https://youtu.be/sj31U_z9MFA*

Figure 5.41: Geocaching Etiquette. URL: *https://youtu.be/GXzIu7p82jg*

Figure 5.42: Hiding a Geocache. URL: *https://youtu.be/aqc8Rnl_fh8*

Figure 5.43: Making a Favorite Point Worthy Geocache. URL: *https://youtu.be/1Is-xEjqwIo*

Figure 5.44: What are Geocaching Trackables?URL: *https://youtu.be/LjSbSsSSTIM*

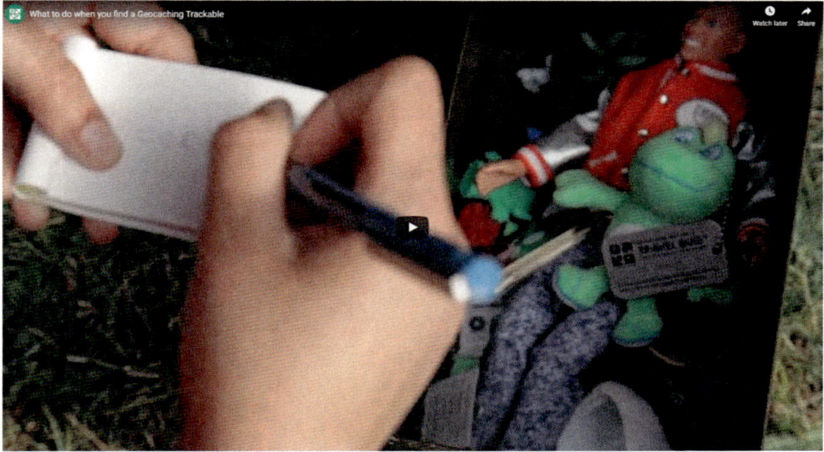

Figure 5.45: What do you do when you find a Geocaching Trackable? URL: *https://youtu.be/nN9jdgM8BDU*

Figure 5.46: How to log a Geocaching Trackable. URL: *https://youtu.be/Un_NSs-HkxY*

Setting up your GPS Receiver

This activity assumes you have access to a GPS receiver. The examples in this book use a Garmin GPSMAP 64. However, less expensive handheld models, such as the Garmin eTrex series, works just as well as long as your computer recognizes them. Take a moment to read the GPSMAP64 Quick Start Manual to become familiar with the menus and options for the GPS receiver. If you are using a different GPS receiver, review the owners manual for that unit.

It is a good habit to clear out old waypoints that you no longer need before starting any project or activity.

Start by turning on the GPS receiver. On the Garmin GPSMAP 64, the power button is on the side of the receiver (**Figure 5.47**). If you are using a different model, locate the power button. Refer to the owner's manual if necessary.

Figure 5.47: On the Garmin GPSMAP 64, find the power button along the side.

Press the *Menu* button one or more times to get to the menu screen (**Figure 5.48**). Then, use the rocker pad to select *Setup* from the menu options. Press the *Enter* button when you are ready.

Figure 5.48: You may need to press the menu button more than once to get to the *Main* menu.

On the *Setup* menu, use the rocker pad to select Reset and press the *Enter* button (**Figure 5.49**).

Figure 5.49: Use the *Enter* button when you want to choose an option.

From the *Reset* menu, choose the *Delete All Waypoints* option. If you are asked to confirm your choice, press *Yes* (**Figure 5.50**).

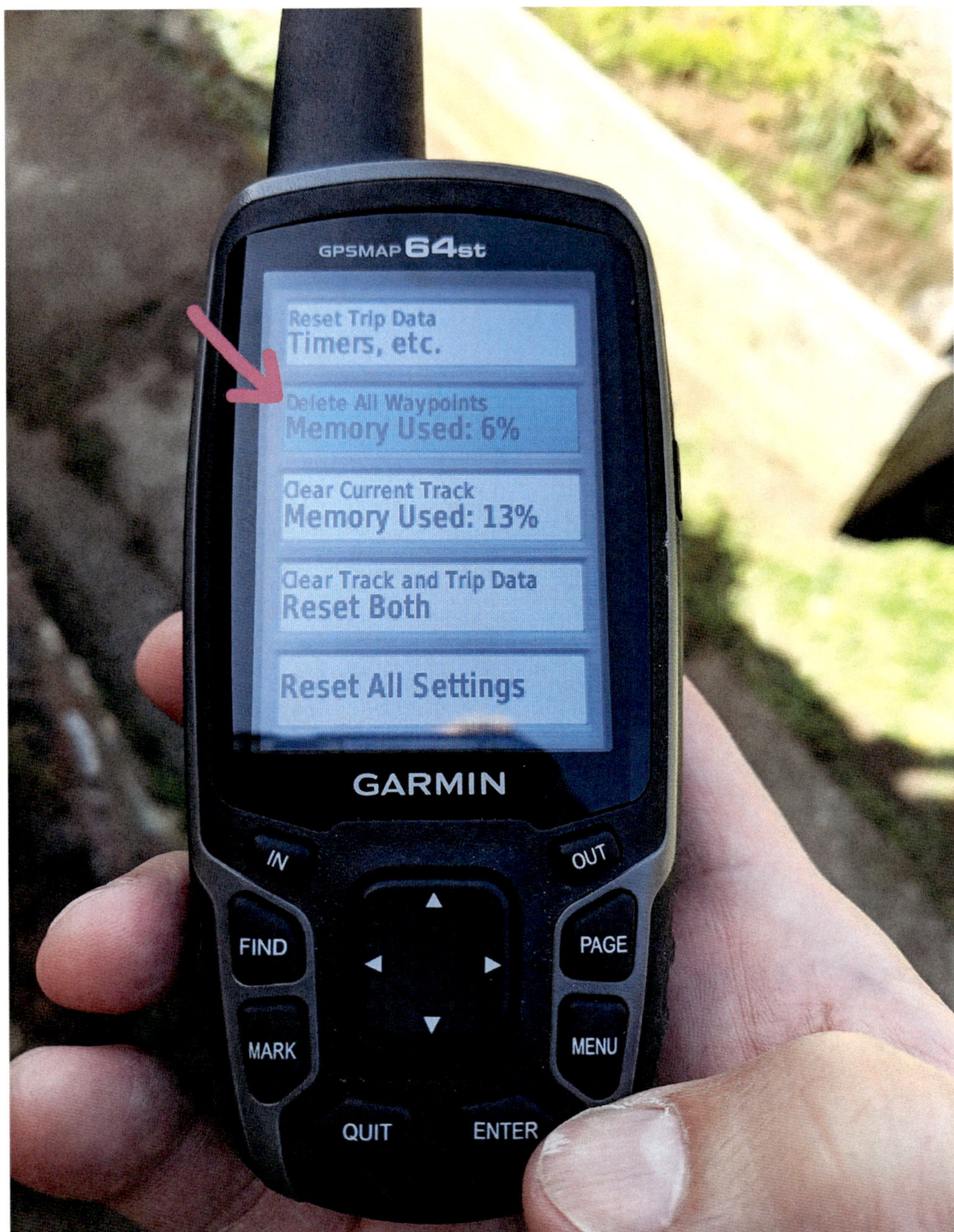

Figure 5.50: Deleting old waypoints saves memory and helps to avoid confusion.

Changing the Position Format

The Geocaching website used to obtain the coordinates uses a specific notation, degrees, and decimal minutes (**hddd°mmmmm'**). Use the *Quit* button to navigate back out to the *Main* menu. Locate and choose the *Position Format* option (**Figure 5.51**).

Figure 5.51: The position format options allow you to change the coordinate system notation.

Choose hddd°mmmmm' as the format (**Figure 5.52**).

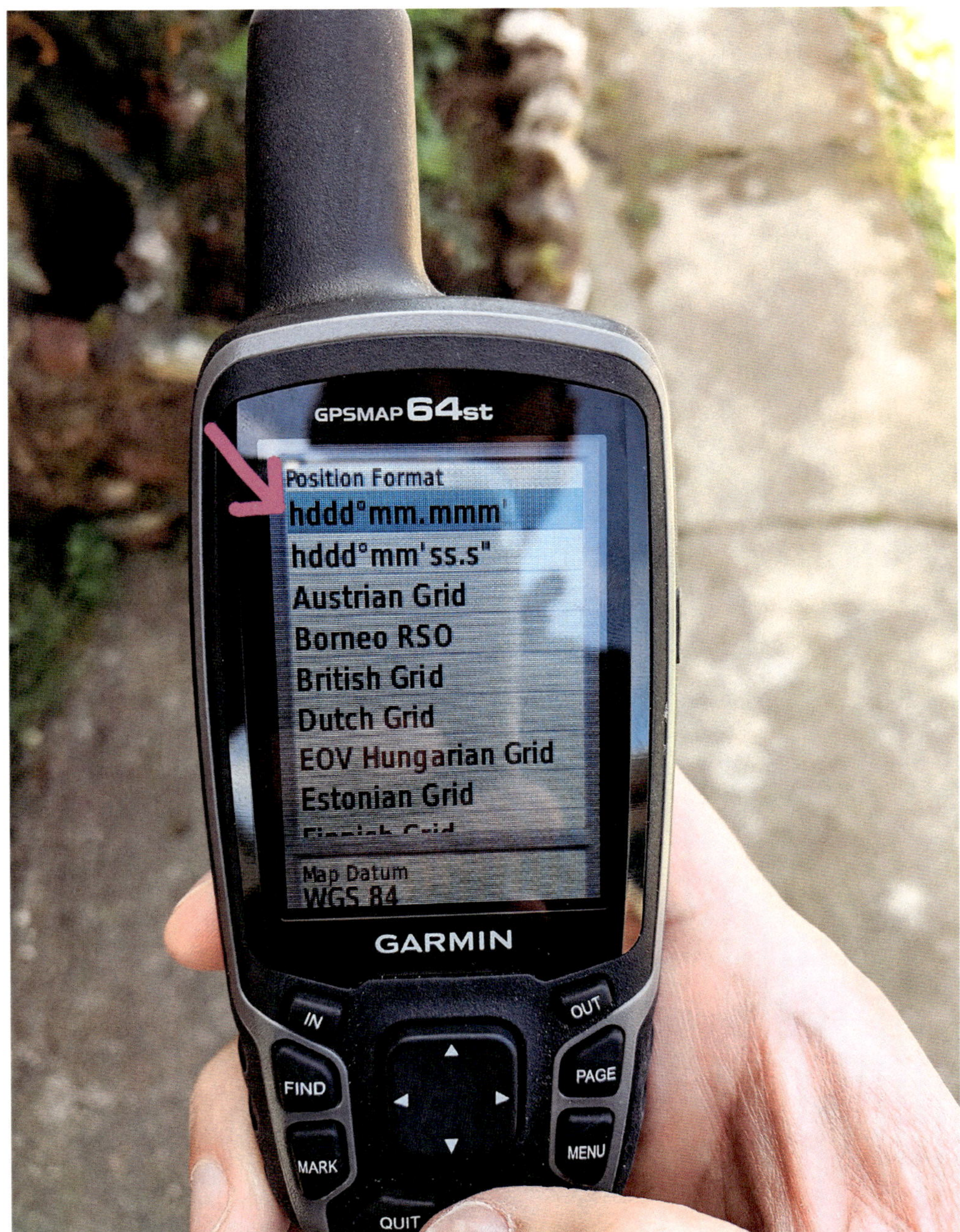

Figure 5.52: The Geocaching website uses degrees and decimal minutes for geocache coordinates.

Use the Quit button to navigate back out to the *Main* menu.

Improving accuracy

Some GPS receivers have access to the **Wide Area Augmentation System (WAAS)**. Enabling WAAS on your device helps to improve accuracy. On the Garmin GPSMAP 64, press *System* (**Figure 5.53**)

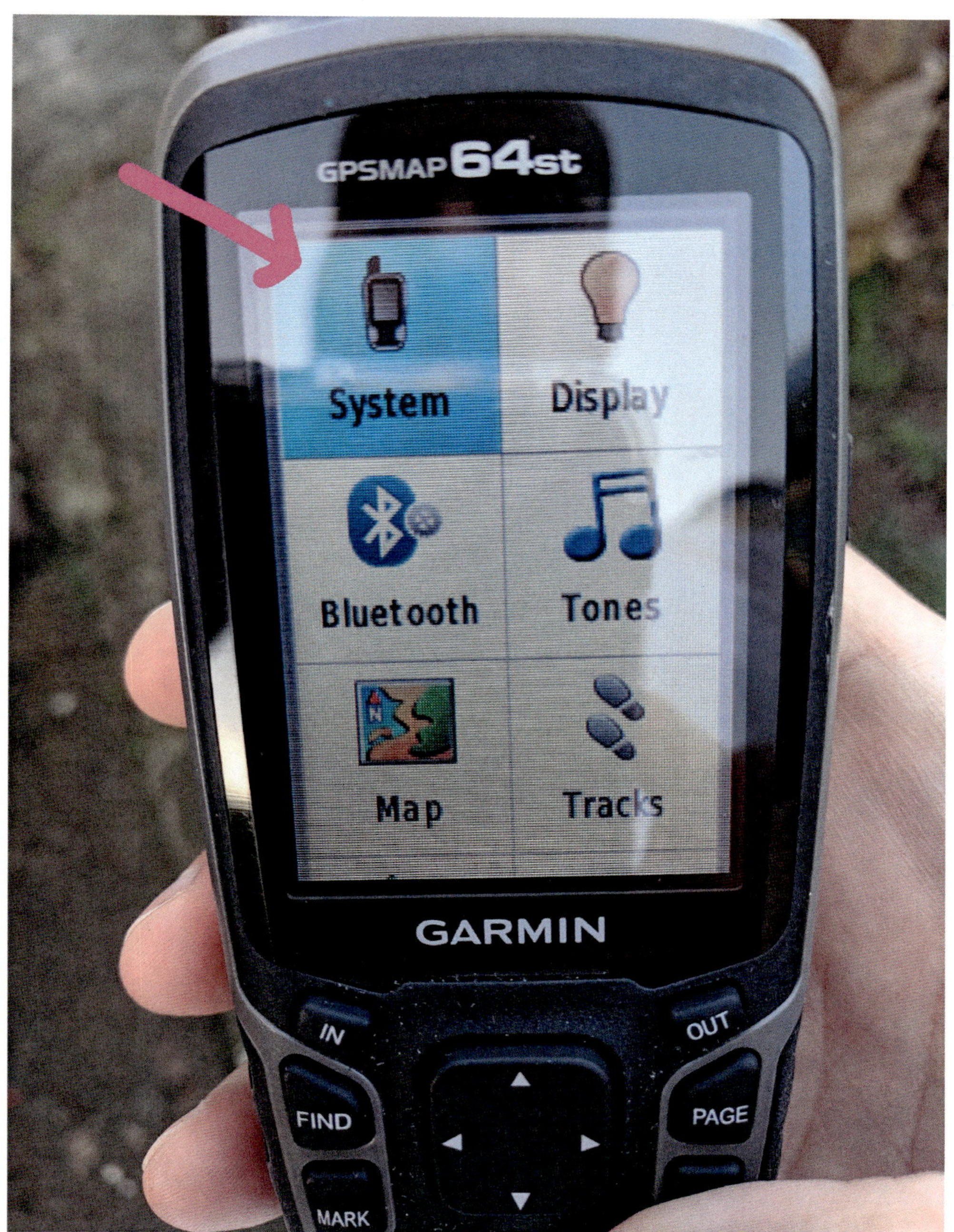

Figure 5.53: Locate the *System* option on the *Main* menu of your GPS receiver.

On the System menu, be sure that the WAAS setting is turned on (**Figure 5.54**).

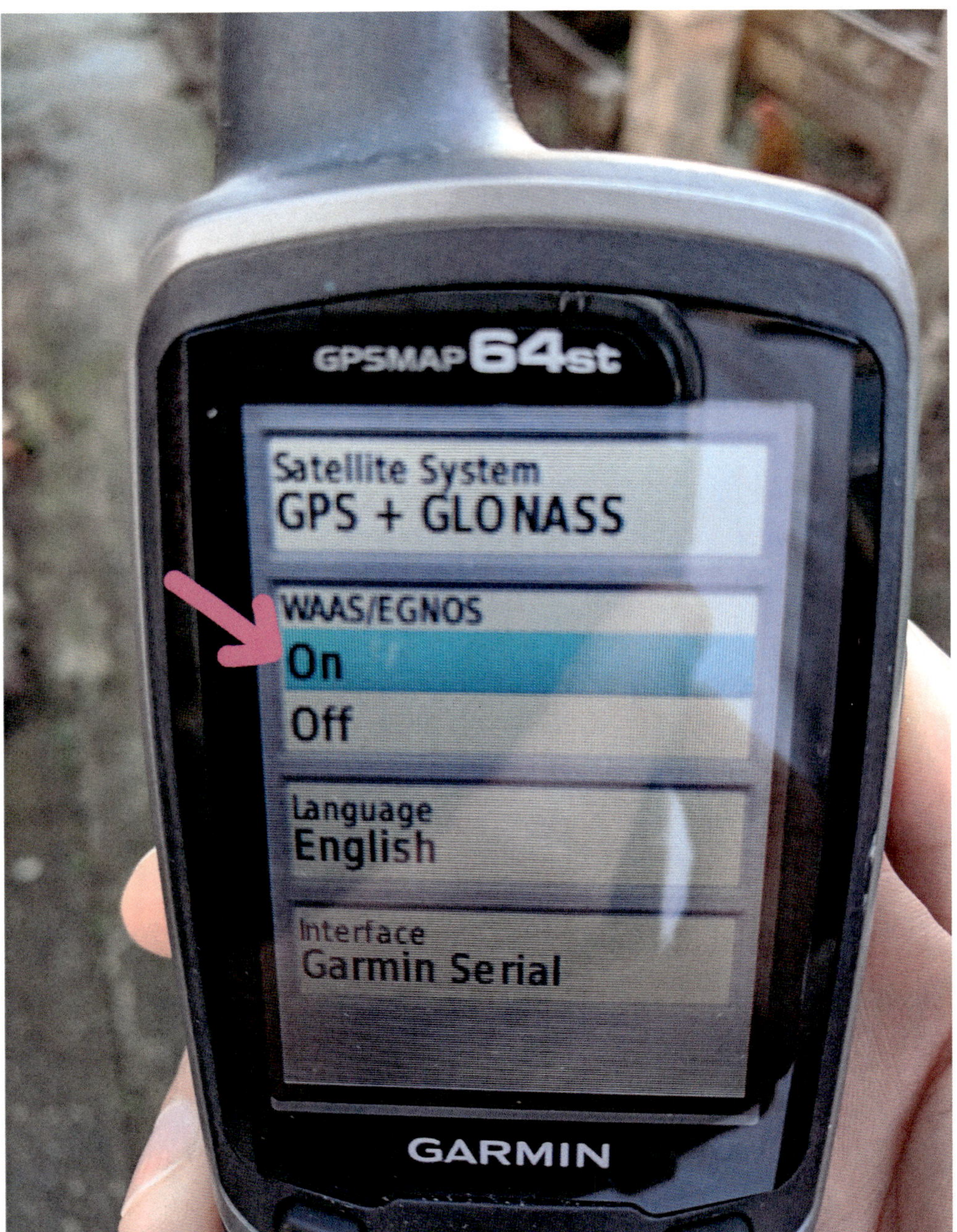

Figure 5.54: Enabling WAAS helps to increase the accuracy of the GPS receiver.

Marking and Editing Waypoints

Now you are ready to enter your geocache coordinates. On the Garmin GPSMAP 64, press the Mark button. If you are using a different model, refer to the owner's manual to learn how to mark a waypoint (**Figure 5.55**).

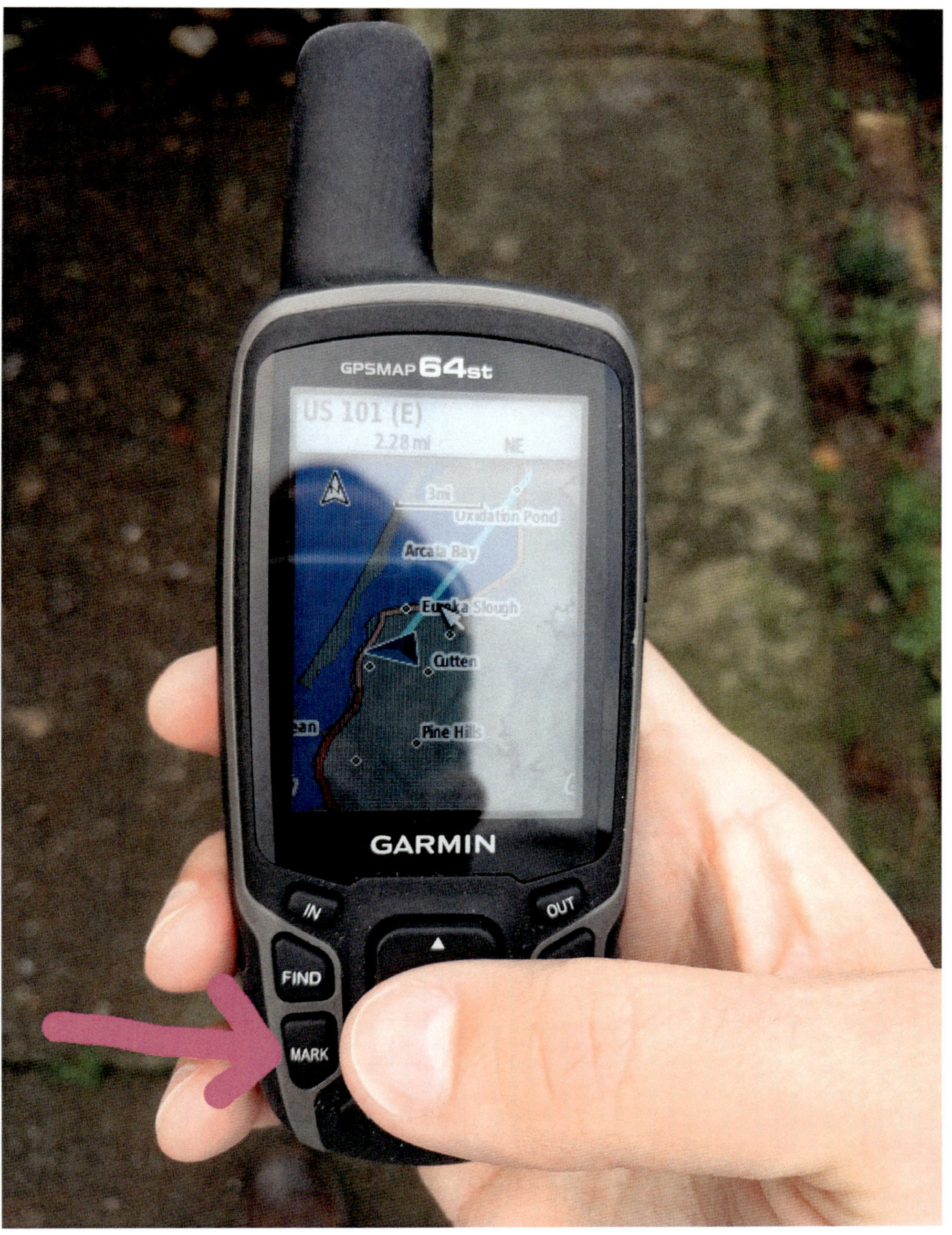

Figure 5.55: Press the Mark button to mark a waypoint.

After marking a waypoint, the GPS receiver takes you to the waypoint edit screen. By default, many GPS receivers assign a series of numbers for waypoint names, such as 001, 002, and 003. Unfortunately, this naming convention sometimes causes problems when working with GIS software. It is a good habit to give your waypoints descriptive names to avoid this issue (**Figure 5.56**).

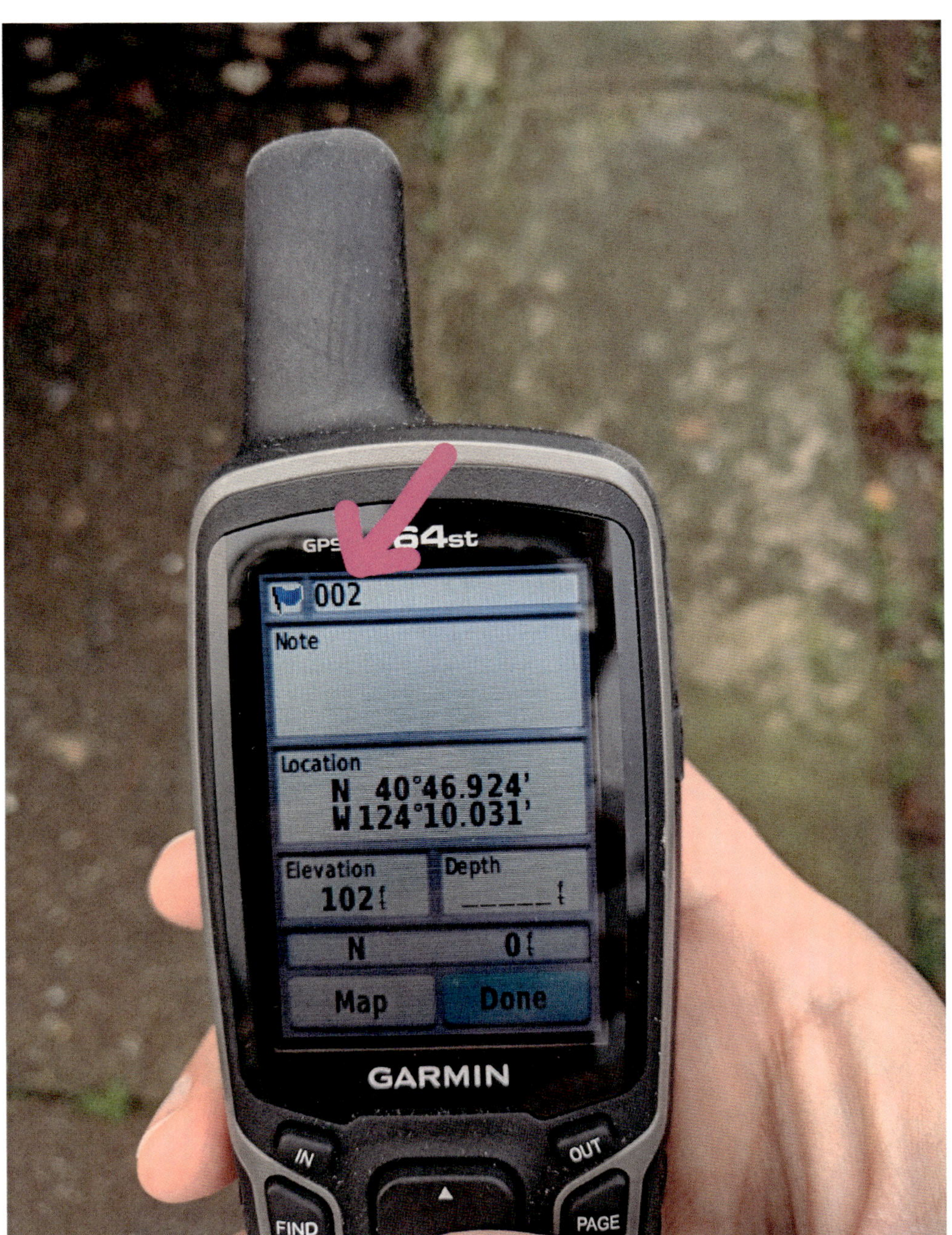

Figure 5.56: Locate the waypoint name at the top. By default, it is a number.

Change the name of your waypoint so that it matches the name of the geocache. On the Garmin GPSMAP 64, use the rocker pad to select the name and press *Enter*. On the edit screen, enter your geocache name (**Figure 5.57**).

Figure 5.57: In this example, the name of the geocache is My Humboldt Degree.

Next, select the location option on the waypoint edit screen (**Figure 5.58**).

Figure 5.58: The location format should be in degrees and decimal minutes format (hddd°mmmmm').

Enter the coordinates for your geocache (**Figure 5.59**).

Figure 5.59: Be sure that the location matches the coordinates provided by the Geocache website.

When you are ready, press enter. Then, select Done to save the waypoint.

Repeat these steps for each of your geocaches. You should attempt **at least three**. You may want to consider adding an extra geocache if you cannot find one or more.

Power down the GPS receiver when not in use to save battery life.

Navigating to Your Geocaches

Start by walking to an open space with minimal obstructions and turn on the GPS receiver. Spend about five minutes standing in the same location to allow the GPS receiver to acquire satellite signals and pinpoint your location.

Choose your first geocache by pressing the Find button. Then, choose Waypoints (**Figure 5.60**).

Figure 5.60: On the Garmin GPSMAP 64, the find button is located on the lower left.

When the list of waypoints appears, select the name of the geocache you want to find (**Figure 5.61**).

Figure 5.61: You should see a list of each waypoint you created.

When you are ready, select, and press GO (**Figure 5.62**).

Figure 5.62: The map indicates the direction of the geocache.

Start walking briskly so the GPS receiver can determine your direction of travel. Use the map to navigate to your geocache. When you arrive at the location, you may have to search carefully for the hidden geocache. If you find one, take a selfie showing your prize (optional).

What if I can't find a geocache?

Don't panic. Some geocaches are hard to find. Others may have been taken or moved by someone else. For this assignment, you are not required to locate each of your geocaches. The purpose of this activity is to become familiar with operating a GPS receiver during a fun, low-pressure activity.

Be Prepared to Write About Your Geocaching Experience

Write a summary of your geocache experience. You may include any photos you took during the activity. Think about the following questions as you write your summary:

» Did you enjoy your geocaching experience? Why or Why not?

» Did you find it challenging to locate the geocaches?

» What kinds of hiding places did you encounter?

» Did your geocaching experience help you to become more familiar with operating a hand-held GPS receiver?

» Do you think you will continue to geocache in the future?

TUTORIAL: MAPPING NOISE POLLUTION DATA USING GPS

The goal of this activity is to learn how to collect data using a GPS receiver and create a map. In this activity, you start by recording noise levels in various locations on campus or near your neighborhood. You then use a GPS receiver to mark a waypoint at each spot. Afterward, you download the data from the GPS receiver using DNRGPS software. After bringing the data into the ArcGIS software and you create a small-sized map to use as a figure in a lab report.

ESTIMATED TIME TO COMPLETE THIS TUTORIAL: 4 HOURS

Learning Outcomes

Readers should be able to accomplish the following outcomes by the end of this tutorial:

- Conduct a data collection project using a GPS receiver
- Practice downloading data from a GPS receiver using DNRGPS software
- Demonstrate how to save one or more GPX files as a single shapefile.
- Perform a table join in ArcMap
- Apply classification methods to quantitative data
- Design a small-sized map for a report

Preparing to Collect GPS Data

Before heading out into the field with your GPS unit, there are several steps you can take ahead of time to increase the efficiency of your fieldwork. These steps include downloading the smartphone application for recording decibel levels and calibrating it if necessary. Then, plan your route before heading out into the field.

Recording Decibel Levels

A popular application for recording decibel levels is the free *Sound Meter* app (**Figure 5.63**). On an Android device, you can locate the application by using the keywords sound meter or decibel.

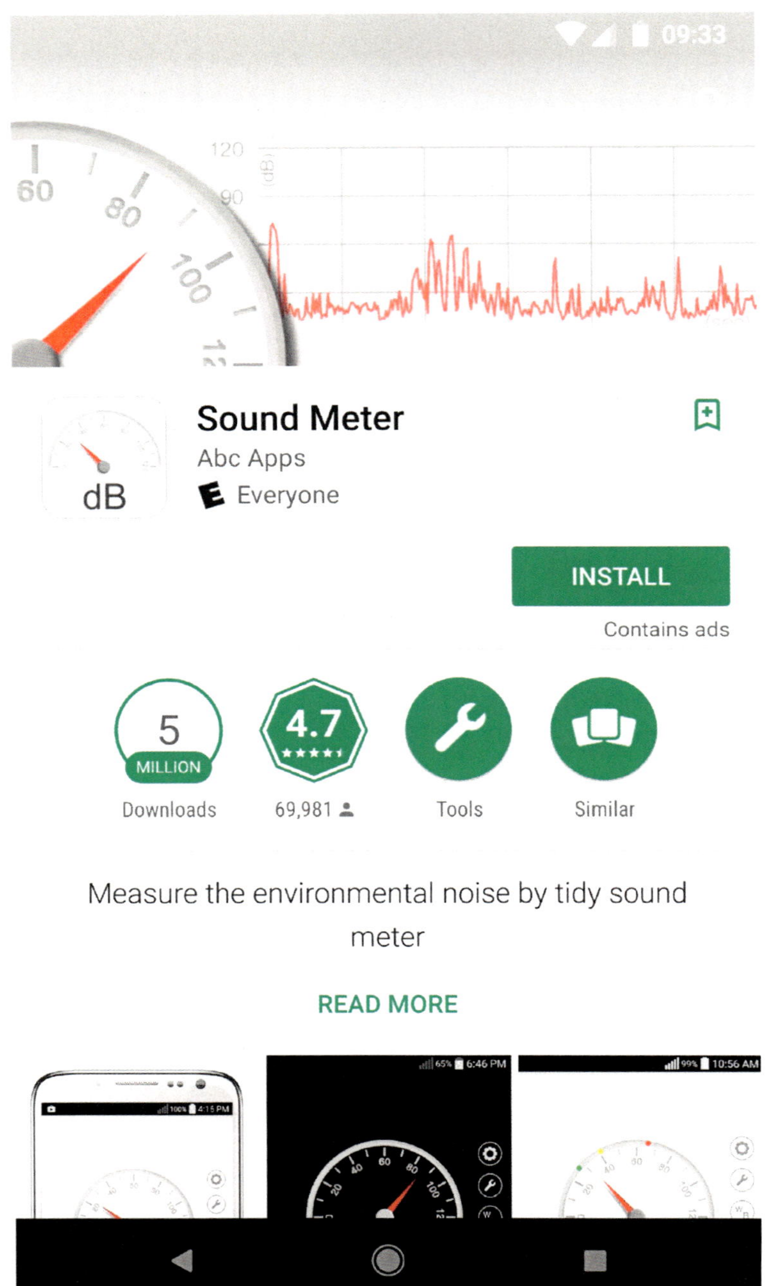

Figure 5.63: Sound Meter is a free application for recording decibel levels.

Related keywords also work for smartphones with other operating systems and app stores. Once you have installed the app, open it and check the ambient sound levels. You may get a message recommending that you calibrate the device (**Figure 5.64**). This step is optional as relative sound levels are more important for the map than accurate decibel readings.

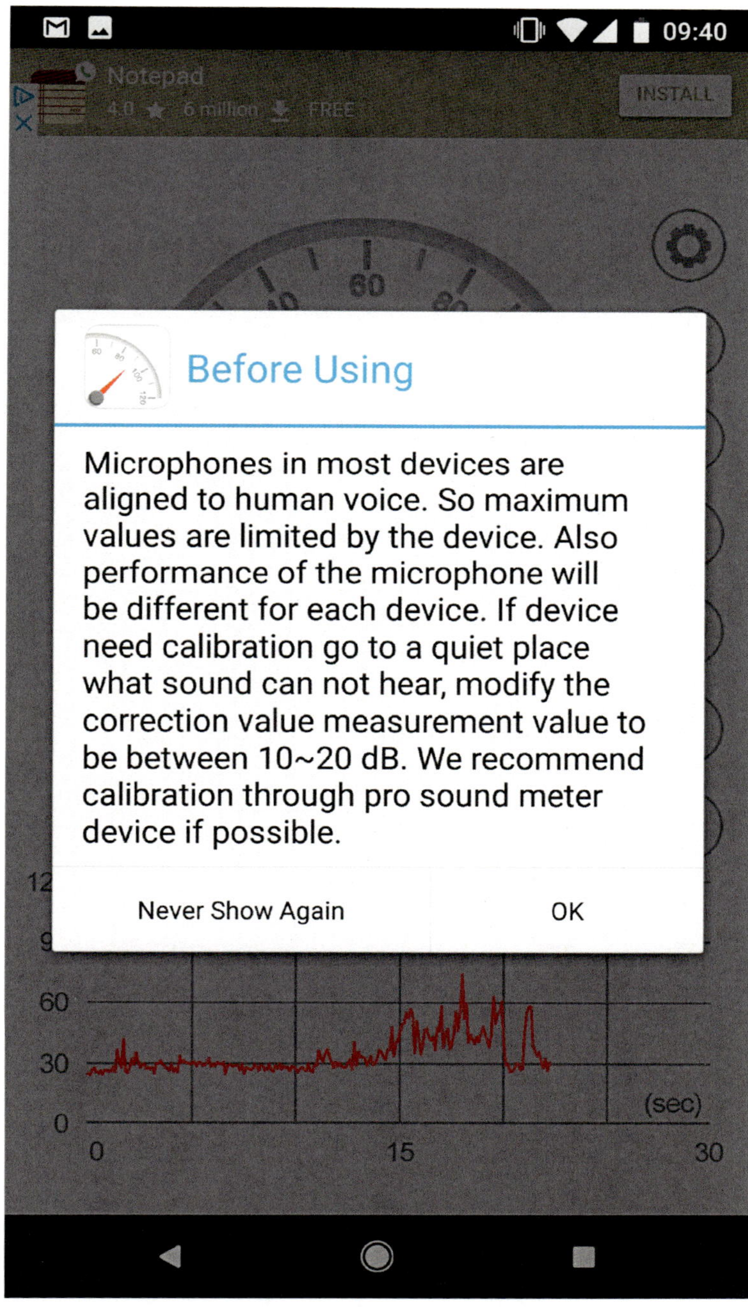

Figure 5.64: Though you may improve accuracy by calibrating your device, this step is optional.

If you want to calibrate your device, use the Sound Meter app to check the ambient noise level for your location (**Figure 5.65**). A quiet library has an approximate value of 20 to 30 dB. A conversation might range from 40 to 60 dB. A lawnmower might read between 80 to 100 dB, and a rock concert can reach as high as 120 dB. A quick Google search for "harmful noise levels" returns quite a few examples. Click the wrench icon and adjust the levels as needed.

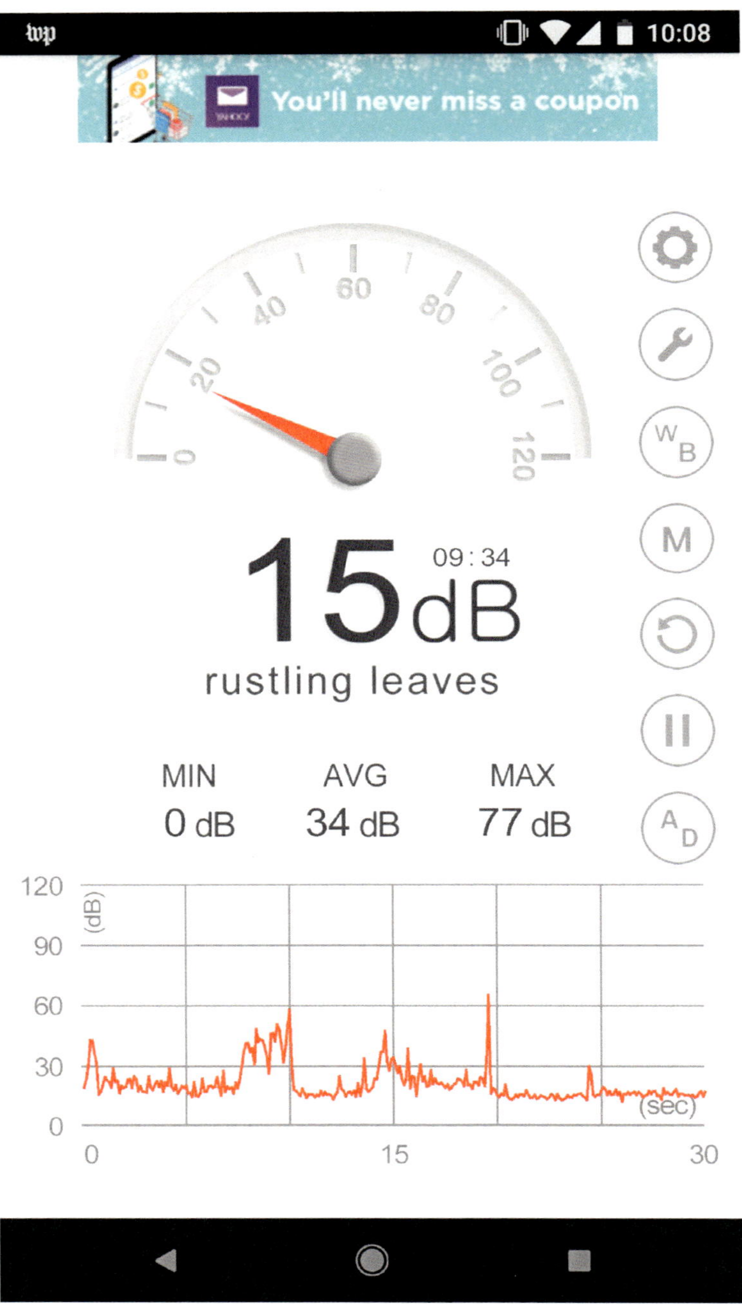

Figure 5.65: The Sound Meter app passively monitors the ambient noise level.

This activity assumes you have access to a smartphone. However, if you do not have access to a smartphone, you can make a chart of average noise levels and make a judgment call for each location you visit. Your range should be from **0 dB** for total silence to **120 dB** for rock-concert levels.

A quick Google search for "harmful noise levels" returns quite a few examples you can use when making your chart.

Planning your Route

Another step that saves time out in the field is to plan your route before heading out to collect data. You can use Google Maps to view your study area (**Figure 5.66**). It might be a good idea to print the map and mark the locations you plan to visit.

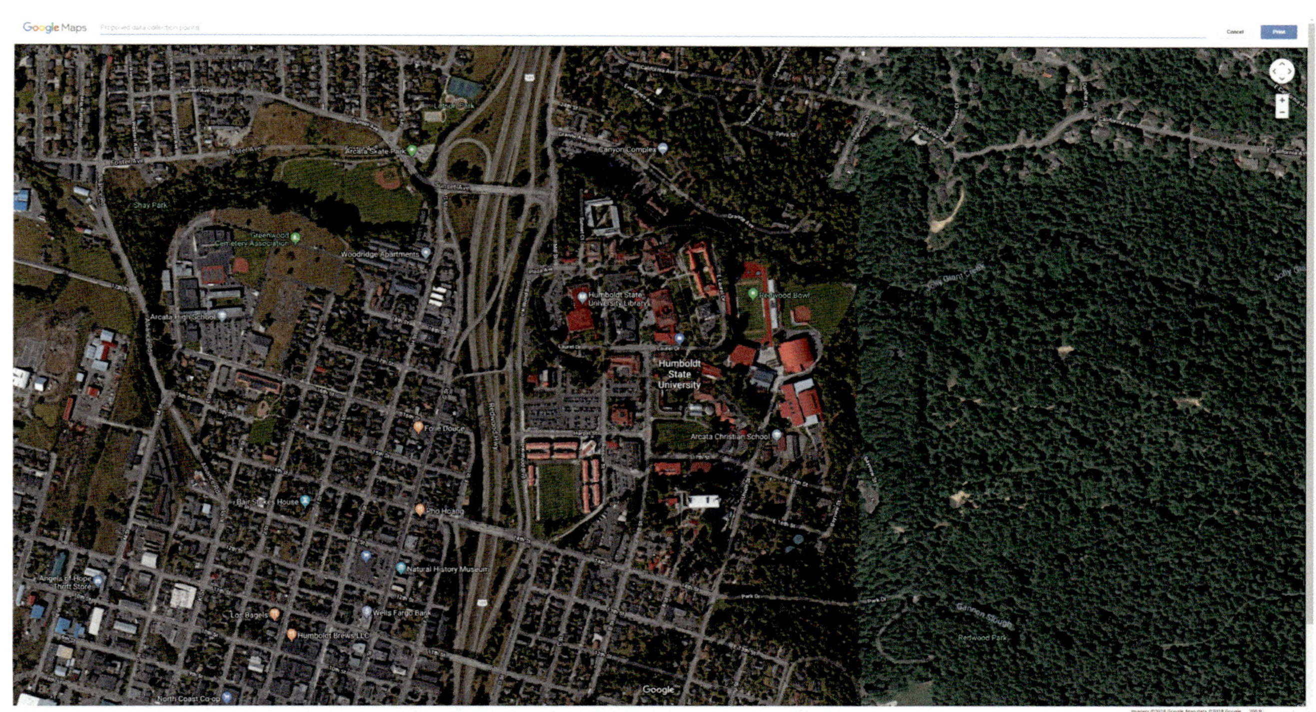

Figure 5.66: You can print out a map of your study area using Google Maps.

Ideally, you want to visit locations with a wide range of noise levels, from quiet to loud. You should also consider the potential for GPS uncertainty and error. Plan to visit sites where you have a good signal coming from as many satellites as possible. Once you mark each location on the map, plan the most efficient route from one waypoint to another.

Setting up your GPS Receiver

This activity assumes you have access to a GPS receiver. The examples in this book use a Garmin GPSMAP 64. However, less expensive handheld models, such as the Garmin eTrex series, will work just as well, as long as your computer recognizes them. Take a moment to read the GPSMAP64 Quick Start Manual to become familiar with the menus and options for the GPS receiver. If you are using a different GPS receiver, review the owners manual for that unit.

It is a good habit to clear out old waypoints that you no longer need before starting any project or activity.

Start by turning on the GPS receiver. On the Garmin GPSMAP 64, the power button is on the side of the receiver (**Figure 5.67**). If you are using a different model, locate the power button. Refer to the owner's manual if necessary.

Figure 5.67: On the Garmin GPSMAP 64, find the power button along the side.

Press the *Menu* button one or more times to get to the menu screen (**Figure 5.68**). Then, use the rocker pad to select *Setup* from the menu options. Press the *Enter* button when you are ready.

Figure 5.68: You may need to press the menu button more than once to get to the *Main* menu.

On the *Setup* menu, use the rocker pad to select Reset and press the *Enter* button (**Figure 5.69**).

Figure 5.69: Use the *Enter* button when you want to choose an option.

From the *Reset* menu, choose the *Delete All Waypoints* option. If you are asked to confirm your choice, press *Yes* (**Figure 5.70**).

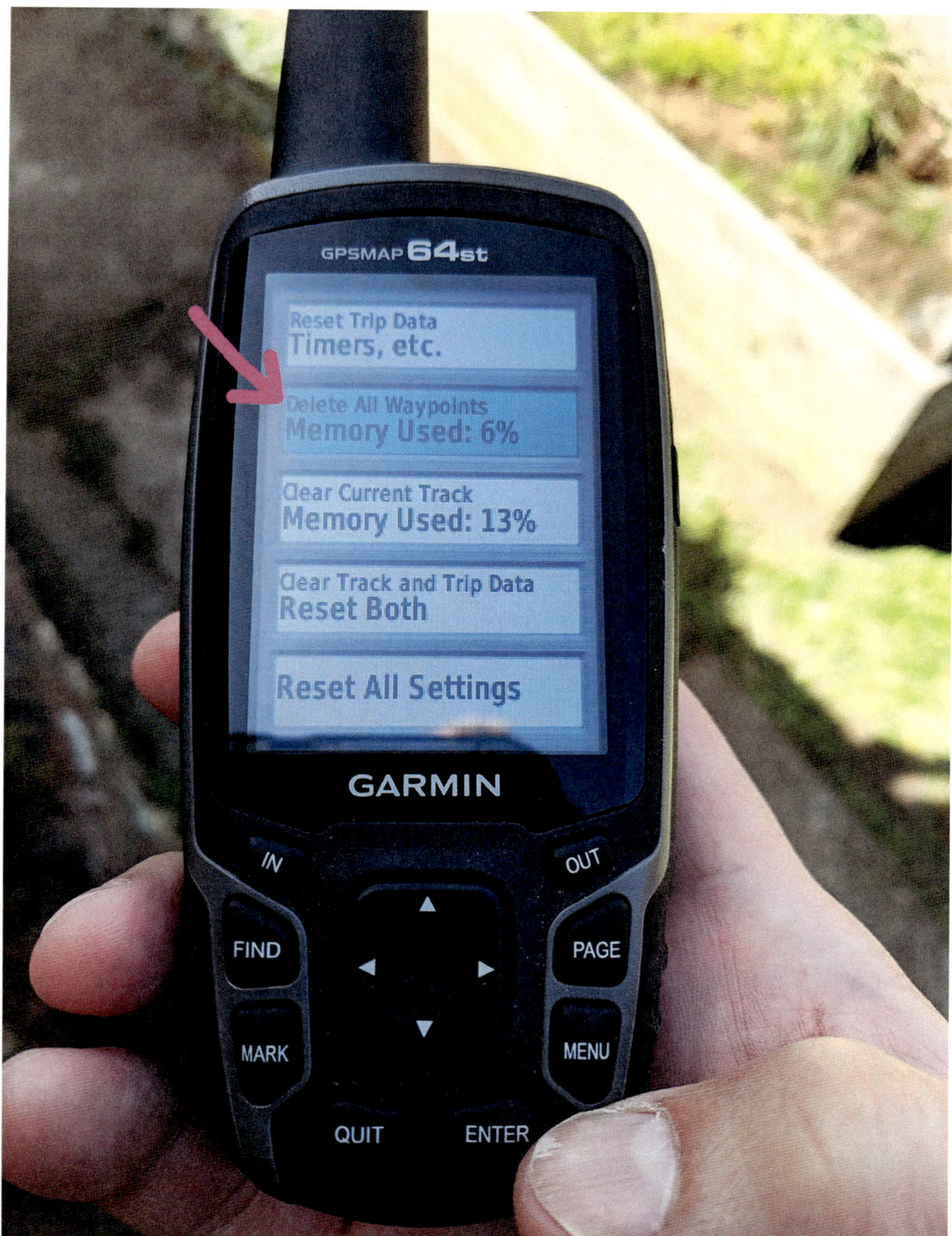

Figure 5.70: Deleting old waypoints saves memory and helps to avoid confusion.

Changing the Position Format

This tutorial uses a specific notation, degrees, and decimal minutes (**hddd°mmmmm'**). Use the *Quit* button to navigate back out to the *Main* menu. Locate and choose the *Position Format* option (**Figure 5.71**).

Figure 5.71: The position format options allow you to change the coordinate system notation.

Choose hddd°mmmmm' as the format (**Figure 5.72**).

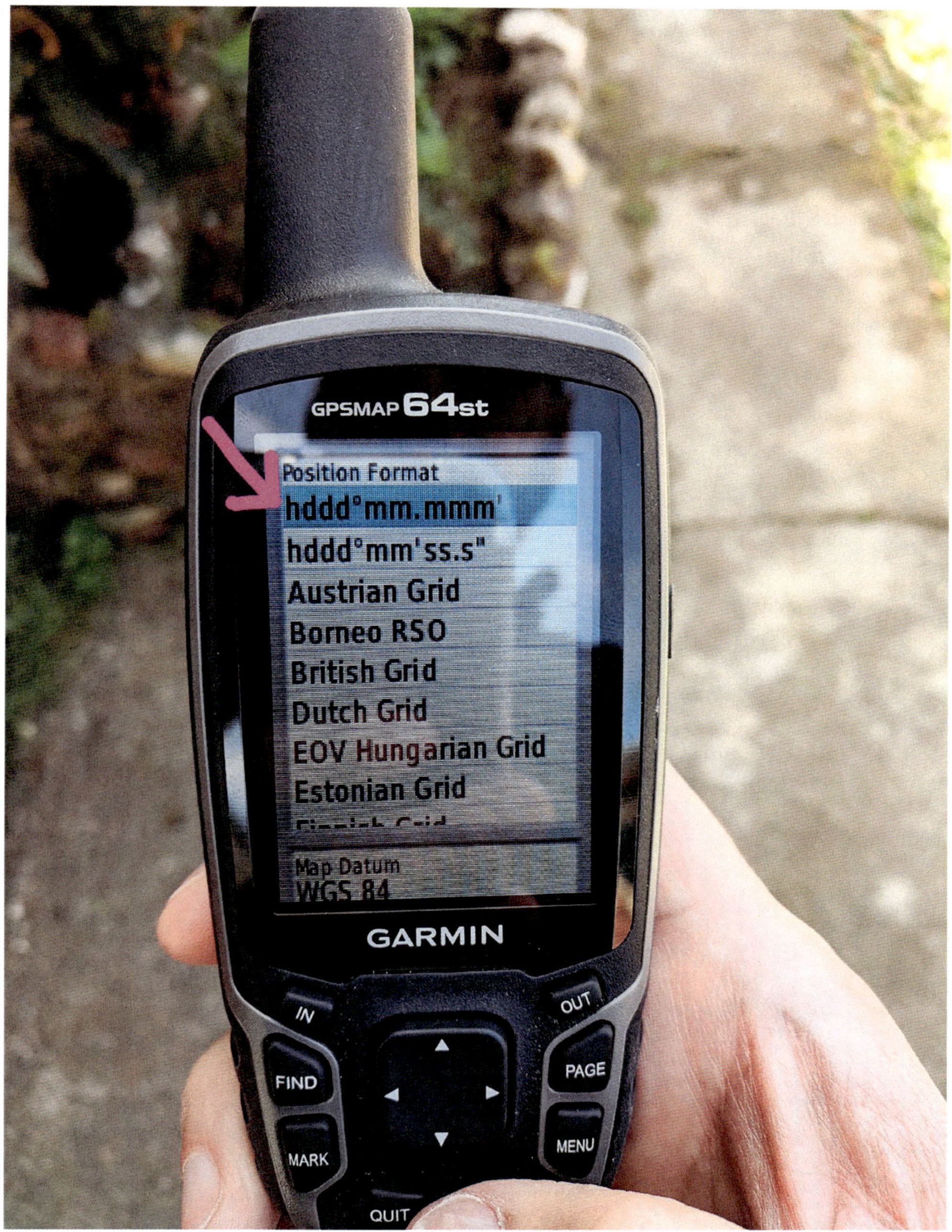

Figure 5.72: The Geocaching website uses degrees and decimal minutes for geocache coordinates.

Use the Quit button to navigate back out to the *Main* menu.

Improving accuracy

Some GPS receivers have access to the **Wide Area Augmentation System (WAAS)**. Enabling WAAS on your device helps to improve accuracy. On the Garmin GPSMAP 64, press *System* (**Figure 5.73**)

Figure 5.73: Locate the *System* option on the *Main* menu of your GPS receiver.

On the System menu, be sure that the WAAS setting is turned on (**Figure 5.74**).

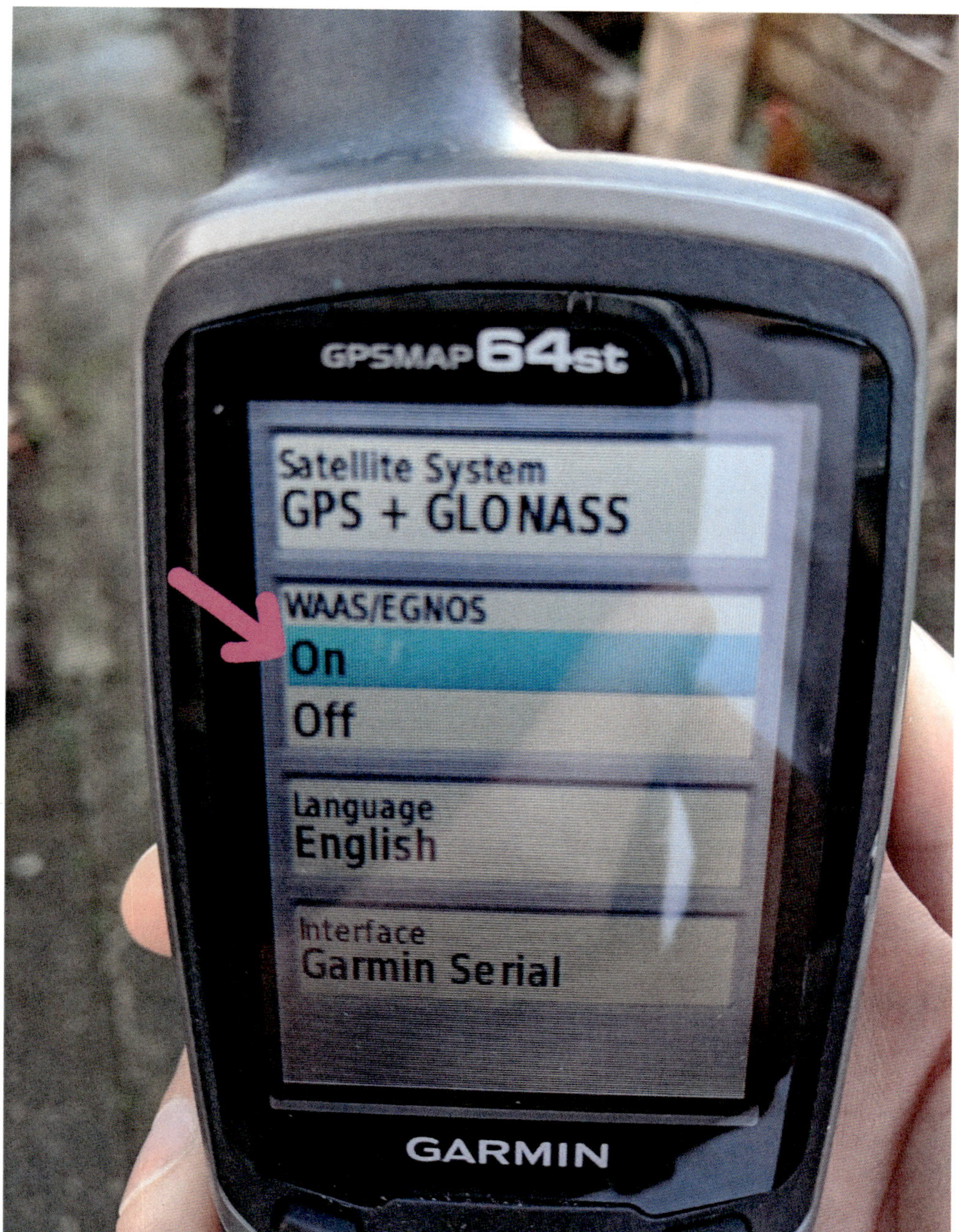

Figure 5.74: Enabling WAAS helps to increase the accuracy of the GPS receiver.

Marking Your Waypoints

You should be prepared to collect data for at least ten waypoints. Before starting your route, you should have the following items prepared:

- A noise level app or a reference chart for typical noise levels
- A map of your planned route
- A pad of paper and a pen for recording noise levels
  - Alternatively, you may use a smartphone application for taking notes, such as Google Keep.
- A GPS receiver
- The owner's manual for your GPS receiver

Walk to the first location on your route. Make a note of the decibel levels by using the smartphone app. If you don't have a smartphone, you should use your reference chart and make a judgment call. When ready, mark your waypoint. On the Garmin GPSMAP 64, press the *Mark* button (**Figure 5.75**). If you are using a different model, refer to the owner's manual to learn how to mark a waypoint.

Figure 5.75: Press the Mark button to mark a waypoint.

Editing Waypoints

For every waypoint you create, you need to edit the name so that it contains a letter. By default, many GPS receivers assign a series of numbers for waypoint names, such as 001, 002, 003. Unfortunately, this naming convention sometimes causes problems when working with GIS software. The GIS software needs to recognize the waypoint names as text, sometimes called a **string**, rather than regarding it as a number.

To ensure that the GIS software views the name as text, edit the name so that it includes a letter.

After marking a waypoint, the GPS receiver takes you to the waypoint edit screen (**Figure 5.76**).

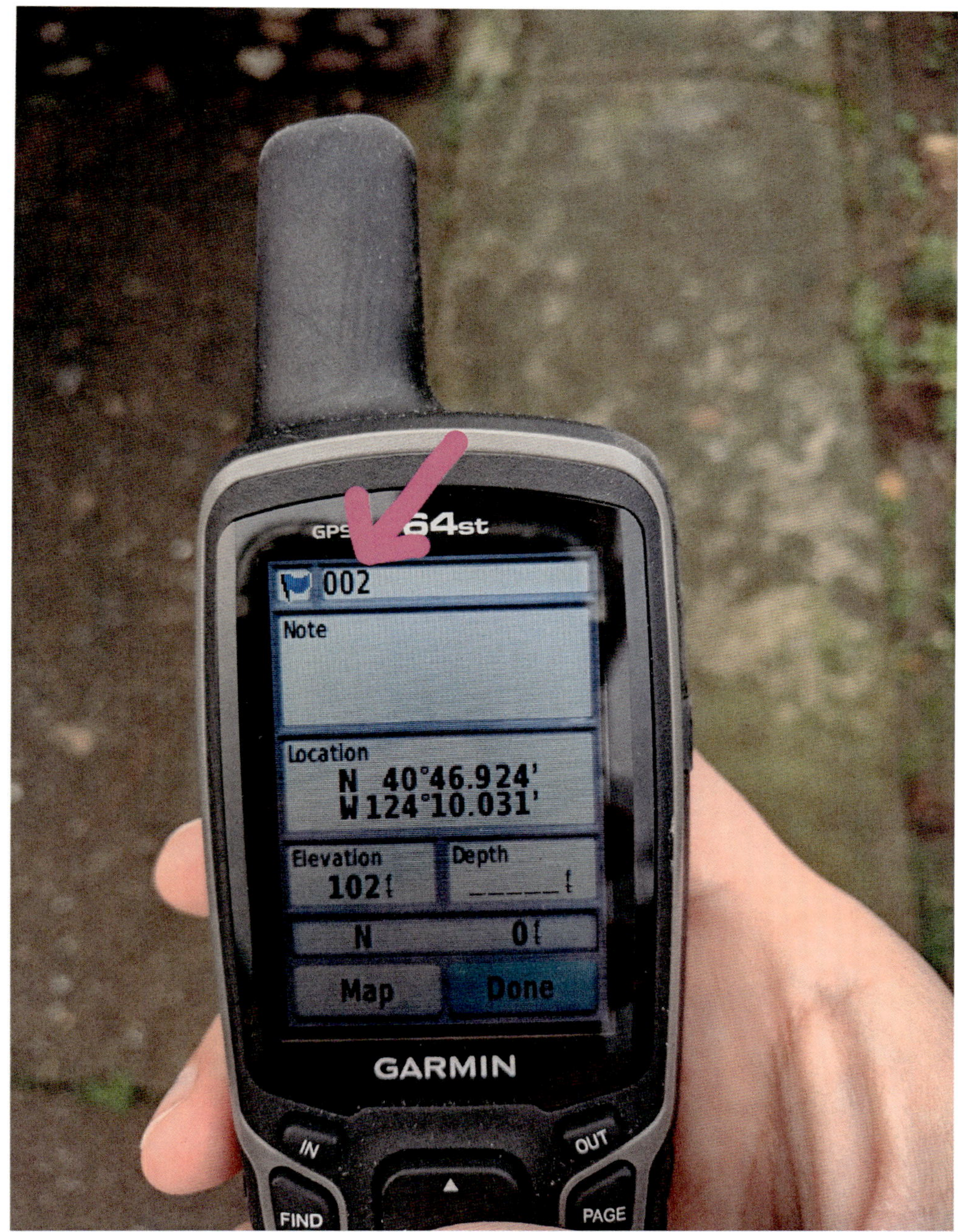

Figure 5.76: Locate the waypoint name at the top. By default, it is a number.

On the Garmin GPSMAP 64, use the rocker pad to select the name and press the Enter. On the edit screen, insert a letter A at the beginning of the name (**Figure 5.77**).

Don't skip this step, or it will cause headaches for you later!

Figure 5.77: The name 002 was changed to A02 to ensure that the GIS software regards the waypoint name as text.

When you are ready, press enter. Then, select Done to save the waypoint.

Either on a paper notepad or using a smartphone app such as OneNote, record the waypoint name and the decibel level at that location. Continue along your route and repeat these steps for at least nine additional waypoints.

Skill Drill: Setting up Your Workspace

When you return to your computer, create a workspace folder for your project on your **desktop**. You learned about creating a workspace in previous activities. Name folder "Noise_Map." Create your three standard subfolders inside, *original*, *working*, and *final*.

Downloading the DNRGPS Application

Navigate to the ***Minnesota Department of Natural Resources***[1] website to download their free GPS application, DNRGPS. You can find this site using a Google search with the keyword DNRGPS. Once there, download the latest version of the software.

> ***At the time this tutorial was published, it is DNRGPS for ArcMap 10.2+. Don't worry if you have a more recent version of ArcMap installed. This version should work just fine.***

Right-click on the link and choose "save link as," then navigate to your workspace folder and save the file (**Figure 5.78**).

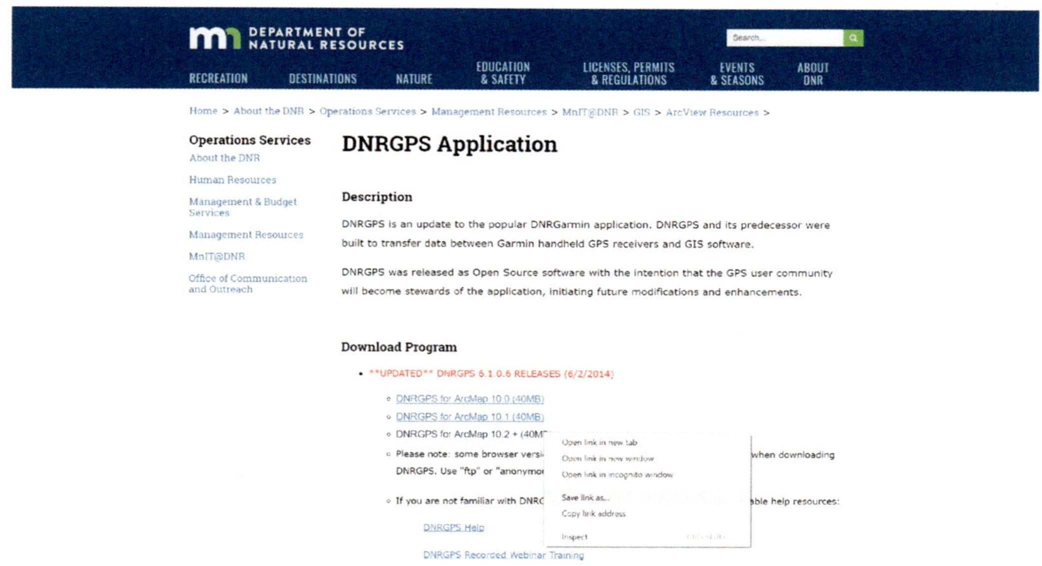

Figure 5.78: The Minnesota Department of Natural Resources provides a free GPS application.

1 https://www.dnr.state.mn.us/mis/gis/DNRGPS/DNRGPS.html

In Microsoft Windows, open your Noise Map workspace folder and locate the downloaded file. The file is compressed, so you need to use 7-zip to decompress it (**Figure 5.79**).

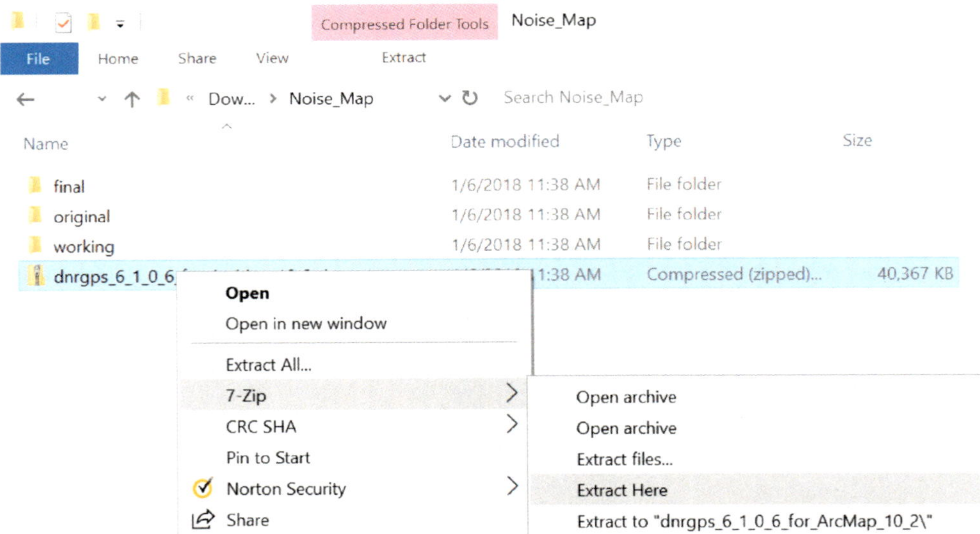

Figure 5.79: You can access the 7zip software by right-clicking on a compressed file.

When done, delete the compressed zip file. Your workspace folder should now contain your three standard subfolders plus a DNRGPS folder (**Figure 5.80**).

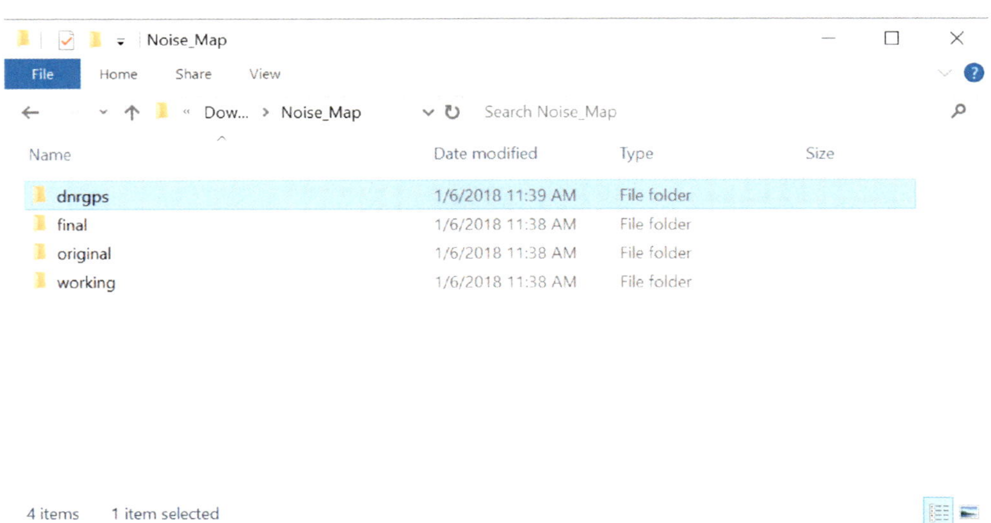

Figure 5.80: The DNRGPS folder can reside in the main workspace folder.

Downloading Your Waypoints

You do not need to install the DNRGPS application on your computer. You can run this application from any directory, even an external flash drive. Before launching the app, make sure you power down your Garmin and plug it into the computer using the USB cord (**Figure 5.81**).

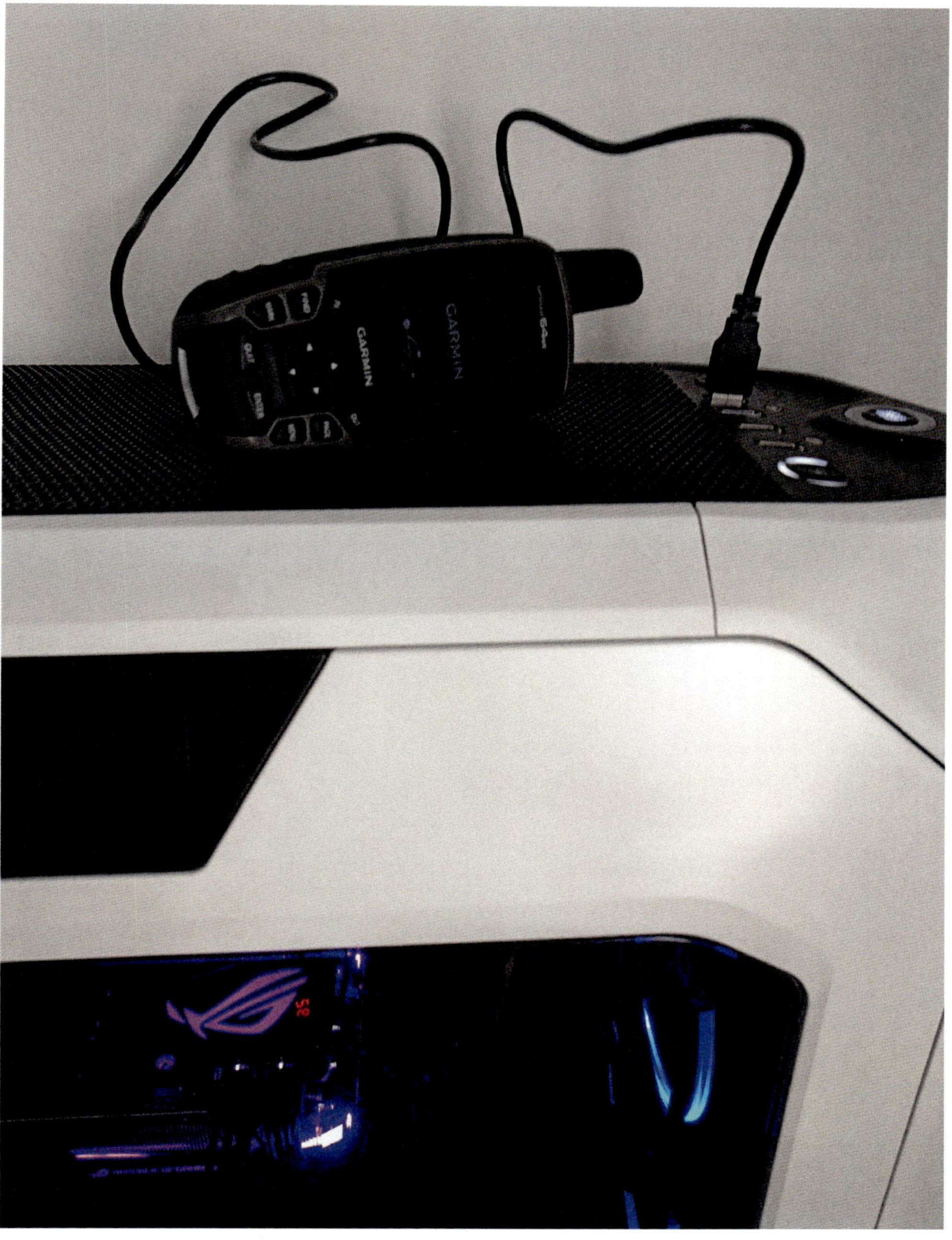

Figure 5.81: You may need to power down your GPS receiver before plugging it into your computer.

Wait a few seconds for the computer to recognize the device. Then launch the DNRGPS application by double-clicking the file "dnrgpgs.exe," located inside the DNRGPS folder (**Figure 5.82**).

Application Tools dnrgps

File Home Share View Manage

Noise... > dnrgps > Search dnrgps

Name	Date modified	Type	Size
documents	6/3/2014 2:46 PM	File folder	
GDAL	6/3/2014 2:46 PM	File folder	
gpsbabel	6/3/2014 2:46 PM	File folder	
dnrgps.exe	6/2/2014 7:09 AM	Application	605 KB
dnrgps.pdf	2/15/2012 3:32 PM	Adobe Acrobat Docu...	1,270 KB
DnrGps_ArcMap.dll	11/13/2013 2:44 PM	Application extension	55 KB
DnrGps_ArcMap_10_0.dll	11/13/2013 9:33 AM	Application extension	55 KB
DnrGps_ArcMap_10_1.dll	4/23/2014 3:31 PM	Application extension	50 KB
ExifLibrary.dll	11/15/2009 2:34 AM	Application extension	80 KB
FileGDBAPI.dll	10/2/2012 4:26 PM	Application extension	4,554 KB
gdal_csharp.dll	11/13/2013 8:45 AM	Application extension	78 KB
gdal110.dll	11/13/2013 8:41 AM	Application extension	9,917 KB
geos_c.dll	4/14/2013 11:41 PM	Application extension	1,580 KB
iconv.dll	10/24/2010 3:55 PM	Application extension	877 KB

31 items 1 item selected 604 KB

Figure 5.82: The dnrgps.exe file starts the app.

Once the application launches, select *Waypoints* from the menu across the top, then click *Download* (**Figure 5.83**)

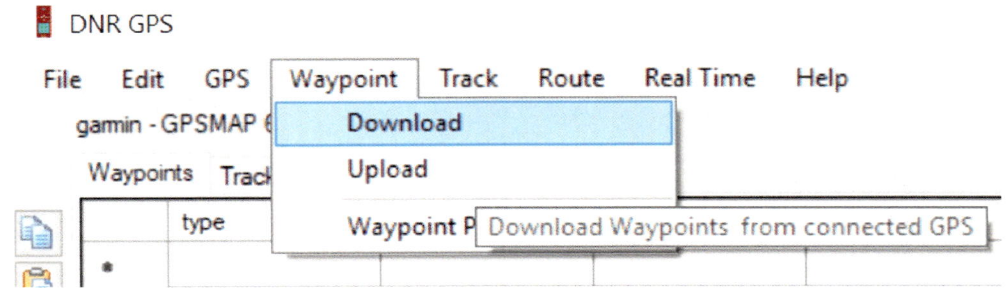

Figure 5.83: The Download option downloads waypoints from the connected GPS receiver.

If it does not appear, you may want to unplug the unit and make sure it is powered off. Then try plugging it in again. Wait a few moments and see if it shows up. Then close and re-open the DNRGPS application.

GPX files are commonly named using the word "waypoint" followed by a date. The Garmin GPSMAP 64 groups all waypoints collected on the same day into one GPX file. If you obtained your data over several days, select all of the files with matching dates

(**Figure 5.84**). You can choose multiple files by holding down the shift key. Once all of your GPX files are selected, click *OK*.

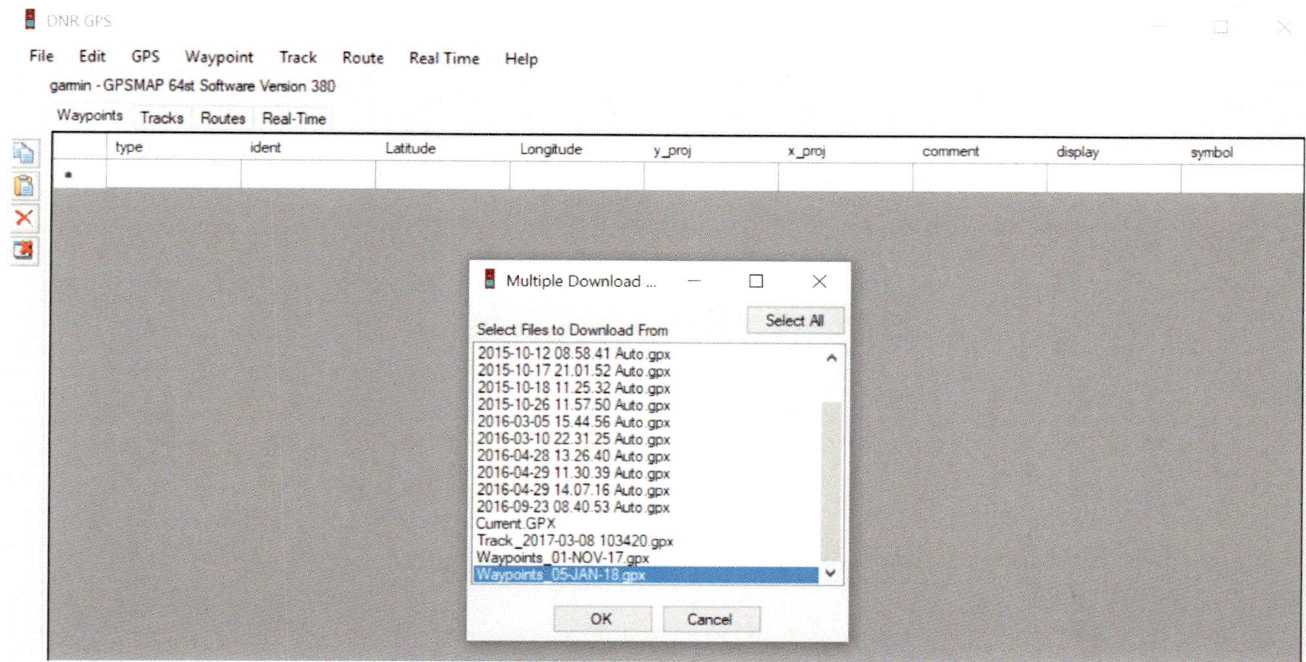

Figure 5.84: Use the dates to locate your waypoints on the list.

Now you should see a table displaying the contents of all of the GPX files downloaded from the GPS receiver. The next step is to save it as an Esri Shapefile. Select the File menu, *Save To*, ArcMap, then *File* (**Figure 5.85**)

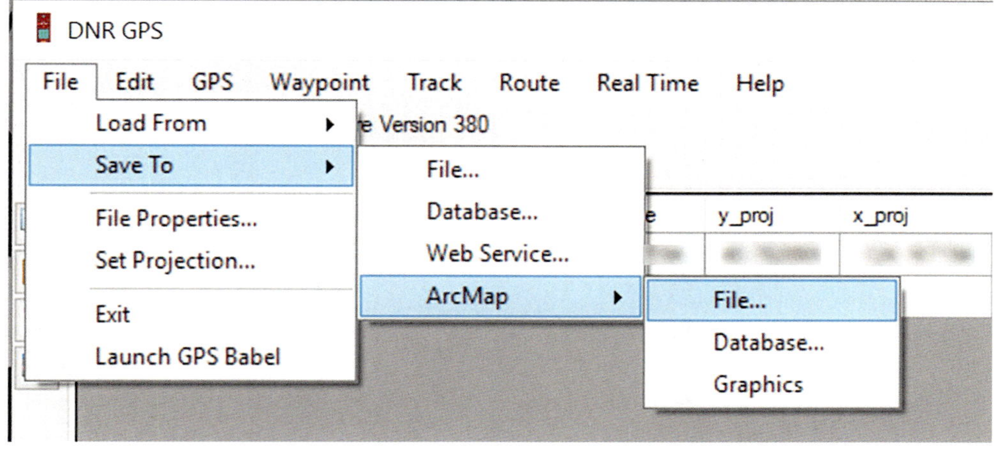

Figure 5.85: Be sure to select ArcMap from the *Save To* menu.

Navigate to your *working* folder. Save it as an Esri Shapefile (3D)(*.shp). You can name the file anything you like, but typically you want to give the file a descriptive name, such as noise_locations.shp (**Figure 5.86**).

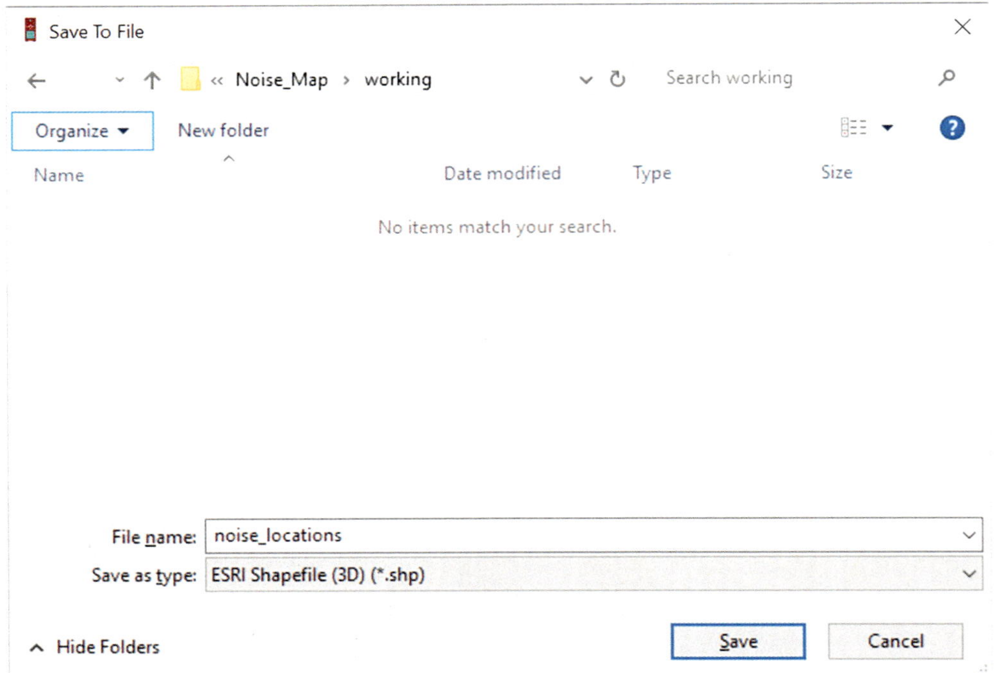

Figure 5.86: Save the waypoints to your *working* folder and provide a meaningful name.

When done, close the DNRGPS application.

Unplug the GPS receiver and make sure the unit is powered down to save battery life.

Skill Drill: Design a Basemap

To give your waypoints context, you need to download layers to create a basemap. If you collected your data on the Humboldt State University campus, the campus map provides an excellent background for your waypoints.

If you chose another study area, use what you learned in previous activities to locate background layers from other sources. Useful background layers include roads, parcels, and city boundaries.

Navigate to the HSU Data Hub on the ***Humboldt Geospatial Website***[2]. Use the keyword *campus* to locate the HSU campus data files. Right-click the *Download* link, select *Save link as*, then save the file to your *original* folder (**Figure 5.87**).

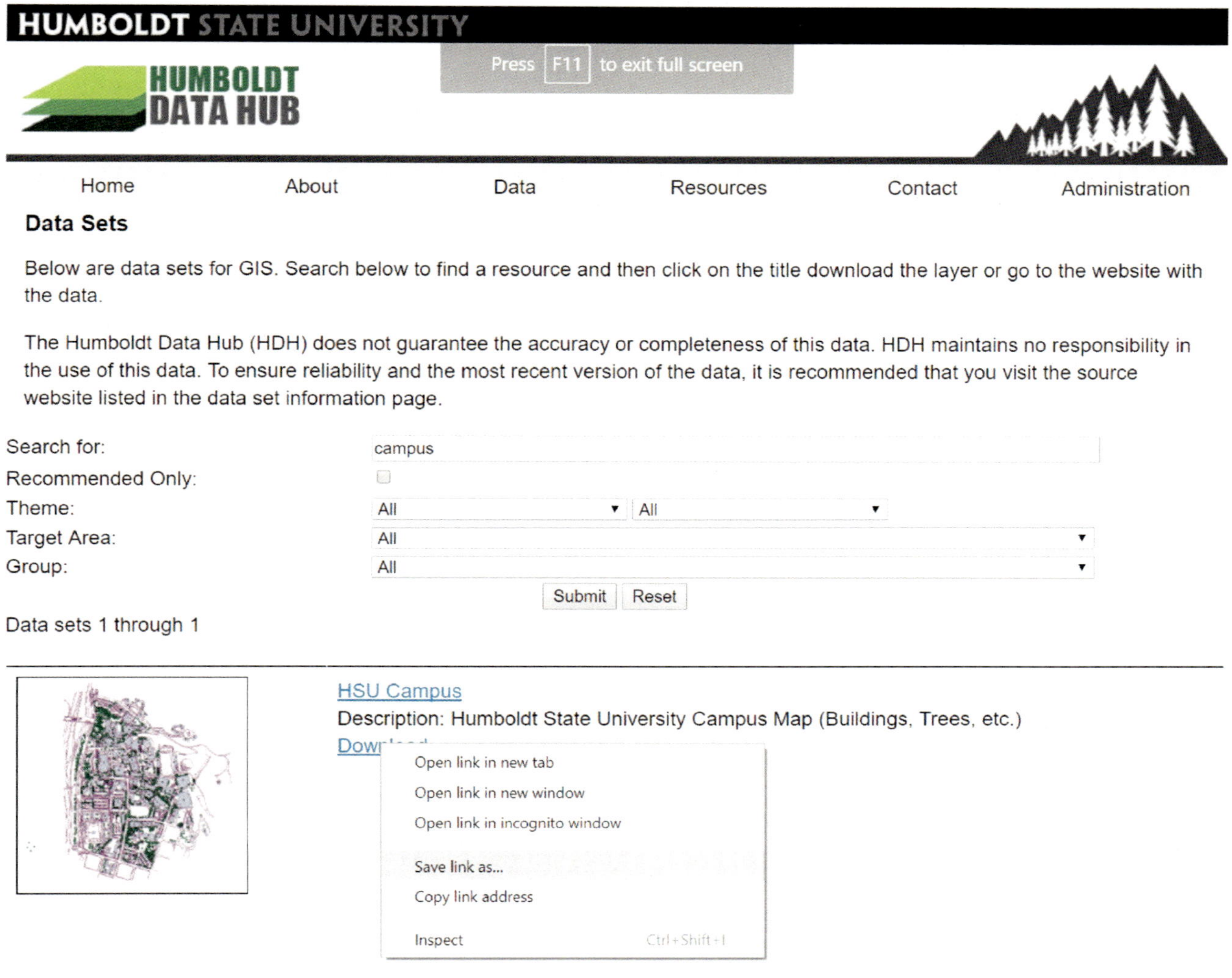

Figure 5.87: The Humboldt Data Hub contains many local datasets.

2 http://gsp.humboldt.edu/cwis438/Websites/HDH/DataSet_List.php

When done, open the *original* folder and decompress the file. You can delete the compressed zip file afterward. Eliminating zip files saves space and prevents confusion later. Launch ArcMap and open a blank map document. From your *original* folder, add the campus map and campus buildings shapefiles to the map. These layers serve as the basemap for the waypoints (**Figure 5.88**).

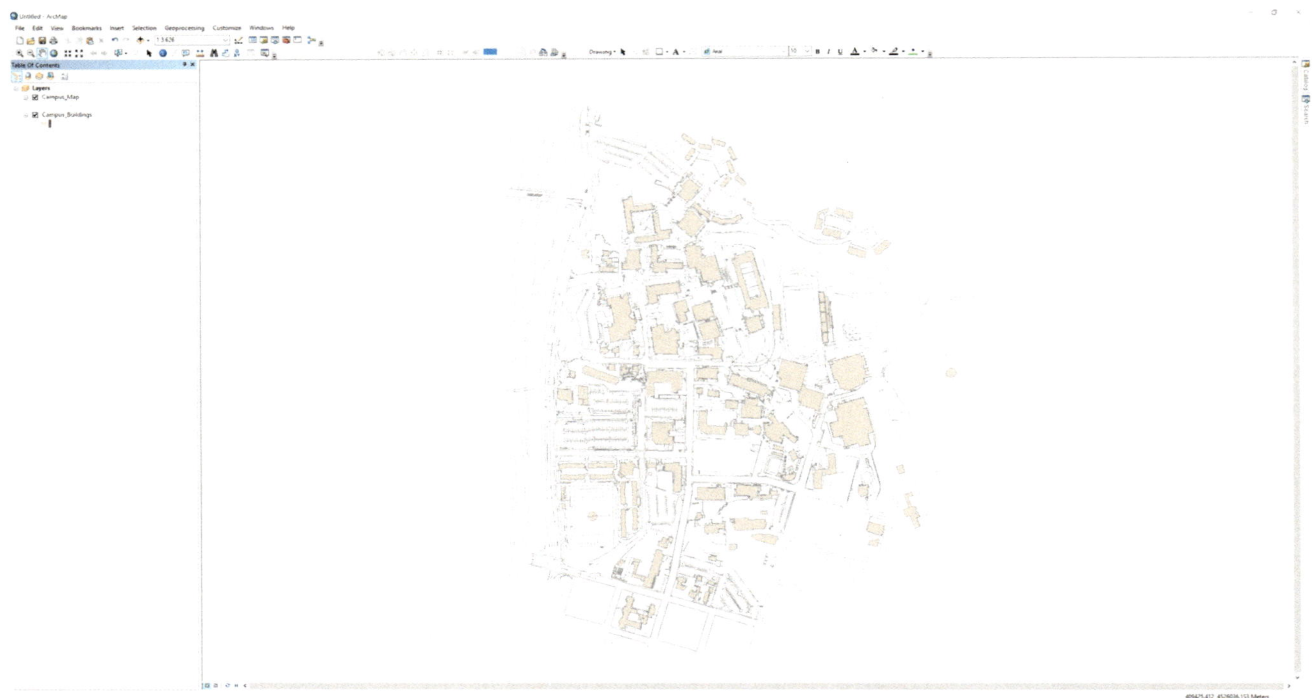

Figure 5.88: ArcMap chooses random colors. Be sure to change the colors to something neutral.

If you chose a different study area, add your local basemap layers instead. Useful background layers include roads, parcels, and city boundaries.

Once your basemap layers are ready, add the noise locations shapefile from the *working* folder. Look over your waypoints and check to see if they are close to the expected locations. Change the point symbol size and color as needed for clarity (**Figure 5.89**).

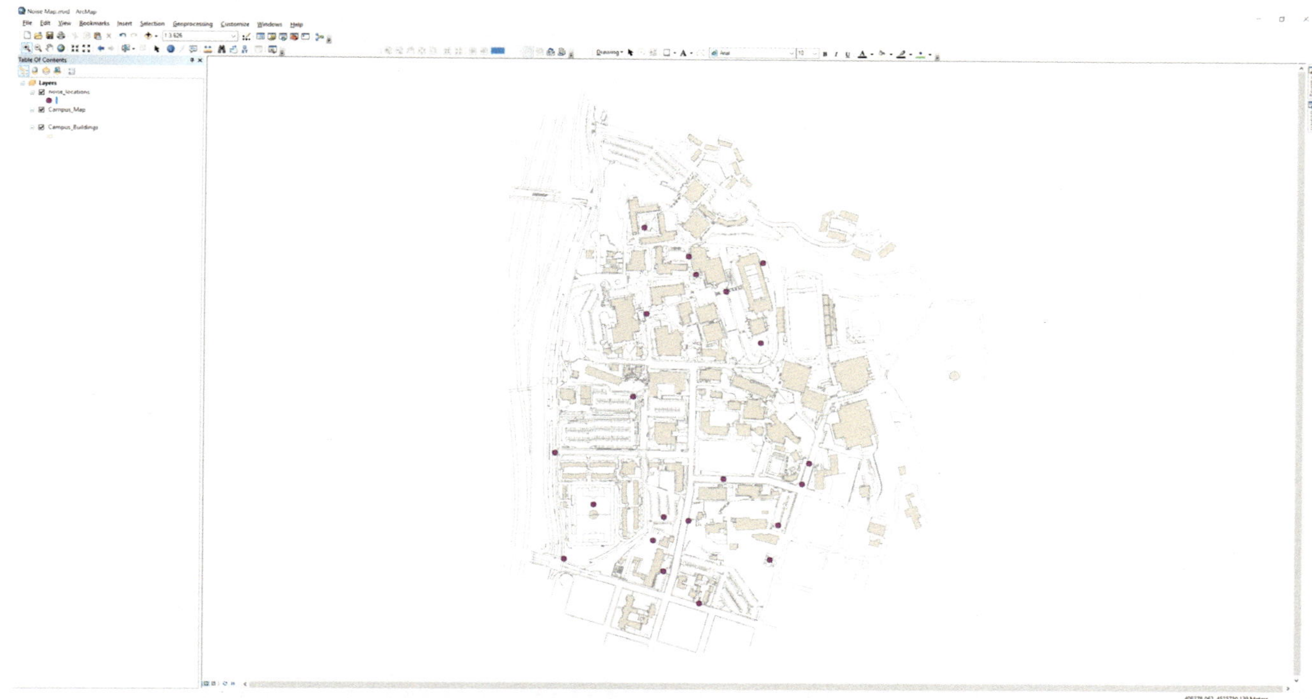

Figure 5.89: The waypoints were changed to a purple color to make them easier to see.

You may notice that some waypoints may be off due to GPS error. Correcting these waypoints is beyond the scope of this activity. For now, just let them serve as a reminder of the potential error and uncertainty present when collecting data using GPS.

Take a moment to save your map document. In a previous activity, you learn about setting the map document properties to store relative pathnames. Be sure to turn on this setting before saving. Give your map document a meaningful name, such as "Noise Map," and save the file to your workspace folder.

Performing a Table Join

Once done inspecting the accuracy of the waypoints, go ahead and open the attribute table by right-clicking on the noise locations layer and select *Open Attribute Table*. If necessary, dock the attribute table to the bottom of the map.

Take some time to look over the information stored in the attribute table. As you can see, the table stores some of the default attributes generated by the GPS receiver. These attributes include the latitude, longitude, altitude, and waypoint names. For the Garmin GPSMAP 64, you can find the waypoint names under a field called *ident* (**Figure 5.90**)

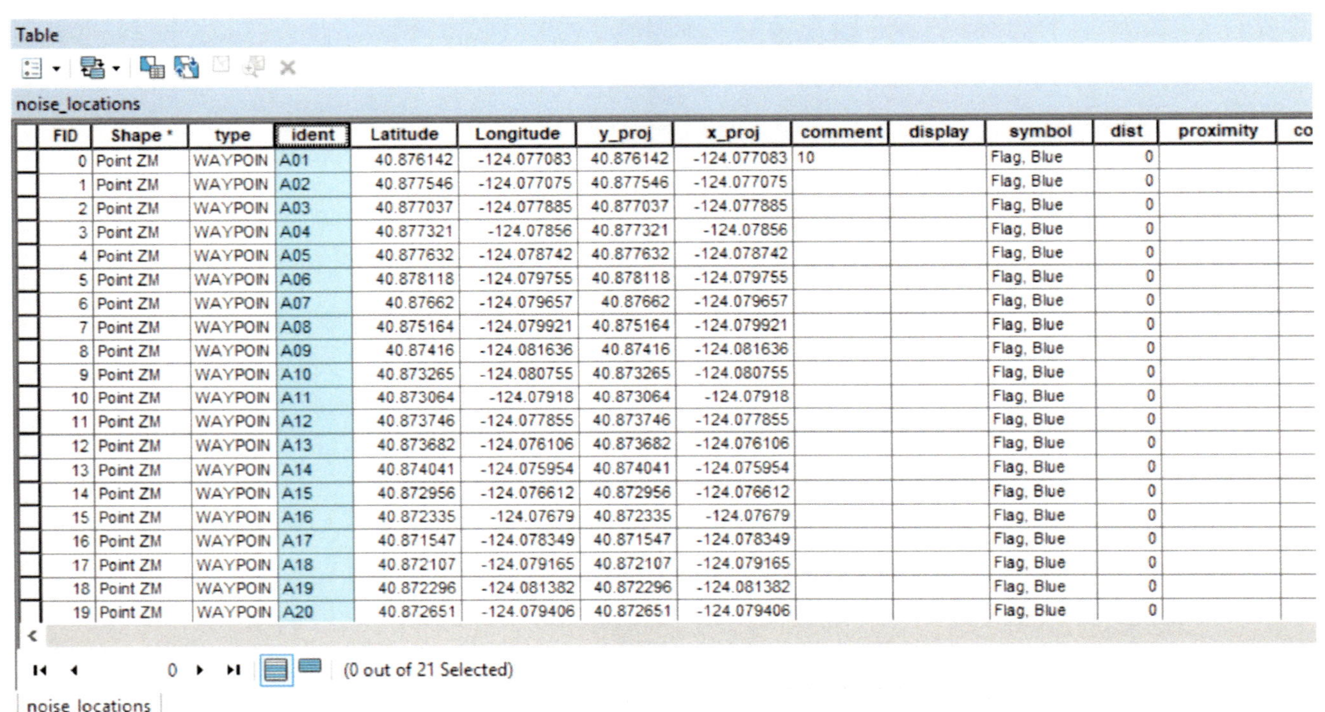

Table

noise_locations

FID	Shape *	type	ident	Latitude	Longitude	y_proj	x_proj	comment	display	symbol	dist	proximity	co
0	Point ZM	WAYPOIN	A01	40.876142	-124.077083	40.876142	-124.077083	10		Flag, Blue	0		
1	Point ZM	WAYPOIN	A02	40.877546	-124.077075	40.877546	-124.077075			Flag, Blue	0		
2	Point ZM	WAYPOIN	A03	40.877037	-124.077885	40.877037	-124.077885			Flag, Blue	0		
3	Point ZM	WAYPOIN	A04	40.877321	-124.07856	40.877321	-124.07856			Flag, Blue	0		
4	Point ZM	WAYPOIN	A05	40.877632	-124.078742	40.877632	-124.078742			Flag, Blue	0		
5	Point ZM	WAYPOIN	A06	40.878118	-124.079755	40.878118	-124.079755			Flag, Blue	0		
6	Point ZM	WAYPOIN	A07	40.87662	-124.079657	40.87662	-124.079657			Flag, Blue	0		
7	Point ZM	WAYPOIN	A08	40.875164	-124.079921	40.875164	-124.079921			Flag, Blue	0		
8	Point ZM	WAYPOIN	A09	40.87416	-124.081636	40.87416	-124.081636			Flag, Blue	0		
9	Point ZM	WAYPOIN	A10	40.873265	-124.080755	40.873265	-124.080755			Flag, Blue	0		
10	Point ZM	WAYPOIN	A11	40.873064	-124.07918	40.873064	-124.07918			Flag, Blue	0		
11	Point ZM	WAYPOIN	A12	40.873746	-124.077855	40.873746	-124.077855			Flag, Blue	0		
12	Point ZM	WAYPOIN	A13	40.873682	-124.076106	40.873682	-124.076106			Flag, Blue	0		
13	Point ZM	WAYPOIN	A14	40.874041	-124.075954	40.874041	-124.075954			Flag, Blue	0		
14	Point ZM	WAYPOIN	A15	40.872956	-124.076612	40.872956	-124.076612			Flag, Blue	0		
15	Point ZM	WAYPOIN	A16	40.872335	-124.07679	40.872335	-124.07679			Flag, Blue	0		
16	Point ZM	WAYPOIN	A17	40.871547	-124.078349	40.871547	-124.078349			Flag, Blue	0		
17	Point ZM	WAYPOIN	A18	40.872107	-124.079165	40.872107	-124.079165			Flag, Blue	0		
18	Point ZM	WAYPOIN	A19	40.872296	-124.081382	40.872296	-124.081382			Flag, Blue	0		
19	Point ZM	WAYPOIN	A20	40.872651	-124.079406	40.872651	-124.079406			Flag, Blue	0		

0 (0 out of 21 Selected)

noise_locations

Figure 5.90: The Garmin GPSMAP 64 uses the field name *ident* for storing waypoint names. Other GPS models might use a different field name.

A **table join** is an operation where tables from two different sources are joined together using a common attribute field. In this activity, you are going to base your table join using the name of the waypoints. The spelling and case must be a perfect match for the table join to work. To ensure that you do not type the names in incorrectly, copy and paste the attribute table into an excel spreadsheet. On the attribute table, click on the gray square on the left side of the top row (**Figure 5.91**).

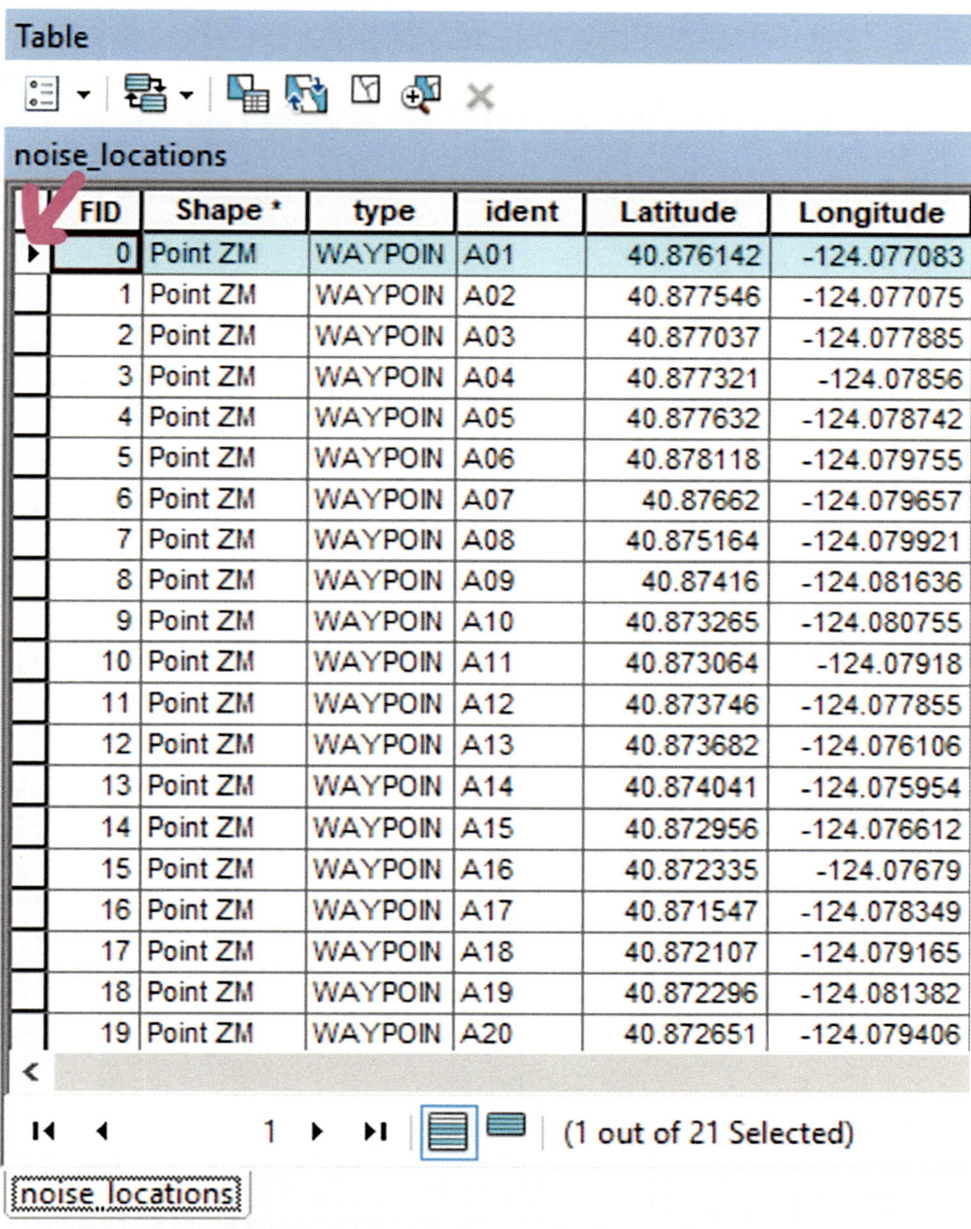

FID	Shape *	type	ident	Latitude	Longitude
0	Point ZM	WAYPOIN	A01	40.876142	-124.077083
1	Point ZM	WAYPOIN	A02	40.877546	-124.077075
2	Point ZM	WAYPOIN	A03	40.877037	-124.077885
3	Point ZM	WAYPOIN	A04	40.877321	-124.07856
4	Point ZM	WAYPOIN	A05	40.877632	-124.078742
5	Point ZM	WAYPOIN	A06	40.878118	-124.079755
6	Point ZM	WAYPOIN	A07	40.87662	-124.079657
7	Point ZM	WAYPOIN	A08	40.875164	-124.079921
8	Point ZM	WAYPOIN	A09	40.87416	-124.081636
9	Point ZM	WAYPOIN	A10	40.873265	-124.080755
10	Point ZM	WAYPOIN	A11	40.873064	-124.07918
11	Point ZM	WAYPOIN	A12	40.873746	-124.077855
12	Point ZM	WAYPOIN	A13	40.873682	-124.076106
13	Point ZM	WAYPOIN	A14	40.874041	-124.075954
14	Point ZM	WAYPOIN	A15	40.872956	-124.076612
15	Point ZM	WAYPOIN	A16	40.872335	-124.07679
16	Point ZM	WAYPOIN	A17	40.871547	-124.078349
17	Point ZM	WAYPOIN	A18	40.872107	-124.079165
18	Point ZM	WAYPOIN	A19	40.872296	-124.081382
19	Point ZM	WAYPOIN	A20	40.872651	-124.079406

Figure 5.91: The gray boxes on the left side of the attribute table allow you to select the record.

Then hold down the shift key and click the gray square on the left side of the bottom row to select all the records. Once the entire attribute table gets highlighted, right-click on any gray square and choose *copy selected* (**Figure 5.92**).

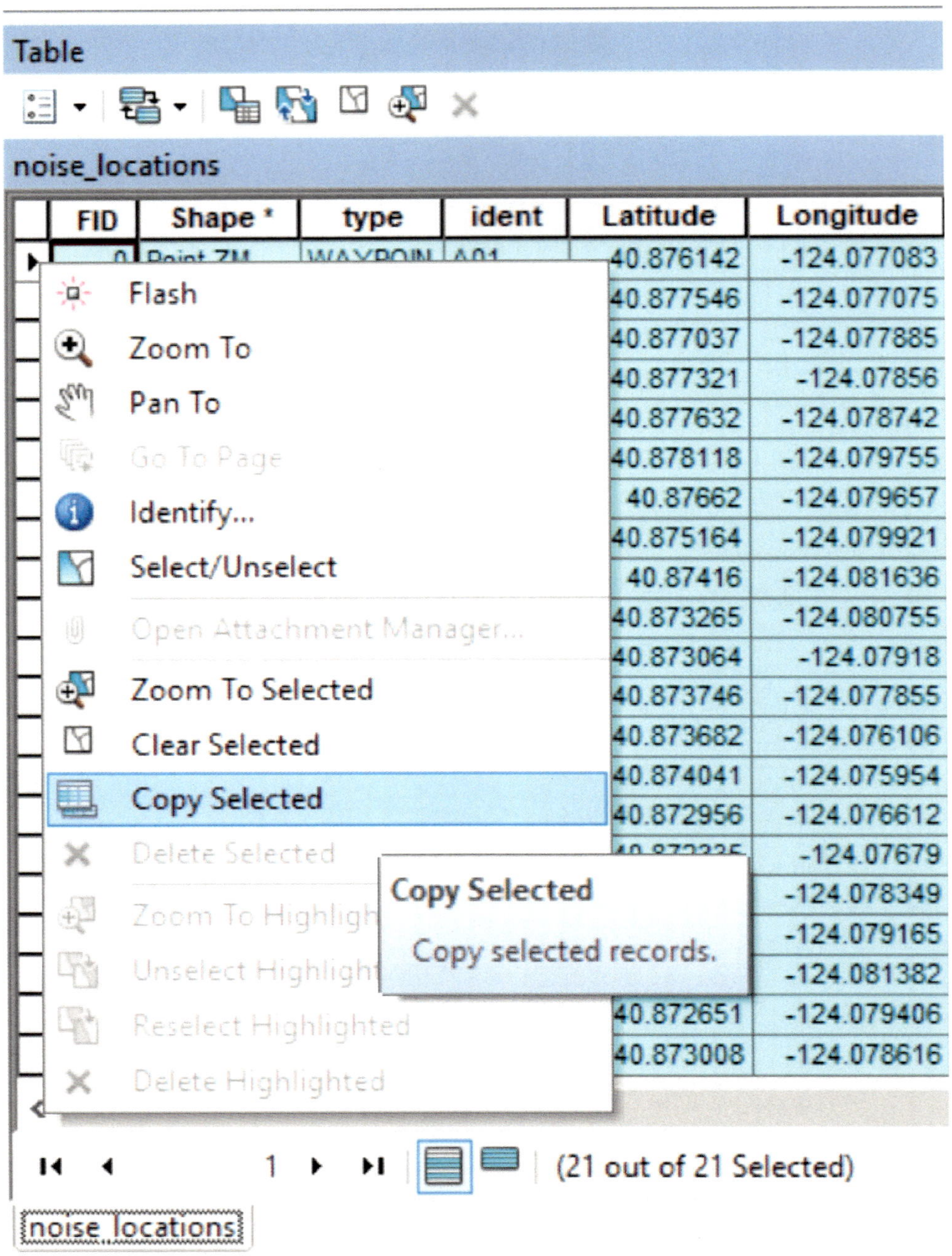

Figure 5.92: Right-clicking on the gray squares gives you access to contextual menus.

Open a blank workbook in Microsoft Excel. Right-click on cell A1 and select *paste*. You only need the names of the waypoints. You should delete all the other columns until just the *ident* column remains. Next to the *ident* field type in the header name *decibel*. Under the word decibel, type in the decibel values you recorded during data collection for each waypoint (**Figure 5.93**).

To ensure that ArcMap recognizes this field as numeric, do not type in the letters dB. Only type the number.

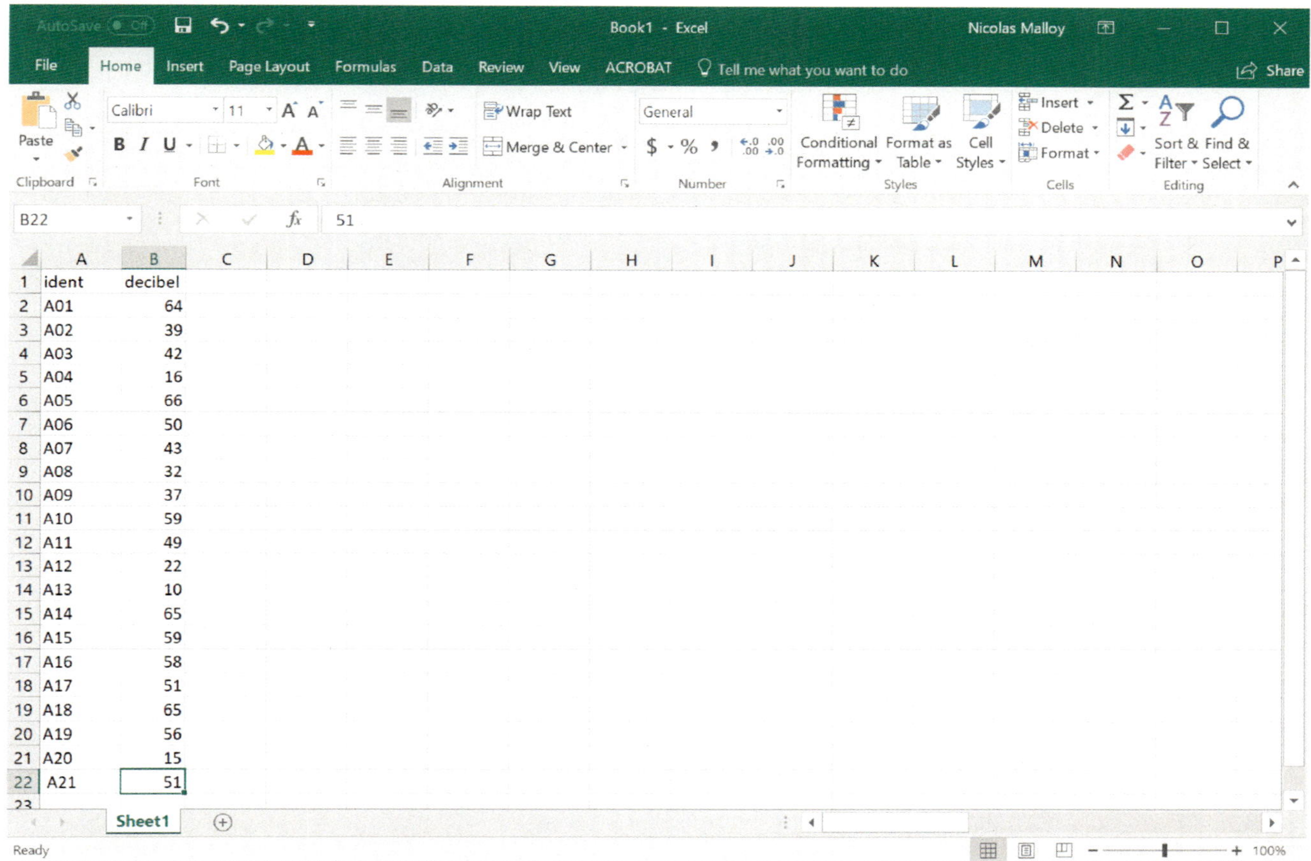

Figure 5.93: The decibel field only contains numbers. Adding letters to this field causes problems.

When ready, save the table as a CSV file (**Figure 5.94**). As you learned in a previous activity, a CSV is one of the simplest forms of geospatial data. When dealing with tables, it is a stable file format for GIS software. Name the file noise levels.csv. Be sure to *close* Microsoft Microsoft Excel after saving the file. ArcMap does not like to have data opened in two places at once. In ArcMap, clear the selected records. In the Table of Contents, right-click on the noise locations layer and select *Joins and Relates*, then *Join* (**Figure 5.95**).

The Join Data window opens up (**Figure 5.96**). Start by choosing which attribute field in the shapefile on which the table join is based. In this case, it is the *ident* field. Then, for step 2, click the yellow file folder icon and navigate to your *working* folder. Choose the noise levels CSV file and click *Add*. For step three, we must find the corresponding field in the CSV file. ArcMap does a pretty good job of finding this automatically, but occasionally you may need to locate it using the drop-down menu. In this case, the corresponding field is also called *ident*. Once you are ready, click *OK*.

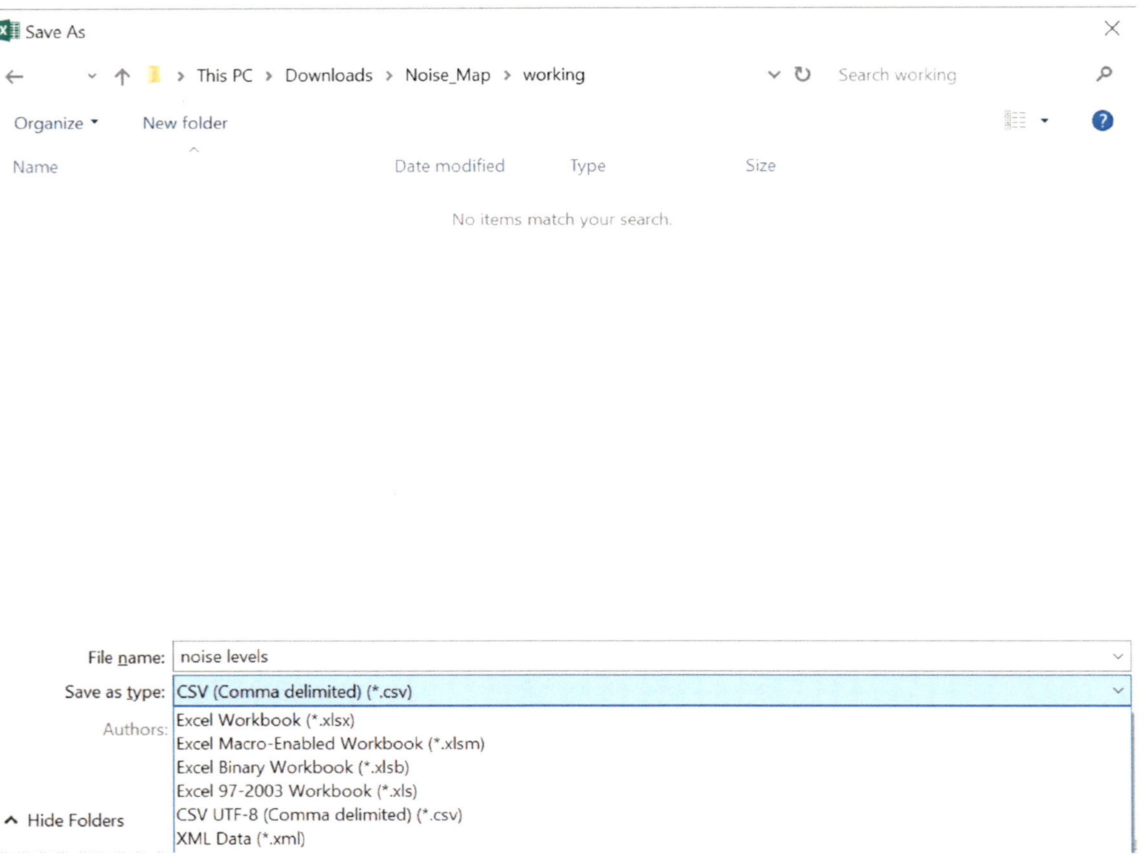

Figure 5.94: You can locate the CSV (Comma delimited) file format by using the drop-down menu under Save as type.

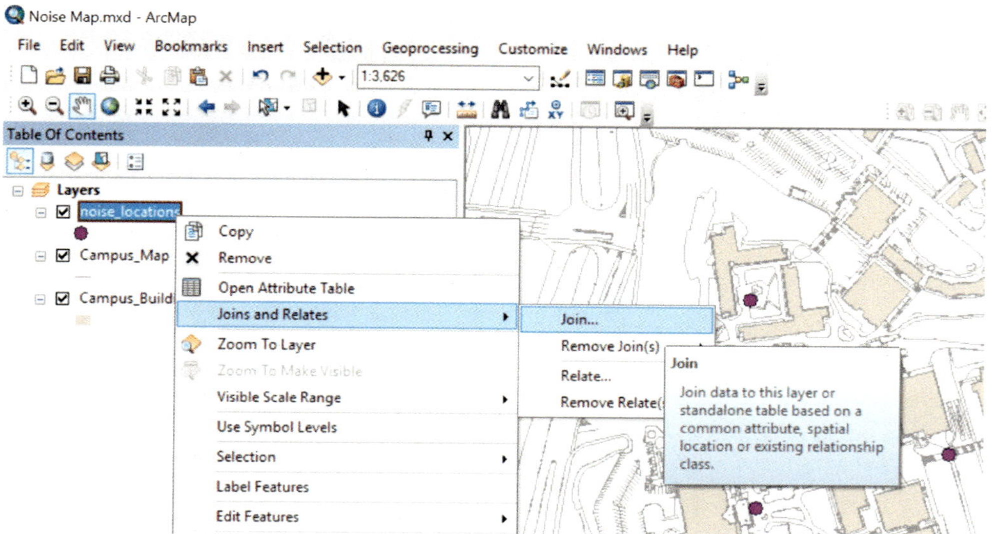

Figure 5.95: Be sure to right-click on the correct layer. In this example, it is the noise locations layer.

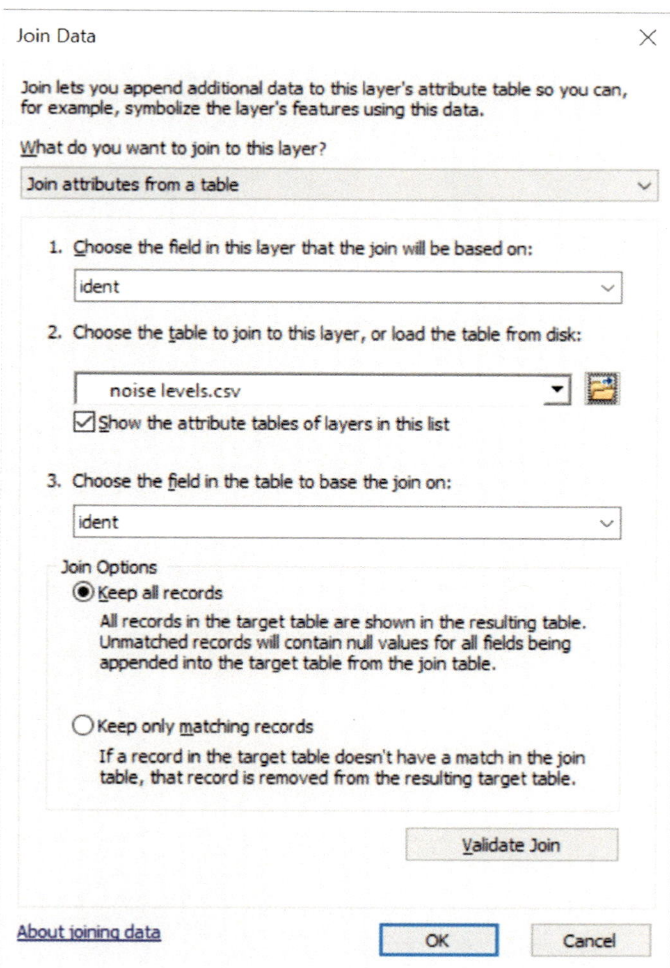

Figure 5.96: Your settings should appear similar to the image shown here.

It may look like nothing happened, but if you open the attribute table, you should see the data from the CSV table appended to the right of the attribute table (**Figure 5.97**). A table join is a temporary state. If you were to load this shapefile into a new map document, you would not see the tables joined together. To make a permanent change, export the data to a new shapefile. In the Table of Contents, right-click noise locations layer and select *Data*, then *Export Data* (**Figure 5.98**) Save the shapefile to your *working* folder and name it decibel_levels.shp. Add the exported data as a map layer.

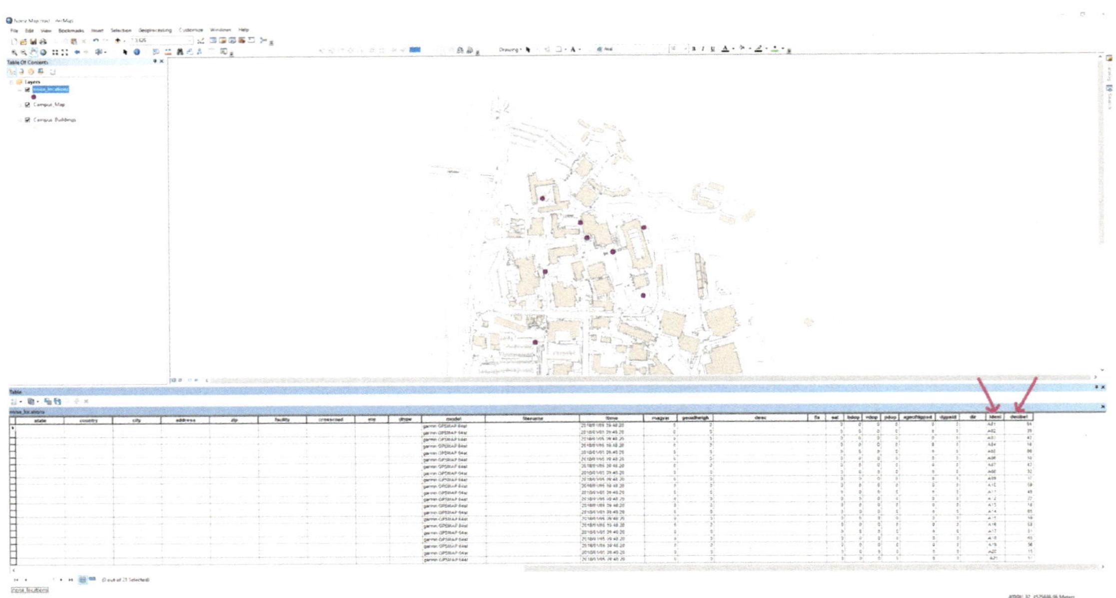

Figure 5.97: The CSV table is attached to the right side.

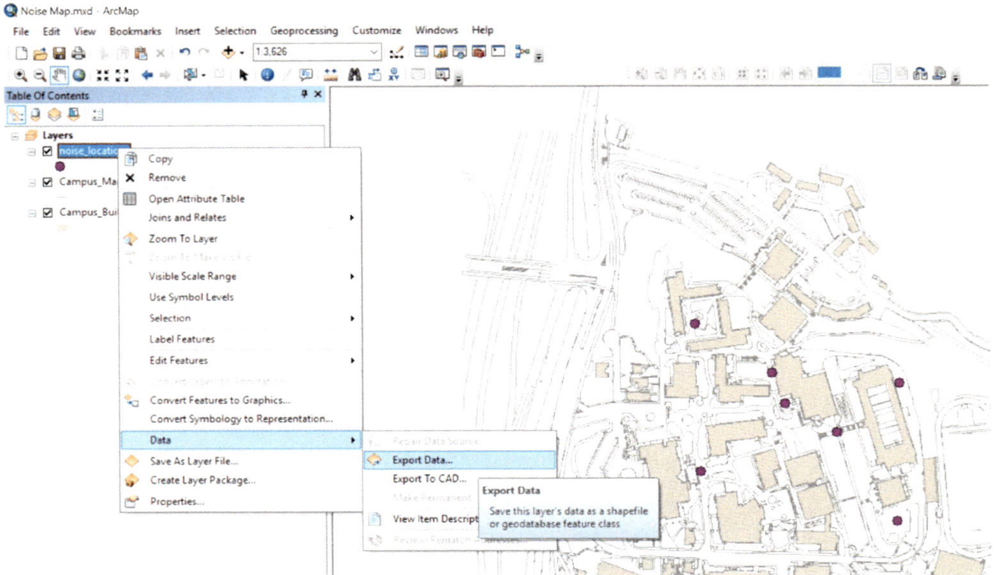

Figure 5.98: You must export the layer with the table join to make it permanent.

When done, remove the original noise locations layer from the table of contents. You should now have three layers in your Table of Contents, the decibel levels, the campus map, and the campus buildings.

Creating a Small-sized Map

Currently, your map only displays the locations of the waypoints. You can provide additional information to your map reader by showing the relative loudness at each site. In a previous activity, you learned how to represent data using graduated symbols. In this activity, you perform similar steps. Using what you learned previously, change the map symbology so that graduated symbols represent the relative loudness at each location (**Figure 5.99**). In this instance, use the *decibel* field for the *Value*. You may leave all other default settings.

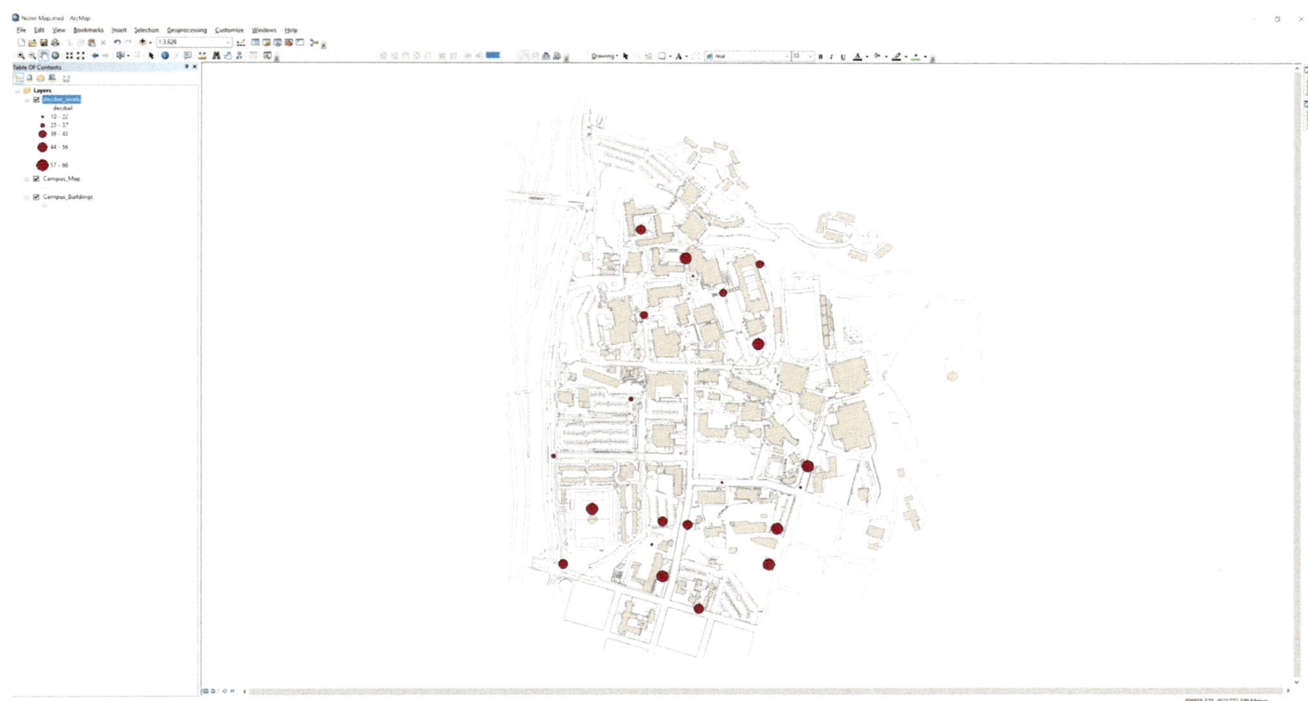

Figure 5.99: Graduate symbols represent each waypoint based on decibel levels.

If you forgot how to represent data using graduated symbols, refer to the Chapter 2 activity, Mapping Earthquakes in Northern California.

Using what you learned in previous activities, create a small-sized map of 6 by 6 inches (**Figure 5.100**). The purpose of this map is for use as a figure in a lab report. When creating small-sized maps, it is necessary to consider the design limitations. They will be more restricted than when you built poster-sized maps. Due to the limited size, you won't have much room with which to work. Only include a **north arrow**, a **scale bar**, and a **legend**. You do *not* need to add a map title directly on the map. Instead, plan to insert a **figure caption** in Microsoft Word as part of a lab report. The caption can take the place of both descriptive text and a map title. For this activity, start by switching to layout view in ArcMap. Then, change the paper size and data frame size to 6 by 6 inches. When the map is the correct size, begin adjusting colors and inserting your map elements.

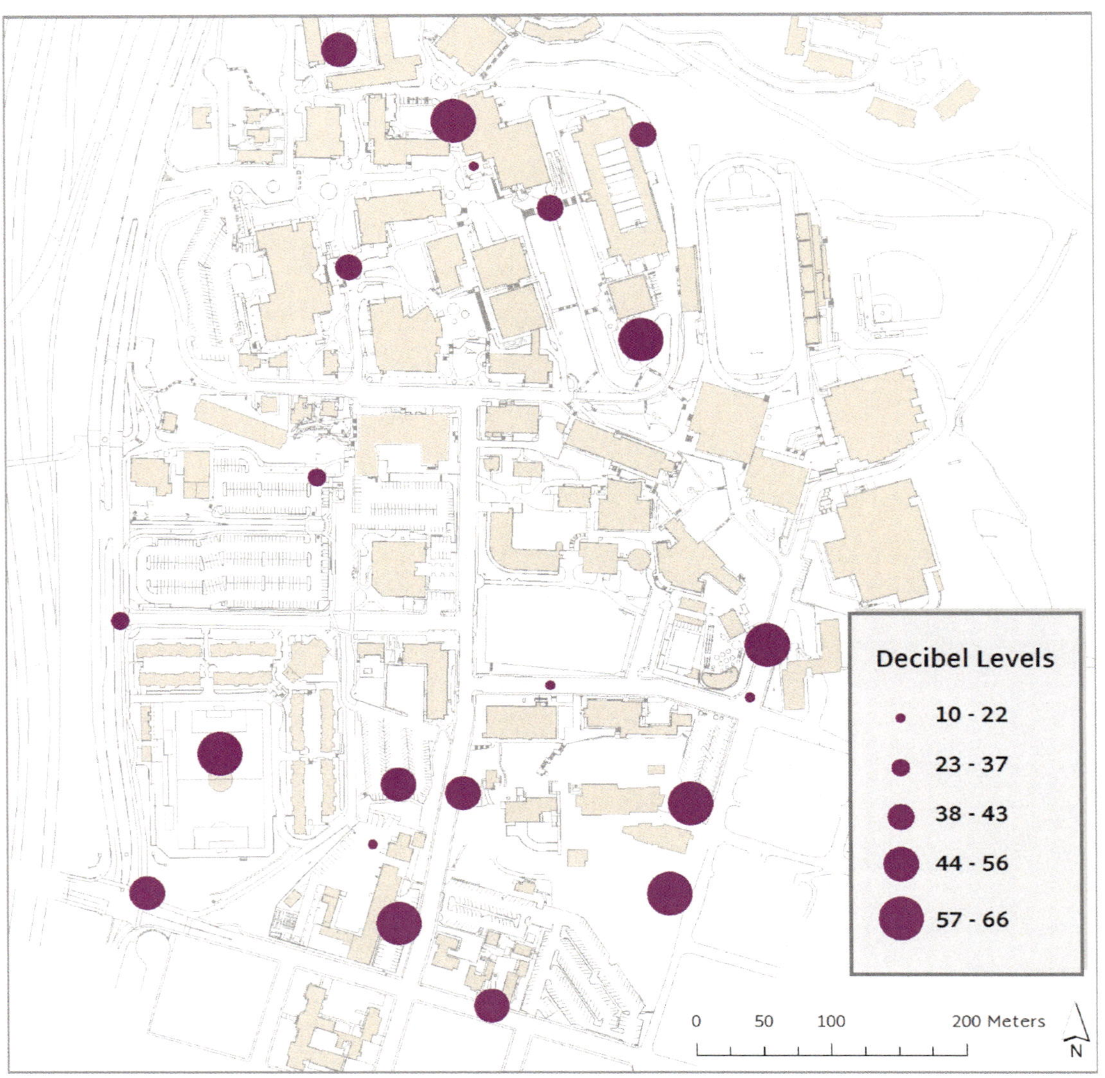

Figure 5.100: This small-sized map was designed at a size of six by six inches before exporting.

Always design and export your map at the correct size for the intended purpose. Never scale down your map after the fact.

When done creating your small-sized map, export the map as a **PNG** file, and save the file to your *final* folder. ArcMap may try to default to a low resolution. Remember to change the image resolution to at least **300 dpi** (**Figure 5.101**).

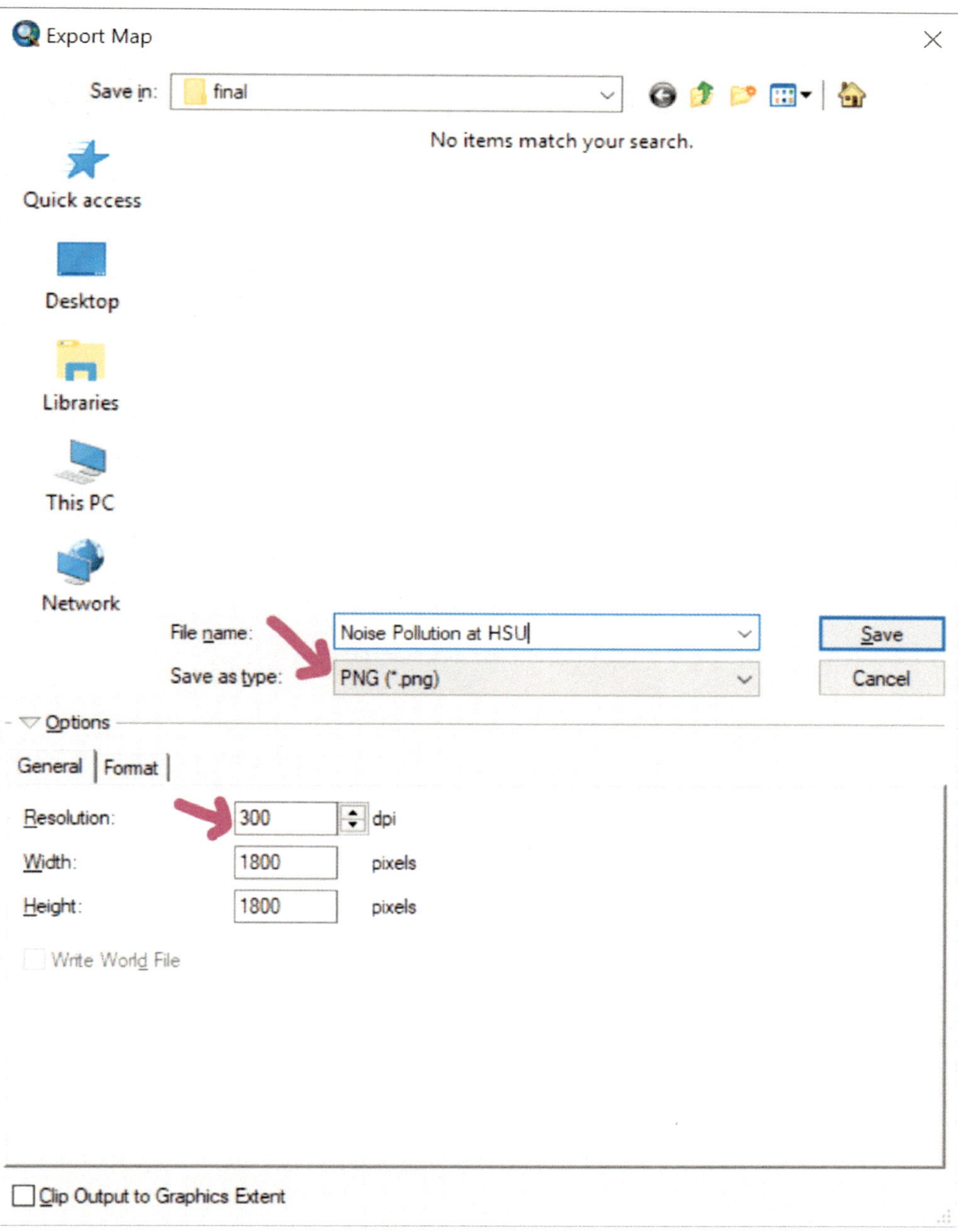

Figure 5.101: For the best results, use the PNG file format and a resolution of at least 300 dpi.

Don't forget to save your map document and back up your workspace folder to Google Drive.

Principal Terms

adjusting key
agonic line
atmospheric error
azimuth
back azimuth
back bearing
bearing
bezel
clock error
closed traverse
control point
control segment
declination diagram
differential correction
differential GPS (DGPS)
directional baseline
electronic distance measurement (EDM)
ephemeris bias
geocaching
geographic north
geometric dilution of precision (GDOP)
geostationary satellites
global navigation satellite systems (GNSS)
Global Positioning System (GPS)
GPS augmentation
GPS time
index pointer
inertial navigation systems (INS)
isogonic lines
isogonic map
lanyard
magnetic declination
magnetic north
mapping-grade GPS
multipath error
Navigation System with Time and Ranging (NAVSTAR) Global Positioning System
open traverse
orienteering arrow
post-mission differential GPS (PMDGPS)
pseudorandom noise code
real-time differential GPS (RTDGPS) correction
receiver clock error
resection
rotating magnetic needle
rovers
satellite ranging
selective availability (SA)
signal masking
space segment
space trilateration
string
survey-grade GPS
survey mark
table join
theodolite
total error budget
total station
transit
traverse
triangulation
trilateration
triple-legged walk
true north
user equivalent range errors (UERE)
user segment
WAAS master station (WMS)
wide area augmentation system (WAAS)
Wide Area Augmentation System (WAAS)

Chapter 6: Image Acquisition and Interpretation

It is possible that many, if not most, readers were not alive before digital aerial imagery became commonplace. Most also have little direct experience with film-based photographs. Today, anyone with an internet connection and a web browser can view images from aircraft and space satellites. With imagery so commonplace and accessible, many might take it for granted. However, there are still new frontiers emerging in the collection, application, and processing of images. The scientific and educational potential of civilian-operated unmanned aircraft systems (UAS) is just one. Chapter 6 presents the phenomenon, concepts, equipment, and methods behind the science of Remote Sensing.

Learning Outcomes

Readers should be able to accomplish the following outcomes by the end of this chapter:

- » Describe the basic principles of electromagnetic energy
- » Summarize spectral bands and their mapping applications
- » Identify different types of imagery resolution
- » Compare the differences between aerial photographs and digital imagery
- » Discuss active and passive sensing
- » List remote sensing platforms
- » Recall historical and current national and commercial programs for image acquisition
- » Explain the principles of image correction
- » Apply image enhancement techniques.
- » Articulate the fundamentals of image interpretation and classification

Electromagnetic Spectrum

Remote sensing is the geospatial science related to obtaining data from a distance using devices that detect emitted or reflected electromagnetic energy. Chapter 1 identified remote sensing as a primary geospatial science because remote sensing applications produce raw geospatial data that are subsequently used in other geospatial sciences. However, recent developments in the field of remote sensing have broadened its scope so that it is more closely enmeshed with geographic information systems (GIS) and mobile mapping technology.

Electromagnetic energy[1] is a form of energy emitted by all matter above absolute zero temperature[2]. Objects can also reflect electromagnetic energy emitted by other objects. Visible light is a familiar form of electromagnetic energy. However, visible light is only a minuscule region of the range of electromagnetic energy called the **electromagnetic spectrum**. Electromagnetic energy travels in waves. Geospatial scientists define it by its wavelength or frequency. A **wavelength** is a distance, measured in nanometers (nm), from one wave peak to the next. The **frequency** is the number of wave peaks passing a fixed point in space per unit time, such as peaks per second. Though scientists assign nominal classifications to narrow regions of the electromagnetic spectrum, called **spectral bands**, it is a continuous phenomenon without clear dividing lines. Nominal classifications, however, do make it convenient with which to work. The following are the broad classifications discussed in this chapter:

- Visible light
    - Blue
    - Green
    - Red
- Infrared
    - Thermal IR
    - Near-IR
- Microwave

1 This term is also referred to as electromagnetic radiation.

2 Absolute zero is 0 Kelvin. To learn more about the Kelvin scale watch *Absolute temperature and the kelvin scale* by the Khan Academy URL: https://youtu.be/eEJqaNaq9v8.

Watch this video By Science at NASA to learn more (**Figure 6.01**).

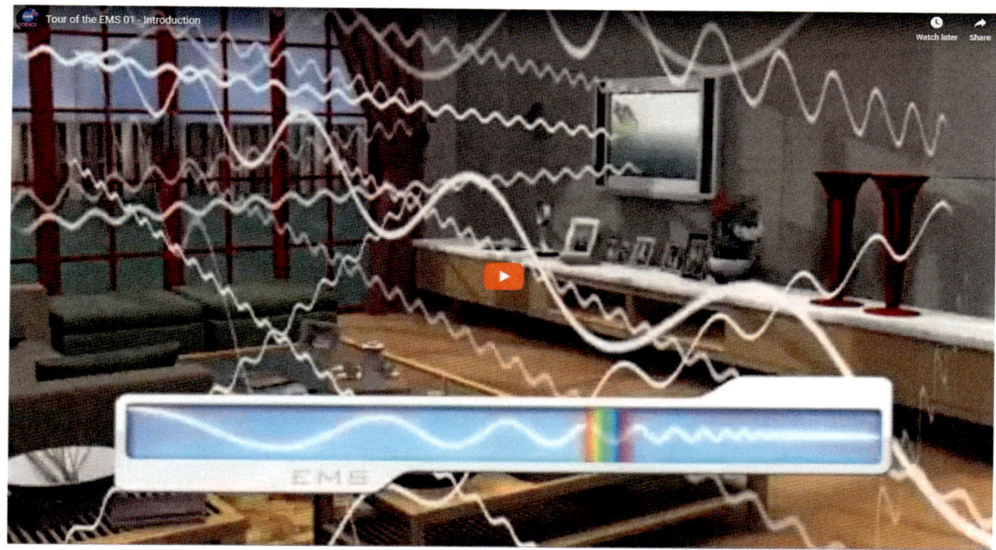

Figure 6.01: This video explains the concepts related to the electromagnetic spectrum. URL: *https://youtu.be/lwfJPc-rSXw*

Passive and Active Sensing

Both passive and active sensing systems are used to collect geospatial data in the visible light portion of the electromagnetic spectrum. **Passive sensing systems** detect naturally emitted or reflected energy (**Figure 6.02**).

Figure 6.02: A photograph taken from aircraft or satellites is a form of passive sensing systems. Source: NASA.

Active sensing systems is a term used in remote sensing when a type of sensor emits its an energy source and records the reflection of that same energy. **Light detection and ranging (lidar)** is an example of an active sensing system using visible light. Lidar uses a laser pulse emitter mounted on aircraft along with the GPS receiver to record the aircraft's vertical and horizontal position. Rapid pulses of laser light strike the surface of the earth and are reflected back to a sensor on the laser pulse emitter. The travel time of the laser pulses indicates distance. The result is a dense point cloud of lidar returns signals. Point clouds are processed to produce high-resolution digital elevation models (DEM) of ground features such as terrain, forest canopy, or man-made structures.

Visible Light

Visible light ranges from 400 nm to 700 nm and is the portion of the electromagnetic spectrum that humans can see. Receptors in the human eye, called cones, perceive color in the visible part of the electromagnetic spectrum. There are three types of cones, and each cone responds to one of three spectral bands, blue (400 nm–500 nm), green (500 nm–600 nm), and red (600 nm–700 nm) (**Figure 6.03**). The pattern in which these three types of cones respond is how the human eye perceives the full range of colors in the visible spectrum.

Figure 6.03: The human eye perceives the full range of colors in the visible spectrum based on the patterns of blue, green, and red wavelengths.

Though listed in reverse order, red, green, and blue (RGB), your computer monitor operates under the same principle, using the RGB colorspace. It emits a combination of blue, green, and red light, which one's eye translates as different colors (**Figure 6.04**).

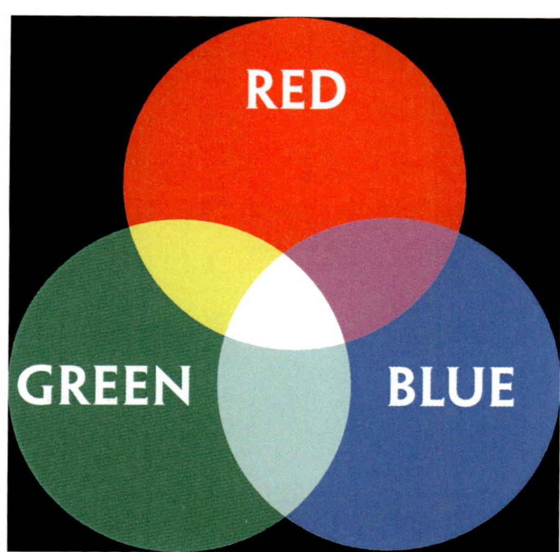

Figure 6.04: The computer has three channels that emit red, green, and blue.

Watch Tour of the EMS 05–Visible Light Waves, courtesy of Science at NASA to learn more (**Figure 6.05**).

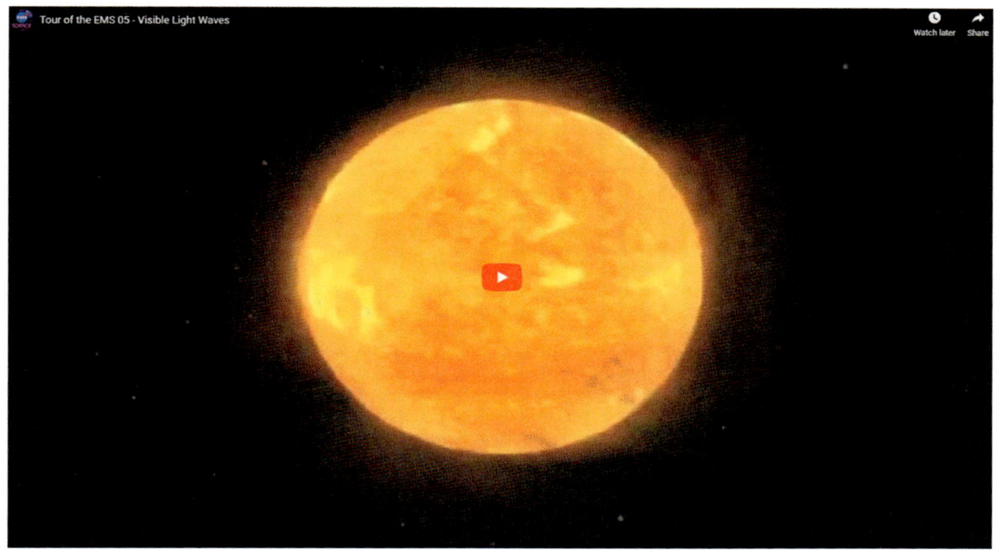

Figure 6.05: This video describes the concepts related to visible light. URL: *https://youtu.be/PMtC34pzKGc*

Near and Thermal Infrared

Infrared energy falls just outside of the red visible light region of the electromagnetic spectrum. Multispectral scanners, sensors that can detect multiple regions of the electromagnetic spectrum at once, are capable of distinguishing between distinct bands within the infrared region. This chapter covers only two in detail, near-infrared (Near-IR)and thermal infrared. **Near-infrared** ranges from 700 to 1300 nm. Remote sensing applications for near-IR include vegetation mapping. Chlorophyll has a high reflectance value in the near-infrared portion of the electromagnetic spectrum (**Figure 6.06**). Analyzing the degree of near-infrared reflectance helps to determine the health of vegetation.

Figure 6.06: An image of a near-infrared false-color composite. Data source: Landsat 8

Manufactured to record green, red, and a portion of the near-infrared band, false-color infrared film captured this phenomenon until the introduction of high-resolution electromagnetic sensors. **Thermal infrared** ranges between 3000 to 14,000 nm and also has some remote sensing applications, including geologic mapping, soil mapping, and fire mapping (**Figure 6.07**). Additional applications include energy efficiency assessments, military applications, and law enforcement applications. Watch the following video by Science at NASA to learn more (**Figure 6.08**).

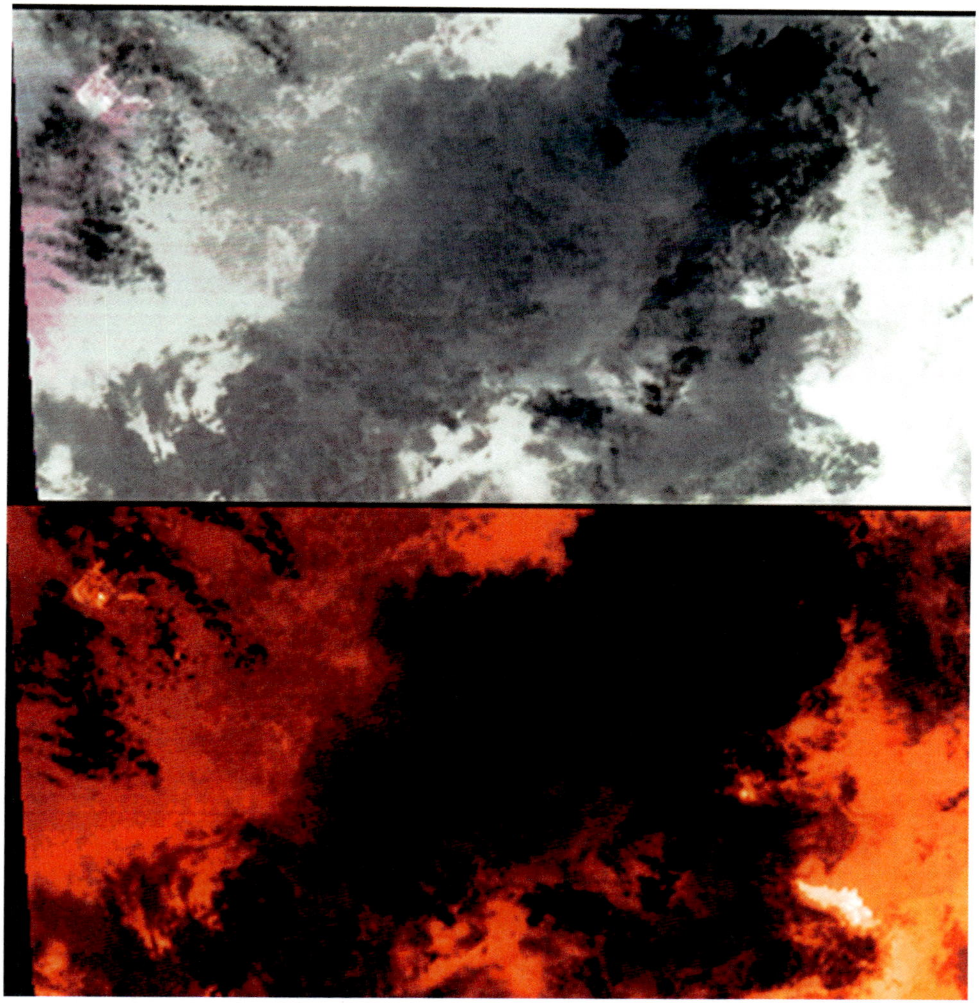

Figure 6.07: A thermal infrared image of Kilauea volcano. Source: NASA.

Figure 6.08: This video provides information about infrared waves. URL: *https://youtu.be/i8caGm9Fmho*

Passive and Active Microwaves

Microwaves with wavelengths between 30 cm to 1 mm are capable of penetrating the atmosphere even under cloudy conditions. This characteristic makes microwaves useful for mapping areas with frequent or perpetual cloud cover. Some natural phenomenon, such as sea ice, radiate passive microwaves. Since 1972, passive microwave sensors have continuously tracked changes in sea ice over time (**Figure 6.09**).

A familiar form of microwave application is **radar** which stands for radio detection and ranging. Radar was developed to detect the presence of objects and to determine their distance. Hollywood has made famous the familiar form of radar called the **plan position indicator (PPI) system** (**Figure 6.10**). A radio sweep indicates the position of radar echoes on a circular display. Radar uses **active microwaves**, and the radar sensor is an *active sensor*.

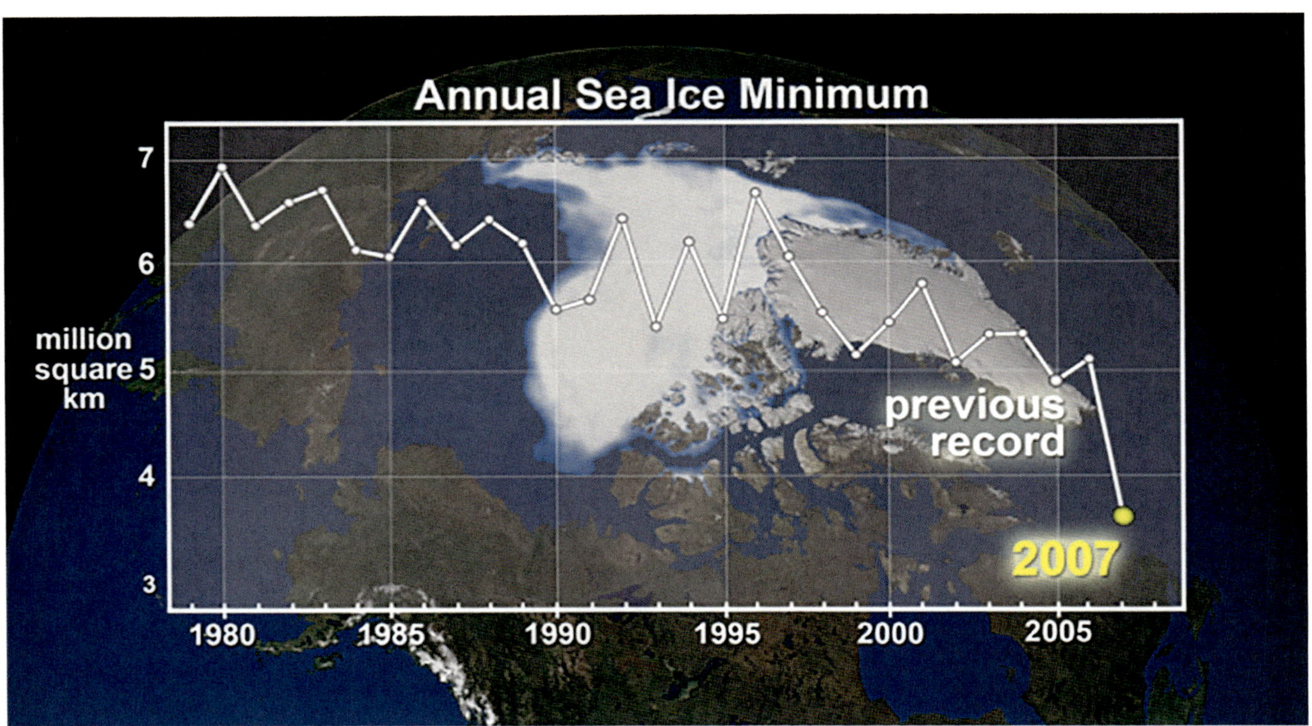

Figure 6.09: This chart by NASA shows the minimum level of sea ice for each year.

Figure 6.10: A screenshot of a plan position indicator (PPI) system from the U.S. Department of Commerce website.

Remote sensing mapping applications using radar include side-looking radar systems mounted on aircraft and used for topographic mapping. **Synthetic aperture radar (SAR)** is an airborne or spaceborne side-looking radar system, which utilizes the flight path to synthesize the effect of a very long antenna. SAR systems are capable of much higher resolution than a shorter antenna would typically allow. Watch the following video by Science at NASA to learn more (**Figure 6.11**)

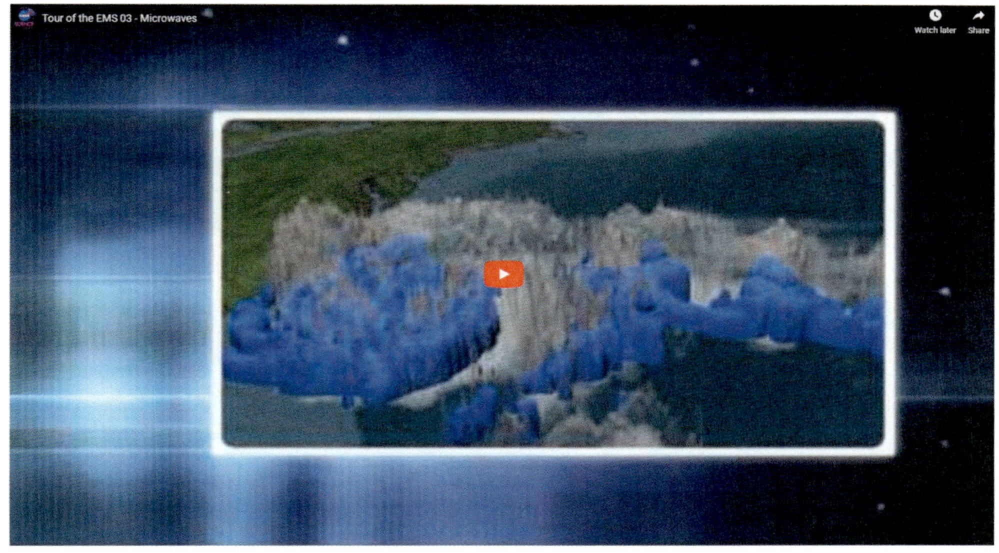

6.11: This video covers the concepts and applications related to microwaves. URL: ***https://youtu.be/UZeBzTI5Omk***

Electromagnetic Interactions

All of the electromagnetic energy interacts with the environment in three major ways. Electromagnetic energy is emitted, reflected, or absorbed (**Figure 6.12**). The instruments one uses and the portions of the electromagnetic spectrum that one studies are all dependent on this interaction with the environment.

Figure 6.12: In this image, electromagnetic energy is emitted by the sun. The energy is then reflected or absorbed by the atmosphere and the environment on earth.

Reflected Energy

One of the ubiquitous environmental interactions with electromagnetic energy comes from solar radiation as it passes through the atmosphere. The atmosphere *reflects* electromagnetic energy. One of the ways the atmosphere reflects energy is **Raleigh scatter**. Tiny atmospheric molecules smaller in diameter than specific wavelengths causes this electromagnetic energy to scatter unpredictably (**Figure 6.13**). This phenomenon has the most significant effect on the shorter blue wavelengths, which is the reason why the sky appears blue instead of black.

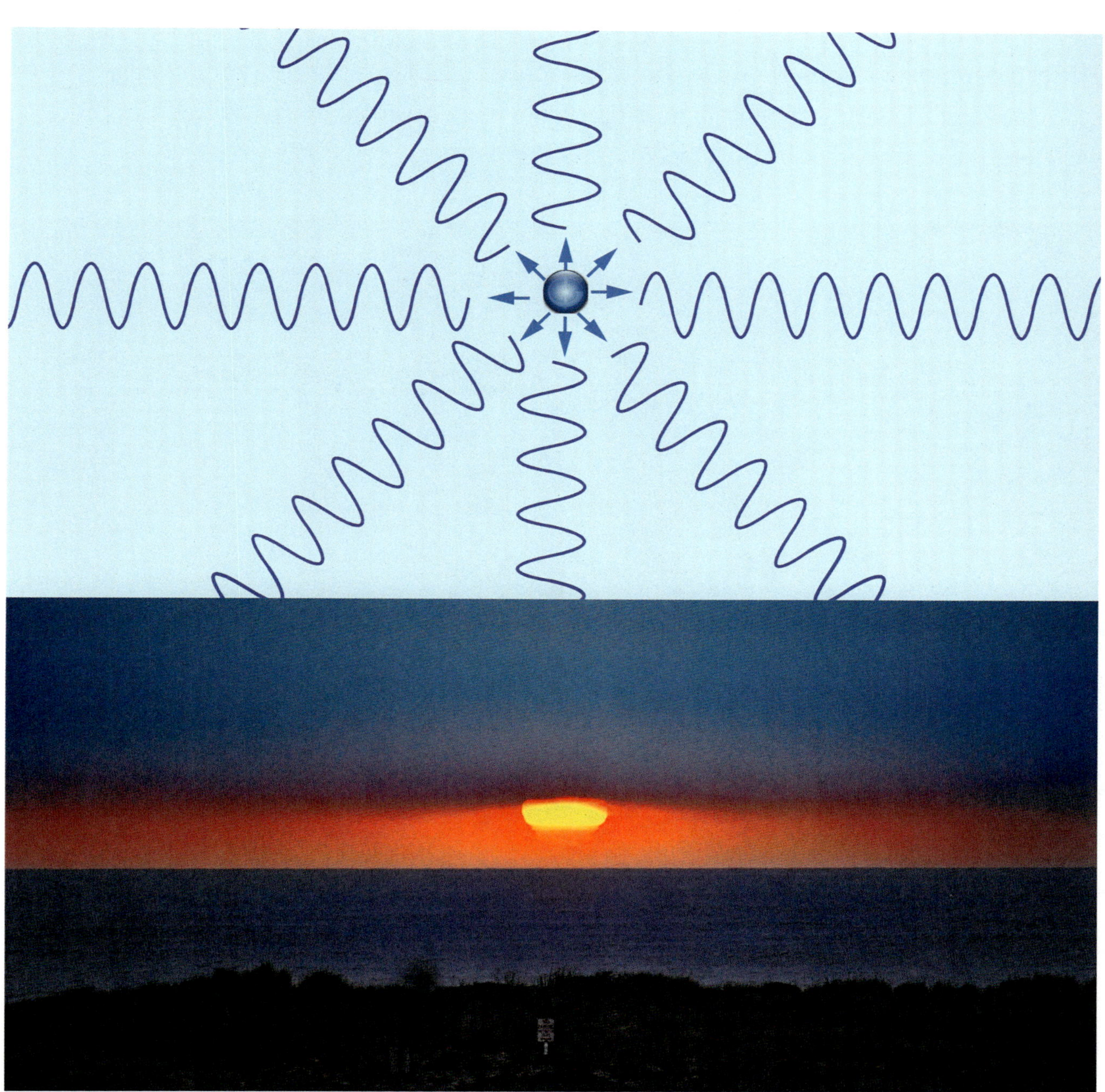

Figure 6.13: When the sun is at a high angle, blue wavelengths are affected by Rayleigh scatter, which causes the sky to appear blue. At Sunset, when the light passes through a thicker portion of the atmosphere, blue wavelengths are filtered out earlier, allowing red wavelengths to pass through cleanly.

Atmospheric Windows

The atmosphere also *absorbs* different wavelengths of solar radiation. As a result, scientists tend to study spectral bands that fall within **atmospheric windows**, ranges of wavelengths unimpeded by the atmosphere (**Figure 6.14**).

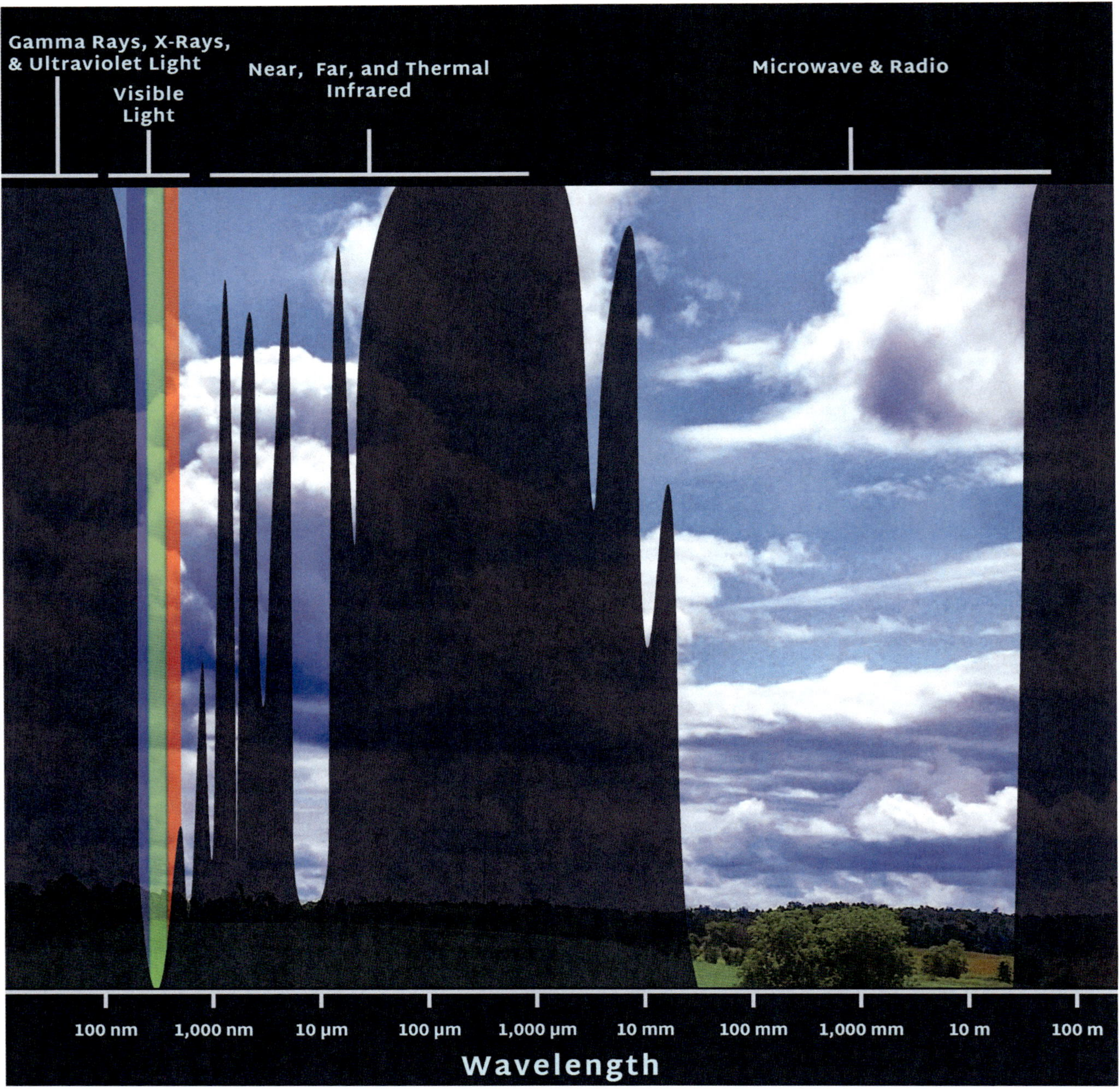

Figure 6.14: Atmospheric windows are regions within the electromagnetic spectrum that passes through the atmosphere. The transparent regions indicate the wavelengths that pass through unimpeded.

Spectral reflectance curves displayed on a graph represent how objects absorb and reflect electromagnetic energy across the spectrum (**Figure 6.15**). The physical and chemical composition of different objects results in distinct **reflectance response patterns**. Sometimes these reflectance response patterns are referred to as **spectral signatures**. This term is a slight misnomer, as *signature* tends to imply a pattern that is absolute or unique. While objects in the natural environment can have very distinct spectral reflective curves, they are not necessarily unique and can also vary with environmental conditions. For example, wet soil will have a slightly different reflective response pattern than dry soil of the same composition. Distinguishing between minute differences in reflectance response patterns and understanding the interaction of electromagnetic radiation and the environment is the basis for interpreting remote sensing data.

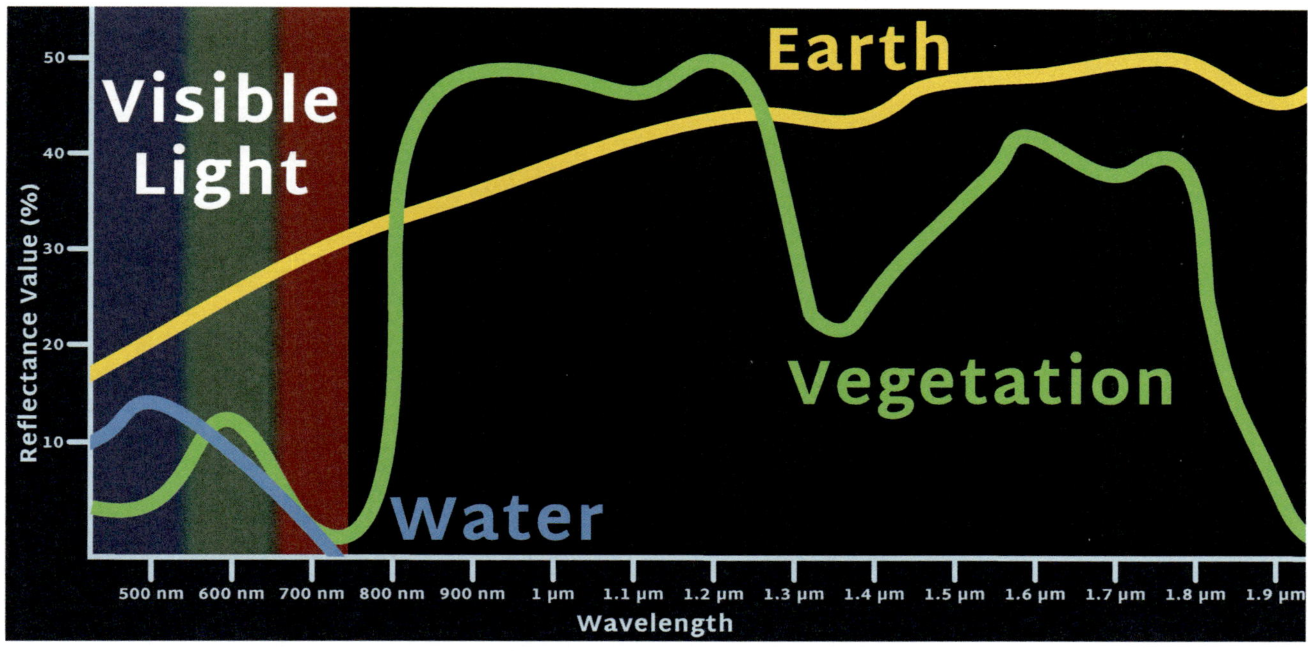

Figure 6.15: Features in the environment will reflect specific spectral wavelengths more than others.

IMAGE ACQUISITION

Historically, specially designed aerial mapping cameras mounted on the bottom of aircraft usually capture images using one of two recording platforms, *low-flight*, and *high-flight*. A **recording platform** describes the type of vehicle and the height from which a sensor operates (**Figure 6.16**). The intended platform dramatically influences the kind of imagery captured and post-processing required. Photographs captured using a **low-flight platform**, from 500 to 3000 meters above the ground, and using film on large 9" x 9" frames, made it possible to capture large-scale, high-resolution photographs. A **high-flight platform**, with photography taken from as high as 3 km above ground, captures images in much smaller scale resulting in less detail. However, these images have the advantage of covering a more extensive area at once. High-altitude photography is also subject to much less *geometric distortion*.

Geometric Distortion

Geometric distortion is the distortion caused by the perspective view of the camera's lens. In a **perspective view**, light enters through the camera lens at a single point. The camera views every position on the ground relative to that perspective (**Figure 6.17**).

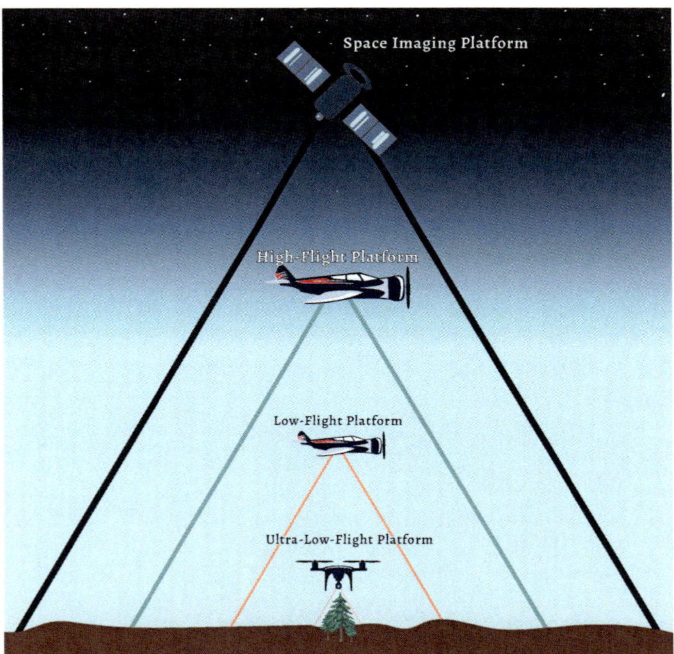

Figure 6.16: The ideal recording platforms used today run from ultra-low-flight platforms to the space imaging platform.

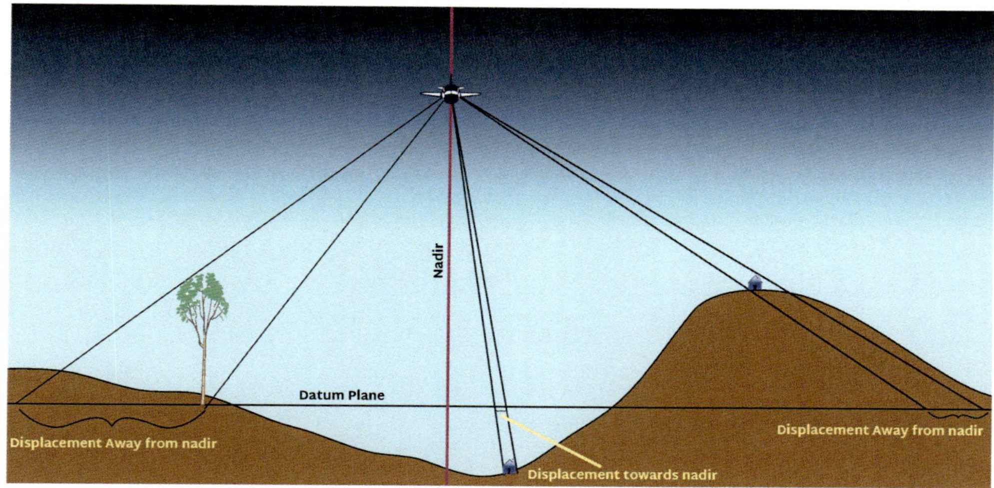

Figure 6.17: The red line indicates the nadir, an area on the ground directly below the camera. All other points are subject to geometric distortion.

This perspective causes this causes both relief displacement and scale distortion. **Relief displacement**, sometimes called **radial displacement**, has the effect of causing tall objects, such as trees or buildings, to tilt outward from the **principal point**, the center, of the photograph (**Figure 6.18**).

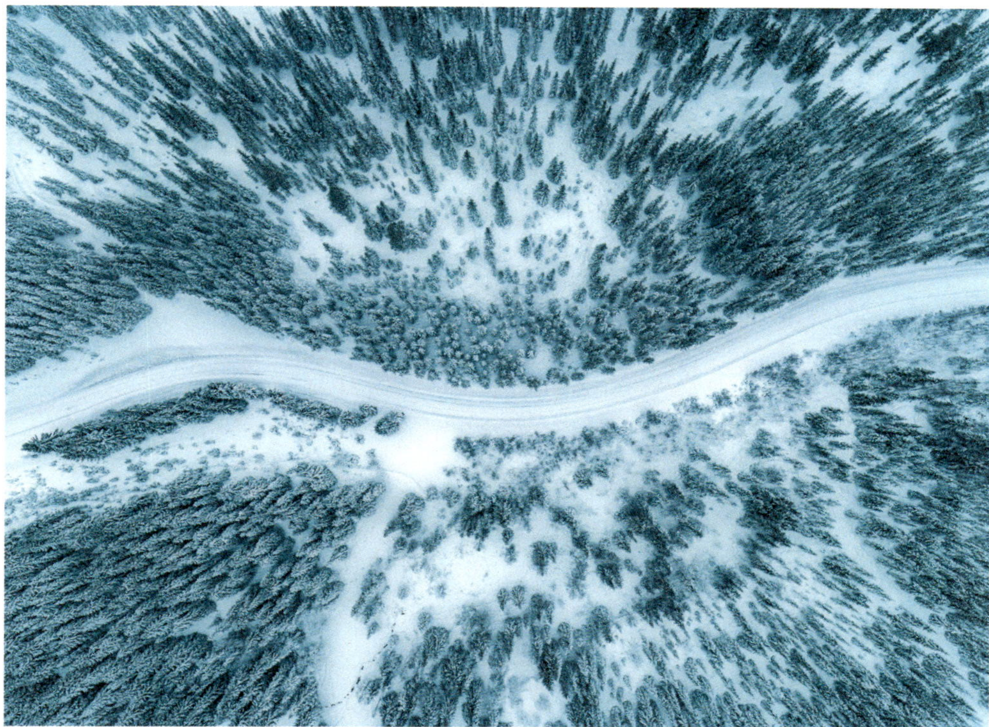

Figure 6.18: Notice the trees near the principal seem vertical, while those far away lean out from the center.

On some photographs, a series of registration marks on the sides called **fiducial marks**, help to determine the principal point (**Figure 6.19**). Lines drawn from opposite pairs of fiducial marks will intersect at the principal point. The area on the ground directly below the camera when a photograph gets taken is called the **nadir**. The nadir point and the principal point will only coincide on a **vertical photograph**, a photograph acquired by a camera aimed directly at the ground from above. Features higher in elevation than the nadir point will be displaced outwards. Features farther away from the nadir point will also vary in scale, resulting in **scale distortion**. The height of the camera above the ground will influence the amount of geometric distortion. Low-altitude photographs experience more geometric distortion, while high-altitude photographs experience less.

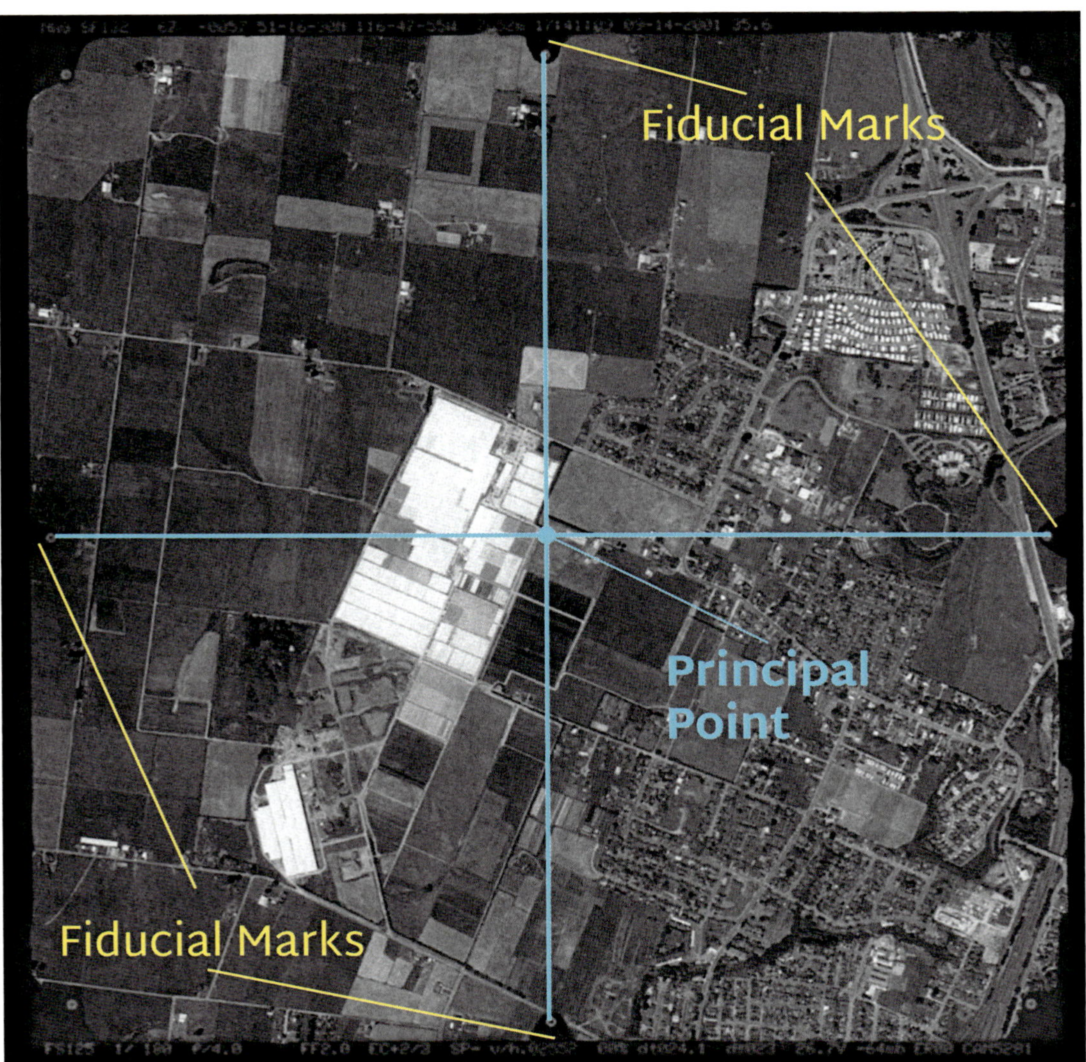

Figure 6.19: On this photograph, the principal point coincides with the nadir point.

Orthorectification and the Orthophotomap

Photo rectification, also called orthorectification, is the process of correcting relief displacement and scale distortion present in aerial photographs so that the photograph becomes planimetrically-correct. In **planimetrically-correct photographs**, the perspective of every position on the ground is from directly above. **Orthophotographs** are planimetrically correct photographs with an accurate scale suitable for mapping (**Figure 6.20**). The USGS produces a series of orthophoto maps at multiple scales. An **orthophoto map is** a map that uses an orthophotograph as its base. USGS also provides digital topographic maps that use orthophotographs as a base, called **digital orthophoto quads (DOQs)**.

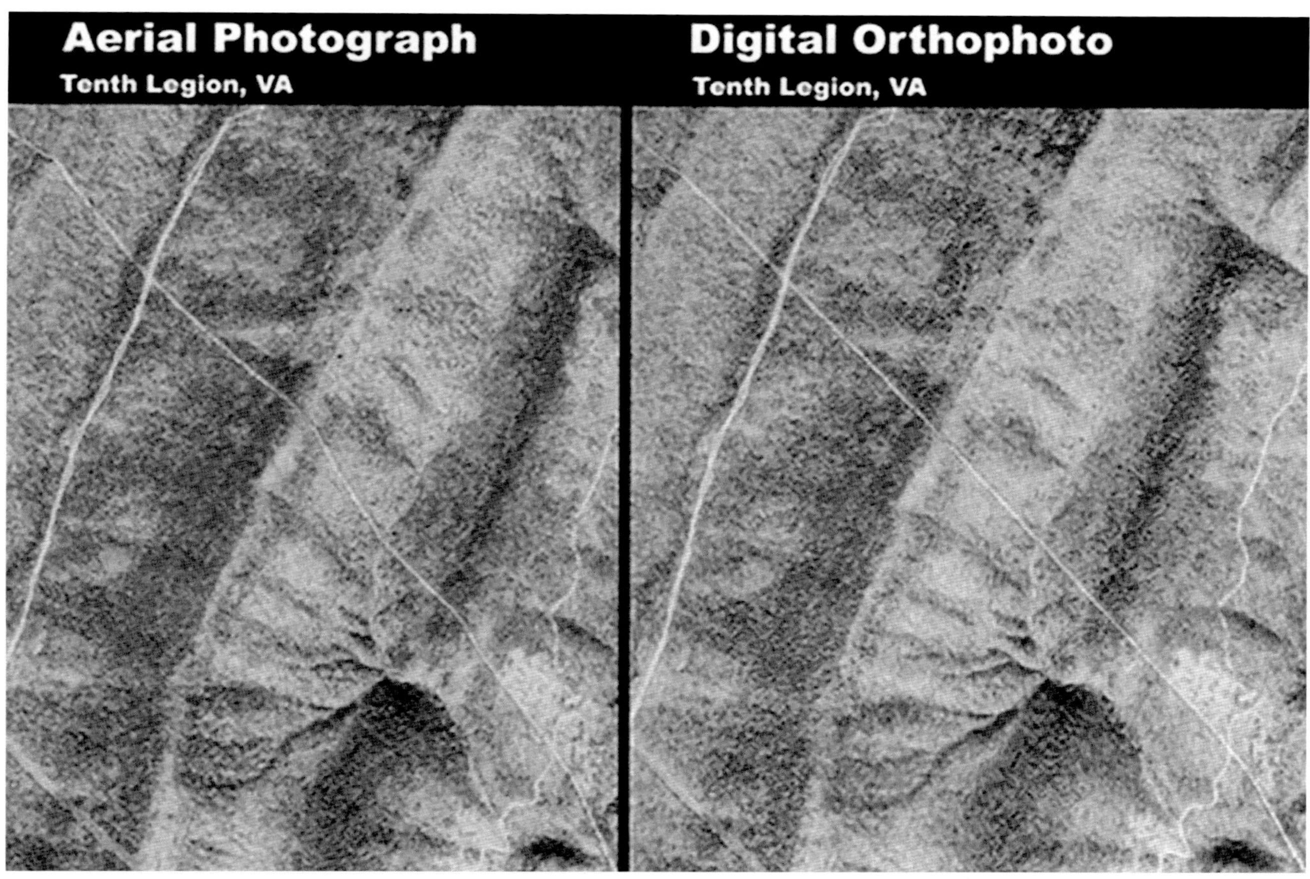

Figure 6.20: Both images show the same pipeline, which is the white line oriented north-west and southeast. The left image is uncorrected, and the pipeline appears curved due to relief displacement caused by elevation changes. The image on the right has been orthorectified. In this orthophotograph, the pipeline looks as it truly is, a nearly straight line. Source: USGS.

Aerial Flight Lines

Typically, one captures aerial photographs by flying the aircraft along a series of overlapping flight lines to ensure complete coverage of an area. **Flight lines** are the path that the aircraft follows while acquiring aerial photographs. **Forwardlap** is the overlap between photographs following a single flight line (**Figure 6.21**). **Sidelap** is the overlap between adjacent flight lines (**Figure 6.22**). Photographs belonging to a specific set of flight lines are managed using a **flight index**, a map with a numbered grid indicating the location and extent of each photograph.

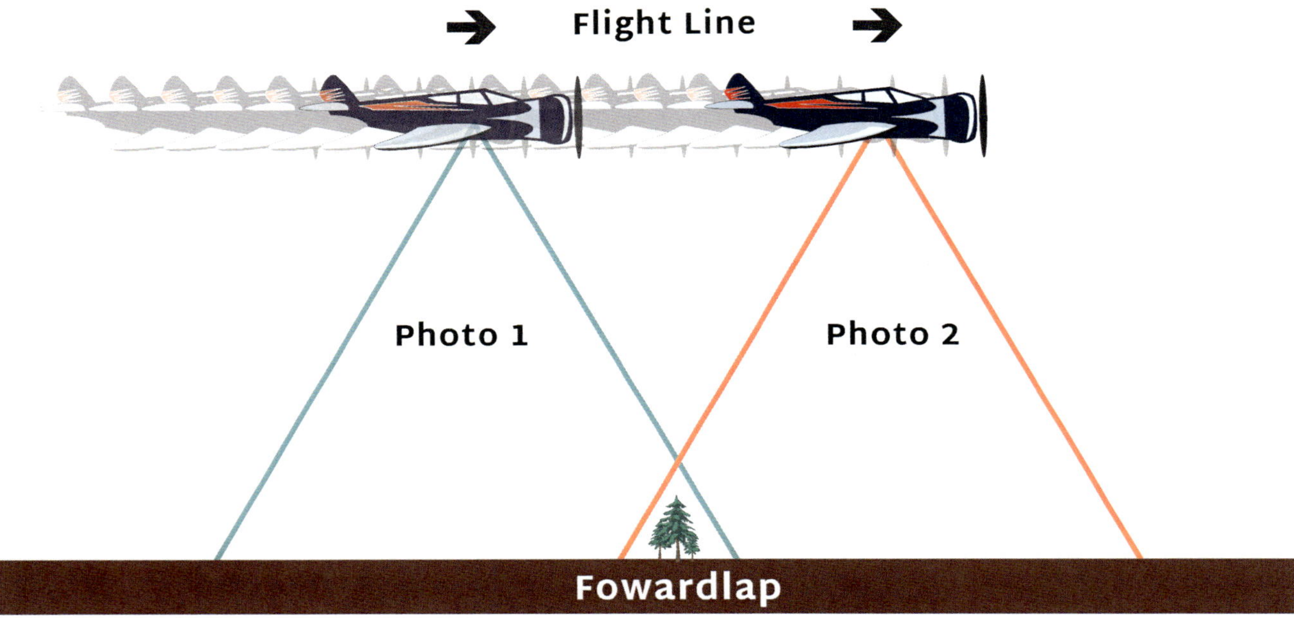

Figure 6.21: This diagram shows the overlap along the flight line called the forwardlap.

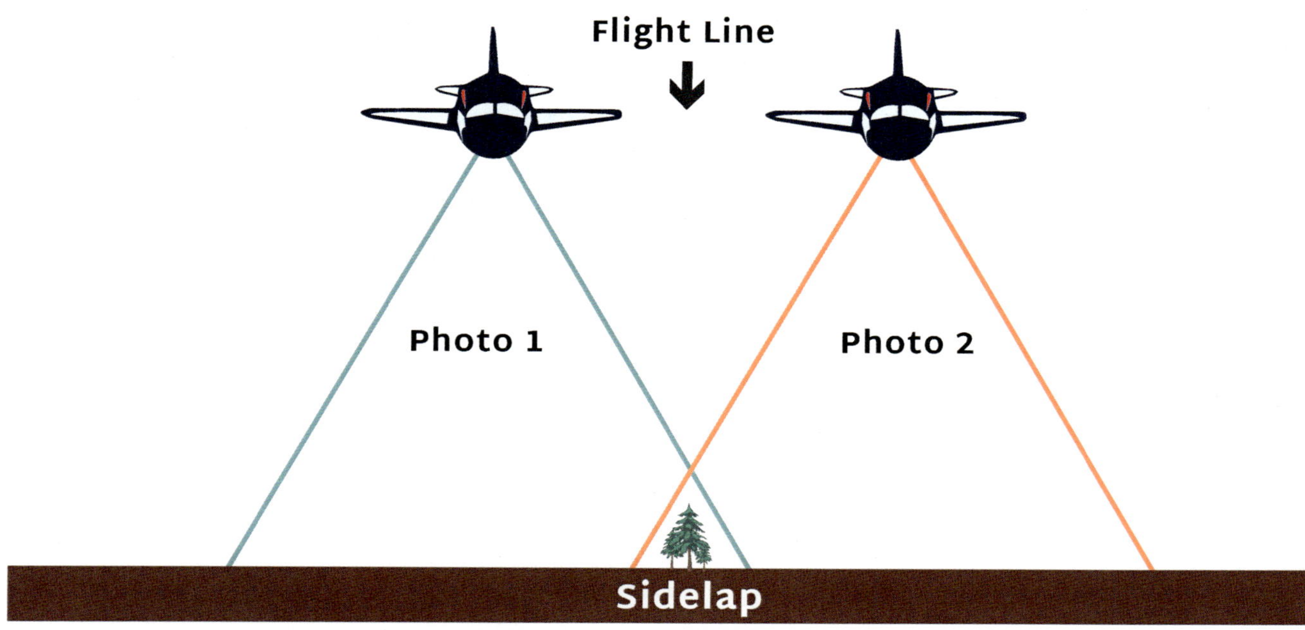

Figure 6.22: This diagram shows the overlap between adjacent flight lines called a sidelap.

Image Formats

When most readers think of image formats, they often consider this to mean image file formats. In this chapter, the term **image format** refers to two broad categories, film-based photographs, and digital imagery. **Film-based photographs** use a film with chemical coatings to capture images. **Digital imagery** relies on electronic instruments, such as sensors or scanners, to capture images. The image format determines the means of storage, distribution, retrieval, and analysis.

Film-based Photographs

Since the 1930s, film-based cameras mounted on aircraft were used to capture aerial imagery with fine spatial detail and positional accuracy. Photographs are the oldest image acquisition format used in remote sensing and archives of photographs produced over the last several decades represent a valuable historical record of landscape change. Many of these photographs only exist in analog form rather than digital, stored in libraries and government archives. Photographs require film with chemical coatings to portray images. **Photographic film** consists of a base or support made of transparent plastic (**Figure 6.23**). One side of the base is coated with one or more layers of light-sensitive emulsions made of silver halide crystals. The size of the film and the coarseness of the crystal grains determine the resolution, also referred to as ground resolved distance (GRD). **Ground resolved distance** is the approximate dimension of the size of the smallest feature that can be reliably determined by a photograph. Coarse grain results in inferior ground resolved distance than fine grain.

Figure 6.23: Credit: Coca-Cola filmstrip, "Black Treasures." (1969)

A single layer of silver halide emulsion produces panchromatic photographs. **Panchromatic photographs** are black and white photographs. Multiple layers of silver halide emulsions separated by light filters can produce both true-color and false-color photographs. **True-color photographs** refer to color analogs of the image captured. **False color-infrared photographs** capture portions of the infrared spectral band nearest to visible light (**Figure 6.24**). They are termed false-color because the color portrayed on the image bears no relationship with the actual color of the infrared light. A single layer of silver halide crystal emulsion can also capture the infrared spectral band. In this instance, one refers to it as a **near-infrared black and white photograph**.

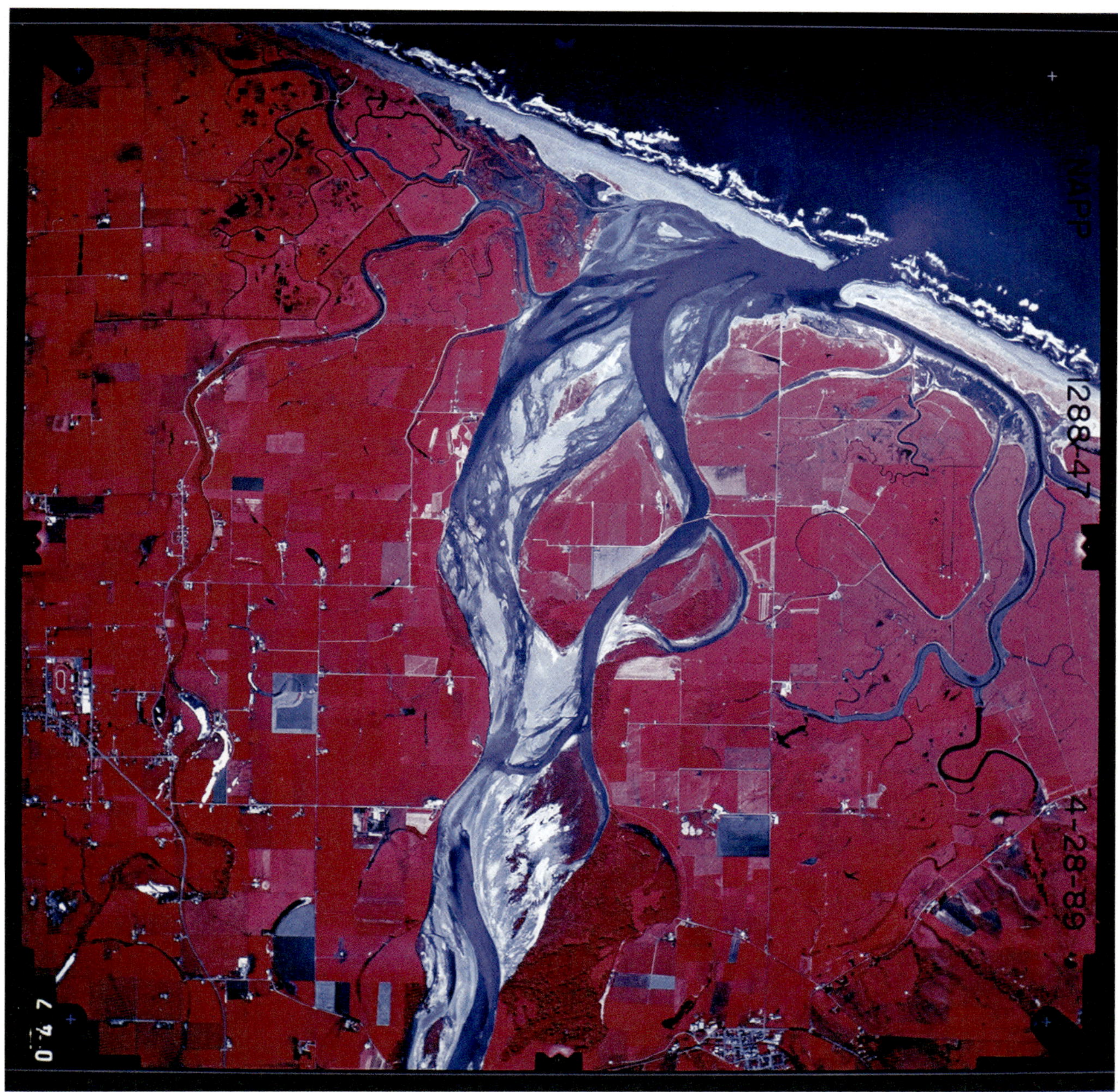

Figure 6.24: This aerial photograph of the mouth of the Eel River, taken with a 9-inch film at approximately 6 kilometers (20,000 ft.) above the ground, used false-color infrared film. It has a scale if 1:40,000. Source: USGS. Double-click or tap twice to view the image in a larger size.

Digital Imagery

Most aerial photographs today are acquired in digital format, which offers several advantages over film-based photographs. Film-based photography presents logistical complications related to storage, distribution, retrieval, and analysis that digital images overcome. Electronic instruments usually referred to as sensors or scanners, capture **digital imagery**. As discussed in Chapter 1, digital images, also called **rasters**, are composed of pixels. Scanners or sensor encode each **pixel** with a number that represents the **reflectance value**, or brightness, of features in the image.

Multispectral sensors, sometimes called **multispectral scanners**, can record separate regions of the electromagnetic spectrum at once. Scanners used in aerial imagery are usually either whiskbroom scanners or push broom scanners. **Whiskbroom scanners**, also called **across-track scanners**, use an oscillating mirror to track the sensor back and forth across a scan line. It tracks at a 90° angle to the flight path. **Pushbroom scanners**, sometimes called **along-track scanners**, contain a linear array of stationary sensors that record multispectral data along a single swath following the flight path.

The figure below depicts three digital images in different parts of the electromagnetic spectrum (**Figure 6.25**). The image on the top is a **true-color composite**, while both images below are **false-color composites**, using portions of the electromagnetic spectrum the human eye cannot see. In the center image, notice how the water in the Eel River is more visible along its entire lengths. Also, the reservoir on the right is easier to see. The third image is a composite of near infrared and shortwave infrared (SWIR). This combination is highly useful for detecting patterns in vegetation, distinguishing wet earth from dry earth, and the differences between rocks and soil that would generally appear the same.

Figure 6.25: Using different spectral band combinations allows one to see features and patterns not ordinarily visible. Data source: USGS.

Image Acquisition Programs

Agencies in the United Stated regularly acquire aerial and satellite imagery and make it available to the public. The three most frequently used sources of imagery in the United States comes from the following three sources:

» National Aerial Photography Program (NAPP)

» National Agriculture Imagery Program (NAIP)

» United States Landsat Program

National Aerial Photography Program (NAPP)

In 1987, the USGS began the **National Aerial Photography Program (NAPP)**. Coordinated by the USGS, the National Aerial Photography Program was an interagency project that captured high-altitude aerial photography for the coterminous United States every five years. The National Aerial Photography Program established systematic protocols for taking aerial photographs with flight paths centered on USGS 1:24,000 scale quadrangles. The program was operational until 2007 and collected more than 1.3 million photographs. The USGS is currently in the process of systematically scanning these photographs at a resolution 14 μ (1,800 dpi). Photographs produced from a film made of cellulose acetate plastic have priority in this process due to the risk of **acetate film base degradation**[3].

3 This phenomenon is also known as *vinegar syndrome*. URL: http://www.filmpreservation.org/preservation-basics/vinegar-syndrome.

Figure 6.26: This 1989 NAPP aerial photograph of Humboldt Bay, taken with a 9-inch film at approximately 6 kilometers (20,000 ft.) above the ground, used false-color infrared film. It has a scale if 1:40,000. Source: USGS.

National Agriculture Imagery Program (NAIP)

USGS began a pilot program called the **National Agriculture Imagery Program (NAIP)** in 2002. The program has expanded since then and is currently ongoing. The goal of the program was to acquire *leaf-on* aerial imagery during the peak growing season throughout the coterminous United States. The National Agriculture Imagery Program began with a five-year cycle using both film-based and digital imagery. It now operates on a three-year cycle and acquires images solely with digital sensors. The National Agriculture Imagery Program collects four-band imagery at a ground sample distance of 1 m or higher. **Ground sample distance (GSD)** is nearly synonymous with the term **cell size** discussed in Chapter 1 and is the length in ground units measured between the center of one pixel to another. The National Agriculture Imagery Program spectral bands include blue, red, green, and near-infrared (**Figure 6.27**). The USGS Earth Resources Observation and Science (EROS) Center manages and distributes free NAIP products on the ***Earth Explorer website***[4].

4 http://earthexplorer.usgs.gov/

Figure 6.27: This NAIP image of San Francisco near Golden Gate Park uses digital imagery. Data source: USGS.

United States Landsat Program

Space imaging platforms have sensors mounted on satellites or spacecraft. Many governmental and commercial programs acquire remotely sensed imagery from space platforms. Applications for satellite imagery include oceanography, watershed management, forestry, agriculture, urban planning, and land use, mineral acquisition, and for other environmental and urban resource management needs. The **United States Landsat Program** has the longest record in the world as a continuous space imaging program. On July 23, 1972, the United States launched the *Landsat 1* satellite mounted with a multispectral scanner system. Sensors on the **multispectral scanner system (MSS)** were capable of scanning four spectral bands between 500 nm and 1100 nm at approximately 80 m resolution. NASA added a thermal-IR band to the MSS sensor with the launch of *Landsat 3* on March 5, 1978. On March 1, 1982, *Landsat 4* was launched, equipped with a sensor called the thematic mapper. The **thematic mapper (TM)** was an advanced multispectral scanning sensor designed to achieve a high resolution and sharper spectral separation. Then on April 15, 1999, *Landsat 7* was launched, equipped with an improved version of the thematic mapper called the **Enhanced Thematic Mapper plus (ETM+)**. The ETM bands include the following spectral ranges:

- » **Band 1**, 450-515 nm blue-green, 30 m GSD, use for bathymetric mapping, distinguishing soil from vegetation, distinguishing deciduous from coniferous vegetation
- » **Band 2**, 525–605 nm green, use for assessing plant vigor
- » **Band 3**, 630–690 nm red, use for discriminating vegetation slopes
- » **Band 4**, 775–900 nm near-IR, use for emphasizing biomass content and shorelines
- » **Band 5**, 1550–1750 nm short-wavelength-IR (SWIR), use for discriminating moisture content of soil and vegetation, penetrates thin clouds
- » **Band 6**, 10,400–12,500 nm thermal-IR, used for thermal mapping, estimating soil moisture
- » **Band 7**, 20,800–20,350 nm Far-IR (FIR), use for mapping hydrothermally altered rocks associated with mineral deposits
- » **Band 8**, 520–900 nm panchromatic, used for high-resolution black-and-white imagery

On February 11, 2013, *Landsat 8* launched from the Vandenberg Air Force Base in California. Landsat 8 is equipped with two types of sensors, the operational land imager, and the thermal infrared sensor. The **operational land imager (OLI)** sensor collects data for the visible, near infrared, and shortwave infrared spectral bands. It also includes two spectral bands used for coastal zone observations and for detecting cirrus clouds. The **thermal infrared sensor (TIRS)** splits the formerly single thermal infrared band into two distinct bands.

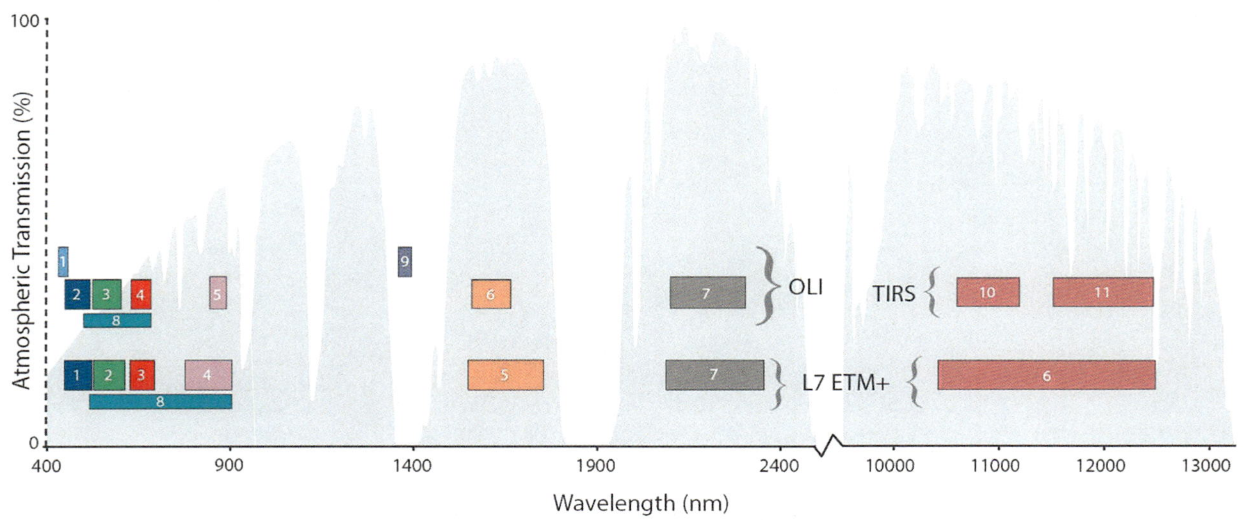

Figure 6.28: A graphic comparing bands from Landsat 7 and Landsat 8. Source: NASA.

Watch the video *Landsat Data Continuity Mission Overview* by NASA to learn more (**Figure 6.29**).

Figure 6.29: This video discusses the Landsat Program in additional detail. URL: *https://youtu.be/mqVKR9OnqqA*

There are other governmental and commercial space imaging platforms. Read the information presented in the following websites to learn the basic facts about a few more:

- Advanced Very High-Resolution Radiometer (AVHRR)[5]
- Systeme Probatoire d'Observation de la Terre (SPOT)[6]
- IKONOS[7]
- OrbView[8]
- DigitalGlobe[9]

5 http://noaasis.noaa.gov/NOAASIS/ml/avhrr.html
6 http://www.esa.int/SPECIALS/Eduspace_EN/SEMIW04Z2OF_0.html
7 http://glcf.umd.edu/data/ikonos/index.shtml
8 https://lta.cr.usgs.gov/satellite_orbview3
9 https://www.digitalglobe.com/

Small Unmanned Aircraft Systems

Small Unmanned Aircraft Systems (sUAS) are an emerging platform for remotely sensed images. A **small unmanned aircraft system (sUAS)** includes the small unmanned aircraft as well as any associated elements required for the safe and efficient operation of the small unmanned aircraft in the national airspace system. The FAA defines a small unmanned aircraft as an unmanned aircraft weighing less than 55 pounds on takeoff, including everything that is on board or otherwise attached to the aircraft (**Figure 6.30**). An **unmanned aircraft** is any aircraft operated without the possibility of direct human intervention from within or on the aircraft. People often refer to a small unmanned aircraft as a **drone**. However, the term drone can have negative connotations as it often references unpiloted military aircraft equipped with a lethal payload.

Figure 6.30: A small unmanned aircraft is one that weighs less than 55 pounds.

This **ultra-low-flight imaging platform** enables unmanned aircraft systems to collect high-resolution imagery using equipment that is relatively inexpensive (**Figure 6.31**). In the wake of this affordable and accessible technology, the field of remote sensing is undergoing a similar transformation, as seen in GIS, cartography, and mobile mapping (GPS). Once the purview of government agencies and large corporations, today nearly anyone can acquire remotely sensed imagery. Affordability and accessibility will no doubt lead to innovation just as it has across the other geospatial sciences. The FAA regulates the use of small unmanned aircraft in the United States under two parts of Title 14 of the Code of Federal Regulations (CFR), Part 101 and Part 107. **Part 101** relates to model aircraft and recreational use. **Part 107** applies to the non-hobby or non-recreational use of small unmanned aircraft systems.

Figure 6.31: Ultra-low-altitude imaging platforms make the acquisition of high-resolution imagery inexpensive.

The term *system* in small unmanned aircraft system applies to both the aircraft and the associated elements. The **associated elements** of a small unmanned aircraft system include, but are not limited to, the control station, the remote pilot in command (PIC), and the visual observer (VO). The **control station** is the interface used by the remote pilot to control the flight path of the small unmanned aircraft (**Figure 6.32**).

Figure 6.32: A control station communicates with the aircraft for flight control. Remote pilots often augment the control station with smartphones or tablets.

A **remote pilot in command (PIC)** is the person that possesses a remote pilot certificate with a small UAS rating and is responsible for the operation and safety of the flight (**Figure 6.33**). A **visual observer (VO)** is a person who is designated by the remote pilot in command to assist the remote pilot in command and the person manipulating the flight controls of the small UAS to see and avoid other air traffic or objects aloft or on the ground. There may be additional elements of a small unmanned aircraft system, including radio communication, cell phones, tablets, and supplementary crew members.

Fixed-wing and Rotorcraft UAVs

UAVs are divided into two broad categories:

» Fixed-wing aircraft

» Rotorcraft

Fixed-wing aircraft for civilian use usually come with wingspans ranging from 3 m to less than 1 m. Larger fixed-wing aircraft are traditionally gas powered, with flight times of around 12 hours. They can usually carry payloads up to 20 kg and require a runway for takeoff and landing. Fixed-wing aircraft with a wingspan smaller than 1 m are typically battery-powered with flight times up to several hours. The smaller aircraft do not require a runway. They are usually, hand launched and belly landed and can carry a payload up to 1 kg. **Rotorcraft UAVs** typically have anywhere between four to eight rotors and have rotor diameters ranging from 2 m to less than 15 cm (**Figure 6.33**). Payloads are in a similar range as fixed-wing aircraft, but this also varies by the number of rotors on the vehicle. It takes more power to sustain flight for a rotorcraft UAV than with fixed-wing aircraft. As a result, rotorcraft UAVs tends to have a much shorter flight time. However, rotorcraft UAVs have the advantage of being more maneuverable, stable, and more flexible regarding takeoff and landing locations.

Figure 6.33: This image depicts a rotorcraft UAV engaged in an agricultural inspection.

UAS Mission Planning

Using unmanned aircraft systems for remote sensing application projects involves three phases:

- Mission planning
- Mission implementation
- Post-mission processing

Mission planning is the most detailed and time-consuming phase of a UAS mission. It includes defining the mission product, assessment of the mission zone, determining the placement of ground control points and line-pairs-per-milliliter (LPM) resolution targets, establishing the delivery schedule, and clarifying the implementation of mission protocol and the responsibilities of the UAS crew members. **Mission implementation** is the execution of the mission plan. When planning a UAS mission, crew members should clearly define the **mission product. The mission product is** the output of the UAS mission, including the imagery resolution and spectral bands. Additionally, the crew should consider the demarcation of the flight paths, image overlap, image scale, and the spatial extent of the mission zone as part of the mission product. Having a clear mission product will ensure the final results will meet expectations.

The next step in mission planning involves careful study of the **mission zone**, the area within and around the flight path, to avoid restricted regions such as airports, sporting events, school campuses, hospitals, and regions with high population density. Violation of federal, state, and local laws can result in steep fines and imprisonment. USGS Digital orthophoto quads and **sectional aeronautical charts**, maps designed for visual navigation of slow to medium speed aircraft, are useful tools to employ when studying the mission zone (**Figure 6.34**).

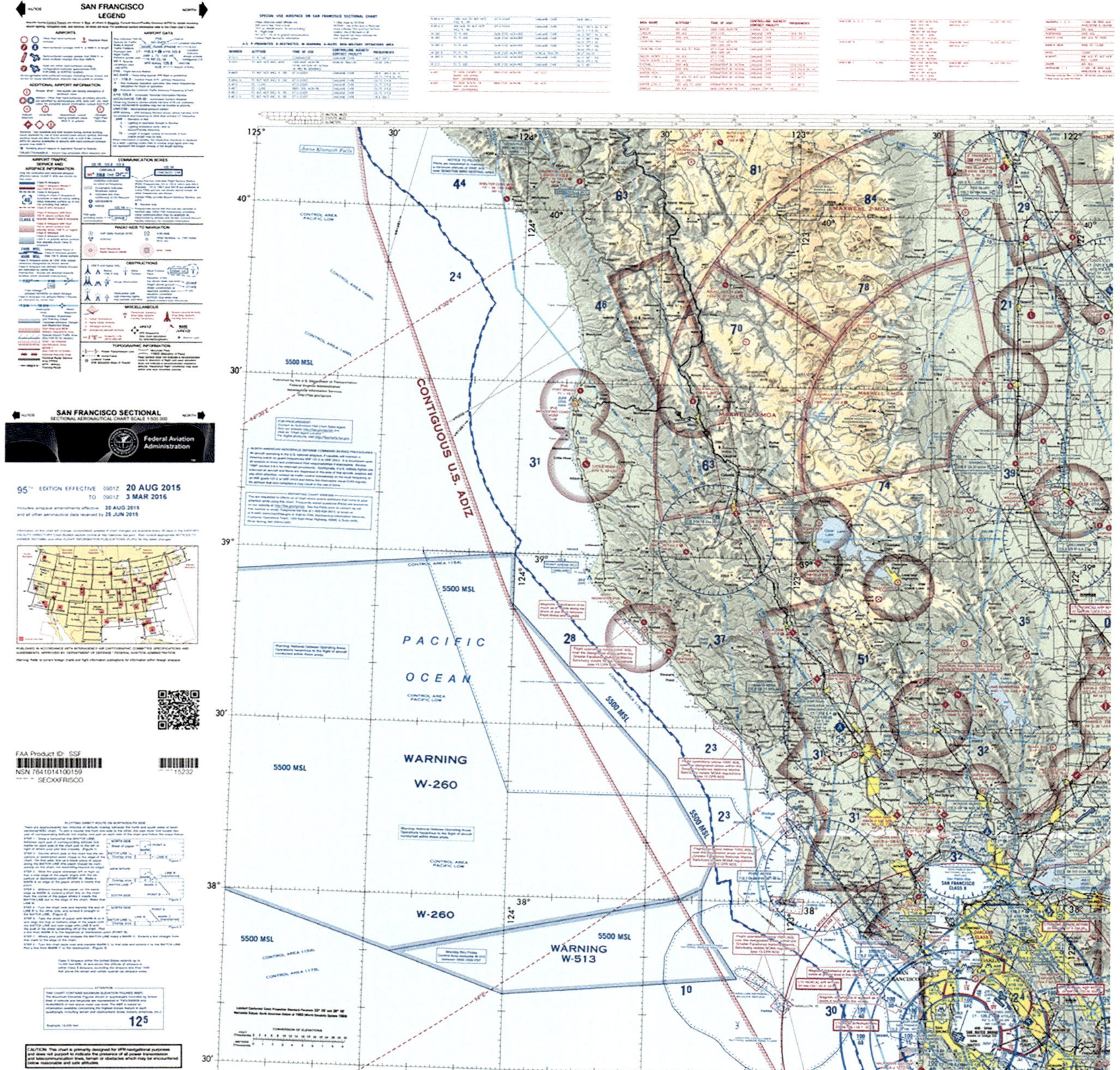

Figure 6.34: A portion of the San Francisco Sectional Aeronautical chart. Source: FAA.

Placement of **ground control points** should be considered to optimize georeferencing in the data processing phase. Ground control points are highly visible targets set into the ground that serve as control points for georeferencing, the establishment of an accurate geometric relationship between the mission zone (ground) and the mission product (image). They should be places that one can locate with a high degree of precision on the ground, on planimetrically correct maps, and in GIS software. The position of ground control points can be established using land survey methods and GPS. Ideally, ground control points should be distributed evenly throughout the mission zone.

The placement of a **line-pairs-per-milliliter (LPM) resolution target** is a standardized way of measuring ground resolution distance (GRD) (**Figure 6.35**). The Line Pairs Per Millimeter (LPM) resolution target was developed by the United States Air Force and has been the standard for a variety of studies over the years. It consists of parallel black lines positioned against a white background. The width of spaces between the lines is equal to that of the lines themselves, and the length of the lines is exactly five times the width. The lines are arranged three bars across with two white spaces in between, form a square. The square patterns form an array which repeats at various sizes. The smallest line pair one can spatially resolve determines the resolution or the ground resolved distance (GRD). Spatially resolved objects are objects that can be visually separated from the other objects around them.

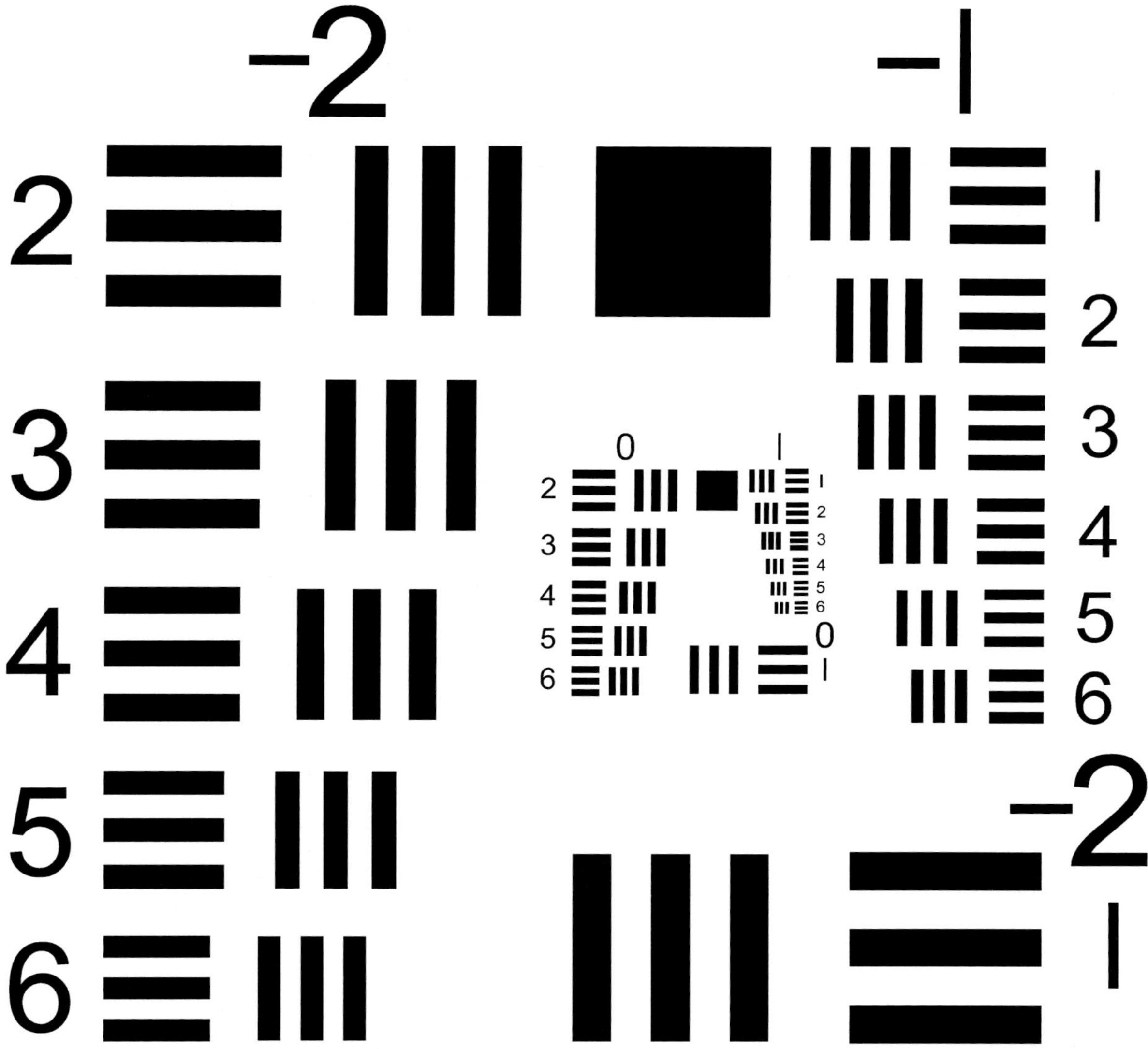

Figure 6.35: The United States Air Force developed the Line Pairs Per Millimeter (LPM) resolution target. A black line and the white space in between make up the "line pair." The smallest line pair one can resolve determines the ground resolved distance (GRD).

A **delivery schedule** is a timetable used for mission implementation. Establishing a detailed delivery schedule will help ensure the right lighting conditions, weather conditions, and logistics of transporting equipment and crew to the mission zone. Included in the delivery schedule should be travel time to and from the mission zone, set up of ground control points and LPM resolution targets, completing safety protocols, UAV launching, landing, and retrieval protocols, and UAV flight time.

Mission protocols describe the responsibilities and the exact procedures executed by the UAS crew during mission implementation. Safety protocols for the prevention of injury and property damage should be the number one priority during UAS mission planning. Some safety conditions to consider are the following:

- The presence of the public within the mission zone
- Overlooked or unexpected features with restricted airspace
- Features or geography that could pose a safety hazard
- Location of UAS crew members during the launch
- Location of UAs crew members during landing and retrieval

Mission protocols should also take into consideration the worst-case scenario, a flyaway. A **flyaway** is the loss of control of a UAV with potential consequences, including a crash landing, property damage, fire, or injury.

Mission product processing is the post-mission procedures and data processing used to develop a useful mission product. Post-mission data processing usually involves georeferencing and orthorectification using GIS and remote sensing image software. Images prepared in this way are suitable for geospatial analysis and the generation of derived image products such as an orthophoto map or a digital elevation model (DEM). The exact details of this process will vary depending upon the quality and size of the images and the software used.

Image Interpretation

Image interpretation is the identification of features seen on an image and the communication of this information to others. Image interpretation is best learned through practice. Successful image interpretation requires a thorough understanding of the phenomenon studied, the geographic region, the imaging device, and platform employed. The study and use of supporting materials, such as topographic maps and field observation reports, should be included in the process when available.

Elements of Image Interpretation

The elements of image interpretation are a combination of eight characteristics of image features interpreters consider:

- Site
- Association
- Size
- Shape
- Pattern
- Tone
- Texture
- Shadows

Site

Site refers to the geographic and topographic location of features. For example, a tree species, such as redwoods, would be located in coastal regions of California, while other vegetation types would be expected in inland desert regions or at high altitudes.

Association

Association refers to context rather than a pattern. On looks at how features associate with others. For example, one may associate a football field with a school campus. Other features visible nearby, buildings, playgrounds, benches, might be more easily identifiable in the same context than on their own.

Size

Size refers to the scale of the features as they relate to the scale of the image and other features nearby. The relative size of an object as they relate to other recognizable features is an intuitive indication of scale without the need for measurements. Features may have similar shapes, such as a tool shed and barn. Size helps to identify the difference.

Shape

Shape refers to the form, dimensions, and outline of features that make them identifiable. For example, the ratio between width and length distinguishes a river from a pond or lake.

Pattern

Pattern refers to the spatial arrangement of features into distinct or repeating formations. For example, an apple orchard would be distinct from a forest because of the regular pattern in which trees are planted.

Tone

Tone refers to the relative brightness, darkness, or color of features. These concepts are similar to the visual variables discussed in an earlier chapter related to the value, saturation, and hue. The tone is arguably the most critical element of image interpretation. Without it, one cannot determine the shape, the pattern, the size, and the texture.

Texture

Texture is the change in tone of an object or feature resulting in apparent roughness for smoothness. The aggregation of features too small to be resolved on their own create texture. The angle of illumination, or shadows, also influence the amount of texture

Shadows

Shadows are the areas of darkness created by the angle of illumination that help to outline features. Shadows are useful for determining the shape and height of objects, but can sometimes also hinder interpretation by obscuring objects within.

Image Interpretation Tasks

The elements of image interpretation are used to accomplish a specific set of image interpretation tasks, which include *classification*, *delineation*, *mensuration*, and *enumeration*. In previous chapters, this book discusses the use of classification and different ways, *data classification*, and *visual classification*. In the context of image interpretation tasks, **classification** is the grouping of features based on the elements of image interpretation. It is the most time-consuming and challenging image interpretation task. The precision of classification groups is based on feature *recognition* and *identification*.

Recognition

Recognition is the discovery of the presence of a feature as a separate object or land cover type from other features on the image. Land cover is the nominal classification of land according to the features or material that covers most of its surface. Identification is the degree to which one group features into a specific class. For example, one might recognize a forest on an image as something different from a field. In this case, one recognizes the land cover type.

Identification

Identification would involve the determination of whether one classifies the forest as conifer or deciduous. The next level of identification would be the determination of the specific species of conifer or deciduous trees that make up the forest. Once one determines the classification, *delineation* follows.

Delineation

Delineation is the process of outlining and defining the boundaries of contiguous classification groups on the image. Using the example from above the areal region of the forest would be separated from the areal region of the field by outlining or defining the boundary between the two. Delineation is necessary to accomplish the following tasks, *mensuration*, and *enumeration*.

Mensuration

Mensuration is the measuring of geometric dimensions, including volume, area, length, and height, as well as the magnitude of the feature or phenomenon. Photographs are often limited to only the measuring of geometric dimensions. Digital imagery can also record the magnitude or brightness reflected by the feature in different parts of the electromagnetic spectrum.

Enumeration

Enumeration is the inventory, tallying, and calculation of discrete features on the image. For example, foresters are routinely interested in the estimation of timber volume as a means of monitoring growth, health, and assessing best management practices. Aerial imagery interpretation assists in these calculations using crown cover, crown diameter, and tree height.

Stereoscopes and Light Tables

There are several standard tools used to aid image interpretation, including stereoscopes, light tables, and image software. A **stereoscope** is a stereovision instrument used to view photographic stereo pairs. A **stereo pair** is a set of two air photos taken from slightly different vantage points, usually produced by the flight path overlap. Using a stereoscope allows one to view stereopairs in **stereovision**, a three-dimensional perspective. A **pocket stereoscope** is constructed by placing two lenses into a frame. The frame can be extended to accommodate differences in pupil distance. The frame is supported by four legs which allow it to be placed over the stereo pair. A **mirror stereoscope** is a more sophisticated version that uses high-quality mirrors attached to the legs of the device that provide a more precise image (**Figure 6.36**). When looking through the stereoscope, each eye is only allowed to see a separate image. The stereo parallax, the change of perspective, between the two images create a three-dimensional view when seen through a stereoscope. Using a stereoscope helps determine differences in height and elevation.

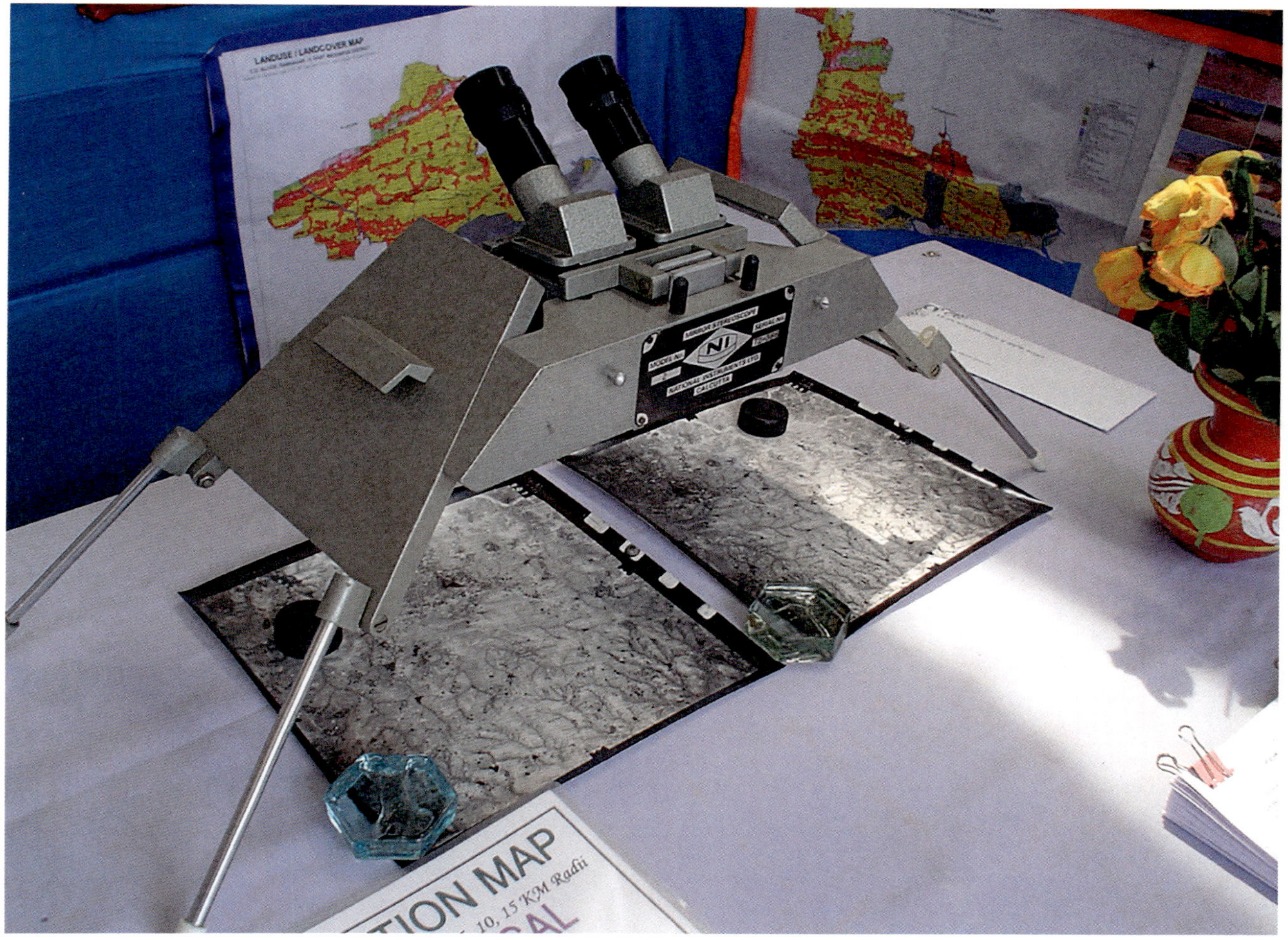

Figure 6.36: This image depicts a mirror stereoscope and photographic stereo pairs.

A **light table** is a flat device with a translucent surface. The light source located underneath the translucent surface illuminates images placed on the light table. Light tables are often used to view film negatives or image transparencies in greater detail.

Image Enhancement

Sometimes image interpretation tasks are assisted with the use of **image enhancement**, the procedure of refining the visual appearance of images using image software. Image enhancement typically applies to digital imagery rather than photographs. Because image enhancement can alter the original characteristics and pixel values of digital images, one should exercise caution when performing an analysis. Unaltered copies of the original image should always be safely archived. There are many techniques used in image enhancement. This chapter discusses two, contrast and stretch. **Contrast** refers to the range of brightness values included in the image. A computer monitor represents contrast using a range of brightness values from 0 to 255. The darkest value, 0, usually appears black while 255, the brightest value, appears white. Remote sensing scanners can often record a more significant range of values reflected in different portions of the electromagnetic spectrum. However, the computer monitor still displays the ranges using values from 0 to 255. If the range of values falls too close together, the result will be an image that is either too bright or too dark to distinguish individual features. This image can be said to have poor contrast. A **contrast stretch**[10] changes the distribution of brightness values across the range from 0 to 255 (**Figure 3.67**). For example, a digital image composed of values falling between 0 and 50 would appear dark. These values can be stretched across the full range of the display system to improve visibility. For example, pixels that had brightness values of 50 could be assigned a brightness value of 150 instead. This modification would have the effect of making those regions in the image appear brighter and increase the contrast against the other pixels in the image.

There are several conventional methods used to produce a contrast stretch, including standard deviation, histogram equalization, and minimum-maximum. **Standard deviation** is a linear function that converts the original brightness values into a new distribution. One accomplishes this method by adding two standard deviations from the mean to the minimum and maximum values. Intermediate values get proportionately distributed between the new minimum and maximum. This distribution has the effect of increasing contrast by spreading the brightness values across a broader range. **Histogram equalization** is a nonlinear function that approaches a uniform distribution of brightness magnitudes, highs, and lows, distributed across the range of output values. Peaks and valleys in the histogram are broadened using this method, which has the effect of enhancing image detail. **Minimum-maximum** increases the distance

---

10 Sometimes the term is also called a stretch.

between tightly grouped values. This method has the effect of enhancing the edges or transitions between regions of contrasting brightness. Minimum-maximum tends to improve the localized contrast between features.

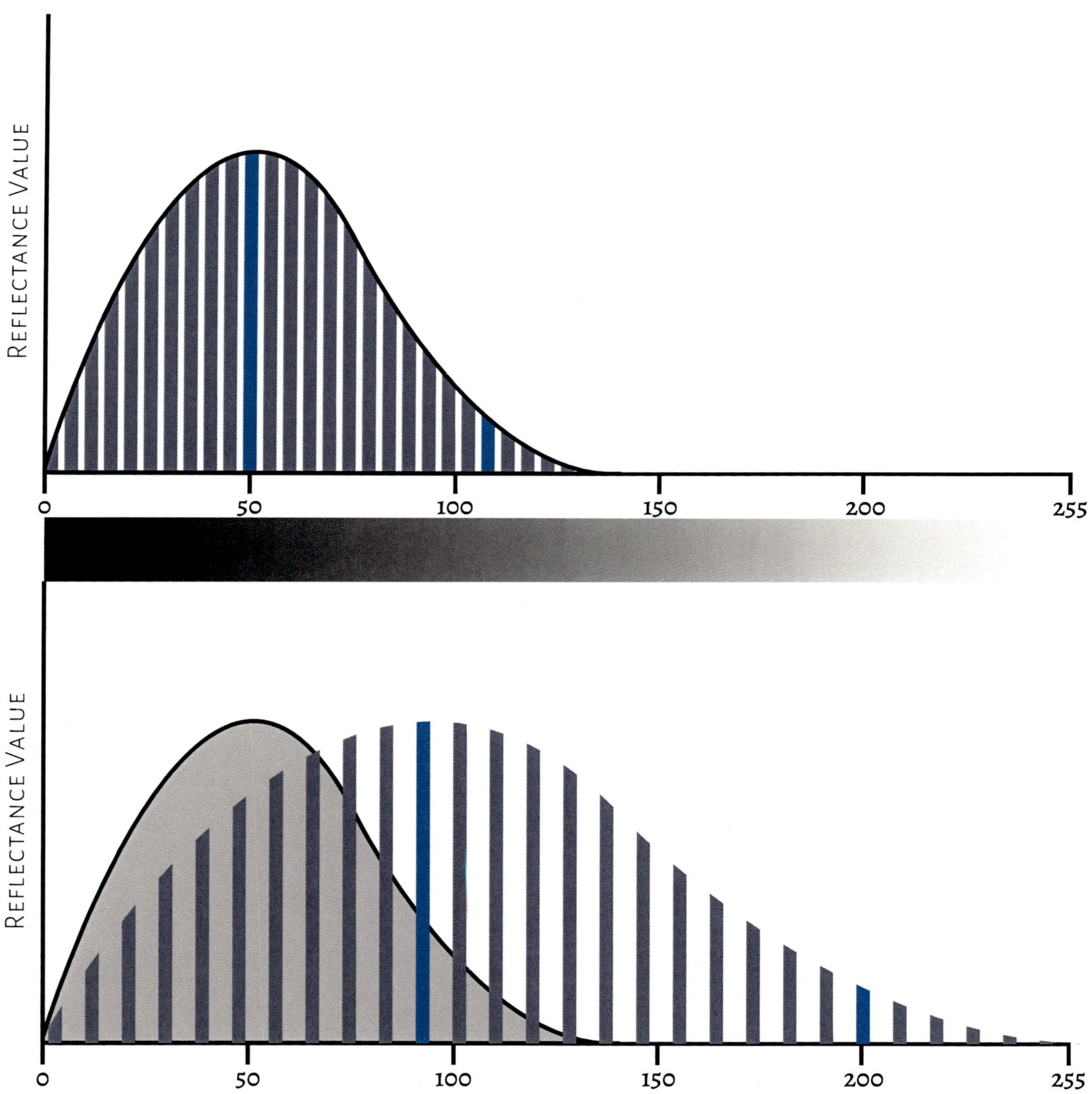

Figure 6.37: The blue bars represent the distribution of pixel values from 0 to 255. 0 is black, and 255 is white. When the values are stretched across the range, a more significant number of pixels are assigned brighter values.

Tutorial: Working with NAIP Imagery

Agencies in the United Stated regularly acquire aerial imagery and make it available to the public. USGS began a pilot program called the **National Agriculture Imagery Program (NAIP)** in 2002. The program has expanded since then and is currently ongoing. In this activity, you learn how to obtain aerial photos, such as NAIP imagery, from the USGS Earth Explorer. After inspecting the images, you then create new raster datasets using the *Project Raster* tool.

ESTIMATED TIME TO COMPLETE THIS TUTORIAL: 2 HOURS

Learning Outcomes

Readers should be able to accomplish the following outcomes by the end of this tutorial:

- » Summarize the steps for creating and organizing a project workspace folder structure
- » Illustrate how to find and download NAIP imagery for an area of interest
- » Describe how to check the spatial reference and resolution of aerial imagery
- » Practice using the *Project Raster* tool in ArcMap
- » Demonstrate how to compress specific folders for backup

Skill Drill: Setting Up Your Workspace

On your desktop, create a new folder and give it a descriptive name, such as *NAIP_Imagery*. Be sure there are no spaces in the name. You may use underscores instead of spaces. Inside this folder, create the following three subfolders: *original, working,* and *final*.

Special Considerations for Remote Sensing Projects

While the standard folder structure works well for most GIS projects, there are some issues to consider when working on remote sensing projects. The file sizes for aerial imagery and satellite data can be quite large. As you modify or derive new imagery data, your workspace folder can reach many gigabytes in size. When it comes to backing up your data in the cloud, the size of your workspace folder can become a problem. Potential trouble can come from your cloud storage limits or from the time it takes to upload your backup files. One way to solve the issue is only back up important folders. For example, it may not be required to keep copies of the original imagery if they are readily available from online sources. In this instance, you may want to only back up files in your *working* folder to save time and storage space. At the end of this activity, you practice this method.

Be sure to pay attention to where you save files and maintain file management discipline.

Downloading Data from the USGS Earth Explorer

In previous courses, you learned how to acquire data from public sources. Here you download digital imagery from the USGS Earth Explorer. Navigate to the Earth Explorer website at ***http://earthexplorer.usgs.gov/***[1]. If you don't already have an account, use the *Register* link located on the upper right of the page. Fill out the User Registration form to create an account (**Figure 6.38**). If you are already registered, click the Login link located on the upper right of the page.

1 http://earthexplorer.usgs.gov/

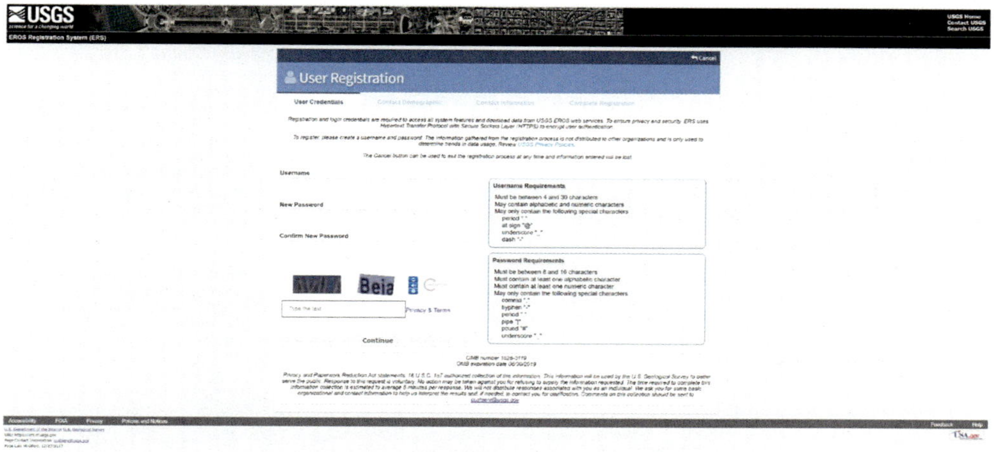

Figure 6.38: Creating an account is free, and it is necessary to have access to high-resolution data.

On the map, use the scroll wheel on your mouse to zoom to the region just north of Humboldt Bay. Place a marker at Humboldt State University by clicking on the map. The geographic coordinates appear on the *Search Criteria* tab on the left, and a corresponding pin gets placed on the map (**Figure 6.39**).

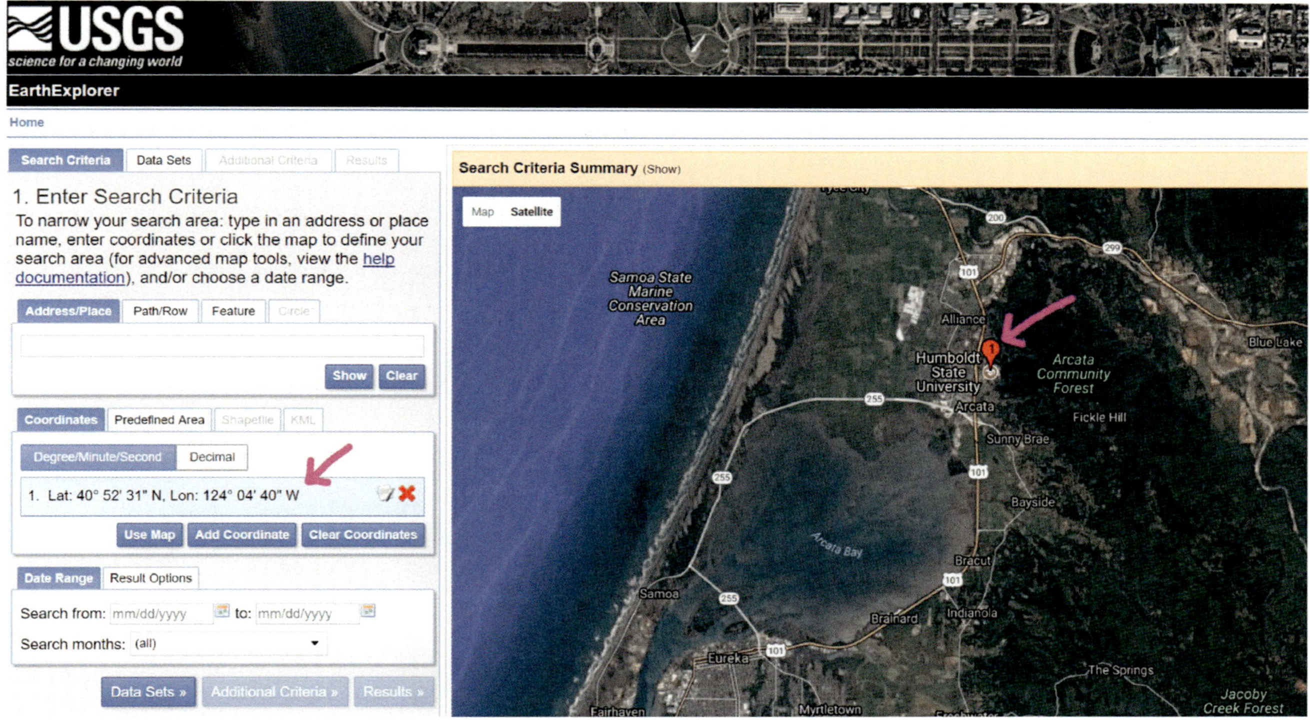

Figure 6.39: You can place a marker in an area of interest.

Next, to Search Criteria, click the tab that says *Data Sets* (**Figure 6.40**). Expand the plus sign next to *Aerial Imagery*. On the list of data sources, check the box for *NAIP GEOTIFF*. When you are ready, click the *Results* button near the bottom.

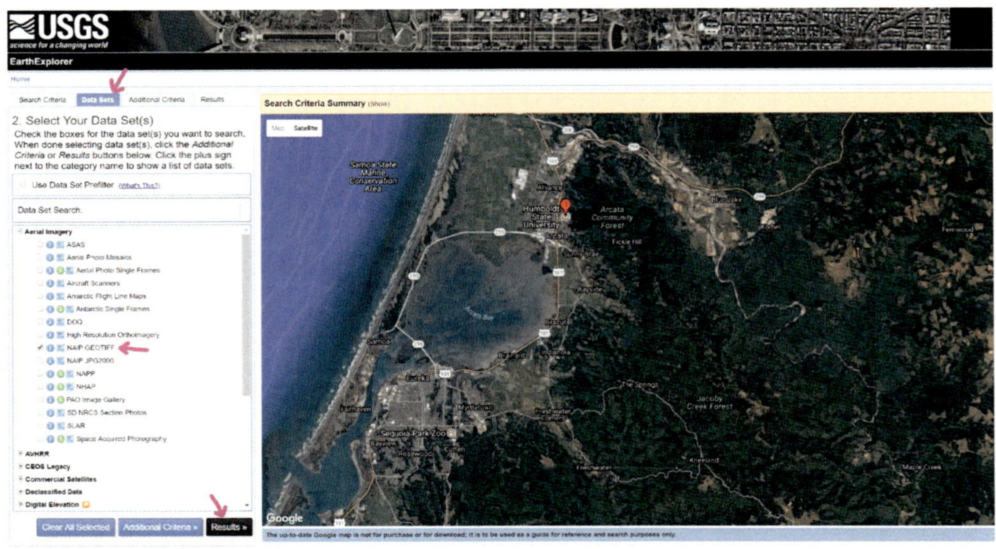

Figure 6.40: The USGS Earth Explorer provides aerial imagery from many sources.

The *Results* tab opens with a list of *NAIP GEOTIFF* images that match the criteria (**Figure 6.41**). The USGS lists the images in chronological order. Click the icon that looks like a foot. The map zooms to the image location, and the image footprint appears on the map.

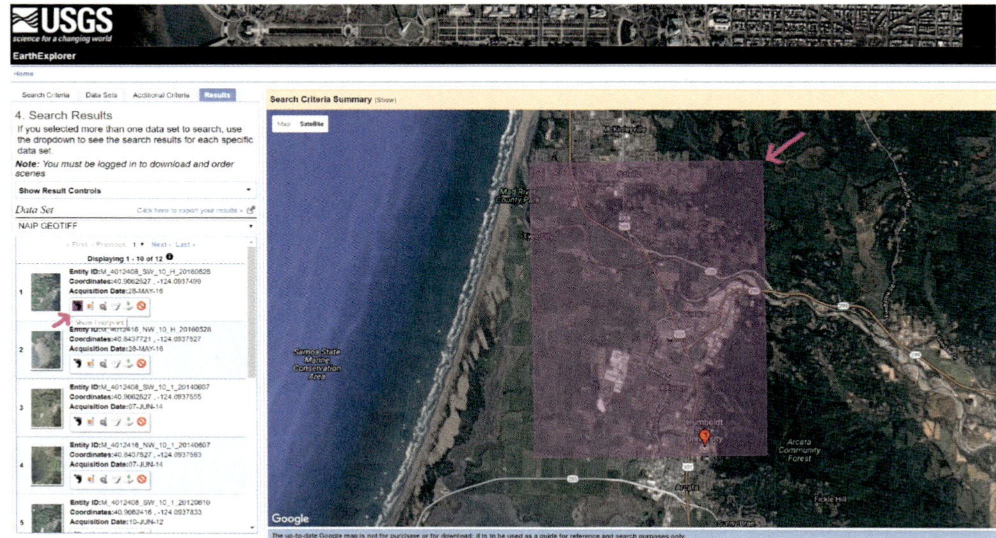

Figure 6.41: You should check the image footprints to make sure they cover your area of interest.

Click the thumbnail of the image to view the metadata. A preview of the image appears along with a metadata table (**Figure 6.42**). Take a moment to read through the metadata. Try to find the answer to the following questions:

» On what date was the image taken?
» Does the image use a map projection? Which one?
» What spatial reference does the image use?
» What is the resolution or cell size of the image?

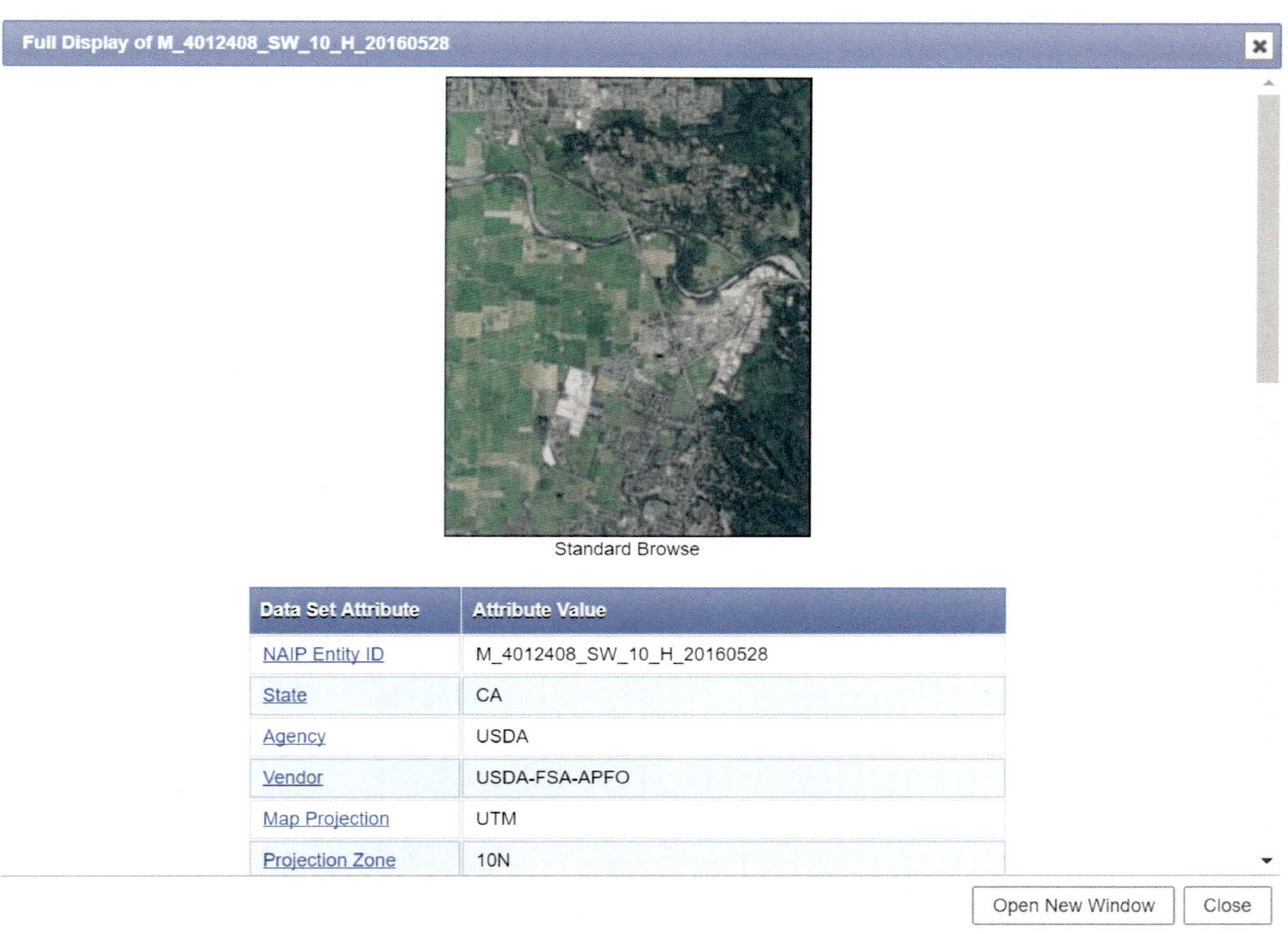

Figure 6.42: The USGS Earth Explorer provides detailed metadata.

When done, close the metadata window. Click the download link for the first NAIP image (**Figure 6.43**).

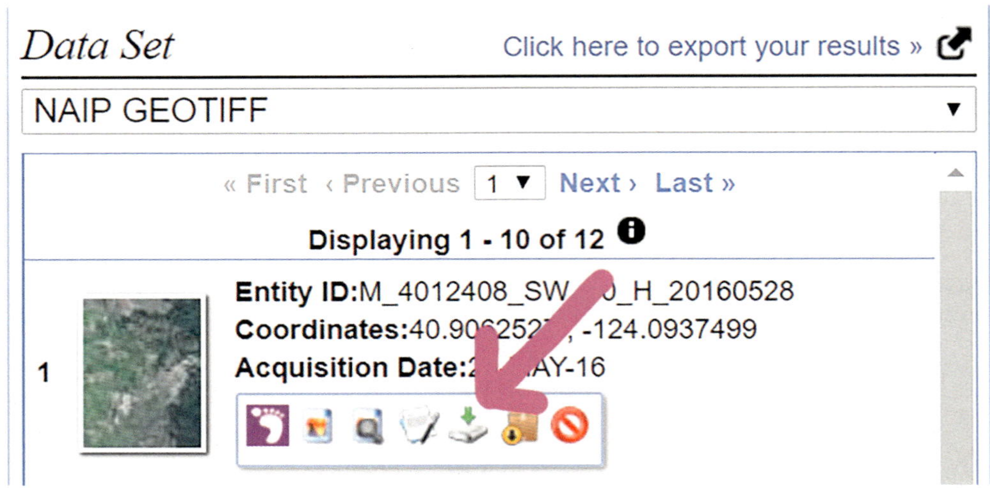

Figure 6.43: The download icon has a green arrow pointing down towards a hard drive.

When the download options appear, click the download button (**Figure 6.44**). In previous activities, you downloaded directly to your *original* folder. In this instance, the file automatically goes to your downloads folder on your computer. Locate the compressed zip file in the downloads folder and manually move it to your *original* folder.

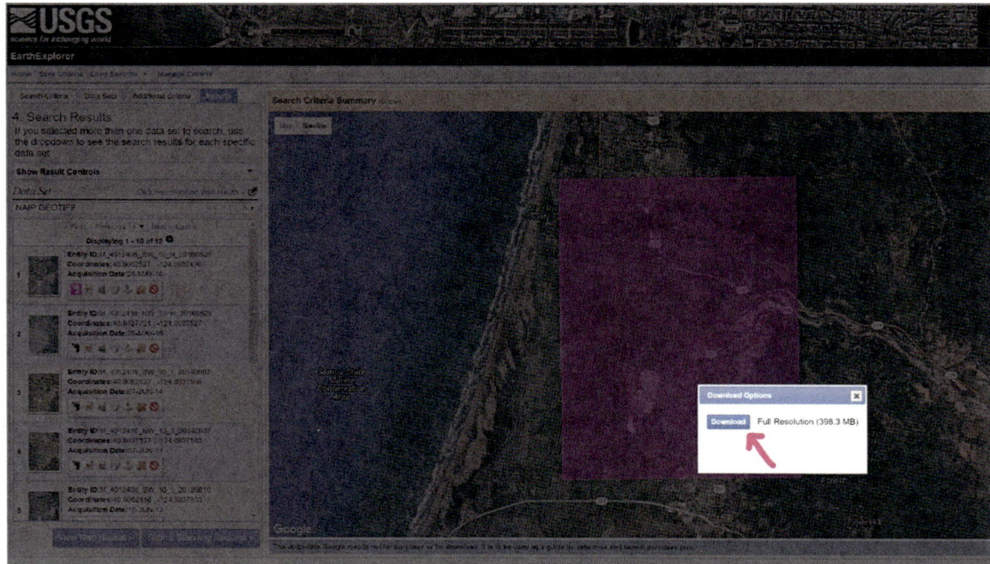

Figure 6.44: If you are logged in, you should have access to the full resolution image when downloading.

Use the 7zip software to decompress the zip file inside your *original* folder (**Figure 6.45**). Remove the compressed zip file when done. This step is especially important when working with imagery because the files are so large. You don't need the zip file anymore. Removing saves space and helps to prevent confusion later.

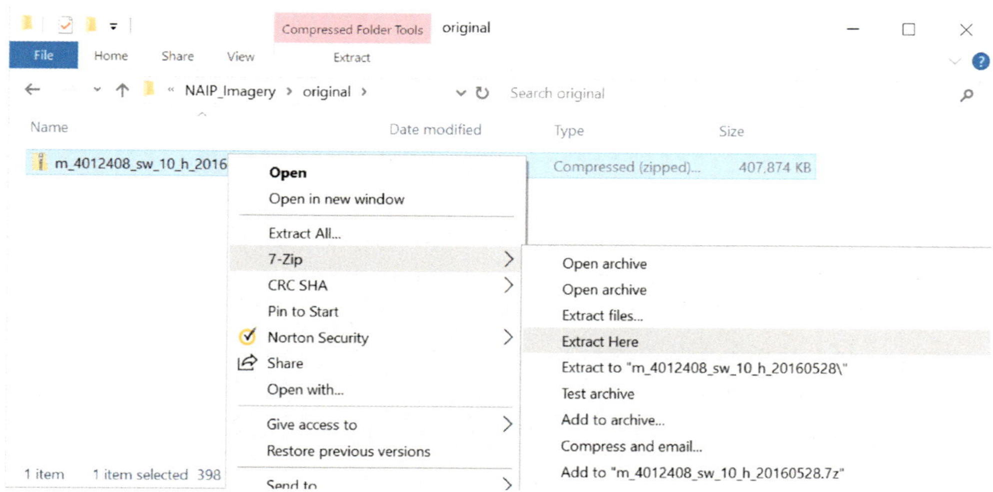

Figure 6.45: You can access the 7zip software by right-clicking on a compressed file

Skill Drill: Download Additional Images

Using what you learned in the previous steps, download the second, third, and fourth NAIP GEOTIFF from the USGS Earth Explorer (**Figure 6.46**).

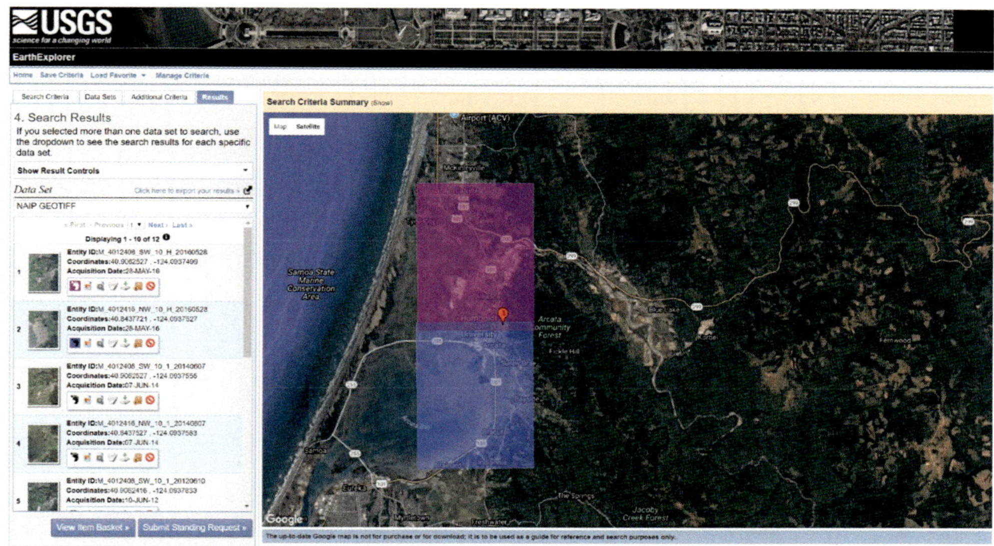

Figure 6.46: As the footprint reveals, there is a slight overlap between images.

Move the files to your *original* folder and decompress them. Remember to remove the compressed zip files when done. Afterward, you should have the contents of the four images inside your *original* folder (**Figure 6.47**). After downloading all four images and decompressing them, launch ArcMap and open a blank map document. Add each of the four images to the map. You may get a message asking if you want to build pyramids (**Figure 6.48**).

In most situations, the answer to this question is always Yes. **Pyramids** are supplementary files for raster datasets. They make copies of the original data in decreasing levels of resolution to enhance the display performance. Though pyramids allow for the rapid display of images, they do increase the overall file size of your workspace folder.

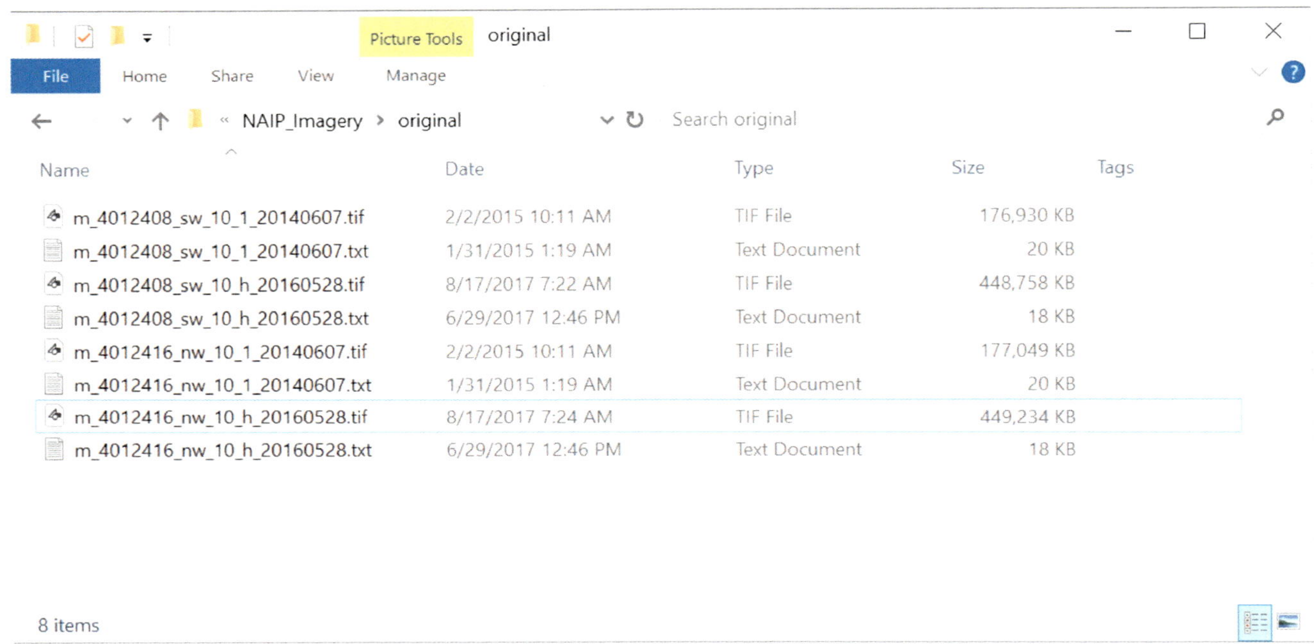

Figure 6.47: Related text files that contain metadata accompany each TIF file.

When done, click the Full Extent button, which looks like a globe, to see all images at once (**Figure 6.49**). In the Table of Contents, uncheck the boxes next to each image. Then, turn on each image one at a time. Compare the differences between overlaying images from one year to the next. Zoom in close to a feature that interests you. Compare the image resolution between images. Try to determine if there are differences in quality from one year to the next.

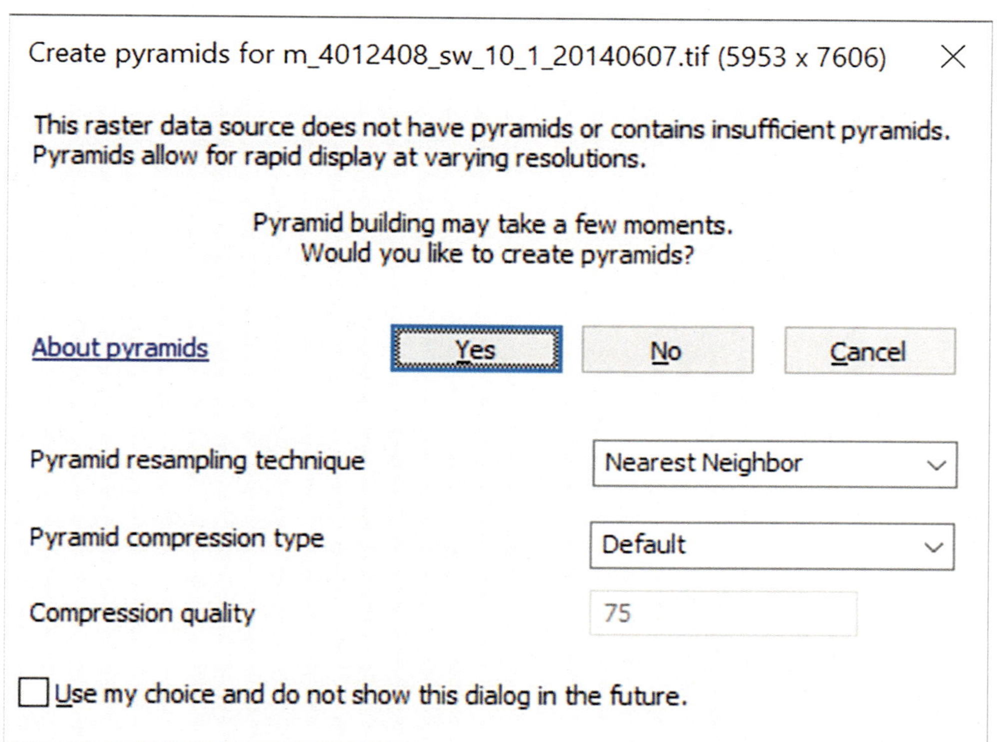

Figure 6.48: Pyramids allow for quick viewing but increase the file size of your workspace folder.

Figure 6.49: Each pair of images covers the northern and southern regions of Arcata.

Creating New Imagery Files Using the Project Raster Tool

Open the layer properties for the first image in the Table of Contents. Then, navigate to the *Source* tab (**Figure 6.50**). Scroll down and locate the spatial reference system. It says NAD 1983 UTM Zone 10N. In a previous chapter, you learned about the UTM spatial reference system.

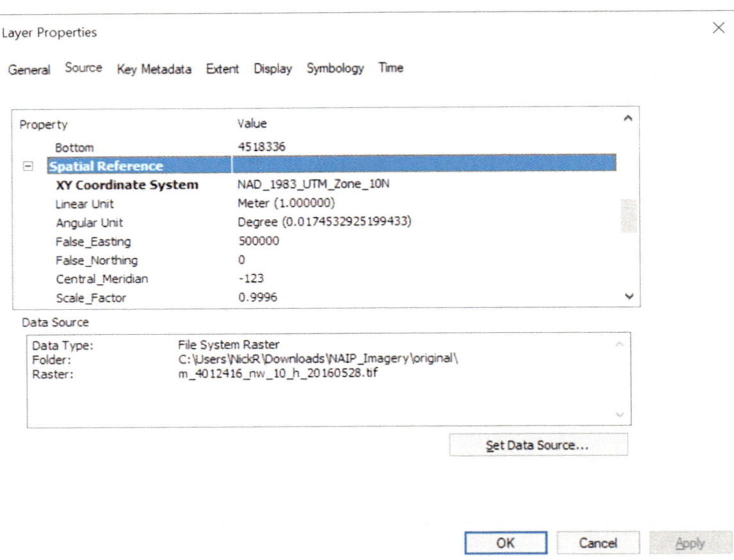

Figure 6.50: The Source tab provides information about a raster dataset, including resolution, file type, and spatial reference.

The UTM system is an excellent spatial reference system, and Zone 10 is optimal for Northern California. However, there may be cases where you need to use a different spatial reference system, such as the State Plane Coordinate System of 1983 (SPC83). In this step, use the *Project Raster* tool to create several new raster datasets in SPC83. In ArcMap, click the toolbox button located on the *Standard* menu. The icon looks like a red toolbox.

The ArcToolbox may appear floating above the map. Dock to the side by dragging it to the left in a similar manner in which you docked the attribute table in previous activities (**Figure 6.51**). Expand the *Data Management Tools*. Next, expand the *Projections and Transformations* toolbox, then *Raster*. Locate the *Project Raster* tool.

To keep track of which image you are currently working on, uncheck all of the images in the Table of Contents except the first image. Then right-click and *Zoom to Layer* (**Figure 6.52**).

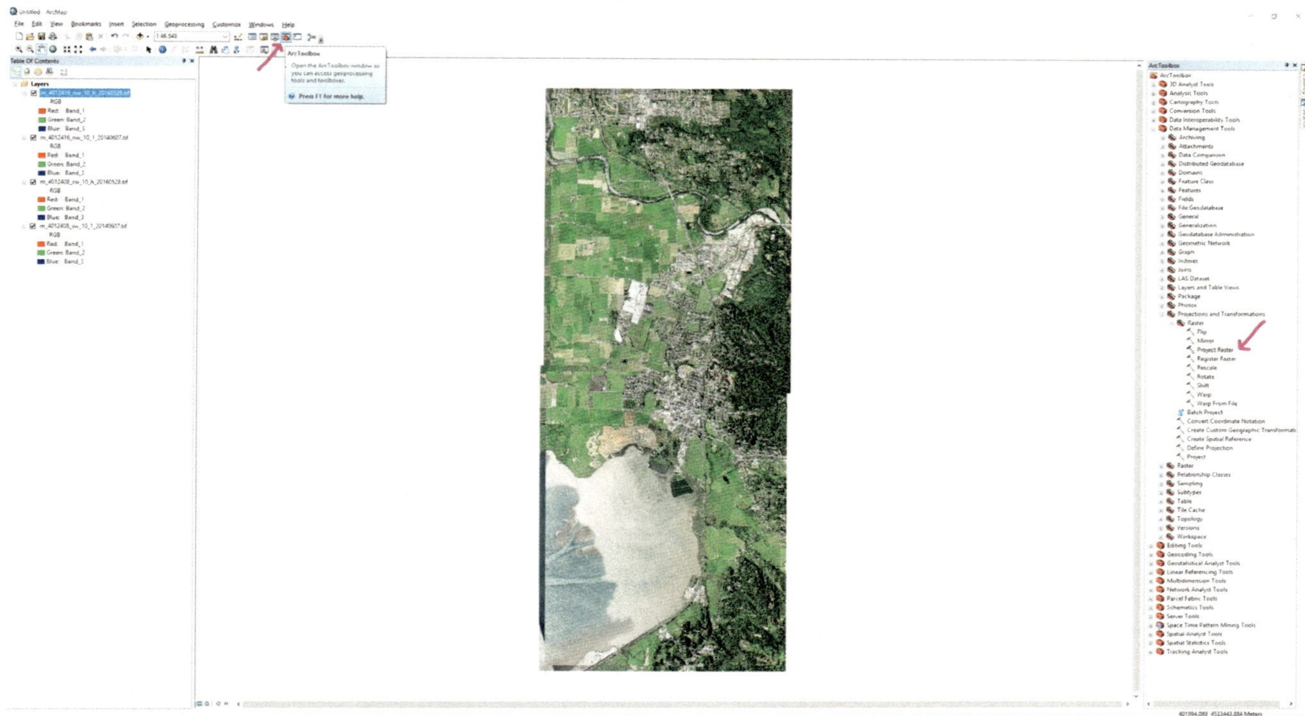

Figure 6.51: The *Project Raster* tool is located within the Raster toolbox inside Projections and Transformations.

Figure 6.52: In this example, the image displays the southern region of the City of Arcata in the year 2016.

Double click the *Project Raster* tool. The tool appears floating above the map. For the *Input Raster*, choose the current raster dataset on the list from the drop-down menu. For the *Output Raster Dataset*, it is essential that you click the yellow file folder icon and browse to your *working* folder (**Figure 6.53**). Name the file ArcataSouth2016.tif and click *Save*. Then, click *Save* a second time.

Always browse to your working folder when saving the output from any tool. Never accept the default location.

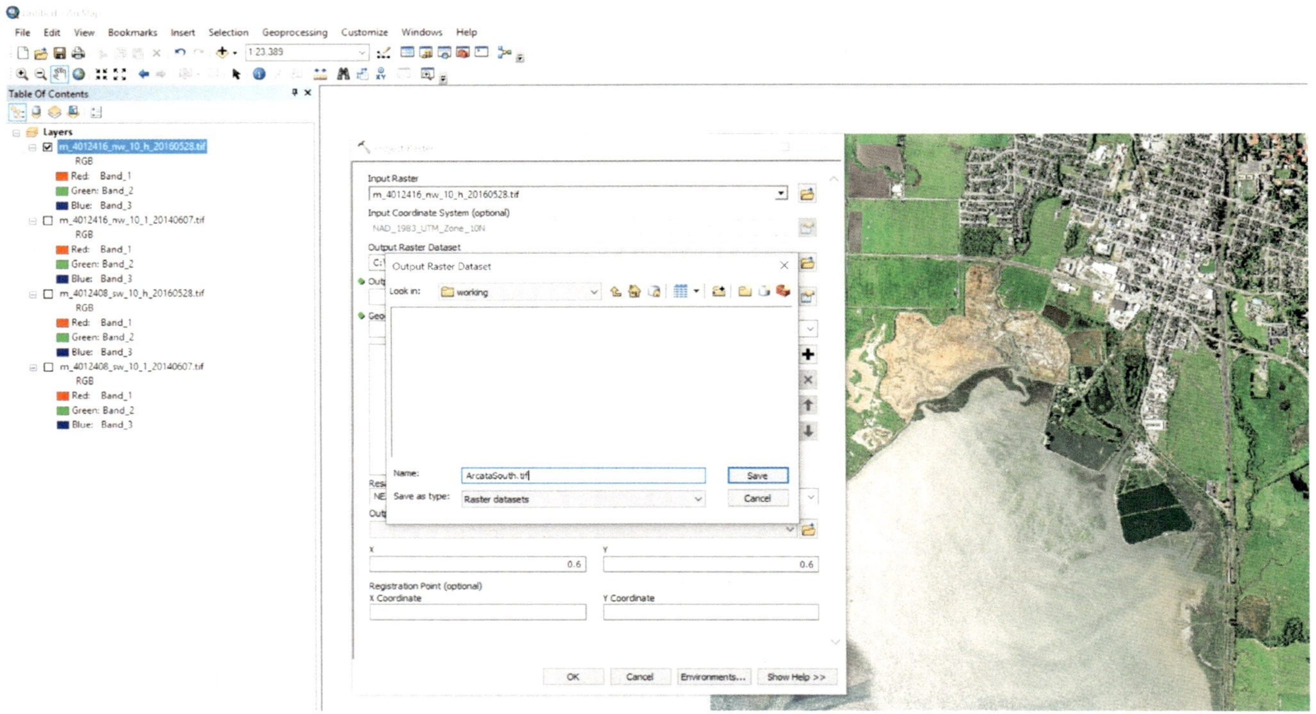

Figure 6.53: Manually adding the **.tif** at the end of the name is necessary.

For the *Output Coordinate System*, click the *Spatial References* button on the right to open the *Spatial References Properties* window (**Figure 6.54**). The icon looks like a hand pointing to a page. Open the Projected Coordinate Systems folder. Expand the State Plane folder, then NAD 1983 (US Feet). Scroll down and select California I as the zone for this spatial reference. When ready, click OK. Leave all other default settings as they are and click OK again (**Figure 6.55**).

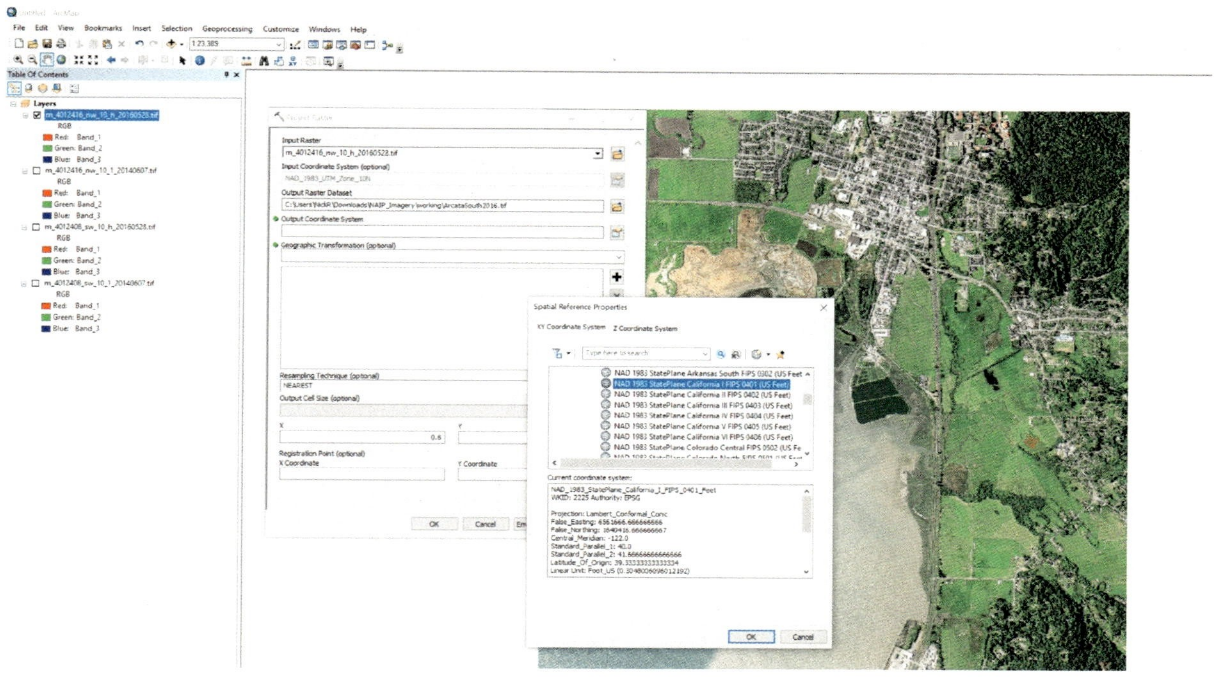

Figure 6.54: The Spatial Reference Properties window allows you to choose a spatial reference system from two main folders, Geographic Coordinate Systems and Projected Coordinate Systems.

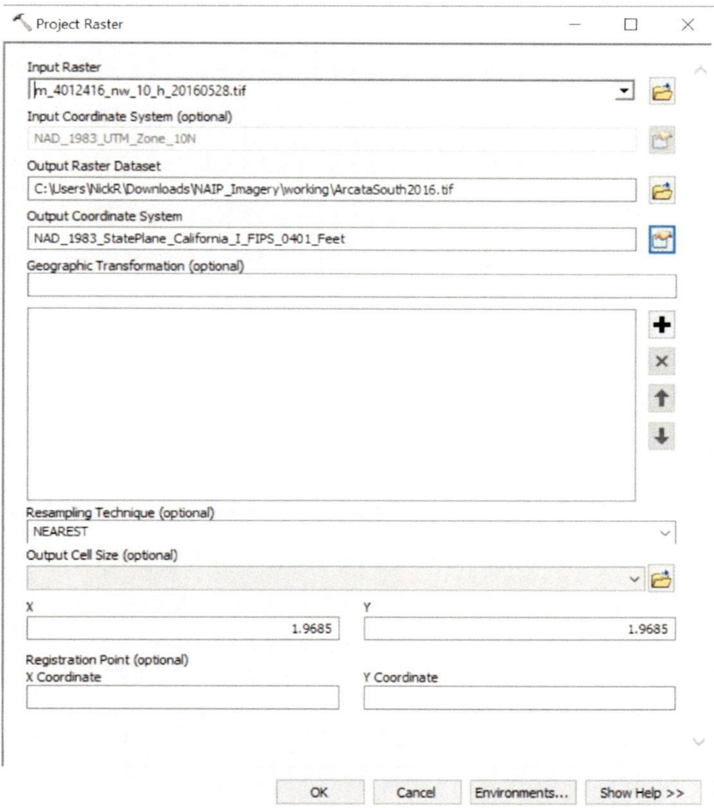

Figure 6.55: Check to make sure that your settings are similar to the ones shown here.

It may take a few minutes for the tool to complete its operation. Once you add the new raster dataset to the map, open the layer properties, and review the *Source* tab (**Figure 6.56**). Compare the following details with the original dataset:

- Cell Size
- XY Coordinate System
- Linear Unit

The new raster dataset uses the SPC83 system, which uses feet instead of meters. The cell size and linear unit reflect these changes.

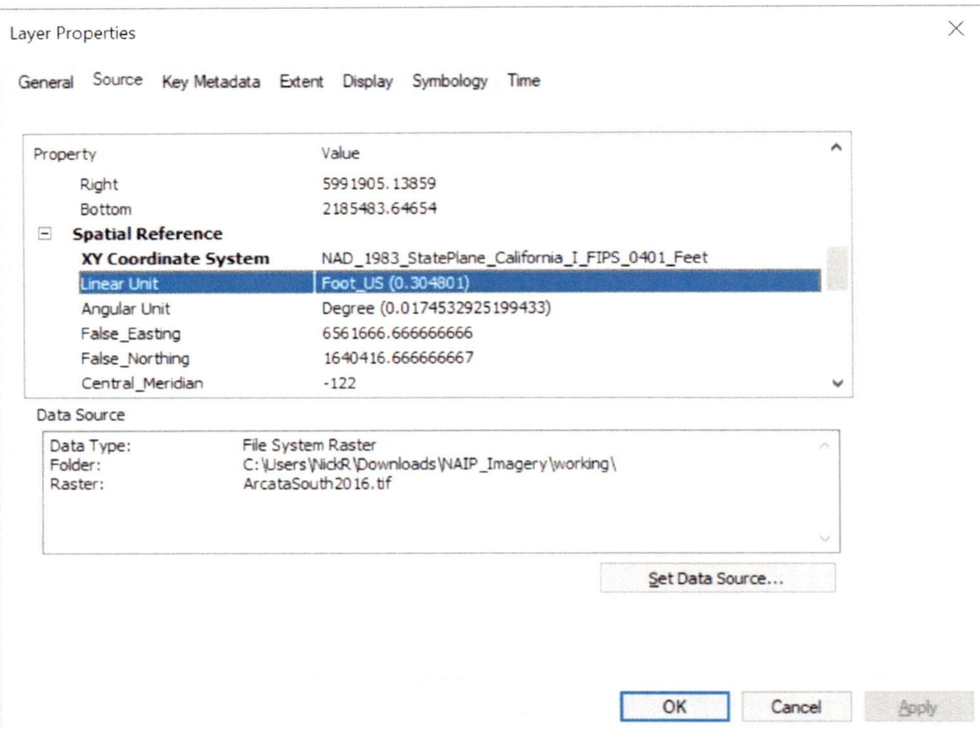

Figure 6.56: The Source tab provides information about a raster dataset, including resolution, file type, and spatial reference.

Skill Drill: Use the Project Raster Tool on the Remaining Datasets

Remove the original image layer for southern Arcata in 2016 from the Table of Contents. Repeat the procedures you learned in the previous steps and run the *Project Raster* tool on the remaining three raster datasets (**Figure 6.57**). Try not to rush the process and make sure that only one raster dataset is visible at a time. Doing so helps to avoid confusion as you name the new files. Be sure to give them appropriate names such as ArcataSouth2014.tif, ArcataNorth2016.tif, and ArcataNorth2014.tif. When done, you should have four new raster datasets saved in your *working* folder.

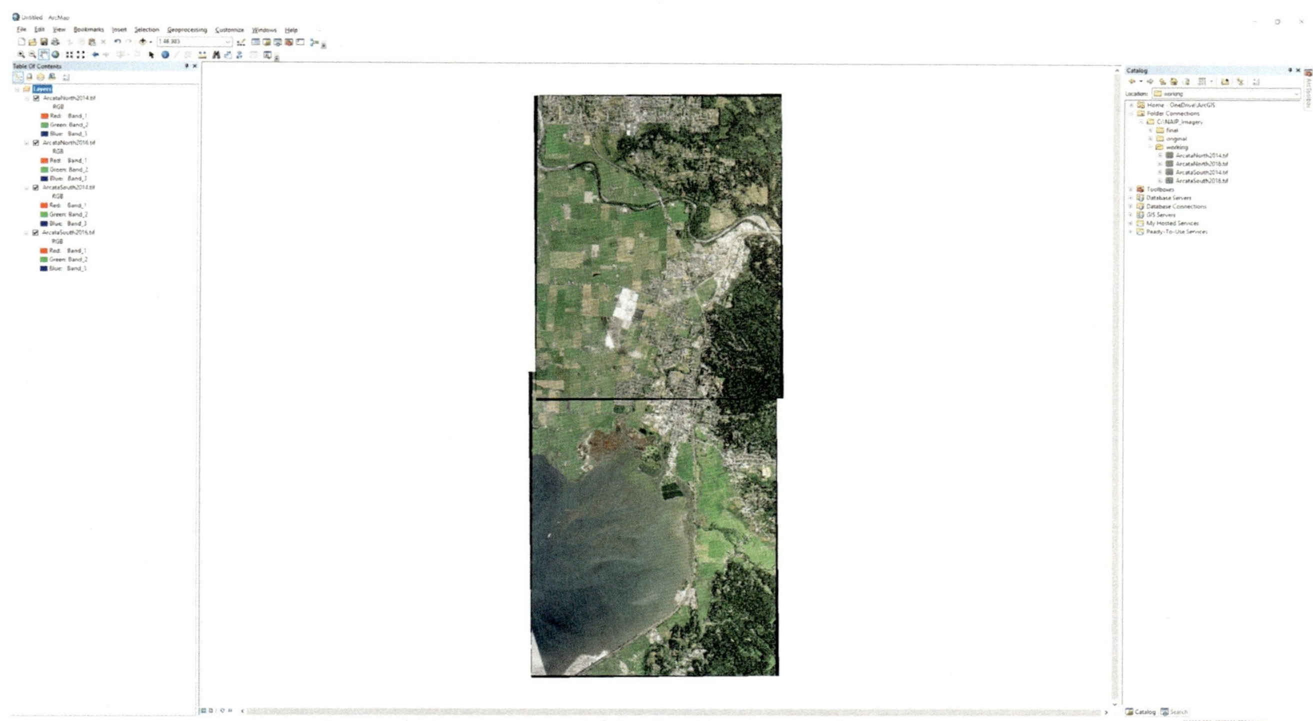

Figure 6.57: Check your *working* folder to make sure it contains the four new NAIP images.

Sharing the Data Using Google Drive

Close ArcMap. You won't need to save the map document for this activity. In this next step, imagine that you need to back up only your *new* raster datasets. Assume that space is limited, and you don't want to save the original files downloaded from the USGS Earth Explorer. Since the *working* folder contains all of the data you need, compress and back it up to Google Drive. Open your workspace folder in Microsoft Windows File Explorer. Right-click on the *working* folder. Select 7-Zip, then Add to working.7z (**Figure 6.58**). It may take a few minutes for the folder to compress.

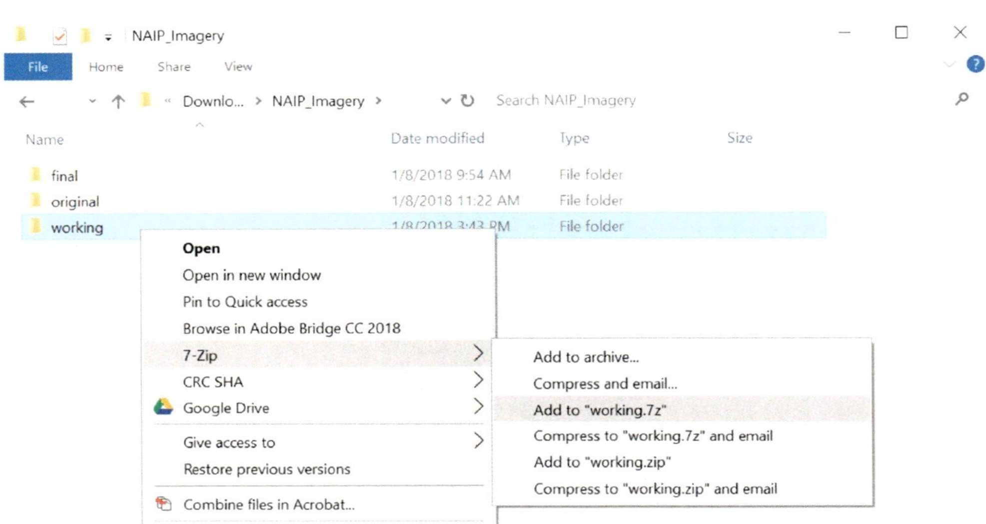

Figure 6.58: You can access the 7zip software by right-clicking on a folder.

The name working.7z is fine when placed in the context of the standard folder structure. However, this compressed file stands alone on your Google Drive. For clarity, rename the compressed 7z file ArcataNAIP.7z (**Figure 6.59**). Right-click on the file and select *Rename*. When renaming the file, be sure it ends with a .7z.

Using the Chrome browser, upload your 7z file to **Google Drive**. Once the file is uploaded, change the settings so that you can share the file using a sharable link. In the Chrome browser, right-click on the name of the file and select *Get shareable link* (**Figure 6.60**).

Toggle the *Link sharing on* button and copy the URL provided (**Figure 6.61**). Be prepared to retrieve the data for later use.

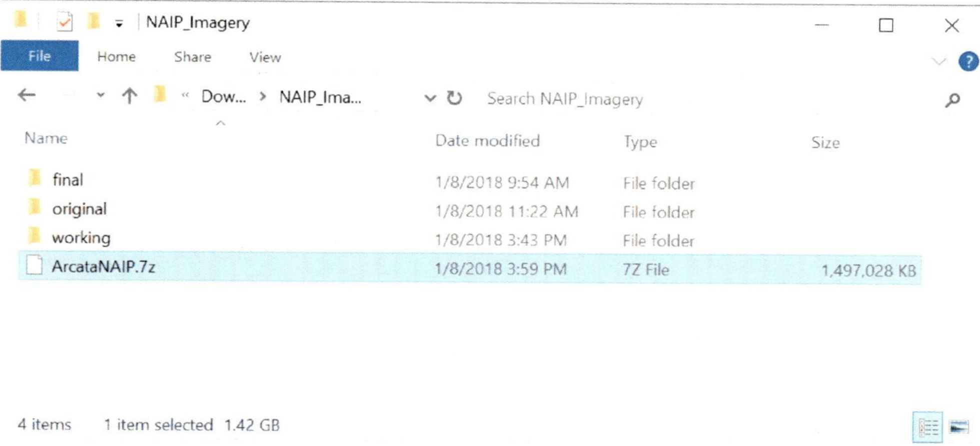

Figure 6.59: As you can see, the file size is nearly 1.5 gigabytes. Be prepared to manage your data for remote sensing projects carefully.

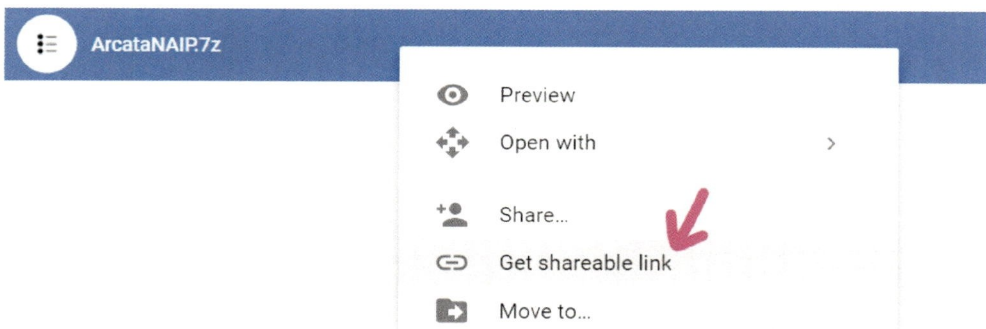

Figure 6.60: You can access the contextual menu by right-clicking on the 7z file in Google Drive.

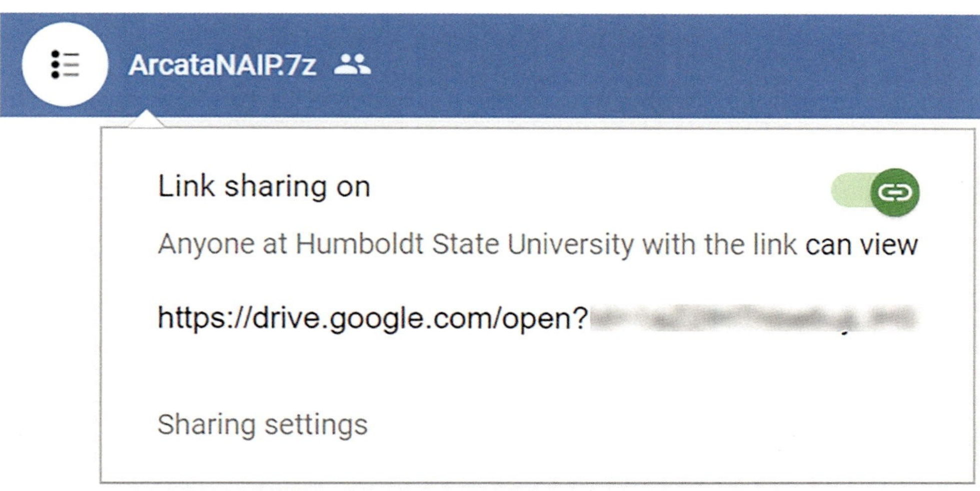

Figure 6.61: For best results, use the Chrome browser.

Tutorial: Georeferencing an Aerial Photograph

Today, most readers have little direct experience with film-based photographs. However, there still may be a high volume of film-based historic photographs stored in basements and file cabinets of federal, state, or municipal government agencies as well as non-profit organizations. When you want to bring these images into a GIS, georeferencing is required. In this activity, you georeference an aerial photo with the help of NAIP imagery downloaded from the USGS Earth Explorer.

ESTIMATED TIME TO COMPLETE THIS TUTORIAL: 1 HOUR

Learning Outcomes

Readers should be able to accomplish the following outcomes by the end of this tutorial:

- » Summarize the steps for creating and organizing a project workspace folder structure
- » Illustrate how to download and decompress a folder stored in the cloud
- » Describe how to check the spatial reference and resolution of aerial imagery
- » Practice using the adding new toolbars in ArcMap
- » Demonstrate how to georeference an aerial photograph

Skill Drill: Setting Up Your Workspace

On your desktop, create a new folder and give it a descriptive name, such as *Georeferencing _Photos*. Be sure there are no spaces in the name. You may use underscores instead of spaces. Don't create any subfolders yet. Read the next section for more instructions.

Special Considerations for Remote Sensing Projects

While the standard folder structure works well for most GIS projects, there are some issues to consider when working on remote sensing projects. The file sizes for aerial imagery and satellite data can be quite large. As you modify or derive new imagery data, your workspace folder can reach many gigabytes in size. When it comes to backing up your data in the cloud, the size of your workspace folder can become a problem. Potential trouble can come from your cloud storage limits or from the time it takes to upload your backup files. One way to solve the issue is only back up important folders. For example, it may not be required to keep copies of the original imagery if they are readily available from online sources. In this instance, you may want to only back up files in your working folder to save time and storage space. In a previous activity, you practiced this procedure by backing up your *working* folder. Unlike previous activities, start by creating only two subfolders: *original* and *final*. Download your compressed 7z file from the previous tutorial, *Working with NAIP Imagery*, and save it to your workspace. The file might be called ArcataNAIP.7z (**Figure 6.62**).

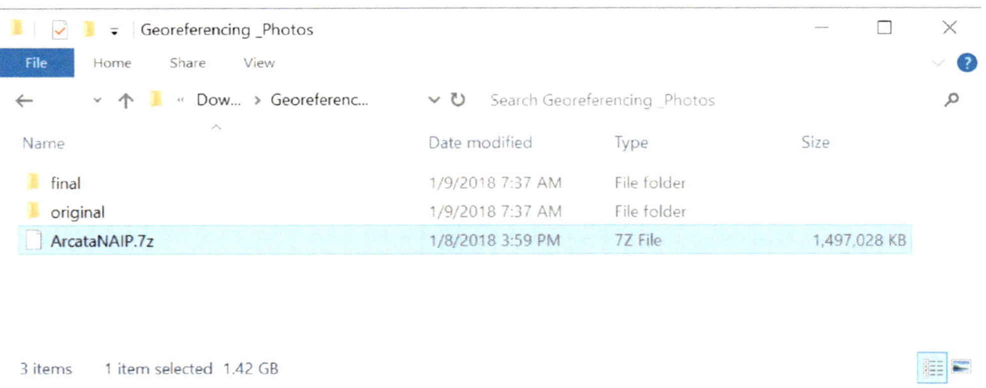

Figure 6.62: The compressed 7z file comes from the previous activity.

Be sure to pay attention to where you save files and maintain file management discipline.

Using the 7zip software decompress ArcataNAIP.7z. It should uncompress the previous *working* folder with four NAIP images inside. It may take several minutes to decompress the file. When done, delete the compressed 7z file from your workspace (**Figure 6.63**). You do not need it anymore.

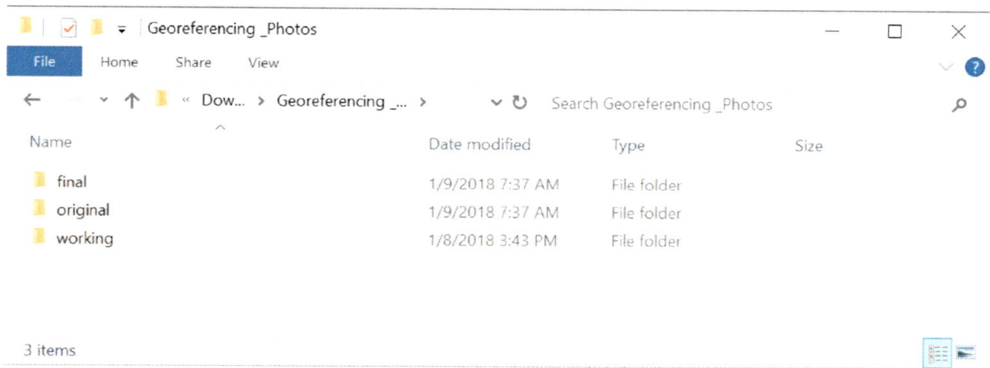

Figure 6.63: The workspace should contain the standard folder structure.

You have one more image to download. The next file is an example of a 9x9 inch aerial photograph digitally scanned and saved as a JPG file (**Figure 6.64**). You can download the file using the following link: ***Aerial Photo with no spatial reference***[1]. When the image appears, right-click on the image, select *Save Link As*, then save the image to your *original* folder.

Figure 6.64: The link takes you to the image page. Once the page opens, download the image by right-clicking.

1 https://www.geospatial.institute/wp-content/uploads/2019/02/nospatialreference.jpg

Working with the Georeferencing Toolbar

Georeferencing is the process of aligning geographic data to a known coordinate system. In this instance, the geographic data is the aerial photograph. You use the NAIP imagery as the known geographic reference. The aerial photograph shows a region north of downtown Arcata and west of Highway 101. The best NAIP image to use here is the one for northern Arcata taken in 2016.

Launch ArcMap and open a blank map document. In the Catalog Window, connect to your workspace folder. Add the ArcataNorth2016.tif (**Figure 6.65**). Try to zoom into approximately the same geographic extent, as shown in the photograph.

Figure 6.65: The map shows roughly the same geographic extent as the aerial photograph.

From the *original* folder, add the aerial photo with no spatial reference. A warning appears telling you that the data is missing spatial reference information (**Figure 6.66**). Usually, this situation poses a serious problem. In this instance, you are expecting this issue. You can close the warning by clicking OK.

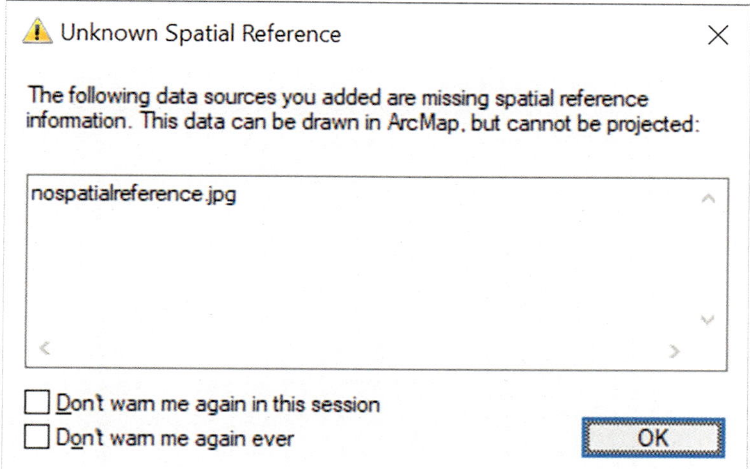

Figure 6.66: One cannot project data that has no spatial reference.

The aerial photo does not appear on the map because ArcMap does not know where it belongs or how large it is. However, the photo does appear in the Table of Contents as a layer.

In a previous activity, you learned how to add toolbars to ArcMap. Right-click on the empty gray area near the top of the ArcMap window to open the contextual menu for toolbars and select *Georeferencing* (**Figure 6.67**). Dock the *Georeferencing* toolbar to the top of the ArcMap window. Notice that the *Georeferencing* toolbar automatically locates the layer with no spatial reference.

Figure 6.67: The *Georeferencing* toolbar locates the first layer in the Table of Contents with no spatial reference.

Double check to make sure the NAIP image has roughly the same geographic extent as shown on the aerial photograph. On the *Georeferencing* toolbar, use the drop-down menu and select *Fit to Display* (**Figure 6.68**).

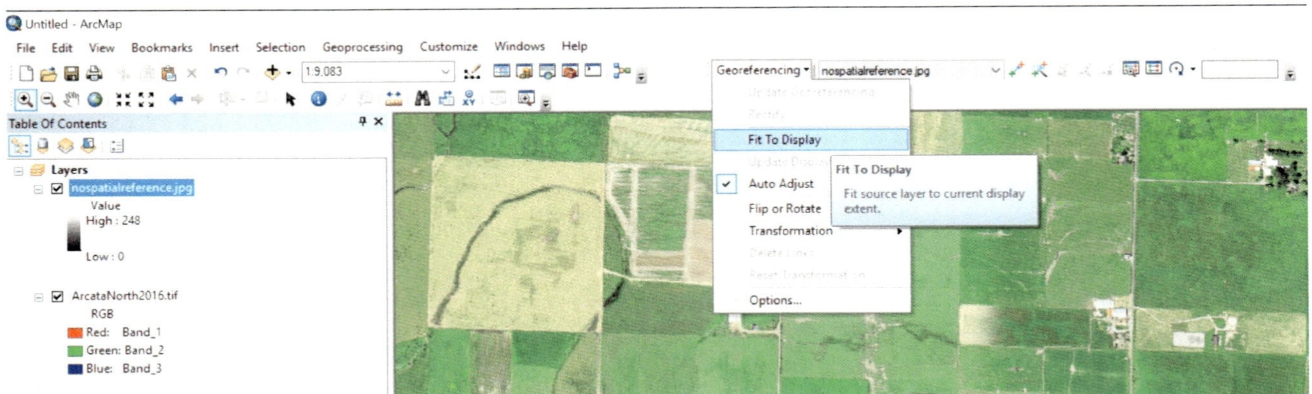

Figure 6.68: Access additional options using the drop-down menu on the *Georeferencing* toolbar.

The aerial photograph fills the data frame (**Figure 6.69**). However, this procedure only roughly georeferenced the photograph. The size and position are still not entirely correct.

Figure 6.69: The Fit to display option gets the photograph roughly into position.

Adding Control Points

To complete the georeferencing, you need to add control points. A **control point** is a location or physical feature with known coordinates used as a reference. Adding control points in ArcMap takes practice. It can be frustrating at times, and you should be prepared to start over if necessary. The order in which you add control points is essential:

1. Mark a point on the unknown spatial reference.
2. Mark the corresponding control point on the known spatial reference.

Choose an easily identifiable feature on the ungeoreferenced aerial photograph that is also visible on the NAIP image (**Figure 6.70**). Zoom in close if necessary.

Figure 6.70: This example uses a road intersection for the first control point and is zoomed closer for accuracy.

On the *Georeferencing* toolbar, click the *Add Control Points* tool (**Figure 6.71**). The icon looks like two plus signs with an arrow in between.

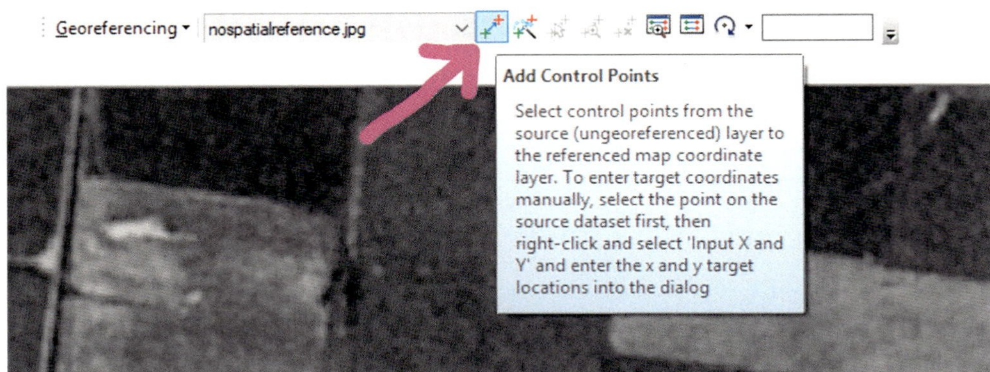

Figure 6.71: The *Add Control Points* tool allows you to define common points between images.

The mouse cursor turns into a cross. Place the cross at the road intersection and click **once** (**Figure 6.72**).

Figure 6.72: The first control point gets added to the unknown spatial reference.

In the Table of Contents, uncheck the box for the aerial photo layer so that only the NAIP image appears in the data frame. Move the cursor over the same road intersection and click **once** (**Figure 6.73**).

Turn the aerial photo on and off to compare. The road intersection on both images should appear in the same place. Repeat these steps for at least two more locations. You should only need about three control points to align this image. Ideally, you want to choose control points that are far away from each other. Clustering control points close together decreases the accuracy. For the second control point, use the intersection between Highway 101 and Giuntoli Lane (**Figure 6.74**). When you add the second control point, the image resizes and rotates.

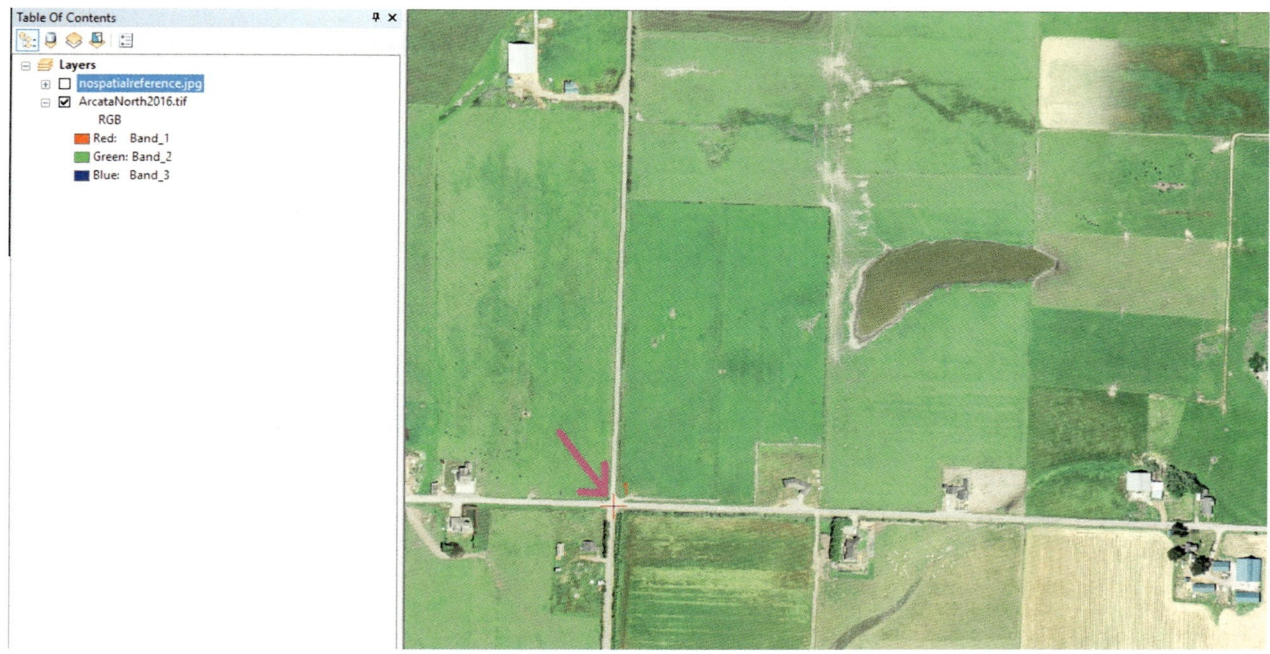

Figure 6.73: A control point is added to the known spatial reference at the same location.

Figure 6.74: The second control point helps to resize the align the image.

Use your judgment and choose a third control point. Ideally, the third point should be located far from the other two. After adding your third control point, return to the *Georeferencing* toolbar. Use the drop-down menu and select *Update Georeferencing*.

Figure 6.75: *Update Georeferencing* saves the spatial reference information.

The aerial photo now has a spatial reference. ArcMap created some additional files to store the spatial reference information. These files are not visible in the Catalog Window because they are considered to be part of the whole image. ArcMap treats this data similar to shapefiles, which also have many parts that do not appear in the Catalog Window. To view these new components, open your *original* folder in Microsoft Windows Explorer (**Figure 6.76**).

To maintain the spatial reference, never separate or delete these extra files.

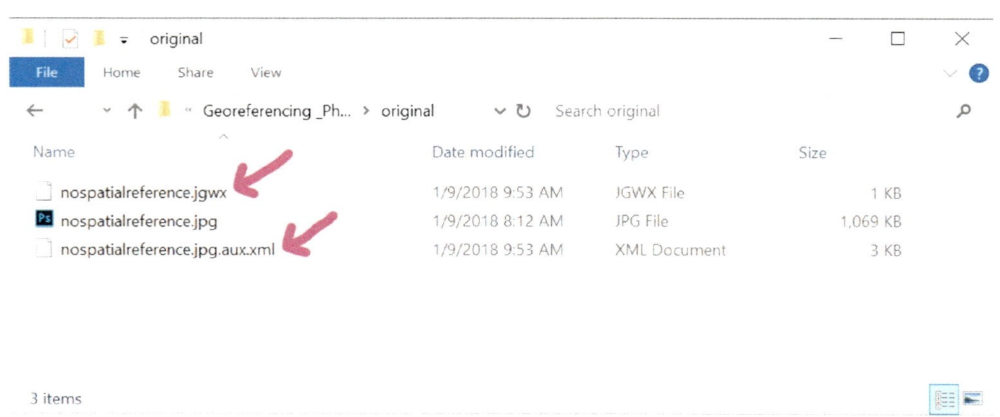

Figure 6.76: The georeferenced aerial photograph now consists of multiple files. Never delete these companion files.

When done, close the ArcMap software. You won't need to save your map document unless you plan to come back and work on this project later. Use the 7zip software to compress your *original* folder as a 7z file. Rename the file georeferencing_project.7z. Upload the compressed 7z file to the cloud using Google Drive. Be prepared to retrieve the data for later use.

TUTORIAL: WORKING WITH LANDSAT IMAGERY

Many governmental and commercial programs acquire remotely sensed imagery from space platforms. Applications for satellite imagery include oceanography, watershed management, forestry, agriculture, urban planning and land use, mineral acquisition, and for other environmental and urban resource management needs. In this activity, you download data from one of the most continuous space imaging programs in the world, the United States Landsat Program. Using Landsat 8 data, you create image composites in natural-color and false-color.

ESTIMATED TIME TO COMPLETE THIS TUTORIAL: 3 HOURS

Learning Outcomes

Readers should be able to accomplish the following outcomes by the end of this tutorial:

- Summarize the steps for creating and organizing a project workspace folder structure
- Illustrate how to find and download Landsat 8 imagery for an area of interest
- Describe the different uses for Landsat 8 spectral bands
- Practice how to create a composite-band image
- Demonstrate how to develop false-color composites

Skill Drill: Setting Up Your Workspace

On your desktop, create a new folder and give it a descriptive name, such as *Landsat_Imagery*. Be sure there are no spaces in the name. You may use underscores instead of spaces. Inside this folder, create the following three subfolders: *original*, *working*, and *final*.

Special Considerations for Remote Sensing Projects

While the standard folder structure works well for most GIS projects, there are some issues to consider when working on remote sensing projects. The file sizes for aerial imagery and satellite data can be quite large. As you modify or derive new imagery data, your *workspace* folder can reach many gigabytes in size.

When it comes to backing up your data in the cloud, the size of your workspace folder can become a problem. Potential trouble can come from your cloud storage limits or from the time it takes to upload your backup files. One way to solve the issue is only back up important folders. For example, it may not be required to keep copies of the original imagery if they are readily available from online sources. In this instance, you may want to only back up files in your working folder or your *final* folder to save time and storage space. Which folder gets backed up depends on your needs. At the end of this activity, you practice this method.

Be sure to pay attention to where you save files and maintain file management discipline.

Skill Drill: Downloading Data from the USGS Earth Explorer

In the previous tutorial, *Working with NAIP Imagery*, you learned how to download data from the USGS Earth Explorer. Navigate to the Earth Explorer website at ***http://earthexplorer.usgs.gov/***. Click the Login link located on the upper right of the page. On the map, use the scroll wheel on your mouse to zoom to the region just north of the United States along the Pacific Coast. For this activity, use the region around Vancouver in British Columbia as the area of interest. Place a marker just north of Vancouver by clicking the map (**Figure 6.77**). The geographic coordinates appear on the Search Criteria tab on the left, and a corresponding pin gets placed on the map.

Next, to Search Criteria, click the tab that says *Data Sets* (**Figure 6.78**). Expand the plus sign next to *Landsat*, then Landsat Collection 1 Level-1. On the list of data sources, check the box for *Landsat 8 OLI/TIRS C1 Level-1*. When you are ready, click the *Results* button near the bottom.

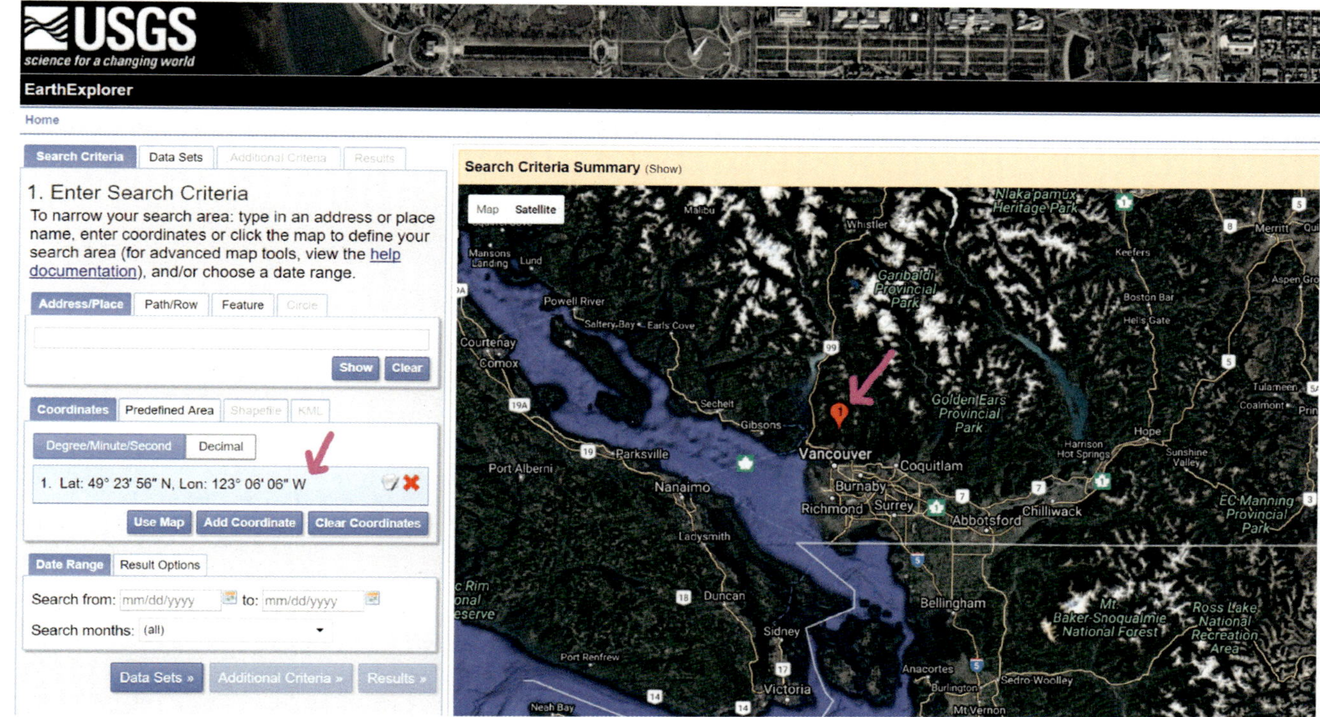

Figure 6.77: You can place a marker in an area of interest.

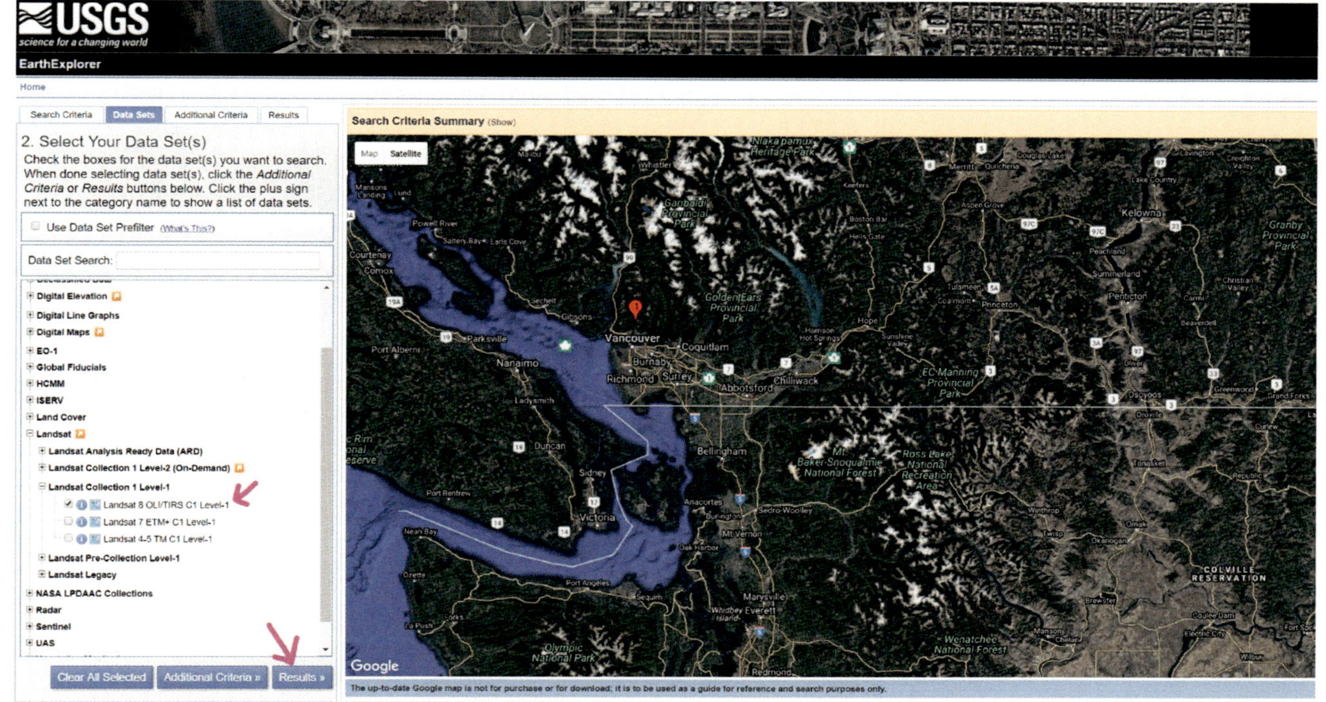

Figure 6.78: The USGS Earth Explorer provides satellite imagery from many sources.

The *Results* tab opens with a list of Landsat 8 images that match the criteria. The USGS lists the images in **chronological order**. In a previous activity, you learned how to browse through the results on Earth Explorer. For this activity, don't worry about the specific location. Any of the images from the search results work fine as an area of interest.

Because the area around Vancouver experiences frequent cloud cover, you need to search for the best image you can find. Ideally, one with little or no cloud cover. Make sure the image is bright green (**Figure 6.79**). If you recall, you can preview the images by clicking on the thumbnail.

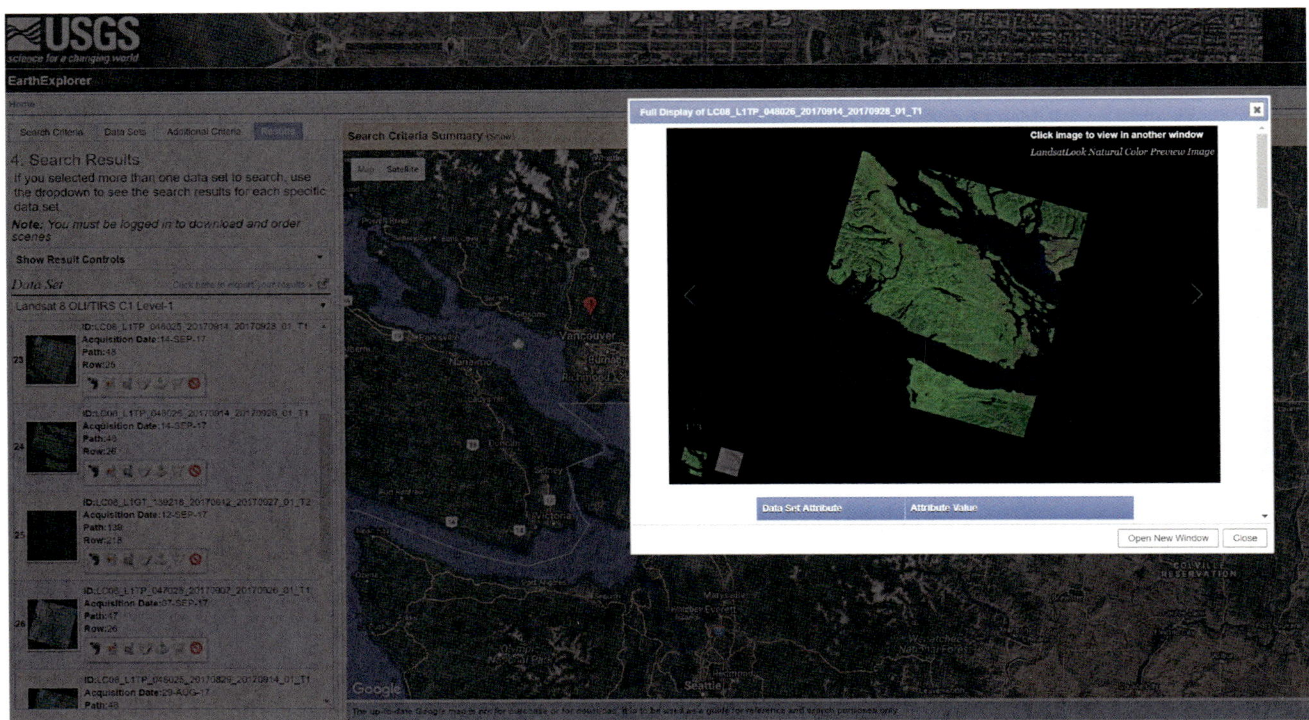

Figure 6.79: This image, dated September 14, 2017, is ideal because the sky is clear.

The preview also provides additional metadata information. Once you have chosen your image, take a moment to find out what kind of metadata is available.

Try to find the answer to the following questions:

» On what date was the image taken?
» Does the image use a map projection? Which one?
» What spatial reference does the image use?
» What is the resolution or cell size of the image?

When done, close the metadata window. Click the download link for the best Landsat 8 image (**Figure 6.80**).

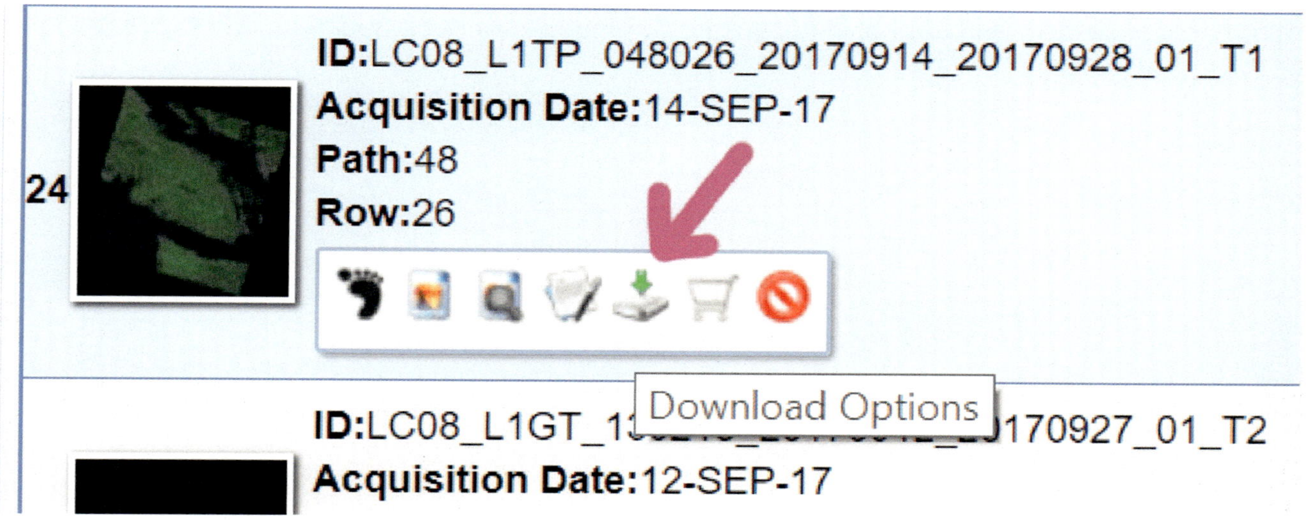

Figure 6.80: The download icon has a green arrow pointing down towards a hard drive.

When the download options appear, click the download button next to Level 1 GeoTIFF Data Product (**Figure 6.81**).

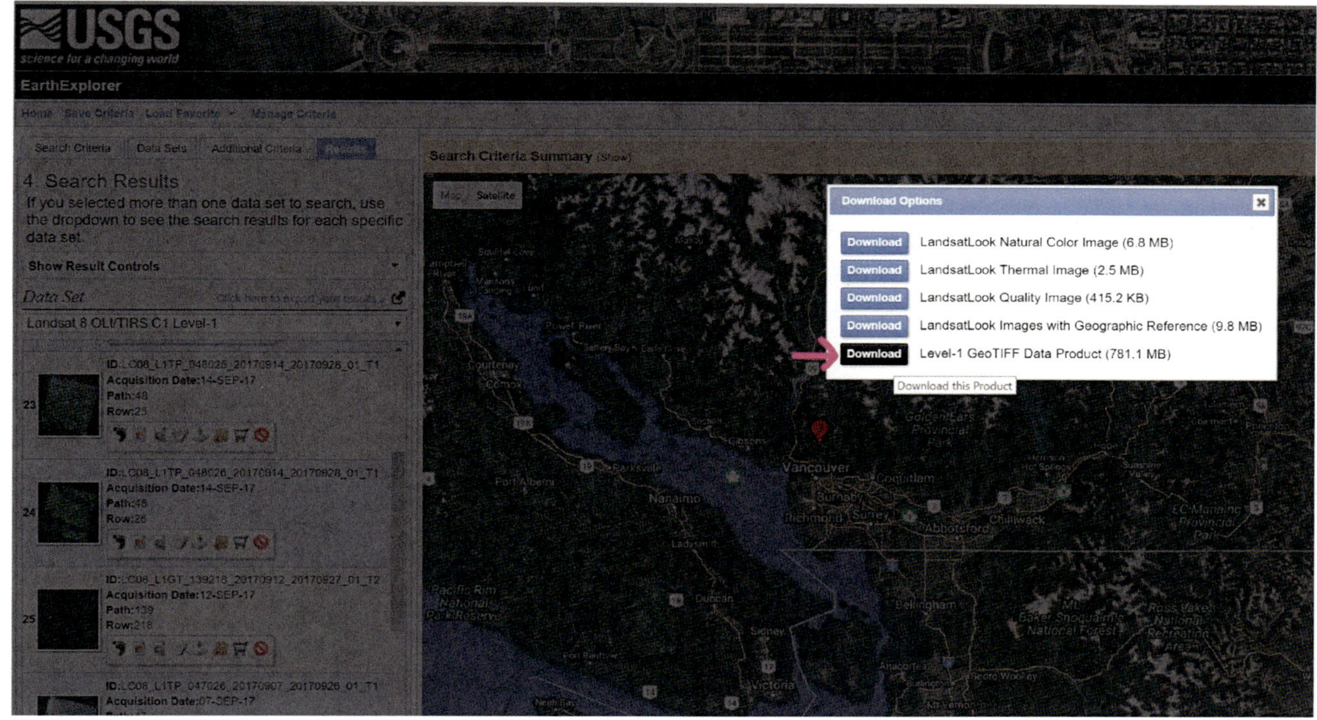

Figure 6.81: If you are logged in, you should have access to the full resolution image when downloading.

The file may take a few minutes to download due to the large file size. In previous activities, you downloaded directly to your original folder. In this instance, the data automatically saves the file to your downloads folder on your computer. You will have to locate the compressed tar.gz[1] file in the downloads folder and manually move it to your original folder.

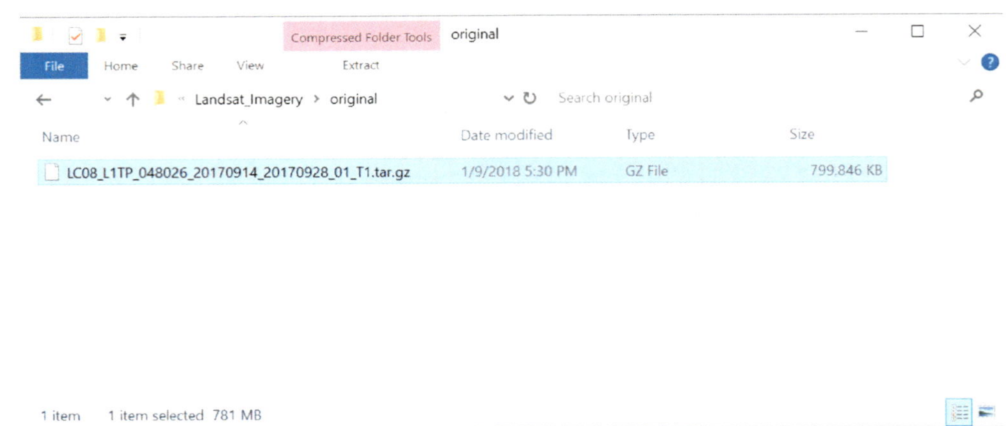

Figure 6.82: The compressed file type ending in tar.gz is frequently used for large datasets.

This type of file, sometimes referred to as a tarball, uses **double compression**, a process where a large file gets compressed two times. You must use the 7zip software *twice* (**Figure 6.83**). Use the 7zip software to decompress the tar.gz file inside your original folder. After decompressing the first time, you should see a file that ends with only the extension *.tar*. Use the 7zip software on this new file that ends only with the extension .tar.

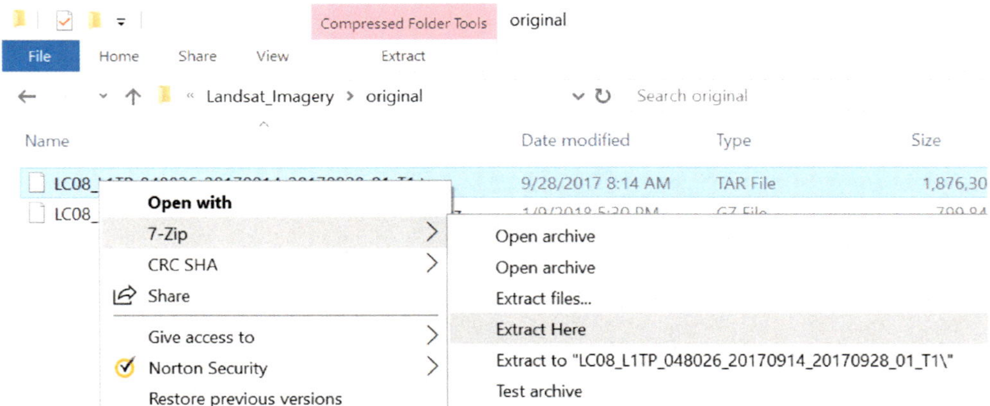

Figure 6.83: After decompressing the first file, you also need to decompress the second file that appears.

1 A compressed file ending in tar.gz is sometimes referred to as a tarball.

Remove both the compressed tar.gz and .tar file when done (**Figure 6.84**). This step is especially important when working with imagery because the files are so large. Removing saves space and helps to prevent confusion later.

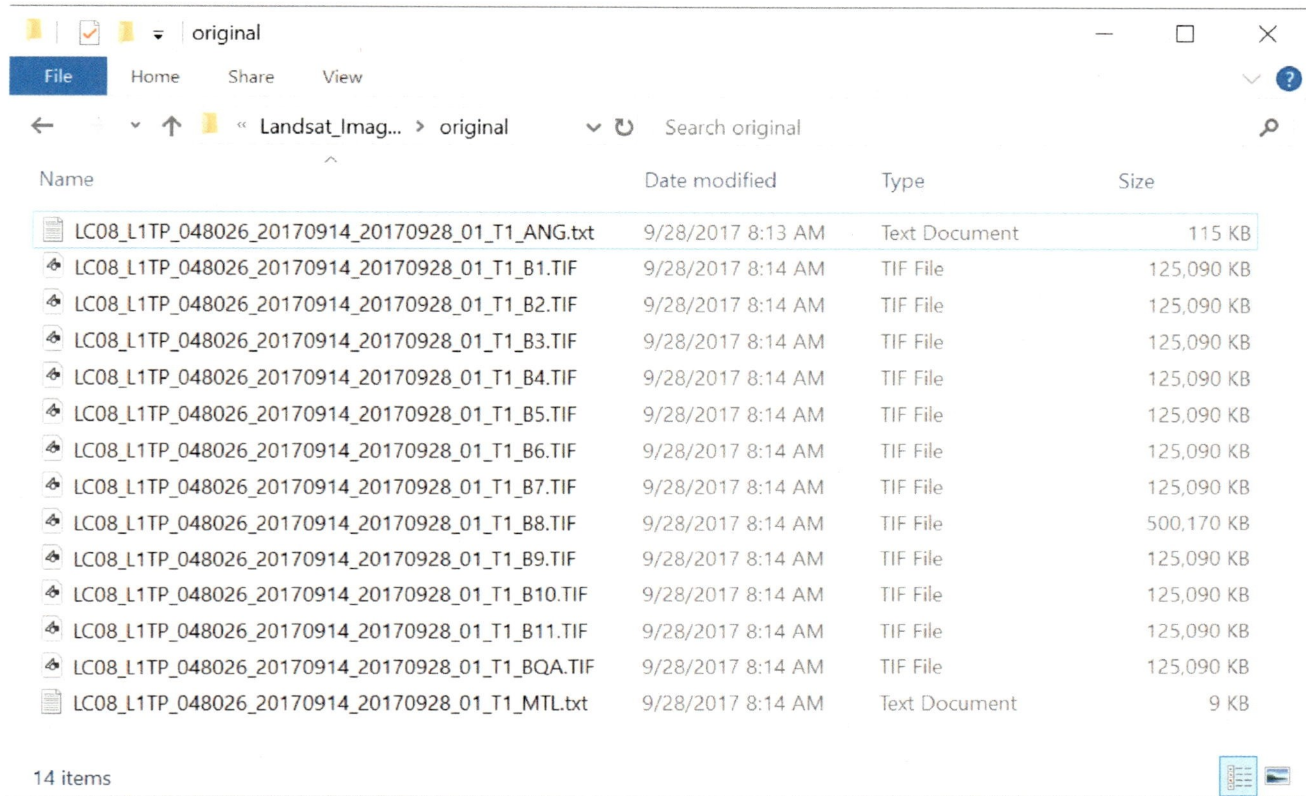

Figure 6.84: After decompressing the tarball, you should have a series of TIF files that represent the Landsat 8 data.

Creating New Imagery Files Using the Composite Bands Tool

Launch ArcMap and open a blank map document.

Disable Background Geoprocessing

In the ArcGIS software, the *Background Geoprocessing* setting is often turned on by default. This setting allows users to continue to work while a tool is running in the background. However, sometimes, this setting stops tools from running or causes other unforeseen problems. To reduce that chances of the ArcGIS software crashing during this exercise, I recommend turning this setting off. After launching ArcMap, open the *Geoprocessing options* from the *Geoprocessing* menu (**Figure 6.85**).

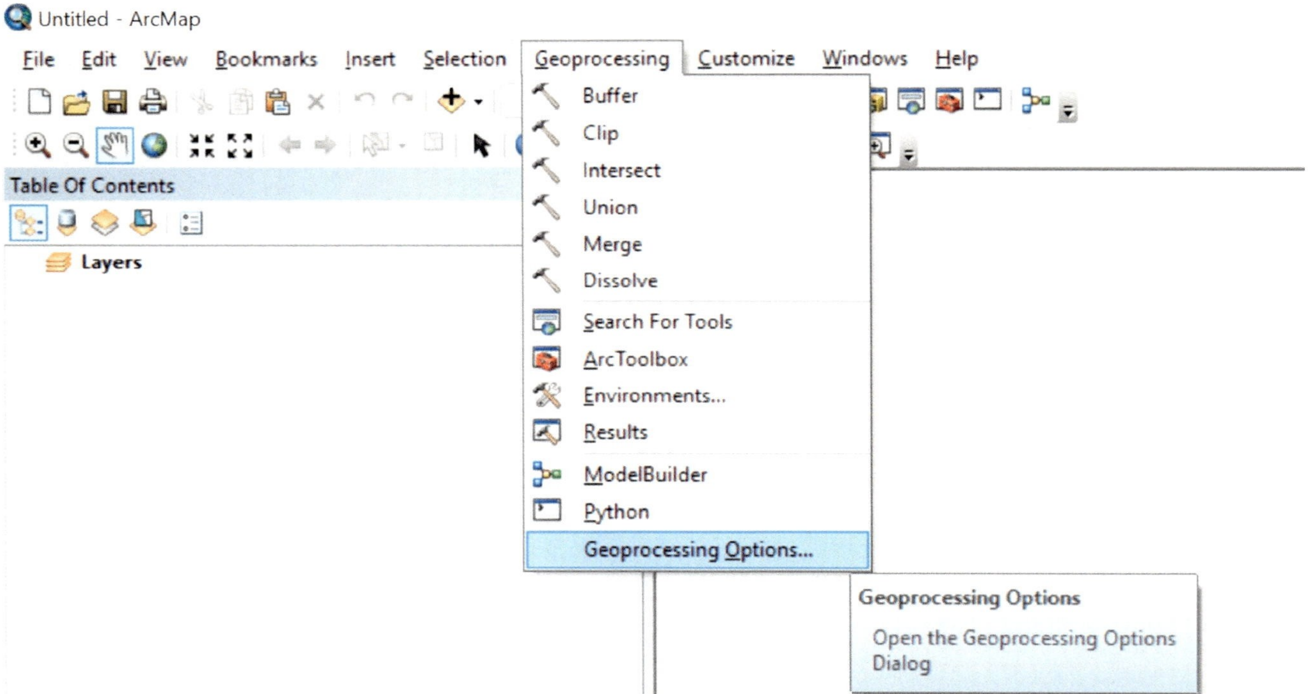

Figure 6.85: The *Geoprocessing* menu contains many useful shortcuts to frequently used tools.

Under *Background Geoprocessing*, uncheck the box next to the word *Enable* (**Figure 6.86**).

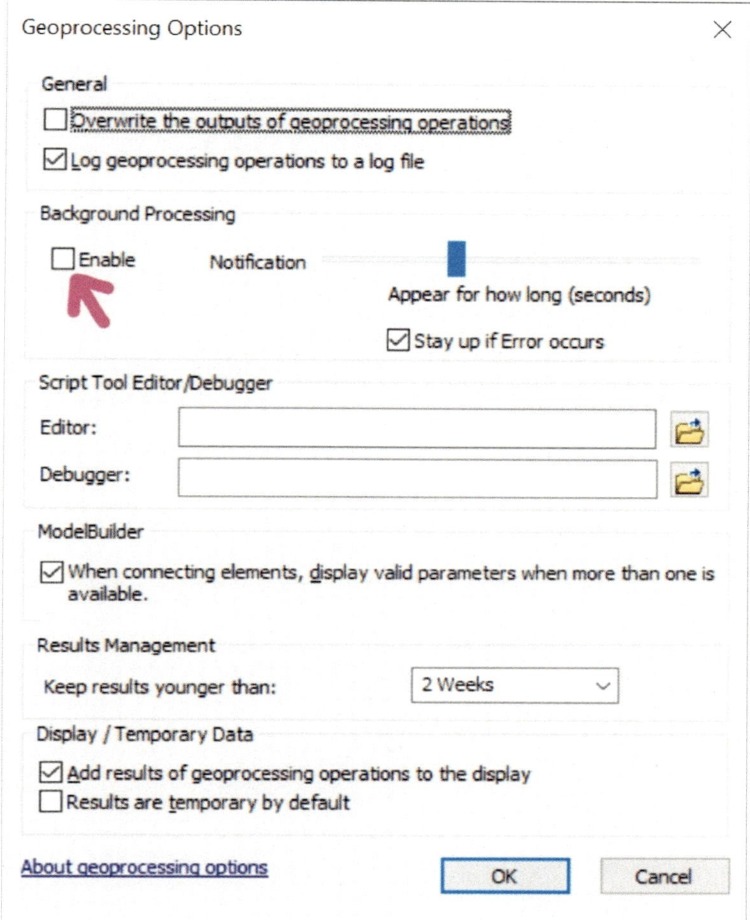

Figure 6.86: Disabling background geoprocessing sometimes makes the ArcGIS software more stable.

Locate the Composite Bands Tool in the ArcToolbox

In the Catalog Window, make a connection to your *workspace* folder.

Do not add any data to the map.

Open the ArcToolbox. In ArcMap, click the toolbox button located on the *Standard* menu. The icon looks like a red toolbox. The ArcToolbox may appear floating above the map. Dock to the side by dragging it to the left in a similar manner in which you docked the attribute table in previous activities. Expand the Data Management Tools. Next, expand the Raster toolbox, then Raster Processing. Locate the Composite Bands tool (**Figure 6.87**).

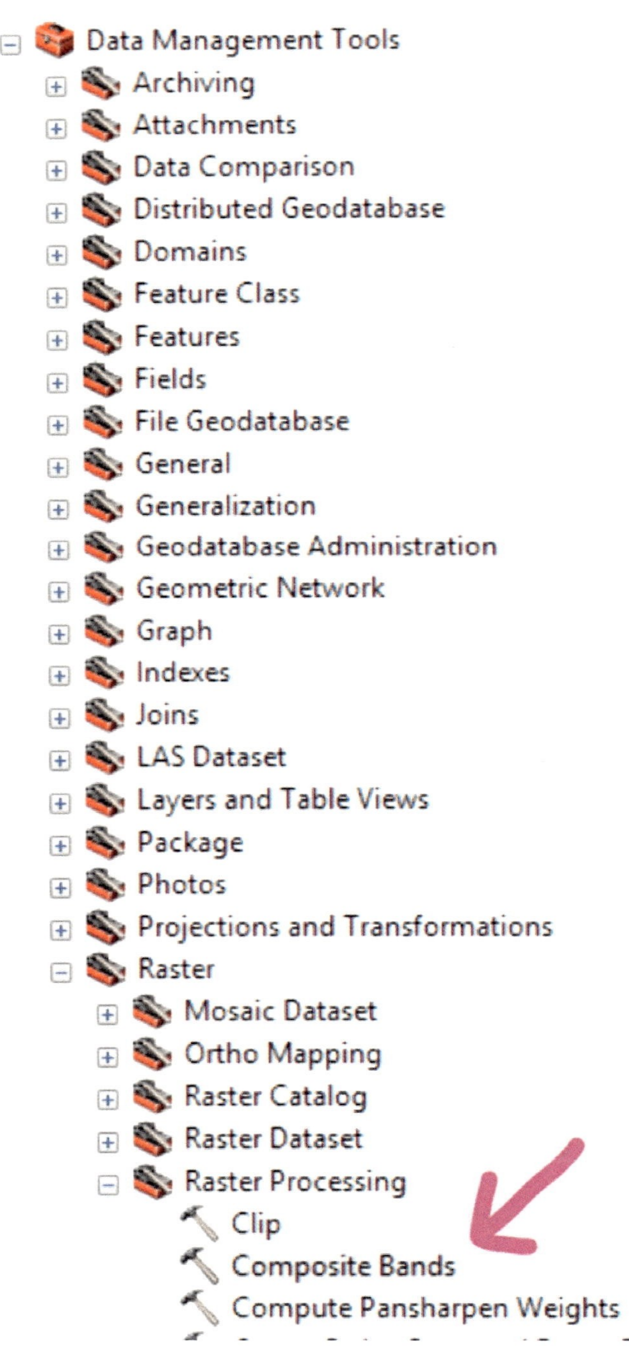

Figure 6.87: The Composite Bands tool is located in the Raster Processing toolbox.

Double click the *Composite Bands* tool. The tool appears floating above the map. For the *Input Raster*, browse to your *original* folder.

In this activity, you only work with bands 1 through 7.

Choose the file that ends in **B1** from the list (**Figure 6.88**). Then click Add.

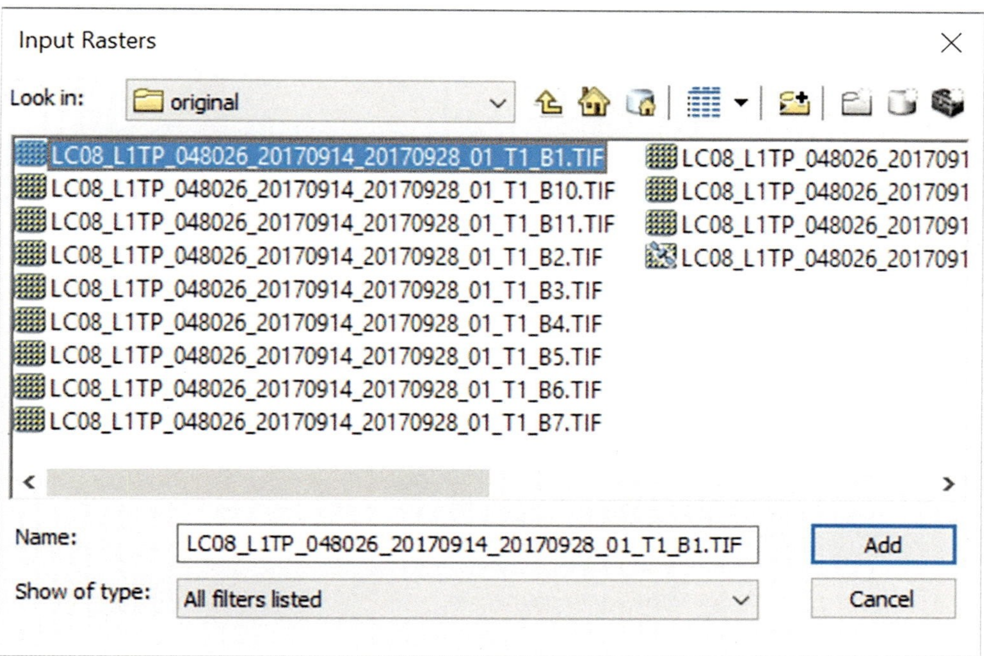

Figure 6.88: B1 stands for Band 1.

Click the yellow file folder icon to browse to the *original* folder again. Select the file that ends in **B2**. Hold down the shift key and select the file that ends in **B7**. All the data in between B2 and B7 should also become selected (**Figure 6.89**). When ready, click *Add*.

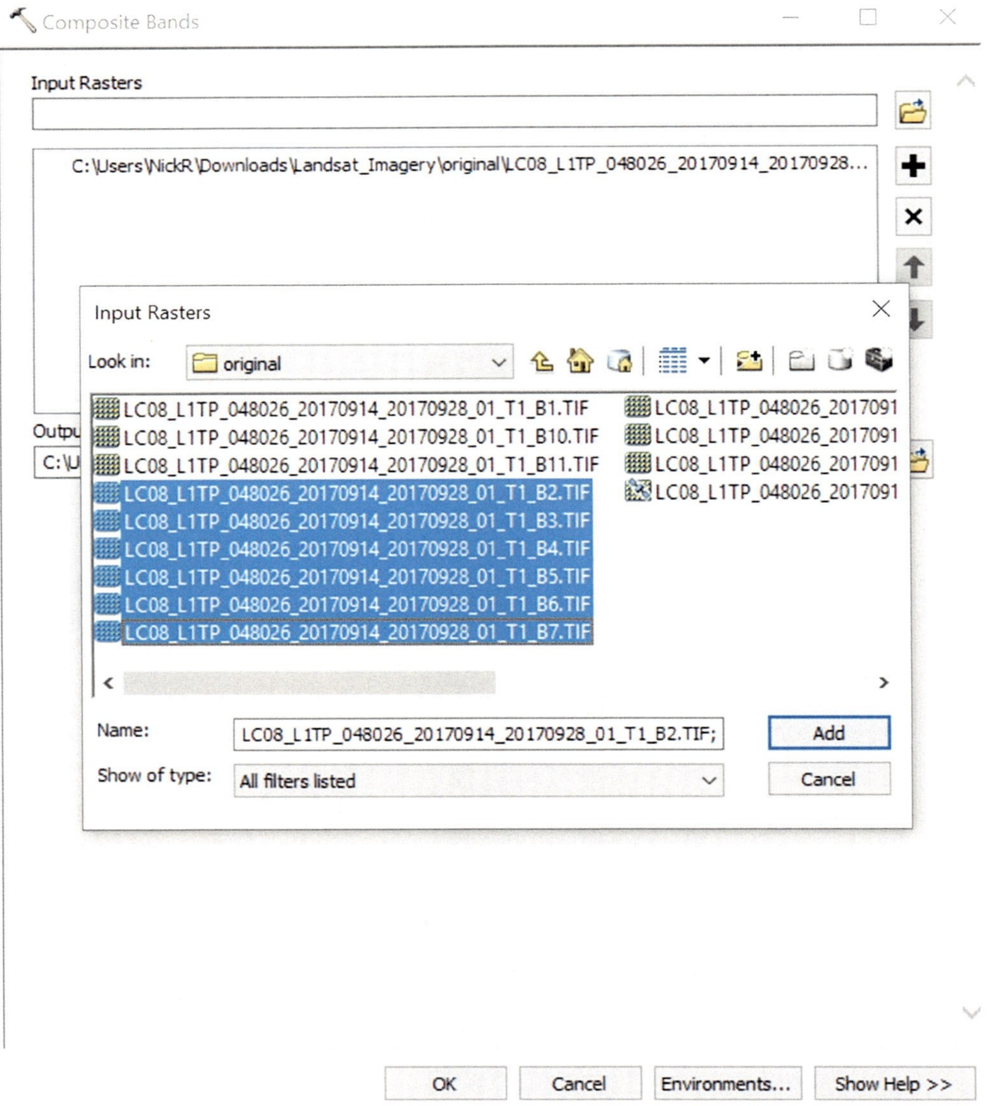

Figure 6.89: Select multiple files by holding down the shift key.

For the Output Raster, browse to your *working* folder. Call the file Vancouver.tif (**Figure 6.90**). When ready, click *OK*.

Manually adding the .tif extension is necessary.

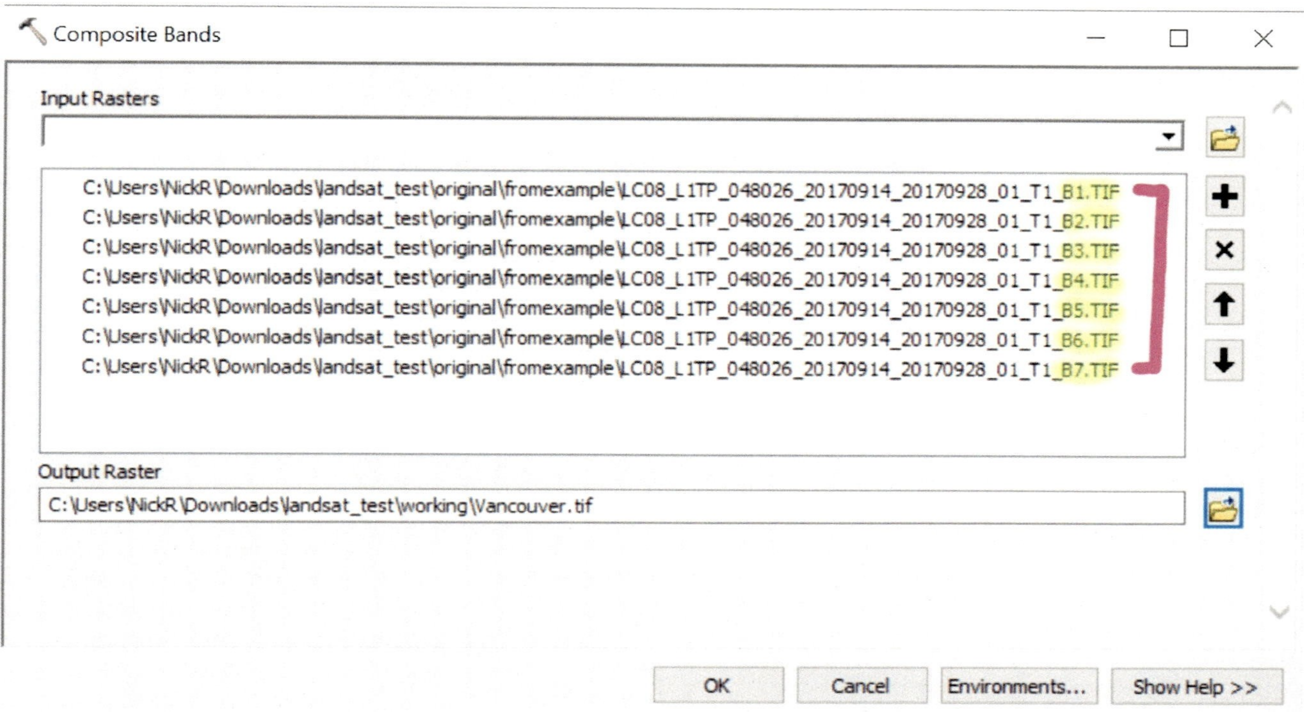

Figure 6.90: Make sure your settings appear similar to the image shown here. Specific file names may vary.

Warning: The order in which you load the bands in the Composite Raster tool is essential. Make sure the bands are listed in the correct order (B1, B2, B3, B4, B5, B6, B7).

It may take several minutes for the tool to complete. Be patient and do not click on ArcMap or the software may crash. When the Composite Bands tool completes, the Vancouver image gets added to the map. At first glance, the image may appear to have strange colors. You make color corrections in the next step.

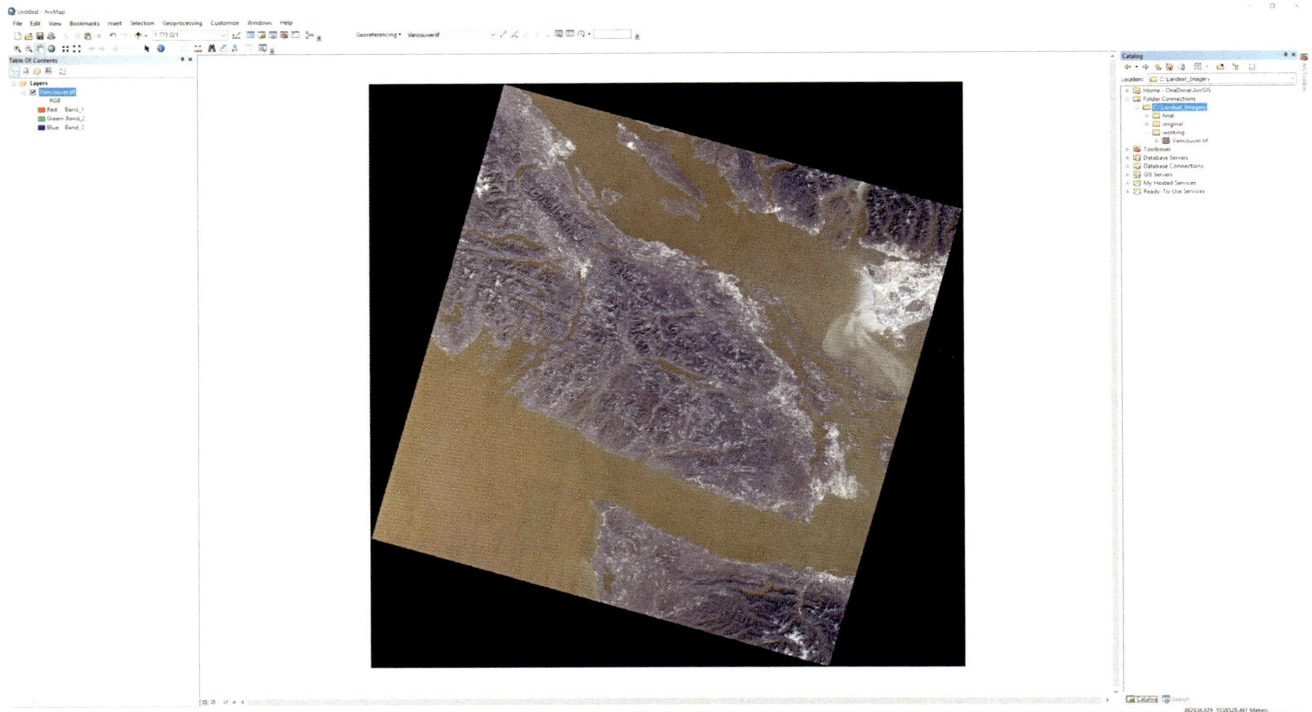

Figure 6.91: The Vancouver composite image is added to the map automatically.

Take a moment to save your map document. In a previous activity, you learn about setting the map document properties to store relative pathnames. Be sure to turn this setting on before saving. Give your map document a meaningful name, such as *Vancouver Map*, and save the file to your *workspace* folder.

Creating True and False-Color Composites

When the layer first appeared in the dataframe, the image may have looked a little strange. The coloration has to do with how the computer monitor displays color using the **RGB colorspace**. The computer monitor emits only three types of colors, red, green, and blue, each assigned to its corresponding channel (**Figure 6.92**).

Receptors in the human eye, called **cones**, perceive color by the way they respond to the combination of blue, green, and red wavelengths. The Landsat 8 composite is made up of many more spectral bands, including portions of the electromagnetic spectrum that the human eye cannot see. To visualize this on screen, we have to replace either blue, green, or red with an alternative spectral band (**Figure 6.93**).

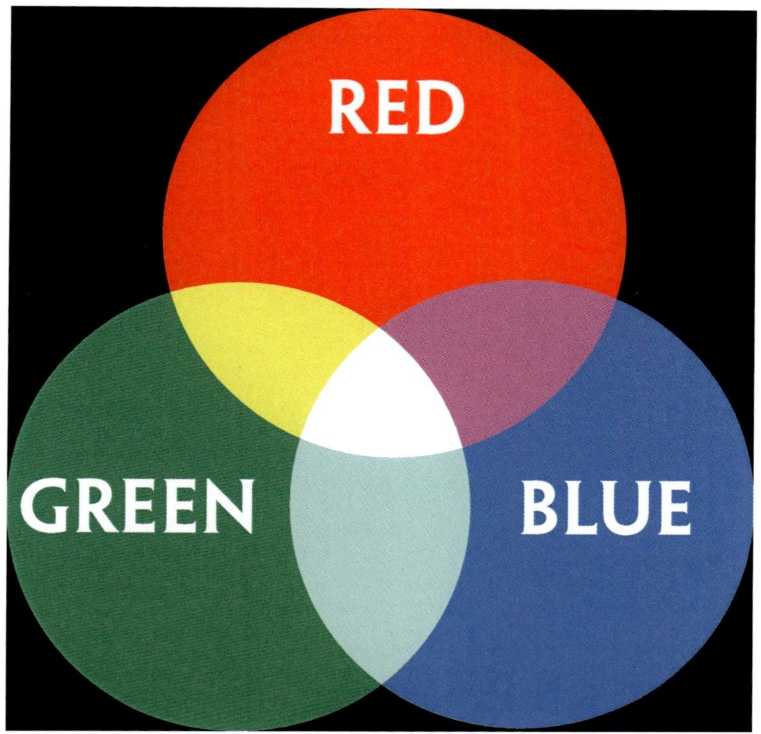

Figure 6.92: The computer has three channels that emit red, green, and blue.

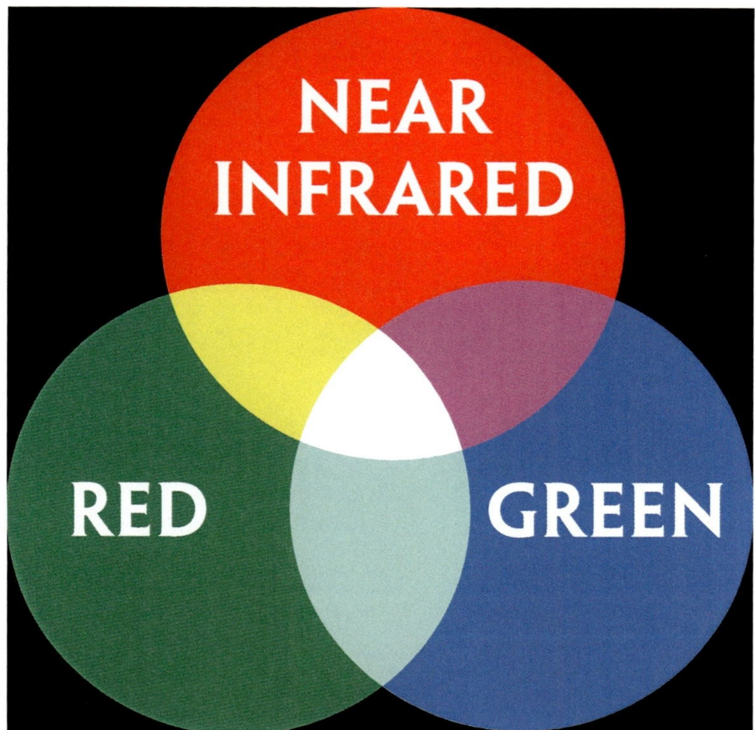

Figure 6.93: It is possible to use a portion of the electromagnetic spectrum you cannot see in either the red, green, or blue channel.

When we first loaded the image in ArcMap, an alternative spectral band is replacing blue, green, and red (**Figure 6.94**). It is the reason why the colors appear strange.

Figure 6.94: To view this image in true-color, assign the Landsat 8 bands representing the blue, green, and red to the right RGB channels.

To learn more about the Landsat 8 bands, navigate to the *NASA website*[2] (**Figure 6.95**). Take a moment to read about each of the Landsat 8 bands. You should become familiar with this information. It lets you know which band combination gives you a true-color image.

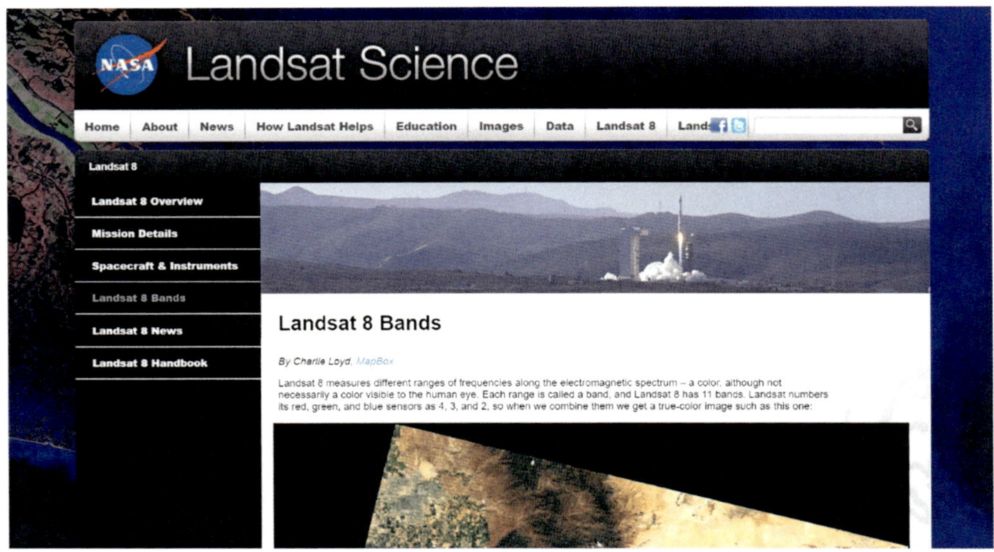

Figure 6.95: The NASA website provides detailed information about the Landsat 8 bands. URL: ***http://landsat.gsfc.nasa.gov/?page_id=5377***

2 http://landsat.gsfc.nasa.gov/?page_id=5377

The table below indicates the Landsat 8 band numbers, the wavelength, resolution, and description.

Band #	µm	Resolution	Description
1	0.433-0.453	30 m	Deep blues and violets
2	0.450-0.515	30 m	Visible blue
3	0.525-0.600	30 m	Visible green
4	0.630-0.680	30 m	Visible red
5	0.845-0.885	30 m	Near-infrared
6	1.560-1.660	30 m	Shortwave infrared (SWIR)
7	2.100-2.300	30 m	Shortwave infrared (SWIR)

When you are done reading about the Landsat 8 bands, go back to ArcMap. Right-click on the *composite raster* layer in the Table of Contents and select *Properties*. Navigate to the *Symbology* tab. Make sure *RGB composite* is highlighted on the left (**Figure 6.96**). You should see the channels, Red, Green, and Blue with checked boxes next to them. One column over you should see a list of bands with drop-down arrows next to them. For the Red channel, choose band 4. For the Green channel, make sure it is band 3. Assign band 2 to the Blue channel. When ready, click *OK*.

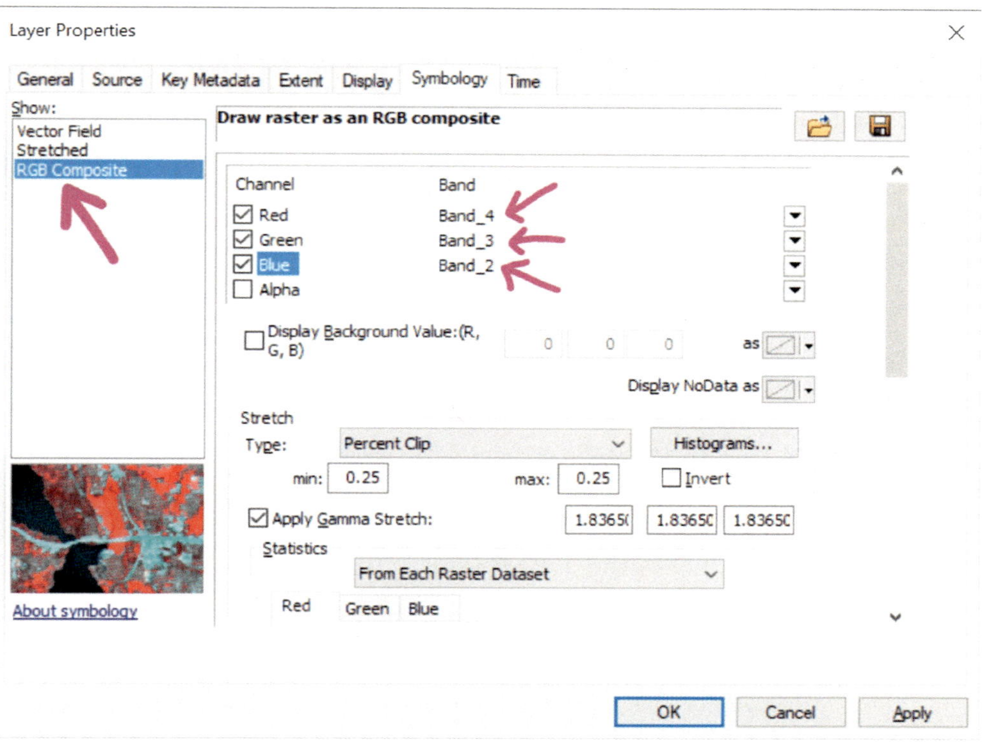

Figure 6.96: According to NASA, Landsat 8 numbers is red, green, and blue sensors as 4, 3, and 2.

The image should now be in true-color (**Figure 6.97**).

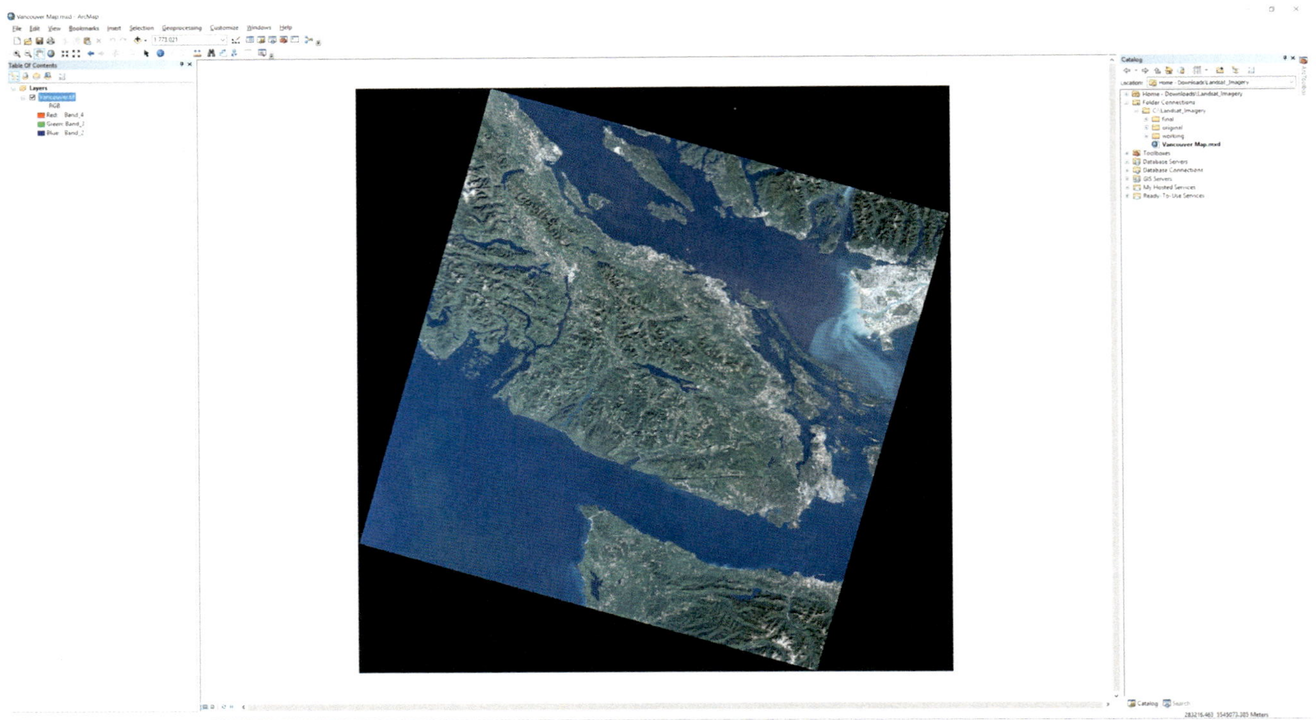

Figure 6.97: By assigning the red, green, and blue bands to the correct red, green, and blue channel, the image now represents true-color.

Because the Landsat imagery captures spectral bands that the human eye cannot see, you can use different band combinations to view these wavelengths. You can accomplish this by creating a false-color composite. One produces a **false-color composite** by replacing either the red, blue, or green bands (4,3,2) with a band that outside of the range of human vision. In this next step, you will create a false-color composite that will allow you to see into the near-infrared. Take a moment to review the ***NASA website***[3] to find out which Landsat 8 band measures the near-infrared and what kind of mapping application for which it is used. When you are ready, return to ArcMap. Create a copy of your *composite* layer in the Table of Contents. You should note that this action is different than creating a copy of the dataset in the Catalog Window. You do not need to duplicate the data. Instead, you will create a duplicate *representation* of the data, or layer, in the Table of Contents. Right-click on the *composite* layer and select *copy* (**Figure 6.98**).

3 http://landsat.gsfc.nasa.gov/?page_id=5377

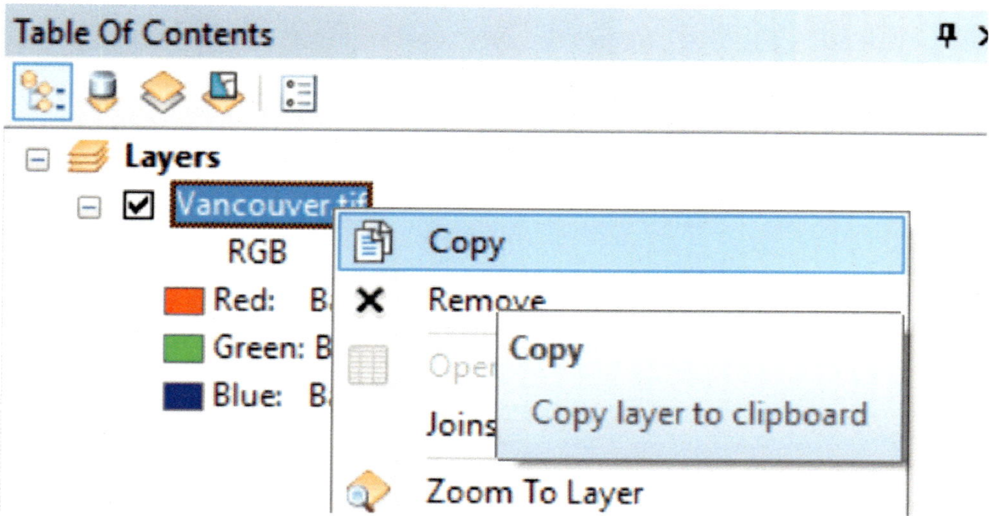

Figure 6.98: Right-clicking on a layer opens a contextual menu with many options.

Then, right-click on the data frame, currently named *Layers*, and select *Paste Layer(s)* (**Figure 6.99**).

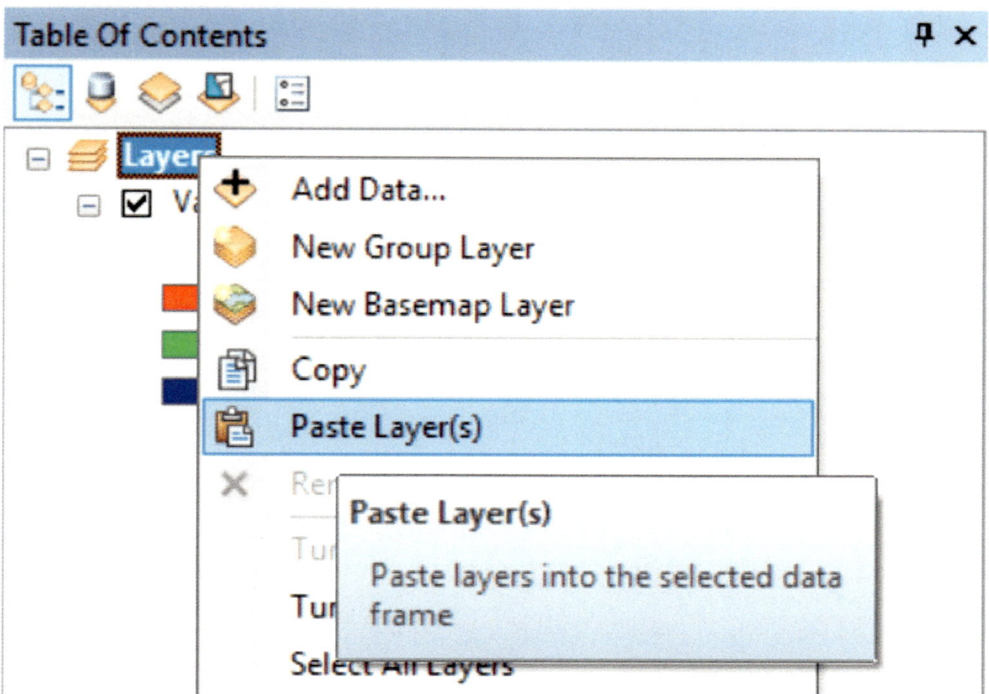

Figure 6.99: You can copy and paste multiple layers in the Table of Contents. This action does not create new datasets, but only representations.

The new layer has the same name and the same symbology settings. To avoid confusion, open the layer properties for the duplicate layer and navigate to the *General* tab (**Figure 6.100**). Name the layer *Near-Infrared False-Color Composite.*

Layer Properties

General Source Key Metadata Extent Display Symbology Time

Layer Name: Near-Infrared False-Color Composite ☑ Visible

Description:

Credits:

Scale Range

You can specify the range of scales at which this layer will be shown:

◉ Show layer at all scales

○ Don't show layer when zoomed:

Out beyond: <None> (minimum scale)

In beyond: <None> (maximum scale)

OK Cancel Apply

Figure 6.100: You can rename a layer in the *General* tab of the layer properties.

When you are ready, uncheck the original layer in the Table of Contents. Navigate to the layer symbology tab for the *near-infrared false-color composite* layer. Assign the near-infrared band to the red channel, the red band to the green channel, and the green band to the blue channel (**Figure 6.101**). Then click *OK*. The result shows a conventional near-infrared false-color composite.

Figure 6.101: In a conventional near-infrared red false-color image, the red and pink areas display healthy vegetation.

Skill Drill: Creating a Custom False-Color Composite

In the Table of Contents, copy and paste your composite image one more time to create a third layer. Name the new layer, "Custom False-color." Take a moment to review the *NASA website*[4] to help decide on a custom band combination. Once you have chosen a custom band combination, make a note on what potential applications the sequence might be used based on the portions of the electromagnetic spectrum made visible.

Once done, create a map layout that is **15 by 15 inches**. You may display the entire Landsat 8 image, or you may zoom into an area on the image that interests you. Insert a north arrow and a scale bar. Don't add any other map elements. To make them visible, you may have to change the colors to white. One by one, turn on each layer in the Table of Contents and export as a PNG file with a resolution of 300 dpi. When finished, you should have three maps, one natural color, one near-infrared false-color, and one custom-made false color (**Figure 6.102**). Save each image to your *final* folder.

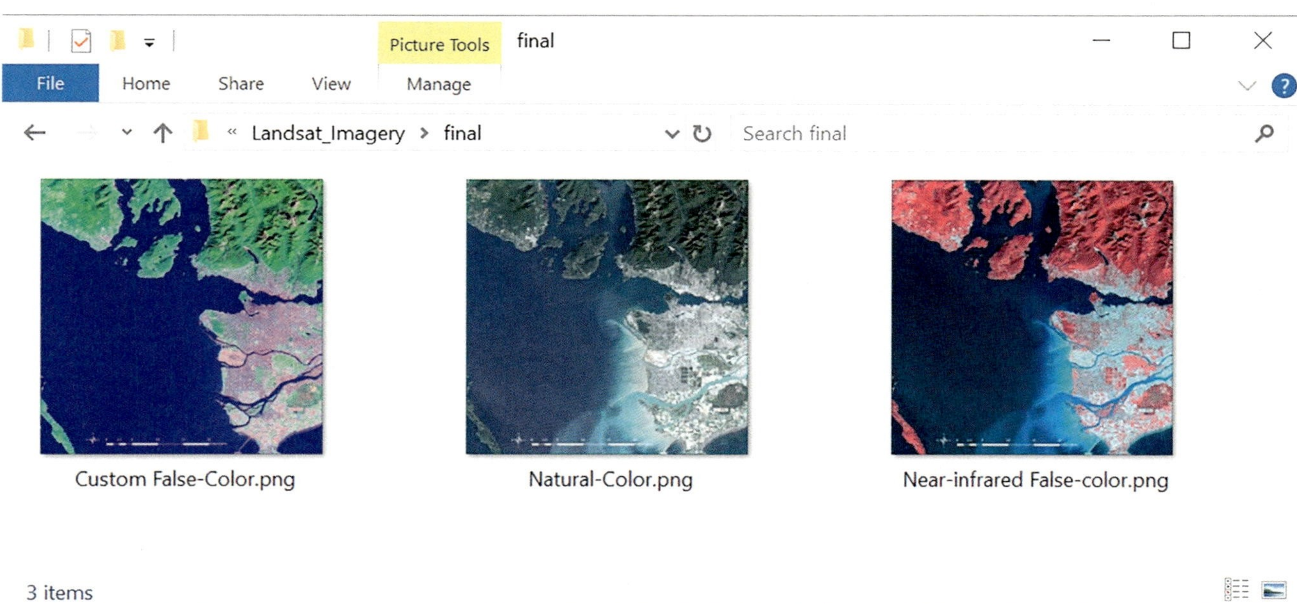

Figure 6.102: At the end of this activity, you should have three 15-inch maps exported to your *final* folder.

4 http://landsat.gsfc.nasa.gov/?page_id=5377

Pricipal Terms

acetate film base degradation
across-track scanner
active microwaves
active sensing systems
along-track scanner
associated elements
association
atmospheric windows
cell size
classification
cones
contrast
contrast stretch
control point
control station
delivery schedule
digital imagery
digital orthophoto quads (DOQs)
double compression
drone
electromagnetic spectrum
Enhanced Thematic Mapper plus (ETM+)
enumeration
false-color composite
false color-infrared photograph
fiducial marks
film-based photographs
Fixed-wing aircraft
flight index
flight lines
flyaway
forwardlap
frequency
geometric distortion
georeferencing
ground control points
Ground resolved distance
ground sample distance (GSD)
high-flight platform
histogram equalization
identification
image enhancement
image format
image interpretation
Infrared energy
light detection and ranging (lidar)
light table
line-pairs-per-milliliter (LPM) resolution target
low-flight platform
mensuration
microwaves
minimum-maximum
mirror stereoscope
mission implementation
mission planning
mission product
mission product processing
mission protocols
mission zone
multispectral scanners
multispectral scanner system (MSS)
multispectral sensors
nadir
National Aerial Photography Program (NAPP)
National Agriculture Imagery Program (NAIP)
near-infrared
near-infrared black and white photograph
operational land imager (OLI)
orthophotograph

orthophoto map
panchromatic photograph
Part 101
Part 107
passive sensing systems
pattern
perspective view
photographic film
pixel
planimetrically-correct photograph
plan position indicator (PPI) system
pocket stereoscope
principal point
pushbroom scanner
pyramids
radar
radial displacement
Raleigh scatter
rasters
recognition
recording platform
reflectance response patterns
reflectance value
relief displacement
remote pilot in command (PIC)
Remote sensing
RGB colorspace
Rotorcraft UAV
scale distortion
sectional aeronautical charts
shadows
shape
sidelap
Site
size
small unmanned aircraft system (sUAS)
space imaging platform
spectral bands
Spectral reflectance curves
spectral signatures
standard deviation
stereo pair
stereoscope
stereovision
synthetic aperture radar (SAR)
texture
thematic mapper (TM)
thermal infrared
thermal infrared sensor (TIRS)
tone
true-color composite
true-color photograph
ultra-low-flight imaging platform
United States Landsat Program
unmanned aircraft
vertical photograph
visible light
visual observer (VO)
wavelength
whiskbroom scanner

Chapter 7: Selection, Proximity, and Overlay

Too often, people conduct a geospatial analysis without consideration for uncertainty and error, map projections, and datums. More often, there is little regard for cartographic convention and communication design goals. A geospatial analysis should consider the properties of geospatial data before applying GIS software tools. Chapter 7 introduces the first steps in learning how to conduct a geospatial analysis. The topics presented within should help to prepare readers for more sophisticated uses of GIS.

Learning Outcomes

Readers should be able to accomplish the following outcomes by the end of this chapter:

- » Recall the components, structure, and organization of a database
- » Discuss types of cardinality between database tables
- » Identify the various types of geospatial queries
- » Explain the function of comparison and Boolean operators
- » Construct an expression using SQL
- » Define the four basic operations used in spatial queries
- » Describe the scope of geospatial analysis operations
- » Execute a proximity analysis
- » Compare different methods of overlay analysis

Database Management Systems and GIS

A significant component of geospatial data resides in a database. Understanding the elements, structure, and organization of a database helps one work with geospatial data more effectively. A **database** is a collection of individual *entities* stored in a highly structured way. **Entities** are unique objects or features represented in the database. For example, a database storing information about Humboldt State University would have entities such as students, instructors, courses, classrooms, and grades. The term database is sometimes used synonymously with the term *database management system*. This use of the term is not entirely correct. In this book, a **database management system (DBMS)** refers to the software used to enter, analyze, search, and report data stored in a database. One of the defining characteristics of a geographic information system (GIS) is that it incorporates the properties of a database management system but has the added functionality of linking a database to a map.

One of the roles of a database is to ensure that each of these entities remains unique and separate from other entities. When working with geospatial data entities are usually made up of geographic units such as cities, counties, states, and countries. Entities can also have relationships with other entities. **Relationships** define the interaction between entities. Using the Humboldt State University example, a student's grade is a relationship between the student, the course, and the instructor.

Most importantly, entities have attributes which the database stores. **Attributes** describe the properties of entities and relationships. For example, an attribute for a student might include the student name, the student address, the student email, and the student identification number. Attributes for a college course might consist of the course title, course number, room number, maximum number of students, and time slot. Attributes for instructors might include name, academic rank, employee ID number, office number, and office telephone number. Geospatial data also has attributes including geographic coordinates, area, length, as well as any number of non-spatial attributes such as population density.

Database Tables

A database stores entities and attributes as tabular data[1]. When discussing databases, one refers to a row in the table as a **record**. Each record represents a *single* entity. For example, in a database table representing students, each row in the table represents an individual student. A single student cannot take up more than one row in the table. When discussing databases, one refers to a column in the table as a **field**. Each field stores a single *attribute type*. An **attribute type** describes the nature of the attribute in the manner in which the database stores it in memory. Attribute types typically include numbers, strings, and dates. A **string** is a data type that represents text. Sometimes it is possible to import non-database tables, such as a Microsoft Excel spreadsheet, into GIS software. The GIS software treats the table as a database table, enforcing the rules and structure of databases. For example, if one imports a table with a field that includes both numbers and text, the GIS software will assign that field the *string* attribute type by default. This assignment means that the database treats the numbers inside that field as text and cannot use them in mathematical calculations.

Along with a distinct attribute type, each field also has a unique name called a **field name**[2], located at the top row of the table called a **header row**. (**Figure 7.01**). A field name generally describes the attributes stored within the field. A field name often has strict limitations such as the maximum number of characters, no spaces, and no special characters allowed. These limitations can cause problems when importing tables such as Microsoft Excel spreadsheets. Best practices for field names include:

» Eliminate any characters that are not alphanumeric or an underscore
» Do not start a field name with a number or an underscore
» Limit the length of the field name to 10 characters

1 Tabular data refers to data in a table format.

2 Sometimes the term *field name* is referred to as a *header*.

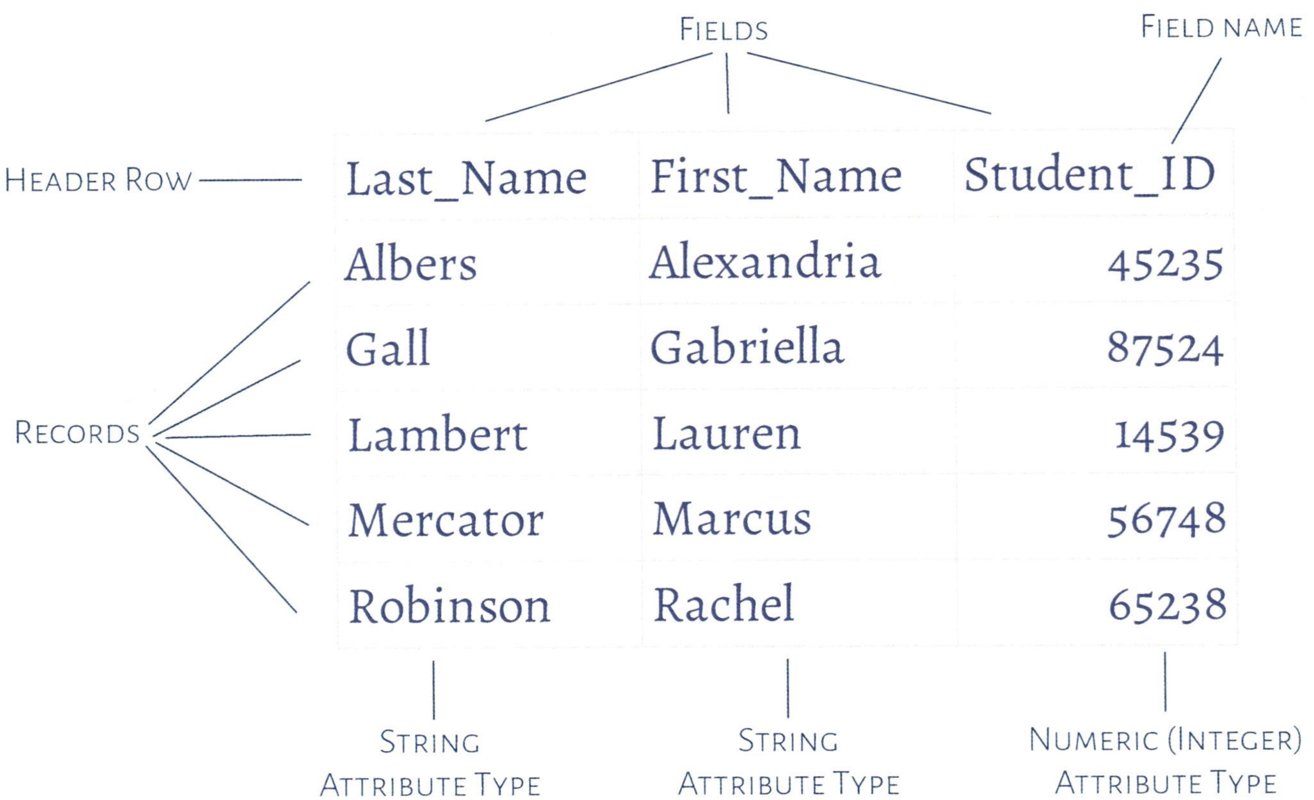

Figure 7.01: It is important to remember that each field can only have one attribute type. Forgetting this limitation is a common mistake that many students make when dealing with data tables.

Primary Keys

Because each entity is a unique object in a database, a database must have the means to enforce the integrity of an entity. Ensuring only one copy of each entity prevents information from being duplicated. A database enforces integrity by assigning each entity a unique identifier called a primary key (**Figure 7.02**). A **primary key** can be composed of one or more attributes. For example, a database table about students might include two students, both with the same first and last name. So that the information between these two students is not confused the database designates a unique attribute, such as a student ID number, to be the primary key.

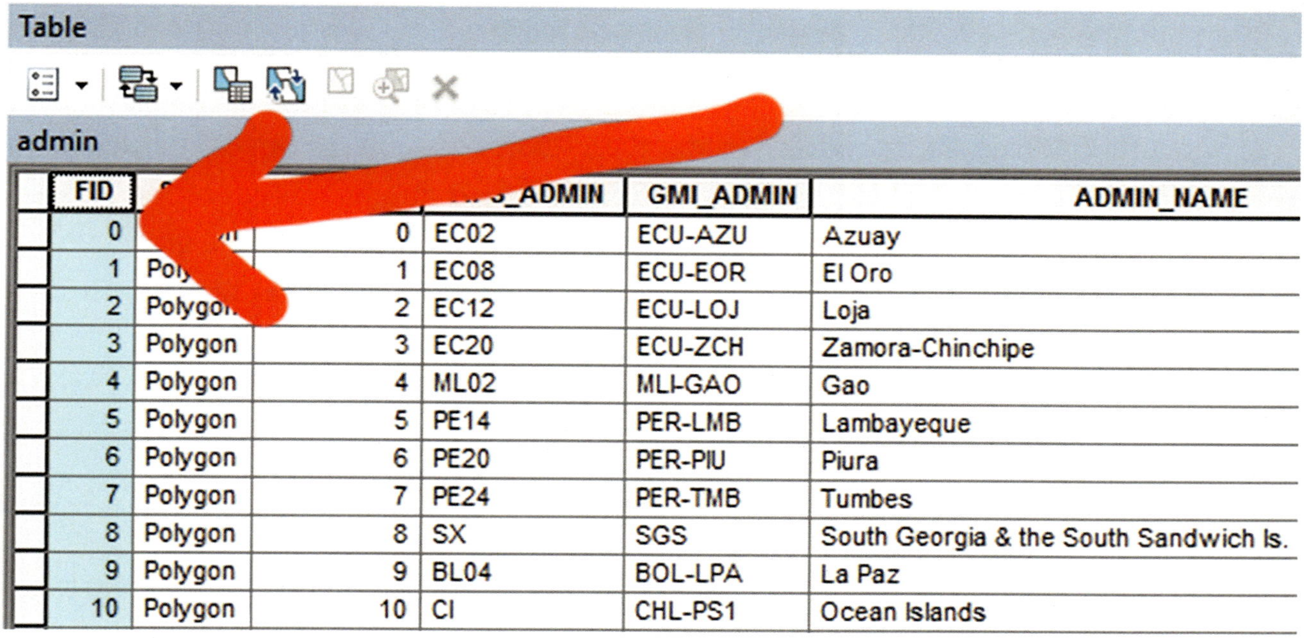

Table

admin

FID	[illegible]	[illegible]	[illegible]S_ADMIN	GMI_ADMIN	ADMIN_NAME
0	[illegible]	0	EC02	ECU-AZU	Azuay
1	Pol[illegible]	1	EC08	ECU-EOR	El Oro
2	Polygo[illegible]	2	EC12	ECU-LOJ	Loja
3	Polygon	3	EC20	ECU-ZCH	Zamora-Chinchipe
4	Polygon	4	ML02	MLI-GAO	Gao
5	Polygon	5	PE14	PER-LMB	Lambayeque
6	Polygon	6	PE20	PER-PIU	Piura
7	Polygon	7	PE24	PER-TMB	Tumbes
8	Polygon	8	SX	SGS	South Georgia & the South Sandwich Is.
9	Polygon	9	BL04	BOL-LPA	La Paz
10	Polygon	10	CI	CHL-PS1	Ocean Islands

Figure 7.02: When working with geospatial data, GIS software often assigns a default numeric index to each record called a Feature ID (FID) to serve as a primary key.

Datasets and Feature Types

A database table is usually a homogenous collection of one type of entity. For example, one database table might only include students and their attributes, while another database table will only include courses and their attributes. The geospatial database table, called a **dataset**, is a database table comprised of a homogenous collection of both entity types and feature types. A **feature type** describes the geometry of the feature, such as a point, a line, or a polygon (**Figure 7.03**). For example, a dataset representing cities in the United States would only include city entities and one feature type to represent them, such as the point feature type. Though a GIS can represent cities using polygons, one would not find polygon feature types mixed with point feature types in the same dataset.

Table

cities

FID	Shape *	CITY_FIPS	CITY_NAME	STATE_FIPS	STATE_NAME	STATE_CITY
1456	Point	55520	Paradise	06	California	0655520
1458	Point	13014	Chico	06	California	0613014
1509	Point	54386	Oroville	06	California	0654386
1593	Point	46170	Marysville	06	California	0646170
1596	Point	81134	Ukiah	06	California	0681134
1601	Point	86972	Yuba City	06	California	0686972
1605	Point	41572	Linda	06	California	0641572
1668	Point	13945	Clearlake	06	California	0613945
1673	Point	73108	South Lake Tahoe	06	California	0673108
1675	Point	51637	North Auburn	06	California	0651637
1686	Point	03204	Auburn	06	California	0603204
1715	Point	62364	Rocklin	06	California	0662364
1733	Point	62938	Roseville	06	California	0662938
1754	Point	13588	Citrus Heights	06	California	0613588
1755	Point	54092	Orangevale	06	California	0654092

Figure 7.03: In this city dataset, notice that the GIS defines the feature type in the Shape field. Each feature is defined as a point. One will not find any other feature types in this dataset.

Database Relationships

Though datasets consist of a homogenous collection of entities and feature types, they can still have relationships with other datasets made up of different types of entities, different feature types, or even with non-spatial database tables. When using GIS software, relationships between database tables are typically established using either table relationships or spatial relationships. A **table relationship** occurs between two or more tables with a common attribute field. A **spatial relationship** occurs between one or more features on a map based on their locations relative it each other.

Establishing a table relationship requires a common attribute field referred to as a key. A **key** is an attribute that exists in two or more tables and which is used to associate them together. For example, one might obtain presidential election results from a website as a simple excel spreadsheet that contains a field with the names of counties across the United States and other fields of the vote tallies for each candidate. One may establish a relationship with any other database table that also has an attribute field of county names.

If one establishes a relationship between one table and another, one has to be sure that the attribute type is the same. So a text field could be associated with another text field, but not to a number field or a date field. Also, the attribute that is establishing the relationship must be precisely the same. So in the previous example, if one of the county names were spelled a different way, no relationship would be established between those two entities. When using GIS software, creating a one-to-one relationship between two tables using a common attribute field is called a **table join**. A table join is one of the most useful and frequently used table relationship operations performed in Geospatial Analysis.

Cardinality

Cardinality defines the type of table relationship between datasets based on the numerical relationship between one dataset's entities and the other. There are three types of cardinality, one-to-one, many-to-one, and many-to-many. In a **one-to-one relationship**, each table only has one matching entity, so only one record corresponds to only one other record (**Figure 7.04**). In a **many-to-one relationship**, a table with one unique entity gets matched with a table containing many references to that same entity. One record on the first table corresponds to many records on the second table (**Figure 7.05**). In **many-to-many relationships**, each table may have a record that corresponds to multiple records on the other table.

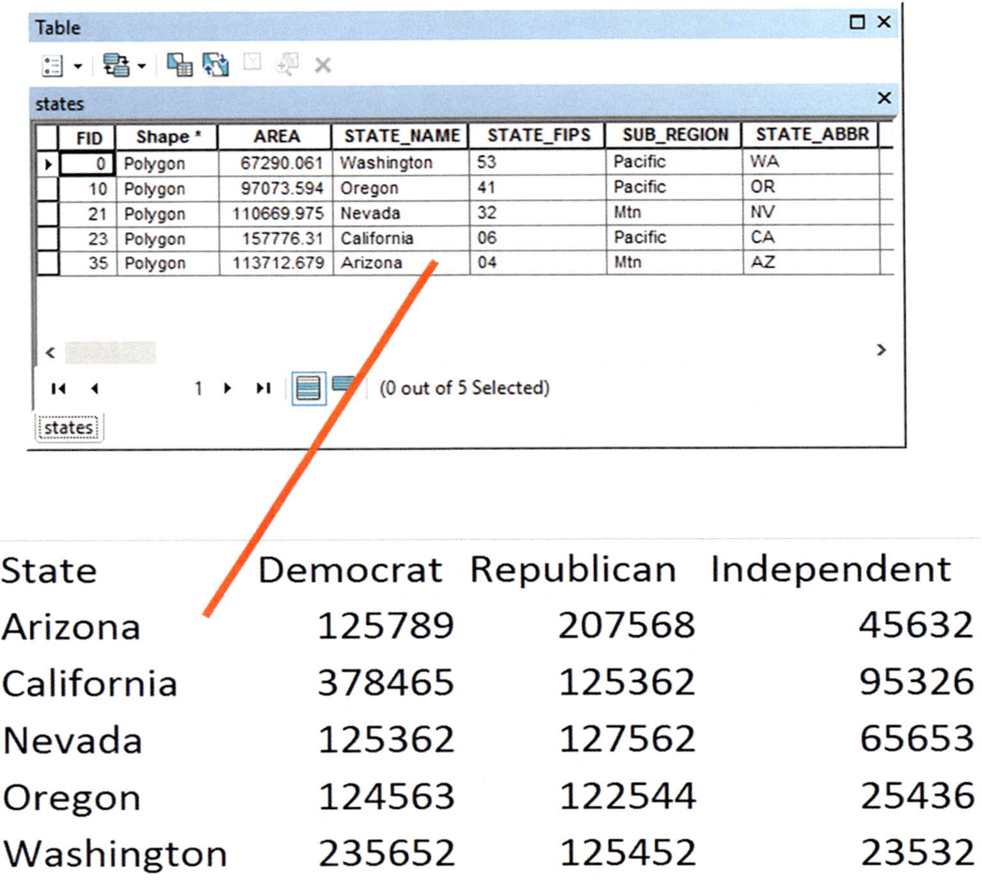

FID	Shape *	AREA	STATE_NAME	STATE_FIPS	SUB_REGION	STATE_ABBR
0	Polygon	67290.061	Washington	53	Pacific	WA
10	Polygon	97073.594	Oregon	41	Pacific	OR
21	Polygon	110669.975	Nevada	32	Mtn	NV
23	Polygon	157776.31	California	06	Pacific	CA
35	Polygon	113712.679	Arizona	04	Mtn	AZ

State	Democrat	Republican	Independent
Arizona	125789	207568	45632
California	378465	125362	95326
Nevada	125362	127562	65653
Oregon	124563	122544	25436
Washington	235652	125452	23532

Figure 7.04: In this image, a table join can be established using a field with common attributes (a key). In this case, it is the "STATE_NAME"field of the first table and the "State" field of the second table. This table join has a one-to-one relationship. One entity in a dataset gets matched with one entity in another.

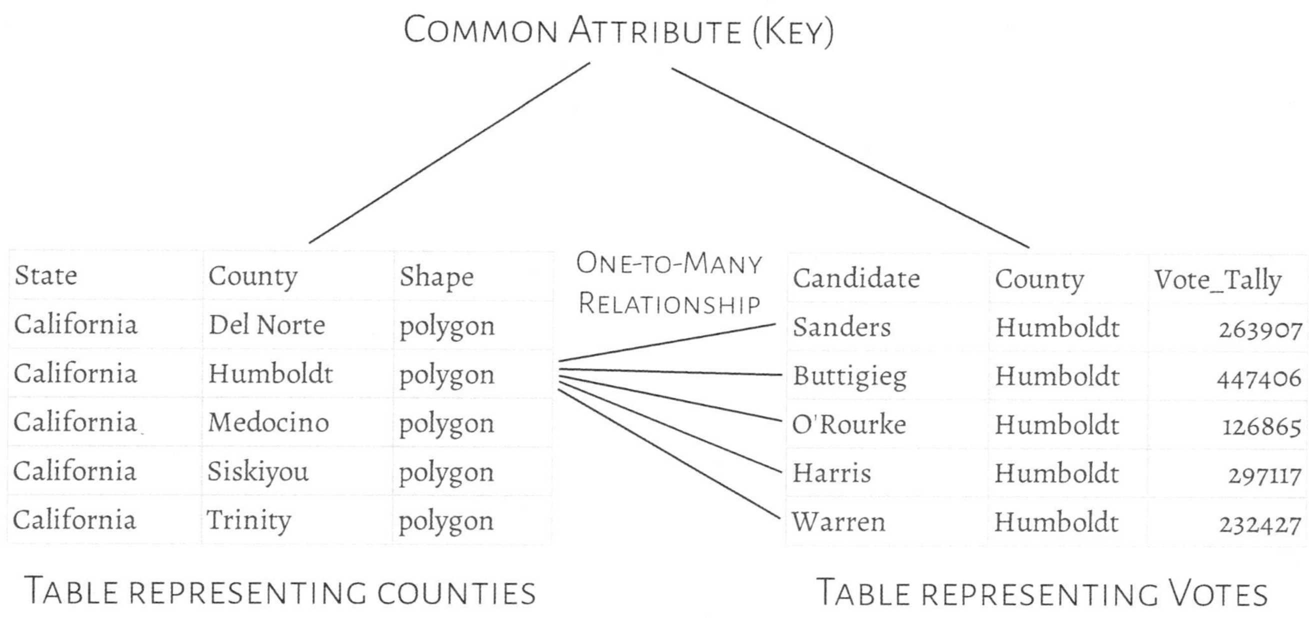

State	County	Shape
California	Del Norte	polygon
California	Humboldt	polygon
California	Medocino	polygon
California	Siskiyou	polygon
California	Trinity	polygon

Candidate	County	Vote_Tally
Sanders	Humboldt	263907
Buttigieg	Humboldt	447406
O'Rourke	Humboldt	126865
Harris	Humboldt	297117
Warren	Humboldt	232427

Figure 7.05: In this example, the tables use the County field as the attribute key to perform a table join. In this instance, the tables have a many-to-one relationship. One record on the first table corresponds to the records on the second table. Note: Vote tallies do not represent actual election results and were randomly generated.

GEOSPATIAL QUERIES

Queries are one of the most frequently used operations performed when working with geospatial data stored in a database. A **query** is a *selection operation*, a procedure that chooses a subset of records based on values of attributes or a set of spatial conditions. There are two primary types of selection operations, *attribute queries* and *spatial queries*.

Attribute Queries

Attribute queries are selections based upon the values of attributes. Attribute selections are commonly performed using a **query builder**, a graphical user interface (GUI) that helps to build selection operations using a **structured query language (SQL)** (**Figure 7.06**). SQL is a standard **syntax**, the structural rules for a programming language, used for retrieving and manipulating data from a database. SQL uses logical expressions to define selection operations. **Logical expressions** are a sequence of *operands* and *operators* that return a value of true or false. An **operand** is the data value in an expression. An **operator** is a symbolic representation of the process for operations performed against one or more operands.

In GIS software, <FIELD_NAME> <OPERATOR> <VALUE OR STRING> is the most basic structure for an expression. The following expression has this basic structure:

"STATE_NAME" = 'CALIFORNIA'

In the expression above, **STATE_NAME** and **California** are *operands*. Notice that 'CALIFORNIA' is enclosed in a pair of primes, mathematical symbols commonly, if inaccurately, referred to as single quotes. In SQL, primes indicate that what is between them is a string. In this expression, "STATE_NAME" is enclosed in a set of double primes. The GIS software recognizes that this value refers to an attribute *field* rather than a specific attribute. The equal sign is a comparison operator. **Comparison operators** test whether two values are the same. The GIS software will check each record for the attribute 'California' under the STATE_NAME field using this expression. If the expression is true, it will select that record.

Other comparison operators include the following symbols:

- > (greater than)
- < (less than)
- >= (greater than or equal to)

» <= (less than or equal to)
» <> (not equal to)

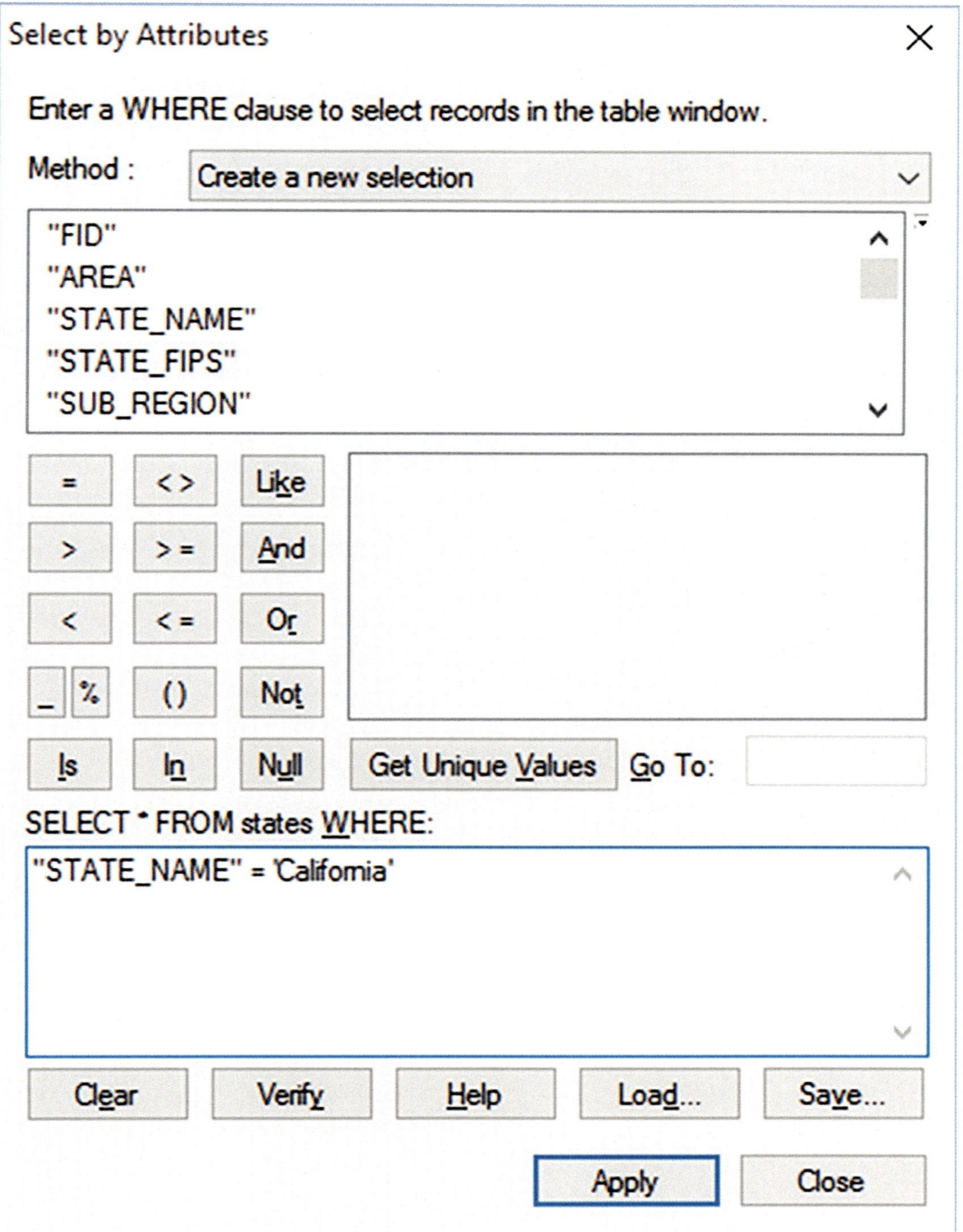

Figure 7.06: Using a query builder can help avoid syntax errors when creating SQL expressions.

Boolean Expressions

A **Boolean expression** results in a true or false condition using a **Boolean operator**. The AND and OR Boolean operators enable expressions to have more complicated structures called **compound expressions**, which compare multiple conditions. The following are Boolean operators:

- AND
- OR
- NOT

The Boolean operator AND is used to limit or reduce the number of records selected. It requires two conditions to be true for the record to be selected. One uses the AND operator when using *multiple attribute fields* as the selection criteria. The Boolean operator OR is used to expand the number of records selected. It is an inclusive operator because either of the two conditions may be true to select a record. One uses the OR operator when using a *single attribute field* as the selection criteria. The Boolean operator NOT negates a condition. The condition must be false for a record to be selected. The following are examples of Boolean expressions:

"STATE_NAME"='California' OR "STATE_NAME"='Oregon'

An OR operator separates the two conditions. The GIS software will check each record for the presence of either the attribute 'California' or the attribute 'Oregon' under the STATE_NAME field using this expression. If either of those conditions is true, the record will be selected. Note that SQL syntax requires a *complete* expression on either side of the Boolean operator. In the first example, each side of the OR operator has a complete expression with the following structure: <Field_Name> <Operator> <Value or String>.

Many readers make the mistake of not constructing a complete expression. Instead, they may write something like the following:

"STATE_NAME"= 'California' OR 'Oregon'

This format is **incorrect SQL syntax, and** no records will be selected.

Another Boolean expression using the AND operator uses the following syntax:

"STATE_NAME"= 'California' AND "POPULATION" > 1000

An AND operator separates the two conditions. The GIS software will check each record for the attribute 'California' under the STATE_NAME field using this expression. It will also review each record to see if the value under the POPULATION field is greater than 1000. The record will only be selected if both conditions are true.

Another common mistake many students make is using an AND operator instead of an OR operator, particularly when trying to make a selection from a single attribute field. For example, a student might use the following expression to select records for both California and Oregon:

"STATE_NAME"= 'California' AND "STATE_NAME"= 'Oregon'

While the syntax is correct, the condition cannot be met. Remember, when using an AND operator, both conditions must be true. Under the STATE_NAME attribute field, no single state has the name California and Oregon. It is either one or the other. In this instance, one should use the OR operator.

Another Boolean expression using the NOT operator uses the following syntax:

NOT ("STATE_NAME"= 'California')

The GIS software checks each record for the attribute 'California' under the STATE_NAME field. The record will only be selected if this condition is false.

Spatial Query Operations

Each record has a one-to-one relationship with a feature on the map. When a record is selected a single corresponding feature on the map will also be selected. This relationship works both ways. When executing a spatial query, one or more features on the map get selected along with the corresponding records in the table. A **spatial query** is a logical expression that selects geographic features by evaluating spatial relationships. There are four broad categories of operations used in spatial queries:

- » Proximity
- » Adjacency
- » Intersection
- » Containment

Proximity operations assess the distance from one feature to another to see if it matches a specific condition. For example, one might want to select land parcels that are within 1 mile of the school using a proximity operation. **Adjacency operations** evaluate whether or not two features share a border along a line segment. For example, one may want to select all states that share a border with Arizona using an adjacency operation. **Intersection operations** are similar to adjacency because they evaluate whether or not to features touch each other. However, intersection includes all features that merely touch in some way. Whereas, adjacency only includes features that share border along a line segment. **Containment operations** use a spatial operation in which polygons, points or line segments in one feature dataset lie within the polygons of another feature dataset. The containment operation evaluates which features lie within the polygon boundaries. For example, one might want to locate all of the fire stations within the city of Eureka using a containment operation. Unlike the attribute query builder, when constructing spatial queries in GIS software, the **spatial query builder** will often not display the SQL syntax. Instead, the query is built using only a graphic interface. However, "under the hood," SQL is still being used.

Geospatial Analysis

Geospatial analysis is the use of one or more spatial operations to solve a problem or answer question. There are hundreds of spatial functions and operations used in geospatial analysis. This book covers only a few of them. In its most basic form, executing spatial operations usually involves a map layer *input* and a map layer *output*.

A **map layer** is the visual representation of geospatial data in GIS software. Solving a problem often requires a chain of spatial operations executed in a sequence. Spatial operations can have the following three kinds of scope:

» Local
» Focal
» Global

Local analysis operations use data at only one input location to calculate the value of the output at the same position. Local analysis operations do not use any values or attributes of adjacent areas. For example, calculating the population density within the state of California is a local analysis operation. It does not include the other states around California in the calculation. **Focal analysis operations**, sometimes called **neighborhood operations**, use the value of the input location as well as values from areas that surround input location. For example, in calculating the total number of bordering states in a U.S. map layer, the output value for any particular state depends upon its neighbors. **Global analysis operations** use all data values from all participating map layers to calculate the value of the output. For example, suppose one wants to rank each state in the U.S. based on population density. A global analysis operation uses values from all states in the map layer in the calculation to evaluate the rank of any particular state.

Spatial Operations

Spatial Operations tend to fall into two primary categories, *proximity* and *overlay*. **Proximity operations** for spatial analysis share similar characteristics as they do for spatial queries in that they evaluate distances from features. An **overlay operation** is a spatial operation in which two or more datasets are superimposed on one another to assess the relationship between features that occupy the same geographic space. Overlay operations also share common elements with the other spatial query operations. The primary difference is that overlay operations use a spatial query to create new datasets.

The following are some basic spatial operations:

» Buffering
» Clip
» Intersect
» Union

Buffering is an operation that creates a buffer and is useful for proximity analysis (**Figure 7.07**). A **buffer** is a zone around the map feature with an area calculated using distance or time. Buffering may be applied to point features, line features, or polygon layers. However, the output will always be a polygon layer.

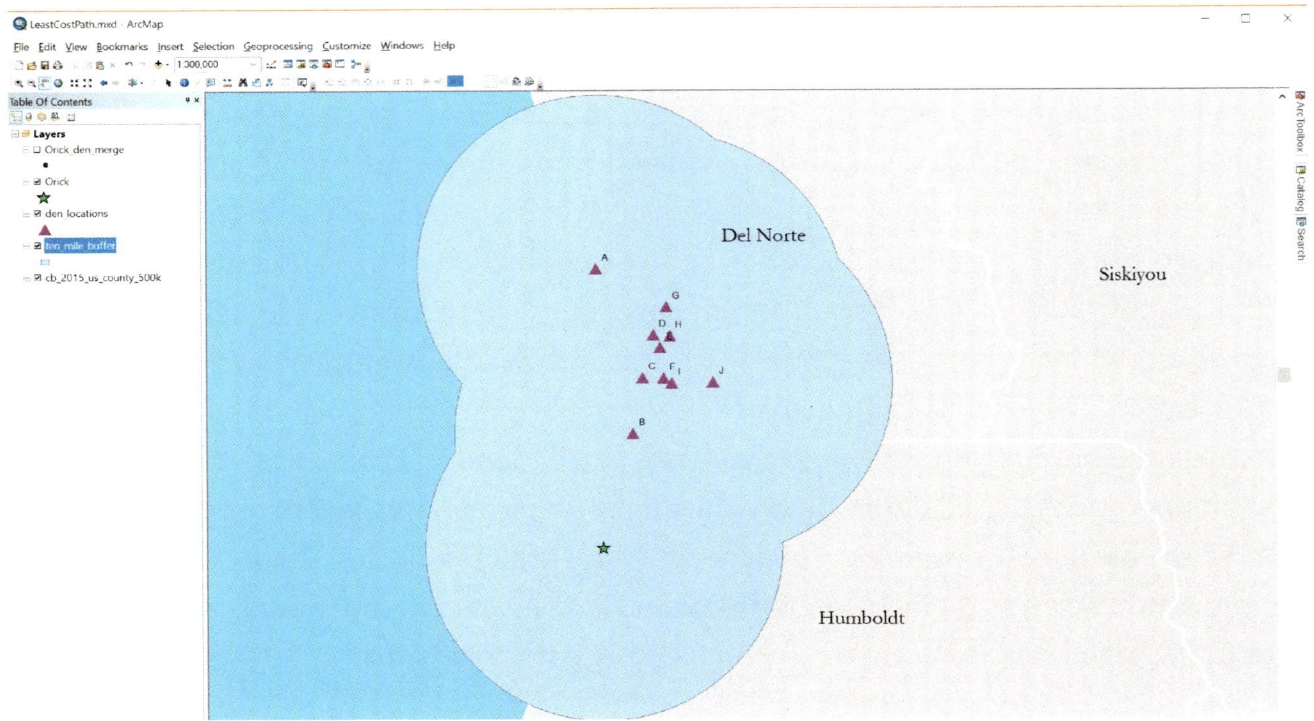

Figure 7.07: In this image, a one-mile buffer was created around a series of point features.

A **clip** is an overlay operation that uses one feature to serve as the outer boundary of another feature (**Figure 7.08**). If one has ever baked cookies, one might be familiar with the concept behind the clip operation. A polygon layer, called the **clip feature**, works like a cookie cutter when overlaid on top of a second layer, the input, which acts like the cookie dough. The output is a copy of the input layer, but with the extent and shape of the clip feature. For example, if one wants to know the length of all roads within the city of Eureka, one can use a polygon representing Eureka to clip a roads layer. The result will be a layer of roads where all the line segments end at the boundary of Eureka. The output of a clip operation will always be smaller than the input.

Figure 7.08: The shapes in this diagram, a star and a pentagon, represent two different map layers. Here, the star acts as the clip feature. The result is a layer that was once pentagon-shaped, but now has a star shape.

Intersect is an overlay operation that preserves overlapping areas (**Figure 7.09**). The output will only contain regions of all included layers that overlap. For example, if one wanted to know which areas in Redwood National Forest intersect with Humboldt County, the output of this operation would be a layer composed of polygons where these two map layers overlapped. The resulting polygons would retain the attribute information from each map layer.

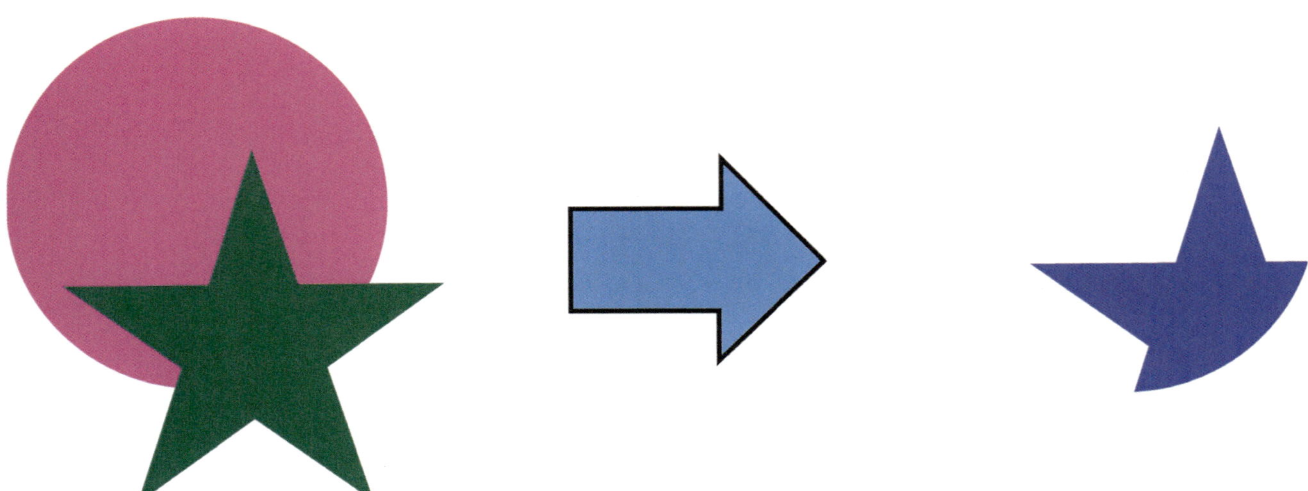

Figure 7.09: The shapes in this diagram, a star and a circle, represent two different map layers. When performing an intersect operation, the output contains only the overlapping areas.

A **union** is similar to an intersect operation in that it combines the attribute information (**Figure 7.10**). The difference is that all features in the participating layers get included in the output. Using the same example, a union of the Redwood national Forest layer and the Humboldt County layer would result in a new layer made up of all polygons from both datasets. The output of a union operation will always be more extensive than any of the inputs.

Figure 7.10: The shapes in this diagram, a star and a rectangle, represent two different map layers. When performing a union operation, the output contains both of the original layers.

TUTORIAL: MAPPING FOOD DESERTS IN SOUTHERN CALIFORNIA

The goal of this activity is to use GIS analysis to map potential food deserts in Southern California. In this activity, you create and organize a workspace folder using a standardized folder structure. You then download and decompress the data from public sources. Using the data, you conduct a GIS analysis using attribute queries, spatial queries, proximity operations, and overlay operations. Once you complete the analysis, you create two small-sized maps using the cartographic design principles learned in **Chapter 2**.

ESTIMATED TIME TO COMPLETE THIS TUTORIAL: 6 HOURS

Learning Outcomes

Readers should be able to accomplish the following outcomes by the end of this tutorial:

- » Summarize the steps for creating and organizing a project workspace folder structure
- » Illustrate the ability to download data from a public source
- » Manage spatial reference systems for an analysis
- » Practice conducting an attribute query
- » Carry out a spatial query
- » Demonstrate how to perform a table join
- » Add and populate fields in an attribute table
- » Show how to implement overlay operations such as clip and erase
- » Exemplify the use of proximity operations such as buffers
- » Practice changing map output size in ArcMap
- » *Apply* symbology and color choices to map features
- » Insert essential map elements using ArcMap
- » Export a high-resolution map

Scenario

A **food desert** is a community, neighborhood, or region where people have limited access to affordable, nutritious food because they live far from a supermarket or large grocery store and do not have easy access to transportation. In this scenario, you are working for a non-profit organization that is interested in identifying potential food deserts in several Southern California counties.

You must use the following criteria in your analysis:

» Limit the study area to several Southern California counties including Los Angeles, Orange, San Bernardino, Riverside, San Diego, and Imperial

» Use Census tracts where people with poverty status is greater than or equal to 25%.

» Find areas more than one mile from a large grocery store or supermarket

Also, you must conduct the analysis using the **Universal Transverse Mercator (UTM) system** along with the **North American Datum of 1983 (NAD 83)**. Southern California lies in **Zone 11** of the UTM system.

Skill Drill: Setting Up Your Workspace

On your desktop, create a new folder and give it a descriptive name, such as "Food_Deserts." Be sure there are no spaces in the name. You may use underscores instead of spaces. Inside this folder, create the following three subfolders: *original, working,* and *final*.

Downloading Data from the National Historical Geographic Information System (NHGIS)

The *National Historical Geographic Information System (NHGIS)*[1] is a website that provides historical data and GIS boundary files from the United States Census. The data is free to use, which makes this site an excellent resource. To access the data, you need to create an account. Watch the *NHGIS Account Registration Tutorial* to learn how to create an account (**Figure 7.11**). Using what you learn in the video, create an account with the NHGIS and log into the website.

1 https://www.nhgis.org/

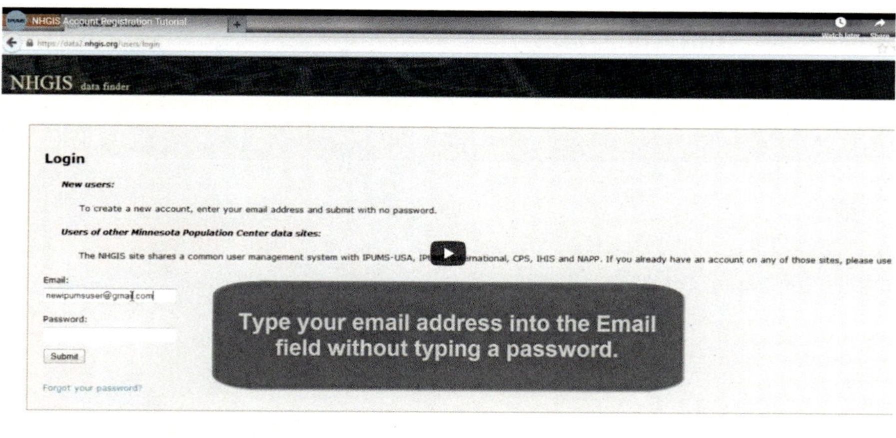

Figure 7.11: This video demonstrates how to create an account. URL: ***https://youtu.be/HoERr8OlpYo***

Applying Data Filters

The NHGIS provides hundreds of unique datasets for years from 1790 through the present day. *Datasets* are aggregated at multiple geographic levels, including Census blocks, Census tracts, counties, and states. The vast number of data files can be overwhelming. To assist you with finding a particular dataset, the NHGIS uses a system of filters to help you narrow down your search. Watch *NHGIS Filter Options* to learn how to use data filters on the NHGIS website (**Figure 7.12**).

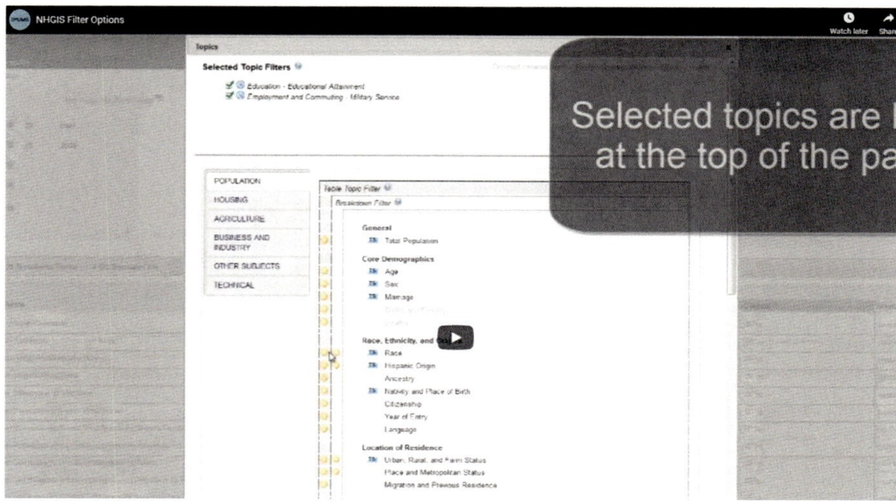

Figure 7.12: This video walks you through the process of applying data filters. URL: ***https://youtu.be/XASUCfhoUdo***

To start, use the first filter to determine the geographic level in which you are interested. Under *Apply Filters*, click on *Geographic Levels* (**Figure 7.13**).

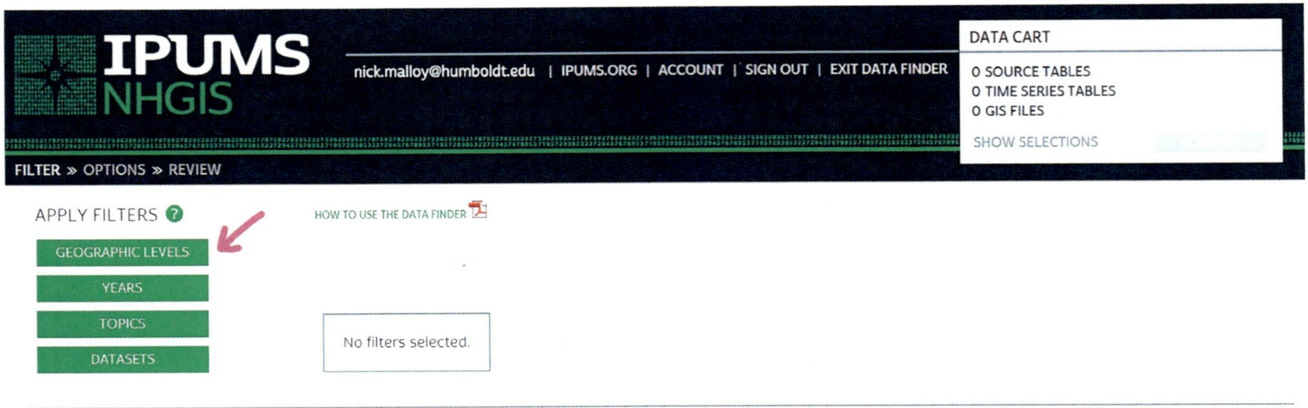

Figure 7.13: The NHGIS website allows you to apply filters to narrow down the search results.

When the *Geographic Levels* options appear, click the plus sign next to *County* and *Census Tract* (**Figure 7.14**). When you are ready, click *Submit*.

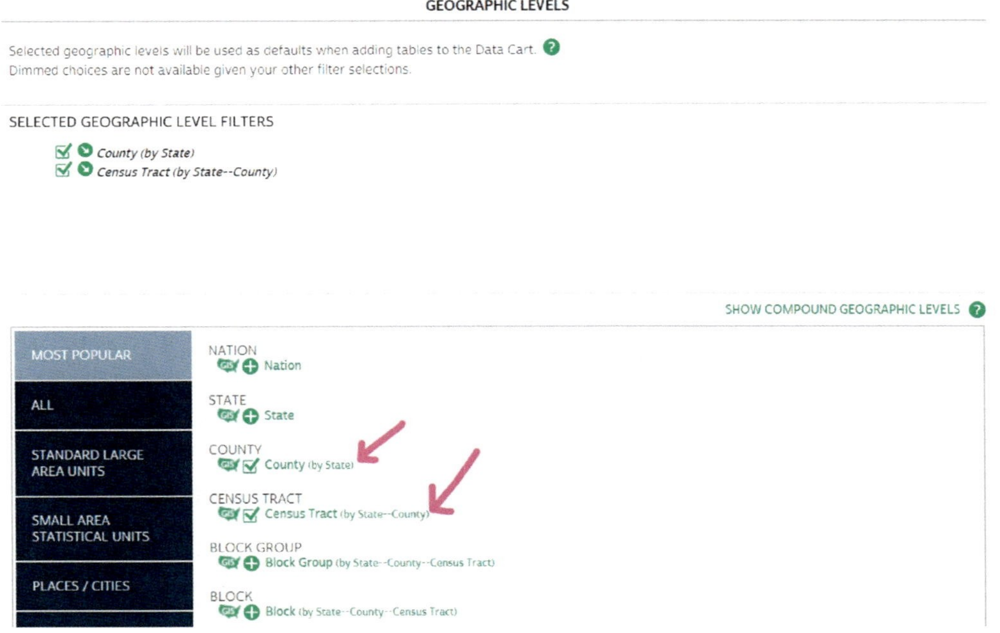

Figure 7.14: The geographic levels define which geographic extents to which the data is aggregated.

Next, click on the *Years* filter. When the *Years* options appear, check the box next to *2016* under *Non-Decennial Years* (**Figure 7.15**). When you are ready, click *Submit*.

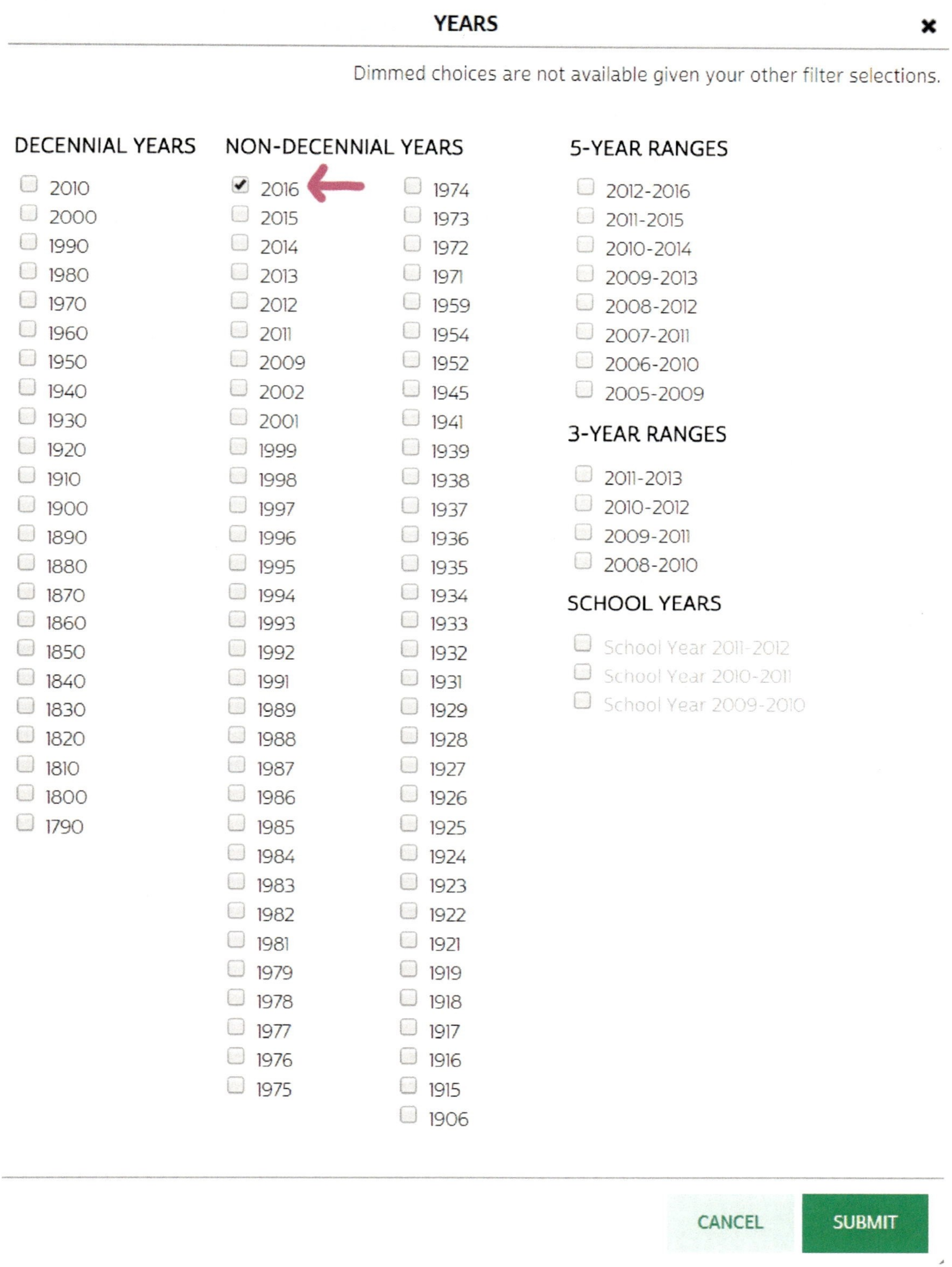

YEARS ×

Dimmed choices are not available given your other filter selections.

DECENNIAL YEARS: 2010, 2000, 1990, 1980, 1970, 1960, 1950, 1940, 1930, 1920, 1910, 1900, 1890, 1880, 1870, 1860, 1850, 1840, 1830, 1820, 1810, 1800, 1790

NON-DECENNIAL YEARS: 2016 (checked), 2015, 2014, 2013, 2012, 2011, 2009, 2002, 2001, 1999, 1998, 1997, 1996, 1995, 1994, 1993, 1992, 1991, 1989, 1988, 1987, 1986, 1985, 1984, 1983, 1982, 1981, 1979, 1978, 1977, 1976, 1975, 1974, 1973, 1972, 1971, 1959, 1954, 1952, 1945, 1941, 1939, 1938, 1937, 1936, 1935, 1934, 1933, 1932, 1931, 1929, 1928, 1927, 1926, 1925, 1924, 1923, 1922, 1921, 1919, 1918, 1917, 1916, 1915, 1906

5-YEAR RANGES: 2012-2016, 2011-2015, 2010-2014, 2009-2013, 2008-2012, 2007-2011, 2006-2010, 2005-2009

3-YEAR RANGES: 2011-2013, 2010-2012, 2009-2011, 2008-2010

SCHOOL YEARS: School Year 2011-2012, School Year 2010-2011, School Year 2009-2010

CANCEL SUBMIT

Figure 7.15: The *Years* option defines the date range for the results.

Next, click on the *Topics* filter. When the *Topics* options appear, scroll down until you see the *Income* category. Click the plus sign next to *Household and Family Income* (**Figure 7.16**). When you are ready, click *Submit*.

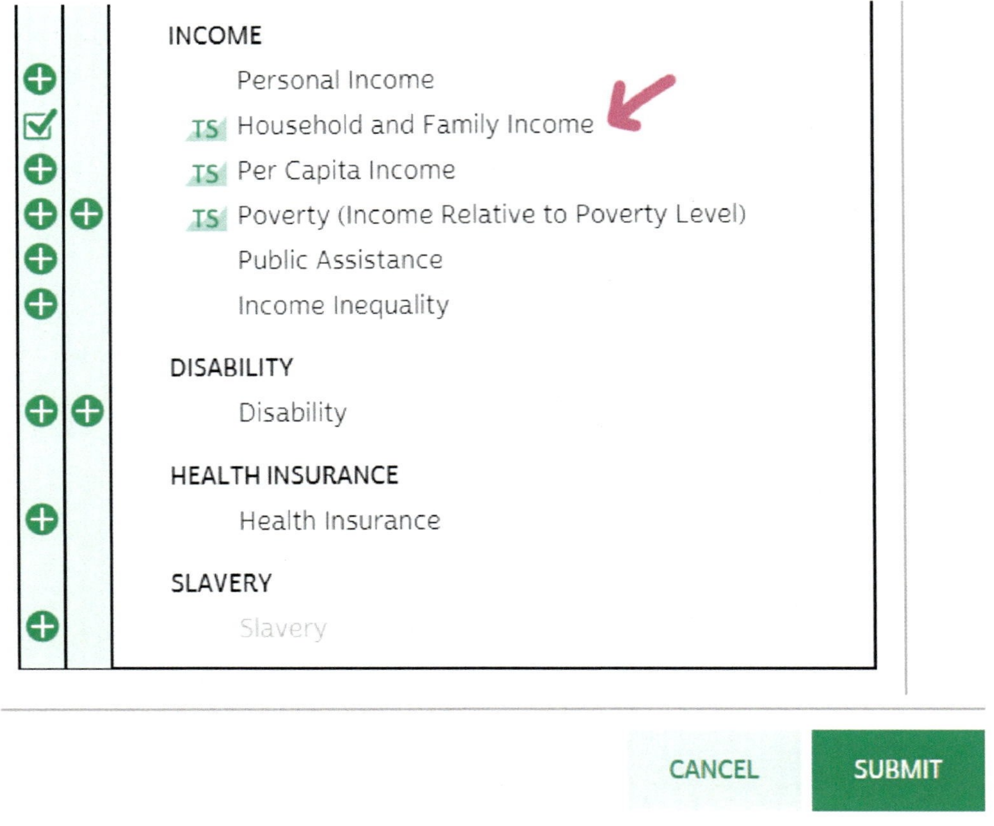

Figure 7.16: The topics options refine your search further by limiting the subject matter of the dataset.

Next, click the *Datasets* filter. Under *2016 American Community Survey*, check the box next to *2012_2016_ACS5b 5-Year Data [2012-2016, Tracts & Larger Areas]* (**Figure 7.17**). When you are ready, click *Submit*.

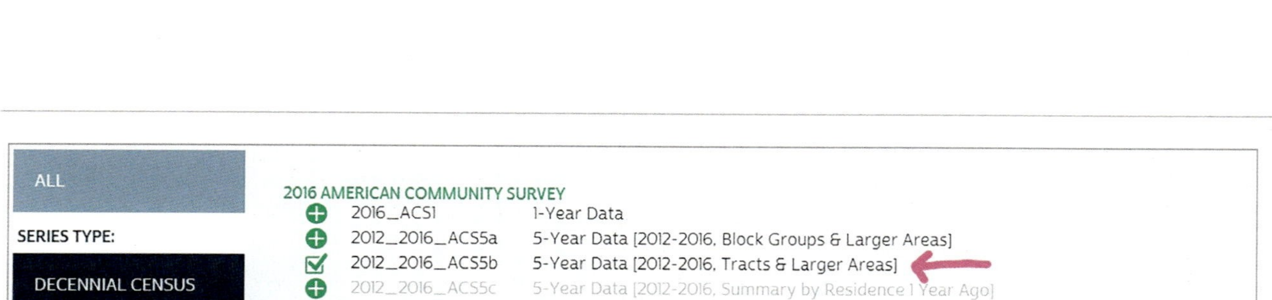

Figure 7.17: With multiple filters active, the NHGIS dims unavailable datasets.

Selecting Data

The results should appear on the page. If necessary, you can use the drop-down menu to select the number of results per page to show all of the results at once. To learn how to choose data and place an order, watch the *NHGIS Data Finder Tutorial* (**Figure 7.18**).

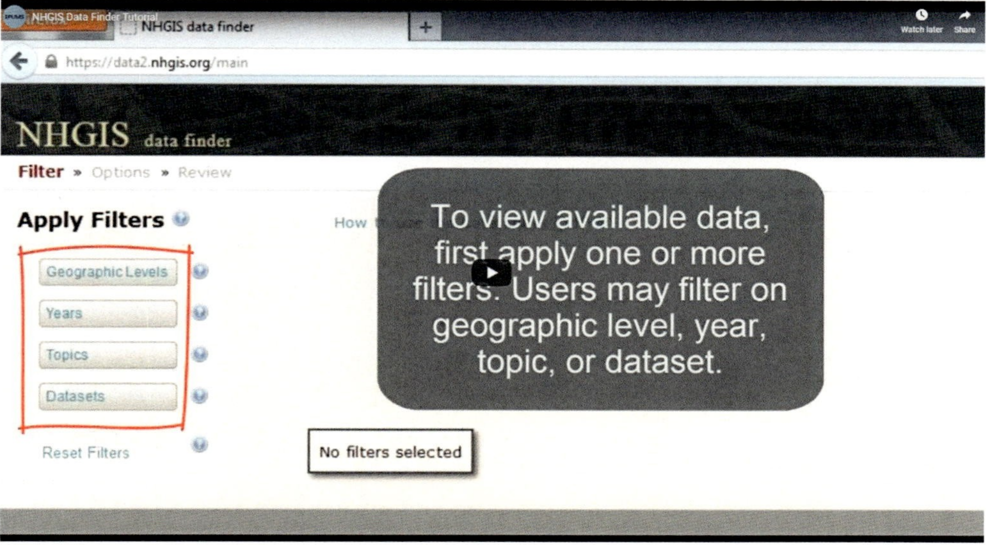

Figure 7.18: This video walks you through how to place an order on the NHGIS website. URL: *https://youtu.be/P1znKKm8vX4*

Under *Source Tables*, locate the table named *Poverty Status in the Past 12 Months of Families by Family Type by Social Security Income by Supplemental Security Income (SSI) and Cash Public Assistance Income* (**Figure 7.19**) When you find it, click the plus sign to add it to your data cart.

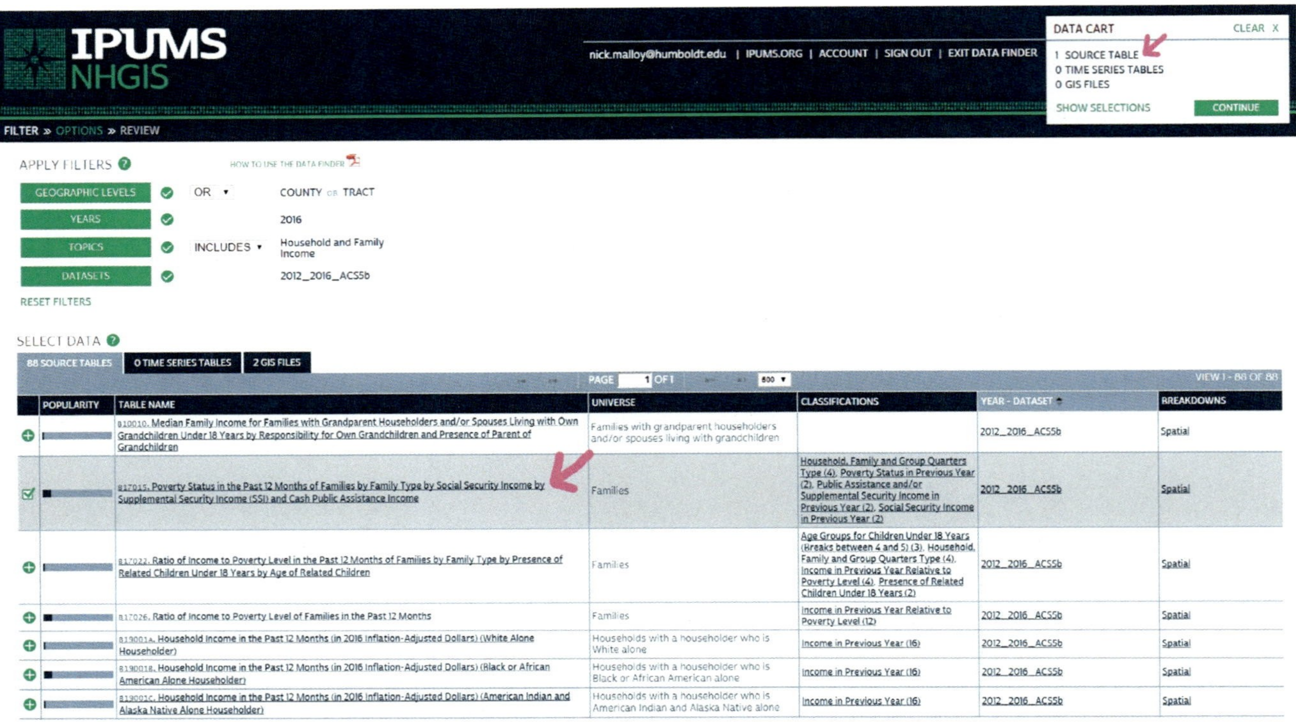

Figure 7.19: When you click the plus sign, the *Data Cart* updates with the number of source tables checked.

Click on the *GIS Files* tab and add the county and Census tract to the *Data Cart* (**Figure 7.20**)

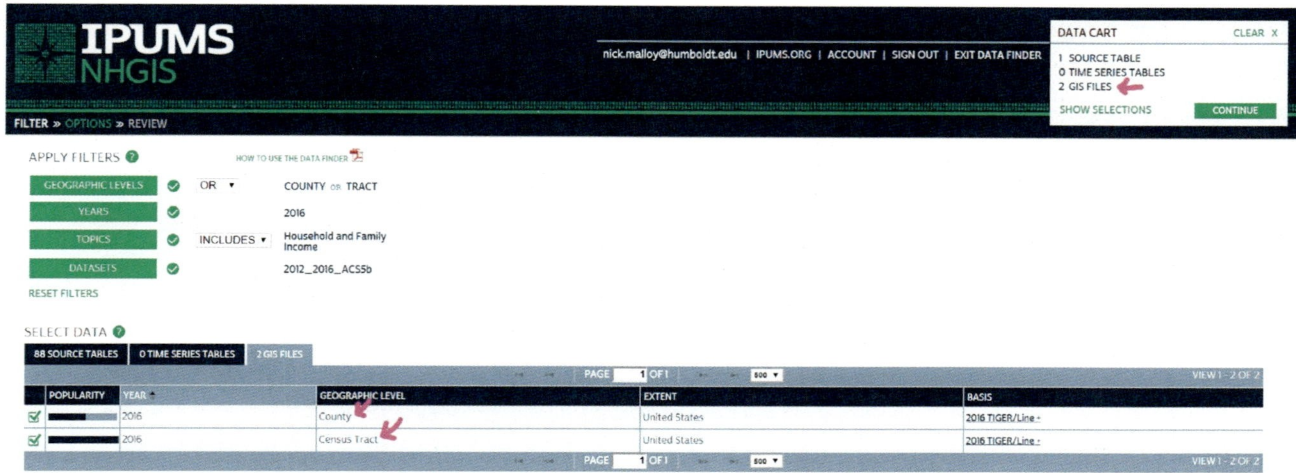

Figure 7.20: The *GIS Files* tab provide County and Census tract shapefiles.

On the *Data Cart,* click *Continue*. On the Options page, you should see a list of your source tables and GIS files (**Figure 7.21**). On the *Data Cart*, click *Continue* again.

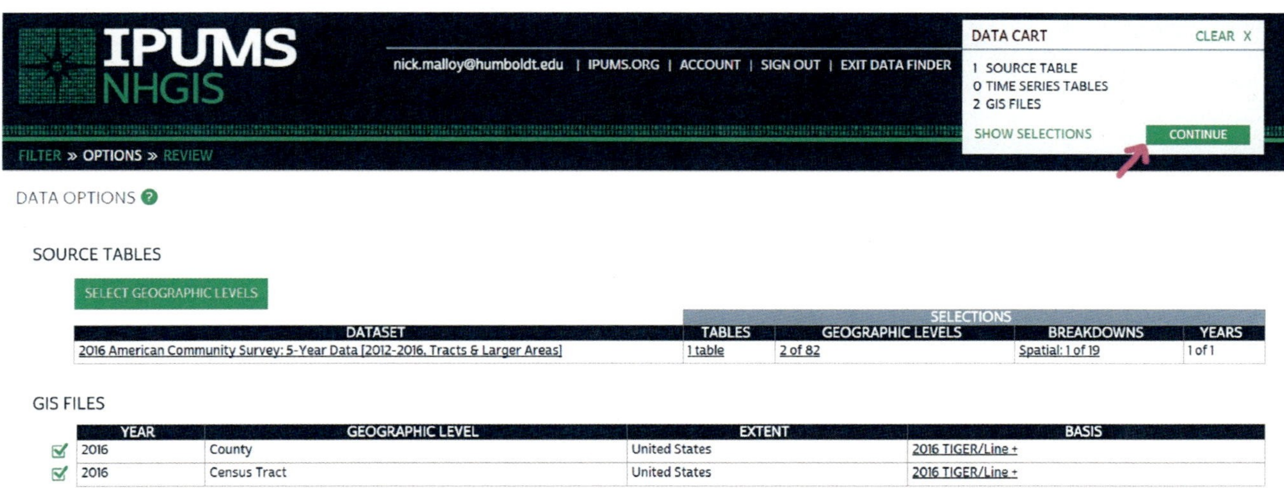

Figure 7.21: You should not need to change anything on the options page.

On the *Review* page, select the radio button next to *Comma delimited* and *Combine different breakdowns/data types into a single data file* (**Figure 7.22**) Leave all the other options unchecked. Write a brief description so that you remember what you requested. When ready, click *Submit*. You will get an email when your data is ready. Click the link in the email.

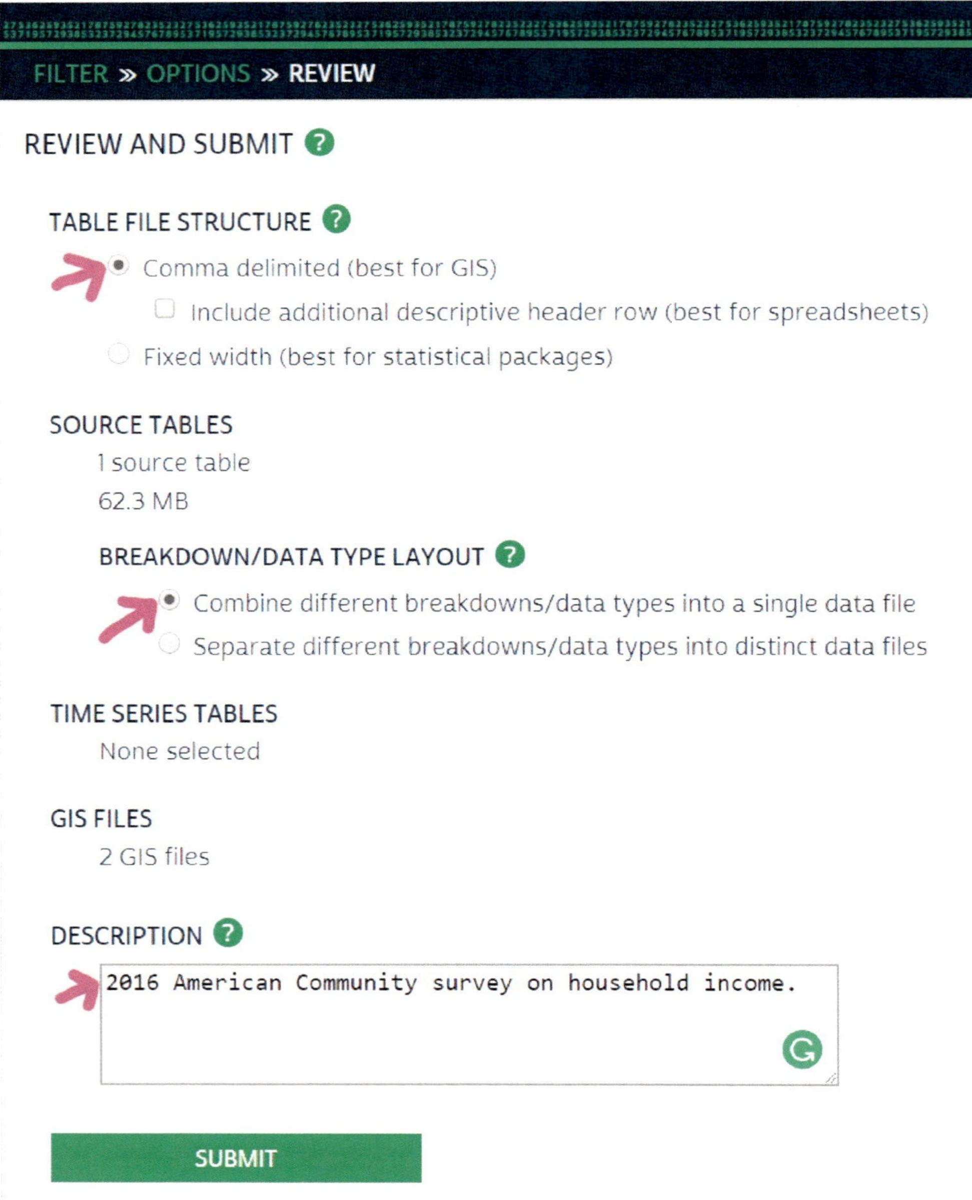

Figure 7.22: The *Review* page provides different data formatting options. A simple comma delimited file is best for GIS.

If you do not get an email within about five minutes, refresh the NHGIS Extracts History webpage.

In a previous activity, you learned how to save files directly to your *original* folder. Use the links under *Download Table Data* and *Download GIS Data* and save the files to your *original* folder. The shapefiles are very large and may take a few minutes to download. Once ready, use the 7zip software to decompress the files. Delete the original compressed zip files when done (**Figure 7.23**). Removing them saves space and helps to prevent confusion later.

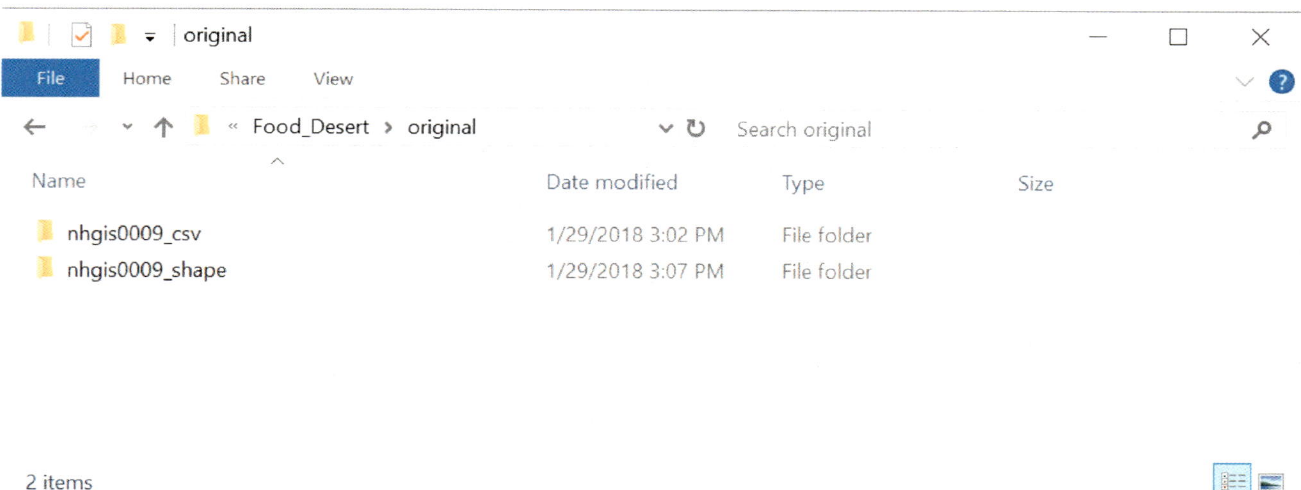

Figure 7.23: The *original* folder contains two subfolders, one for the shapefiles and one for the CSV tables. In this instance, both folder names end in 0009. The specific number may vary.

The Census tract and county shapefiles are **double compressed**, meaning you should find a zip file within a zip file. It is a common source of confusion when working with NHGIS data. You need to decompress the contents of the shape subfolder. If you fail to do so, you will not be able to use the files in ArcMap. Open the shape subfolder and decompress the contents (**Figure 7.24**).

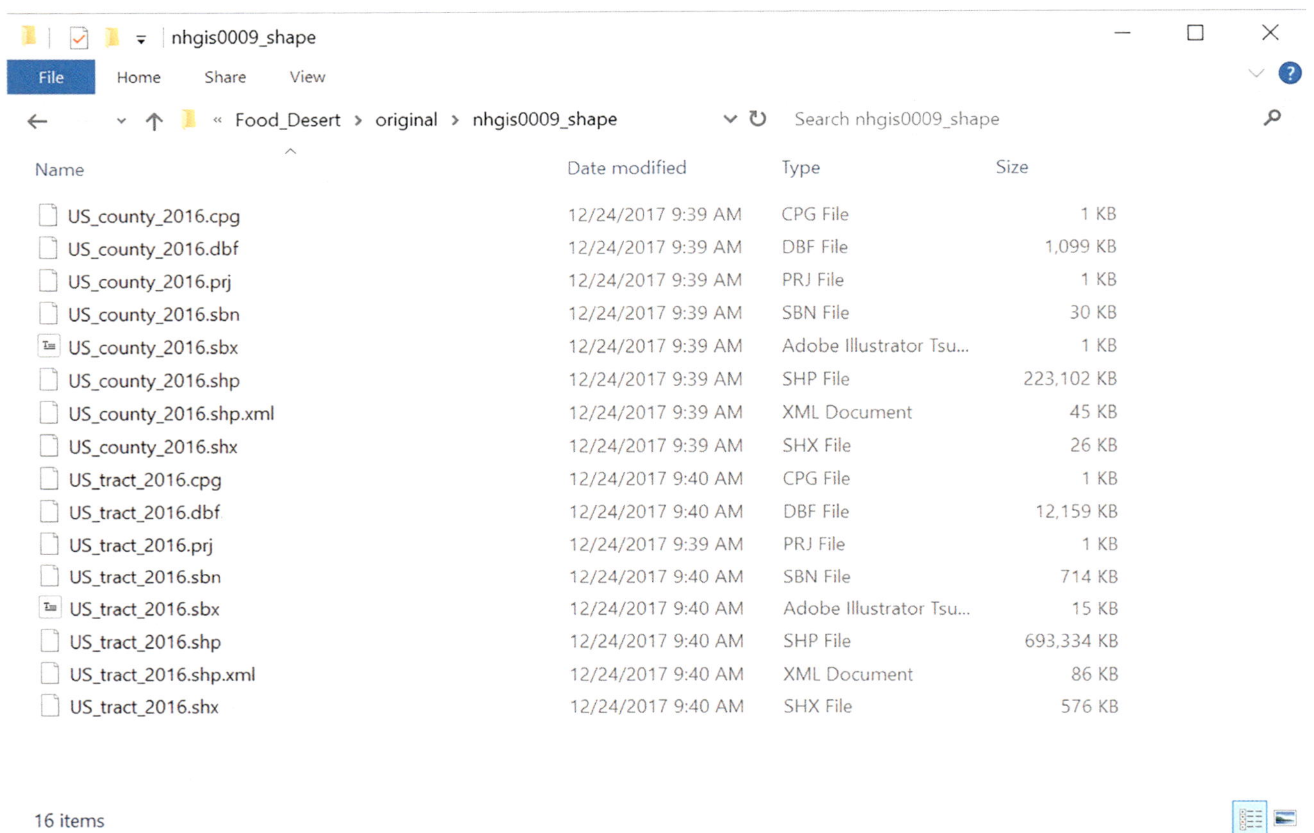

Figure 7.24: The contents of the shape subfolder are extracted, and the zip files removed.

The NHGIS uses a series of numbers and letters that they generate when you acquire the data. Unfortunately, these numbers are not very user-friendly. In this example, the household and family income csv files were assigned the file names " nhgis0009_ds226_20165_2016_county.csv,", and nhgis0009_ds226_20165_2016_tract.csv." To prevent confusion, be sure to rename these files to the human-friendly names County_Level_Income.csv, and Tract_Level_Income.CSV (**Figure 7.25**).

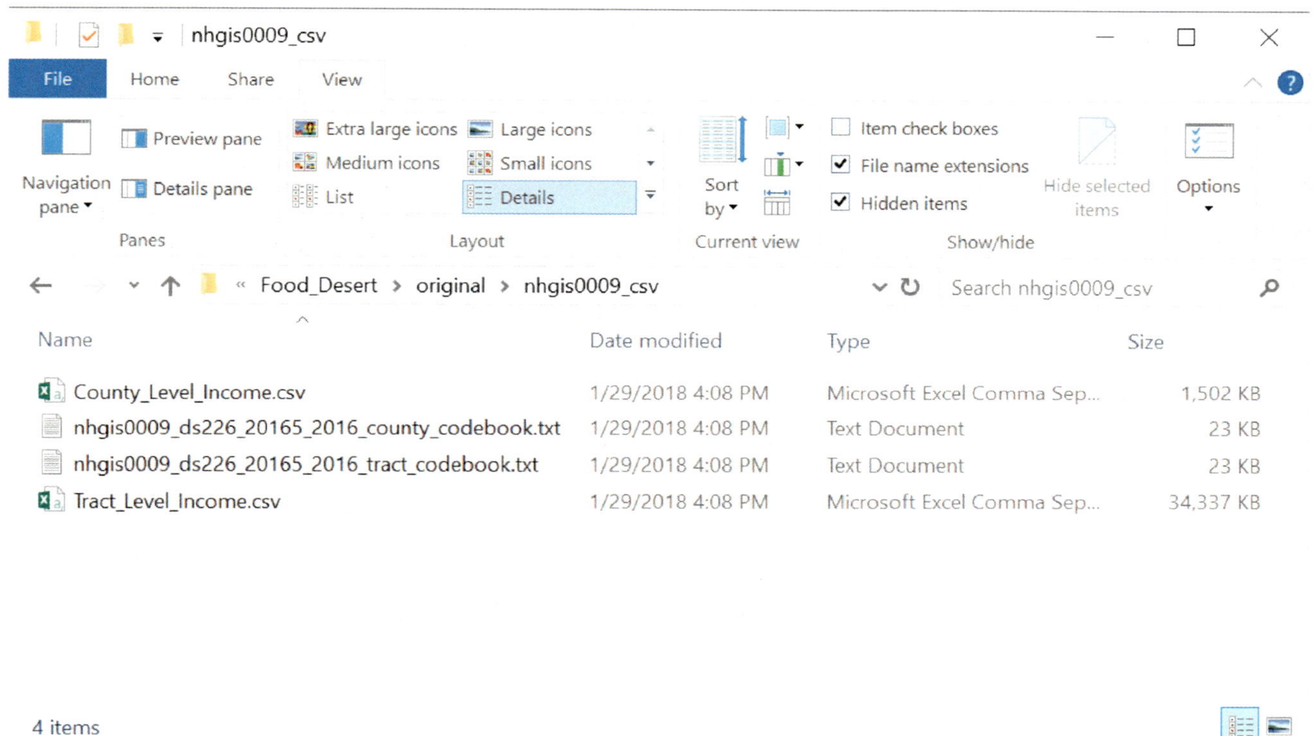

Figure 7.25: Renaming the files to something more meaningful helps to prevent confusion.

Skill Drill: Downloading Population Data from the NHGIS

Using what you learned in the previous step, return to the NHGIS website and download total population data at the Census tract level for the year 2016. Use the following filters:

» *Geographic Levels:* Census Tract

» *Years:* 2016

» *Topics:* Total Population

The source table is named *Total Population*. You *do not* need a GIS file because you already downloaded the census tract shapefile in the previous step. After a few minutes, the table should appear on the Extracts History page (**Figure 7.26**).

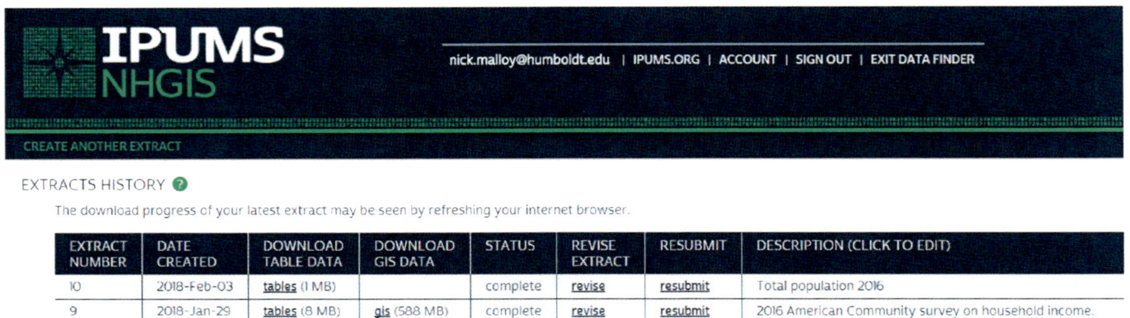

Figure 7.26: The NHGIS Extracts History page displays your order history and provides links for downloading data.

Save the table to your *original* folder and decompress the file. Delete the original compressed zip file when done. Removing it saves space and helps to prevent confusion later. The *original* folder should now contain three subfolders (**Figure 7.27**).

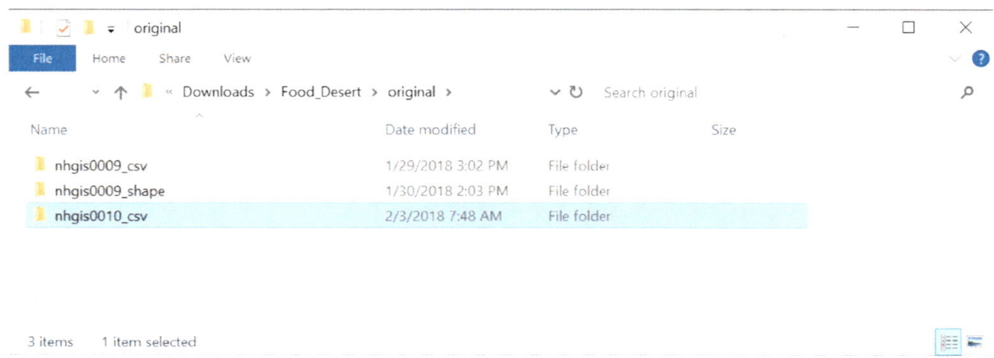

Figure 7.27: The two CSV datasets are assigned a different number. Keep track of which number belongs to which dataset.

In this instance, the total population CSV includes the number 0010. Over time, this number may vary. Take a moment to rename the CSV file to something more meaningful, such as Tract_Level_Population.csv (**Figure 7.28**).

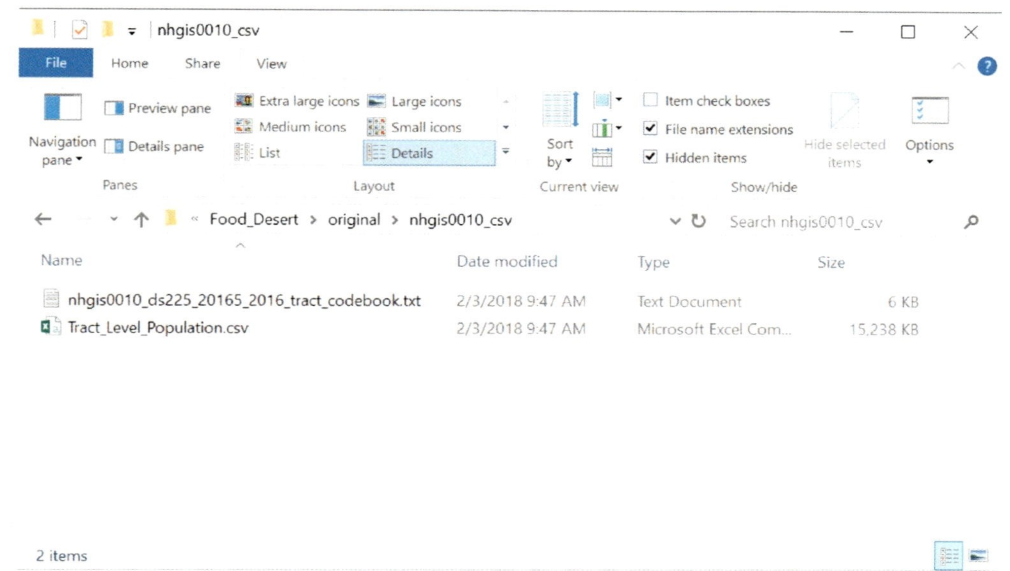

Figure 7.28: Renaming the files to something more meaningful helps to prevent confusion.

Using the Project Tool

It is essential that all of the layers in a geospatial analysis use the same spatial reference system. Also, it is crucial to use the optimal spatial reference system for the study area. Recall from Chapter 4 that a **spatial reference system** defines a geographic location and includes the map datum, if applicable the coordinate reference system, and if applicable, the map projection. For this analysis, you must use a projected coordinate system optimized for Southern California, the **Universal Transverse Mercator (UTM) Coordinate System** along with the **North American Datum of 1983 (NAD 83)**.

Launch ArcMap and open a blank map document. In the Catalog Window, connect to your workspace folder. Open the *original* folder and locate the shapefile for U.S. counties. Add the layer to the map (**Figure 7.29**).

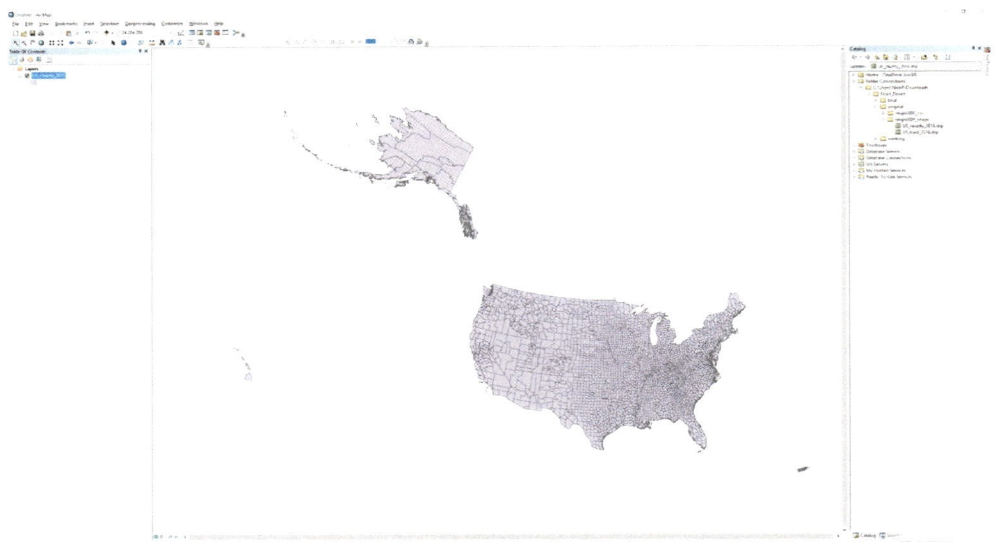

Figure 7.29: The U.S. counties were added from the shape subfolder.

Take a moment to check the spatial reference information for the counties layer. Right-click on the counties layer in the Table of Contents and open the layer properties. Navigate to the *Source* tab. Find the answers to the following questions:

» What map projection is used for this dataset?

» What datum is used for this dataset?

You will find that the map projection and datum is one that is optimized for the United States as a whole. However, there are better choices for Southern California. For your study area, you will use the UTM system. In Chapter 4, you learned that the UTM system contains a series of 120 sections called **UTM zones**. For Southern California, you use **Zone 11 N**.

In a previous activity, you used a tool called *Project Raster* to create a new raster dataset in the desired spatial reference system. *Project Raster* only works on raster datasets. The *Project* tool provides similar results, but only works on vector datasets, like shapefiles. In this step, you use the *Project* tool to create a new shapefile in the desired spatial reference system.

Note: You cannot change the spatial reference of an existing dataset. Once created, the data is immutable. The Project and the Project Raster tools create new datasets using the desired spatial reference.

In ArcMap, click the toolbox button located on the *Standard* menu. The icon looks like a red toolbox. The ArcToolbox may appear floating above the map. Dock it to the side by dragging it to the left in a similar manner in which you docked the attribute table in previous activities. Expand the Data Management Tools (**Figure 7.30**). Next, expand the Projections and Transformations toolbox. Locate the *Project* tool. Double-click the tool to open it.

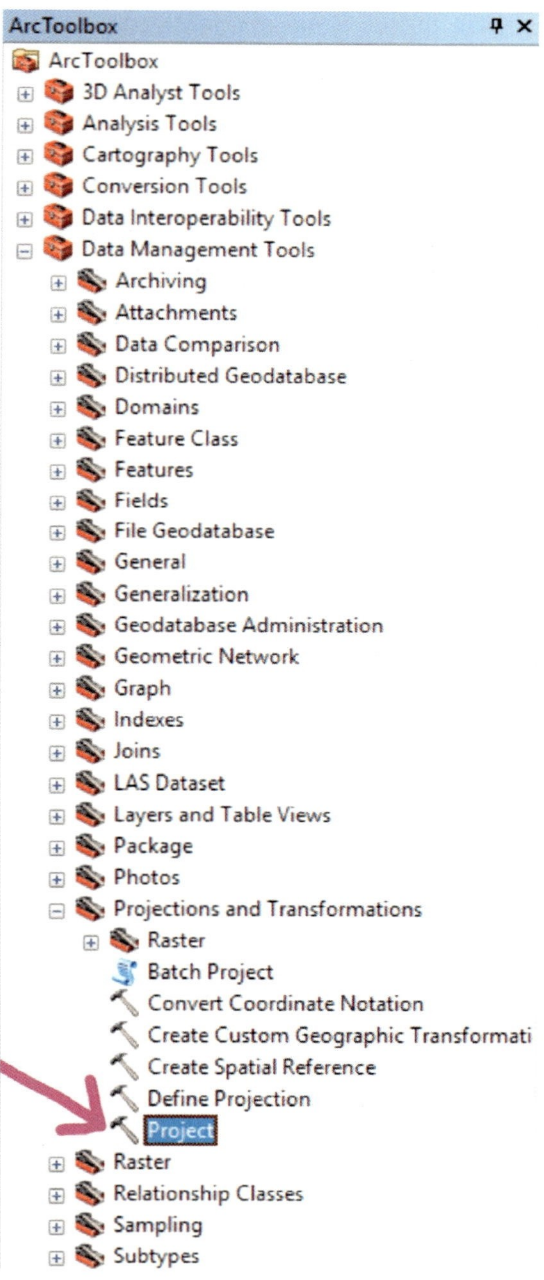

Figure 7.30: The *Project* tool is used for vector data.

For the *Input Dataset or Feature Class*, use the drop-down menu to select the layer representing U.S. counties. For the *Output Dataset or Feature Class*, click the yellow file folder icon and browse to your *working* folder. Name the file US_Counties_NAD83_

UTM11N (**Figure 7.31**). For the *Output Coordinate System*, click the button on the right. When the *Spatial References Properties* window opens, click the plus sign next to *Projected Coordinate Systems* to expand the folder. Scroll down until you see the *UTM* folder. Expand it, then open the *NAD 1983* folder. Select *NAD 1983 UTM Zone 11N* and click *OK*. You may leave all other default settings and click *OK* (**Figure 7.31**).

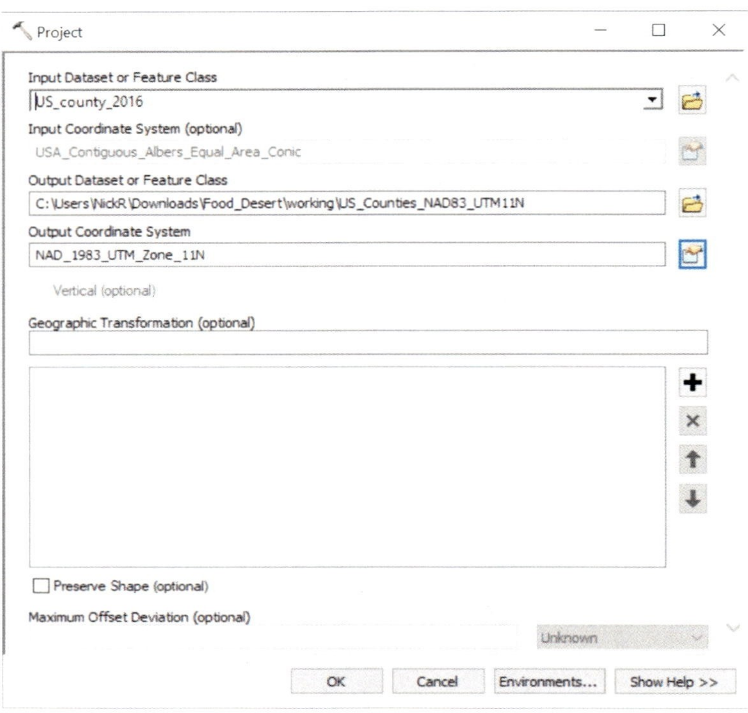

Figure 7.31: Check to make sure your settings match those in this image.

When the *Project* tool completes, ArcMap may not automatically add the new layer to the map. In the Catalog Window, open the *working* folder and locate the new shapefile. Add the data to the map. Since you do not need it anymore, remove the original U.S. County layer from the Table of Contents. Eliminating unnecessary layers helps keep the Table of Contents organized and prevents confusion.

Resetting the Data Frame Coordinate System

As you can see, the newly created shapefile looks identical to the original, even though it is using a different projection. You might be wondering why? As you learned in Chapter 1, the data frame adopts the spatial reference of the first layer you add to the Table of Contents. The data frame window displays the map using **project-on-the-fly**. Even though the layers in the Table of Contents might use different projections and coordinate systems internally, ArcMap tries to line them up on the screen using the coordinate system defined in the **data frame** properties. In this instance, the data frame currently uses the spatial reference from the *original* U.S. County layer. In this step, you must change the **display projection** of the data frame to match the new county layer. Open the data frame properties and navigate to the *Coordinate System* tab. Expand the *Layers* folder and select *NAD 1983 UTM Zone 11N* (**Figure 7.32**). When ready, click *OK*.

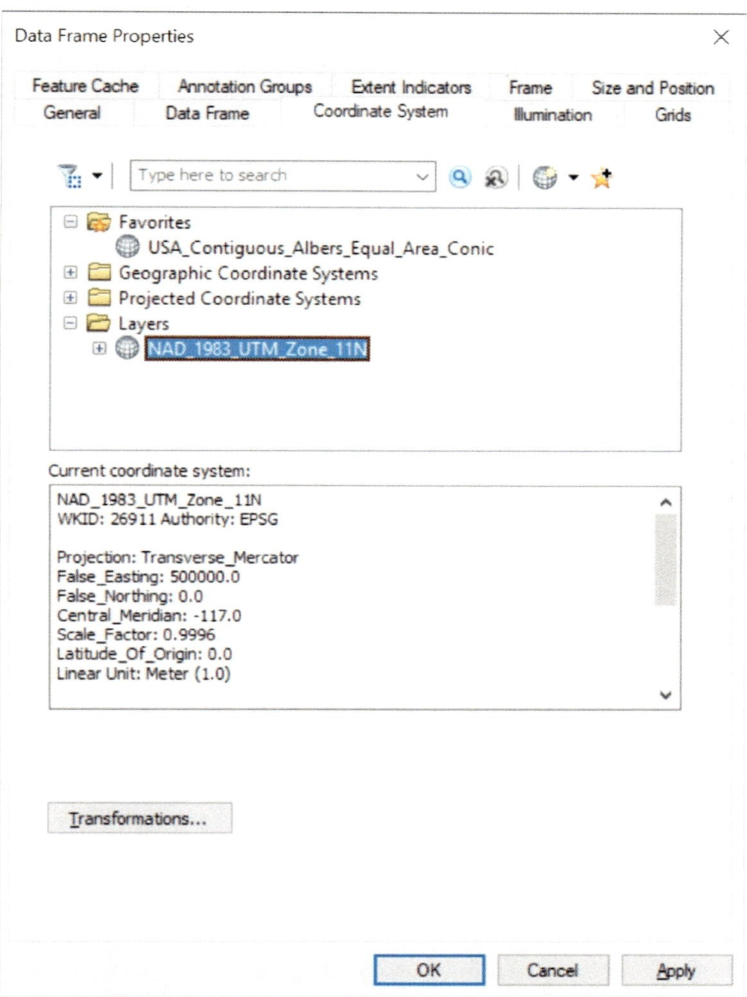

Figure 7.32: The *Layers* folder provides a shortcut to spatial references for layers currently in the Table of Contents.

The data frame visibly updates the display projection with the new spatial reference system, which is centered on Southern California (**Figure 7.33**). Take a moment to set your map document properties to *store relative paths*. Then, save the map document in your workspace folder. Name the file food desert Map.

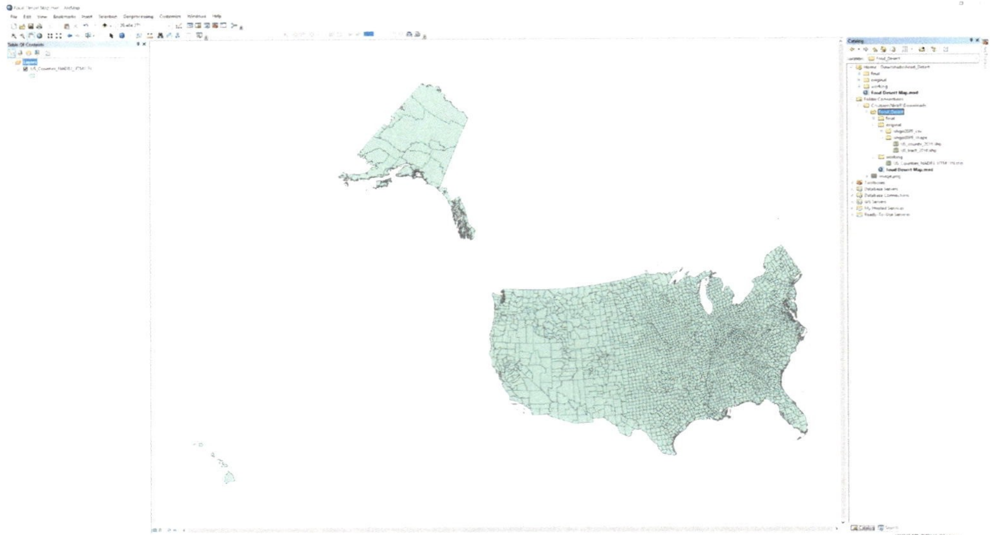

Figure 7.33: You may notice that the eastern and western extent of the United States is cut off. If you recall, the Transverse Mercator projection displays a wedge-shaped region called a gore.

Skill Drill: Use the Project Tool on U.S. Census Tracts

Using what you learned in the previous step, use the *Project* tool on the shapefile representing the U.S. Census tracts. The new shapefile should also use NAD 1983 UTM Zone 11N for the spatial reference system. Be sure to save the output to your *working* folder. Add the new Census tract layer to the map (**Figure 7.34**).

Warning: The Census tract layer is an extensive dataset. It may take a few minutes for the Project tool to work.

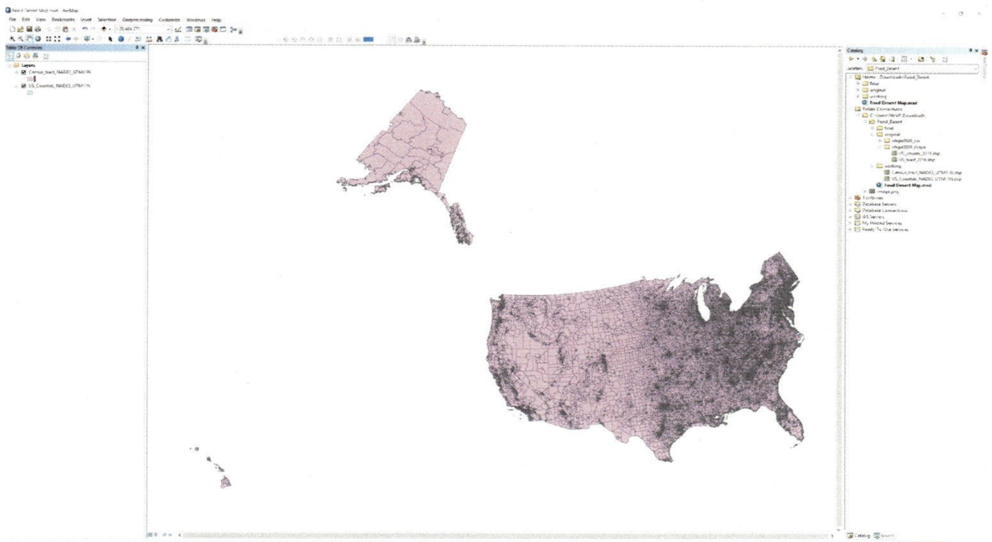

Figure 7.34: The U.S. Census tract layer is a large file. You may want to consider unchecking the layer in the Table of Contents until you need to use it.

Defining the Study Area Using an Attribute Query

For this analysis, you must limit your study area to several counties in Southern California:

- » Los Angeles
- » Orange
- » San Bernardino
- » Riverside
- » San Diego
- » Imperial

To start, use an attribute query to create a layer for California counties. In this chapter, you learned that **attribute queries** are selections based upon the values of attributes. In the Table of Contents, uncheck the box next to the Census tract layer so that only the counties layer is visible. Open the attribute table for the counties layer. To save space, you may want to dock the attribute table to the bottom of the ArcMap window (**Figure 7.35**). Take a moment to read through the attribute table to find out what kind of county-level information is available in the shapefile.

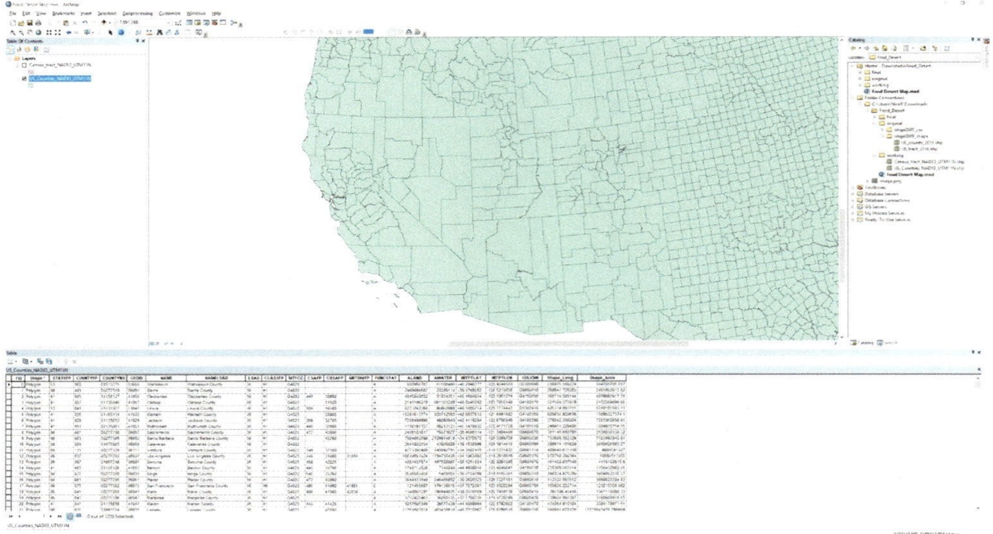

Figure 7.35: The U.S. counties shapefile comes with some basic information about the county polygons, including, area, feature type, and a state and county identification code.

Among the attributes is an identification code for each state. Right click on the field name STATEFP and select Properties from the contextual menu (**Figure 7.36**).

FID	Shape *	STATEFP	COUNTYFP	COUNTYNS	GEOID	NAME	NAMELSAD
0	Polygon	53				Wahkiakum	Wahkiakum County
1	Polygon	06				Sierra	Sierra County
2	Polygon	41				Clackamas	Clackamas County
3	Polygon	41				Clatsop	Clatsop County
4	Polygon	53				Lewis	Lewis County
5	Polygon	41				Klamath	Klamath County
6	Polygon	41				Jackson	Jackson County
7	Polygon	41				Multnomah	Multnomah County
8	Polygon	06				Sacramento	Sacramento County
9	Polygon	06				Santa Barbara	Santa Barbara County
10	Polygon	06				Calaveras	Calaveras County
11	Polygon	06				Ventura	Ventura County
12	Polygon	06				Los Angeles	Los Angeles County
13	Polygon	06				Sonoma	Sonoma County
14	Polygon	41				Benton	Benton County
15	Polygon	06				Kings	Kings County
16	Polygon	06	061	00277295			County
17	Polygon	06	075	00277302			ancisco County
18	Polygon	06	041	00277285			County
19	Polygon	06	043	00277286			sa County
20	Polygon	41	047	01135859			County
21	Polygon	06	035	01693324	06035	Lassen	Lassen County

Figure 7.36: You can access the contextual menu for attribute fields by right-clicking on the field name.

The *Field Properties* window provides information about the attribute field, including the name, the alias[2], and the type (**Figure 7.37**). It also shows you any special rules that may apply to the field, such as maximum character length. The *Field Properties* window reveals that the STATEFP filed has a data type of *string* with a maximum length of 2 characters.

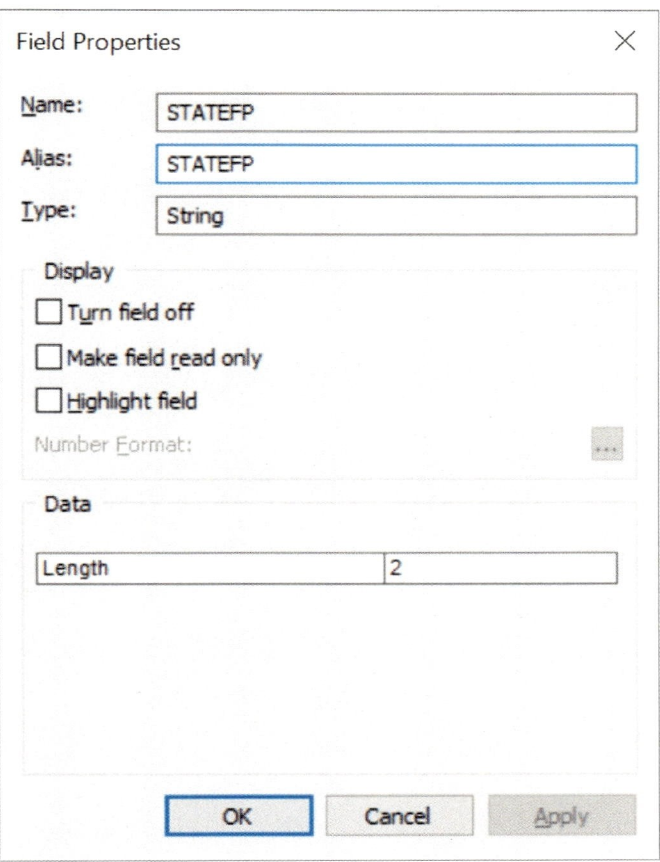

Figure 7.37: Checking the field properties is a good idea when conducting an analysis.

Though the code looks like a number, the shapefile's database stores it as a string data type. In this chapter, you learned that a **string** is a data type that represents text.

The fact that the code is stored as a string will have implications when you create your SQL statement in the following steps.

Use the code for California, which is '06', in the attribute query to select California counties and create a new layer. On the attribute table, click the *Select by Attributes* button, which is the third icon from the left, to open the *Select by Attributes Query Builder* (**Figure 7.38**).

2 An Alias is where you can provide a human-friendly name.

Table

US_Counties_NAD Select By Attributes

	FID	Shape *	STATEFP	COUNTYFP	COUNTYNS	GEOID	
▸	0	Polygon	53	069	01513275	53069	Wahkia
	1	Polygon	06	091	00277310	06091	Sierra
	2	Polygon	41	005	01155127	41005	Clacka

Figure 7.38: Using the S*elect by Attributes* button opens the *Select by Attributes* window for the current layer.

In this chapter, you learned that a **query builder** is a graphical user interface (GUI) that helps to build selection operations using a **structured query language (SQL)**. As the name implies, SQL is a programming language with structural rules, called **syntax**. SQL is used for retrieving and manipulating data from a database. At the top of the *Select by Attributes* window, you have a choice of several different methods (**Figure 7.39**). Take a moment to read through them. They will give you an idea of what you can do with the Select by Attribute tool. For this step, use the method, Create a new selection.

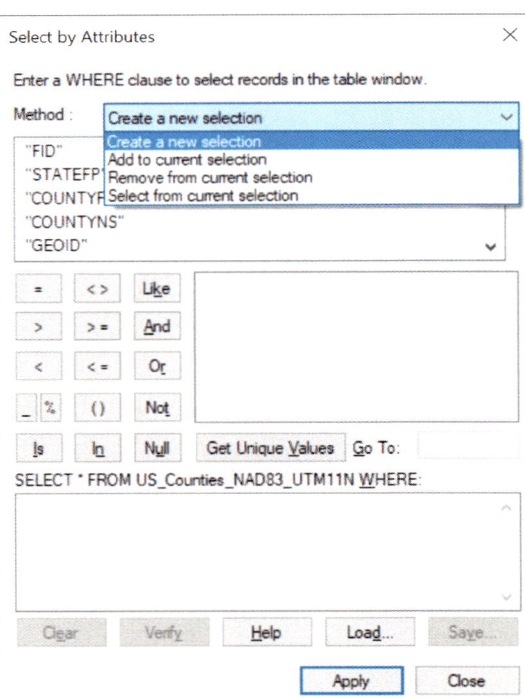

Figure 7.39: The S*elect by Attributes* Query Builder provides different ways for refining your selection.

Below the word *Method*, you should see a list of field names. Double-click on STATEFP. Notice that "STATEFP" appears in the box at the bottom (**Figure 7.40**). As you work, the *Select by Attributes* tool will create an SQL statement to query the attributes in the county layer's database. Next, click the = sign button. An equal sign appears after the field name. Click on the button that says 'Get Unique Values." A list of all the STATEFP values appears. From the list, double-click on '06' to add it to the SQL statement. Your SQL statement should say "STATEFP" = '06'. When you are ready, click *Apply*.

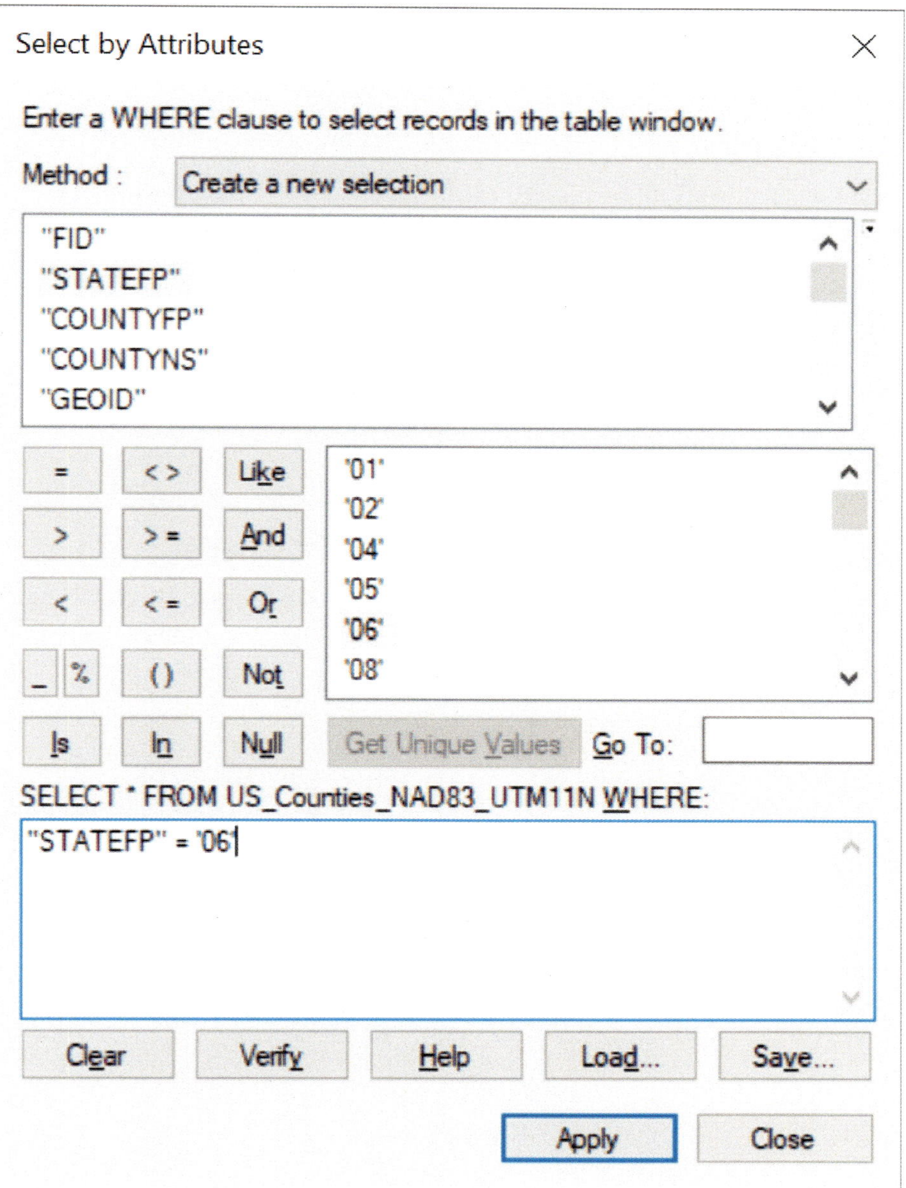

Figure 7.40: Because you used the tool to build the SQL statement, you know the syntax and spelling will be correct.

Notice that some of the counties on the map are now selected (**Figure 7.41**). These are counties that have a state code of "06."

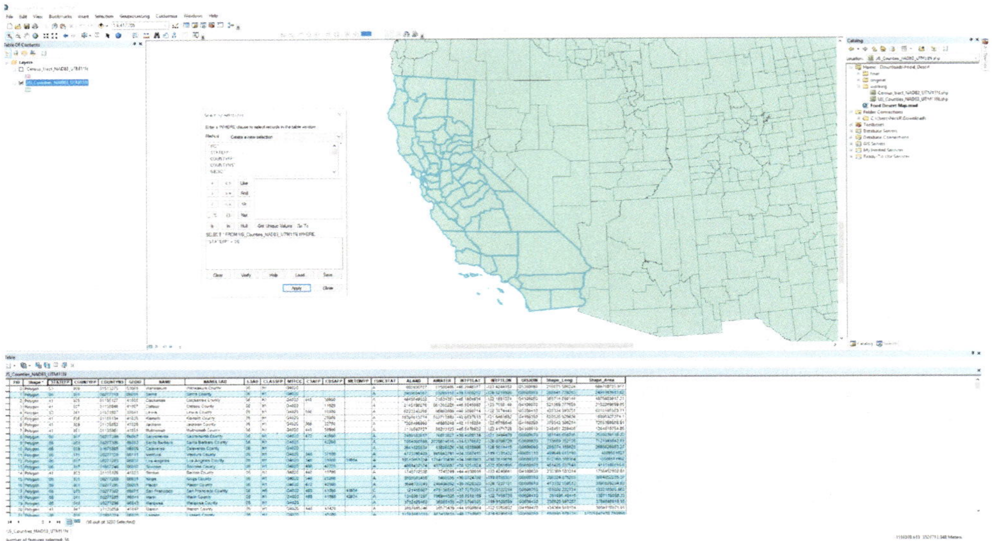

Figure 7.41: The counties in California are selected on the map and in the attribute table.

Go ahead and close the *Select by Attributes* window. Right-click on the counties layer and select *Data*, then *Export Data*. Click the yellow file folder icon and navigate to your *working* folder. Call the file California_counties and save it as a shapefile (**Figure 7.42**). After saving the new shapefile, add the layer to the map.

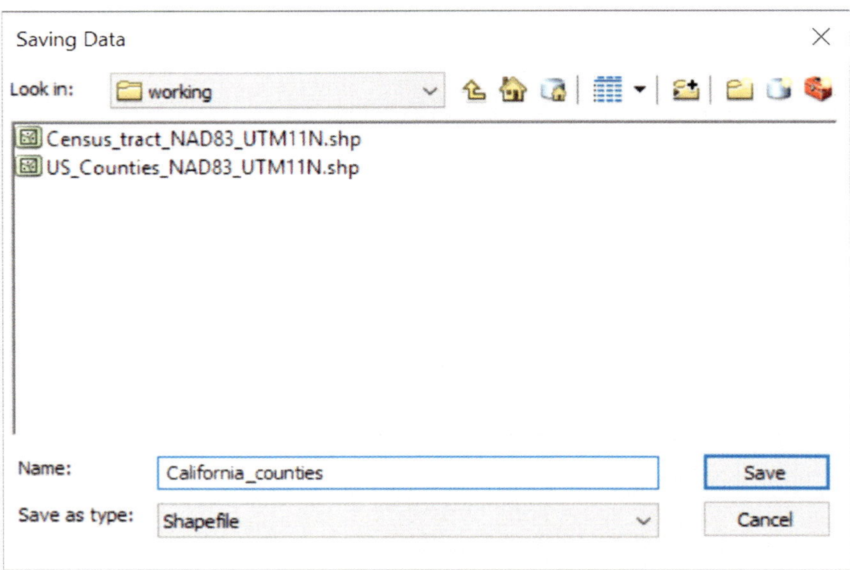

Figure 7.42: Always save data to the desired folder. Never accept the default location.

The next few steps are essential to staying organized. Try not to rush.

Clear your selected records by clicking on the *Clear Selected Features* button. It is the fifth button from the right on the attribute table and looks like a white square. You no longer need the original counties layer, so remove it from the Table of Contents. Right-click on the counties layer and select *Remove*. Now what remains are only the counties in California (**Figure 7.43**).

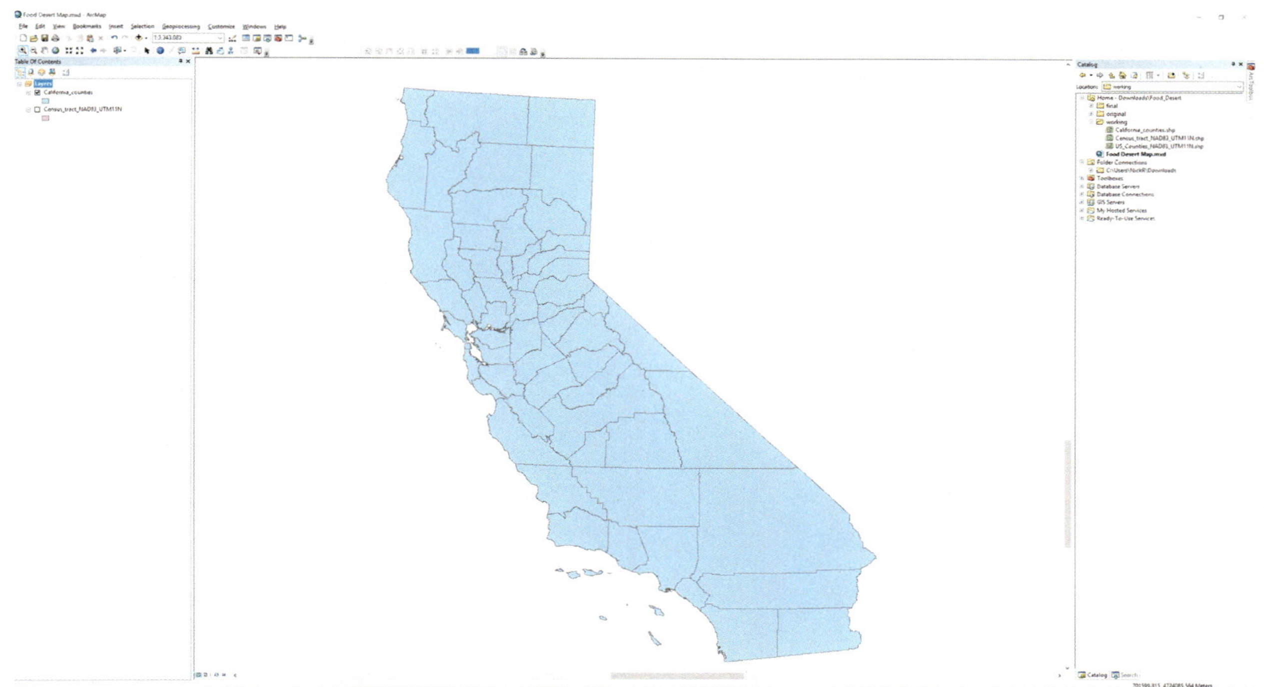

Figure 7.43: Eliminating unnecessary layers keeps the Table of Contents organized and can help to prevent confusion.

Skill Drill: Defining the Study Area Using an Attribute Query

In the previous step, you used an attribute query to select specific records from the counties layer. In this step, you must perform an attribute query on the **U.S. Census tract** layer to define the study area for this analysis. The study area includes the counties of Los Angeles, Orange, San Bernardino, Riverside, San Diego, and Imperial. Refer to the list below for the county codes.

» *Los Angeles:* 037
» *Orange:* 059
» *San Bernardino:* 071
» *Riverside:* 065
» *San Diego:* 073
» *Imperial:* 025

In this step, you need to perform an SQL query using a Boolean operator and a compound expression. In this chapter, you learned that **Boolean operators** compare multiple conditions, enabling you to create expressions with complicated structures. In this instance, you must use the OR operator to produce a compound expression that selects the desired counties.

> ***This skill drill might feel challenging if it is your first time creating a compound expression. Review the text earlier in this chapter related to Boolean expressions to see an example of a compound expression along with explanations.***

Create a new shapefile from the selected records. Call the new file Queried_Census_tracts. When done, clear the selection and organize the Table of Contents by removing the original U.S. Census tract layer from the Table of Contents (**Figure 7.44**). The results may not appear as expected. As you can see, county codes are not unique across different states. Many states in the U.S. use the same county codes. As a result, counties outside of California were also selected. In the next step, you must use a spatial query.

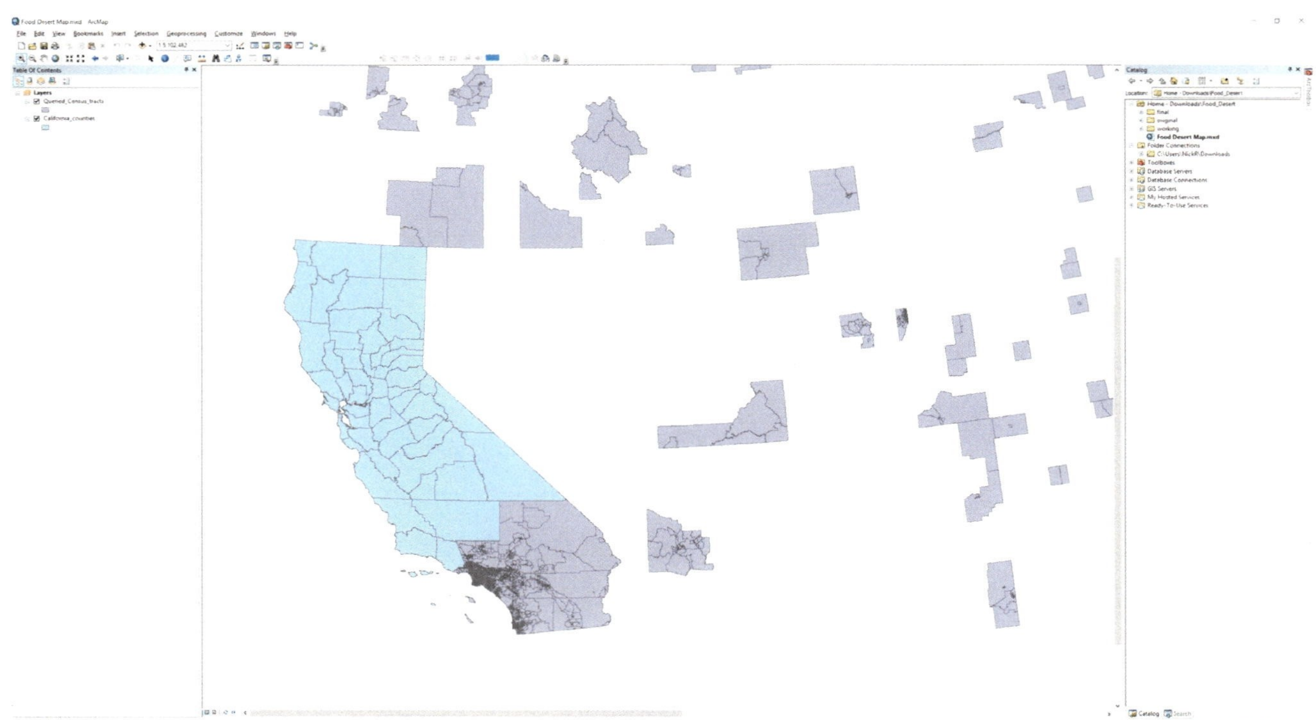

Figure 7.44: Eliminating unnecessary layers keeps the Table of Contents organized and can help to prevent confusion. After completing this step, you should have only two layers in the Table of Contents.

Refining the Study Area using a Spatial Query

In this chapter, you learned that a **spatial query** is a logical expression that selects geographic features by evaluating spatial relationships. In this instance, the spatial relationship is the location of the census tracts relative to the California counties. You need to use a spatial query to find out which census tracts fall within California. From the *Main* menu, choose Selection, then *Select by Location* to open the *Select by Location* query builder. At the top of the *Select by Location* window, take a moment to read the tool description. It says, "Select features from one or more target layers based on their location in relation to the features in the *source layer*" (**Figure 7.45**) When using this tool, you define the target layer, the *source layer*, and the selection method.

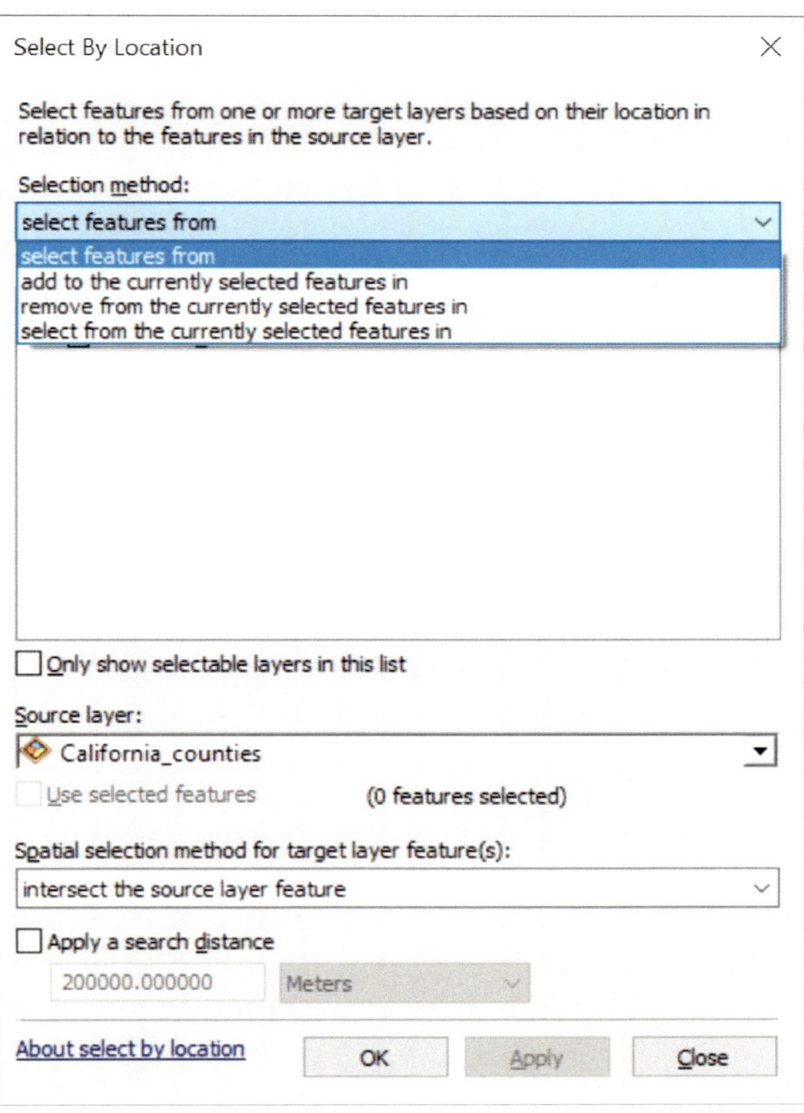

Figure 7.45: The *Main* menu provides shortcuts to attribute and spatial query builders.

Near the top, you have a choice of several different selection methods (**Figure 7.46**). Take a moment to read through them. They provide you with an idea of what you can do with the *Select by Location* tool. For this step, use the method, *select features from*.

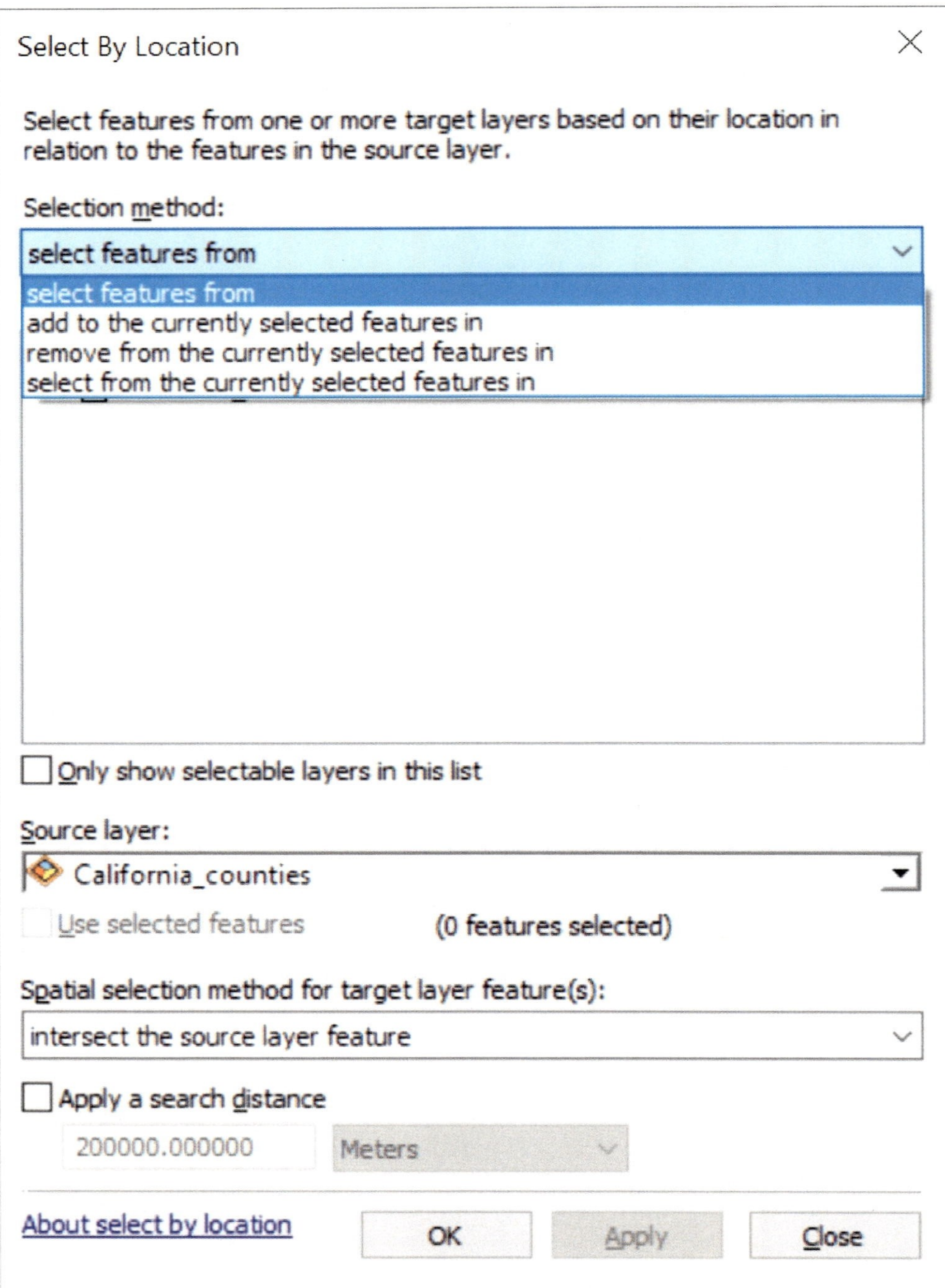

Figure 7.46: The *Select by Location* query builder provides different ways for refining your selection.

In the *Select by Location* query builder, the selection method applies to the *target* layer. In this instance, you want to select features from the Census tracts. Under the words, *Target layer(s)*, check the box next to the *Queried Census Tract* layer (**Figure 7.47**).

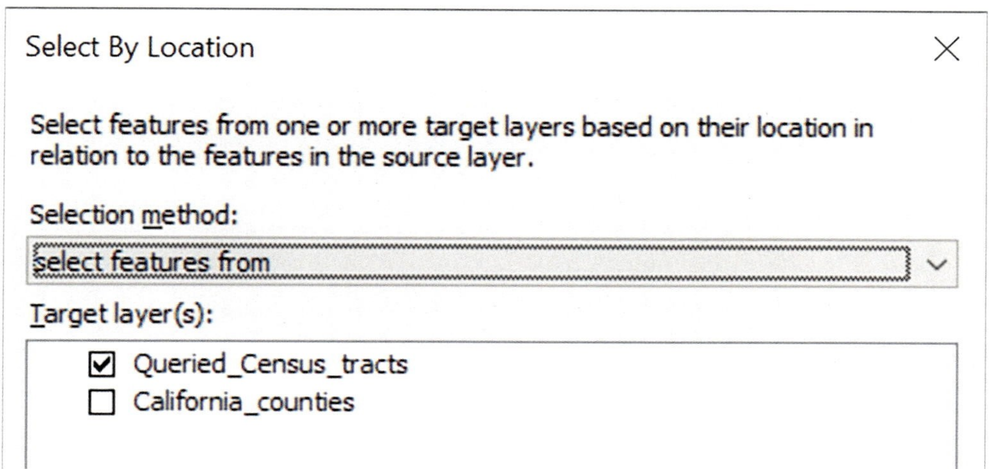

Figure 7.47: The target layer is the layer with features from which you want to select.

With this setting, the tool selects features from the *Queried Census Tract* layer based on their location in relation to the features in the *source layer*. In the next step, you define the *source layer*. Use the drop-down menu to select the California counties layer (**Figure 7.48**). It may appear automatically.

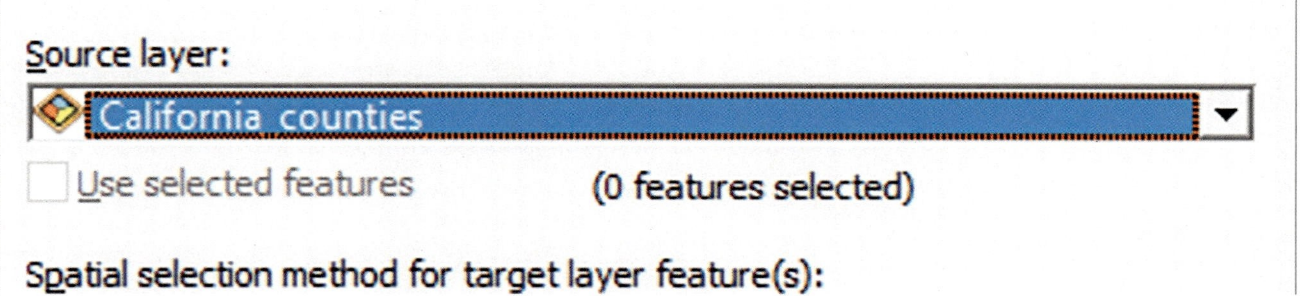

Figure 7.48: The *source layer* is the basis for the spatial relationship with the target layer.

The next setting defines the spatial relationship between the target layer and the *source layer*. The drop-down menu under *Spatial selection method for target layer feature* has many options. Take a moment to read through them to get an idea of the types of **spatial relationships** you can define. There are two very similar selection methods you may want to compare:

» Are within the source layer feature
» Are completely within the source layer feature

Try each method by clicking *Apply* and viewing the results on the map. The *completely within* option excludes Census tract polygons that are touching the outer boundary of the California counties layer (**Figure 7.49**). This result is *not* what you want, but it is useful to understand the difference between the two methods.

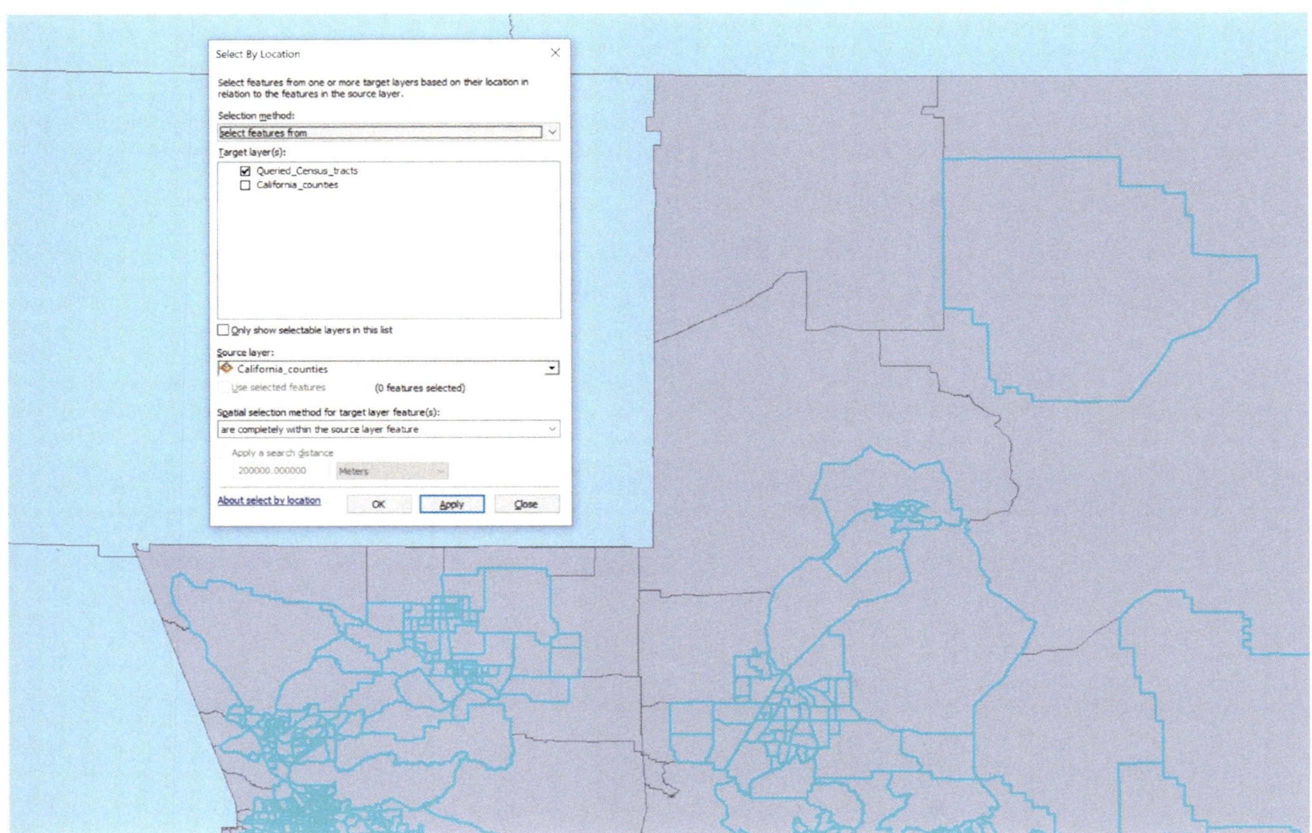

Figure 7.49: The *completely within* method excludes any polygons touching the boundary of the *source layer*.

In this instance, you want to choose, *are within the source layer feature* so that you include the Census tract polygons touching the outer boundaries of the California counties layer (**Figure 7.50**). When ready, click *OK*.

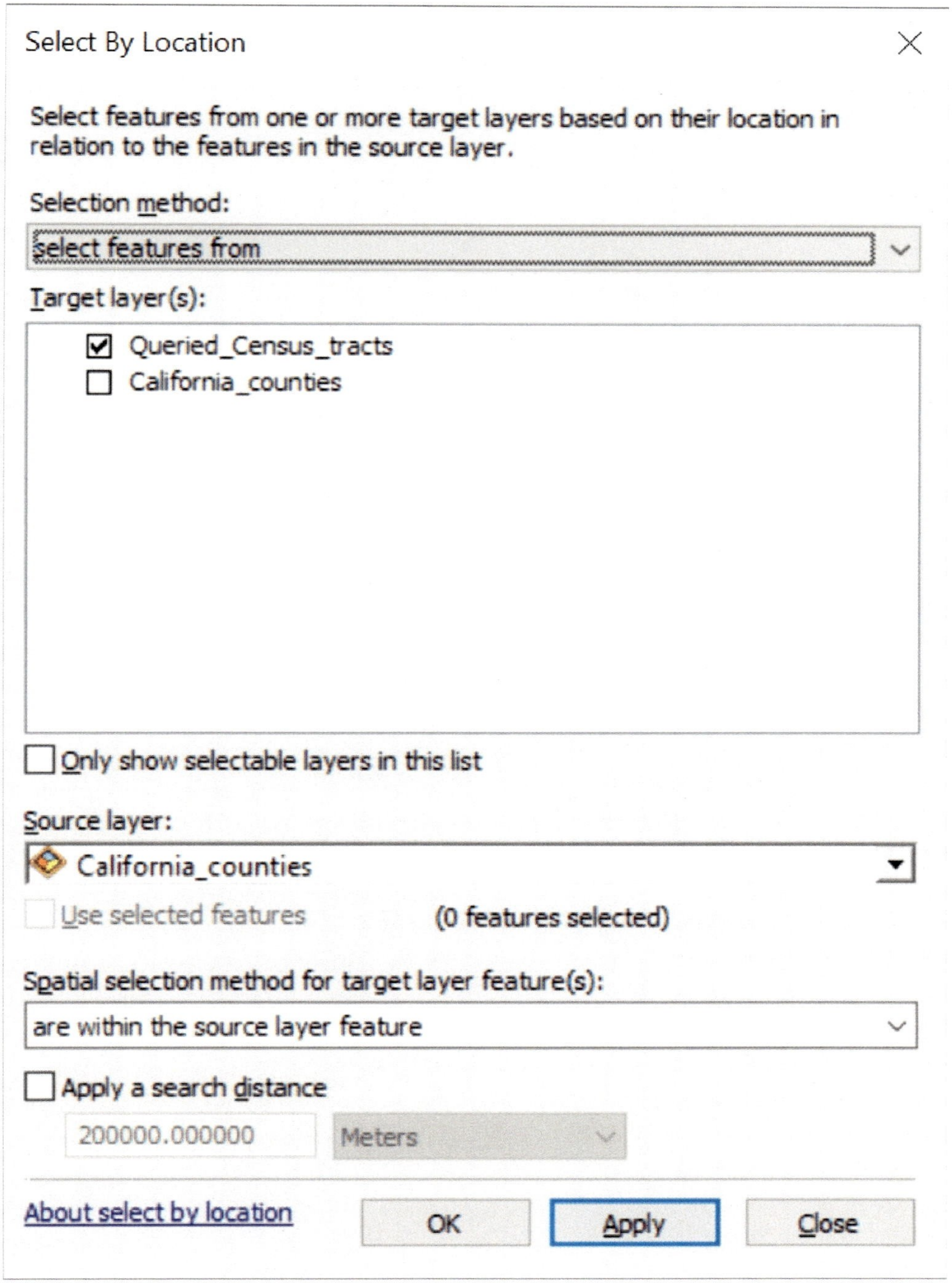

Figure 7.50: Check to make sure your settings match the options on this image.

All of the Census tracts within California are selected. Right-click on the *Queried Census Tract* layer and select *Data*, then *Export Data*. Click the yellow file folder icon and navigate to your *working* folder (**Figure 7.51**). Call the file Study_Area and save it as a shapefile.

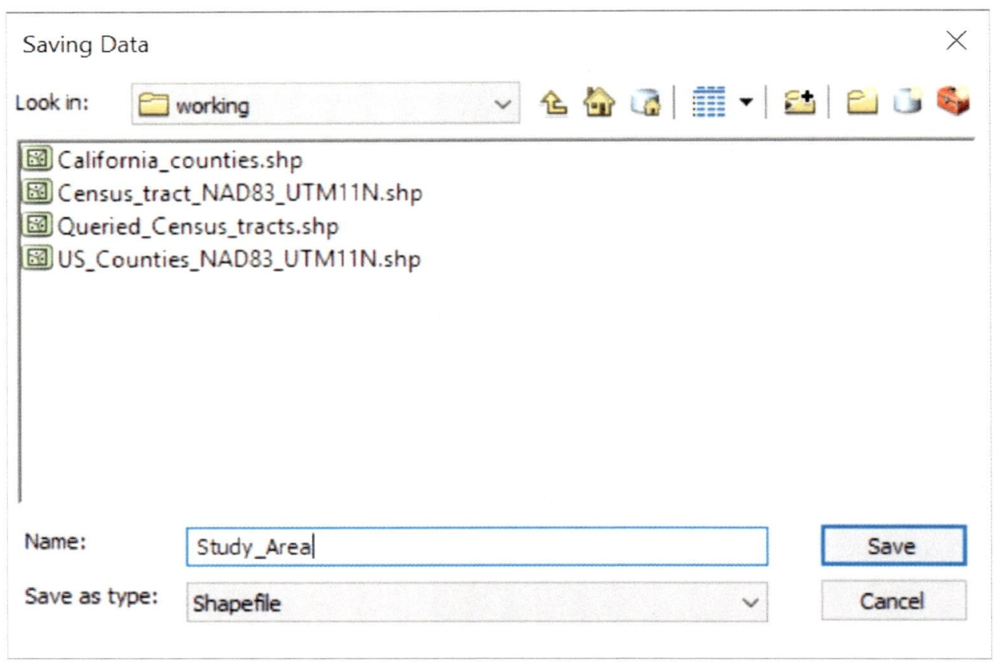

Figure 7.51: Always save data to the desired folder. Never accept the default location.

After saving the new shapefile, add the layer to the map. The next few steps are essential to staying organized. Try not to rush. Clear your selected records by clicking on the *Clear Selected Features* button. It is the ninth button from the right on the Tools toolbar and looks like a white square. You no longer need the *Queried Census Tract* layer, or the California Counties layer, so remove them from the Table of Contents (**Figure 7.52**). Now what remains is only the study area.

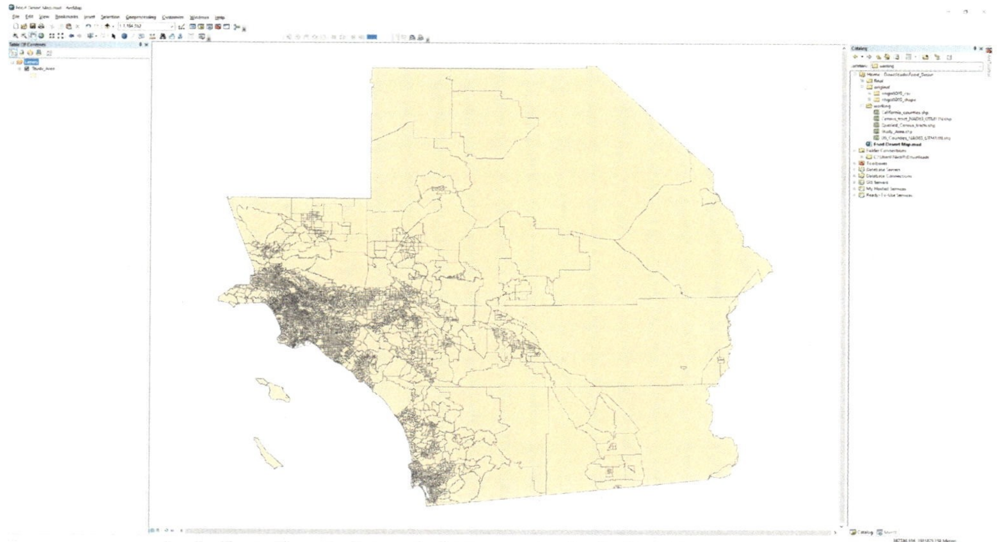

Figure 7.52: Eliminating unnecessary layers keeps the Table of Contents organized and can help to prevent confusion.

Performing a Table Join

In this chapter, you learned that a **table join** is when you establish a one-to-one relationship between two tables using a common attribute field. In this step, you must create a table join between the study area layer and the Census tract CSV table for household and family income. To help you understand how this works, open the attribute table for the study area layer (**Figure 7.53**).

Table

Study_Area

FID	Shape *	STATEFP	COUNTYFP	TRACTCE	GEOID	NAME	NAMELSAD	MTFCC	FUNCSTAT	ALAND	AWATER	INTPTLAT	INTPTLON	GISJOIN	Shape_Leng	Shape_Area
0	Polygon	06	037	204920	06037204920	2049.20	Census Tract 2049.20	G5020	S	909972	0	+34.0175004	-118.1974975	G0600370204920	3900.872736	909975.273699
1	Polygon	06	037	205110	06037205110	2051.10	Census Tract 2051.10	G5020	S	286960	0	+34.0245059	-118.2142985	G0600370205110	2807.460463	286960.74301
2	Polygon	06	037	205120	06037205120	2051.20	Census Tract 2051.20	G5020	S	1466130	0	+34.0187546	-118.2117956	G0600370205120	6117.387099	1466128.05922
3	Polygon	06	037	206010	06037206010	2060.10	Census Tract 2060.10	G5020	S	1418138	20551	+34.0682177	-118.2320356	G0600370206010	6425.056453	1438691.05772
4	Polygon	06	037	206020	06037206020	2060.20	Census Tract 2060.20	G5020	S	843243	30290	+34.0571230	-118.2311021	G0600370206020	3666.418123	873536.485466
5	Polygon	06	037	206050	06037206050	2060.50	Census Tract 2060.50	G5020	S	1643426	94606	+34.0299036	-118.2244531	G0600370206050	7791.830805	1738032.99445
6	Polygon	06	037	207400	06037207400	2074	Census Tract 2074	G5020	S	862957	6458	+34.0562223	-118.2466420	G0600370207400	4625.328814	869413.768552
7	Polygon	06	037	192410	06037192410	1924.10	Census Tract 1924.10	G5020	S	683899	0	+34.0798577	-118.3181008	G0600370192410	3307.841793	683900.131066
8	Polygon	06	037	192510	06037192510	1925.10	Census Tract 1925.10	G5020	S	340750	0	+34.0798690	-118.3068568	G0600370192510	2480.081499	340749.909379
9	Polygon	06	037	192520	06037192520	1925.20	Census Tract 1925.20	G5020	S	307802	0	+34.0799255	-118.3024972	G0600370192520	2385.984295	307804.165808
10	Polygon	06	037	192610	06037192610	1926.10	Census Tract 1926.10	G5020	S	415408	0	+34.0813650	-118.2961539	G0600370192610	2785.841417	415406.555273
11	Polygon	06	037	192700	06037192700	1927	Census Tract 1927	G5020	S	531205	0	+34.0800134	-118.2881064	G0600370192700	2932.449053	531203.624909
12	Polygon	06	037	194500	06037194500	1945	Census Tract 1945	G5020	S	685342	0	+34.0781753	-118.3695958	G0600370194500	4550.445232	685339.694154
13	Polygon	06	037	195100	06037195100	1951	Census Tract 1951	G5020	S	1566297	329063	+34.1022274	-118.2669741	G0600370195100	6626.231816	1895360.00969
14	Polygon	06	037	195300	06037195300	1953	Census Tract 1953	G5020	S	580349	0	+34.0992506	-118.2836893	G0600370195300	3366.183373	580351.189252
15	Polygon	06	037	137501	06037137501	1375.01	Census Tract 1375.01	G5020	S	2514410	0	+34.1649344	-118.5768036	G0600370137501	7372.945431	2514411.79716
16	Polygon	06	037	137502	06037137502	1375.02	Census Tract 1375.02	G5020	S	2090222	0	+34.1634518	-118.5976927	G0600370137502	5866.444322	2090219.50585
17	Polygon	06	059	011711	06059011711	117.11	Census Tract 117.11	G5020	S	1131330	0	+33.8856931	-117.8692764	G0600590011711	4199.12258	1131326.97649
18	Polygon	06	059	011712	06059011712	117.12	Census Tract 117.12	G5020	S	1431776	0	+33.8766552	-117.8702302	G0600590011712	4787.728369	1431779.06978
19	Polygon	06	059	011715	06059011715	117.15	Census Tract 117.15	G5020	S	2476274	0	+33.8823899	-117.8528754	G0600590011715	7102.129608	2476275.36514
20	Polygon	06	059	011716	06059011716	117.16	Census Tract 117.16	G5020	S	1548521	0	+33.8721868	-117.8529250	G0600590011716	5439.971541	1548521.96577
21	Polygon	06	059	042114	06059042114	421.14	Census Tract 421.14	G5020	S	3773683	0	+33.4194943	-117.5931457	G0600590042114	10861.88224	3773683.05827

1 (0 out of 4405 Selected)

Study_Area

Figure 7.53: An image of the study area attribute table.

The shapefile currently stores some basic information about the Census tracts, including area, state codes, and county codes. However, the attribute table lacks any demographic data, such as population, income, or poverty level. Recall from a previous step that you downloaded information from the NHGIS about household and family income in the United States. The shapefile *does not* currently store this data. The *CSV tables* you downloaded store the household and family income data.

To place that information on the map, you need to join the CSV table to the study area layer. The joining of the two tables requires a key. As you learned in this chapter, a **key** is a common field, one that exists on both tables, and is used to associate them together. The NHGIS provides an attribute field that serves as a key called GISJOIN (**Figure 7.54**). In the Table of Contents, right-click on the study area layer and select *Joins and Relates*, then *Join* (**Figure 7.55**). When the *Join Data* window appears, use the drop-down menu to choose the option *Join attributes from a table* (**Figure 7.56**)

INTPTLAT	INTPTLON	GISJOIN	Shape_Leng	Shape_Area
+34.0175004	-118.1974975	G0600370204920	3900.872736	909975.273699
+34.0245059	-118.2142985	G0600370205110	2807.460463	286960.74301
+34.0187546	-118.2117956	G0600370205120	6117.387099	1466128.05922
+34.0682177	-118.2320356	G0600370206010	6425.056453	1438691.05772
+34.0571230	-118.2311021	G0600370206020	3666.418123	873536.485466
+34.0299036	-118.2244531	G0600370206050	7791.830805	1738032.99445
+34.0562223	-118.2466420	G0600370207400	4625.328814	869413.768552

Figure 7.54: Both the CSV table and the study area attribute table have a field called GISJOIN.

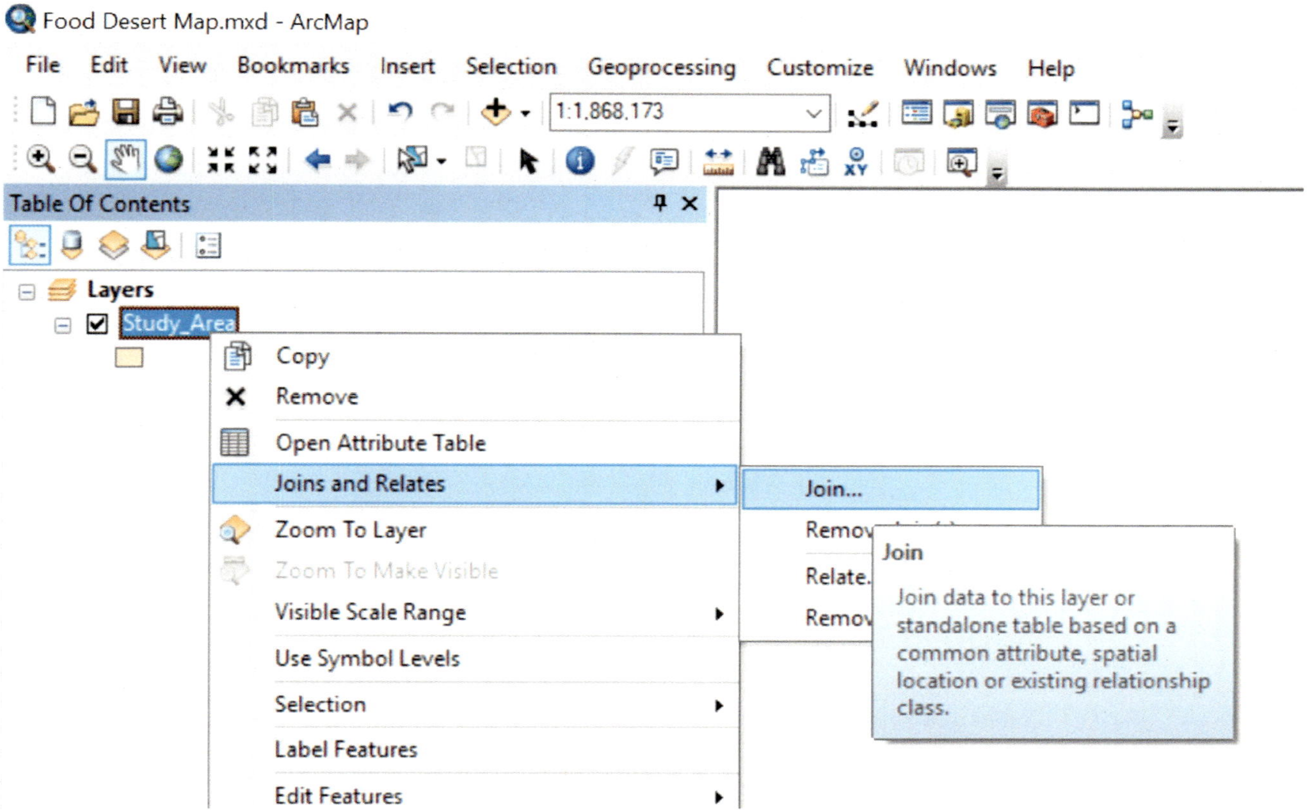

Figure 7.55: You can access contextual menus by right-clicking on a layer in the Table of Contents.

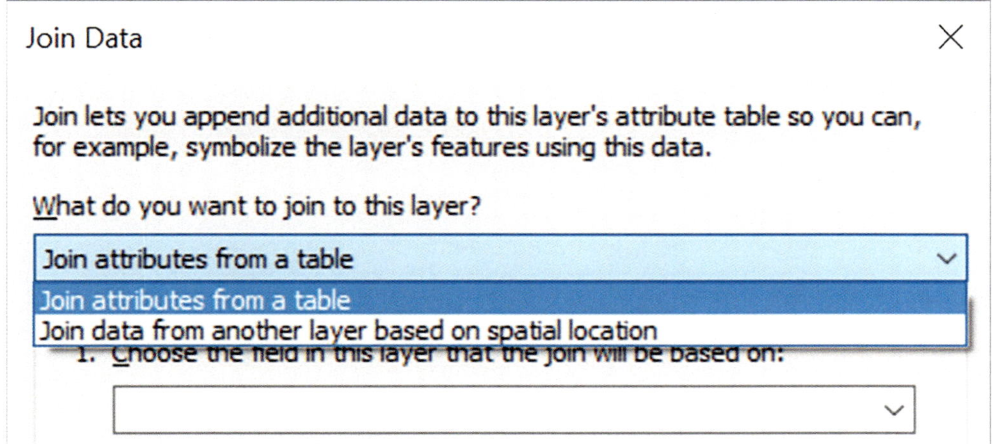

Figure 7.56: When joining data, you have options to base the join on a table or a location.

Next to option 1, *Choose the field in this layer that the join will be based on*, use the drop-down menu to select the *GISJOIN* field (**Figure 7.57**).

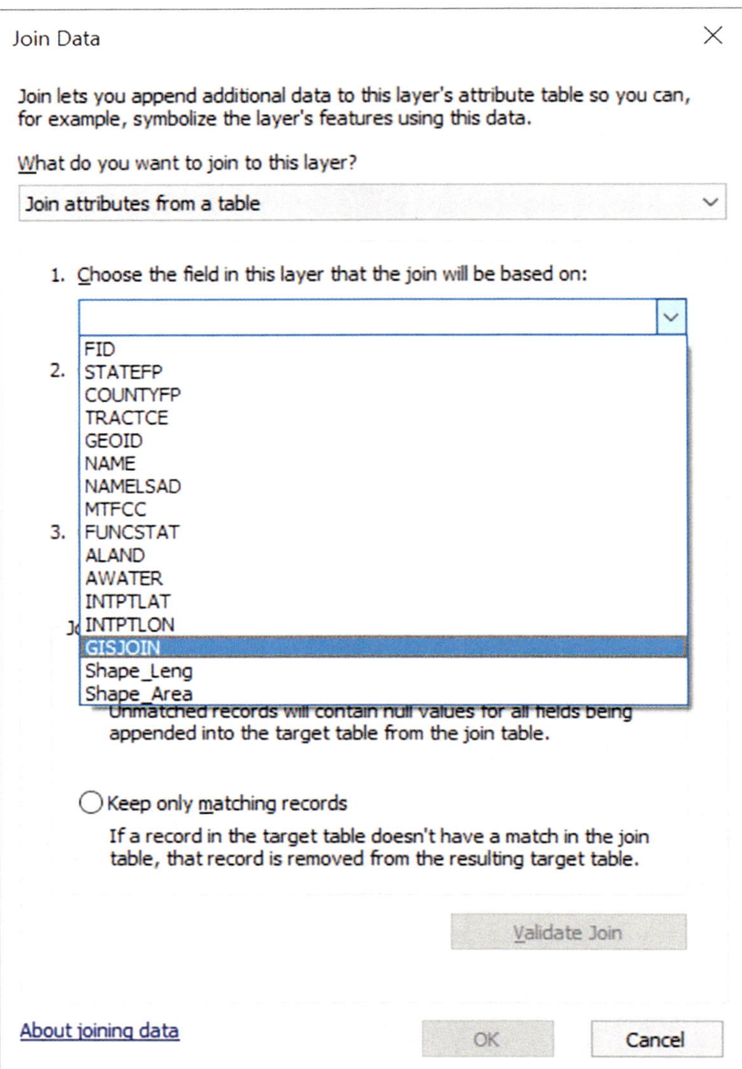

Figure 7.57: Option 1 applies to the layer in the Table of Contents.

Next to option 2, *Choose the table to join to this layer, or load the table from disk,* use the yellow file folder icon to browse to the *original* folder (**Figure 7.58**). Then, open the CSV subfolder and locate the CSV file for the 2016 tracts that relate to household and family income. Be sure you are in the correct folder for household and family income data and not for the total population. When you are ready, click Add.

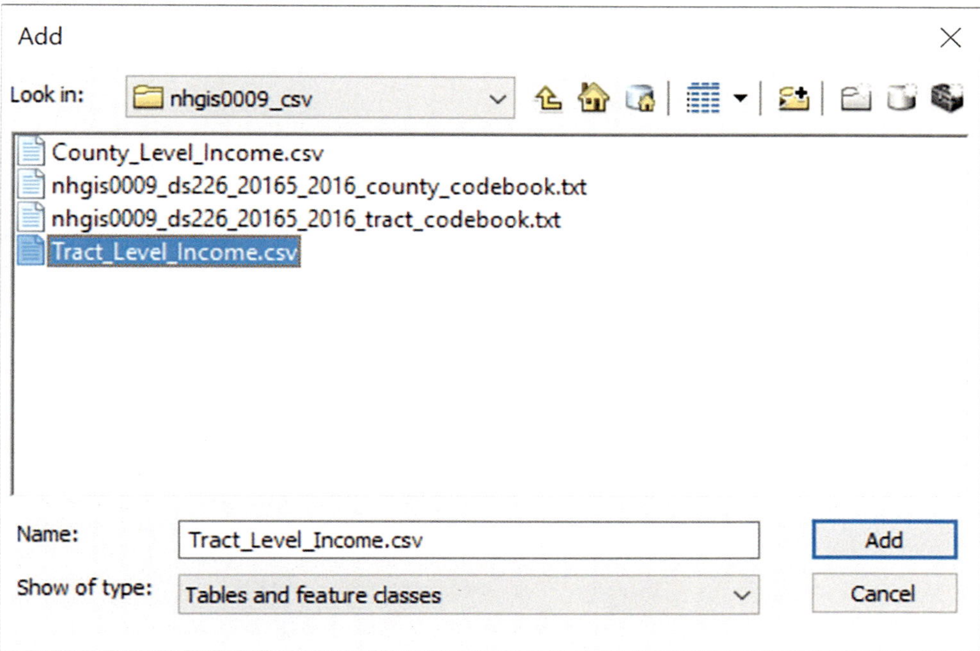

Figure 7.58: The NHGIS provides csv files and codebooks for each dataset. The CSV stores the data table while the codebook stores the metadata.

Next to option 3, *Choose the field in the table to base the join on*, make sure to choose the GISJOIN field (**Figure 7.59**). The ArcGIS software usually locates the attribute field that serves as a key automatically. Leave all of the other default settings. When you are ready, click *OK*.

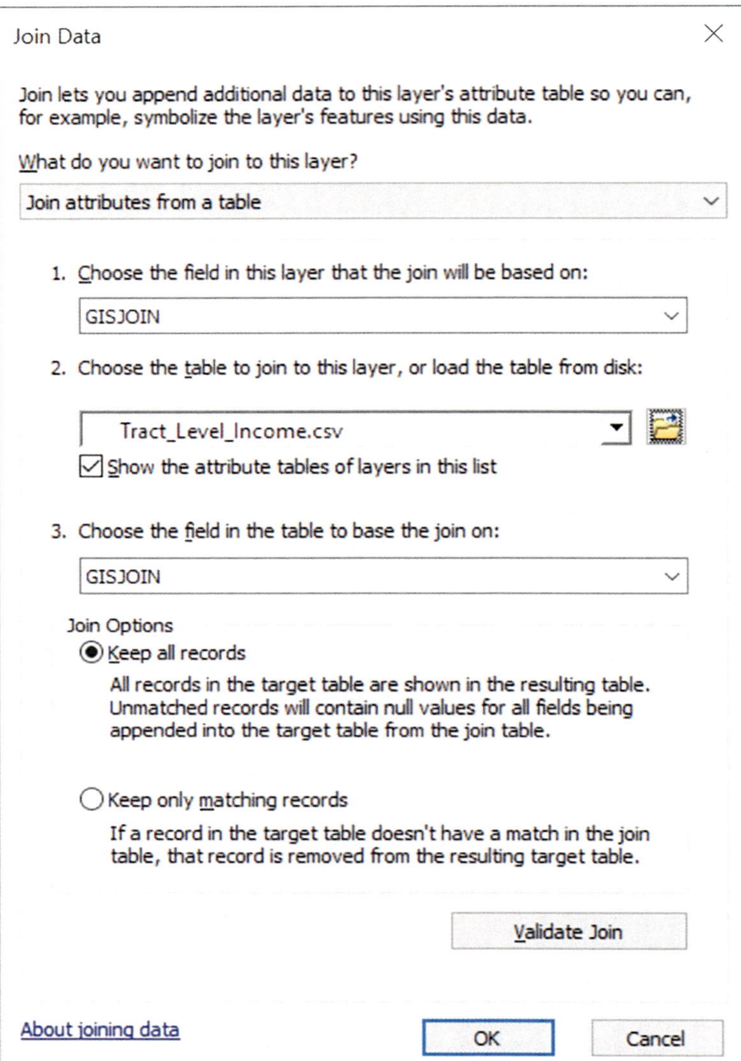

Figure 7.59: Check to make sure your settings match the options on this image.

The contents of the two tables are now merged. To view this, open the attribute table for the study area layer. Sometimes the column widths are set too wide after a table join. You can fix this issue by clicking the table options button and selecting *Restore Default Column Widths* (**Figure 7.60**)

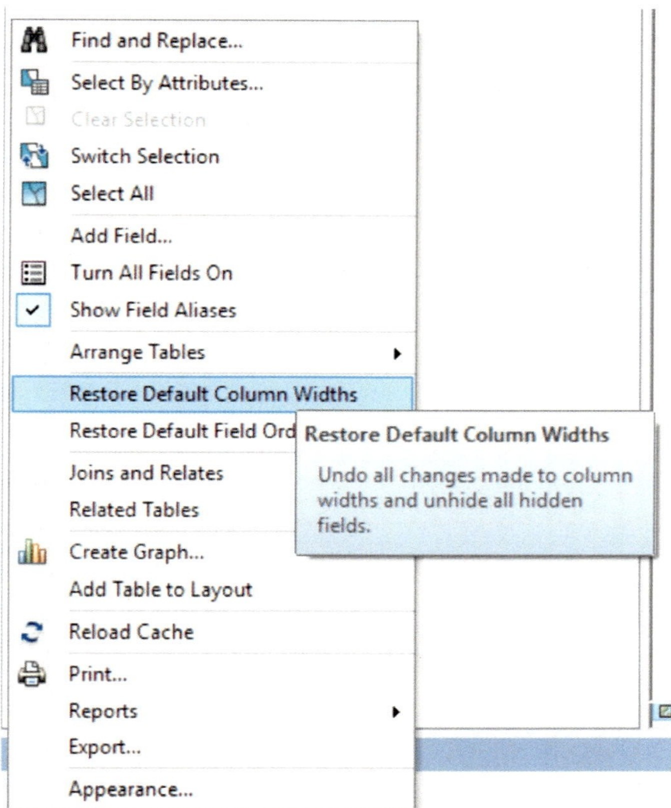

Figure 7.60: The table options button is located on the upper left side of the attribute table.

You may need to scroll to the right to see the contents of the CSV table joined to the layer (**Figure 7.61**).

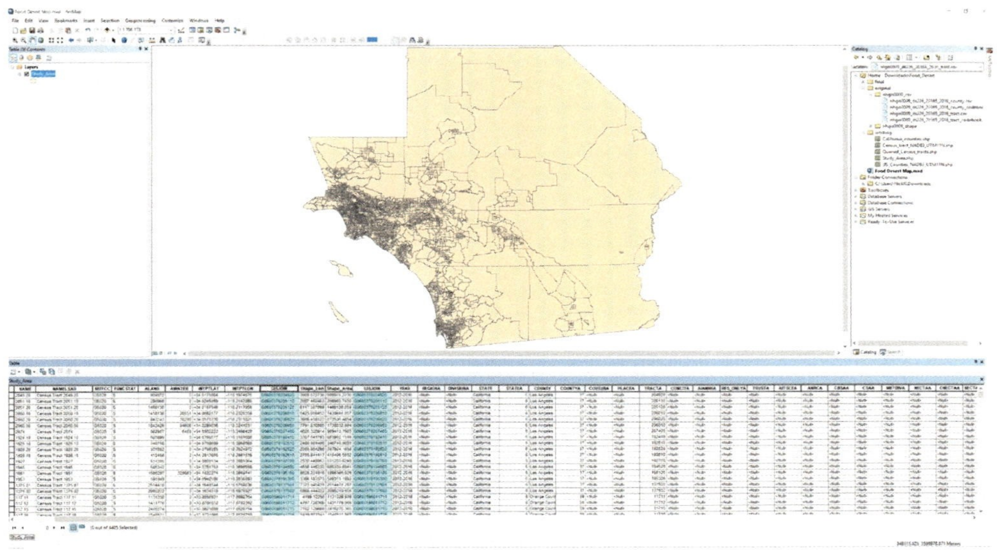

Figure 7.61: The records from each table are matched using the GISJOIN field.

A table join is a temporary condition. You must export the layer to make the join permanent. Make sure there are no features selected. If necessary, use the *Clear Selected Features* button. In the Table of Contents, right-click on the study area layer and select *Data*, then *Export Data*. Click the yellow file folder icon and navigate to your *working* folder (**Figure 7.62**). Call the file Poverty_Status and save it as a shapefile.

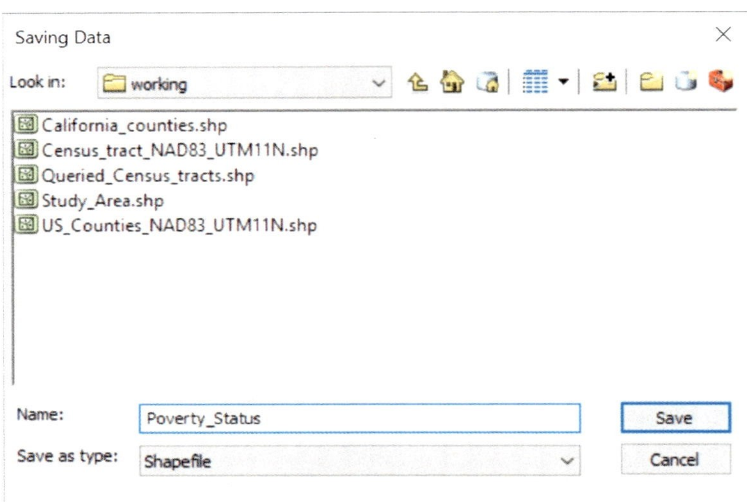

Figure 7.62: Always save data to the desired folder. Never accept the default location.

After saving the new shapefile, add the layer to the map. You no longer need the study area layer, so remove it from the Table of Contents. Right-click on the study areas layer and select *Remove*. Now only the poverty status layer remains in the Table of Contents.

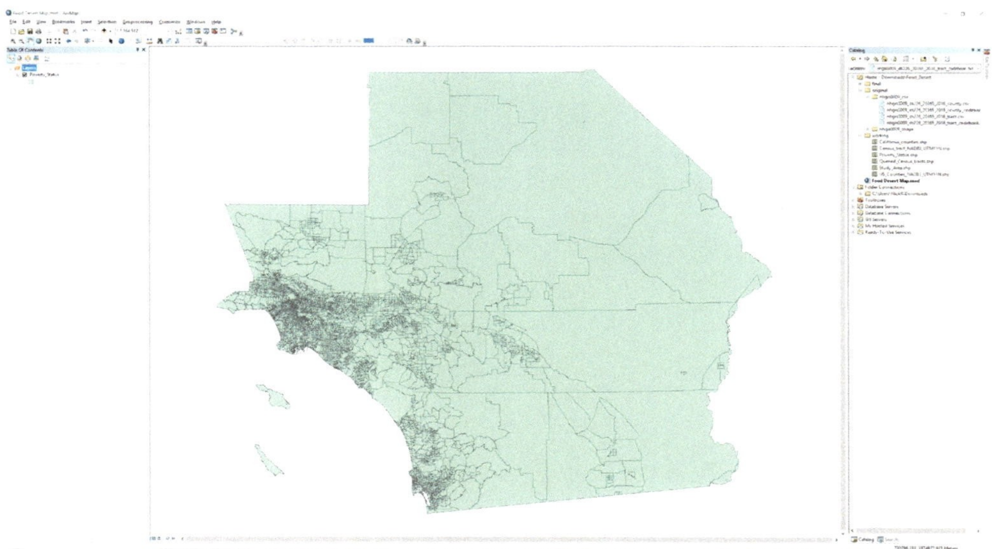

Figure 7.63: Removing unnecessary layers keeps the Table of Contents organized and can help to prevent confusion.

Skill Drill: Perform a Table Join for Total Population

Using what you learned in the previous step, perform a table join between the poverty status layer and the CSV table for the *total population* (**Figure 7.64**) Try not to confuse the file with the household and family income data.

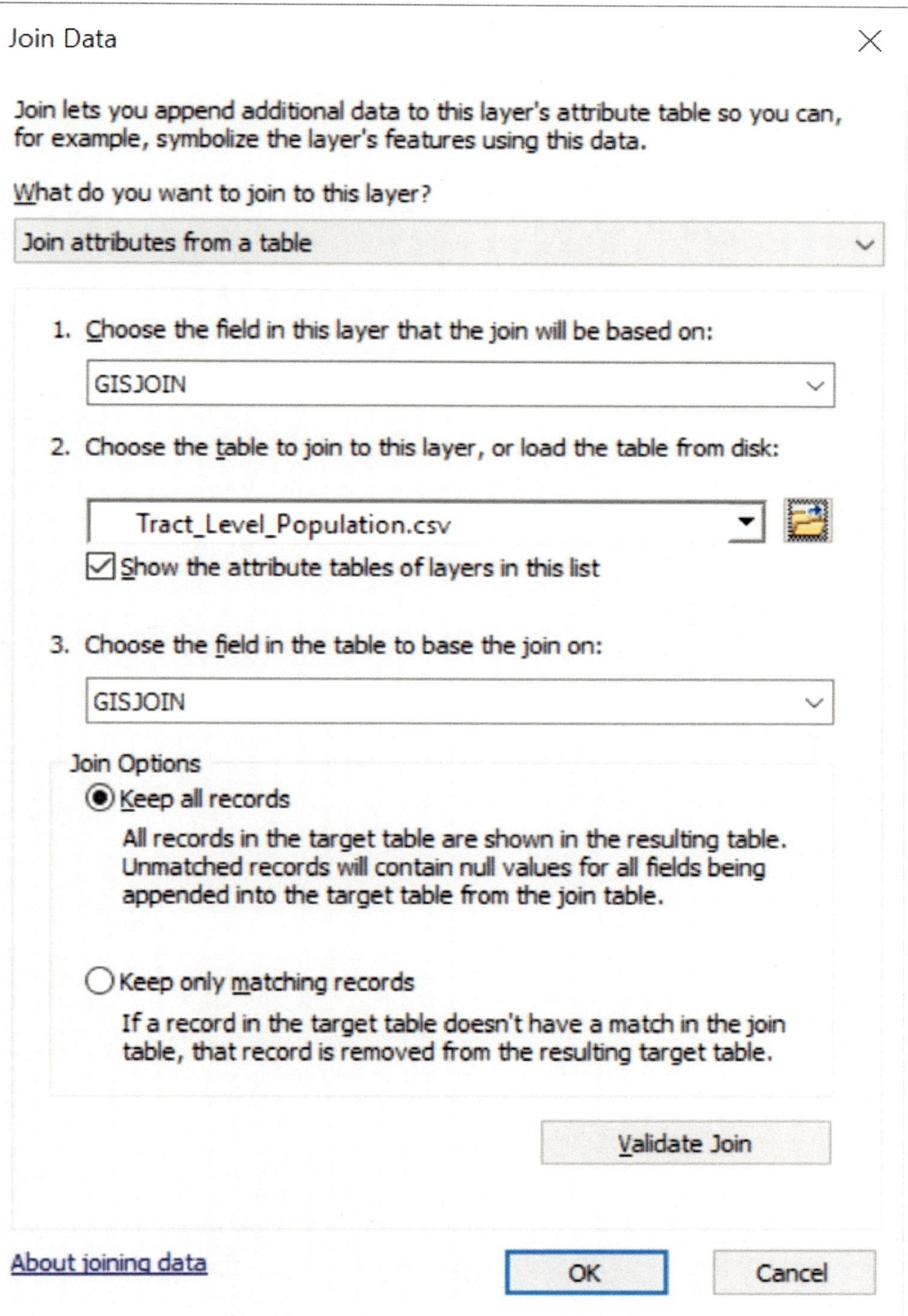

Figure 7.64: Check to make sure your settings match the options on this image.

After performing a table join, export the data. Save the new shapefile to your *working* folder (**Figure 7.65**). Call the file Population_and_Poverty. Add the new layer to the map and remove the poverty status layer.

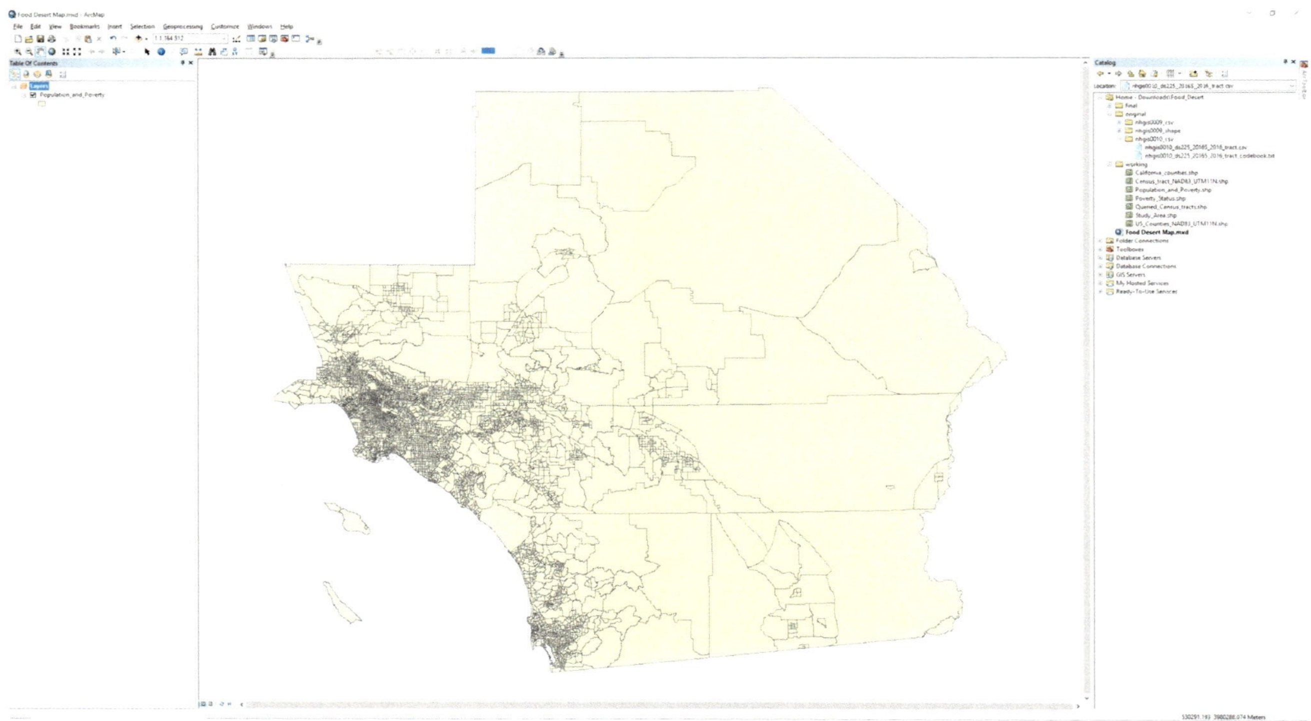

Figure 7.65: Eliminating unnecessary layers keeps the Table of Contents organized and can help to prevent confusion.

Adding a New Field to an Attribute Table

Now that you have a layer with both total population and poverty data, the next step is to create a field in the attribute table that records the percentage of the population that is below the poverty level. Open the attribute table for the population and poverty layer. Click the *Table Options* button located on the upper left of the attribute table. Then select, *Add Field* (**Figure 7.66**).

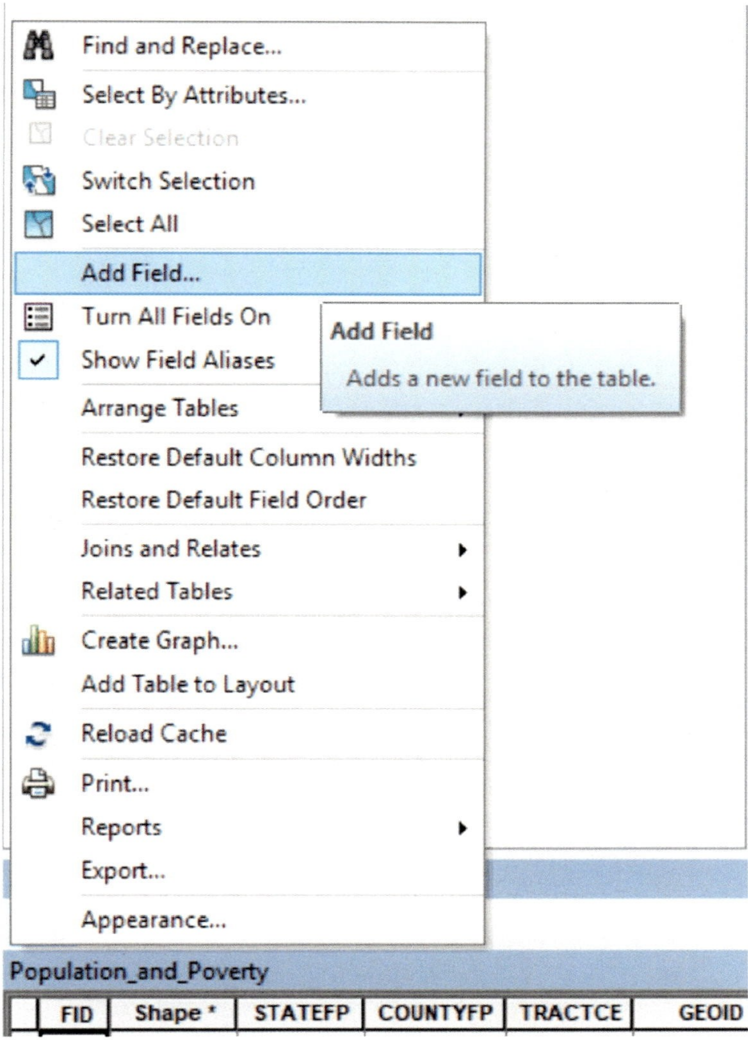

Figure 7.66: The table options button is located on the top left side of the attribute table.

As you learned in this chapter, a column in a database is called a **field**, and each store a single attribute type. You also learned that each field has a unique name called a field name and that the field name has strict limitations. In this case, you should use the following best practices when naming fields:

- Eliminate any characters that are not alphanumeric or an underscore

» Do not start a field name with a number or an underscore
» Limit the length of the field name to 10 characters

Use Pct_Pov for the field name (**Figure 7.67**). This name is short for *percent poverty*. Using the drop-down menu, change the field type to *Double*. Leave all of the remaining default settings and click *OK*.

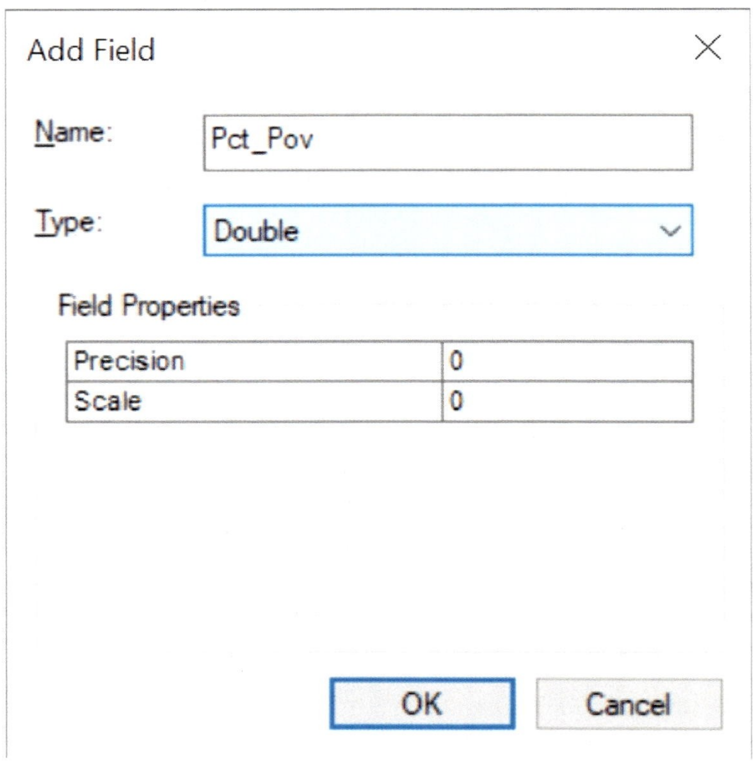

Figure 7.67: With limitations on field name length, abbreviations or codes are often used.

On the attribute table scroll all of the way to the right side to locate the new field. Right click on the field name and open the properties (**Figure 7.68**).

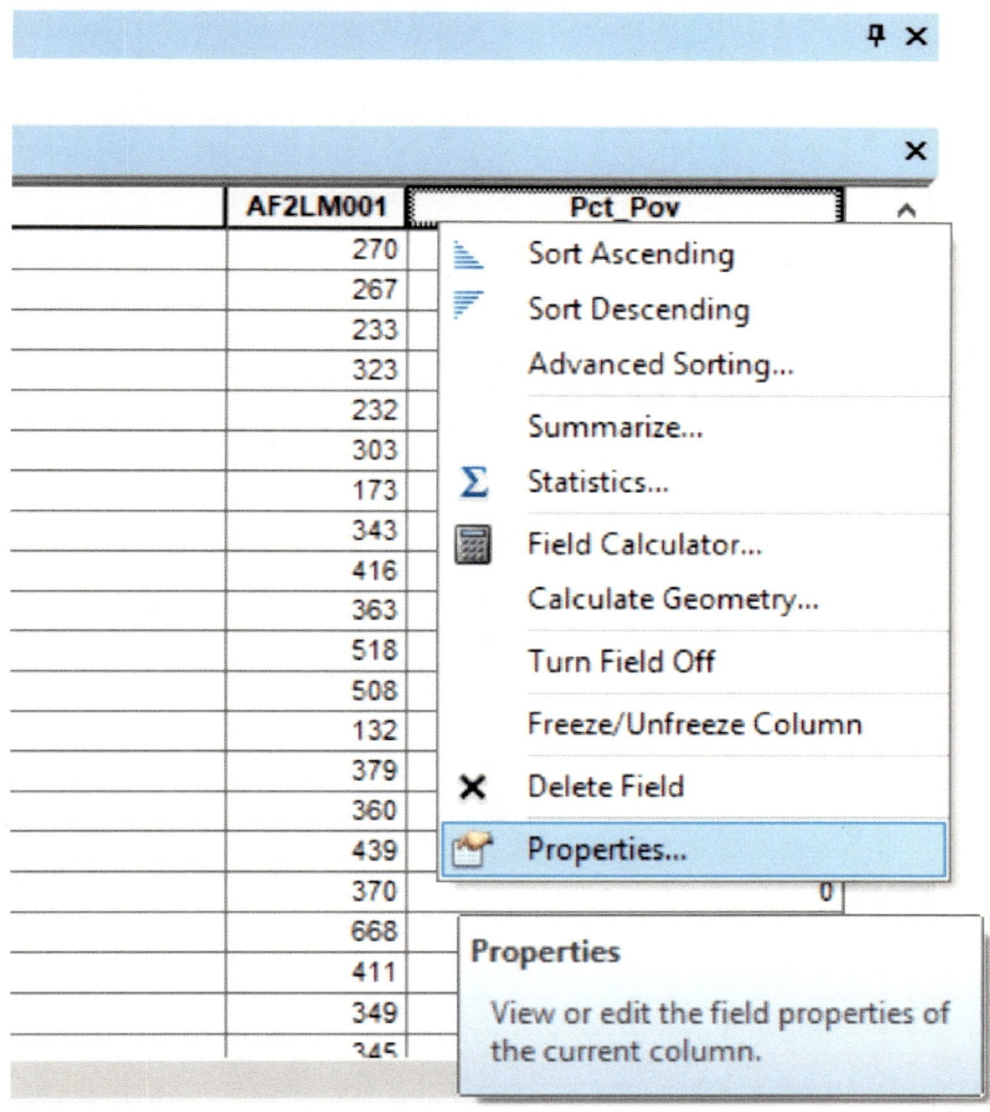

Figure 7.68: You can access the contextual menu for attribute fields by right-clicking on the field name.

The *Field Properties* window provides information about the attribute field including the name, the alias (where you can provide a human-friendly name), and the type. An alias does not have the same strict limitations as the field name. Use the alias option to give the field a human-friendly name such as *Percent Poverty*. When ready, click *OK*. You may notice that the percent poverty field is filled with zeros. Zero is the default value for a new field of this type. In the next step, you must populate the field with the percentage of persons in poverty using the *Field Calculator*.

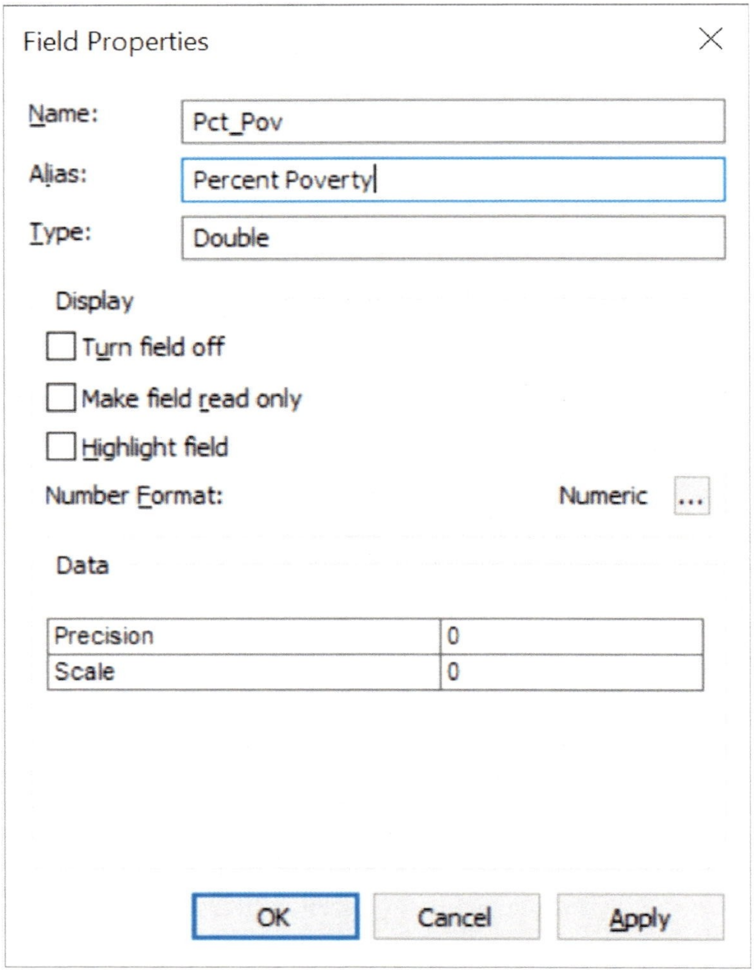

Figure 7.69: An alias is for display purposes only. The database still uses the original field name.

Using the Field Calculator

The *Field Calculator* is a tool that allows you to perform mathematical calculations to set the value for records in a field. The *Field Calculator* can use the Python scripting language or VB Script.

Don't worry! You won't have to understand programming languages. The Field Calculator has a graphic user interface to help you, and the calculation you will use in the next step is straightforward.

In this step, you must use the *Field Calculator* to divide the total number of people with poverty status by the total population to get the percentage of people with poverty status.

» PERCENT POVERTY = TOTAL POVERTY/TOTAL POPULATION

Before you begin the calculation, it is important to exclude records where the total population is zero. Dividing by zero will result in an error. To do this, you must select the records that do not have a total population of zero. As you learned in this chapter, field names have strict limitation. The people at NHGIS used a series of codes for the field names. The meaning behind the codes can be found in the *codebook files* that accompanied the CSVs. In this instance, the code for the total population is **AF2LE001**. On the attribute table, click the *Select by Attributes* button, which is the third icon from the left, to open the *Select by Attributes Query Builder* (**Figure 7.70**). Enter the following expression:

» NOT("AF2LE001" = 0)

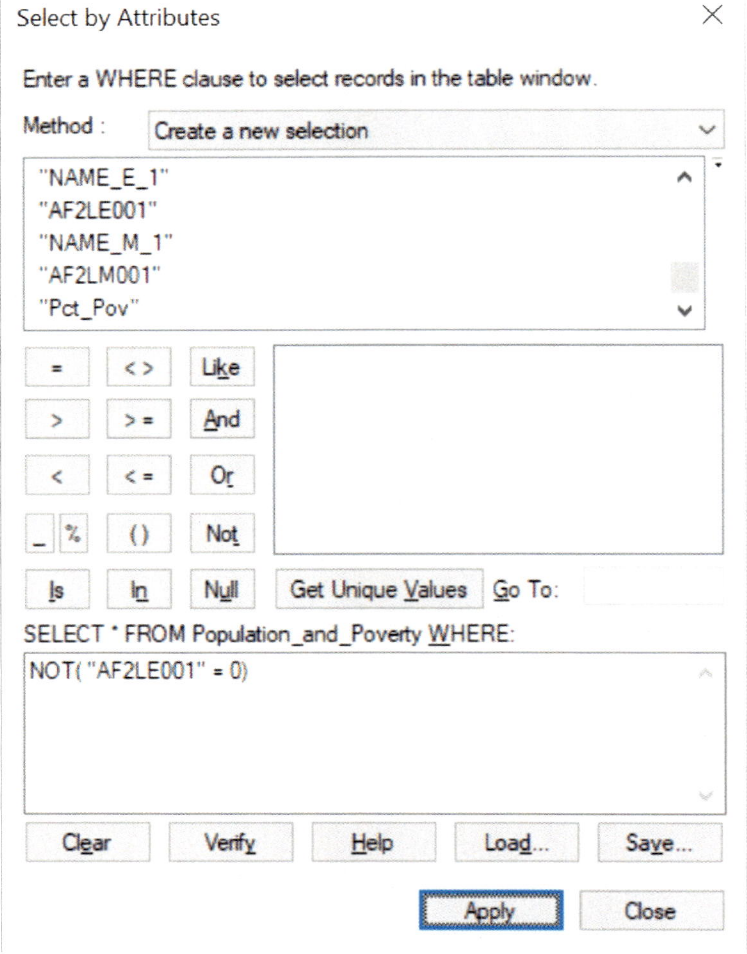

Figure 7.70: The Boolean operator NOT negates a condition. The condition must be false for a record to be selected.

Most of the records on the attribute get selected. When using the *Field Calculator* tool with records selected, the calculations only work on the selected records. Because you the records where the total population is zero are not selected, they will be excluded from the calculation. This step prevents you from dividing by zero. Right-click on the field name for the percent poverty field, then choose *Field Calculator* (**Figure 7.71**)

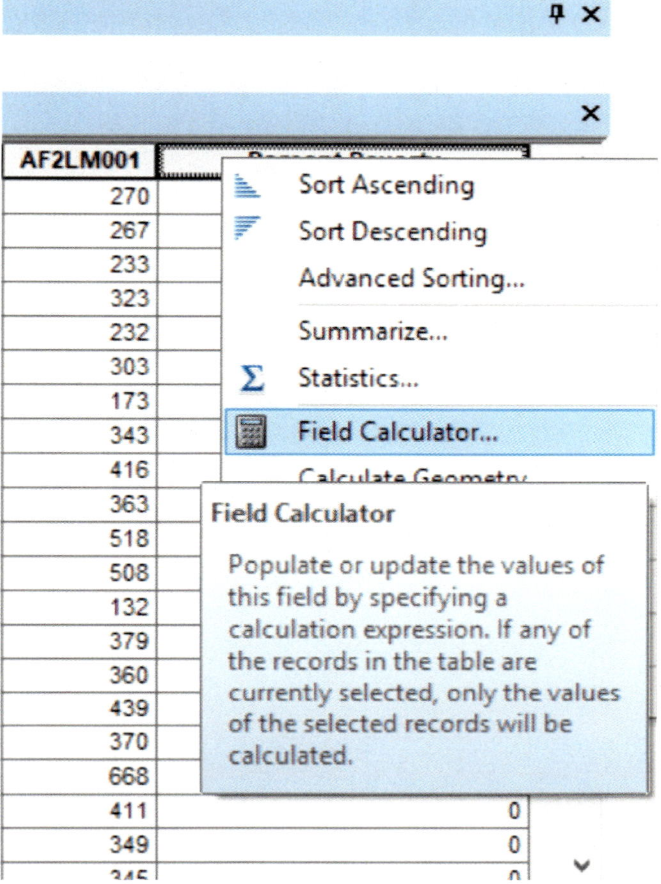

Figure 7.71: You can access the contextual menu for attribute fields by right-clicking on the field name.

A warning message may appear (**Figure 7.72**). Take a moment to read what it says. After reading the warning, click *Yes*.

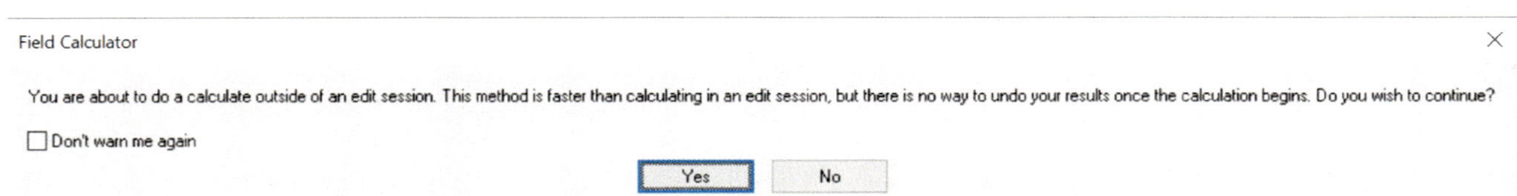

Figure 7.72: The ArcGIS software is warning you that you are about to make a permanent change to the shapefile.

The *Field Calculator* has several options. At the top, near the word *Parser*, you have a choice between VB Script or Python. You don't need to change anything here because these options mainly apply to the functions available. In this instance, you will use simple arithmetic. Under the word *Fields*, you should see a list of field names. The meaning behind the codes can be found in the *codebook files* that accompanied the CSVs. In this instance, the code for the total number of people with poverty status is **AGJZE001**. Locate **AGJZE001** on the list and double-click on it (**Figure 7.73**). Then, click the button with the backslash symbol.

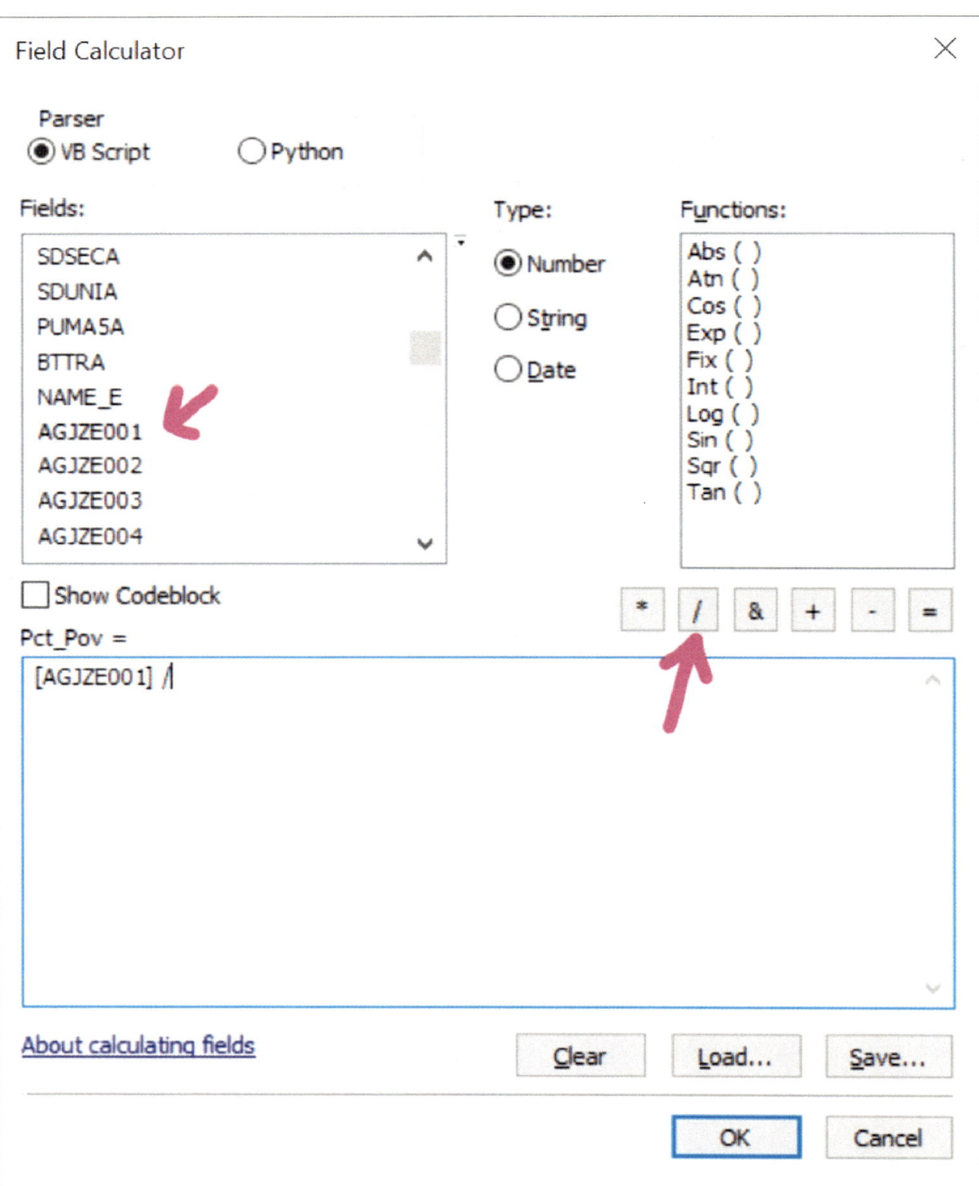

Figure 7.73: When using VB Script, the field calculator places brackets around field names.

The field name and the division symbol appear in the white text field on the bottom half of the *Field Calculator*. However, the expression begins with the words above, "Pct_Pov =." The content in the text field completes the expression. The code for the total population is **AF2LE001**. Locate **AF2LE001** on the list and double-click on it (**Figure 7.74**).

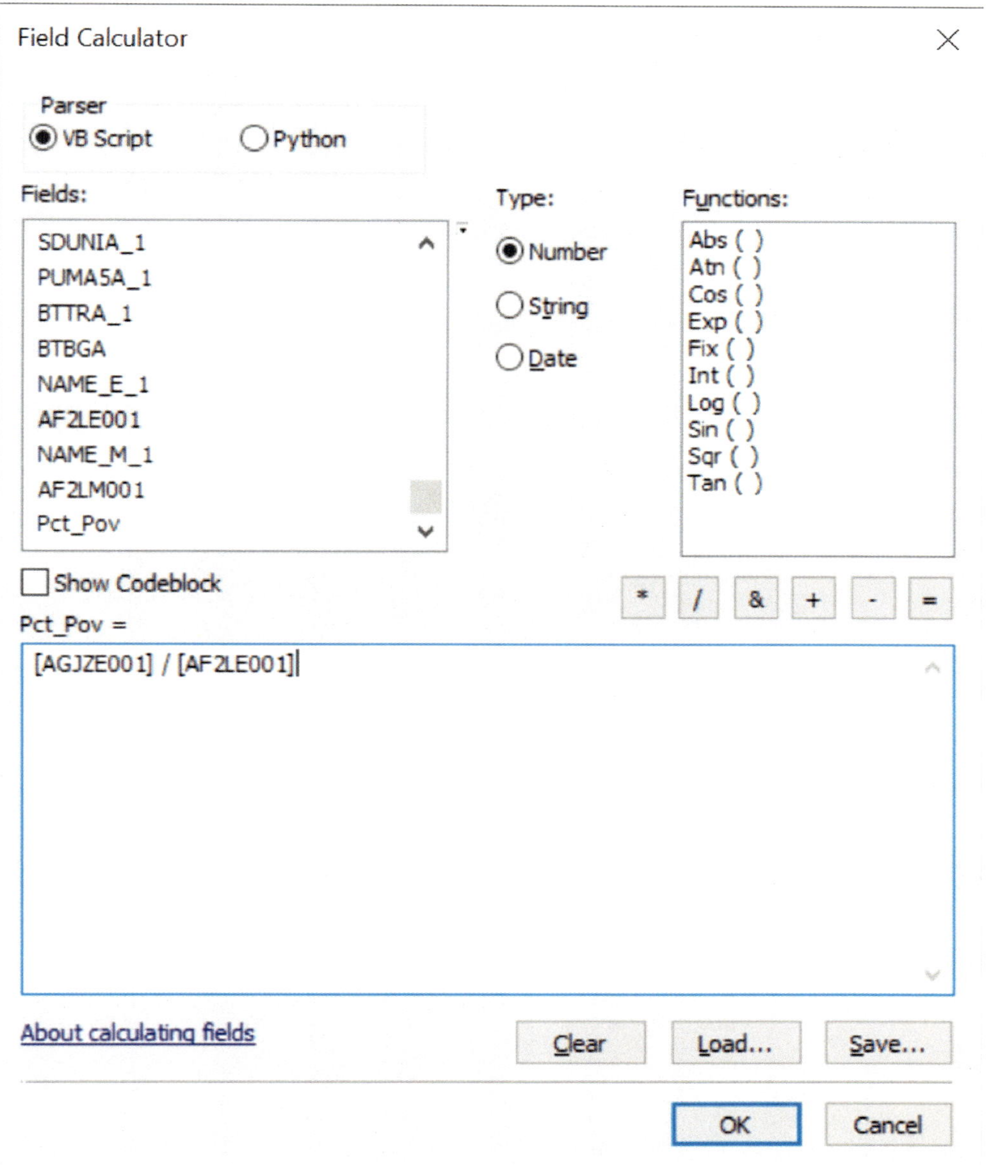

Figure 7.74: The field name for the total population, AF2LE001, is located near the bottom of the list.

The VB Script expression now includes the following structure:

» Pct_Pov = [AGJZE001] / [AF2LE001]

Just remember, only the part of the expression *after* the = sign should appear in the text box. When you are ready, click *OK*. If the correct records were selected, you should not receive an error message. On the attribute table, you should see the percent poverty field updated with the new values (**Figure 7.75**). Go ahead and clear the selected records.

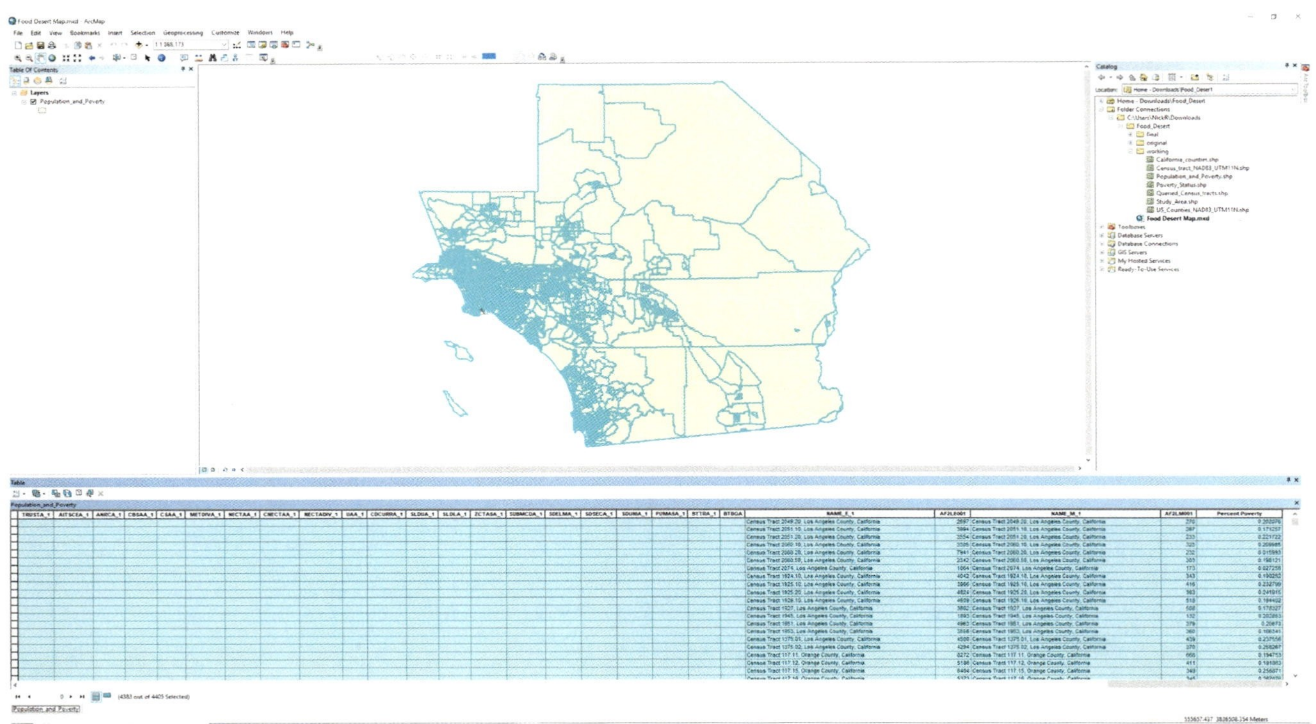

Figure 7.75: The *Field Calculator* will only work on the selected records.

Skill Drill: Using an Attribute Query to Identify Percentages of Poverty

Using what you learned in previous steps, perform an attribute query using the updated percent poverty field. Select the Census tracts where the percentage of persons with poverty status is greater than 0.25 (25%) (**Figure 7.76**).

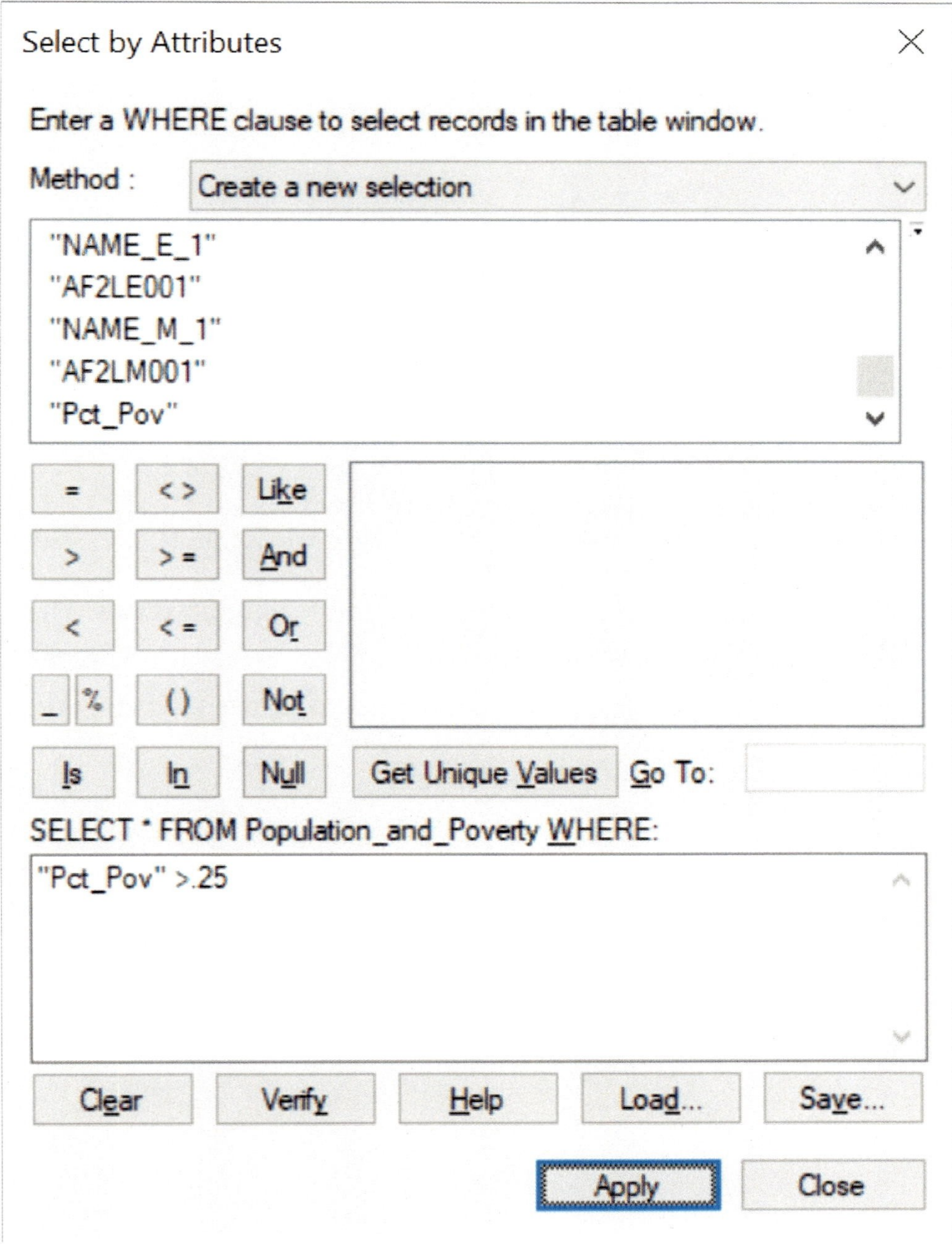

Figure 7.76: Be sure to use .25 to represent 25%.

Export the selected records as a new shapefile. Be sure to save the file to your *working* folder. Call the file, At_risk_communities (**Figure 7.77**).

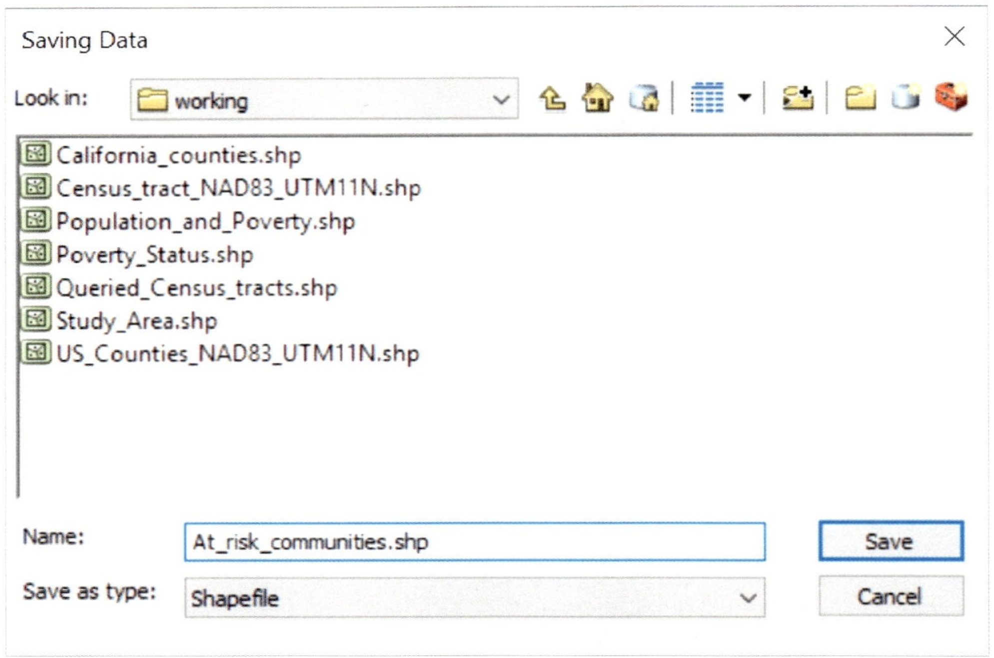

Figure 7.77: Always save data to the desired folder. Never accept the default location.

When done, add the new layer to the map and clear the selected records. This time, you may keep the original percent poverty layer in the Table of Contents (**Figure 7.78**). You will use this layer again in a later step. Take a moment to resave your map document.

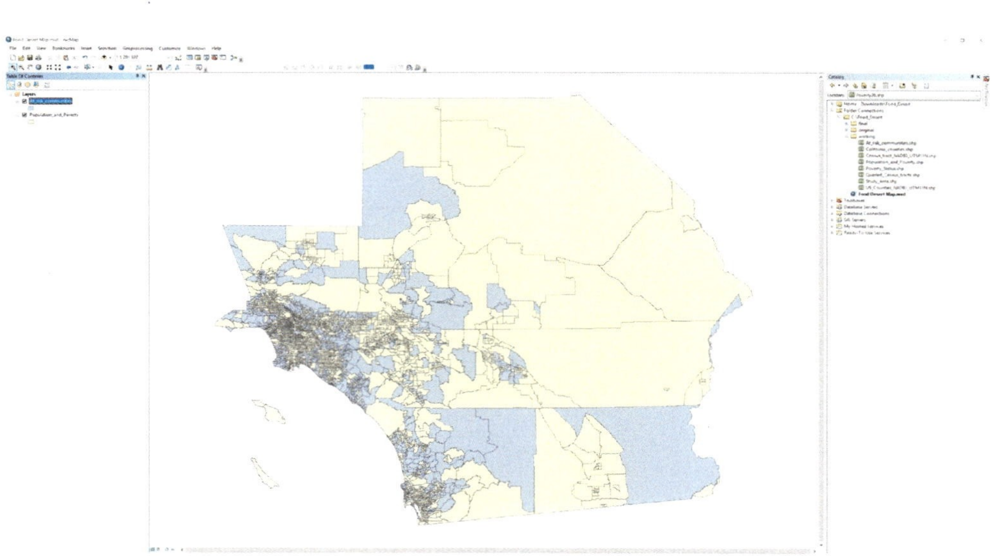

Figure 7.78: The blue polygons represent areas where the percentage of persons with poverty status is greater than 25%.

Downloading Data from ArcGIS Online

Esri curates a large number of datasets and provides them to the public via ArcGIS Online. Though Esri allows access to these datasets, they are proprietary and may not be used for profit. In this step, you must download grocery store data from ArcGIS Online using the ArcMap interface. From the *Main* menu, choose *File*, then *Add Data*, then *Add Data from ArcGIS Online* (**Figure 7.79**).

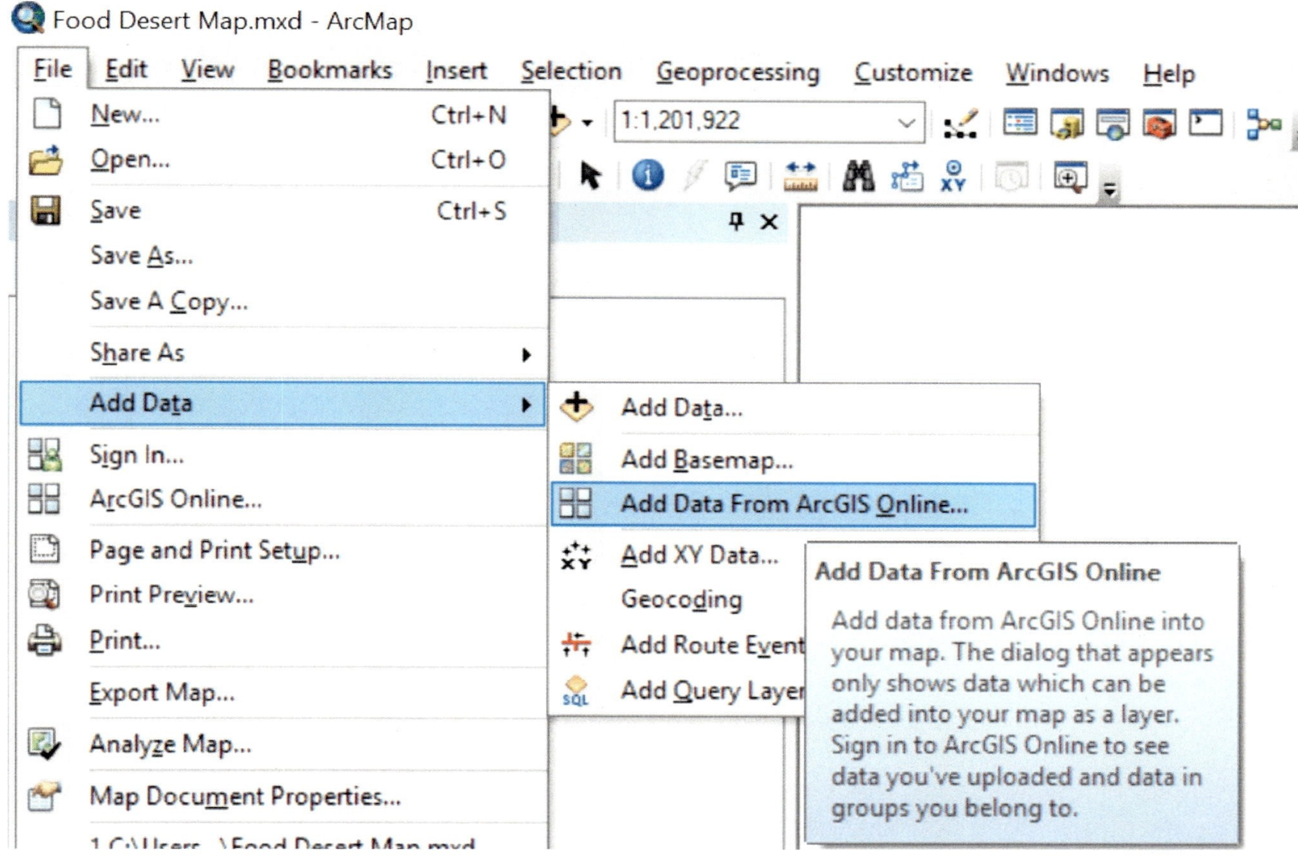

Figure 7.79: The Add Data flyout menu provides several options for adding data.

When the ArcGIS Online data search window appears, use the keyword, "US Grocery Stores Esri," to find the data (**Figure 7.80**). Locate the data with the same name as the keyword and click Add.

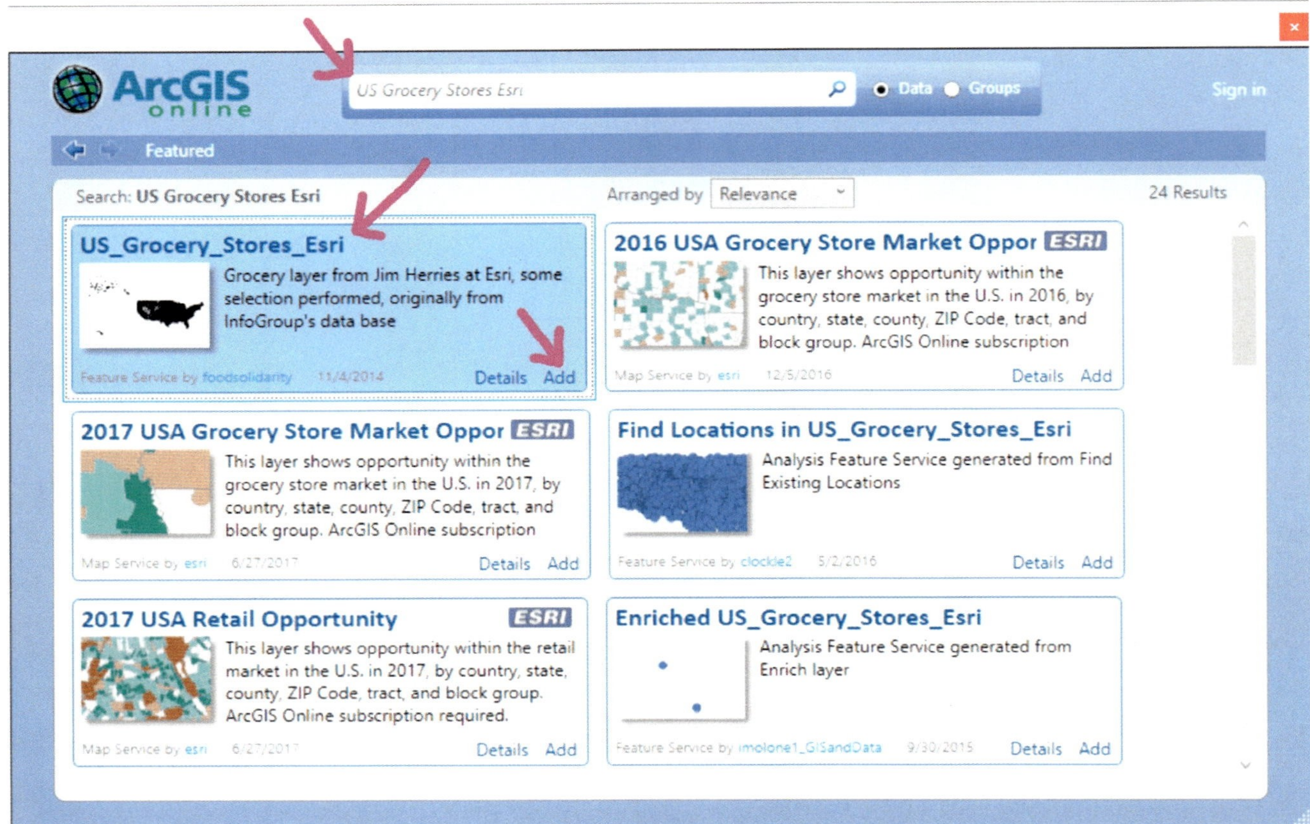

Figure 7.80: Esri provides a wide range of datasets via ArcGIS Online. These datasets are proprietary and may not be used for profit.

When the data downloads and is added to the map, the *Geographic Coordinate Systems Warning* may appear because the grocery data is not in the UTM coordinate system. You will correct this issue in the next step. Click *OK* to close the warning.

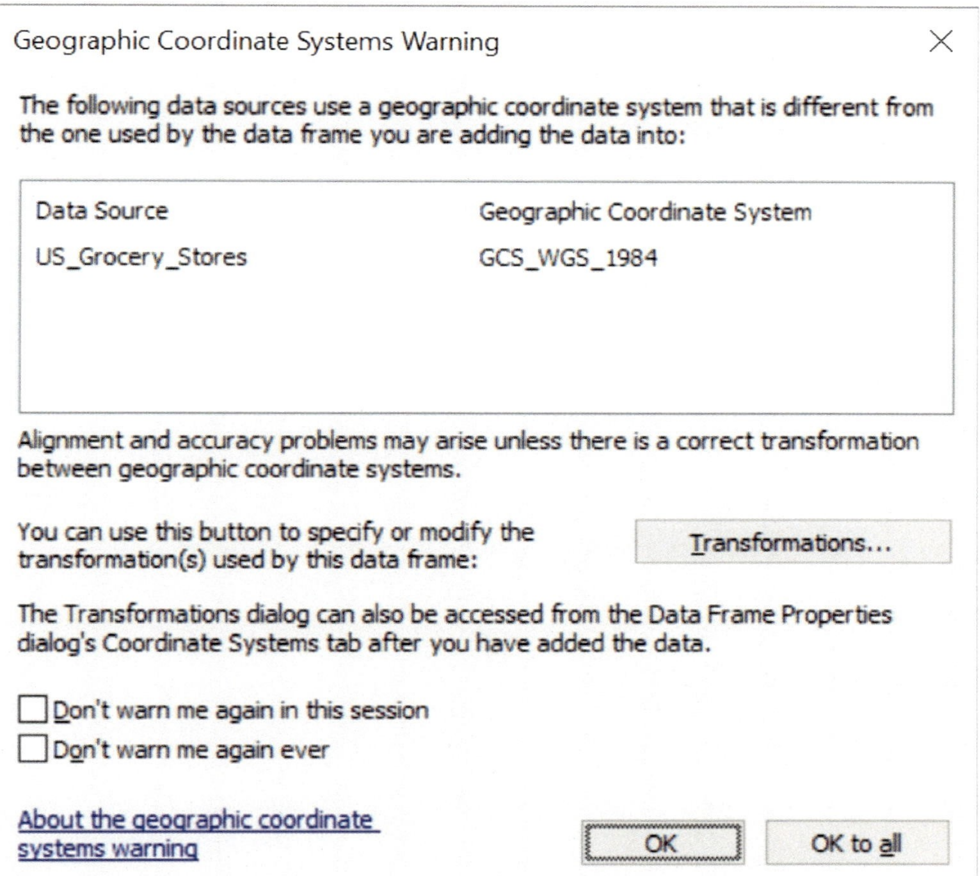

Figure 7.81: ArcMap warns you when the spatial reference system for the layer you add does not match the data frame.

Skill Drill: Using the Project Tool to Match All Layers

Using what you learned in a previous step, open the *Project* tool. The US Grocery Store layer is located in a layer group with a similar name. For the *Input Dataset or Feature Class*, choose the *US Grocery Stores* layer inside the *US Grocery Stores* Esri group layer (**Figure 7.82**).

Figure 7.82: Group Layers can sometimes be confusing. Be sure to choose the correct layer, US_Grocery_Stores.

For the Output Dataset for Feature Class, make sure you click the yellow file folder icon and browse to your *working* folder. Name the file, Grocery_NAD83UTM11N (**Figure 7.83**)

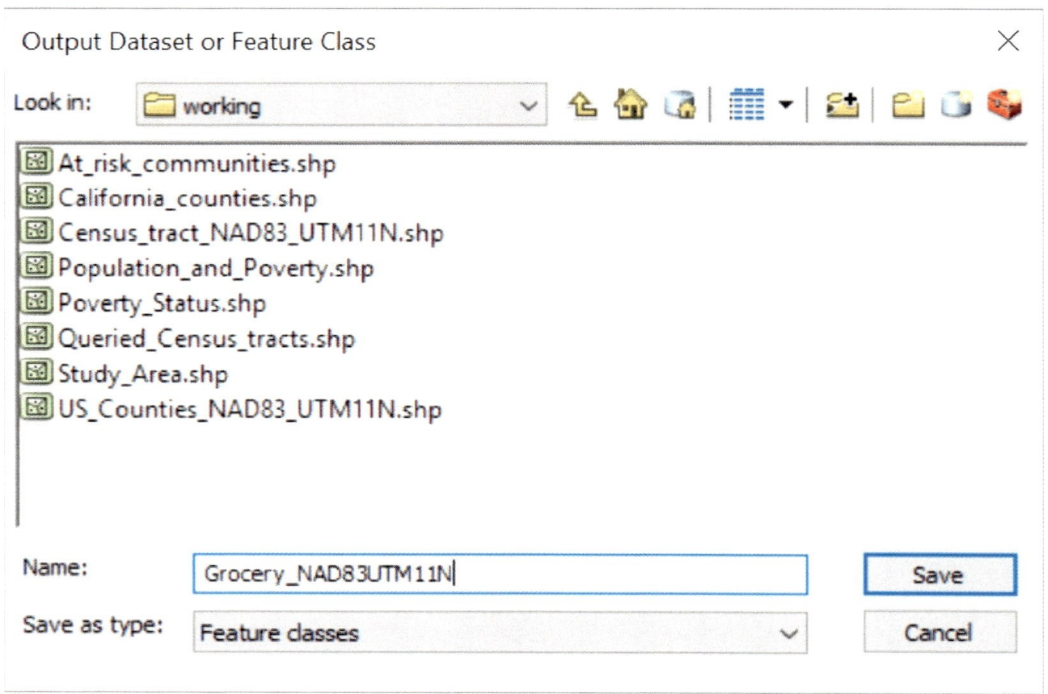

Figure 7.83: Always save data to the desired folder. Never accept the default location.

For the *Output Coordinate System*, match the other layers by using **NAD 83 UTM Zone 11 N**. Leave all of the other default settings as they are and click *OK* (**Figure 7.84**)

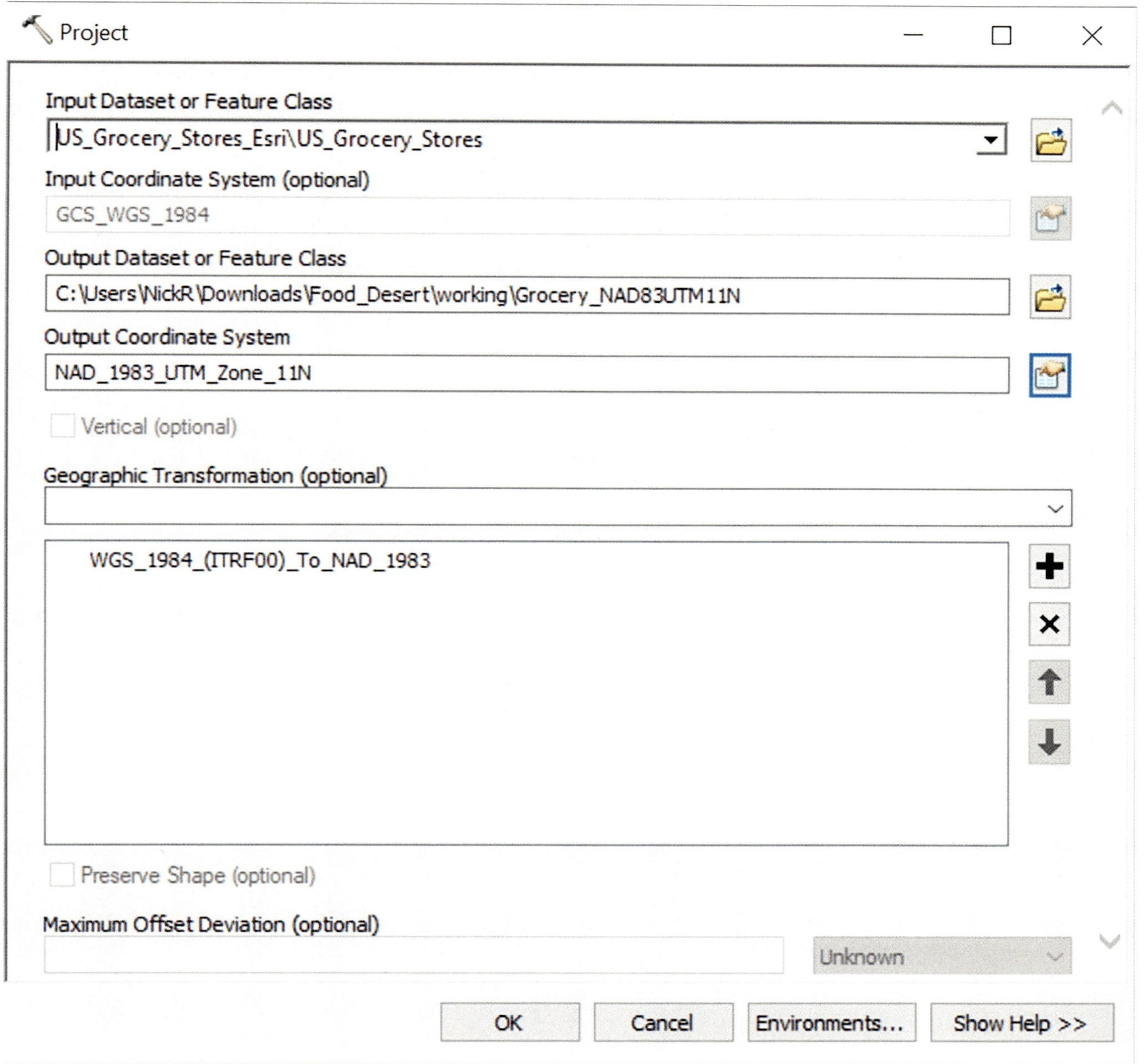

Figure 7.84: Check to make sure your settings match those in this image.

The grocery store layer contains many points. As a result, it may take a few minutes for the *Project* tool to run. When the Project tool finishes, remove the layer group downloaded from ArcGIS Online. Expand your *working* folder in the Catalog Window. Add the projected grocery store layer to the map. Click the Full Extent button on the Tools toolbar to view the extent of the grocery store data in relation to the study area (**Figure 7.85**). As you can see, the projected grocery store layer contains tens of thousands of data points outside the study area. In the next step, you will make this dataset more manageable using the *Clip* tool.

Figure 7.85: The projected grocery stores are added to the map.

Using the Clip Tool

In this chapter, you learned that an **overlay operation** is a spatial operation in which two or more datasets are superimposed on one another to evaluate the relationship between features that occupy the same geographic space. In this step, you must perform an overlay operation called a clip. When performing a **clip operation**, a polygon layer, called the clip feature, acts like a cookie cutter. In this instance, you must use the projected grocery store layer as the input feature and the population and poverty layer as the clip feature. The result will be a grocery store layer with points only within the counties of Los Angeles, Orange, San Bernardino, Riverside, San Diego, and Imperial. Using what you learned in a previous step, open the ArcToolbox. Expand the *Analysis Tools*, then Extract. Locate the *Clip* tool (**Figure 7.86**). Double-click the tool to open it.

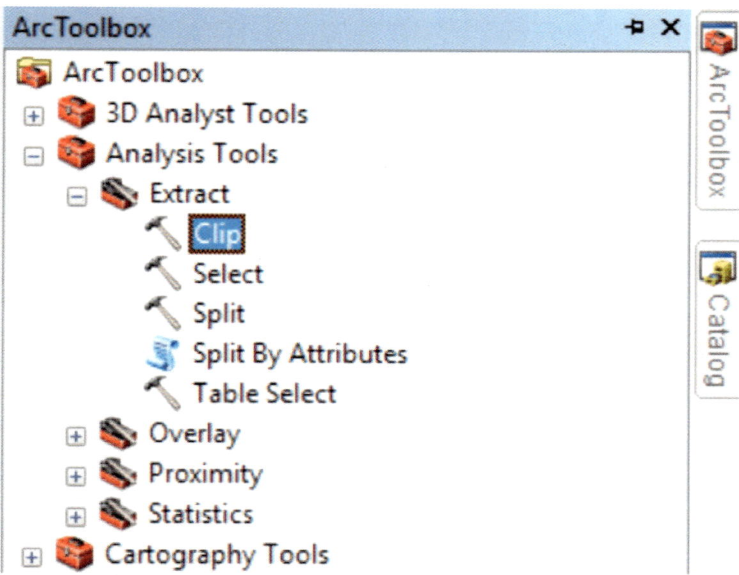

Figure 7.86: The *Analysis* toolbox contains many frequently used tools.

The *Input Features* is the layer to be clipped. For the *Input Features,* use the drop-down menu to select the grocery store layer. The *Clip Feature* is the shape used to clip the input. For the *Clip Features,* choose the population and poverty layer, which represents the study area. For the Output Feature Class, be sure to browse to your *working* folder. Call the file, SoCal_Groceries (**Figure 7.87**)

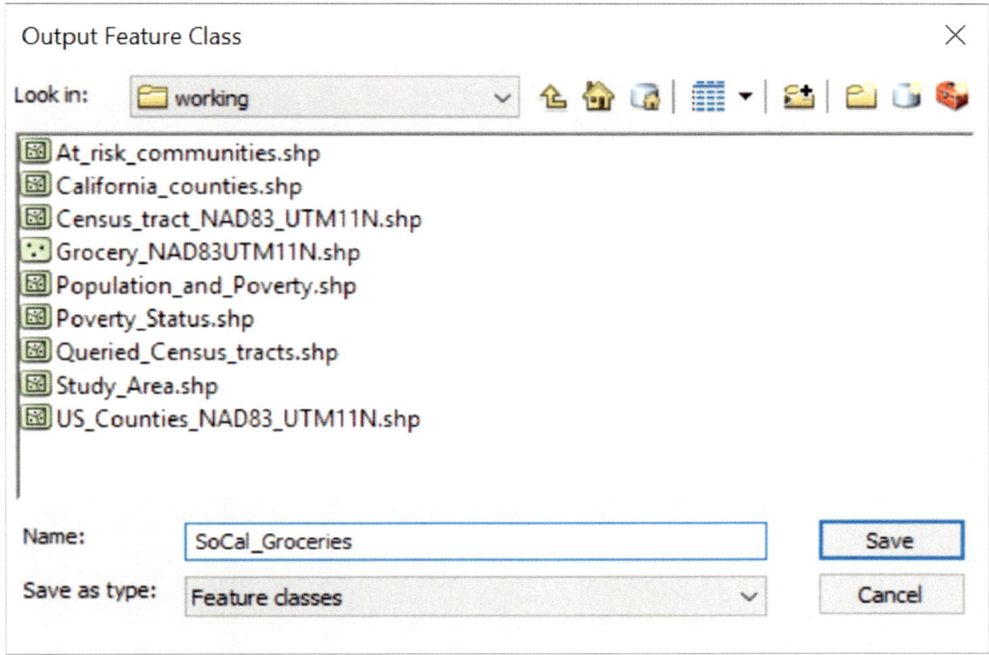

Figure 7.87: Always save data to the desired folder. Never accept the default location.

Leave all other default settings as they are (**Figure 7.88**). When you are ready, click *OK*.

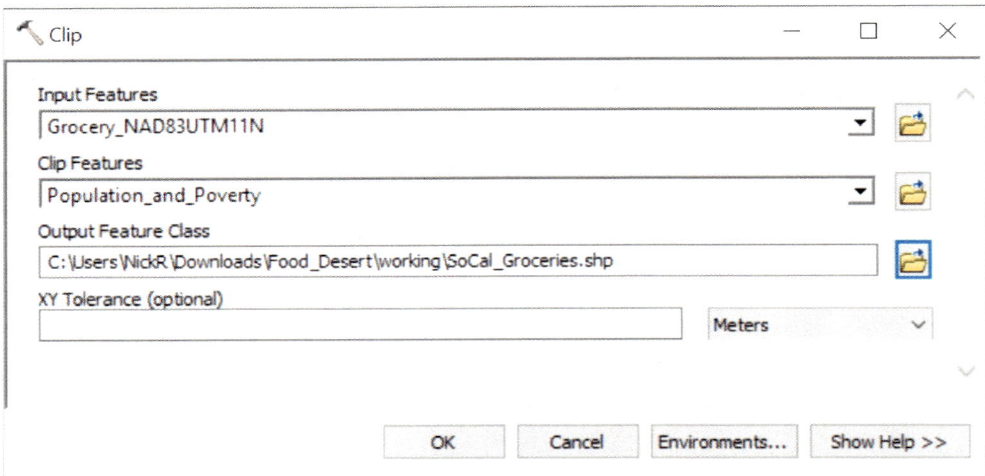

Figure 7.88: Check to make sure your settings match those in this image.

When the clip operation is complete, remove the original grocery store layer from the Table of Contents. You should now have a grocery store layer in Southern California (**Figure 7.89**).

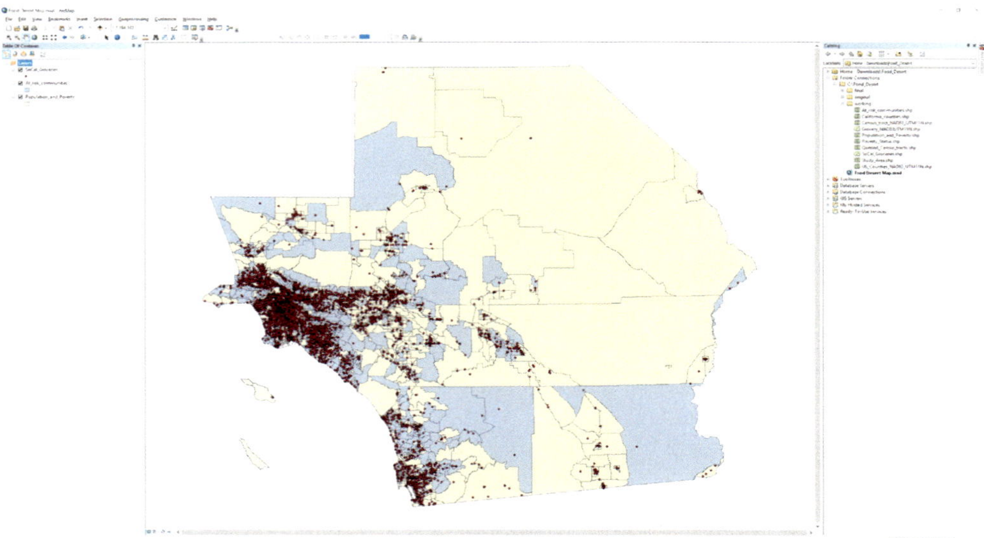

Figure 7.89: The clip operation is useful for extracting a subset of data.

Take a moment to consider the following question:

» What other method did you recently learn about that could have been used to generate the grocery store layer for Southern California?

You may find that in GIS Analysis, there are often several ways to meet the same objective.

Skill Drill: Using an Attribute Query to Locate Large Grocery Stores and Supermarkets

The problem with the Southern California grocery store layer is that it contains information about all types of grocery stores, large and small. Recall from earlier that a **food desert** is a community, neighborhood, or region where people have limited access to affordable, nutritious food because they live far from a supermarket or large grocery store. In terms of food access and quality, larger stores often provide access to a variety of food departments, including fresh produce, baked goods, dairy, seafood, and meats. According to a 2009 report to Congress by the United States Department of Agriculture (USDA), a store should have an annual sales volume of at least $2 million to be considered a *large* grocery store or supermarket. Using what you learned in a previous step, perform an attribute query on the Southern California grocery store layer (**Figure 7.90**). Select the stores with a daily sales volume greater than or equal to $5480.

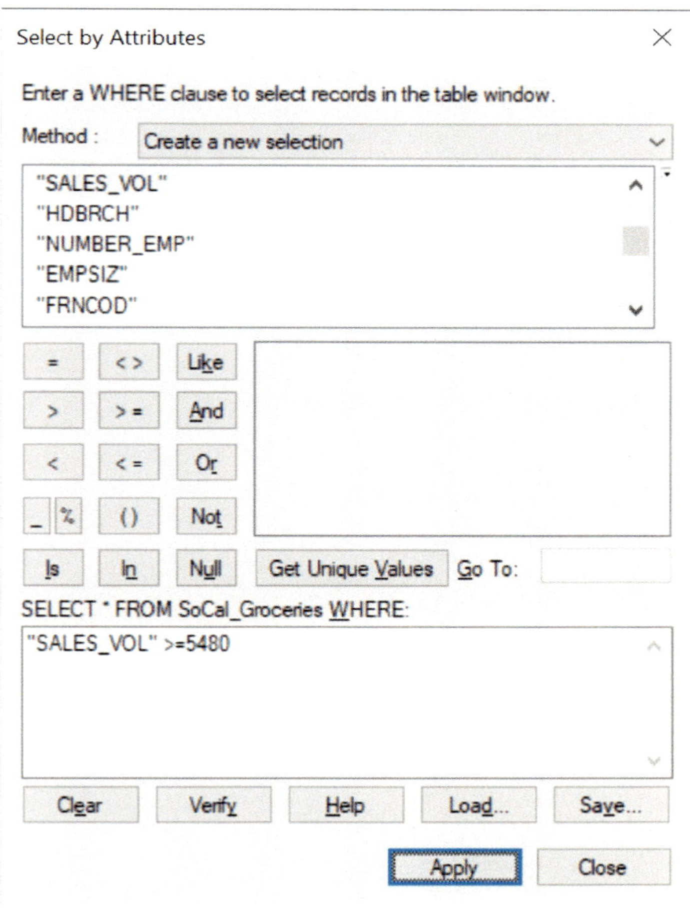

Figure 7.90: Use an attribute query to select stores with a daily sales volume of at least 5480.

Export the selected features as a new shapefile and save to your *working* folder. Call the file Large_Grocery_Stores (**Figure 7.91**)

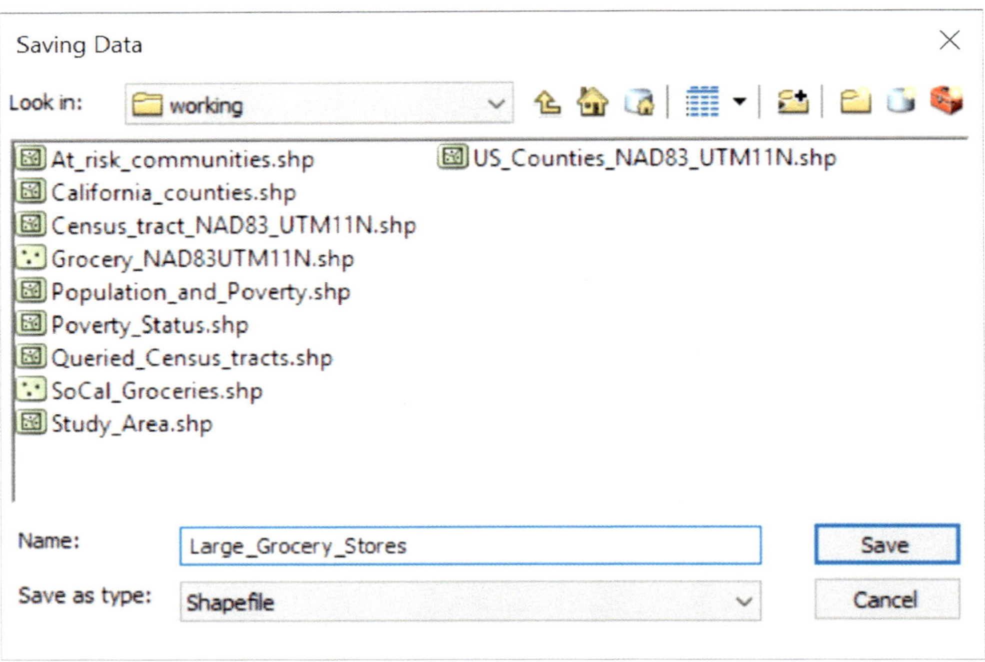

Figure 7.91: Always save data to the desired folder. Never accept the default location.

When done, add the new layer to the map. Clear the selected records and remove the Southern California grocery store layer from the Table of Contents (**Figure 7.92**).

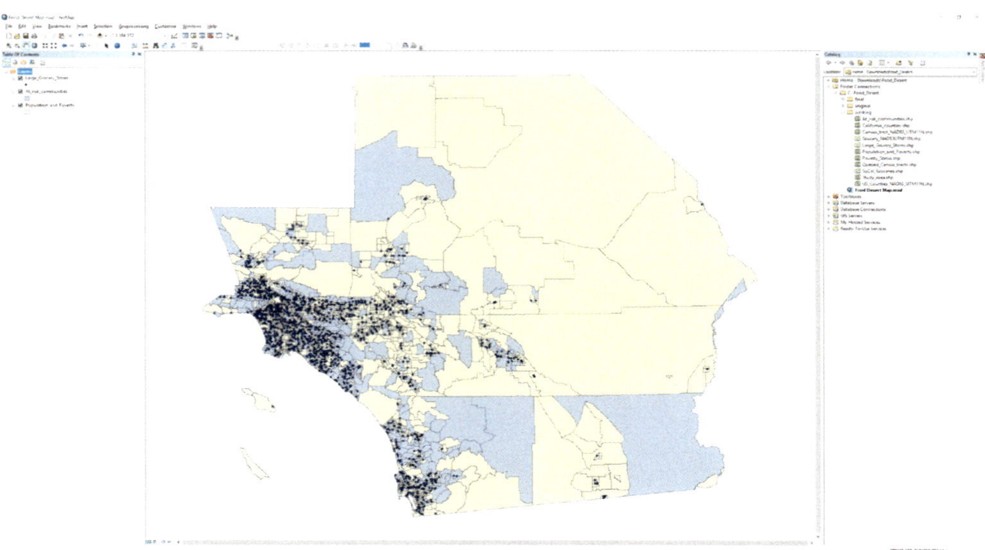

Figure 7.92: The large grocery stores and supermarkets are added to the map.

Using the Buffer Tool

As you learned in this chapter, a **buffering operation** creates a buffer, which is a zone around a map feature based on distance or time. In this analysis, you are interested in areas more than one mile from large grocery stores or supermarkets. To define this area, you must use the *Buffer* tool to create a one-mile buffer around the large grocery store layer. Using what you learned in a previous step, open the ArcToolbox. Expand the *Analysis Tools*, then *Proximity*. Locate the *Buffer* tool (**Figure 7.93**). Double-click the tool to open it.

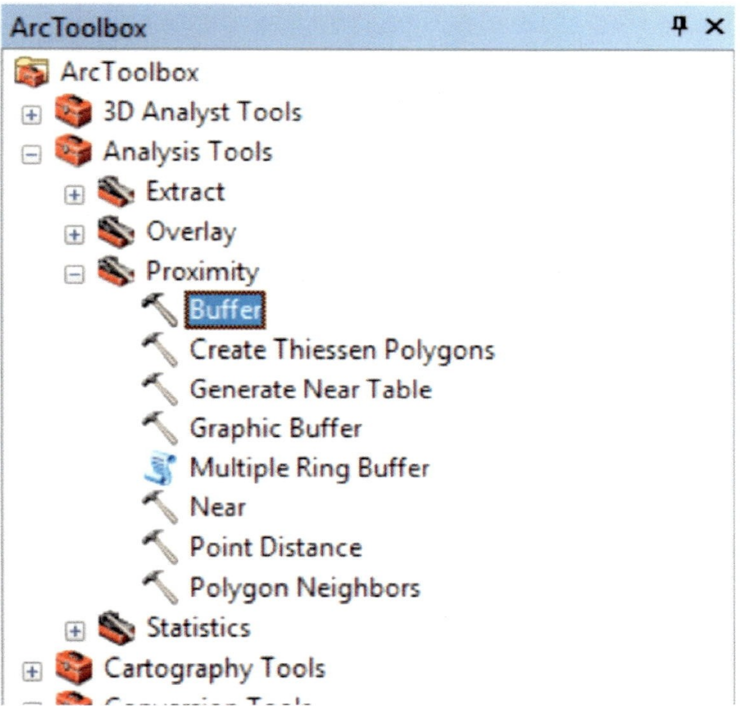

Figure 7.93: The *Analysis* toolbox contains many frequently used tools.

For the *Input Features*, use the drop-down menu and select the large grocery store layer. For the Output Feature Class, use the yellow file folder icon and browse to your *working* folder (**Figure 7.94**). Name the file, One_Mile_Buffer.

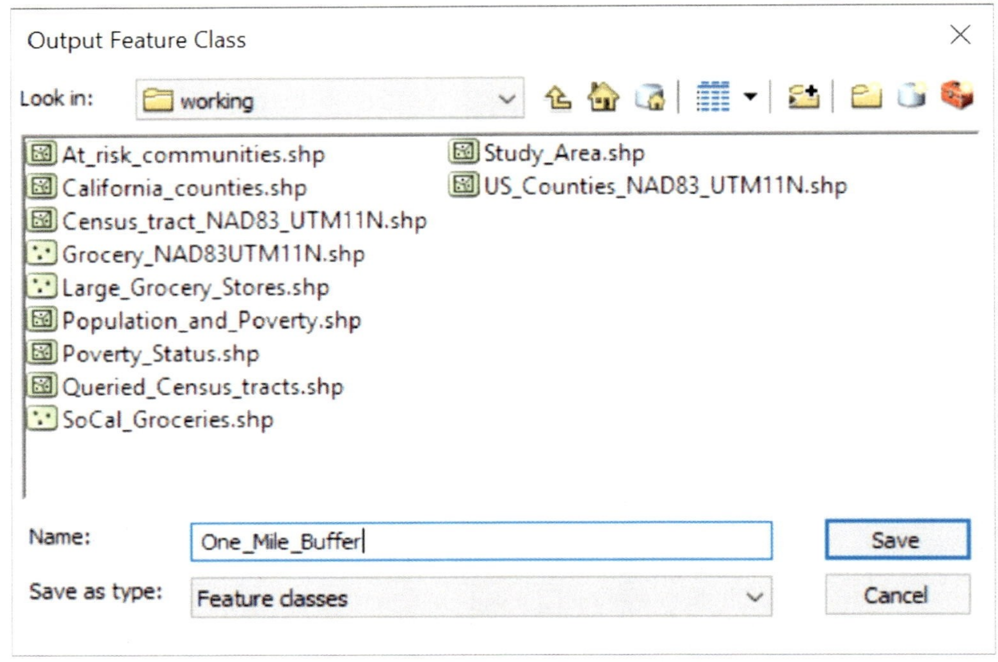

Figure 7.94: Always save data to the desired folder. Never accept the default location.

Under *Linear Unit*, enter the number 1. Use the drop-down menu to change the units to *Miles*. For clarity, any overlapping buffers should be represented as a single polygon. To do this, change the *Dissolve Type* to *ALL*. When this option is selected, all the buffers are dissolved together into a single feature, removing any overlap. Leave all other default settings as they are and click *OK* (**Figure 7.95**).

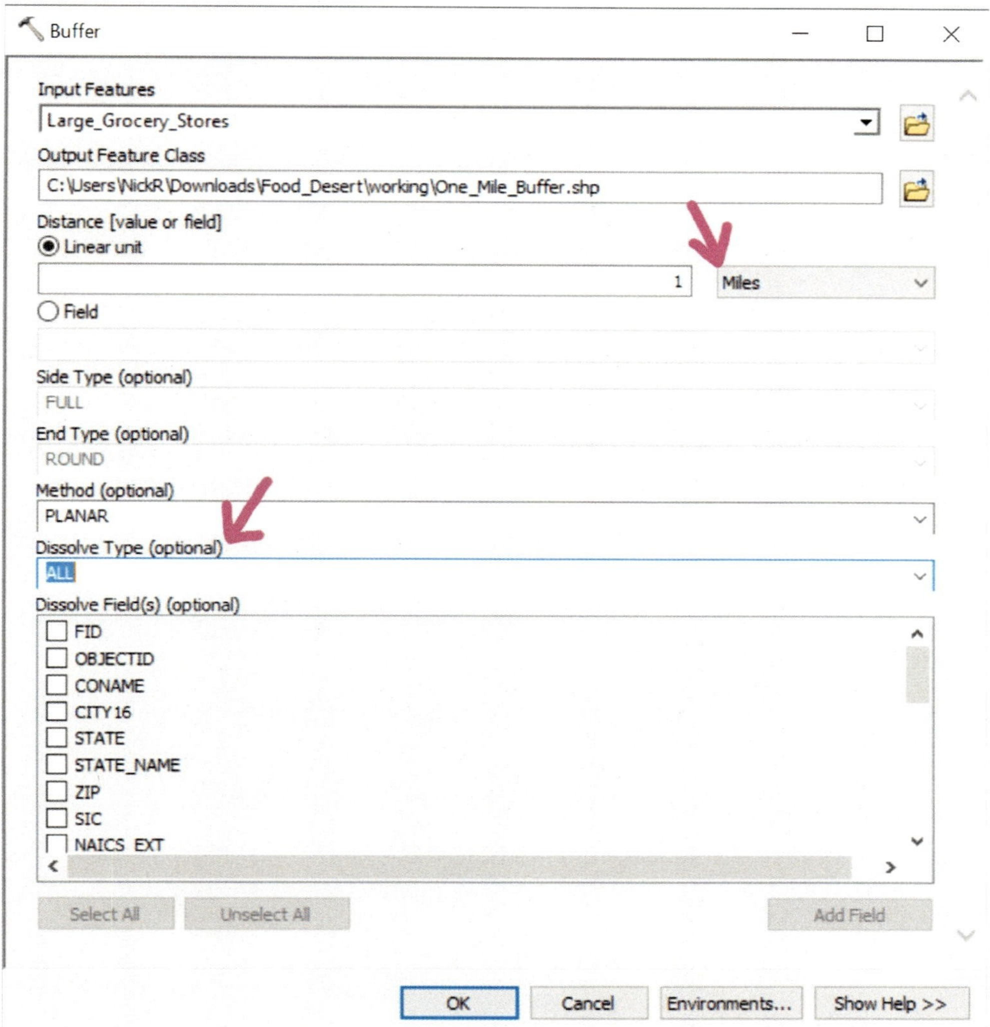

Figure 7.95: Be sure to check the linear unit and the dissolve type.

When the *Buffer* tool completes, a polygon with a distance of one mile should appear around each larger grocery store (**Figure 7.96**).

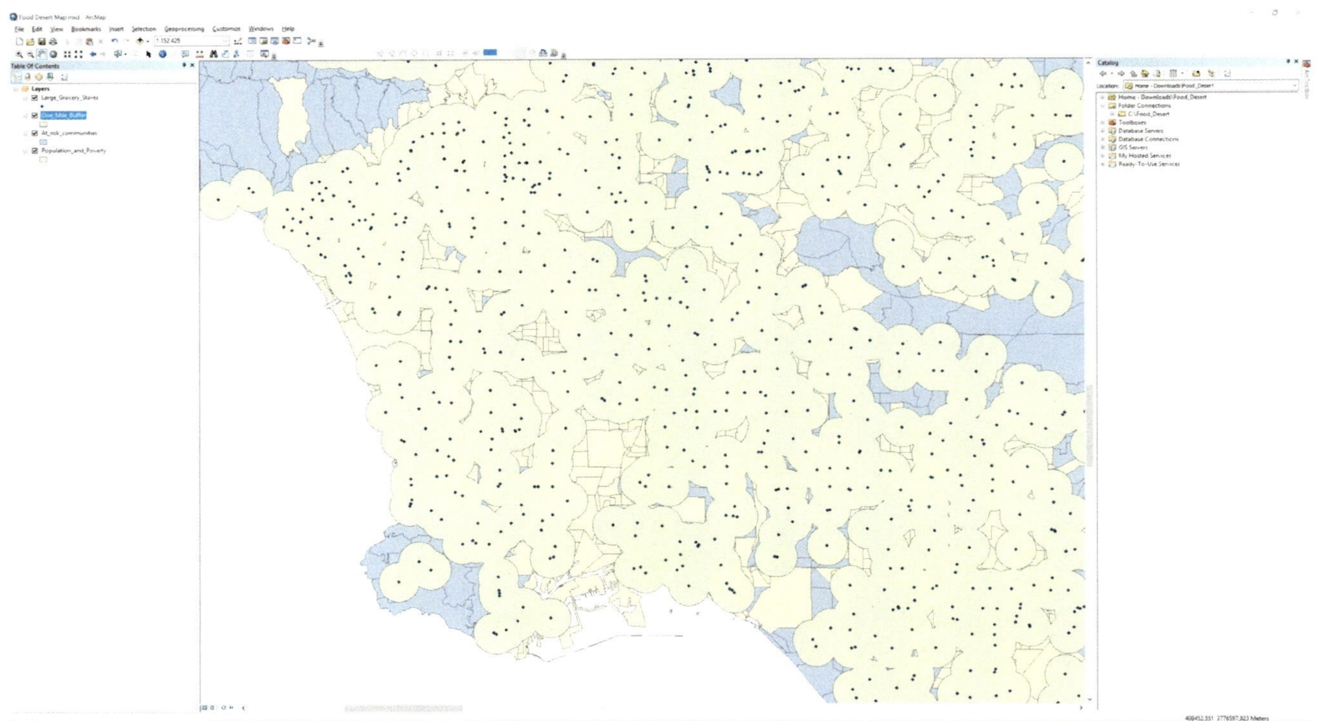

Figure 7.96: Upon closer inspection, you will see a buffer extending one mile from each grocery store.

Using the Erase Tool

With the one-mile buffer around large grocery stores and supermarkets, you can identify areas that are potential t. These are the areas with communities at risk that fall outside the buffer zones. To create a layer with regions of communities at risk that is outside the buffer zones, you must use the *Erase* tool. **Erase** is an overlay operation where the shape and areas of one layer are used to erase the areas of another. Only the non-overlapping regions are kept. When compared to the *Clip* tool, the *Erase* tool works in an opposite manner. Using what you learned in a previous step, open the ArcToolbox. Expand the *Analysis Tools*, then Overlay. Locate the *Erase* tool (**Figure 7.97**). Double-click the tool to open it.

For the *Input Features*, use the drop-down menu and select the at-risk communities layer. For the *Erase Features*, use the drop-down menu to choose the *One-mile buffer* layer. For the *Output Feature Class*, use the yellow file folder icon and browse to your *working* folder (**Figure 7.98**). Name the file, Food_Deserts.

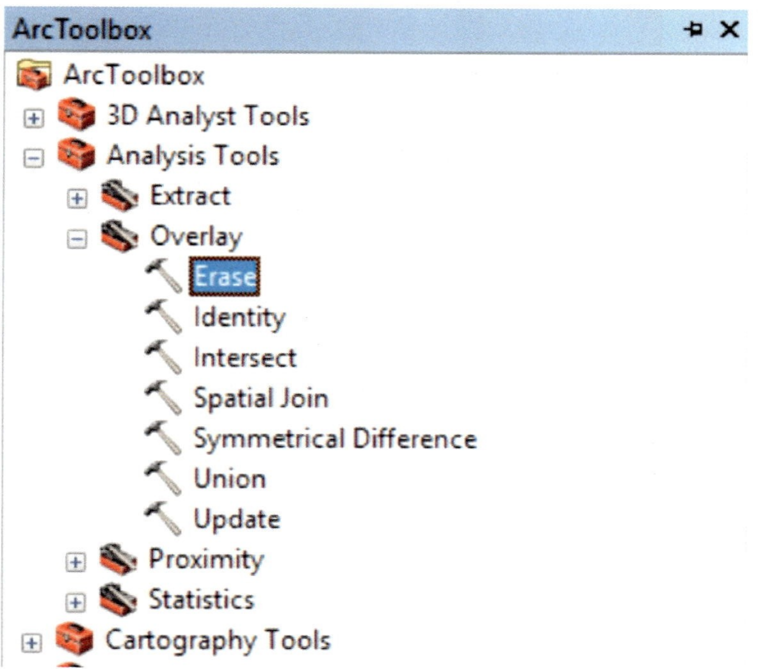

Figure 7.97: The Analysis toolbox contains many frequently used tools.

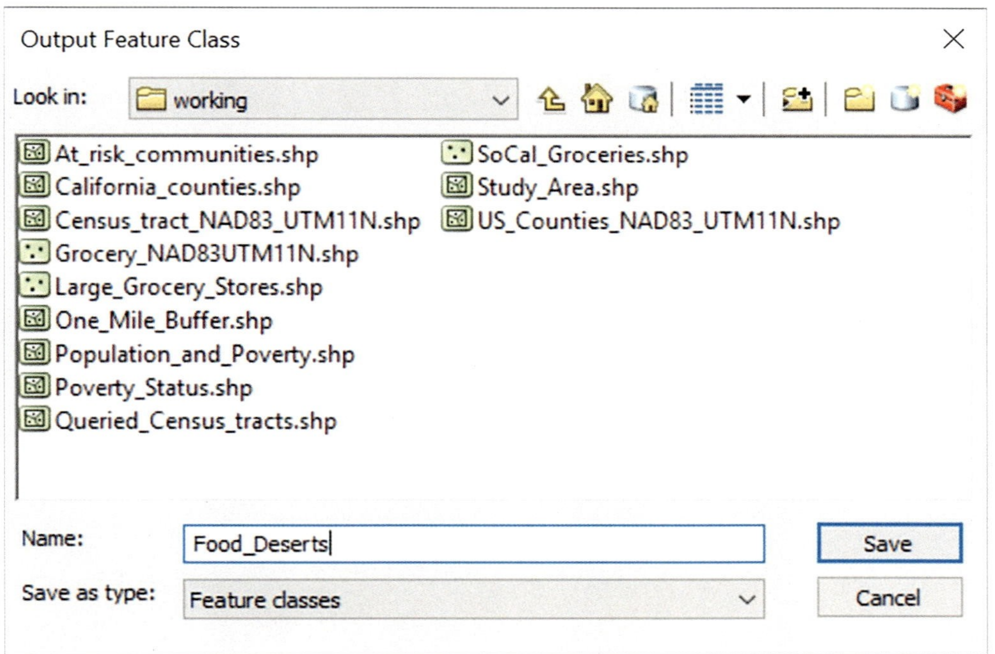

Figure 7.98: Always save data to the desired folder. Never accept the default location.

Leave all other default settings as they are and click *OK* (**Figure 7.99**).

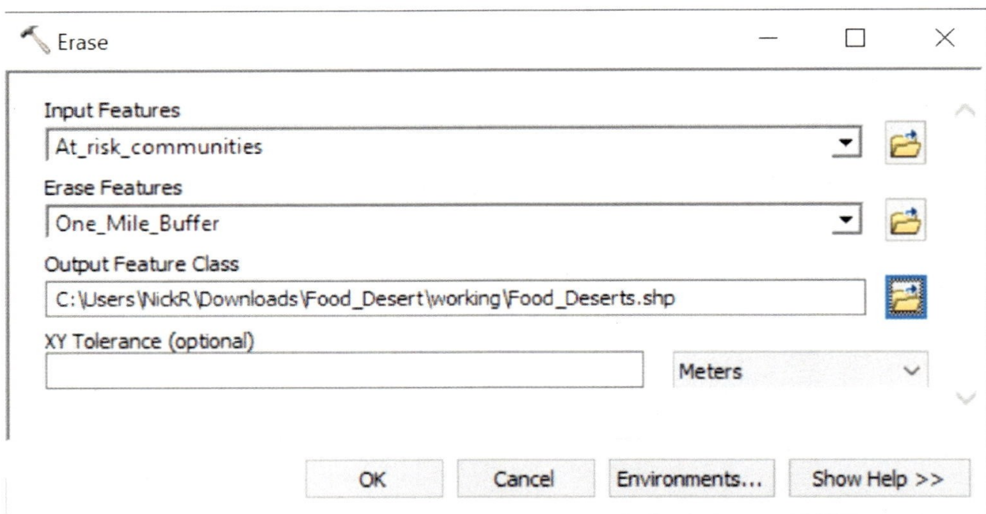

Figure 7.99: Check to make sure your settings match those in this image.

When the *Erase* tool completes, you should have a layer that meets all of the criteria for the analysis. These areas represent potential food deserts in Southern California (**Figure 7.100**). Take a moment to re-save your map document before moving on to the next step.

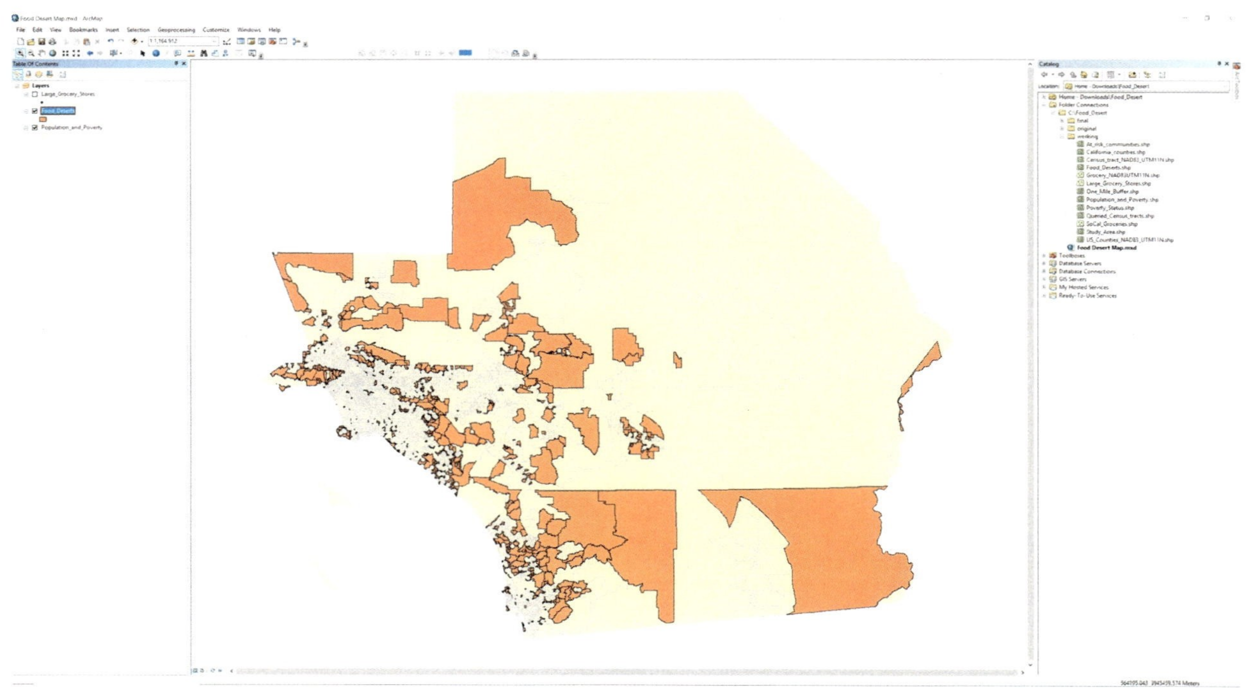

Figure 7.100: For clarity, the buffer layer and the at-risk layer were removed from the Table of Contents and the grocery store layer was turned off.

Skill Drill: Creating Small-Sized Maps for a Report

Using what you learned in previous activities, create two small-sized maps of 6 by 6 inches. The purpose of these maps is for use as figures in a lab report. When creating small-sized maps, it is necessary to consider the design limitations. They must be more restricted than poster-sized maps. Due to the limited size, you do not have much room with which to work. Only include a north arrow, a scale bar, and a legend. You do not need to add a map title directly on the map. Instead, plan to insert a **figure caption** in Microsoft Word as part of a lab report. The caption can take the place of both descriptive text and a map title. You can also add acknowledgments to the NHGIS and Esri in the report, so you do not need to add them to the map. Remember, when designing the small-sized maps for this activity, start by switching to layout view in ArcMap. Then, change the paper size and data frame size to 6 by 6 inches. When the map is the correct size, begin adjusting colors and inserting your map elements.

Map Purpose and Audience

For each map, you may assume that the audience is the board of directors of the non-profit organization requesting the analysis. You may also assume that they are familiar with the location of the study area and the Southern California counties. Thus, your maps should focus on displaying the results of the analysis. The first map must illustrate the study area as a choropleth map[3] that communicates the percentage of persons in poverty for each Census tract (**Figure 7.101**).

3 In ArcMap, choose *Graduated Colors* as the option in the *Symbology* tab to create a chloropleth map.

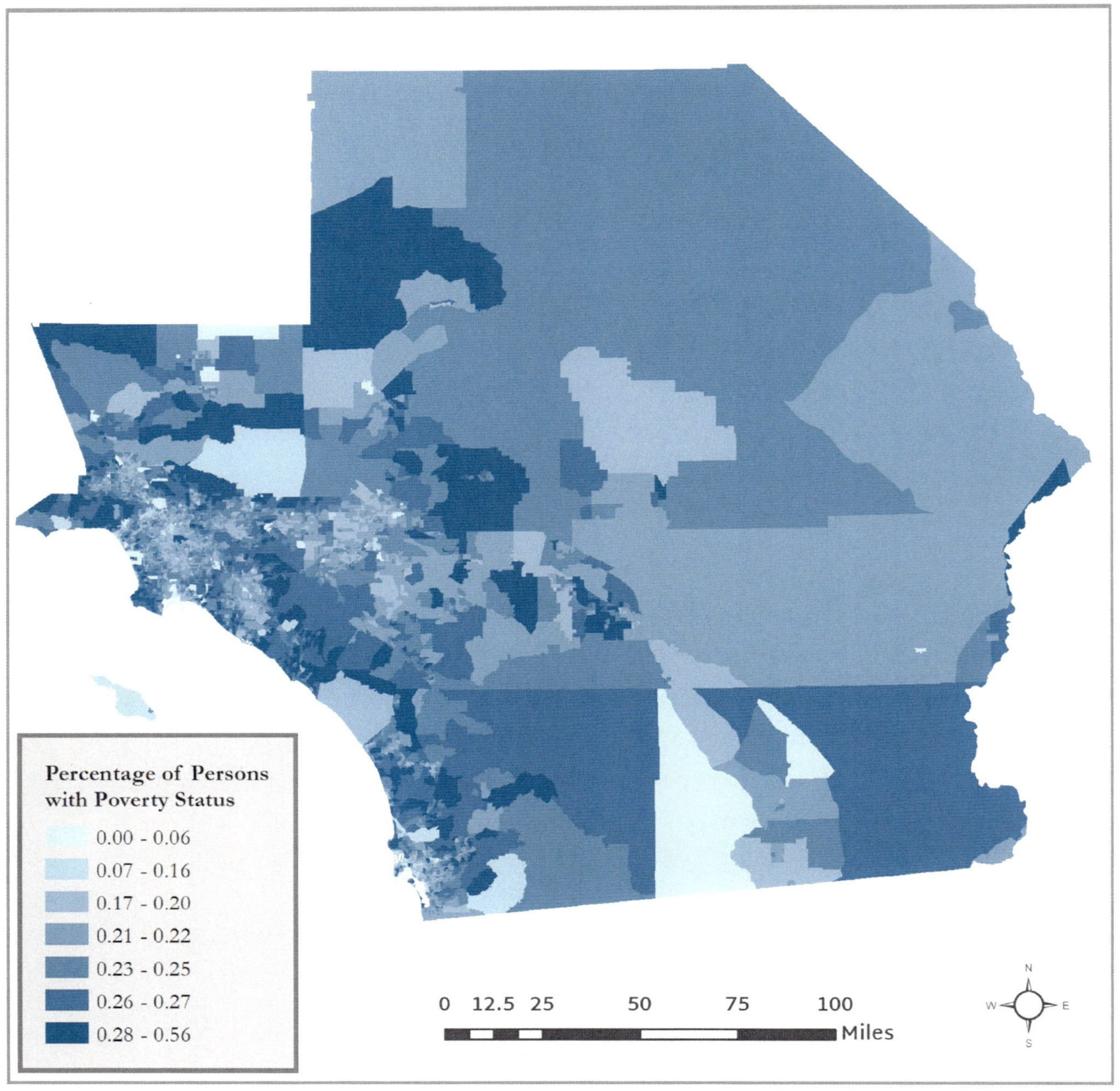

Figure 7.101: In this example, the Natural Breaks classification method with seven classes was used. Polygon outlines were removed for clarity.

The second map should be designed to communicate the location and extent of food deserts throughout the study area (**Figure 7.102**).

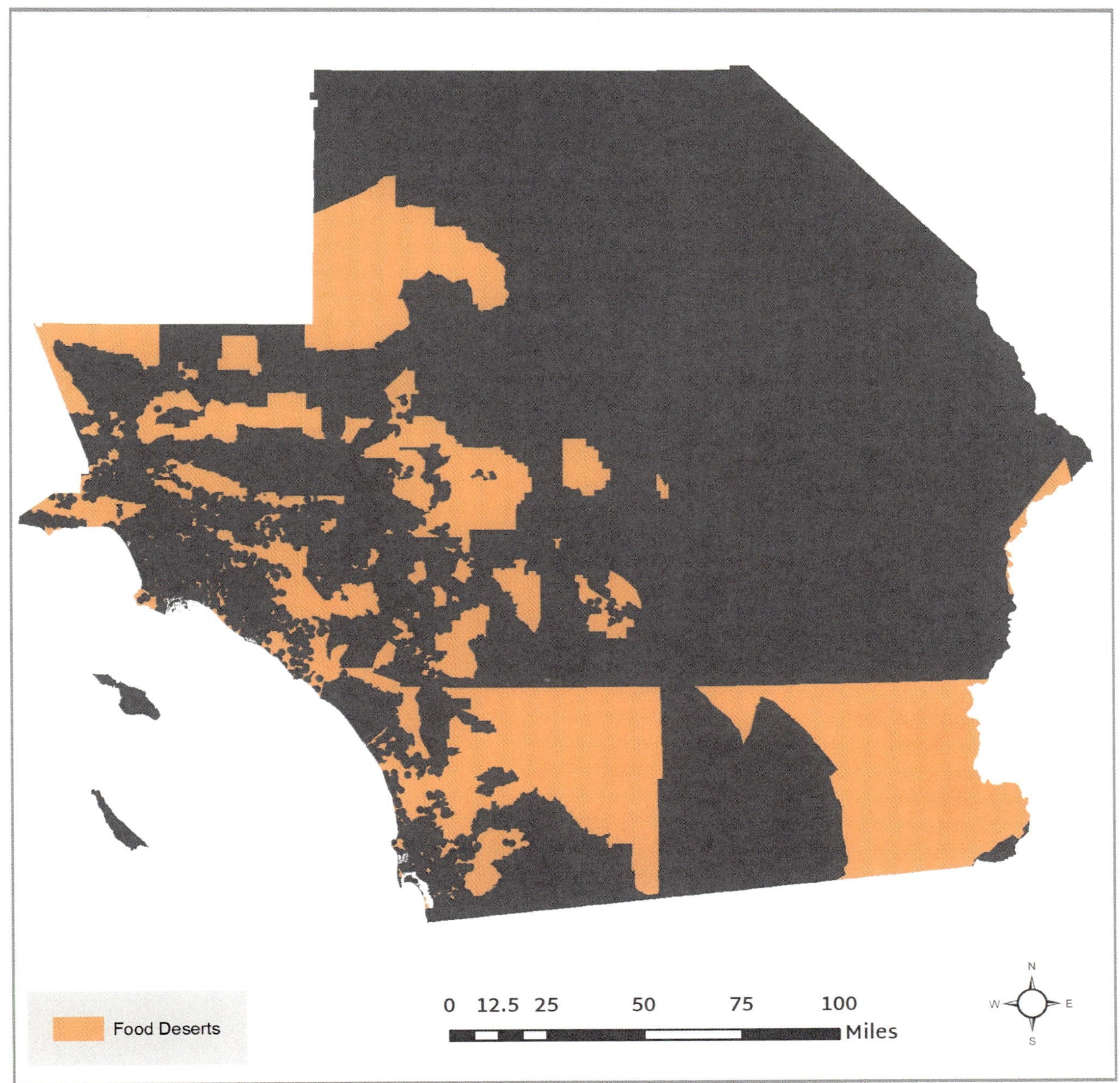

Figure 7.102: In this example, a solid color was used to represent food deserts. Polygon outlines were removed for clarity.

When you are done creating your small-sized maps, export the maps as a PNG, and save the files to your *final* folder. ArcMap may try to default to a low resolution. Remember to change the image resolution to at least 300 dpi.

Principal Terms

adjacency operation
attribute queries
attribute query
attributes
attribute type
Boolean expression
Boolean operator
buffer
buffering
buffering operation
Cardinality
clip
clip feature
clip operation
comparison operator
compound expression
containment operation
database
database management system (DBMS)
data frame
dataset
display projection
entities
Erase
feature type
field
field name
focal analysis operation
food desert
geospatial analysis
global analysis operation
header row
immutable
intersect
intersection operation
key
local analysis operation
Logical expressions
many-to-many relationship
many-to-one relationship
map layer
neighborhood operation
North American Datum of(NAD)
one-to-one relationship
operand
operator
overlay operation
primary key
project-on-the-fly
proximity operation
query
query builder
record
relationships
spatial query
spatial query builder
spatial reference system
spatial relationship
string
structured query language (SQL)
syntax
table join
table relationship
union
Universal Transverse Mercator (UTM) Coordinate System
Universal Transverse Mercator (UTM) system
UTM zones

INDEX

D

E

F

G

H

I

N

O

P

Q

R

S

T